Energy and Society

[An Introduction]

Harold H. Schobert

Taylor & Francis
New York • London

Denise T. Schanck, *Vice President*
Robert H. Bedford, *Editor*
Brandy Mui, *Production Editor*
Catherine M. Caputo, *Assistant Editor*
Thomas Hastings, *Marketing Manager*
Mariluz Segarra, *Marketing Assistant*

Published in 2002 by
Taylor & Francis
29 West 35th Street
New York, NY 10001

Published in Great Britain by
Taylor & Francis
11 New Fetter Lane
London EC4P 4EE

Library of Congress Cataloging-in-Publication Data

Schobert, Harold H., 1943–
   Energy and society : an introduction / by Harold H. Schobert
     p. cm.
   Includes bibliographical references and index.
   ISBN 1-56032-767-7 (alk. paper)
     1. Power resources.   2. Power (Mechanics)   I. Title.

HD9502.A2 S3975 2001
333.7—dc21   2001027963

CONTENTS

1. Introduction — 1
2. Energy, Work, and Power — 9
3. Human Energy — 17
4. The Energy Balance — 33
5. Fire — 39
6. Firewood — 49
7. Combustion for Home Comfort — 65
8. Waterwheels — 81
9. Wind Energy — 101
10. The Steam Engine — 113
11. Heat and Thermal Efficiency — 129
12. An Introduction to Electricity — 153
13. How Electricity is Generated — 169
14. Impacts of Electricity on Society — 181
15. Electricity from Falling Water — 195
16. Electricity from Steam — 211
17. Boilers and Heat Transfer — 229
18. Electricity from Coal — 241
19. Energy for Transportation — 257
20. Petroleum and its Products — 281
21. Gasoline — 297
22. Cars and their Impact — 311
23. Jet Engines and Jet Fuel — 327
24. Diesel Engines and Diesel Fuel — 343
25. Atoms and 'Atomic' Energy — 355
26. Radioactivity, Fission, and Chain Reactions — 367
27. Nuclear Power Plants — 387
28. The Nuclear Controversy — 403
29. Energy and the Environment — 427

30. Acid Rain   441
31. Vehicle Emissions and Emissionless Vehicles   463
32. The Greenhouse Effect   477
33. Fossil Energy: Reserves, Resources, and Geopolitics   501
34. Renewable Energy from Biomass   531
35. Electricity from Wind   565
36. Energy from the Sun   581
37. Nuclear Fusion: Bringing the Sun to Earth   605

Glossary   619
Bibliography: Resources for Learning More   633
Index   637

ACKNOWLEDGEMENTS

Two people made major contributions to the preparation of the manuscript of this book. Aubrey Dick provided a great contribution in locating sources of illustrations and securing the permissions for them; she even took a number of the photographs herself. The inimitable Abby Bitner assisted in numerous ways with correspondence, filing, invoices, and the many related clerical chores. Their help was crucial in bringing this book to completion. In addition, Elizabeth Young provided excellent assistance in 'translating' many of the rough sketches from my old lecture notes into illustrations, sometimes on fairly short notice. Many of my colleagues, at Penn State and elsewhere, helped by providing specific information, lending materials from their files, or suggesting other sources. Assistance, suggestions and, especially, remarkable patience were provided by several people at Taylor & Francis and at Keyword Publishing Services, most notably Rob Bedford and Sue Nicholls. It is a pleasure to thank all of these people for everything they've done. Of course, any blame for errors of fact or of omission is solely mine.

PERMISSIONS

Figure 1.1   Reprinted, by permission, from Philadelphia Printing, Willow Grove, PA

Figure 2.1   Reprinted, by permission, from Peter J. Pappano

Figure 3.1   Photo by Aubrey Dick

Figure 3.4   Photo by Aubrey Dick

Figure 3.5   Photo by Aubrey Dick

Figure 3.6   Reprinted, by permission, from Art Resource, New York, NY

Figure 4.2   Photo by Aubrey Dick

Figure 4.3   Photo by Aubrey Dick

Figure 5.1   Reprinted, by permission, from Visual Image Presentations, Bladensburg, MD

Figure 6.1   Photo by Simon Fraser. Reprinted, by permission, from Photo Researchers, Inc., New York, NY

Figure 6.2   Photo by Simon Fraser. Reprinted, by permission, from Photo Researchers, Inc., New York, NY

Figure 6.3   Reprinted, by permission, from Good Time Stove Company, Goshen, MA

Figure 6.4   Reprinted, by permission, from U.S. Geological Survey, Reston, VA

Figure 6.5   Reprinted, by permission, from Jim Whitmer Photography, Wheaton, IL

Figure 7.1   Photo by Aubrey Dick. Reprinted, by permission, from Valerie P. Dick

Figure 7.2   Reprinted, by permission, from *The Power of the Machine: The Impact of Technology from 1700 to the Present Day* by R.A. Buchanan, © Penguin USA, New York, NY

Figure 7.5   Reprinted, by permission, from *Victorian London Street Life in Historic Photographs* by John Thomson, © 1994 Dover Publications, Inc., Mineola, NY

Figure 8.4   Reprinted, by permission, from Jim Whitmer Photography, Wheaton, IL

Figure 9.1   Reprinted, by permission, from PhotoSphere Images Ltd, Canada

Figure 9.2   Reprinted, by permission, from PhotoSphere Images Ltd, Canada

Figure 9.4   Reprinted, by permission, from Visual Image Presentations, Bladensburg, MD

Figure 10.1   Reprinted, by permission, from Visual Image Presentations, Bladensburg, MD

Figure 10.3   Reprinted, by permission, from *Modern Steam Car and its Background*, Revised Edition, by Thomas S. Derr, © 1998 Lindsay Publications, Bradley, IL

Figure 13.14   Reprinted, by permission, from Electric Power Generation Association, Harrisburg, PA

Figure 15.6   Reprinted, by permission, from Hagley Museum and Library, Wilmington, DE

Figure 15.9   Reprinted, by permission, from Steve Dewey, Department of Plants, Soils and Biometeorology, Utah State University, Logan, UT

Figure 16.2   Reprinted, by permission, Elizabeth Young, The Energy Institute, The Pennsylvania State University, University Park, PA

Figure 16.3   Reprinted, by permission, from Robert Hazen, Geophysical Laboratory, Carnegie Institution of Washington, Professor of Earth Science, George Mason University, Washington, DC

Figure 16.4   Reprinted, by permission, from *A History of Mechanical Inventions*, 2nd Edition, by Abbot Payson Usher, © 1998 Dover Publications, Inc., Mineola, NY

Figure 16.5   Image originally appeared in *Elementary Treatise on Physics: Experimental and Applied*, 17th Edition, edited by A.W. Reinhold, © 1905 William Wood and Company

Figure 18.1   Reprinted, by permission, from Reliant Energy, Texas

Figure 19.2   Reprinted, by permission, from Bygone Designs, Newport, MN

Figure 19.3   Reprinted, by permission, from *Modern Steam Car and its Background*, Revised Edition, by Thomas S. Derr, © 1998 Lindsay Publications, Bradley, IL

Figure 19.4   Reprinted, by permission, from *Engineering in History* by R.S. Kirby, S. Withington, A.B. Darling, and F.G. Kilgour, © 1990 Dover Publications, Inc., Mineola, NY

Figure 19.6   Reprinted by permission, from *The Construction of a Gasolene Motor Vehicle* by C.C. Bramwell, Lindsay Publications, Bradley, IL

Figure 19.7   Reprinted, by permission, from National Air and Space Museum, Smithsonian Institution, Washington, DC

Figure 19.8   Reprinted, by permission, from National Air and Space Museum, Smithsonian Institution, Washington, DC

Figure 20.10   Reprinted, by permission, from The Energy Institute, The Pennsylvania State University, University Park, PA

Figure 20.11   Reprinted, by permission, from *Chemistry in Context* by A. Truman Schwartz, Diane M. Bunce, Robert G. Silberman, Conrad L. Stanitski, Wilmer J. Stratton, and Arden P. Zipp, © 1997 The McGraw-Hill Companies, New York, NY

Figure 22.2   Reprinted, by permission, from Bygone Designs, Newport, MN

Figure 22.3   Image originally from the General Electric Company, New York

Figure 22.4   Reprinted, by permission, from Bygone Designs, Newport, MN

Figure 22.6   Reprinted, by permission, from Bygone Designs, Newport, MN

Figure 22.7   Reprinted, by permission, from Bygone Designs, Newport, MN

Figure 22.8   Reprinted, by permission, from Bygone Designs, Newport, MN

Figure 23.1   Reprinted, by permission, from National Air and Space Museum, Smithsonian Institution, Washington, DC

Figure 23.6	Reprinted, by permission, from *The Romance of Engines* by Takashi Suzuki, © 1997 Society of Automotive Engineers, Warrendale, PA
Figure 26.6	Reprinted, by permission, from Photo Researchers, Inc., New York, NY
Figure 28.3	Reprinted, by permission, from Photo Researchers, Inc., New York, NY
Figure 29.1	Reprinted, by permission, from Photo Researchers, Inc., New York, NY
Figure 30.3	Reprinted, by permission, from National Atmospheric Deposition Program, Champaign, IL
Figure 30.5	Reprinted, by permission, from Photo Researchers, Inc., New York, NY
Figure 30.6	Reprinted, by permission, from Phototake, New York, NY
Figure 30.7	Reprinted, by permission, from *The Acid Rain Sourcebook*, edited by Thomas C. Elliot and Robert G. Schwieger, © 1994 The McGraw-Hill Companies, New York, NY
Figure 32.7	Reprinted, by permission, from Photo Researchers, Inc., New York, NY
Figure 34.5	Reprinted, by permission, from the Whole-in-One Catalog
Figure 34.7	Reprinted, by permission, from the Whole-in-One Catalog
Figure 34.8	Reprinted, by permission, from Photo Researchers, Inc., New York, NY
Figure 35.1	Reprinted, by permission, from Phototake, New York, NY
Figure 35.2	Reprinted, by permission, from Photo Researchers, Inc., New York, NY
Figure 35.3	Reprinted, by permission, from Photo Researchers, Inc., New York, NY
Figure 36.3	Reprinted, by permission, from Photo Researchers, Inc., New York, NY
Figure 36.6	Reprinted, by permission, from Photo Researchers, Inc., New York, NY

INTRODUCTION

ENERGY USE AND US

Consider the beginning of a typical day. As we get out of bed, we're probably concerned with whether we feel warm enough, with turning on lights so we can see, and with cooking something for breakfast.

All of these activities depend on *energy*. Many of us live in homes with electric heating; if not, natural gas or heating oil are other popular choices. We rely on electricity for lighting. Most of us would use electric stoves or electric microwave ovens for cooking; the rest of us probably rely on gas stoves. *Energy* is the central topic of this course. Although we are introducing the word now, we are going to defer its formal definition until Chapter 2. For now, we'll rely on an intuitive definition or understanding of energy.

Of course, cooking breakfast means that we have to have something to cook. Not many of us these days subsist entirely on what we can raise or catch ourselves. Having food available first requires planting, cultivating, and harvesting on farms. Most farmers use a variety of agricultural machinery that operates with gasoline or diesel fuel. The harvested food has to be transported to plants where it is processed or prepared and then packaged. Then the packaged food is transported to warehouses and stores for sale to us, the consumers. Transportation from farm to processing plant, and from processing plant to stores will use gasoline or diesel fuel. Various kinds of energy might be used in the processing of foods, including electricity, natural gas, and heating oil.

As we get dressed and begin to go about our daily routine, we depend on a huge variety of manufactured articles. Very, very few of us weave our own cloth, turn logs into boards to make wooden articles, or make any of the other items we use throughout the day. Manufacturing begins with the production of raw materials, such as making metals from their ores. Many kinds of energy might be used in such operations, including electricity, heating oil, or coal. Then the raw materials have to be fabricated into useful articles. Fabrication might include molding, casting, machining, or weaving (as but a few examples). Many of these operations use electrically powered machinery, and coal, heating oil, or natural gas as heat sources. The manufactured articles then have to be transported to stores.

Then on most days we must get ourselves out and around. If we walk or bicycle, we use energy from our own muscles for transportation. Cars or light trucks have gasoline engines or possibly small engines that use diesel fuel. Electric cars, or hybrid electric/gasoline cars are coming onto the market too. If we take the bus, we are probably relying on diesel fuel or natural gas. Trains may have diesel locomotives or operate using electricity. Small airplanes may use gasoline as fuel; others use jet fuel.

Our manufacturing and transportation, in addition to the many uses of energy in the home for warmth, cooking and lighting, together consume prodigious quantities of energy (Figure 1.1). The key idea that should come from thinking about how we get through the day is this:

ENERGY IS UBIQUITOUS IN OUR LIVES AND, IN FACT, IS SO COMMON THAT WE SELDOM EVEN THINK ABOUT IT.

For most of us, it's likely that we rely most often on two kinds of energy: electricity and gasoline.

We surround ourselves with electrical appliances and gadgets. Most of us own a majority of the items on this list: television set, microwave oven, stereo system, personal computer, electric razor, hair dryer, refrigerator, lamps for reading or study, coffee maker, electric clock, electric pencil sharpener, electric tooth brush, power tools, and radio. In the kitchen, for example, most people now expect to find a stove, a refrigerator, and, often, a dishwasher and a microwave oven. What else can we find in the 'modern kitchen'? A coffee maker, an espresso machine, an electric can opener, a pasta maker, a bread maker, a crock pot, an electric carving knife, a toaster or toaster oven (or

FIGURE 1.1   This photograph of the Philadelphia skyline illustrates the prodigious use of energy in our society—the consumption of electricity in the many lights of the buildings, and the use of gasoline in the cars speeding down the road on the left side of the picture.

both), and a blender or food processor. If there's a problem, it's how to find counter space to use all this stuff, or some place to store it, not whether or not there will be enough electricity to operate these gadgets.

The idea of even asking whether there would be 'enough' electricity to operate a gadget we're buying when we got it home might sound silly. Almost invariably we simply assume that we can purchase and plug in a limitless number of electrically operated items. (Admittedly, there are occasional exceptions, particularly with large electrical appliances such as clothes dryers or electric stoves, or possibly with some hobby items such as electrically operated machinery for a home workshop.) Usually, if we think of any limit at all, it's that we are limited by the number of electric outlets we have available for plugging items into. We can even solve that problem if we remember to buy some 'cube taps' (that let us plug three items into a single outlet) or outlet strips (that provide five or more electrical outlets from an original single outlet). Though it's very unwise, and perhaps even illegal in some localities, some of us probably even plug cube taps into outlet strips, or connect two or more outlet strips together.

Our assumption about the eternal availability of essentially unlimited quantities of electricity is tested when there is a power failure. When the electricity supply suddenly fails, we might have a momentary bit of panic until we assure ourselves that we're OK, but then many of us react to a power failure with a feeling of annoyance or anger. We were watching that TV show, or cooking that meal, or reading that book, feeling that we could do those sorts of things as much as we wanted, any time we wanted, and now, suddenly— no electricity.

It's sometimes helpful to remember that not everyone in the world gets to enjoy a lifestyle based on assumptions we take for granted. There are places in the world where, if electricity is available at all, it is only 'on' for a certain period each day. In some places the use of electricity has been strictly rationed. For example, during the last years of the horrible dictatorship of Nicolai Cauciescu in Romania, each apartment or home was allowed *one* 25-watt light bulb. In other places, electricity does not exist at all, for instance, in rural regions of many of the impoverished nations in Africa and Asia.

We usually have the same cavalier attitudes toward gasoline. We expect that we can drive around without ever once worrying about whether we will be able to buy gasoline whenever and wherever we need it, and as much as we want. (Probably only a few of those who recall the gasoline shortages during the oil embargoes of the 1970s, or perhaps the gasoline rationing during the Second World War, may challenge this attitude.) Even at 2 a.m. on Christmas morning we can find someplace to buy gasoline. Gasoline seems to be as easily and widely available as water.

> *Many commodities which are most useful to us are esteemed and desired but little. We cannot live without water, and yet in ordinary circumstances we set no value on it. Why is this? Simply because we usually have so much of it that its final degree of utility is reduced nearly to zero. We enjoy every day the almost infinite utility of water, but then we do not need to consume more than we have. Let the supply run short by drought, and we begin to feel the higher degrees of utility, of which we think but little at other times.*
> —Jevons[1]

But, just as with electricity, it is helpful to remember that the assumptions we make about the availability of gasoline are not valid everywhere in the world. They apply only to the so-called developed or industrialized nations. Even within that small subset of the world's nations, gasoline is cheaper in the United States than in virtually any other nation. In many industrialized nations, even those such as Japan or the western European countries with strong economies and modern industries, gasoline costs $3.50 per gallon. About the only countries in the world where gasoline is cheaper than it is in the United States are Venezuela, which has enormous domestic supplies of petroleum, and Iraq, where the national economy has been grossly distorted by the United Nations sanctions following the 1990–91 Persian Gulf War.

Another way of illustrating our dependence on energy is to consider it from the other perspective: how we would live if electricity and fuels (especially petroleum products and natural gas) were not available any more.

- What would we eat? Probably the best we could do would be foods raised by ourselves or foraged in the woods. Many people living in large cities would probably starve.
- How would we get around? We could use horses, if we had them. We could ride bicycles, until they broke or wore out. Most of us would be confined to an area that would be accessible by walking. Perhaps if we were lucky enough to live near the coast or near a navigable river, we could travel by boat—one that we paddled or rowed, or that used the energy of the wind to sail, or a steamboat that used wood as fuel.
- How would we stay warm? We could use firewood, if we had access to it and for as long as the wood lasted. A few clever persons might rig up solar energy collectors, or figure out how to use windmills or water wheels to operate electrical generators.
- What would we use in our daily lives? When our clothes, tools, and utensils broke or wore out, we would have no replacements, except for things that could be made of wood (assuming we hadn't burned it to stay warm), or wool or cotton cloth.
- How would we regulate our days? Most of us would rise at dawn and go to bed at dark, because there would be little artificial light other than fires.

In essence, the vast majority of people—especially city dwellers—would quickly freeze and starve in the dark. Of the survivors, most would be reduced to a fairly brutish existence not unlike that experienced by the poor during medieval times. A very few, those who were competent at subsistence farming and at manufacturing or repairing of small tools and machinery (the Amish, for example), might 'make it.'

ENERGY USE AND NATIONAL WELL-BEING

So far, the principal point of discussion has been one person's reliance on energy. Most of us take advantage of different kinds of energy during the day, and we expect energy to be so abundant, and so available at low cost, that we seldom even pay attention to it. Now we shift focus, from the impact of energy use on one person's life (and specifically one person living in a reasonably wealthy, developed nation) to the role of energy on a country's well-being.

Energy consumption is an absolutely necessary component of industrial society. Before the Industrial Revolution of the 18th century, most energy use relied on two sources: human and animal muscles (Chapter 3), and the energy of wind and water available in nature (Chapters 8 and 9). The chemical energy in firewood was the main source of heat and light (Chapter 6). Since then, there have been two major historical transitions in the way we use energy. The first was the steam engine (Chapter 10), for which the principal fuel was coal. The second involved a major diversification of energy use and energy sources. This transition was based on two major components. One was the development of electricity (Chapters 12–18). Electricity was a double benefit: it was the first form of energy in which the place at which it is produced could be separated by miles—indeed, hundreds of miles—from the place where it is used. It is also the only energy source that is easily converted into light, heat, or mechanical work wherever it is used. The other major component was the internal combustion engine, which introduced an enormous mobility into society (Chapters 22–25). Petroleum and its products (Chapters 20–22 and 26) gradually took over the position of being the dominant global energy source by the end of the 20th century.

At any particular time, there is a tendency among the countries of the world for a correlation between income growth and growth in energy consumption. This same relationship holds for many countries if we examine their income and energy use over a period of years. The correlations are not perfect. Different countries may have different ways of reporting economic statistics. Some factors, such as fuel supplies obtained by bartering, may not even be counted in official statistics. Over a period of years, a country may change the way it collects and uses statistics. Even though these uncertainties exist, until about 1975 there seemed to be fairly good correlation between the economic well-being of a country, as measured by its total output of goods and services (**gross domestic product**, GDP) and its energy consumption. The countries with the largest GDP tended to use the most energy, while undeveloped economies used less energy.

The ratio of energy consumption to GDP is defined as the **energy intensity** of the economy. Regardless of whether we consider the present energy situation, the history of energy use, or predictions of energy use in the future, we find that population growth, economic development, and technological progress are the major considerations affecting economic development. As a country develops over a period of time, a first approximation, the growth of GDP occurs through an increase in population—with its consequent demand for housing, transportation, consumer goods, and services—and results in an increase in energy consumption too. During that time of development, the energy intensity of that country's economy is nearly constant. The growth of GDP virtually parallels that of energy consumption.

This was true in the United States beginning in the mid- to late 1930s (as the Great Depression was coming to an end) and continuing until the early 1970s. During that time, the GDP seemed to grow at a fixed ratio to energy consumption. In Britain a quite similar relationship between energy use and GDP existed in same time period.

Since the mid-1970s, though, there has been a remarkable change in the relationship between GDP and energy use in the developed countries. The rate of energy consumption per dollar of GDP generated has been falling steadily in the major developed nations for the last quarter-century. The period of greatest change came between 1975 and 1985, when, in the United States, the energy intensity (energy used per dollar of GDP) dropped by one-fifth while the GDP grew by 30%. In recent years the energy intensity is continuing to drop by about 2% per year.

Several factors are responsible for this change, which is not unique to the United States, but is generally experienced by most of the developed nations. First, from the end of the Second World War until the early 1970s the cost of energy (referenced to one specific time, so as to adjust for inflation) actually dropped. During that period, many utilities encouraged consumers, even via cash incentives, to use as much electricity or natural gas as possible. The oil price shock in the early 1970s caused industries to develop ways of operating their various processes in ways that would reduce energy consumption (there had been no incentive to do this in an era when the cost of energy was dropping). A second major factor that 'decouples' the use of energy from growth in GDP is the transition of the developed countries to the so-called post-industrial society. This change represents a transition from an industry-based economy to a service-based one, requiring much less energy. In addition, those industries that do remain tend to be ones, like making computers, that add considerable value to the raw materials or components but consume relatively little energy; the heavy, energy-intensive industries such as steel or cement move out.

We must recognize that these comments do not apply to all the countries or all the people of the world. There are enormous differences in the levels of economic development, standards of living, and access to energy around the world. For example, the richest one-fifth of the world's population is responsible for consuming four-fifths (80%) of the world's goods and services, and in doing so use more than half the world's energy. In stark contrast, the poorest one-fifth of the population consumes only 1% of goods and services, and accounts for about 5% of energy consumption. Differences in per capita GDP from one country to another are correlated with differences in economic structure. Low per capita GDP is usually associated with countries in which agriculture has a large share of the economy. As industrialization takes place, GDP and energy use increase, approximately in parallel, with increasing incomes until high levels of per capita GDP are reached. Then there is a transition to the post-industrial economy, with a decreasing reliance on energy for generating GDP.

WHERE WE'RE GOING

Hopefully this book will help dispel concerns that gaining an understanding of 'energy' is either mysterious or difficult; and there is no need to feel overwhelmed or intimidated by politicians, salesmen, hucksters, or demagogues. What if—for example—an electric utility were to propose to construct a nuclear power plant in your town? Likely, you would be bombarded with an entire range of arguments, pro and con, ranging from the assurance that nothing can ever possibly go wrong to the assertion that if this reactor is built you and your families will be exposed to so much radiation that you'll glow in the dark. How can we figure out where the truth lies? (Perhaps it's already obvious to most readers that in this specific example the truth lies somewhere between these extremes—but where, exactly?) In fact, rather than being at their mercy, you can become very frightening to the politicians, salesmen, hucksters, and demagogues, because the most dangerous person in the world is someone equipped to think for herself or himself.

Furthermore, because our society is changing at such an incredible pace, it is vital to be able to continue to learn throughout our lives. Consider a couple of very common items that are nowadays displayed at the checkout stands of discount stores: digital wristwatches, which can be as cheap as about $2.98, and simple four-function electronic calculators that are about $10. A quarter-century ago the watches were about $200 and the calculators well over $300. Compact disks of music have replaced vinyl records in about two decades. Gasoline containing lead additives has been wiped off the market in about the same time.

It's hoped that this book will help the reader feel confident to continue reading elsewhere (e.g., other books, magazines such as *Scientific American* or *Discover*, or the science section of *The New York Times*), continue learning, and continue probing. When we visit another country, we can better appreciate its culture and customs if we can understand something of the language. Unfortunately, it seems to many people that science itself has now become another country, or another culture. The technology of energy and the issues surrounding energy use are all around us. Many energy-related terms are in common use: acid rain, greenhouse effect, cold fusion, the China syndrome, melt-down, and superconductors are a few examples. It helps to know what these terms mean. This is especially so because, as we will soon see (Chapter 2), a scientist sometimes uses a word in a slightly different way or with a more restricted meaning than when the same word is used in everyday conversation.

⚛ The concept of a society divided into two cultures, of which one is science, has been argued eloquently and most famously by the British writer and physicist C. P. Snow, in his appropriately titled book, *Two Cultures and the Scientific Revolution* (1959).

A second point is that there are limits as to what can be accomplished with energy. We can't create energy out of nothing. The best we can hope to do is utilize the amount of energy available to us and convert it from one form to another. Ideally, we would hope to be able to convert energy completely from one form to another, with no waste or losses. As we'll see, we'll never really be able to attain this ideal of a device that is 100% efficient in converting energy from one form into another. Common electric motors come pretty close, changing about 90% of the electric energy we supply to them to doing the work we want done. An electric power plant that burns coal is much worse; only about 35% of the coal's energy comes out of the plant as usable electricity. Steam locomotives, that at their zenith represented mighty examples of brawn and brute force, are terrible, with less than 10% of the energy of the coal fuel becoming available to drive these behemoths.

A third point that will be made is that there are limits to what scientists and engineers can accomplish, and what they can tell us. A current environmental issue is the problem of acid rain (to be discussed in Chapter 30). Acid rain is known to be caused by emissions of sulfur and nitrogen oxides into the air. One possible source of sulfur oxide emissions is power plants that burn fuels having high concentrations of sulfur. Scientists can develop methods for removing sulfur from fuels before they're burned, and for capturing the sulfur oxides before they can be released to the environment. Engineers can design plants for actually realizing these processes on a large scale. Economists can calculate the increase in the electricity cost that results from an electric utility having to pay a premium for low-sulfur coal, or for installing and operating equipment to capture sulfur oxides. But none of these people can tell you, or calculate for you, what your optimum choice would be between increased electric bills and increased loss of forests due to acid rain. It is up to the individual citizen (or a group of citizens voting together) to make the choice between increased electric costs and the forest destruction that might come from acid rain. Some people might be willing to accept substantially higher electric bills to increase protection of the environment. Others might adopt the 'if you've seen one tree, you've seen 'em all' attitude. The choice is not something that is subject to scientific calculation—the choice is something that each person must make individually by weigh-ing many factors.

As another example, petroleum is used both as a source of fuels (such as gasoline, diesel fuel, and home heating oil) and as a source of many of the synthetic materials that are ubiquitous in our daily lives. It's possible that petroleum may dwindle in supply sometime in the mid-21st century. If that happens, should we continue to burn scarce petroleum, or should we save it to make plastics? There is no correct answer to this question (though of course zealots on one side or the other will claim there is) and scientists cannot answer it unequivocally. This is an issue persons will have to think through carefully and then express an opinion by voting, or by attempting to persuade elected representatives to vote in some particular way.

A NOTE ABOUT THE CHAPTER NOTES

In most of the chapters of this book you will encounter three kinds of notes. One is called *supplemental information*. This is a series of notes that occur throughout the text, flagged by an icon: ⚛. The purpose of these notes is to supply some extra discussion pertaining to specific topics within the text. This supplemental information is not crucial to understanding the discussion presented in the main body of the text, but is provided to amplify some issues for the reader wishing to push a little deeper into the subject. The second kind is *citations*. These are the sources from which very specific information in the text has been obtained; they are listed at the end of the chapter, and are flagged in text by superscript numbers. The third is *for further reading*, listed below the citations. This is a bibliography of books and magazine articles that the reader can use to pursue a particular topic further and in greater depth. For those wishing to learn more, a bibliography of useful sources is presented at the back of the book.

CITATION

1   Jevons, William Stanley. *The Theory of Political Economy*. Penguin Classics: New York, 1970; Chapter III. Also cited in Heilbroner, Robert. *Teachings from the Worldly Philosophers*. Norton: New York, 1996; Chapter V.

FOR FURTHER READING

Cassedy, Edward S.; Grossman, Peter Z. *Introduction to Energy*. Cambridge University: Cambridge, 1998. Chapter 4, on the demand for energy, provides a thorough economic discussion of the relationship, or lack of one, between energy consumption and gross domestic product.

Goldemberg, José. *Energy, Environment, and Development*. Earthscan: London, 1996. This book explores the three-way relationship among energy, economics, and the environment, with much emphasis on the impacts on the less-developed nations. Chapter 3 discusses relationships between energy use and economic development.

Hill, Robert; O'Keefe, Phil; Snape, Colin. *The Future of Energy Use*. Earthscan: London, 1995. This book covers both technical and economic issues of energy policy, including economic costs of different forms of energy and a look to the future. Chapter 1 includes a discussion of relationships between energy use and gross domestic product.

Nakic'enovic', Nebojsa; Grübler, Arnulf; McDonald, Alan. *Global Energy Perspectives*. Cambridge University: Cambridge, 1998. This book examines the balance, and trade-offs, among fossil fuels, alternative fuels, efficiency increases, and energy conservation in meeting increasing consumer demands for clean, convenient energy. Chapter 3 discusses issues of energy intensity, and gross domestic product.

2

ENERGY, WORK, AND POWER

The last chapter introduced some examples of how dependent we are on energy. That discussion assumed that everyone has a rough idea of what is meant by the word 'energy.' As we progress, though, we will find that we need to be more careful of the way in which we use certain terms. Many words used casually in everyday conversation have more restricted or specialized definitions when used in a scientific context. This is very important to keep in mind when reading scientific material or when having discussions with scientists. For effective communication, it's crucial to make sure that everyone is using words in the same way.

> How hard one must work in order to acquire his language—words by which to express himself! ... With the knowledge of the name comes a distincter recognition and knowledge of the thing. ... My knowledge was cramped and confined before, and grew rusty because not used—for it could not be used. My knowledge now becomes communicable and grows by communication. I can now learn what others know about the same thing.
> —Thoreau[1]

That's especially the case in today's world, because scientists use a variety of jargon and acronyms for efficient communication with colleagues in the same scientific discipline. In fact, scientists trained in one discipline often have difficulty reading professional papers concerning a different discipline unless they first learn the 'lingo.' Though this understanding of the meaning of terms is crucial for reading and learning about science and technology in our modern world, the importance of meaning in this regard has been recognized at least since the time of Aristotle. ⚛ He argued, some 2,400 years ago, that scientific knowledge cannot be communicated if the terms that are being used are not used in their correct senses.

⚛ Aristotle, who died in 322 B.C., was a scientist and philosopher; he was one of the first persons (and indeed one of the few in all of history) of whom it might reasonably be said that he had a command of the entire range of human knowledge. Indeed, some would argue that Aristotle, more than any other thinker, is responsible for providing the form and the original content of Western civilization. His writings covered medicine, law, art, biology, metaphysics, rhetoric, natural history, and many other subjects. Various anthologies of his basic works are available, including, as examples, *The Pocket Aristotle* (The Pocket Library, 1958), *Aristotle on Man in the Universe* (Gramercy, 1971), and *Aristotle* (Easton Press, 1995).

THE RELATIONSHIP OF WORK, ENERGY, AND POWER

Three terms will be of special importance to this book: energy, work, and power. All three are quite commonly used in informal, everyday conversation and writing, often to mean different things. Consider these examples:

- I should mow the lawn today, but I just don't have the energy.
- If you eat this candy bar, it will give you some quick energy.
- That band's performance was just packed with energy.
- That test was a lot of work!
- I have to go to work today at 4 o'clock.
- That outfielder can really hit a ball with a lot of power.
- The Corvette is a really powerful car.

There are plenty of other examples. The definitions we will use throughout the remainder of this book are more restricted and have their basis in science. We begin with the scientific definition of work (that is, by the meaning scientists use to define the word):

WORK IS CAUSING AN OBJECT TO MOVE INTO, OR OUT OF, SOME POSITION, ESPECIALLY WHEN IT MOVES AGAINST A RESISTANCE.

In places where it's important to emphasize that a common word is being used in its more restricted definition, we'll write it with capitals, as WORK. Eventually, it will be possible to infer from the context whether we're using a word in its restricted sense or in its more common, informal usage.

Some examples of WORK include:

- Carrying boxes of belongings or furniture up stairs, against the resistance of gravity.
- Driving down the road, against the resistances of friction caused by tires moving against the road surface and the mechanical parts of the car against each other.
- Converting sheets of metal or plastic into useful objects, against the resistance of the natural stiffness or strength of the material.
- Sending electricity from the generating station to our homes, against the electrical resistance of the wires.

In fact, just about everything we try to accomplish in our daily lives—manufacturing items, transporting ourselves or our goods, using electricity—represents doing WORK. The first real concept of WORK as we now use the term derives from a German philosopher, Gottfried Leibniz.⊛ A general understanding of WORK dates from the early decades of the 19th century. To accomplish the things we do, and obtain the things we use, requires doing WORK. In order to do this WORK, we need ENERGY.

ENERGY IS THE CAPACITY FOR DOING WORK.

⚛ Gottfried Leibniz, born in Leipzig, Germany in 1646, was the sort of universal genius whose life work influenced many fields, including, in his case, mathematics, logic, geology, linguistics, and several others. He is perhaps best known nowadays for his invention of the calculus, made apparently independently of the almost simultaneous development of the subject by Isaac Newton. A controversy raged for centuries on the issue of which of these two great men had priority of invention and whose system of the calculus was better. Leibniz died in 1716.

ENERGY, therefore, is the key to accomplishing, or having the capacity for doing, all of the activities that form the basis of our modern industrial society. The word 'energy' is only about two centuries old. Thomas Young proposed it in the early nineteenth century.⚛

'Then idiots talk,' said Eugene, leaning back, folding his arms, smoking with his eyes shut, and speaking slightly through his nose, 'of Energy. If there is a word in the dictionary under any letter from A to Z that I abominate, it is energy. It is such a conventional superstition, such parrot gabble!...'

—Dickens[2]

⚛ Like Leibniz, Thomas Young also contributed significantly to many different fields. He was born in 1773, could read by age two, and is said to have read the Bible completely through twice by age four. His contributions include an early decipherment of the Rosetta Stone, which provided valuable information on how to understand Egyptian hieroglyphics, and contributions to the theory of color that are used today in both color television and color photography. His name survives in Young's modulus, which describes the elasticity of a material, its ability to resist changing its length when subjected to being compressed or stretched.

In several of the following chapters we'll also see that it's helpful to think of some concepts related to energy by the using the analogies related to money. It's obvious—sometimes painfully so—that you can't *spend* unless you have *money*. Put in a different way, *money* is the capacity or ability to *spend*. *Money* and *spending* have the same relationship as ENERGY and doing WORK. Why do we do WORK? It's to get something done, to accomplish something, to make something happen. We need ENERGY to do this. Why do we *spend*? It's to acquire things, to obtain some service, to get what we need (or want) for our daily lives. We need *money* to do this.

Let's push the money–energy analogy a bit further. What happens when we buy something? We transfer money from ourselves, or our bank accounts, to the seller. The act of *spending* involves a transfer of *money*. In exactly the same way, the act of doing WORK involves a transfer of ENERGY. As we explore different applications of ENERGY in the coming chapters, we will also occasionally stop to examine what kinds of ENERGY are being transferred, and how the transfer takes place.

There's another important aspect of what happens when we buy something. The total amount of money does not change. True, the amount of *your* money drops, and the amount of money that the seller has goes up. However, the total amount of money in circulation in the country doesn't change. In other words, *spending* involves a transfer of *money* but not a change in the total money supply. Similarly, doing WORK involves a transfer of ENERGY, but doesn't change the total amount of ENERGY available in the world.⚛ The concept of POWER addresses the question of how quickly we can do

FIGURE 2.1   A Corvette is an example of a powerful car, in both the everyday and 'scientific' senses of the word. Because it can do a great deal of WORK in a short time, it indeed has significant POWER.

WORK. A powerful car (Figure 2.1), for example, is one that can do a great deal of WORK and, usually, consumes a great deal of ENERGY, in a short time.

POWER IS THE TIME RATE OF DOING WORK.

⚛ We have to be careful not to push this analogy too far. It is true, and indeed is one of the most fundamental laws of science, that the amount of energy in the universe is constant. Regardless of what we might want to do, we can neither create new energy, nor destroy the energy that now exists. All we can do is move it around into different forms. However, governments do have the ability to increase or to decrease the amount of money in circulation as part of their financial policies. Whether doing so is a good idea or not is an issue of fierce debate among economists, and would get us way too far afield from the present discussion.

We can express this relation mathematically as

$$\text{Power} = \text{Work/Time}$$

When we say that a particular item (for example, a car) is more powerful, that is, has more POWER, than another, what we mean is that either it can do more WORK in a given amount of time, or it takes less time to do the same amount of WORK.

To recapitulate, WORK is required for accomplishing virtually all aspects of our industrial society (manufacturing, transportation, and so on), ENERGY gives us the ability to do that WORK, and POWER tells us how quickly (or slowly) we can get it done. *Anything* that we attempt to accomplish in our daily lives—doing WORK with our own bodies, transporting ourselves or the goods that we consume, manufacturing items, or distributing electricity, involves doing WORK. *Everything* that we want to do, or want to have to have done for us, in our daily lives involves doing WORK. But, we need ENERGY to do that WORK. The reason that ENERGY is important therefore springs directly from its definition—it is what we need to do WORK. POWER tells us how fast, or how slowly, we can accomplish a given amount of WORK (or, from the other perspective, how much WORK we can do in a given amount of time). To sum up, doing WORK lies at the heart of every important activity that provides the goods and services we enjoy; ENERGY is what is required to be able to do WORK; and POWER indicates how fast we can do it.

HOW WORK GETS DONE

What does the doing of WORK actually involve? We've described the concept *qualitatively* as moving an object against some resistance, of which gravity is a familiar example. The concept of WORK is usually defined *quantitatively* as the application of a force of some sort through a distance. That is,

$$\text{Work} = \text{Force} \times \text{Distance}$$

There are two implications in this relationship. First, if there isn't a force being exerted, or if there isn't a distance being traversed, then WORK isn't being done. Imagine helping a friend move into a new apartment. You're handed a heavy box of books, or a heavy suitcase with the instruction, 'Here, hold this,' while your friend fumbles around to find the door key. You may be holding that box or suitcase till it feels like your arms are ready to pop out of their sockets, but if you're not moving it, you're not doing WORK. All of us have had the experience of being faced with a very difficult examination question, and having to think as hard as we can for some time to come up with an answer. Sitting still in a classroom seat, exerting no force, nor moving anything through a distance, is not WORK. Of course it would be fully justified in both instances to exclaim, 'That was a lot of work!' in our everyday connotation of the word. But, in our restricted definition of the term, it really wasn't any WORK at all.⚛

⚛ **This point refers to the human body as an entity, and in this example the body is not doing WORK. At a different organizational level—that is, in the individual organs of the body—WORK is being done.**

As for the second implication, let's consider a force of 500 pounds that has to be lifted 2 feet. We can immediately figure out the amount of WORK involved:

$$\text{WORK} = \text{Force} \times \text{Distance}$$
$$= 500\,\text{lb} \times 2\,\text{ft} = 1000\,\text{ft-lb}$$

However, many combinations of numbers can be multiplied to give a product of 1000 foot-pounds. Consider a force of 50 pounds moved 20 feet:

$$\text{WORK} = \text{Force} \times \text{Distance}$$
$$= 50\,\text{lb} \times 20\,\text{ft} = 1000\,\text{ft-lb}$$

Why should we care about this mathematical quirk, that $50 \times 20 = 500 \times 2$? Let's generalize: a large force moved through a small distance can be equivalent in WORK to a smaller force moved through a larger distance. Suppose you had to move a 500-pound object—a big boulder, a heavy piece of furniture, a safe. Very, very few of us are capable of lifting and carrying an object that weighs 500 pounds. Practically all of us, though, can exert a force of 50 pounds. The secret of moving so heavy an object through a short distance is to apply a much smaller force, but through a bigger distance. How do we do that? In this instance, with a lever.

The simple tools that our primitive ancestors developed and that we still use—the lever, the wedge or inclined plane, the pulley—derive from the concept that we can do an enormous amount of WORK simply by applying a small force through a relatively large distance. Almost certainly the unknown individuals who figured out how to use sturdy sticks to help move heavy rocks or logs had no scientific concept of WORK, but likely an intuitive understanding quickly developed. This was of enormous importance to those early ancestors when the WORK to be done far exceeded the ability of relatively puny human muscles.

Many of us don't even think of activities like climbing or lifting as WORK. Many others, though, still do a large amount of WORK every day, especially the manual laborers who work on farms, on construction jobs, or in factories. In our modern industrialized society the proportion of people who do a great deal of WORK is steadily decreasing. However, for the overwhelming portion of human history, the only way that humankind had of accomplishing WORK was to rely on the muscles of the human body.

A BRIEF HISTORY OF THE HUMAN USE OF ENERGY

In the chapters that follow, we will develop the concept of energy in a historical context, progressing from the kinds of ENERGY available to the earliest humans up through the most modern developments, and even looking a bit into the future. To set the stage for that development, we will consider briefly a 'time line' of human development and energy use.

Hominids, the family of primates of which we are a member, are about 5,000,000 years old. Modern humans (*Homo sapiens*), people that would be essentially similar to us in appearance, thought patterns, behavior, and genetics, appeared about 240,000 years ago. Even if we restrict our attention only to so-called modern humans, we still have a long time period to consider.

Archeological evidence suggests that hominids have been making use of fire for 500,000 years (which is only one-tenth the time that hominids have been in existence). Some evidence suggests that hominids may have been using fire for much longer, up to 1,400,000 years. The earliest uses of fire required that the fire be started naturally, such as by lightning, and then kept going. If the fire went out, there was no way of restarting it. This fact may be responsible for various religious rituals and folk rites concerning guarding the fire.

If we take a more restricted view of the use of fire—the ability to start a fire at will and to control the fire as desired—that skill is much more recent and may have occurred only with Neolithic humans around 9,000 B.P. (before the present).

Some few animals may have been domesticated for probably as long as there have been hominids. Domestic animals may originally have been kept as pets; archeological evidence suggests that domestication of animals may have occurred 9,000 to 11,000 years

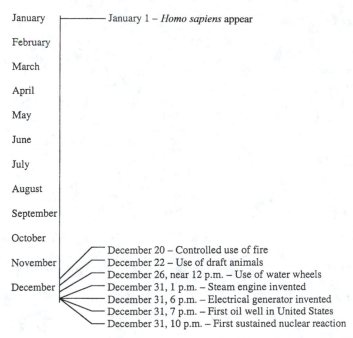

January   January 1 – *Homo sapiens* appear

February

March

April

May

June

July

August

September

October

November

December

December 20 – Controlled use of fire
December 22 – Use of draft animals
December 26, near 12 p.m. – Use of water wheels
December 31, 1 p.m. – Steam engine invented
December 31, 6 p.m. – Electrical generator invented
December 31, 7 p.m. – First oil well in United States
December 31, 10 p.m. – First sustained nuclear reaction

FIGURE 2.2 If all of human history were compressed into a single year, the use of ENERGY sources by humans would not have begun until December 20.

ago. However, the use of animals to do WORK for us (so-called draft animals) may have occurred about 7,000 B.P. with reindeer that were used in northern Europe to haul sledges. If we accept the origin of *Homo sapiens* to date from 240,000 B.P., then for the first 233,000 years of human existence we relied on our own muscles to do WORK. We then figured out how to have animals help us, but they too rely on the action of their own muscles.

Waterwheels used to do WORK, such as grinding grain, first came into use about 3,000 B.P. In other words, it is only in the past 3,000 years—a scant 1/80th of human existence—that we have had any source of ENERGY other than human or animal muscles.

The first practical windmill in western nations that can be dated with certainty was put into operation in 1185. Steam engines for doing practical WORK were developed about 300 years ago. Crude engines, for pumping water from coal mines, were built in 1698; an improved version was developed in 1721. The first clear demonstration of the sustained generation of electricity happened in 1831. The first oil well in the United States was drilled in 1859. The first sustained nuclear chain reaction was achieved in 1942.

To put these dates into perspective, we can imagine what it would be like if the whole 240,000 years of *Homo sapiens* were condensed into a single year (Figure 2.2).

This shows that we humans have had to rely on our own muscles in order to do WORK for almost the entire span of human history—all but nine 'days' of the human 'year.' All but about four and a half 'days' of the human year relied either on human or animal muscles.

Another way of interpreting this compression of human use of ENERGY into a single 'year' considers that for 97% of the entire time that humans have existed, we have relied on our own muscles for doing WORK. In the next chapter we will examine the source of the ENERGY that gives our bodies the ability to do WORK.

CITATIONS

1   Thoreau, Henry David. *Journals*. Volume XI, 1858; p. 137. Also conveniently found in Walls, Laura Dassow (Ed.) *Material Faith: Henry David Thoreau on Science*. Houghton Mifflin: Boston, 1999; pp. 82–83.

2   Dickens, Charles. *Our Mutual Friend*. Penguin Classics: New York, 1997; pp. 29–30. Many other editions of this book are also readily available.

FOR FURTHER READING

Asimov, Isaac. *The History of Physics*. Walker: New York, 1966. A history of the subject from the earliest times to the early 1960s, written for the layperson. Chapter 7 discusses the concepts of work and energy.

Aubrecht, Gordon J. *Energy*. Prentice-Hall: Englewood Cliffs, NJ, 1995. A very useful introductory text on the subject of energy. Chapter 3 deals with the topics of this present chapter.

Berry, R. Stephen. *Understanding Energy*. World Scientific: Singapore, 1991. Chapter 1 of this book provides an introduction to the concepts of force and work.

Highsmith, Phillip. *Physics, Energy, and Our World*. Saunders: Philadelphia, 1975. This elementary physics text requires very little mathematical background. Chapter 6 covers the topics of energy, work, and power.

Purrington, R.D. *Physics in the Nineteenth Century*. Rutgers: New Brunswick, 1997. As the title implies, this book provides a history of the development of physics in the 19th century. It will be quite interesting to anyone curious about the topic, but assumes considerably more scientific background from the reader than do the other books listed here. Chapter 5 covers topics relating to energy and work.

3

HUMAN ENERGY

We saw in the last chapter that if we could compress the 240,000 years of *Homo sapiens* into a single year, then for the first 11 months and 22 days of that year, all the way up to December 22, we relied on our own muscles to do WORK. We also saw, in the definitions we established, that in order to do WORK, we have to have the capacity or ability to do WORK—in other words, we need ENERGY. Where does that ENERGY come from?

Even prehistoric peoples probably appreciated that there is some correlation between the intake of food and ENERGY. In times of famine or of deliberate starvation (as, e.g., sometimes happens to prisoners of war) or of extended fasting (e.g., for religious purposes) human ENERGY decreases and can reach a point at which a person is no longer able to fend for, or even care for, himself. The world's history is full of dreadful tales of human suffering and death during famine or deliberate starvation: the Irish potato famine of the 1840s, Stalin's 'liquidation of the kulaks' in the Soviet Union in the 1930s, the starvation of concentration camp inmates during the Holocaust of World War II, extensive famine in China as a result of Mao Tse Tung's 'great leap forward' in the 1950s, and, in recent years, the inability of North Korea to provide adequate nourishment for its citizens.

Our appreciation for the processes by which food is actually utilized in our bodies is probably only about 200 years old. However, it has been recognized for a long time that food taken into the body has three roles: some of it provides ENERGY, some is used structurally for making bones and tissues or is stored as fat, and some is excreted as waste material. As scientists began to learn how to do accurate chemical analyses, and to understand the physiology of the body, three key observations were made:

- The major components of foods are carbon, hydrogen, and oxygen.
- Many foods contain small amounts of nitrogen and sulfur, and various other chemical elements in still smaller quantities.

■ We inhale air, which is essentially a mixture of nitrogen and oxygen. When we exhale, we find that the amount of oxygen in the air is depleted, and our breath contains measurable amounts of carbon dioxide and water vapor.

At about the same time in history, other scientists were studying what happens when common fuels, such as wood or coal, are burned. They too made three key observations

■ The major components of fuels are carbon, hydrogen, and oxygen.
■ Some fuels contain small amounts of nitrogen and sulfur, and various other elements in still smaller quantities.
■ A fuel needs air to burn. The air surrounding the fire becomes depleted in oxygen, and contains carbon dioxide and water vapor.

Clearly, there are some significant similarities between foods and fuels!

We recognize intuitively that ENERGY of some sort is liberated in a fire, simply because we can feel the heat. (Actually, establishing this relationship, that heat is a form of energy, is a very crucial point in our understanding of energy. For now, we are glossing over this relationship, but we will investigate it in some detail in later chapters.) In a very simplistic way, we can say that

$$Fuel + O_2 \rightarrow CO_2 + H_2O + Energy$$

and that

$$Food + O_2 \rightarrow CO_2 + H_2O$$

Since there are so many similarities between foods and fuels, in terms of chemical composition and the chemical processes that seem to occur when each is consumed, it seems reasonable to assume, by analogy, that the consumption of food also liberates ENERGY. In other words, we may amend our statement for food by writing

$$Food + O_2 \rightarrow CO_2 + H_2O + Energy$$

An abundance of observations indicate that there is an analogy between burning a fuel to get ENERGY from a fire and converting food to ENERGY in our bodies. (We should understand, though, that the analogy is by no means exact. We will revisit this point later in this chapter to examine why.) In addition, we observe from personal experience that different foods provide different amounts of ENERGY, an example being eating candy or other sugar-rich food to obtain a quick burst of ENERGY. For these reasons, it would seem that we should be able to rate, or measure, the amount of ENERGY that a food can provide in the same way that we can compare the amounts of ENERGY in different kinds of fuels.

The experimental apparatus that allows us to determine that amount of ENERGY in foods or fuels is a **calorimeter** (Figures 3.1 and 3.2). A sample of food or fuel is weighed very accurately and placed in the sample container with air or pure oxygen. The container holding the food or fuel sample is surrounded by an accurately weighed amount of water. The sample is then ignited—for example, by using an electric spark. The heat liberated from the burning food or fuel passes into the water. Thanks to the thick insulation around the calorimeter, the heat cannot escape from the water, and consequently, the temperature of the water rises, and is measured by a thermometer (or other temperature-measuring device). The amount of heat that will cause an increase in temperature in a known quantity of water is known very precisely (it's called the

FIGURE 3.1 A laboratory calorimeter, used for measuring the heat released when food or fuel is consumed.

specific heat⊛ of water). Having measured the weight of water before starting the experiment and having observed the temperature rise during the experiment, we can then figure out the amount of heat (ENERGY) that was produced from our sample of food or fuel.

⊛ Specific heat is a characteristic physical property of *all* substances. Water has one of the highest known specific heats, higher than virtually all other common substances. Since the specific heat of water is numerically 1 in either the British system or the metric system, the actual values of specific heats can be used to determine how much heat is needed to raise the temperature of a given amount of some material relative to the same amount of water. As an example, many kinds of steel have specific heats of about 0.1. It takes only one-tenth (i.e., 0.1/1) as much heat to raise the temperature of steel by some desired amount as would be required to obtain the same temperature rise of the same amount of water.

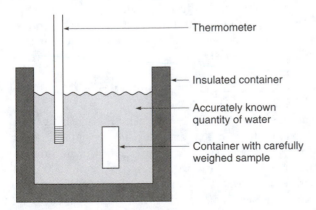

Thermometer

Insulated container

Accurately known quantity of water

Container with carefully weighed sample

FIGURE 3.2 The essential components of the calorimeter.

In the United States, along with one or two undeveloped countries, we still cling to the British system of units (even the British have abandoned the system). In this system, the ENERGY is expressed in **British thermal units**, which are commonly known simply as Btus. One Btu is the amount of heat that will raise the temperature of one pound of water one degree Fahrenheit.⚛

> ⚛ Strictly speaking, the Btu is defined in this way only for water at its point of highest density, which is 39.1°F. However, the variation with temperature is so slight that for our purposes we can consider that the definition holds for liquid water at any temperature.

The rest of the world uses the metric system. A convenient measurement of energy in the metric system is the **calorie**. One calorie is the amount of heat that will raise the temperature of one gram of water one degree Celsius.⚛ Unfortunately, our discussion is complicated by the fact that nutritionists have defined their own 'calorie.' It is called the **nutritional calorie**, or sometimes the 'large calorie.' It should be—but usually isn't— written with an upper case C: Calorie. One nutritional calorie (Calorie) is the amount of heat that will raise the temperature of 1000 grams of water one degree Celsius. Thus we have the relationships that

1 nutritional calorie (Calorie) = 1000 calories = 1 kilocalorie

> ⚛ Strictly speaking again, the calorie is defined for water at a temperature of 15°C. But, like the Btu, the variation with water temperature is so slight that we can consider that the definition holds for liquid water at any temperature.

An additional important point must be recognized. The amount of energy available to our bodies will depend not only on the kind of food we eat, but also on how much we eat. We will certainly derive more energy from eating an apple than, for example, allowing a single tiny grain of sugar to dissolve on our tongue. When we make a comparison of the amounts of energy available from different foods (or from different fuels) we need to make the comparison on the basis of the same quantity. We should make comparisons on the basis of British thermal units per pound (Btu/lb) or calories per gram (cal/g). In nutrition, unfortunately, there are all sorts of zany units, such as Calories per ounce (mixing metric and British units) or Calories per cup (mixing the metric system with arbitrary cooking units). The important thing to remember is that whatever the 'per' refers to—cups, ounces, pints, or whatever— must always be the same for the comparison to be fair.

Some examples of the energy contents of different foods are shown in Table 3.1. In the next chapter we will see how to use these data.

The Calorie intake for a dietary ENERGY allowance varies with the age and gender of the person, the type of WORK he or she is doing, and the climate. It takes about 1000 Calories per day simply to sustain life. An adult in a temperate climate doing a typical range of daily activities (but not heavy labor) would require about 2000–2500 Calories per day. (This is equivalent to a 100-watt light bulb being on continuously.) A worker who spends his or her day in heavy manual labor will require about 4000 Calories per day.

TABLE 3.1 THE ENERGY CONTENTS OF SOME COMMON FOODS, IN UNITS OF CALORIES PER OUNCE

Almonds	141	Lamb chop	89
Apples	9	Leeks	6
Apricots, raw	6	Lentils	26
Avocados	55	Liver	63
Bacon	77	Lobster	30
Bananas	21	Mackerel	47
Beans, white	29	Margarine	182
Beans, green	6	Melon, honeydew	5
Beef	54	Milk, skim	9
Beets	11	Milk, whole	16
Blackberries, raw	7	Mushrooms	3
Brazil nuts	154	Mussels	22
Bread, white	58	Nectarines	16
Bread, wheat	54	Oatmeal	100
Broccoli	7	Oil, vegetable	225
Brussels sprouts	4	Onions	9
Butter	185	Oranges	12
Cabbage	6	Parsnip	12
Carrots	5	Pasta	91
Cauliflower	5	Peaches	9
Celery	9	Peanuts	142
Cheese, cheddar	104	Pears	15
Cheese, cottage	24	Pepper, red	5
Cheese, Edam	79	Pineapple	12
Cherries	17	Pork chop, broiled	82
Chicken	35	Potato chips	129
Chocolate	127	Potatoes, baked	21
Cod	24	Potatoes, boiled	19
Cookies	128	Prunes	34
Corn	23	Raisins	62
Crab	32	Raspberries	6
Cream, heavy	112	Rice	32
Cucumber	4	Rutabaga	4
Dates	53	Salmon	49
Eggplant	3	Spinach	6
Egg, boiled	41	Strawberries	9
Figs	53	Sugar	99
Graham cracker	99	Tomatoes	3
Grapefruit	10	Tuna, canned in water	39
Grapes	17	Turkey	35
Haddock	24	Walnuts	131
Ham	42	Yogurt	15
Honey	72	Zucchini	3
Jam	65		

ENERGY IN FOODS AND ENERGY IN FUELS

So far we've implied that we can use a calorimeter to measure the energy released from foods or from fuels, and we've talked about foods or fuels as being the sources of, say, a certain number of Btus or of Calories. However, in this discussion we have neglected the essential question of what the energy difference is between a food and a fuel. Actually, we have been 'neglectful' for one very simple reason:

THERE IS NO DIFFERENCE IN THE ENERGY RELEASED FROM A
SUBSTANCE WHETHER IT IS USED AS A FOOD OR BURNED AS A FUEL.

As two examples: Suppose we could be so extravagant that we could burn sugar in our home furnaces or in automobile engines. We would find that *exactly* the same amount of energy is released when we burn sugar as a fuel as when we consume the same amount of it in our bodies as food. Granted, the specific chemical pathways by which sugar is converted to carbon dioxide and water in our bodies or is converted to carbon dioxide and water in a furnace are quite different (remember, we don't set ourselves on fire by eating), but the energy output is exactly the same. Next, consider the hypothetical case that our bodies would let us metabolize coal (or, put crudely, that we could eat coal). If that were actually true, the energy that would be released to our bodies by eating some amount of coal would be *exactly* the same as we would get by burning that amount of coal in some combustion device. As an example, a standard slice of cheese pizza contains 145 Calories (or 145 kilocalories). The energy in 24 slices of pizza is equivalent to that in one pound of coal.

We consume food: the food goes through that strange set of vessels and organs within us, and is brought into various parts of the system, into the digestive parts especially; and alternately the portion which is so changed is carried through our lungs by one set of vessels, while the air that we inhale and exhale is drawn into and thrown out of the lungs by another set of vessels, so that the air and the food come close together, separated only by an exceedingly thin surface: the air can thus act upon the blood by this process, producing precisely the same results in kind as we have seen in the case of the candle. The candle combines with parts of the air, forming carbonic acid,⚛ and evolves heat; so in the lungs there is this curious, wonderful change taking place. The air entering, combines with the carbon (not carbon in a free state, but, as in this case, placed ready for action at the moment), and makes carbonic acid, and is so thrown out into the atmosphere, and thus this singular result takes place; we may thus look upon the food as fuel . . . the oxygen and hydrogen are in exactly the proportions which form water, so that a sugar may be said to be compounded of 72 parts of carbon and 99 parts of water; and it is the carbon in the sugar that combines with the oxygen carried in by the air in the process of respiration, so making us like candles; producing these actions, warmth, and far more wonderful results besides, for the sustenance of the system . . .

. . . A candle will burn some four, five, six, or seven hours. What then must be the daily amount of carbon going up into the air in the way of carbonic acid? What a quantity of carbon must go from each of us in respiration! . . . A man in twenty-four hours converts as much as seven ounces of carbon into carbonic acid; a milch cow will convert seventy ounces, and a horse seventy-nine ounces, solely by that act of respiration. That is, the horse in twenty-four hours burns seventy-nine ounces of charcoal, or carbon, in his organs of respiration to supply his natural warmth in that time. All the warm-blooded animals get their warmth in this way, by the conversion of carbon, not in a free state, but in a state of combustion. And what an extraordinary notion this gives us of the alterations going on in our atmosphere. As much as 5,000,000 pounds, or 548 tons, of carbonic acid is formed by respiration in London alone in twenty-four hours. And where does all this go? Up into the air. . . . As charcoal burns it becomes a vapour and passes off into the atmosphere, which is the great vehicle, the great carrier for conveying it away to other places. Then what becomes of it? Wonderful is it to find that the change produced by respiration, which seems so injurious to us (for we cannot breathe air

twice over), is the very life and support of plants and vegetables that grow upon the surface of the earth.

—Faraday[1]

⚛ Faraday is speaking of carbon dioxide. A solution of carbon dioxide in water is nowadays sometimes referred to as carbonic acid.

The mythical average person, who needs about 2500 Calories per day, could obtain this ENERGY—in principle—by eating about 12 ounces of good quality coal.

WHY CAN'T WE EAT COAL?

The amount of energy released when, say, sugar is eaten is the same as it would be if we burned it in the fireplace at home. The amount of energy released when we burn coal in the fireplace is the same as would be released if coal were used as food. So, why can't we eat coal? Or why can't we drink oil?

He gave out an oily-alcoholic odour; it was said of him that he drank petroleum.

—Mann[2]

The answer derives from the way that we digest food.

There's an old saying, 'you are what you eat.' This may be true in a certain sense, in that the kinds and amounts of food we eat affect our weight, stamina, and health. But on the other hand, we are not made of pieces of beef or broccoli or biscuits, and milk does not course through our veins.

You may be an undigested bit of beef, a blot of mustard, a crumb of cheese, a fragment of an underdone potato. There's more of gravy than of grave about you, whatever you are!

—Dickens[3]

Something happens to the food that we eat to transform it into the components of our bodies. The chemical components of foods include proteins, carbohydrates, and fats. Though these represent unique kinds of chemical families, they all undergo one particular kind of chemical reaction after they are eaten. In each case, the first step in the digestion process involves the specific proteins, carbohydrates, or fats reacting with water. These molecules are broken down into smaller components by this process. The breaking apart of molecules by reaction with water is called **hydrolysis**. The hydrolysis of proteins produces amino acids; hydrolysis of carbohydrates, sugars; and hydrolysis of fats, fatty acids. Many subsequent chemical reactions inside our bodies then convert these amino acids, sugars, and fatty acids into the new chemical components of our cells or tissues. Some of these simple compounds may be used immediately by our bodies as a source of ENERGY. Others are stored as a reserve supply of ENERGY for future needs.

Enzymes facilitate the chemical processes inside our bodies. The family of enzymes represents a collection of biological catalysts, substances that speed up chemical processes inside the body, and also allow those processes to operate at lower temperatures than might otherwise be needed if no catalyst were present. A subfamily of enzymes is responsible for the hydrolysis reactions of fats, proteins, and carbohydrates. An important characteristic of enzymes is that they interact only with very specific chemical structures. The way a key fits into and opens a lock is a common analogy for the way that enzymes facilitate chemical reactions. To unlock a door, a very specific pattern of notches or grooves on the key must fit exactly into the corresponding parts of the mechanism of the lock. Otherwise, the lock can't be opened. In the same way, an enzyme-catalyzed reaction requires that the reacting molecule—a protein, for example—fit exactly into the geometric structure of the enzyme molecule. If it does not, the molecule won't react. Because the enzyme-catalyzed reactions depend so very highly on specific aspects of molecular geometry, many enzymes can catalyze the hydrolysis of only one particular linkage in the food molecule. Thus a host of different enzymes might be required to digest a single protein molecule.

Starch is a common carbohydrate (Figure 3.3). A molecule of starch consists of many subunits of simple sugar molecules connected to each other via chemical bonds to oxygen atoms. The linkage sugar–oxygen–sugar has a very specific geometrical arrangement. Without worrying about details, let's call this a linkage type *a*. Cellulose is another carbohydrate. It too consists of many subunits of sugar molecules connected via oxygen linkages. However, the sugar–oxygen–sugar linkage in cellulose has a different geometrical arrangement than the one in starch. Let's call this a *b* linkage. Our bodies have enzymes that assist in the hydrolysis of the *a* linkage, but we lack the kind that will hydrolyze *b* linkages. Thus we can eat and digest starches, but not substances in which cellulose is a major constituent, such as wood or grass.

To get back to our question of why can't we eat coal or drink oil, even though coal and petroleum have elemental compositions similar to some foods and even though foods liberate the same energy when eaten or burned, we can't eat coal because we don't have the enzymes that will allow us to digest coal.⚛⚛ For the same reason, we can't drink oil.

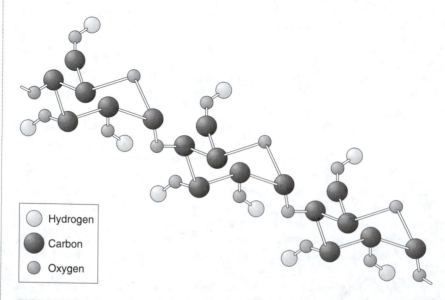

Hydrogen

Carbon

Oxygen

FIGURE 3.3   A portion of the molecular structure of starch. In this diagram, the circles represent atoms. For ease of illustration, some of the hydrogen atoms have been left out.

⚛ There is now a research effort underway in many nations to develop bacteria that will have enzymes to allow the partial digestion of coal or petroleum. In particular, there is great interest in having bacterial enzymes attack carbon–sulfur chemical bonds in the fuel, as a way of removing sulfur before the fuel is burned.

⚛ Unfortunately, not all humans necessarily possess a complete set of enzymes needed for digestion of the many things we normally eat. As an example, the sugar that occurs in milk, lactose, is split apart in our bodies by an enzyme called lactase. However, many people—especially including many people of African, Asian, or Jewish ancestry—do not have lactase and are unable to digest milk or products made from milk. This situation, which is called lactose intolerance, can lead to various medical problems, including diarrhea and stomach cramping. A solution that works for some folks is to take a lactase supplement as pills, or added to foods.

WHY DON'T WE CATCH ON FIRE WHEN WE EAT?

The ENERGY available from a particular substance—whether it be sugar or coal—is the same whether we eat that substance or burn it. We all know that spectacular fires can occur in the burning of fuels. Some of us have direct experiential evidence, from outdoor barbecue grills or kitchen stoves, that foods can also produce remarkable flames if accidentally ignited. Taking those facts into consideration, we might then ask why, when the energy is liberated from food in our bodies, we don't just catch on fire ourselves.

The chemical pathways of burning fuel and digesting food lead to the very same products, the important ones being carbon dioxide and water vapor. The details of these two processes, examined at the molecular level, are quite different. The burning of a substance usually involves a reaction with oxygen molecules from the surrounding air. In some cases, the fuel being burned will be partially decomposed by the heat of the fire, and the actual combustion reaction itself will involve the chemical interaction of those decomposition products with oxygen. This is the case for chemically complex fuels such as wood or coal. With chemically simpler fuels, such as natural gas, molecules of the fuel itself will react with oxygen. As we've seen, the digestion of a substance as food proceeds through a very complicated sequence of biochemical reactions, facilitated by a family of enzymes. Different chemical pathways, even though they lead to the same products, proceed at different rates. Digestion is much slower than burning. A bit of fat accidentally ignited on your stove top burns quickly; that same amount of fat taken into your body may require hours to digest. The total energy released may be the same, the products may be the same, but the *rates* of these processes are very different.

THE LIMITATIONS OF DIGESTION

The difference in rates of burning and digestion has an important implication for human activity. ENERGY is the capacity to do WORK. The relatively slow release of ENERGY means, in effect, that the amount of WORK we can do in a given time—per second, per minute, per hour—is limited. The amount of WORK done in a given time is POWER. The comparatively slow rate of digestion means that ENERGY release is slow, hence the

WORK we can do in a given time is limited, and thus so is our POWER. Put another way, we humans are not very powerful.

Imagine having the job of clearing a field of rocks. Let's assume that it's a bit of a strange field, so that all the rocks are the same size, say 20 pounds. Suppose that, to get rid of the rocks, you need to pick them up and carry them a hundred yards (the length of a football field). Allowing time for trudging back and forth, by the end of a day you'll have shifted a hundred of these rocks. A hundred 20-pound rocks are a ton of rock. Your POWER output for the day is a ton per eight hours (which is 480 minutes). Few of us would need to be on this job very long before getting the idea that there must be a whole lot easier way to move these rocks than picking them up one by one and carrying them a hundred yards—for instance, we could load them all in the back of a truck of appropriate size, and drive the truck a hundred yards (Figure 3.4). Suppose it takes the truck a minute to go the hundred yards (this is equivalent to an average speed of only three and a half miles per hour). In comparison to what we could do with our bodies, the truck takes only 1/480th of the time; from the other perspective, the truck is 480 times as powerful as a human (Figure 3.5).

Notice the relationship: A human is not as powerful as a truck. From our fundamental definition of POWER (Chapter 2) a human therefore does not use ENERGY as fast as a truck. In a human, the ENERGY source, food, is digested (converted to ENERGY) rather slowly, and we of course do not catch on fire when we eat. In a truck the ENERGY source, gasoline or diesel fuel, is consumed relatively quickly and does burn. Fast consumption of ENERGY leads to greater POWER.

Of course this example is contrived. What kind of field has rocks of identical size? None. Wouldn't it make more sense to carry several rocks at a time? Of course. Why carry every single rock exactly a hundred yards? No good reason. The value of this contrived example is that it illustrates a process of arriving at a rough solution to a problem, or a process of reasoning through a situation by making up a simplistic set of assumptions and using some numbers that are easy to manipulate, even mentally. This kind of approach to solving problems is sometimes referred to as a 'back-of-the-envelope' calculation. The

FIGURE 3.4 These large trucks moving rocks from a quarry represent an easier and faster way of doing WORK than relying on our own muscles.

FIGURE 3.5 The comparative sizes of the person and the truck in this photograph also indicate their relative amounts of POWER.

term implies that we can work out a solution without needing a calculator or computer, but simply by jotting a few numbers on the back side of an old envelope.

We needn't do a back-of-the-envelope calculation to realize that we humans are not as powerful as a large truck (Figure 3.5). Long, long before there were trucks and other sorts of machinery, early humans learned to harness (literally) many kinds of animals to use the energy of animal muscles rather than our own. We are not as powerful as the various kinds of animals that were pressed into service for doing WORK. For heavy, slow jobs, the primary animal of choice was the ox. Depending on the specific job, an ox can exert about ten times as much force as a human. Horses were preferred for lighter work, but situations in which speed was important. The horse can exert about the same amount of force as an ox—at least for a short time—but can do the work much faster. In terms of POWER, the ox may be some seven to eight times as powerful as a human, but the horse is some ten to twelve times as powerful.

We have made use of plows for at least 6,000 years. Probably the earliest plows were pulled by humans, but eventually systems of harnesses were developed to allow animals to pull the plow. Another crucial contribution of animal ENERGY to early agriculture was the use of oxen, donkeys, or camels hitched to waterwheels to provide irrigation water for fields. At one time or another, in various parts of the world, elephants, llamas, and buffalo have also been used.

Until about 1700, human and animal muscles remained the dominant way of doing WORK in Western society (Figure 3.6). A major drawback of this ENERGY source is that the ENERGY produced, or the POWER output, must be applied within a very short distance of its source (that is, of the human or animal doing the work). A society of this kind is sometimes referred to as a 'low-energy society.' Low-energy societies are often very stable societies, but also tend to be resistant to change.

Probably the most spectacular example of social solutions to doing WORK is the construction of the pyramids of ancient Egypt. They represent the most remarkable example of engineering construction due almost entirely to human ENERGY. No one knows for sure how the pyramids were actually constructed. Although some ludicrous

FIGURE 3.6 For almost all of the time that humans have been in existence, WORK was accomplished by using the ENERGY of muscles.

suggestions have been made that the pyramids were constructed by ancient astronauts from outer space, or other 'New Age' nonsense, it seems clear that the pyramids were built with no other ENERGY source than human muscles. A recent estimate for the Great Pyramid of Khufu (a ruler also known as Cheops) suggests that the pyramid took 23 years to build, and the construction involved about 1,250 workmen working for the entire 23 years. The stone blocks also had to be quarried and hauled on sledges—pulled by men—to the construction site. Some of the blocks weigh 50 tons. Overall, the combined effort in quarrying, transporting the stone, and building the pyramid may have involved 10,000 men. The pyramids remain an enduring testimonial to the ability of the human body to do WORK.

Slavery represents the negative aspect of social solutions to the problem of finding ways to do WORK. Slaves have been used throughout most of human history to provide a variety of basic services, some amenities, and even luxuries. Ancient Rome was one of the societies most dependent on slaves, including for such fundamental activities as mining and maintaining the water supply. In the relatively recent history of the United States, slaves were used as the foundation of the agrarian society of the southern states. Sadly, the institution of slavery, though poorly and incompletely documented, is believed to exist even at the start of this new century in some isolated spots of the Arabian Peninsula, southeast Asia, and Africa.

The limitations of human POWER have some additional implications. Suppose that no trucks were available. Is there a quicker way to clear the field than by lugging these rocks, one by one? One way would be to organize a group of your friends. You could set them all to work clearing rocks, or use them to form a 'human chain' in which one person picks up a rock, passes it to the next, and so on, until the person at the other end of the chain adds it to the growing pile. Suppose that instead of a hundred 20-pound rocks, there was only one one-ton rock to be moved. Nobody is going to pick that up and move it. Perhaps, though, if you had a sufficient number of ropes that could be tied onto the rock and had all of your friends pulling on the ropes, you could get it to move. In each of these instances, what you have done is to arrive at 'social solutions' to the problem.

Now suppose that you did have a truck of the right size, but you were still confronted by the one-ton rock. How could you get a rock that big into the back of a truck?

One way might be to construct a long, gently sloping ramp, and slowly roll the rock up the ramp into the truck. (The ENERGY required to move the rock is exactly the same whether we lift it, in one mighty gorilla-like yank, or whether we slowly roll it, inch by inch, up a gentle slope. However, the POWER needed is much, much less in the second case—because it takes a greater amount of time to use that energy.) Another way might be to rig a set of pulleys that we could use to hoist the rock into the truck. In each of these instances, what you have done is to arrive at *mechanical solutions* to the problem.

These simplistic examples illustrate two strategies that are available to us when we are confronted with a job that is beyond the limitations of ENERGY or POWER of our bodies: the social solution or the mechanical solution. The first case involves organizing people to work collectively on a common task. The second involves inventing tools or machines to help us do the job.

WHERE DOES THE ENERGY IN FOOD COME FROM?

Throughout this chapter we have talked about how we as humans have used food as a source of ENERGY. In that discussion, we've neglected completely the issue of where the energy in food comes from in the first place. Most of us eat a varied diet from both plant and animal sources. Vegetarians or vegans eat only plant products. Perhaps some few individuals restrict themselves to a diet of only meat and other foods (milk and eggs, for example) from animals. The animals that we normally consume as food obtained their nourishment by eating plants directly or eating products of plants, such as nuts, fruits, or berries. Since animals and their milk and eggs derive from the animals having eaten plants, it doesn't matter for this analysis whether we have a mixed diet, a vegetarian diet, or an all-meat diet. They all derive ultimately from plants. Thus the question of where the energy in food comes from can be narrowed to asking where the energy in plants comes from. It comes from the sun.

The crucial process in the growth and development of plants is **photosynthesis**. Plants absorb carbon dioxide from the atmosphere and, by using energy from sunlight, combine it with water to form simple sugars. The overall process can be represented by the equation

$$6CO_2 + 6H_2O \rightarrow C_6H_{12}O_6 + 6O_2$$

where $C_6H_{12}O_6$ is the molecular formula of a simple sugar molecule. The plant is then able to utilize the sugar molecules in various biochemical reactions to form all of the many chemical substances needed for the growth and life processes of the plant. The prefix 'photo-' signals us that light is somehow important in the photosynthesis process. Indeed, photosynthesis cannot take place in the absence of sunlight.

The ability of plants to remove carbon dioxide from the atmosphere is a vital component in the movement of carbon around and through our planet's atmosphere, hydrosphere, and biosphere. We will revisit this important role later, when we discuss the build-up of carbon dioxide in the atmosphere and the possible serious implications that build-up may have for global climate.

Although photosynthesis can be represented simplistically by this equation, the detailed chemical pathway by which carbon dioxide is converted to sugar is extremely complicated. Unraveling the complex chemistry of photosynthesis resulted in the Nobel Prize in chemistry being awarded (1961) to the American chemist Melvin Calvin.

Photons, packets of energy from the sun, drive the photosynthesis process. Photosynthesis produces the chemical energy source that plants need for their life processes. We humans eat plants (or we eat the animals that ate the plants) to obtain the energy for our bodies to function. In other words,

HUMAN ENERGY ULTIMATELY DERIVES FROM SOLAR ENERGY.

Our muscles operate on energy stored up from the sun. As we will see, *all* the ENERGY that we make use of ultimately derives from one of only two possible sources: solar energy or nuclear processes.

Unfortunately, we humans are not very efficient solar energy converters. The atmosphere absorbs much of the solar energy reaching the Earth. Then, plants themselves are not very efficient at using solar energy, only about 0.2%. Finally, we consume only a tiny fraction of the plant or animal population on Earth. Overall, our human ENERGY represents about two millionths of the solar energy received on Earth.[4]

Since our human ENERGY derives from the foods we eat, then changes in diet can have a potentially significant impact on society. For example, the late Middle Ages—particularly the 12th century—were characterized by a tremendous flourishing in many areas of human activity, including architecture, politics, mechanical invention, and other areas of intellectual development. Some historians have related this flourishing to the increased consumption of beans (and similar vegetables of the legume family) about that time. The same change is believed to have affected the civilization of the Anasazi people✿ around the year 600. Beans provided an excellent supplement to a diet that was based largely on cereals, because beans are good sources of the B vitamins and the essential nutrient lysine,✿ neither of which are available in high concentrations in the cereals. The improved nutrition led directly to improvements in health and longevity.

✿ The Anasazi people were a civilization that flourished in what is now the southwestern part of the United States for about 1,200 years—roughly from 100 A.D. to 1300 (about six times longer than the United States itself has been in existence!)—and held on until the coming of the Spanish occupancy around 1600. Their name derives from a Navajo word meaning 'Ancient Ones.' Their modern descendants are the Hopi and Zuni. An excellent essay on this remarkable culture, 'Searching for Ancestors' can be found in the collection *Crossing Open Ground* by Barry Lopez (Vintage, 1989).

✿ Eight vitamins constitute the family known as the B vitamins, most of which have various roles in the metabolism of proteins, fats, and carbohydrates. Lack of B vitamins in the diet can result in eye problems, dermatitis, and more serious diseases such as beriberi and pellegra. Lysine is one of nine so-called essential amino acids ('essential' meaning that it *must* be included in the diet because it cannot be synthesized from other chemicals by the body); the amino acids are the fundamental molecular building blocks of proteins. In addition to the book by Gebelein listed below, an excellent introductory book on the science of food is *On Food and Cooking* (Harold McGee, Scribner's: New York, 1984).

CITATIONS

1   Faraday, Michael. *A Course of Six Lectures on the Chemical History of a Candle.* Chatto and Windus: London, 1861.
2   Mann, Thomas. *Buddenbrooks.* Vintage: New York, 1952.
3   Dickens, Charles. *A Christmas Carol.* McElderry: New York, 1995.

There are many other editions of these three excellent books available, including relatively inexpensive paperback editions.

4   Houwink, R. *The Odd Book of Data.* Elsevier: Amsterdam, 1965; p. 58.

FOR FURTHER READING

Butler, Samuel. *Erewhon.* Heritage Press: Norwalk, CT, 1934. Originally published in the 1870s, this book is a Utopian satire that includes the remarkable idea of a society operating not in terms of money as medium for reckoning people's wealth and social status, but rather of using the amount of ENERGY or POWER as a measure of a person's capital. This book is generally kept in print by one publishing house or another, often in relatively inexpensive paperbacks.

Caneva, Kenneth L. *Robert Mayer and the Conservation of Energy.* Chapter 3, on physiology and medicine, discusses Mayer's work on respiration and 'animal heat.' Mayer was a ship's doctor, born in a small town in Germany, whose observations on the color of seamen's blood led him to develop relationships between heat, motion, and force.

Caret, Robert L.; Denniston, Katherine J.; Topping, Joseph J. *Principles and Applications of Inorganic, Organic, and Biological Chemistry.* Brown: Dubuque, 1997. This is an introductory textbook in college-level chemistry mainly—but not exclusively—intended for students majoring in health sciences. Many of the chapters of this book, notably 22 and 23, provide a much more detailed view of the chemistry respiration, metabolism, and human energy production.

de Camp, L. Sprague. *The Ancient Engineers.* Barnes and Noble: New York, 1993. This book provides an easily readable history of technology from ancient times through the Renaissance. Chapter 2 includes a discussion of the construction of the Pyramids.

Derry, T.K.; Williams, Trevor I. *A Short History of Technology.* Dover: New York, 1993. A relatively inexpensive paperback edition of a book that treats the history of technology from ancient times through about 1900. Chapter 8 discusses the use of humans and animals for work.

Faraday, Michael. *The Chemical History of a Candle.* Collier: New York, 1962. This is one of the classic examples of scientific explanation for persons who are not professional scientists—one of the great books of science. It is seldom out of print, and available in editions other than the specific one cited here. Lecture VI discusses the analogy between the burning of a candle and the 'burning' of food in the body.

Gebelein, Charles G. *Chemistry and Our World.* Brown: Dubuque, IA, 1997. This excellent introductory chemistry textbook has two useful chapters on the chemistry of food and of digestion processes. Chapter 15 provides a discussion of the chemistry of food.

Heinrich, Bernd. *Bumblebee Economics.* Harvard: Cambridge, 1979. Any book by this author is a delight. This one discusses the use of energy supplies by bees in maintaining their body warmth and in flying. It is, in a sense, the 'bee analog' of the human energy discussed here.

Matthiessen, Peter. *The Wind Birds.* Chapters: Shelburne, VT, 1994. Some shore birds migrate thousands of miles, for example making two journeys per year between Argentina and the Canadian Arctic. This book touches (e.g., Chapter 3) on issues of ENERGY storage and use by these remarkable birds in the making of these extraordinary journeys.

Salvadori, Mario. *Why Buildings Stand Up*. Norton: New York, 1990. This book intersperses chapters on the technology of architecture with chapters on examples of buildings or other structures illustrating those principles. Chapter 2 provides an excellent discussion on the construction of the Pyramids.

Strandh, Sigvard. *The History of the Machine*. Dorset: New York, 1989. A very well illustrated history of machines, from the use of human muscles through computers. Chapter 4 provides information on muscle 'power.'

4

THE ENERGY BALANCE

In the last chapter we introduced several ideas: Foods and their consumption have many points in common with fuels. The source of ENERGY for the human body is food. Different foods contain different amounts of ENERGY. We can measure the 'energy content' of different foods (or of different fuels) with a calorimeter. In this chapter we will build on these ideas to develop two new concepts. The first is that we can keep track of ENERGY in the same way as we keep track of money, and follow changes in ENERGY similarly to balancing a checkbook or bank account. The second is that, when ENERGY is used, some of it is wasted rather than being converted to useful WORK.

We'll start with an analogy. Suppose that you get $100 (as your paycheck, a gift, or some other source), and suppose that you buy an item that costs $50. Now—what happened to the other $50? Of course it didn't disappear or 'evaporate'; this is money that you have saved, whether by keeping it on your person, or by actually putting it into your bank account. You have stored that remaining $50 someplace. If we express this relationship mathematically, we could write

$$\text{Money } in - \text{Money } used = \text{Money } stored$$

But now let's take another example. Suppose this time once again you get $100 from some source, and you want to buy a new winter coat that costs $125. In this example, the 'money stored' is actually a negative number ($-$25). What does the negative number mean? This transaction is not possible unless you somehow can provide the 'missing' $25 by removing it (as implied by the minus sign) from the money you have saved and stored earlier; in other words, the total amount of money you have in storage goes down.

Since many of us develop some intuitive understanding of the relationships among the amount of money we receive, the amount that we use, and what we have saved as soon as we are old enough to be entrusted with money of our own, why is it worth considering such apparently simple examples? The reason is that the situation with human ENERGY is exactly analogous. Every day (excepting those occasions when we can't eat because of illness or we are fasting) we consume (eat) a certain number of Calories. Also, each day we do WORK that requires ENERGY in the form of Calories. We can therefore write:

Food energy *in* − Food energy *used* = Food energy *stored*

Figure 4.1 illustrates one way in which food energy is stored, in fat cells. Before we examine the possible solutions to this new equation, we might ask how food energy is stored in our bodies. This storage is effected in several ways. Short-term energy storage (analogous to money we might keep in a checking account for almost immediate use) is accomplished by storing carbohydrates in the liver. Our long-term reserves of energy, akin to investments of money, are established because our bodies store up a reserve supply of ENERGY as fatty tissue, or fat.

There are three possible general solutions to this equation. First, consider the case where the food energy in is exactly equal to the food energy used. Regardless of the actual numerical values of these terms, subtracting two identical numbers invariably leads to the difference being zero. In other words, in this case food energy stored = 0. In practical terms, we would see that we have neither increased nor decreased the amount of fat stored, and therefore we would neither gain nor lose weight.

Second, suppose that the amount of food energy taken in is greater than the amount of energy used. (As a specific example, a person might consume 2000 Calories and use up 1500 in doing WORK.) In this case, the 'food energy stored' term will be a positive number (specifically, 500 Calories). That means we are increasing the amount of ENERGY stored in our 'food energy bank' in the same way (but unfortunately not in the same form!) as we store money in our financial banks. We see this storage manifested as a gain in body weight. Put succinctly, if we take in more Calories than we use, our weight goes up.

The third type of generic solution to the food energy equation is the case in which the ENERGY that we use is greater than the amount we take in. (A specific example might be a situation in which a person uses 2400 Calories in doing WORK but only obtains 1800 Calories by eating.) Now the 'food energy stored' term is a negative number (−600 Calories). The only way that we can account for doing this much WORK with so few Calories taken in is by 'withdrawing' some ENERGY from our 'energy bank.' This is exactly analogous to the example in which in order to purchase a $125 coat with a $100 paycheck, we must withdraw $25 from the money stored in our financial (savings) bank. (In fact, the analogy goes further. There are people in the world who are in such poverty that this sort of financial transaction is not possible, because they simply do not have the additional $25 saved some place. Tragically, people who are in the last stages of starvation may no longer have any extra food energy stored in their bodies; for them, it would be impossible to do more WORK than they can 'pay for' with food energy taken in

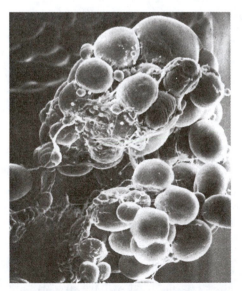

FIGURE 4.1 A high-magnification view of adipose tissue in the human body— our long-term storage of ENERGY in the form of fat.

that day. When the WORK to be done means getting up from bed to obtain food and even that seemingly little bit of WORK is impossible, then death is inevitable.) We see the withdrawal of ENERGY from our body's energy bank as a loss of weight. In fact, this is the basis for the approach to losing weight by exercising. If we adopt a vigorous exercise regimen so that the WORK of exercise requires more ENERGY than we consume as food Calories, we lose weight.

These relationships tell us that we can keep track of the human energy that we use, and the human energy that we take in as food, in exactly the same way that we keep track of our personal finances. Indeed, sometimes people who are concerned about losing excess weight, or maintaining a particular body weight, will speak of 'counting Calories,' and this works in just the same way as counting our money.

We have seen previously that there is really nothing special about food energy; that is, burning sugar as a fuel would liberate the same amount of energy as we obtain when we eat sugar. In fact, the importance of understanding the balance of food energy is that it is perfectly generally applicable to any kind of energy.

Before we explore this point in more detail, it is helpful to understand the concept of **system**.

> A SYSTEM IS WHATEVER PORTION OF THE UNIVERSE IS UNDER OUR SPECIFIC OBSERVATION OR STUDY.

For example, when we are concerned about balancing human energy and counting Calories, the SYSTEM is our body. In a calorimeter, the SYSTEM is the sample and its container, the water, the thermometer, and the insulated box. In *exactly* the same way as we can speak of balancing our checkbooks or bank accounts, or counting Calories to see whether we've gained or lost weight, we can speak of the **energy balance** for a SYSTEM. Let's use the symbol E to represent any type of energy. Then we can speak of the energy balance as being

> FOR ANY SYSTEM, $E_{in} - E_{used} = E_{stored}$

If more energy is put into a SYSTEM than is taken (or comes) out, then some has to be stored inside the SYSTEM. On the other hand, if we take more energy out of a SYSTEM than we put into it, then some of the energy that had previously been stored in the system must have been used. What if there is no energy previously stored in the system? This transaction is impossible—and analogous to a person dying of starvation. This leads to a general, and extremely important, statement:

This statement has great implications, which we will revisit from time to time through the rest of the book.

> IF NO ENERGY IS STORED OR 'HIDDEN' IN A SYSTEM, A PROCESS THAT PRODUCES MORE ENERGY THAN WAS ORIGINALLY PUT IN IS *IMPOSSIBLE*.

We will end this chapter by examining one more concept concerning the energy balance, and that is ENERGY which cannot be used to WORK. Many of us have experienced being outdoors on a cold winter day, and find it comfortable to do a bit of exercising, because doing so helps us to feel warmer. We also know that in the summertime we can become uncomfortably hot when doing a lot of WORK outdoors. (Perspiring—Figure 4.2—is one of our body's mechanisms for trying to get rid of this excess heat. In extreme cases, our body can become so overheated that it 'shuts itself off,' that is, we can pass out from heat stroke.) These same effects occur with inanimate objects. After you've been driving your car for a while, the hood will feel perceptibly warm. An electric motor that's running steadily will also feel slightly warm.

FIGURE 4.2 Perspiration is one of the body's ways of dealing with the waste heat generated as ENERGY is used to do WORK.

THE PROBLEM OF WASTE HEAT

What does your body's getting hot have to do with helping you do WORK? What does the hood of a car getting warm have to do with the WORK of moving the car down the road? What does the casing of an electric motor getting warm have to do with helping it turn a fan (or whatever WORK it's doing)? The answer to every one of these questions is exactly the same: absolutely nothing.

Let's consider the cases of doing some exercise to stay warm outdoors in the winter, or of getting 'overheated' when working outdoors in the summer. Where does this heat come from? It doesn't come from the sun, since these phenomena can certainly occur on cloudy days or at night. The only source of this heat is the ENERGY of our bodies, which we know comes from food. So far we've emphasized that food ENERGY is converted to WORK. But, this common observation of becoming warm or overheated actually suggests a process that we could write as

$$\text{Energy} \rightarrow \text{Work} + \text{Waste (as heat)}$$

A simple example is illustrated in Figure 4.3. This relationship is one of the most profound observations we can make about the world and how it works, so profound and so thoroughly pervasive in everything we do that rarely do we even think about it. One way that we can express this relationship is by the following:

WHENEVER ENERGY IS CONVERTED INTO WORK, A PORTION IS WASTED (USUALLY AS HEAT).

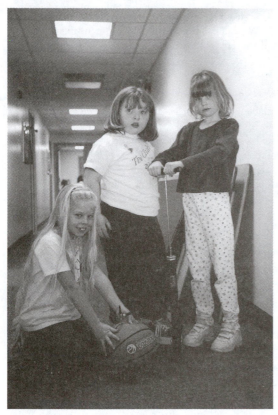

FIGURE 4.3 The generation of heat, as in the barrel of this air pump, is often a signal that ENERGY is not being converted completely to WORK.

An alternative way of expressing the very same thing is this:

> ENERGY CANNOT BE CONVERTED COMPLETELY TO WORK (I.E., WITH 100% EFFICIENCY). SOME ENERGY IS INEVITABLY WASTED.

The ENERGY of our bodies is obtained from the food we eat. As we've seen in the last chapter, the energy in food ultimately derives from solar energy, through the vital process of photosynthesis. When we use our human ENERGY, a portion can be converted into WORK, and the remainder appears as waste heat. Where does that energy that is apparently wasted go? A portion of the heat can be absorbed by nearby objects, but eventually it becomes radiated away from Earth into outer space.⚛ There's no way for us to get it back or to make any further use of it. Essentially the waste heat that we generate becomes a part of an enormous reservoir of heat energy in the universe—but, the crucial point is that such energy is no longer available to us.⚛

⚛ Think of the thousands of power plants and factories, the millions of motor vehicles, the billions of people . . . all generating heat every day. If there were not some way to get rid of the excess heat, we would have made Earth uninhabitable long ago. Nowadays there is great concern that we may finally be on the brink of doing just that, because we are changing the composition of the atmosphere in such a way as to trap more of this heat on

Earth, rather than allowing it to radiate away from the planet into space. This problem is popularly known as the greenhouse effect. We will examine it in detail later in this book, and will find that an understanding of the concept of the energy balance is one of the keys to understanding the greenhouse effect.

The fact that the energy is no longer available to us is not because we lack the appropriate kind of spaceship to fly somewhere and bring it back. Rather, the reason is much more fundamental. The kind of energy associated with heat can only be made to move (in other words, used) if there is a difference in temperature between two objects. The waste heat radiated away from the Earth becomes part of an unimaginably vast reservoir of heat energy in the universe, which happens to be at uniform temperature. There is no way it can ever be used again.

Recall another aspect of our analogy of energy and money: ENERGY is the capacity to do WORK, and money is the capacity to spend. Spending involves the transfer of money, and doing WORK involves the transfer of ENERGY. The ENERGY that can easily be used to do work is ordered, while heat, which is not so easy to use, is disordered. What does this mean? Imagine lifting a heavy box of books from the floor to a table. As you do WORK on the box, the box and its contents are moving in a single direction (up). Imagine hitting a baseball. As you do WORK on the bat, it is moving in a single direction (ideally, straight at the ball). Both are examples of ordered energy. If we had a microscope so powerful as to let us watch the motions of isolated atoms and used it to examine a hot object, we would see that the individual atoms are also all moving—but not in any concerted way all in the same direction; they would be jumping all around in all sorts of directions. For this reason, we speak of heat as being disordered energy. We all have abundant information from daily experience that it's easy to disorder something. It's easy to make a mess. We also know that it's hard to put order into something; given a messy room, or house, or lawn, or car, it takes effort to convert the disorder in the mess back into order. In other words, if we've created a mess, it's hard to undo what we have just created. It is relatively easy to convert ordered energy into disordered energy (e.g., waste heat). It is difficult to convert disordered energy—heat—into ordered forms of energy.

FOR FURTHER READING

Luria, Salvador E.; Gould, Stephen Jay; Singer, Sam. *A View of Life*. Benjamin/Cummings: Menlo Park, CA, 1981. A well-written introductory college-level biology textbook. Chapter 5, on energy metabolism, is particularly relevant here.

5

FIRE

As mentioned in Chapter 2, fire was probably the first ENERGY source used by humans, other than the muscles of the human body itself. Fire was either the first or second chemical reaction to be exploited by humans.⚛ The discovery of fire ranks with the invention of stone tools as one of the crucial steps forward in the development of early humans. The history of the relationship of humans and our hominid ancestors with fire is divided into three phases. First, of course, was a complete ignorance of fire. This period began with the appearance of our hominid ancestors some five million years ago. The second period was a time when there was knowledge of the existence of fire for warmth and cooking, but no knowledge of how to create fire. The discovery and first use of fire probably occurred sometime between 500,000 and 1,500,000 years ago, in East Africa. There is some evidence of fire having been used in cave settlements about a million years old. However, our ancestor *Homo erectus* (often called Peking Man, because the first archeological discoveries were made in China) is the first creature we know for certain used fire, about 400,000 years ago. Archeological remains of hearths found in the Choukoutien cave in northern China clearly show that Peking Man left traces of charred bones of animals. Finally, the ability was developed to create fire as desired and to use it to meet our needs. This probably happened in the Neolithic era, about 9,000 years ago.

⚛ Besides fire, the other chemical process exploited by early humans must have been fermentation.

No one knows how or where our ancestors first discovered fire. Very likely it was discovered independently at many places around the world at different times. Some anthropologists suggest that the knowledge of fire may have been discovered, lost or forgotten, and then rediscovered. Peking man had certainly discovered fire. Every culture since that one has used fire, although it is not clear that all knew how to make fire. Remarkably, the pygmies in the tropical rain forest on the Andaman Islands (off the south coast of Burma) have, until recent years, carefully tended fires ignited by lightning or other natural sources, because they knew no methods for making a fire when they needed it.

Evidence supporting the idea that fire was discovered independently around the world is that virtually all the peoples of the world have a myth or legend about how they first acquired fire. Examples are the Prometheus legend of the ancient Greeks and the similar legend of the ancient Vedics about Pramantha. In the Greek legend, Prometheus was the person who stole fire from the gods for the benefit of mankind. Prometheus did this because he took pity on the sorry state that humans were in without fire.

The importance of fire to primitive humans is reflected in many ways. Fire is used in the ceremonies of many religions. All of us, sooner or later, discover for ourselves that a fire can cause a painful burn. Thus, in some religions fire is a symbol of divine punishment. If not fire itself, then some symbol of fire, such as candles, might be used. Various religious and ceremonial rituals involve the concepts of the 'sacred flame' or the 'eternal flame.' For a long period of human history, it was of utmost importance that a fire, once started, be kept going. This importance is also reflected in many cultures. According to legend, when the Greeks destroyed Troy, the hero Aeneas (who was supposedly the son of Aphrodite and the greatest of the Trojan heroes after Hector) carried the sacred fire from Troy to Italy, where it was then preserved in Rome. The ancient Romans worshipped the goddess Vesta, who was an adaptation of the earlier Greek goddess, Hestia. Worship of Vesta led to the cult of the Vestal Virgins, women who were consecrated to the service of Vesta and who were charged with keeping the flame perpetually burning. In the Jewish religion, the lighting of one candle from another in the eight-day celebration of Hanukkah also indicates the importance of perpetual fire. Vestiges of these concepts survive in our modern times, as for example in the 'eternal flame' around graves (such as for the tombs of unknown soldiers, or at President John Kennedy's grave, and even in the Olympic torch).

The importance and religious significance of an eternal flame may derive from the considerable difficulties that primitive humans must have experienced in igniting fires at will.

EARLY USES OF FIRE

Though nowadays we think of fire as something that destroys, that burns things up, in many primitive cultures fire was regarded not as a destroying element, but as a transforming element—something that causes important changes. Virtually all primitive cultures worldwide have used fire for the same purposes: to keep warm; to drive off animals, especially predators; to clear woods and forests for agriculture; to cook; to dry and harden wood and clay; to heat and split stones to make them easier to move when clearing land or digging wells or mineshafts.

It is likely that humankind's first contact with fire was with fires ignited by natural phenomena, such as lightning or volcanoes. In prehistoric Japan, for example, the Ainu people worshiped a fire goddess who, they thought, lived in a volcano. The steps in learning how to use fire could have come from observing natural phenomena. Reducing the amount of fuel available can control fire, or it can be invigorated and spread by blowing on it (as the wind does). Fire can be put out by water, as the rain does. Perhaps even the technique of lighting fire by rubbing pieces of wood together may have been learned by watching tree branches being rubbed together by wind.

The first use of fire might have been for safety, such as to frighten wild animals away from caves or other shelters. Probably the second use of fire was to provide warmth. It may be that the gathering of humans, or our pre-human ancestors, around the fire to keep warm and safe would have increased bonding among the group. This bonding would be essential to get the group working cooperatively to kill larger, stronger animals for food, to organize migrations, or to make plans for planting crops. Such bonding around the fire could also have contributed to the development of language.

In temperate and cold climates, the heat from the fire was vital. We humans are much more susceptible to cold than are most animals. (Consider animals that survive outdoors through harsh winter conditions; few of us could do that.) It is likely that simply having a warm, dry shelter improved the health of early humans and our hominid ancestors. There appears to be a collective subconscious urge in the human psyche to get warm, or stay warm, when we are ill. Usually when we are sick (unless with a very high fever) many of us feel better—emotionally or psychologically if not physically—wrapping in blankets or putting on warm clothes. Small wonder, then that our name for the most common ailment that afflicts us is a *cold*.

The importance of cooking and warmth to human well-being was reflected in many early religions. Prehistoric Japanese actually worshipped cooking ovens. In ancient Greece the goddess Hestia was the patroness of fires in homes (and in temples). The Romans took Hestia into their religion as Vesta. Even nowadays we sometimes hear the expression 'hearth and home' describing the comforts and pleasures of home and family life; in this expression it's the hearth—the floor of the fireplace or cooking area—that gets precedence.

Fire also provides light. Archeologists have found primitive lamps, small bowls made of soft stone in which animal fat could be burned. Some of the remarkable wall paintings in caves dating from the last Ice Age were done in parts of the caves that daylight could not reach. This fact suggests that lamps must have been used to help make these paintings.

The next major use of fire probably discovered by humans is cooking. For example, a legend of the aborigines of Queensland, Australia, is that the bodies of kangaroos killed in a tribal hunt had been put in a pile. Lightning ignited the surrounding grass. Some brave soul then discovered that cooked kangaroo tasted better than raw kangaroo. It's possible that cooking must have appeared to be magical to primitive peoples. Some nutritionists recommend eating much of our food, especially fruits and vegetables, raw. However, many ancient cultures have stories, admonitions, or proscriptions about the virtues of cooked food relative to raw food. Cooking of food can destroy harmful germs or other parasites. Cooking food certainly provides a greater range of tastes. Cooking makes more substances potentially available to us as foods, because some plants are inedible or poisonous if used raw. In addition, cooking provides ways of preserving food. The smoking of cooked food, especially meat, is a preservative technique. Examples are the jerky and pemmican prepared by Native Americans. This smoking technique may have been one of the first benefits of early humans' use of fire in food preservation. Cooked food can also be preserved if sealed away from air. People still do this in home canning. Primitive techniques involved storing cooked food in the bladders of animals that had been killed or in hollow sections of bamboo that were then sealed with tallow.

The ability to preserve food represented a major step forward for humankind. Storing preserved food eliminates the cycle of 'feast or famine.' Before there were ways of storing food, people had to eat as much food as they could while it was available (and before it spoiled) and go hungry in those times when none was available. Preserving food frees people from the constant foraging and hunting and provides time for other activities, such as making tools, utensils, and other artifacts.

Directly related to cooking itself is the use of fire for hardening wood. A shaped piece of wood heated carefully in the fire can be used as a weapon or as a tool. Humans do not have the strong canine teeth and jaw muscles of carnivorous animals, so we are not as good at chewing up raw or partially cooked carcasses of food. Sharpened and fire-hardened sticks help to tear up raw or partially cooked meat. Probably it did not take much for someone then to recognize that the same sharpened and fire-hardened sticks could be used to tear up carcasses on the hoof—that is, as hunting weapons.

Two other discoveries, both probably accidental, arose from the use of fire. Someone (or, again, most likely more than one person at different times in different parts of the world) noticed that various kinds of clays in the soil around the fire, or that had fallen into the fire, were converted into fairly hard, watertight material. The various minerals of the clay family are plentiful in most parts of the world. The ability of a fire to harden clay could have been discovered independently in many parts of the world whenever a fire happened to be built on a patch of ground containing abundant clay. These primitive ceramics could be used for making cooking vessels, containers for storage of foods, or even containers for transporting preserved foods when the group moved from one area to another. The prehistoric clay vessels from Japan, found on the island of Honshu, are the oldest known pottery, produced about 13,000 years ago. Even 4,000 years later, the peoples of the Near East (in the region of modern-day Iran) were making pottery that was sun-dried rather than being heated in fire.

Somewhere, sometime, ores of various metals, such as tin, lead, and copper, fell into the fire and pieces of metal were found in the ashes. These observations led to the deliberate production of metals and then of alloys (for example, bronze is an alloy of copper and tin). Metals could be used as tools for manufacturing things, as farming implements, and, of course, as weapons. The importance of the discovery of metals and their use marks the transition in human history from the Stone Age to the Bronze Age. Metals are today so common in our everyday lives that we lose sight of the fact that this discovery was one of the greatest of technical leaps forward in human history.

It became appreciated that fire was not only something for domestic applications of cooking, lighting, keeping animals at bay, and warmth. Fire also was a medium whereby raw materials—clay minerals in the soil and ores of metals—could be changed into useful new products.

Where rock was an obstacle in clearing land for farming, for digging wells, or in early efforts at mining, it could be broken apart by a technique called fire-setting. A fire was started on, or around, the rock to heat it. Pouring cold water on it could then splinter the hot rock. The rapid cooling of the rock caused a contraction that set up very severe internal forces that would shatter the rock. In medieval Europe, miners would leave fires burning over a weekend to get rock hot enough to shatter by pouring cold water on it when the workweek began.

THE PROCESS OF BURNING

Clearly, fire was of great use to early humans. It appears from the archeological record that a very long time—millennia—must have elapsed between humans' discovery of fire and the discovery of how to ignite or control a fire at will. We will not try to trace here all the steps in the long, complex struggle that led to a scientific understanding of the processes that occur in a fire. Rather, let's consider some common qualitative observations that can easily be made about fire. Fires require air to burn. (This is useful to us both in kindling a fire, because we need to make sure abundant air is available and in trying to stop a fire, because we can put a fire out by smothering it.) Something is consumed (i.e., burned up) in the process. In some fires the burning material appears to be totally consumed, as in the case of burning gasoline, and in others some small amount of an incombustible residue, ash, remains behind. Certainly we notice that heat is given off, as is light. Sometimes sound is also given off, as the fire 'crackles' as it burns.

The 'something' that is consumed in the fire is the **fuel**. A fuel is something that is consumed to produce energy, most commonly, a material that is burned to produce heat. Since the two halves of this definition of fuel are equivalent, we need to notice that hidden in the definition is an extremely important fact that we will revisit later:

HEAT IS A FORM OF ENERGY.

The evidence of our senses is that a fire destroys or consumes something. This is especially the case when the fuel is a solid or liquid, such as wood, coal, or petroleum products. If the fire is allowed to burn until all the fuel has been consumed, we see quite clearly that the fuel is gone and nothing, except perhaps for some incombustible ash, remains. Various ideas were developed over the years to explain the action of a fire. However, beginning with a few careful experimental observations in the 17th century, and culminating with the studies of the French scientist Antoine Lavoisier⚛ (Figure 5.1) in the late 18th century, scientists established that, in fact, nothing is truly destroyed in a fire.

⚛ Antoine Laurent Lavoisier (1743–1794) is considered by some historians of science to be the father of modern chemistry. In other words, the detailed experimental work of Lavoisier represented a turning point between the scattered, empirical testing of earlier workers and a modern science based on theories and the detailed understanding of chemical processes. In the mid-1780s Lavoisier demonstrated clearly that there is no change of mass in a system undergoing chemical reaction. In 1789 he published a definitive work on the distinction between elements and compounds. Although other chemists had prepared oxygen before Lavoisier's work, he recognized that oxygen is an element, and that the process of combustion is that of chemically combining a fuel with oxygen. In fact, it was Lavoisier who coined the name 'oxygen,' on the mistaken belief that oxygen is a component of all acids (the name comes from the Greek words for 'acid' and 'to form'). Because Lavoisier had held some government positions during the reign of Louis XVI, he was sent to the guillotine during the French Revolution. His international reputation as one of the most illustrious scientists of the 18th century was not enough to save him.

Very precise measurements showed that the weight of the fuel plus the weight of the air involved in the fire equaled *exactly* the weight of the products of combustion (including the leftover air). This relationship holds to the highest degree of precision we can possibly measure. Further, Lavoisier showed that fires don't require *air* to burn—strictly speaking, they require the oxygen that's in the air. When a solid or liquid fuel is burned, matter appears to be destroyed in the fire because the principal products of combustion, carbon dioxide and water vapor, are invisible gases.

Nothing is really destroyed in a fire, in the sense that the weight of the fuel burned, and the weight of the oxygen used to consume the fuel are equal to the weight of all of the chemical products of combustion. But something seems to be produced: ENERGY. We can feel it as heat, see it when the flame is luminous, and sometimes hear it in the crackling of the firewood. Let's consider in more detail this observation that ENERGY is given off from a fire.

First, we must select a SYSTEM for study. We can take one of the simplest fuels, natural gas. The main ingredient of natural gas is methane, CH_4. The other component of

FIGURE 5.1 Antoine Lavoisier (1743–1794), the French scientist who is generally regarded as the founder of modern chemistry.

the SYSTEM will be oxygen. We can imagine a container that holds a mixture of methane and oxygen in the exact proportions needed to combust the methane completely to carbon dioxide and water.⚛ This SYSTEM of methane and oxygen can be kept for years with nothing happening inside it. However, if we choose to supply a small amount of ENERGY to the SYSTEM, as, for example, by inserting a lighted match, or from an electric spark,⚛ we would find that the methane–oxygen mixture burns with great rapidity and provides a large quantity of energy.

The important parts of the SYSTEM are the fuel (in this case, methane) and oxygen. Recall our energy balance:

$$E_{in} - E_{out} = E_{stored}$$

In a fire, the term E_{out} usually will be quite large. On the other hand, the term E_{in} is usually very small, indeed trivial. For example, if we ignited the methane fire with a match, the E_{in} from the match is much, much smaller than E_{out} from the burning methane. We've also said that that a process that produces more ENERGY (i.e., E_{out}) than was put into the SYSTEM is impossible, *unless* ENERGY was somehow stored or 'hidden' in the SYSTEM to begin with. Certainly a fire is not impossible. Then since E_{out} is much, much greater than E_{in}, we need to inquire what this E_{stored} term is.

⚛ When there is exactly enough oxygen present to completely burn the fuel to carbon dioxide and water, we speak of this condition as being one of stoichiometric combustion. The term 'stoichiometric' derives from a Greek word meaning 'element.' It is not necessary for stoichiometric conditions to exist in order for a fuel to burn. Situations can occur in which a system has more oxygen than is needed for complete burning, or has less oxygen. The latter case is undesirable, since, instead of producing carbon dioxide, the carbon in the fuel might be converted to soot or, worse, the potentially lethal carbon monoxide. We will return to this concept when we discuss the burning of gasoline in automobile engines.

⚛ Many processes that appear spontaneous will proceed readily, and often rapidly, provided that we give a 'nudge' to get the process started. A physical example is provided by a brick or block standing on its small end. A block standing in this position can topple over, and once it begins falling, will continue to do so unless we somehow intervene (by catching it, for example). However, the block will stand in the upright position essentially forever unless it is somehow provided with a small nudge to get it started. In the example of combustion discussed in this chapter, the match or electric spark provides the nudge. We call this 'nudge' the activation energy.

We write the chemical equation✧ for the combustion of methane as

$$CH_4 + 2O_2 \rightarrow CO_2 + 2H_2O \text{ (+ heat!)}$$
methane + oxygen → carbon dioxide + water

> ✧ We saw in the previous chapter that photosynthesis could be represented by a deceptively simple chemical equation. However, the exact chemical pathways involved are so complex that working them out was an achievement deserving of a Nobel Prize. In the same way, the equation for the combustion of methane appears to represent an extremely simple chemical process. But, in an analogous sense, the combustion of even a small, five-atom molecule like methane is a devilishly complicated business that may involve over a hundred intermediate steps on the route from methane and oxygen to carbon dioxide and water.

This equation tells us that one molecule of methane reacts with two molecules of oxygen to produce a molecule of carbon dioxide and two molecules of water. We can write this equation in a slightly different, but completely equivalent, form, showing the way the atoms are connected to each other (Figure 5.2). Regardless of which of these two equations we look at, we should be able to see that it satisfies the requirement that no matter is actually destroyed. On the left-hand side there are one carbon atom, four hydrogen atoms, and four oxygen atoms. On the right-hand side there are still one carbon atom, four hydrogen atoms, and four oxygen atoms. What has changed? Only the way in which the atoms are linked (that is, chemically bonded) together.

The collection of a carbon atom, four hydrogen atoms, and four oxygen atoms represents the components of our methane–oxygen SYSTEM. In the combustion process, this system has undergone a change. Originally, the four hydrogen atoms are bonded to the carbon atom, and the oxygen atoms are bonded to each other.

At this point, we introduce a new bit of jargon: **state**. A STATE is the condition in which a SYSTEM happens to exist.

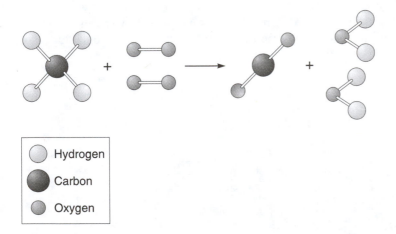

Hydrogen

Carbon

Oxygen

FIGURE 5.2 The molecular structural changes accompanying the complete combustion of methane in oxygen. In this diagram, the circles represent atoms.

We can arbitrarily call the assemblage of atoms in which hydrogen is bonded to carbon in methane, and the oxygen atoms are bonded to themselves, STATE 1 of the SYSTEM. At the end, the carbon atom is now bonded to two oxygen atoms, and each of the remaining two oxygen atoms is bonded to two hydrogen atoms. Let's call this STATE 2. Therefore, in very general terms, the burning of methane in oxygen to make carbon dioxide and water represents a change of this SYSTEM from STATE 1 to STATE 2.

When this change happens, heat is given off. Since heat is a form of energy, we could speak more broadly to say that energy is given off, or is released from the SYSTEM when it changes from STATE 1 to STATE 2. There is one other important consideration. We know from repeated experience that the conversion of methane and oxygen into carbon dioxide and water will proceed spontaneously once we have supplied a small amount of ENERGY—all we need do is strike a match. We also know from experience that the reverse change, the conversion of a mixture of carbon dioxide and water into methane and oxygen, does not happen spontaneously.

Let's represent this with a diagram (Figure 5.3). This diagram applies to the particular case of methane combustion. We can change this diagram to make it much more general (Figure 5.4). This generalized diagram is a pictorial representation of an extremely important principle:

WHEN A SYSTEM UNDERGOES A SPONTANEOUS CHANGE FROM ONE STATE TO ANOTHER, ENERGY IS RELEASED.

We'll see later how we can build on this diagram and this principle to extend our understanding of various ways in which ENERGY is obtained and used. For now, recall that we need ENERGY in order to be able to do WORK, and it is WORK that provides all of the things that we rely on in our daily lives: transportation, manufacturing goods, and transmitting electricity, for example. Getting the ENERGY to do the WORK therefore means that we need to figure out how to take advantage of spontaneous changes in SYSTEMS.

But we have glossed over one other issue—where does the ENERGY of a fire come from? We have said, on the basis of the energy balance equation, that ENERGY is somehow hidden or stored in the SYSTEM (methane + oxygen) before the fire is started. We say this because the value of E_{out} in the energy balance (the energy released by the fire) is much greater than the value of E_{in} (provided by the match). As we saw in Chapter 4, a situation in which E_{out} is greater than E_{in} is impossible unless some ENERGY was stored in the SYSTEM. Most certainly a methane fire is not impossible. Where can the stored ENERGY be? There is only one place it can be. It must be stored in the only thing that

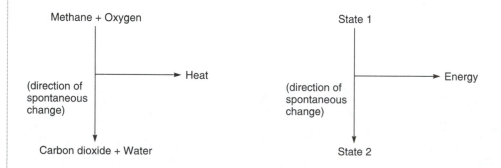

FIGURE 5.3 An energy diagram for the combustion of methane.

FIGURE 5.4 A generic energy diagram for any SYSTEM.

changes as the methane burns: the chemical bonds between atoms. Since ENERGY comes out of a burning SYSTEM, then, continuing our example of methane, it must be that more ENERGY is stored in four C—H and two O—O bonds than is stored in four H—O and two C=O bonds. Generalizing from this specific example,

> THE ENERGY RELEASED DURING COMBUSTION COMES FROM ENERGY STORED IN THE CHEMICAL BONDS OF THE FUEL AND THE OXYGEN.

FOR FURTHER READING

Bronowski, Jacob. *The Ascent of Man*. Little, Brown: Boston, 1973. This splendid book traces the history of humankind largely in terms of scientific and technological progress. Chapter 4 deals with fire.

Derry, T.K.; Williams, Trevor. *A Short History of Technology*. Dover: New York, 1960. This book provides a discussion of the history of technological discoveries and inventions, and their impact on human society. The first chapter of the book describes some of the early uses of fire. Available as a relatively inexpensive paperback.

Hazen, Margaret Hindle; Hazen, Robert M. *Keepers of the Flame*. Princeton University: Princeton, 1992. This book discusses many aspects—good and bad—of fire in American culture in the period 1775–1925. Chapter 6, on 'understanding fire,' is particularly relevant to the material covered in this chapter.

Lyons, John W. *Fire*. Scientific American Library: New York, 1985. A well-illustrated discussion of many aspects of fire, including the fundamentals of combustion, uses of fire, and its dangers.

Poirier, Jean-Pierre. *Lavoisier*. University of Pennsylvania: Philadelphia, 1996. A biography, available as reasonably inexpensive paperback, of the person often regarded as the founder of modern chemistry.

Rossotti, Hazel. *Fire*. Oxford University: Oxford, 1993. The first three chapters of this well-written book provide a discussion of the principles of combustion and ignition.

6

FIREWOOD

Historically, the first fuel that humankind relied on in large amounts was wood. There are several reasons for this. In many places of the world, wood is abundant. It flourishes on the Earth's surface (Figure 6.1), so we do not need to drill or dig into the Earth to find it. Wood is relatively easy to handle and store, unlike, for example, natural gas, which requires pipes, valves, and storage tanks. Wood—unless it's wet—is relatively easy to ignite, unlike some kinds of coal, for example. Wood is easy to burn in open fires. Most anyone can figure out how to get a wood fire started, but other fuels can require some fairly sophisticated devices—gas stoves, for example. Wood was the key fuel that sustained humanity through most of our history. It was probably the dominant fuel well into the mid-19th century, when it was supplanted by coal.

ENERGY CRISES MADE OF WOOD

Wood as a fuel is often considered to be a 'renewable' energy source, because, in principle, we can grow more fuel (wood) to replace the amount we have harvested for immediate use. However, this principle only works with very careful management of the wood supply. It is very easy for an increasing population, with increasing energy demands, to outstrip the ability of the wood supply to keep up with demand. Even in the time of ancient Rome, fuel wood got to be in scarce supply, and by the Middle Ages people were already mining coal to alleviate their dependence on wood. In the 17th century, Britain was still almost entirely dependent upon wood. Because of increasing shortage of wood in Britain, some of the first colonists of America were astonished when they saw the apparently limitless forests of New England.

> *Here is good living for those that love good Fires. Though it bee here somewhat cold in the winter, yet here we have plenty of Fire to warme us, and that a great deale cheaper*

FIGURE 6.1 This flourishing forest in Washington state illustrates how abundantly wood occurs in many parts of the world.

than they sel Billets and Faggots in London. Nay, all Europe is not able to afford so great Fires as New-England.

—Higginson[1]

Likewise, many Mediterranean countries were so dependent on wood that the hillsides were stripped of their trees to provide firewood and lumber (Figure 6.2), causing a range of environmental problems of the kinds that we will discuss later in this chapter.

Although it seemed as if the forests of America were limitless, it didn't take long for reality to overcome perception. By 1637, only seven years after Francis Higginson's claim of 'good living for those that love fires,' poor people in Boston were already suffering from a scarcity of firewood.

The issue of the ownership of the wood is so crucial that it has left a mark on our language yet today. In medieval England, serfs originally were limited to whatever firewood they could pick up from the ground, such as fallen trees or dead limbs that had come down. The land, the trees, and the wood all belonged to the local lord. Eventually, the scarcity of wood resulted in this rule being relaxed. The feudal lords realized that dead serfs (who had died as a result of not having fuel for warmth and cooking) were of no use to them. Shepherds were allowed to knock or pull down dead branches using their crooked staffs. Farm workers could do likewise, using a billhook, an implement that's used for rough pruning or for clearing brush. In other words, these peasants were allowed to have all the wood they could collect, *by hook or by crook*.

> *In their Country Plantations, the wood grows at every Man's Door so fast that after it has been cut down, it will in seven Years time grow up again from seed to substantial Fire Wood.*
>
> —Trefil[2]

Actually, there is an interesting counterargument to the concept that dead serfs are useless to their master. Nikolai Gogol eloquently expresses this argument in his book *Dead Souls*. (Numerous editions of this book, including inexpensive paperbacks, are available.)

It did not take long—less than a century—for the young, expanding cities to get trapped in America's first energy crisis. As the population grew, so too did the energy (wood). But as more and more of the wood was cut, fresh wood had to be transported for longer distances to reach the consumers in the cities. Eventually a point came at which supply could no longer adequately match demand. In the middle of the 18th century, the price of wood in New York City quadrupled in a few years.[2]

The situation became so serious in some places that even the gathering of driftwood from beaches became subject to legislation. One New England town passed legislation in 1765 'to choose four persons to take care of the town's beach, in order to prevent other towns' people carrying off sand, rockweed, and clams, and trash [*driftwood*], or other rubbish of said beach.'[3] Throughout the colonial period and well into the 19th century, this problem affected the great cities of the eastern seaboard. Philadelphia, for example, suffered firewood shortages during the war of 1812. From 1635 to 1850, the amount of wood burned in Concord, Massachusetts was equivalent to having burned every tree in the town three times over.[4] In three centuries (up to 1930) Americans burned approximately 12.5 billion cords⚛ of wood.[5]

> ⚛ A standard measure of quantity of firewood is the *cord*. A cord is a stack of wood 4 × 4 × 8 feet, or 128 cubic feet. However, only about 80 cubic feet of this volume is solid wood, the remainder being air space. Also, measuring wood by volume doesn't tell us very much about how much energy we are buying, because the amount of heat released on combustion varies widely for different types of wood, and also depends on whether the wood is wet or dry. As a rule, hardwoods have more heating value than softwoods. Heating value per cord of wood ranges from 24,600,000 Btu for hickory to 12,500,000 Btu for aspen.[3] There are two other measures of wood sometimes used for the purchase of smaller quantities. A *face cord* is a 4 × 8-foot stack of wood cut into any desired lengths, rather than standard four-foot lengths. A *rick* is a 4 × 8-foot stack of wood cut into 16-inch pieces.

People responded to this energy crisis in ways that were remarkably like the better-known energy crisis of the 1970s–early 1980s. When any energy source becomes in short supply, one of the first, and indeed most effective, responses is to find ways of conserving the resource, or using it more efficiently. In the 1970s, the responses included switching away from the large 'gas guzzlers' to smaller cars with much greater fuel efficiency, and to take much greater care with energy consumption in the home. Benjamin Franklin, with his invention of the so-called Franklin stove (Figure 6.3) contributed one of the 'efficiency responses' in the 1740s and 1950s. (Franklin himself christened his invention 'The Pennsylvania Fire-Place.') One of the improvements made by Franklin was an opening near floor level that allowed cold air from the room to enter an 'air box' around the stove, be warmed there, and then be returned into the room via a second, higher opening. A second response to an energy crisis is to find an alternative fuel, to replace the one in short supply. In the 1970s, there was considerable interest in switching away from expensive, imported petroleum to less expensive, domestic coal. In the 1740s and after, the thrust was away from expensive, scarce wood to use instead—coal. Aside from the improvements in technology made in the two centuries between the mid-1700s and late 1900s, the other difference in these situations was that, in the 1700s, some coal was imported from Europe, mainly from Britain. The enormous supplies of North American coal had scarcely begun to be

FIGURE 6.2 The over-use of forested land for firewood leads to the environmental problem of deforestation, shown here. Compare this ruined landscape with Figure 6.1.

FIGURE 6.3 The Franklin stove was a major innovation in the use of firewood in the home.

tapped. However, the Revolutionary War, followed by developments in transportation systems in the early 19th century, resulted in coal from the interior of the original United States being shipped to the seacoast cities.

WOOD IN THE INDUSTRIALIZED WORLD TODAY

For most of us living in industrialized nations, wood has been almost totally abandoned since coal, and especially oil and gas, became available. The main exception would be fireplaces, which many people seem to find to be more of a decoration than a component of the home heating system. Without a doubt, electric heat, or oil or gas fuels are much more convenient for the homeowner than is wood. However, with proper care for the design and uses of wood burners or fireplaces, wood can provide supplemental heating and may offer some cost savings in areas where it is less expensive than gas or oil.

Since most of us who use wood nowadays rely on it only as auxiliary heat in fireplaces or small woodburners, it is difficult to appreciate the immense amount of labor that was involved in running an entire household using wood as the only source of energy. Before there could even be a fire, there was lots of work in preparation. This preparatory work involved cutting down the trees, hauling them home, chopping the wood into fireplace- or stove-sized pieces, stacking it, and finally bringing it into the house and putting it on the fire. Everybody in the family had a job. Men cut the trees and chopped and split the wood, children gathered kindling wood and carried wood to keep woodbox full in the house, and women transferred the wood to the fire as more was needed. Starting at the fireplace, the wood had to be arranged properly on the hearth. Usually this began with a foundation of large logs, including an exceptionally large log— the backlog⚛—at the rear of the fireplace. Additional wood could be piled onto this foundation of logs as it was needed. This fire was the source of energy for heating the home, cooking food, and boiling water. Tending to this fire was a constant job in the household. It could not have been pleasant to work right over the hot fire, and to try to manipulate hot cooking utensils in the fire. The fireplaces in some medieval dwellings were so huge that the cook worked *in* the fireplace! A wood fire is almost certain to produce some soot and ashes, and these are almost certain to get on the floors and furniture. Thus, cleaning was a permanent chore.

⚛ The original meaning of the word *backlog* was simply a very large log rolled into position at the back of a fireplace. There, it could provide a large, reserve supply of fuel that would take a long time to consume. Our modern usages of the word, as an accumulation of work not yet done, or as a reserve supply of something, derive from this original meaning.

Wood heating in the home can be achieved in two broad classes of devices: open and enclosed heaters. Fireplaces are examples of open heaters. Enclosed heaters are the closed stoves, usually made of metal. The enclosed heaters are much superior to open heaters in terms of providing heat. For one thing, it's easier to control the combustion process in an enclosed heater, such as by opening and closing air vents. In fact, enclosed heaters are sometimes referred to as 'complete combustion' burners. Also, enclosed heaters do a better job of radiating heat into the room. However, few, if any, of the enclosed heaters can provide the aesthetic pleasure that comes from watching an open fire in a fireplace.

A well-built metal stove can provide about three times as much heat to a room as can a fireplace, using the same amount of wood in each. This is the importance of the invention of the Franklin stove. It helped to alleviate our first energy crisis because it could generate much more heat from a given amount of wood (or, from the other, perhaps more important, perspective, could generate the same required amount of heat using less wood). By using less wood to meet the needs of a given amount of heat in a home, the Franklin stove, in effect, meant that dwindling wood supplies could last much longer. For all the enjoyment we might get from having a fireplace in the home and watching the fire, fireplaces are actually not very good means of providing heat. On cold winter evenings it's even possible that a fireplace in operation can *cool* the rest of the house, by pulling cold air in from outside.

When efficiency of wood use and providing useful heat are the main concerns, then an airtight stove (a complete combustion burner) is unquestionably the best choice. The best models available provide numerous features to obtain the most efficient use of wood. These features include the control of air intake by thermostats; a so-called secondary intake system, which brings unburned gases back through the stove for a second attempt at combustion; and water coils to generate hot water using the hot gases from the fire. Such stoves represent the best approach to using wood effectively. But, they have no way of providing the aesthetic pleasure that many people derive from watching a wood fire in a fireplace. There is something about an open, woodburning hearth that strikes a powerful chord in many people. In many areas of the United States, a fireplace is one of the few improvements that one can make to a home that increases its market value by more than the cost of the added fireplace itself.

Aside from their aesthetic appeal or pleasure, and the value they may add to the home, there is otherwise little good to be said about a fireplace. Assuming a tightly sealed house, many of the products of burning the wood remain in the room and must be breathed by the people in it. (Some of the emissions from wood combustion are discussed later in this chapter.) At the same time, the fire is using up oxygen from the room. An open fire is also a potential safety hazard.

For the individual householder relying on wood, the ideal source of firewood is a managed woodlot. In favorable cases, the simple acts of pruning and picking up broken limbs can yield a cord of wood per acre every year.[6] Harvesting trees for firewood, as well as for timber, requires careful management. The management practice is generally referred to as 'sustained-yield.' The principle of sustained-yield management is that of allowing nature each year to grow back the amount of wood that has been removed. Of course, trees cannot immediately be replaced on a one-for-one basis. That is, if we were to cut down a mature tree for firewood this fall, planting a seed or even a replacement sapling will not result in the replacement of that mature tree in only one year. Because of this, at least several acres of sustained-yield forestland are needed to supply firewood for a typical home. For example, a house of modest size, 1,000–1,200 square feet, in a state such as Pennsylvania, that has cold but not subzero winters, would need eight to ten acres to produce enough wood for one winter's heating. In turn then, this means that planning to have all of the population of a large industrialized nation (like the United States) rely completely on firewood home heating is hopelessly unrealistic—there just isn't enough land to grow sustained-yield forests for all of us.

A managed woodlot of hardwood trees will provide roughly a cord per acre. A rough rule is that the 'average' house will need five cords of wood per winter.[2] Of course, there is no such thing as an 'average' house, and the actual need varies with the size of the house, the quality of its insulation, the type of wood being burned, and the temperature that the occupants expect to maintain. However, if we use this rough rule as a guideline, then each household needs five acres of *renewable* woodlot. Other estimates suggest more—about 7.5 acres.[7] Not all of us live in houses. Apartment buildings would require much more available wood. A 30-story apartment building might require the equivalent of a square mile of woodlot.[2]

Since so many of us now live in cities, and in apartment buildings, the idea of a large-scale return to wood as an energy source in most industrialized nations would mean that much land now used for agriculture or forestry would have to be converted to plantations, or managed forests, of fast-growing trees used exclusively for firewood. (We should take care to remember that all of this wood would have to be cut, dried, and transported into the cities. Each of these steps would itself have a significant requirement for energy.) Speaking of wood as a renewable resource does not mean that trees would be cut and forests *as we know them today* would be regenerated. Forests might exist, but they would be managed to provide maximum yields of firewoods. In many places, the specific kinds of trees in the forests would be different. This change would in turn result in a significant change in the **ecosystem**, that is, the organisms of a natural community—the forest—with their environment.

There are several drawbacks or limitations to wood as a fuel. Fireplaces are not very efficient. Cutting, drying, and transporting wood all have associated energy requirements. Maintaining apartment buildings or cities on wood as an energy source would likely be an insurmountable challenge. Nevertheless, wood has enjoyed a resurgence in popularity in the United States. Many people perceive that there is something 'natural' or 'organic' about wood that cannot be said of other fuels. The burning of wood is by no means pollution-free, as we discuss later in this chapter. In some rural areas, the individual householder can obtain quite adequate supplies of wood, often at low cost. As we've seen, the best of the woodburning stoves are much more efficient than fireplaces. When all factors, pro and con, are considered, it still seems very unlikely that wood can displace the fossil fuels or electricity as a major source of energy in most industrialized nations.

WOOD IN THE THIRD WORLD

Even today, wood is the dominant fuel in many parts of the Third World. Thirty to forty percent of the world still relies on wood (Figure 6.4).[8] The so-called 'Third World' usually refers to those countries other than the industrialized democracies and the former Soviet Union and allied socialist states. The Third World includes all of Africa and South America, parts of Asia, and many island nations, such as Indonesia. As examples of the importance of wood in the Third World: Kenya still today obtains 68% of its energy from wood; Nepal, 94%; Ethiopia, 96%, and Mozambique, 98%. In 1980, wood accounted for nearly 60% of the total energy consumption in Africa.[8] The heavy reliance of Third World countries on wood has several important consequences.

Broadly, the Third World countries have two kinds of energy needs. One is the energy that will help the country develop, for example, energy needed for industrial production, for transportation, and for mechanized agriculture. The other is energy needed for simple survival of populace, energy for cooking, and heating, for example.

We can balance the energy needs against the possible sources of energy that could meet these needs. Energy sources can be put into three categories. *Commercial* energy sources include oil, natural gas, coal, and electricity. *Traditional* energy sources, sometimes also called non-commercial sources, include wood; charcoal, which is usually made from wood; agricultural crop wastes (an example being bagasse, the straw-like material left over after sugar cane has been processed to extract its sugary juices), and animal wastes. In addition, some countries may be developing *new* energy sources, solar energy being one example.

Commercial energy sources are the ones that can meet the energy demands for development. However, there is a 'catch.' The nation must have sufficient financial capital resources, either internally or supplied by foreign aid, in order to develop the commercial energy sources. These are expensive. Drilling an oil well can cost thousands of dollars per foot drilled. An electric power plant of modest size can easily cost hundreds of millions of dollars. The situation in Indonesia is an example. Indonesia is richly endowed with energy—coal, petroleum, and natural gas. It has enough energy supplies not only to meet its own needs for many years to come, but also possibly to be

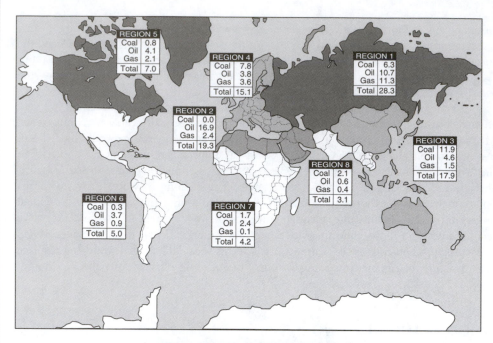

TOTAL ENERGY PRODUCTION	
Coal	31.0
Oil	47.0
Gas	22.0

FIGURE 6.4   Wood is still a dominant ENERGY source in many developing nations.

a significant exporter. Yet in many parts of Indonesia, wood is still the major energy source, because of the lack of capital to invest in the extensive development of the coal, gas, and petroleum.

Usually, the traditional energy sources are the ones used to meet the needs of energy for survival. Often this is because they are not bought and sold as commodities, but rather can be gathered freely from the local environment. In many parts of the world, wood falls into this category, and indeed wood is the most important source of energy for most of the Third World. Wood is the classic example of a traditional energy source. It is the energy source that supplies most people's needs for cooking, heating, and domestic comfort. An example is Kenya, where the average consumption of wood is one ton per year per person.

Many of the Third World countries are faced with problems of burgeoning populations. As more and more people need more and more wood, several factors come into play regarding wood availability. The first consideration is the distance people have to travel to collect wood. The second is having access to the wood—who owns it, or owns the land on which the wood can be found. A third consideration that may also be important is whether any public- (government) or private-sector constraints exist concerning use of the wood.

In Africa, population growth and the increasing amount of urbanization drive a developing energy crisis. That is, not only are there more people, but more of them are moving to cities. As we have seen in considering America's first energy crisis, one possible response is switching to a different kind of fuel. Unfortunately, this is not an attractive option in some African countries because of the comparatively high price of petroleum products.

Firewood availability is also a gender issue. In most Third World nations, collecting wood is considered to be 'women's work' (Figure 6.5). Again, as wood becomes increasingly scarce, more and more time has to be devoted to gathering it. In Kenya, for example, women may spend as much as 24 hours per week gathering wood. In some cases, the bundles of wood to be carried back home may weigh in excess of a hundred pounds. This time is taken away from family activities, such as child rearing. It also takes away from assisting with farming, from planting, through watering and weeding, to harvesting. The loss of time also severely impacts and restricts the women's ability to have time for improving their lot, such as by working outside the home or by obtaining education.

As wood is used up in the vicinity of a village and becomes scarce, several negative aspects of this situation become apparent. People have to go farther distances, and therefore spend much more time, to get wood. The time spent doing this is taken away from other essential tasks. People may be forced to switch to poorer quality fuels, such as agricultural wastes. The fuel quality will affect the performance of stoves or burners used for heating and cooking. People may instead wind up buying fuels, such as kerosene, instead of collecting

FIGURE 6.5 In some developing countries the gathering of firewood is considered 'women's work.' So much time can be expended on this chore that the women have almost no time for other important activities, such as getting further education or working outside the home.

and using free wood. Their limited financial resources may be used up, at least partially, to buy fuel rather than to purchase other necessities such as food or clothing.

There are environmental problems associated with the use of firewood. Often, all the available trees might be cut down over wide areas, for example in circles of 10–12 miles' radius around a village. This extreme devastation is called **deforestation**. Also, later in this chapter we will discuss some of the problems resulting from the burning process.

Deforestation has a number of consequences. It can result in the loss of habitat for a wide variety of plants and animals, essentially, the destruction of an ecosystem. In fact, even a small stand of trees, which one might not think of as a 'forest,' can still have significant benefit to the local **ecology**, the interrelationships of organisms with one another and with the environment in which they live. Some kinds of tough conifers can be planted on land that was ruined by erosion resulting from the clearing of previous hardwood forest. The soil can be restored by these hardy trees to a point at which other forest species can again thrive. Besides controlling erosion and rejuvenating soil, trees can have an important influence on the microclimate of land by providing shade, helping to moderate temperatures and humidity, and serving as windbreaks. When used as a windbreak, trees can also help to reduce heating needs for a home by as much as 30%.[6] Destruction of the trees means that their root systems, which held the soil in place, are gone too. In rainy periods the fertile topsoil can be washed away. This loss of soil can result in mudslides, and, if it's washed into rivers or lakes, can cause pollution of these bodies of water. Alternatively, destruction of the ecosystem can result in the spread of deserts (a process called desertification). Deforestation also contributes to the problem of global warming, which we will discuss in detail in Chapter 32.

The environmental problems of deforestation and desertification should not be minimized. However, cutting trees for firewood is not the principal cause of deforestation in most places. The major source of the problem is the cutting of forests to open up land for agriculture, either for growing crops or for grazing animals. Even commercial logging is not so severe as the cutting of forests for agriculture. In fact, in many Third World countries, firewood does not come from cutting forests; rather, it often comes from the practice of 'agroforestry.'⚛ In some cases, the wood burned for fuel is the leftover scraps of wood remaining from other applications.

> ⚛ Agroforestry is a land management practice in which trees are combined with crops, animals, or both, in a comprehensive approach to land use. The land yields a whole range of products rather than a single product (such as, say, wood, or grain, or meat). Ideally, agroforestry would rely on trees that are well suited to the region, and that have multiple uses—providing shade for animals or crops, yielding fruits or nuts, and eventually being harvested for wood. Experience in some countries, like Kenya, indicates that, given help and support, farmers will actually work to increase the amount of wood growing on their farms—not, as sometimes imagined, ruthlessly chop it all down.

Numerous human factors impact deforestation. Overpopulation is an obvious and severe one. The national priorities in many impoverished Third World nations might favor agriculture (especially for raising crops or animals that can be exported for hard currency). In some regions, especially Africa, national boundaries imposed quite artificially by long-departed colonial powers make it difficult for nomadic peoples to migrate across borders.⚛ And, we should never downplay the impact of utter managerial incompetence, no matter how well intentioned. In Brazil, for example, probably the dominant cause of deforestation in the Amazon basin are schemes of resettling of groups that have already failed once in agriculture because soil conditions were not adequate. Production of charcoal may also cause deforestation in some places. For example, in Brazil, trees are cut to produce charcoal for use in the steel industry. The exploitation of the resource by humans is made all the worse by the fact that many forest ecosystems are very fragile. Once the ecosystem has been damaged, it has difficulty recovering.

⚛ **Contrary to what is often asserted about nomads or the nomadic way of life, a true nomad does not wander aimlessly about, but tends to follow a relatively fixed pattern of seasonal migration. A splendid explanation is provided by Bruce Chatwin's article, 'Nomads,' in his collection of essays *What am I Doing Here*, Penguin, 1989.**

To sum up,

THE POOR PEOPLE OF THE THIRD WORLD HAVE AN ENERGY CRISIS—THE LACK OF AVAILABILITY OF FIREWOOD—THAT THREATENS THEIR SURVIVAL AND THEIR ENVIRONMENT.

HOW DOES WOOD BURN?

In the last chapter we introduced the first example of a combustion reaction, the burning of methane. In the coming chapter we will revisit methane combustion to illustrate the effects of different amounts of oxygen (or air) on the combustion process. Chemically, methane is the simplest fuel in common use. Wood is a much more complex material, so it should come as no surprise that the combustion of wood is, in turn, a much more complex process.

To begin with, virtually all wood contains some amount of water in the form of free water molecules. Freshly cut, 'green' wood, is wet, and can contain appreciable amounts of water. As much as 50% of the weight of a piece of green wood can be water. This high moisture content can lead to a smoky fire, and, since water has no value as a fuel, reduces the amount of heat released when compared to burning the same weight of dried wood. When a piece of wood is put on the fire, these molecules of water boil off. Then the thoroughly dried wood begins to break down chemically as the molecules of the components of wood are heated to higher and higher temperatures. The principal chemical components of wood are cellulose and lignin. As their molecules are broken down by heat, two kinds of product are formed. One kind appears as vapors of various compounds, and collectively these vapors are referred to as *volatiles*. As the volatiles make their way out of the solid wood, they are responsible for the cracking and hissing noises that sometimes accompany a wood fire. The other product is a solid material very rich in carbon, called *char*. At this point no burning has yet occurred; all that has happened is that the components of wood have been decomposed by heat into volatiles and char. Ignition of the volatiles produces the flame that we see apparently dancing over the fire. Ignition of the solid, carbon-rich char produces the glowing embers or 'coals.'

Consider what happens when you light a gas burner (on a gas stove, or a propane torch, for example), and what happens when you light a candle. Touching a lighted match to a source of gaseous fuel causes instantaneous ignition. A flame appears at once. Lighting a candle is not quite so easy. It's necessary to hold a match or lighter to the wick for several moments before the candle is lit. The difference lies in the fuel. A gaseous fuel consists of small molecules (typically with one to four carbon atoms) that mix easily with air and ignite readily. The wax in a candle contains molecules with more than 20 carbon atoms. These molecules have to be broken down by heat from the match and form a vapor. Because it is harder to do this than it is to ignite a small, gaseous molecule, there is an evident time delay in lighting a candle. The cellulose and lignin molecules in wood are even larger and harder to break apart than are the wax molecules in a candle. As a result, wood is even harder to ignite. Furthermore, some heat is needed simply to vaporize the volatiles from the wood.

The heat that is released from burning wood comes from the same source of 'stored energy' as we have discussed for the burning of methane. Many of the chemical bonds in wood are carbon–hydrogen bonds. The interaction of the volatiles from wood with oxygen follows essentially the same process as with methane; the production of new carbon–oxygen and hydrogen–oxygen bonds releases some energy. The energy supplied to the wood by the match, or by other pieces of burning wood, is enough to begin to break down the carbon–hydrogen bonds to allow the new carbon–oxygen and hydrogen–oxygen bonds to form.

With perseverance, we can supply enough heat to wood to break apart the cellulose and lignin molecules and drive off volatiles. Ignition of the volatiles produces a luminous flame accompanied by abundant soot. Once the fire is started, it continues because some of the heat of the fire breaks down more wood, producing more volatiles. This is the case only if the pile of burning is well constructed, so that heat can be utilized in this way. If too much heat is lost, the unburned wood cannot break down chemically, and the fire will go out. Over centuries, many useful rules-of-thumb have evolved for building and maintaining good wood fires. As examples: Three logs in good physical contact with each other burn better than if the same three logs were spaced apart. This occurs because each of the logs in contact can share its heat with the others. It's generally good to allow the ashes to stay underneath the fire. Ashes are very good at trapping heat (notice how long the ashes in a charcoal grill stay hot), and can radiate this heat back to the fire.

Woods differ greatly in the amount of heat produced per pound of wood burned (Figure 6.6). There are also considerable differences in moisture content and the yield of ash. Two premium woods for combustion are hickory and oak. A cord of hickory or oak provides as much heat as a ton of coal or 200 gallons of fuel oil. Softwoods, such as pine and poplar, are not nearly so good, taking up to twice as many cords to provide the same heat as one cord of hickory or oak. Of course, other factors need consideration in selection of wood for a fire: ease of igniting, amount of smoke produced, and how quickly the wood is consumed are all important.

PROBLEMS ASSOCIATED WITH WOOD COMBUSTION

In comparison with coal and many fuel oils, wood contains virtually no sulfur, so produces no harmful sulfur oxides on combustion.⊛ However, wood fires can often produce a fair amount of smoke, and the potential pollution caused by smoke from these fires needs to be considered. Some of the smoke and particulate matter in the smoke comes from the resins in the wood. Many people consider the smoke from wood-burning fireplaces or outdoor bonfires with some nostalgia. However, that may be the case for a single, isolated, small wood fire. Adding the wood smoke from thousands of wood-burning households to a city atmosphere that is already seriously polluted with motor vehicle exhaust fumes, power plant emissions, and factories can greatly multiply the environmental problems. The smoke comes from the burning of moist wood, or from burning wood that contains large amounts of resins.

⚛ The formation of sulfur and nitrogen oxides, and their effects on the environment, are considered later in this book, particularly in the chapters on electricity generation from coal and on acid rain.

In addition to being an atmospheric pollution problem, excessive smoke is also a potential safety problem, because it can cause a fire hazard by leaving soot and creosote ⚛ deposits in a flue. These deposits can ignite, causing a dangerous chimney fire that could in turn ignite the rest of the house. A wood fire generally leaves only a small amount of ash for disposal. If one wants to be very self-reliant, it's possible to use the ash in making soap and as a fertilizer. Harvesting wood often can result in much less damage to the surrounding environment (except for the actual cutting of the trees, of course) than coal mining or oil drilling.

⚛ So-called wood creosote is a liquid of fairly high boiling temperature (above 200°C) that is produced when the components of wood are partially broken apart chemically by the heat of the fire, rather than being burned. It contains a variety of chemical compounds. (The partial decomposition of molecules by heat is called *pyrolysis*.) Wood creosote actually has some valuable uses, serving, for example, as an antiseptic. In many chimneys, particularly in home fireplaces, the vapors of creosote that are leaving the actual region of the fire condense to form a tar-like liquid on the chimney walls. This sticky condensed creosote can trap and accumulate particles of soot, which is produced by the incomplete burning of the wood. Soot is nearly pure carbon. The mixture of creosote and soot is a superb fuel. If the upper part of the chimney becomes hot enough, and if enough oxygen is available, the creosote–soot mixture can ignite inside the chimney. This of course makes the chimney even hotter, and in serious cases the hot chimney can in turn ignite floors or walls of the house, resulting in a disastrous house fire.

Wood smoke contains a variety of chemicals, called carcinogens, which are known or suspected to cause cancer. ⚛ Smoke particles may increase susceptibility to respiratory illness. Atmospheric emissions from many small wood-burning stoves and fireplaces, localized in individual homes, are much more difficult to trap than emissions from large, centralized power plants. Towns in New England, the Pacific Northwest, and some of the Rocky Mountain states have begun to put limits on wood stove use. Because of the difficulty of controlling emissions from many small units vis-à-vis a single, large, centralized unit, if a large fraction of residential heating could be converted back to wood burning (ignoring issues of wood supply, for example), the pollution due to particles in smoke and to potential carcinogens would be serious. Right now there are areas in the United States where wood burning is popular, and concentrations of carcinogens are ten to a hundred times higher than the average concentration in cities.[9] This problem could be a major detriment to a return to large-scale use of wood.

⚛ These chemicals include benzo(a)pyrene, dibenz(a,h)anthracene, benzo(b)fluoranthene, benzo(j)fluoranthene, dibenzo(a,l)pyrene, benz(a)anthracene, chrysene, benzo(e)pyrene, and indeno(1,2,3-cd)pyrene. They are all known or suspected carcinogens. Of course, their formation by pyrolysis is not limited to wood. Many kinds of plant material that pyrolyze as part of the burning process produce these cancer-causing substances. An excellent source is tobacco.

White Ash, *Fraximus americana*
Calorific value when green: 4,600 Btu/lb
Calorific value when air dried: 5,400 Btu/lb

American Beech, *Fagus grandifolia*
Calorific value when green: 3,900 Btu/lb
Calorific value when air dried: 5,400 Btu/lb

American Chestnut, *Castanea dentate*
Calorific value when green: 2,600 Btu/lb
Calorific value when air dried: 5,800 Btu/lb

Eastern Cottonwood, *Populus deltoids*
Calorific value when green: 3,000 Btu/lb
Calorific value when air dried: 6,000 Btu/lb

American Elm, *Ulmus americana*
Calorific value when green: 3,600 Btu/lb
Calorific value when air dried: 5,700 Btu/lb

FIGURE 6.6   Not all kinds of wood provide the same amount of heat when burned. This chart illustrates the differences among some common kinds of wood.

WHERE DOES THE ENERGY IN WOOD COME FROM?

Finally, we should keep in mind the obvious facts that wood comes from trees, and trees are members of the plant kingdom. As we've seen in the last chapter, the crucial step in the life processes of plants is photosynthesis. This process requires the light of the sun—solar energy—to drive the conversion of carbon dioxide and water to the simple sugars that plants then use as energy sources. Since trees require this sunlight, wood ultimately derives from the energy of the sun.

WOOD ENERGY IS A FORM OF STORED SOLAR ENERGY.

Bitternut Hickory, *Carya cordiformis*
Calorific value when green: 4,000 Btu/lb
Calorific value when air dried: 5,400 Btu/lb

Sugar Maple, *Acer saccharum*
Calorific value when green: 4,100 Btu/lb
Calorific value when air dried: 5,600 Btu/lb

Red Maple, *Acer rubrum*
Calorific value when green: 3,700 Btu/lb
Calorific value when air dried: 6,000 Btu/lb

FIGURE 6.6 (continued)

CITATIONS

1   Higginson, Francis. In: Hazen, M.H.; Hazen, R.M. *Keepers of the Flame*. Princeton University: Princeton, 1992; p. 3.
2   Trefil, J. *A Scientist in the City*. Doubleday: New York, 1994; pp. 103–107.
3   Stilgoe, J.R. *Alongshore*. Yale University: New Haven, 1994; pp. 91–92.
4   McGregor, Robert Kuhn. *A Wider View of the Universe*. University of Illinois: Urbana, 1997; p. 15.
5   Hazen, M.H.; Hazen, R.M. *Keepers of the Flame*. Princeton University: Princeton, 1992; p. 159.
6   Dahlin, D. 'Home heating with wood.' In: *Energy Primer*. Portola Institute: Menlo Park, 1974; pp. 154–160.
7   Tenner, E. *Why Things Bite Back*. Knopf: New York, 1996; pp. 92–93.
8   Goldemberg, J. *Energy, Environment, and Development*. Earthscan: London, 1996; pp. 68–70.
9   McFarland, E.L.; Hunt, J.L.; Campbell, J.L. *Energy, Physics, and the Environment*. Wuerz: Winnipeg; 1994; pp. 374–375.

FOR FURTHER READING

Cohen, I. Bernard. *Benjamin Franklin's Science*. Harvard: Cambridge, 1990. An excellent description of the Franklin stove, what it really is, and how well it fared both as a device and commercially, is provided in a supplement 'The Franklin Stove' by Samuel Edgerton, p. 199ff.

Dahlin, Dennis. 'Home heating with wood.' In: *Energy Primer*. Portola Institute: Menlo Park, 1974; pp. 154–160. Although parts of the *Energy Primer* are now somewhat dated a quarter-century after publication, Dahlin's article remains a very useful source of information on the pros and cons of using wood as a fuel in the modern home.

Freeman, Castle, Jr. *Spring Snow*. Houghton Mifflin: Boston, 1995. This book of short essays on nature in New England includes several discussions on firewood and its gathering.

Hill, Robert; O'Keefe, Phil; Snape, Colin. *The Future of Energy Use*. Earthscan: London, 1995. Describes the history and uses of various forms of energy, their impacts on society and on the environment, the associated economic costs, and future energy supplies. Available as a relatively inexpensive paperback. Chapter 3 provides a discussion of the firewood problem in developing nations, and the concept of agroforestry.

Stoner, Carol Hupping (Ed.) *Producing Your Own Power*. Rodale: Emmaus, PA, 1974. Part III of this book discusses applications for heating and cooking with wood.

Tillman, David A. *Wood as an Energy Resource*. Academic Press: New York, 1978. Chapter 1 covers the history of the use of wood as a fuel, and Chapter 3 discusses wood combustion.

7

COMBUSTION FOR HOME COMFORT

Most of us use fuels in our daily lives in two ways: for transportation, and for domestic heating (and sometimes cooking). In our personal transportation, most of us use a single product, gasoline. In contrast, many kinds of fuels are used for domestic heating. These include liquefied petroleum gases (LPG), fuel oil, and kerosene, all of which are made from petroleum; natural gas; coal; and wood. Some people live in 'all electric' homes with electric heating, but, as we'll see in later chapters, most of the electricity that we use is generated in power plants that burn coal, petroleum products, or natural gas.

CENTRAL HEATING IN HISTORY

Central heating is a method of heating homes in which one central furnace provides the heat, which then is distributed around the house by heated air, hot water, or steam. Central heating offers many conveniences compared to the alternative of having a separate fireplace or stove in every room. It may seem as if central heating is a relatively modern development, but in fact the idea goes back at least 2,000 years, to the Han Dynasty (202 B.C.–A.D. 220) of ancient China. In northern China homes were built with a raised floor, for sleeping, over the furnace or oven. Pottery models that show how this system worked have been found in archeological excavations of tombs from the Han Dynasty.

At about the same time (around 100 B.C.), a concept of central hot air heating was developed in Rome. The person credited with the invention is Caius Sergius Orata, who was in the business of fish and oyster farming. To keep his livestock warm and healthy, Orata used a furnace to produce hot air, which passed underneath the bottoms of the fish or oyster tanks. This was accomplished by keeping the fish or oyster tanks raised off the floor, giving space underneath to allow passage of the heated air. The original concept, which was intended simply to help the fish and oyster business, was quickly extended to keep tanks of water comfortably warm for humans—that is, to provide the heating for public baths. Another short step in development extended the concept further, to heat an entire house rather than to heat only the bath. As the Roman Empire expanded, particularly into the north and west of Europe, where the climate makes heating for comfort very desirable, so too did the expansion of hot-air central heating.

The largest application of central heating in ancient times, at least in terms of size, was the use of central heating for public baths. In Constantinople (modern-day Istanbul), a naturally occurring petroleum-like material, perhaps somewhat similar to gasoline, was burned in open rooms underneath the baths. The heat warmed both the floors and the water baths themselves. This system worked too well, in a sense, because in some baths the bathers had to wear sandals with thick soles to avoid getting burns on the feet. The largest of Constantinople's public baths was used by more than 2,000 people a day. Operation of all of the public baths must have involved a prodigious consumption of fuel.

In the United States central heating began to be incorporated in homes toward the end of the 19th century. Many of the earliest central heating systems were of similar design to boilers used to generate steam for various processes in factories. The improvement in comfort, relative to earlier heating systems, was tremendous. Before the advent of central heating, those rooms with a fireplace or stove might seem to be blazingly hot, yet other rooms in the house would, at the same time, be miserably cold. A central heating system allowed heat sources, such as radiatiors, to be located at many points throughout the house, resulting in a much more even heating. The central furnace was not so wasteful of fuel as a fireplace, which often sent much of the heat straight up the chimney. By the early 20th century, a centrally heated American home, even a small and modest dwelling owned by people of limited financial means, was far more comfortable in the wintertime than the grandest of European manors and palaces with their scattered fireplaces.

A SECOND LOOK AT WOOD

In the previous chapter we saw that wood was the dominant fuel for a very long time in human history, and that it still remains very important in much of the Third World, despite the tremendous societal and environmental problems arising from relying on wood. Wood has a number of virtues as a fuel: It is abundant in many places in the world. It is easily gathered. It is relatively easy to handle and store. And, it is generally easy to ignite and burn without sophisticated hardware. In other words, wood is an excellent 'low tech' fuel.

After serving as the main home heating fuel for centuries, wood was largely abandoned when coal, oil, and gas became available. Fireplaces have continued to be a decorative feature in homes, but their role in heating has largely been neglected. Still today the open wood-burning fireplace seems to provide powerful emotional or psychological comfort, as well as some nostalgic link to simpler (and, as often presumed, better) time. Wood fires in the home, in fireplaces or so-called 'woodburners' can be a useful supplement to central heating methods. Properly managed, wood combustion as an auxiliary source of home heat can cut the consumption of oil or natural gas. It's questionable whether wood would ever experience a resurgence that would make it again a dominant choice for central heating in the home. One major reason for this is that wood-fueled units simply can't provide the convenience of oil and gas heaters, which require no significant amount of time or work on the part of the householder.

THE SMOKE PROBLEM

One of the most persistent problems in home heating is what to do about the smoke that is often produced, especially from open fires (Figure 7.1). Anyone who has sat too close to a campfire or to a smoky fireplace in the home may have experienced sore eyes; damaged clothing; food tasting of smoke; clothes or hair smelling of smoke; smoke and ashes on clothes, in the hair, or around the house; even damaged furniture. Benjamin Franklin complained about 'skins almost smoked to bacon.' Persons who already have respiratory problems can find them made all the worse by exposure to smoke.

The problem has been with us since the Stone Age. The first solution was a fairly simple and direct one—simply to cut a hole in the roof to let the smoke escape. This of course has some rather obvious disadvantages (especially when it's raining). A much better solution was the invention of the chimney. A chimney is mentioned in the writings of Theophrastus (a pupil of Aristotle) in Greece in the early fourth century B.C. However, chimneys only became common some 1800 years later. By the 12th century (A.D.), building construction had advanced to a point where large rooms in which the feudal aristocracy held court were now being installed on to the second story of buildings. To accommodate this change in design, it was also necessary to move the central fireplace or hearth from its customary position in the center of the room so that, instead, it was up against an outside wall. A central hearth would have been quite inconvenient in a two-story house except

FIGURE 7.1 Smoke is an unpleasant by-product of many fires, including those used for home heating.

FIGURE 7.1   (continued)

perhaps in a very lofty main room or hall that provided sufficient space for the smoke to escape. Even worse, on an upper level a central hearth would have been positively dangerous without a solid flooring of non-combustible material such as stone.

At the same time, the fireplace or hearth was also covered with a canopy or hood in an effort to collect the smoke. Because matches did not exist at that time (matches were not invented until the mid-19th century) it was easier to keep the fire lit continuously, rather than to have to light a new fire each time one was wanted. Keeping the fire lit continuously was not safe in a central hearth at night without some means of protecting the inhabitants from the fire. The protection was provided by a so-called fire cover placed on the fire. Many of these covers were made of pottery, a circular lid with holes for ventilation. In many of the large houses a bell would be rung to signal the inhabitants that the fire was about to be covered, and further signaled that the time had come to cease household activities for the night. After the transition from central hearths to fireplaces mounted against an external wall, the ringing of the fire-cover bell itself, rather than the covering of the fire, was the signal that activities were over for the night. In French the fire cover, and the associated bell, were known as the *couvre-feu*. The Anglicized version is *curfew*.

Smoke remained a chronic problem for homemakers for many more centuries (Figure 7.2). In the 18th century, 600 years after chimneys began to become commonplace in the large buildings of northern and western Europe, people were still struggling with smoke in the home. A significant part of the overall problem was recognized by Benjamin Franklin. It was simply the poor design of many of the fireplaces and chimneys. Franklin addressed this problem by inventing a particularly effective type of stove, as we will discuss

FIGURE 7.2 Buildings clustered closely together, often with multiple fireplaces and multiple chimneys, can contribute to enormous emissions of smoke into the local environment.

in the next section. Benjamin Thompson (Count Rumford) redesigned the common fireplace to prevent downdrafts (currents of air) from blowing smoke into the room (Figure 7.3). (As we will see in later chapters, both of these Benjamins have had several important roles in the development of energy.)

Another inventor of the time, Charles Peale, developed the so-called 'smoke eater.' Smoke contains a great deal of soot, formed by essentially carbonizing the fuel rather than burning it completely. Under the right conditions, soot can be burned, since it is mainly carbon. Peale's smoke eater forced smoke back through the fire to remove the soot by burning it.

Improving the way the fireplace was designed and built addressed only part of the problem. Numerous other factors also had a role. For example, wet or green wood can be counted on to produce smoke, an observation that many a camper has verified. An excessive build-up of soot in the chimney could cause smoke (or, worse, could contribute to a chimney fire). Even the prevailing weather was important, because sometimes winds could affect the way a chimney would 'pull' smoke. The situation was helped somewhat as cooking and heating stoves began to replace fireplaces as the principal source of heat. Stoves had the advantages that the fire was enclosed, and that smoke could be controlled by regulating the amount of air allowed into the stove via dampers. However, the use of stoves instead of fireplaces was not a complete solution. For example, smoke production and emission was especially hard to regulate when several stoves throughout the house were connected into a single chimney. For example, sometimes a gust of wind can produce enough air pressure at the chimney to prevent smoke from escaping; some of the smoke can be forced back down the chimney into the fireplace. When multiple stoves are connected into a common chimney the problem is worsened, because smoke can blow from a stove in one room partway up the chimney and then be forced out into a different room.

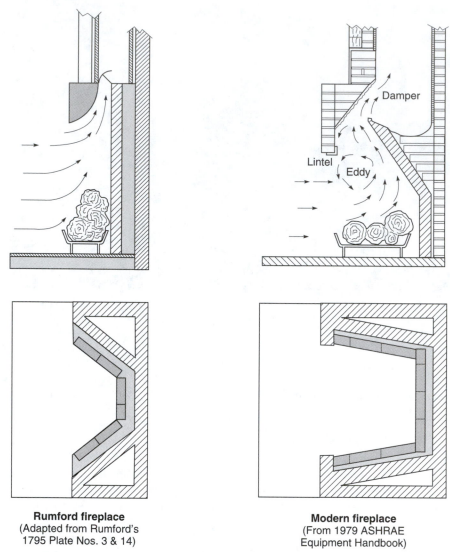

Rumford fireplace
(Adapted from Rumford's
1795 Plate Nos. 3 & 14)

Modern fireplace
(From 1979 ASHRAE
Equipment Handbook)

FIGURE 7.3 The fireplace as redesigned by Benjamin Thompson to reduce the amount of smoke produced.

STOVES

Probably the first real improvement in domestic heating and cooking was made by the first person in American history to be recognized as a 'world class' scientist and inventor. He later gave up his career in science for one in politics and diplomacy. His name was Benjamin Franklin. The first so-called Franklin stove was developed in the 1740s (Figure 7.4).

The Franklin stove did not achieve immediate popularity. Probably the most serious drawback was that it could not be used for cooking, because early designs called for it to be installed in the fireplace, replacing the open fire and taking up enough room to make cooking on or around it difficult. Furthermore, with the relatively primitive metallurgy of the time, the plate iron of which the stove was made tended to crack. The market for the

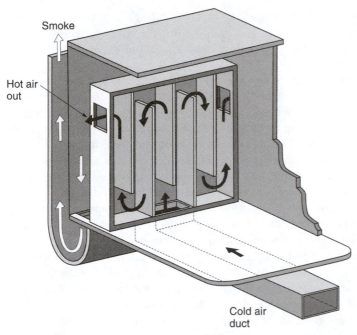

FIGURE 7.4 A schematic diagram of the Franklin stove, showing the cold air intake, air circulation, and the exit for the heated air. A picture of the stove is shown in Figure 6.3.

Franklin stove was limited to the rich, who could afford to own both a fireplace for cooking and a separate stove for heating. An iron stove is heavy, and that meant that it would be expensive to transport, especially so in the days before railroads, where freight transport was limited to horse-drawn or ox-drawn wagons. The transportation problem effectively limited a given market to a very small one that could be served by small manufacturers selling only to customers in the immediate locality.

Solutions to these two problems came in the middle of the 19th century. First, a rapidly expanding system of canals made iron (and coal) available over a wide area, but at still relatively low prices. Canals made the stoves themselves readily available to consumers over a wide geographical area, and helped in a second way by lowering the price of the fuel (coal) that would be used in them. Around the same time, it was learned how to 're-smelt' iron to produce parts that wouldn't crack. An added benefit was that the improved manufacturing allowed making a stove that could be used for both heating and cooking. The mid-19th century was also a time of the great westward emigration in the United States. These new iron stoves were ideal for the pioneers heading west by wagon, because they could be taken apart, transported in the wagon, and then reasonably easily reassembled and got working.

THE FUELS

Coal

Cast iron stoves for cooking and heating with coal were developed in the 1830s and became fairly common by the 1850s. Like the use of wood, the domestic use of coal also required an immense amount of labor. The coal had to be cleaned (to remove rocks, for example), the pieces sorted by size, and hauled from the coal bin to the stove or fireplace. The fire had to be kept going by periodic additions of more coal. Ashes had to be raked out and taken out of the house.

With both coal and wood there was also the problem of cleaning the chimney to remove accumulations of soot and creosote that could catch on fire there. A fire in the chimney could ignite the rest of the house. Various methods were used for cleaning the chimney, such as a brush on a long handle, or a chain lowered down the chimney and rattled to knock soot and creosote deposits loose, or, in what must have been a memorable day in the household, stuffing a live goose up the chimney. In human terms, though, the worst method was the use of 'climbing boys' by professional chimney sweeps (Figure 7.5).

In the days long before occupational health and safety laws, and before laws regulating child labor, young boys would be put to work either to climb upward through the chimney or to be lowered downward on a rope. They would scrape away the accumulated soot and creosote. This job must have been about one step removed from actually being in Hell. The first disease ever traced to occupational exposure was discovered in chimney sweeps and their climbing boys: cancer of the scrotum. Some of the chemicals that occur in the creosote or tar that accumulates in the chimney are carcinogenic. Steady exposure to such chemicals on a daily basis lead to this dreadful form of cancer. Worse, sometimes boys might be put into the chimney *while it was still hot*. This horrific testimony was recorded in England in hearings of the Parliamentary Committee on Climbing Boys (1817):

THE LONDON SWEEP.

[From a Daguerreotype by BEARD.]

FIGURE 7.5 Being a chimney sweep was a dirty, dangerous, and poorly paid occupation.

On Monday morning, 29 March 1813, a chimney sweeper of the name of Griggs attended to sweep a small chimney in the brewhouse of Messrs. Calvert and Co. in Upper Thames Street; he was accompanied by one of his boys, a lad of about eight years of age, of the name of Thomas Pitt. The fire had been lighted as early as 2 o'clock the same morning, and was burning on arrival of Griggs and his little boy at eight. The fireplace was small, and an iron pipe projected from the grate some little way into the flue. This the master was acquainted with (having swept the chimneys in the brewhouse for some years), and therefore had a tile or two broken from the roof, in order that the boy might descend the chimney. He had no sooner extinguished the fire than he suffered the lad to go down; and the consequence, as might be expected, was his almost immediate death, in a state, no doubt, of inexpressible agony. The flue was of the narrowest description, and must have retained heat sufficient to have prevented the child's return to the top, even supposing he had not approached the pipe belonging to the grate, which must have been nearly red hot; this however was not clearly ascertained on the inquest, though the appearance of the body would induce an opinion that he had been unavoidably pressed against the pipe. Soon after his descent, the master, who remained on the top, was apprehensive that something had happened, and therefore desired him to come up; the answer of the boy was, 'I cannot come up, master, I must die here.' An alarm was given in the brewhouse immediately that he had stuck in the chimney, and a bricklayer who was at work near the spot attended, and after knocking down part of the brickwork of the chimney, just above the fireplace, made a hole sufficiently large to draw him through. A surgeon attended, but all attempts to restore life were ineffectual. On inspecting the body, various burns appeared; the fleshy part of the legs and a great part of the feet more particularly were injured; those parts too by which climbing boys most effectually ascend or descend chimneys, viz., the elbows and knees, seemed burnt to the bone; from which it must be evident that the unhappy sufferer made some attempts to return as soon as the horrors of his situation became apparent.

—Carey[1]

Coal certainly required a great deal of domestic labor, generally well over an hour each day to sift the ashes, lay the fire, add coal to it, carry coal and empty the ashes, and put 'blacking' on the stove. (Blacking was a paste applied to the stove to give it a uniform, glossy black polish.) However, using coal introduced a split in the way labor is applied to energy supply. With wood, all of the necessary labor could be supplied by the household, even beginning with cutting the trees if need be. As coal became important, things began to change. Some of the work relating to the use of coal had to be supplied not by the householder, but by specialists—specifically, coal miners and those who are employed in distributing and selling coal.

Nowadays coal is rarely used as the primary source of heat in homes, at least in most industrialized countries of the world. There are a number of perceived problems with coal that diminish its popularity and usefulness for domestic heating. It has the lowest calorific value, per unit weight, of the fossil fuels (though it is higher than wood). Solids, like coal, cannot be conveniently metered or controlled, at least on the small-scale furnaces typical in homes. Coal is perceived to be a dirty fuel. Compared to the other fossil fuels, it produces a substantial amount of ash residue on combustion, which is undesirable because it must be collected and hauled away, and a fine dust of ash particles can get through the house. Handling lumps of coal can produce some coal dust, from lumps knocking or grinding against each other, and this too can be a nuisance inside the house. Some coals burn with a smoky flame, again adding to pollution in, and outside, the home. In contrast to oil and natural gas, it takes some effort to kindle a coal fire, and once it is going, it burns best if kept at a steady heat output. That is, it is not easy to 'turn down' or 'turn up' a coal fire, and coal lacks the quick on/off response of gas or oil. An abundant amount of storage space, usually in the basement, must be set aside for a coal bin. One very positive aspect of coal is that, at least in some places, it can be by far the least expensive fuel in terms of dollars per unit heat produced.

Petroleum Products

Oil products have tremendous advantages relative to coal or wood. These include the facts that oil furnaces offer essentially instantaneous on/off control (that is, eliminating the laborious task of 'building' a coal or wood fire and waiting some time for it to give off abundant heat), no dust or dirt in handling the fuel, and no ashes to be collected and disposed of. The rise of oil heating in homes was helped greatly by the invention of thermostats in the 1920s. Thermostats allowed totally automated home heating, controlled to a chosen temperature. The cleanliness of oil relative to coal and the ease of controlling an oil furnace helped oil displace coal for home heating, especially shortly after the Second World War, as oil pipelines spread through the United States and heating oil became readily available. Even in the heart of coal territory, such as the anthracite-mining region of northeastern Pennsylvania, it became a status symbol in the 1950s to have one's old coal furnace torn out and replaced by an oil burner.

Several fuels derived from petroleum are in use for domestic heating. Fuel oil is the most popular of these. Kerosene is a useful fuel, especially for auxiliary space heaters, such as might be used to heat a garage or basement area.

> . . . if a stove was not burning satisfactorily you poured a gallon or two of kerosene on the flames. Women were burned bald, and children were consumed like little phoenixes (although they never rose again from their ashes) and even fathers of households were Called to Their Reward blazing like heretics in the fires of the Inquisition, and shrieking to be extinguished. But with the cost of kerosene nowadays only the very rich would be able to afford this form of spectacular demise.
>
> —Davies[2]

Liquefied petroleum gas (LPG) is a useful gaseous fuel, particularly in areas not served by natural gas pipelines. Fuel oil shares many of the desirable features of natural gas. It has a high calorific value (though not as high as gas). It is a fluid, and is relatively easy to monitor, control, and automate. It generally provides a quick on/off response. On the other hand, fuel oil has some slight disadvantages. Many fuel oils contain some small amount of sulfur, which contributes to air pollution. Fuel oils may also contain small amounts of incombustible material that accumulates in the furnace as ash, requiring a periodic cleaning and maintenance of the furnace. Fuel oil does require some space to be set aside for its storage, and some arrangements to be made with the local fuel company for delivery.

Natural Gas

Gas heaters were introduced early in the 20th century. Natural gas is now the most popular fuel for domestic heating in the United States. Gas shares many of the advantages of oil as a source of domestic heat: it can be automated and thermostatted, there is no dust or ashes, and there is an instant on/off capability. If the furnace is working properly, there is no soot produced. Gas also has two additional advantages that oil does not have. First, oil burners are complex. The oil has to be 'atomized' into droplets and sprayed into the air in the furnace. Gas mixes freely with air, so generally a gas furnace is simpler and more trouble-free than an oil furnace. Second, oil has to be stored at the home, in tanks in the basement or outdoors near the house. Gas, though, is supplied straight from the distribution system, so there is no storage problem and no wasted space in the home to store fuel (such as fuel oil tanks or a coal bin). Also, there's no need for the householder to have to make arrangements for delivery of the fuel. Finally, natural gas has the highest calorific value, per unit weight, of any common fuel.

Gas-fired appliances came on the market around 1900. By the 1930s, it was predicted that there would be rapid growth in the use of gas once pipelines came into service. Indeed, this prediction proved quite accurate. Pipelines began to be built in 1925. By 1952, consumption of natural gas was double that at the outbreak of the Second World War (1939). Twenty years later, about 40 million homes were using gas. In addition, individual consumers were using more gas each year.

If cost were the only consideration, gas could not compete with coal. In terms of cost per useful amount of energy, coal is the cheapest fuel. Rather than compete on the basis of price, the gas industry stressed many of the advantages of gas that we have listed above. In terms of price, coal is the easy choice. In terms of convenience and cleanliness, gas is the easy winner. Gas remained the preferred fuel even after price increases in the 1970s.

The switch from coal to oil and gas as preferred fuels for home heating also meant an essentially complete shift in the application of labor, from the householder to specialized labor. With oil and gas, essentially all of the labor is supplied by specialists, the oil well drillers, the people who work in refineries, those who work with distribution and supply systems, as examples. For wood, all of the energy supply of the household could be obtained by those in the home, starting with the cutting of the trees. Though few people would actually try to mine or collect their own coal, it can nevertheless be done. For example, some very impoverished people managed to heat their homes by walking along railroad tracks and collecting coal that had fallen off railroad cars. *Nobody* is likely to drill his or her own oil well and set up a backyard oil refinery. In the course of the transition from wood to coal to oil or gas, individual homeowners have changed from energy producers to energy consumers.

A century ago, heating, cooking, and lighting of homes was a matter of hard physical labor for the homeowner or for servants. With the advent of oil, and especially gas, the homeowner no longer had to chop wood, break coal, bring wood or coal into the home, and collect and haul ashes. Now all that was required was hitting a switch or twisting a thermostat. But—because of the shift in the application of labor, the homeowner could no longer provide energy for himself or herself. The homeowner now had to rely entirely on the work of specialists in the oil and gas industries. The hard physical work of the 19th century also meant that a home was essentially 'energy autonomous' because, if need be, the whole energy supply could be provided by the homeowner. The physical labor was traded for the great convenience of gas and oil. The price was surrendering autonomy and relying on specialized labor—and, of course, the monthly utility bills.

COMBUSTION—THE STORY SO FAR

Since natural gas is relatively simple chemically (we've seen that the dominant component is methane, CH_4), we will use it as a basis for introducing some of the chemistry of combustion. First, there are several lines of evidence suggesting a role for air in the combustion process. As our bodies function, converting food to obtain energy, we breathe in air but exhale air that is depleted in oxygen and that contains carbon dioxide. Air is important for fires. Primitive people recognized this by facilitating the ignition and expansion of a fire by blowing on it or by fanning it. We can put out a fire by smothering it, with a wet blanket or sand for example, to exclude air.

The principal components of air are nitrogen and oxygen. Oxygen is the substance that reacts with the fuel in the combustion process. Depending on the amount of air (oxygen) available, there are three possible combustion processes. We will illustrate this by assuming that natural gas is methane, CH_4. (In fact, most natural gas delivered to the consumer contains small amounts of some other gases, but generally is indeed more than 90% methane.)

With abundant oxygen, we obtain complete combustion:

$$CH_4 + 2O_2 \rightarrow CO_2 + 2H_2O$$
$$\text{methane} + \text{oxygen} \rightarrow \text{carbon dioxide} + \text{water}$$

(recall Figure 5.2). This is the reaction we introduced in the previous chapter to discuss the ENERGY changes during combustion. Notice that the carbon in the fuel has become carbon dioxide, CO_2. Of the three reactions we discuss here, this one provides the greatest amount of heat per unit of methane burned.

If not enough oxygen is available for complete combustion, we obtain incomplete combustion (Figure 7.6):

$$CH_4 + 1.5O_2 \rightarrow CO + 2H_2O$$
$$\text{methane} + \text{oxygen} \rightarrow \text{carbon monoxide} + \text{water}$$

Notice that in this case the carbon-containing product is carbon monoxide, CO. Carbon monoxide is a deadly poison, and its production in some combustion processes can lead to accidental death. This may happen with stoves or furnaces that are not properly adjusted to allow sufficient air for the amount of fuel present, or sometimes may happen when an amateur car mechanic is working on a car with the engine running

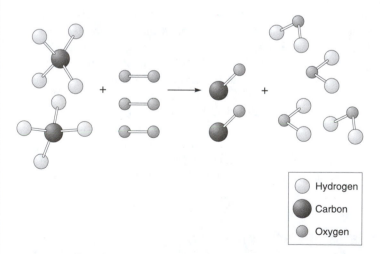

FIGURE 7.6 The molecular structural changes accompanying the incomplete combustion of methane in a limited amount of oxygen. In this diagram, the circles represent atoms. The carbon monoxide produced in this reaction is a dangerous poison. (Compare Figure 5.2.)

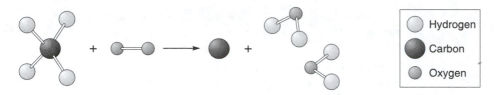

FIGURE 7.7 The molecular structural changes accompanying the combustion of methane in very limited oxygen (oxygen-starved combustion). In this diagram, the circles represent atoms. The carbon forms as soot. (Compare Figures 5.2 and 7.6).

in a closed garage, not obtaining adequate ventilation. Aside from the very serious issue of the hazards of carbon monoxide, this incomplete combustion reaction yields less heat per unit of methane burned than does complete combustion.

In very limited amounts of oxygen, we obtain oxygen-starved combustion (Figure 7.7):

$$CH_4 + O_2 \rightarrow C + 2H_2O$$
methane + oxygen → carbon (soot) + water

Carbon usually forms as soot. As a minimum, soot formation is aesthetically undesirable. In addition, soot can contribute to air pollution or, if inhaled, to human respiratory tract problems. This reaction produces the least amount of heat per unit of methane burned.

A WORD ABOUT THE HUMBLE MATCH

Through most of this time the actual creation of fire was still a very difficult job. Fires were started by using flint and steel, rotating 'fire sticks,' or simply by resorting to 'borrowing' fire from a neighbor. The ordinary match was developed, in a reasonably practical form, in 1836. It was so remarkable an invention, in terms of giving people the ability to create fire as needed, that some surveys of 19th century technology listed the match as one of the three greatest inventions of the century (the others being the steam boat and the cotton gin).

Matches were first invented in 1827. To be useful, matches need to be able to be ignited very easily. At first, a very inflammable form of phosphorus❀ was the chief ingredient of matches. However, this material can be very hazardous, leading to severe problems with the gastrointestinal tract and liver, circulatory problems, and death. Its use was outlawed by many countries during the 19th century. Matches are of two types: the 'strike-anywhere' match, which can be ignited on any handy surface, and the 'safety' match, which requires a special surface on the matchbook or box for use.

❀ Phosphorus, like a number of other elements, can exist in very different forms. Many times the different forms of an element, called allotropes, display very different physical properties and chemical reactivities. Perhaps the best-known allotropes occur with the element carbon: diamond and graphite. The form of phosphorus used in the original matches was the allotrope commonly called white phosphorus. It is extremely poisonous, as little as about 1/250th of an ounce being fatal. Even if just handled with unprotected fingers, it can produce deep, slow-to-heal skin lesions. White phosphorus will ignite spontaneously in air at temperatures as low as 30–35°C (85–95°F). Suffice it to say that a material this nasty and dangerous to handle has found a variety of military uses, and in rat poison.

Strike-anywhere matches typically contain a material that is easily ignited by friction, a material to provide that friction, an oxidizer to help the burning take place, and glue, to hold the other ingredients in place. Often phosphorus sulfide is the material to be ignited, finely ground glass helps provide the friction, and potassium chlorate serves as the oxidizer. When the match is rubbed on any surface, the phosphorus sulfide is ignited and the potassium chlorate helps to sustain a vigorous burning. Although the combustion of the phosphorus sulfide takes place in a very short time, it usually generates enough heat to ignite the wood of the matchstick. In some matches the stick will have been dipped in paraffin to help it burn more easily.

Safety matches contain the ground glass for friction and a material to assist ignition (usually a different, somewhat less dangerous form of phosphorus⊛) on the dark-colored strip that is found on the side of the box or at the bottom of the matchbook. The tip of the match itself contains the oxidizer, potassium chlorate, and a substance to initiate the burning of the match, such as antimony sulfide. In this case, the most combustible substance in the system, the phosphorus, is separated from the oxidizer. When the match is struck, friction produces enough heat to ignite a bit of the phosphorus on the side of the box. It in turn quickly ignites the antimony sulfide–potassium chlorate mixture on the match itself. The burning of a bit of the phosphorus on the igniting strip each time a match is struck is why, after such a strip has been used many times, it becomes difficult to light other matches on it. Safety matches get their name from being relatively safer than the strike-anywhere matches (in part because of the separation of the phosphorus and oxidizer). Either kind can of course lead to property destruction, injury, or loss of life if misused.

⊛ The friction strip used to ignite safety matches contains a different allotrope of phosphorus, called red phosphorus. It is not nearly so reactive nor dangerous to handle as white phosphorus—unless it contains traces of white phosphorus as an impurity, it is quite safe to handle and work with.

CITATIONS

1   Carey, John (Ed.) *Eyewitness to History*. Harvard University Press: Cambridge, MA, 1987; pp. 280–281.
2   Davies, Robertson. *The Papers of Samuel Marchbanks*. Totem, 1985, p. 369.

FOR FURTHER READING

Black, Newton Henry; Conant, James Bryant. *New Practical Chemistry*. Macmillan: New York, 1943; and Kruh, Frank O.; Carleton, Robert H.; Carpenter, Floyd F. *Modern-life Chemistry*. Lippincott: Chicago, 1941. These two books are examples of introductory chemistry texts published in the 1930s–50s. They are treasure-troves of information on practical aspects of the chemistry of things in our daily lives: steel, fertilizers, glass, ink, textiles, and, of course, matches. An interested reader can often obtain such books at very little cost at used book sales. In these two specific books, relative passages (on matches) are Chapter XXV and Unit 8, respectively.

Dahlin, Dennis. 'Home heating with wood' in: *Energy Primer*. Portola Institute: Menlo Park, 1974; pp. 154–160. Though parts of this book are dated a quarter-century after publication, Dahlin's article remains a very useful source of information on the pros and cons of using wood as a fuel in the modern home.

Davies, Robertson. *The Papers of Samuel Marchbanks*. Totem Press: Don Mills, Ont., 1987. This delightful book, by one of Canada's finest authors, provides an excellent glimpse of the struggles of a homeowner (Marchbanks) with his coal furnace.

Hazen, Margaret Hindle; Hazen, Robert M. *Keepers of the Flame*. Princeton University: Princeton, 1992. This book discusses many aspects—good and bad—of fire in American culture in the period 1775–1925. Chapters 5 and 6 are particularly relevant to the material covered in this chapter.

Hotton, Peter. *Coal Comfort*. Little, Brown: Boston, 1980. Devoted to ways of heating the home with coal, including clear conceptual drawings and examples of coal stoves and furnaces suitable for home use.

Hunter, Christine. *Ranches, Rowhouses & Railroad Flats*. Norton: New York, 1999. An excellent and well-illustrated book on the architecture of American homes. Chapter 3 includes a discussion of space heating.

James, Peter; Thorpe, Nick. *Ancient Inventions*. Ballantine: New York, 1994. This is an extensive compilation of ideas from the dawn of history through 1492. Chapter 10 provides a discussion of early concepts of central heating.

Rossotti, Hazel. *Fire*. Oxford University: Oxford, 1993. Several chapters of this well-written book provide a discussion of the uses of fire for warmth, heating, and cooking in the home.

Tenner, Edward. *Why Things Bite Back*. Knopf: New York, 1996. This book examines the 'revenge of unintended consequences' in technology—how a technological advance sometimes leads to consequences no one had ever foreseen. Chapter 4 includes a discussion of the unintended indoor air pollution arising from stoves and furnaces.

Wood, Margaret. *The English Mediaeval House*. Studio Editions: London, 1994. The evolution of English houses up to about 1540. Chapters 17–19 particularly relate to arrangements for cooking and heating.

8

WATERWHEELS

It is likely that the use of water was the second major energy source to be used by humans (after fire). For that reason, waterwheels are important in the history of the development and use of energy. Also, waterwheels are relatively simple mechanical devices that we can use to understand some important concepts.

KINETIC ENERGY

We start with a simple observation that most of us first make as children: If we are wading in a stream of water, we can feel the water pushing against our legs. In very rapid streams, we have to be careful that we don't get knocked down by the rushing water. Similarly, if we're wading on the coast we can feel the push or pull of the tide on us. Sometimes we can get knocked off our feet by a big wave coming in. Certainly if we were to stand under a waterfall (Figure 8.1) we would feel the rush of water over our bodies.

Suppose we envision making a simple wheel with paddles on it (Figure 8.2). If we partially immersed this wheel into a flowing stream, then the water will push against it, just as it pushes against us (Figure 8.3). By pushing against only one paddle (in this illustration) the water will cause the wheel to turn. Then, if we have connected the shaft of the wheel to some piece of machinery, we can do useful work, such as the grinding of grain. This concept is the basis of the waterwheel.

Now let's consider a related observation: When standing or wading in a swimming pool or a pond, we don't worry about being knocked down by the water. Similarly, we should envision that sticking our waterwheel into a swimming pool isn't going to cause it to move. What's the difference between this situation and the ones previously discussed? In a swimming pool or pond, the water isn't moving (at least, not very much). Only moving water can push or pull us around, or turn the waterwheel. Pushing us, or pushing a waterwheel, represents doing WORK. Only moving water has the capacity for doing WORK, which means that moving water must have some kind of ENERGY.

FIGURE 8.1   We can often observe for ourselves the force in rapidly moving water.

> THE ENERGY ASSOCIATED WITH MOVING BODIES IS CALLED KINETIC ENERGY.

As mentioned earlier, if we connect the waterwheel to some sort of mechanical device, we can use the device to do WORK, such as grinding grain, or sawing wood. A generalization of this statement is that:

> KINETIC ENERGY CAN BE USED TO DO WORK.

POTENTIAL ENERGY

Let's now consider a somewhat more ambitious way of using a water wheel (Figure 8.4). We can take advantage of a waterfall (Figure 8.4). The kinetic energy of the water falling down the waterfall turns the wheel and thus—with appropriate machinery—does WORK.

Consider a volume of water just at point **A** of a waterfall, as indicated in Figure 8.4. Can that bit of water do WORK? No (at least not yet), because it does not have kinetic energy in the direction downward along the waterfall. We do know, though, that the water at **A** could *potentially* do WORK. All we need to do is let that bit of water acquire kinetic energy, by falling off the cliff over the waterfall. In other words, at **A** the water has no kinetic energy, but *potentially* it could acquire some. To describe this situation we say that the water at **A** has potential energy.

What makes the water fall? It is pulled, toward the center of the Earth, by the force of gravity. To be complete, we would say that the water at **A** has **gravitational potential energy**, because its potential is due to its position in Earth's gravitational field. Gravitational potential energy is the ENERGY associated with an object by virtue of its position in a gravitational field.

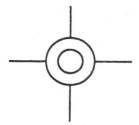

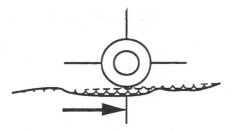

FIGURE 8.2 Attaching a set of paddles to a wheel provides a simple way of making a primitive waterwheel.

FIGURE 8.3 Dunking a simple wheel like that shown in Figure 8.2 into a stream allows the kinetic energy of the moving water to turn the wheel.

Now let's add three more points. First, potential energy can be converted into kinetic energy. Second, point **B** is the point at which the water is hitting the blades of the waterwheel. The kinetic energy of the water is used here to impart kinetic energy to the wheel; in turn, the motion of the wheel can be used to do WORK. Put colloquially, point **B** is 'where the action is.' Third, consider what has happened to our volume of water at **C**. We might be tempted to conclude that it no longer has potential energy. However, that's not quite true, because it could—in principle, at least—encounter a further waterfall somewhere down the stream. The water still has some gravitational potential energy at **C** because it retains the potential of falling still further. (In fact, it will always have some gravitational potential energy unless it happens to be at the center of the Earth.)

To recap, water at **A** had high gravitational potential energy. In passing over the waterfall, some of that potential energy was converted to kinetic energy. The kinetic energy of the falling water was able to do WORK, turning the waterwheel. Then, the water was at a new, *lower* gravitational potential energy.

FIGURE 8.4 A conceptual diagram (right) of a waterwheel being operated by a waterfall (example above). At point **A** the water has high potential energy. At point **B**, some of the kinetic energy of the water is transformed to WORK. At point **C**, the water has lower potential energy than it did at **A**, and—because it is flowing away from the wheel—still has some kinetic energy.

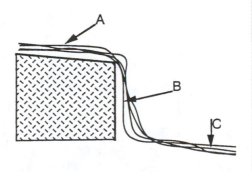

SPONTANEOUS CHANGE AND THE ENERGY DIAGRAM

We know from plenty of our own observations that things always fall *down*. That includes us, when we stumble or trip. Anything that is able to move will spontaneously fall from a position of high gravitational potential energy to a new position of low gravitational potential energy. We also know from abundant observations that things never spontaneously fall *up*. (Sometimes we see an effect on television or in the movies where a film is run backwards and people or things 'fall up.' Usually this is done for its humor. We find it funny because it is so contrary to our usual, expected experience.)

> For every natural event, there is an allowed direction and a forbidden direction. Apples fall; they don't jump up to their branches. Sugar lumps dissolve in your coffee; they don't form by themselves in a cup of sweetened coffee.
>
> —de Duve[1]

So—objects move spontaneously from high gravitational potential to low gravitational potential. We've also seen that we can extract kinetic energy—in this case, by using a waterwheel—when that change from high to low potential energy occurs.

That being the case, we can simplify our SYSTEM (water, a waterfall, and a waterwheel) into a simple diagram (Figure 8.5). This diagram looks suspiciously like, but not identical to, the diagram we've introduced earlier (Figure 5.5) for another SYSTEM—a carbon atom, four hydrogen atoms, and four oxygen atoms—from which we could also extract ENERGY. In the case of these chemical changes (combustion), we can speak of the SYSTEM as being in a state of high 'chemical potential' energy, analogous to a SYSTEM in a state of high gravitational potential energy. In fact, these two diagrams are so similar that we can create a 'generic energy diagram' (Figure 8.6) that applies to both SYSTEMS. This a crucial diagram that we will use to explain the behavior of a large number of systems. The key concept that goes along with this diagram is this:

WHEN A SYSTEM UNDERGOES A SPONTANEOUS CHANGE FROM HIGH POTENTIAL TO LOW POTENTIAL, WE CAN EXTRACT ENERGY (OR WORK) FROM THE SYSTEM.

TYPES OF WATERWHEEL

The most basic design (and simplest concept) in waterwheels is the undershot wheel. The earliest design was a simple paddle wheel that was immersed in the stream or river. The undershot wheel was in use in places like northern Denmark and China (where it was used for milling rice) by about A.D. 30. A waterwheel has been discovered in the ruins of Pompeii, where it was buried in the volcanic eruption of A.D. 79. The undershot waterwheel also has some practical problems. The WORK done by the undershot wheel derives almost entirely from the kinetic ENERGY of the water, which in turn is dependent on the velocity of the water. The WORK and POWER of the wheel are directly affected by seasonal changes in the rate of flow of the stream or river. In particular, they are subject to seasonal fluctuations in the water supply. In hot weather, or in times of drought, the stream level may drop so low that there is not enough water to turn the wheel. In times of flood, there may be so much water that the wheel can't turn, because it is totally sub-

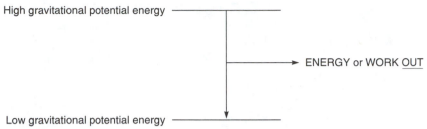

FIGURE 8.5 The energy diagram for a waterwheel. Compare Figures 5.3 and 5.4.

merged. Despite these potentially serious drawbacks, the undershot wheel was used in many places around the world for centuries. The longevity of the undershot wheel very likely is due to the ease, simplicity, and relatively low cost of its construction.

In an undershot waterwheel, particularly of the paddle-wheel design, there is a significant loss of ENERGY because of the shock and turbulence as the water hits the paddles. If turbulence occurs in the water because of its hitting the blades, some ENERGY is, in essence, wasted in causing the water to swirl and mix with itself. (Think of how we must use ENERGY—applied to the spoon—to cause turbulence deliberately when we want to mix cream with coffee.) An improvement in this design did not occur until about 1800. In hindsight, the improvement was quite simple: the shape of the paddles was changed. In the improved design, called a Poncelet undershot waterwheel, or simply a Poncelet wheel (Figure 8.7), the paddles or vanes are curved so that they allow the water to enter the wheels with a minimum of the turbulence or shock that causes energy loss. The curved shape of the paddle forces the water to 'run up' the paddle, all the while exerting a force on the wheel, and transferring its kinetic energy to the wheel. As the energy of the water is transferred to the wheel and the wheel rotates, the water falls off the wheel with nearly zero velocity. The Poncelet wheel is 70–85% efficient at converting the kinetic energy of the water to WORK. In contrast, the flat-bladed paddle wheel might be only 20–25% efficient.

In an overshot wheel, the water flows down over the wheel (Figure 8.8). The overshot wheel is superior to the undershot in cases where the stream flow is scanty. In fact, the overshot wheel can work in any stream, provided that we build a dam upstream of the mill. Doing so creates a millpond. This lets us keep a constant flow of water the year round. In a dry season, we have a reserve in the millpond; in a wet season we can let the excess water past the dam via a sluice and sluice gates. The earliest historical record of the overshot wheel is a picture of one in a mural in the Roman catacombs, from the third century A.D.

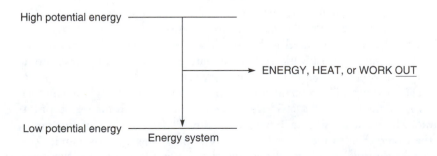

FIGURE 8.6 The generic energy diagram, fleshed out from that in Figure 5.4.

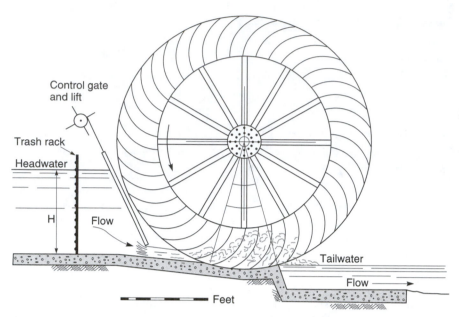

Control gate
and lift

Trash rack

Headwater

H

Flow

Feet

Tailwater

Flow

FIGURE 8.7 A Poncelet undershot waterwheel.

The overshot wheel is also vertical on a horizontal axle. The blades of the wheel serve to divide it into bucket-like compartments. Water pours into these compartments or buckets from above. In an overshot wheel, only a small part of the WORK is a result of impact of the water hitting the blades. Most derives from the weight of the water as it descends in the 'buckets' as the wheel turns. The efficiency of the overshot wheel is 65–80%.

When installed with a dam upstream, the overshot wheel is mounted between a trough or channel bringing water in at a high level (i.e., at high gravitational potential energy). This trough is known as a 'head race.' There needs to be a second trough or channel that allows the water to flow away from the wheel at a lower level (at lower gravitational potential energy). This second trough is known, in the jargon of the business, as a 'tail race.' The diameter of the wheel is about the same as the difference between the two water levels—that is, so it fills almost completely the space between the head race and tail race. Water flows out of the head race into the buckets at the top of the wheel. Its weight causes the bucket to fall, turning the wheel at the same time. As the wheel turns, each succeeding bucket is filled and adds to the moving weight until it is just past half-way down. At that point, the water begins to spill out of the wheel into the tail race as the bucket tilts over. By the time the bucket has reached the lowest level, it is empty.

A variation of the overshot wheel is the so-called breast wheel. This wheel rotates against a small dam (called a breast work) so that the water is retained in the buckets for a greater distance than would happen in the ordinary overshot wheel.

The so-called floating mill is an illustration of the cliché that 'necessity is the mother of invention.' In 537 the Ostrogoths, a Germanic people moving from the region of the Black Sea, invaded Italy and laid siege to Rome. At that time all that remained of the Roman empire was the eastern half, whose capital was Byzantium (modern-day Istanbul). The Byzantine general Belisarius was assigned the task of defending Rome during the siege. Most of the watermills were supplied from the extensive system of aqueducts that were used to bring water into Rome. The besieging Goths thought it would be possible to starve Rome

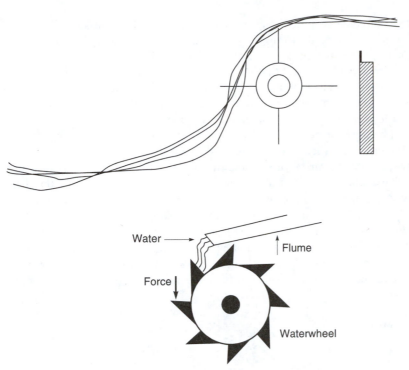

Water

Flume

Force

Waterwheel

FIGURE 8.8   An overshot waterwheel.

into surrender by cutting off the water supply from the aqueducts. Without water to operate the mills, the Romans could not grind grain to make flour, and hence could not bake bread. Belisarius had paddle wheels mounted in boats that were then moored in the Tiber River. The Tiber flows through Rome, and the Goths had no way of halting its flow. By connecting these wheels with grinding mills, the supply of flour was assured. This design slowly spread over the next thousand years, first apparently to Venice and Baghdad. Three floating mills were built under bridges in the Seine River in Paris. By the end of the Middle Ages, many floating mills were in use, mainly for grinding grain or pumping water. Floating mills built in the Rhine River at Cologne in the 15th century were maintained in use for at least three centuries. Victor Hugo's immortal novel *Notre Dame of Paris* mentions in several places pedestrians being splashed by water from the bishop's mills set in the Seine.

WATERWHEELS IN THE ANCIENT WORLD

The waterwheel is a device that allows us to extract energy from a system—falling water—that is undergoing a spontaneous change from high potential energy (in this specific case, high gravitational potential) to low potential energy. If we could connect some sort of device to the shaft or axle of the waterwheel, we could then use that kinetic energy to do WORK.

In fact, the waterwheel was the first practical device that freed humans from reliance on their own muscles or on draft animals for WORK. We don't know where or when waterwheels were first used, or who developed them, to harness the kinetic energy of flowing water.

Around 300 B.C. there may have been some use of waterwheels connected, via a chain drive, to a set of buckets that could scoop up water and deliver it for household use or for irrigation. This is simplest, and almost certainly the earliest, application of waterwheels. The wheel itself was a simple paddle wheel that would have been driven by the current of the stream. This type of device is called a noria. Its origin is most likely the Mediterranean region of the ancient world. The use of the noria did not spread so quickly, at least in the ancient Mediterranean world, as was the case with other wheels that we will describe below. This may possibly have been because the noria needs an abundant supply of rapidly flowing water.

By about 100 B.C., waterwheels were in use in several scattered regions of the world—northern Turkey (probably the first), Greece, and India—for grinding grain and extracting oil from olives. These wheels were so-called horizontal wheels mounted so that the shaft stuck up vertically and drove the millstone directly (Figure 8.9). These horizontal wheels did have several practical problems. It's necessary to build some sort of a channel so that the water hits only one side of the wheel, to get it to turn. Because of that, the horizontal wheel only works in fast-flowing streams, such as are found in the mountainous regions of northern Turkey and northern Greece. In addition, the speed of the millstone is exactly equal to the speed of the waterwheel, which in turn depends on the speed of the stream.

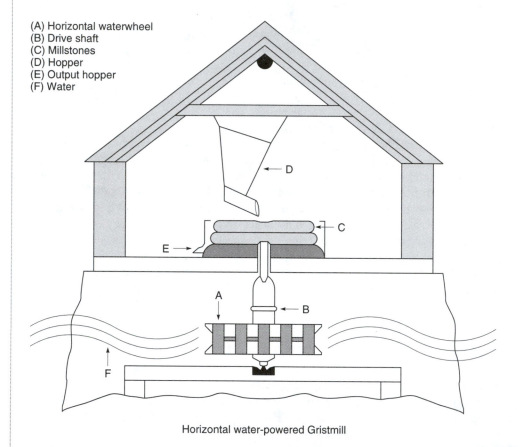

(A) Horizontal waterwheel
(B) Drive shaft
(C) Millstones
(D) Hopper
(E) Output hopper
(F) Water

Horizontal water-powered Gristmill

FIGURE 8.9 As this diagram of a horizontal waterwheel demonstrates, there are many difficulties with horizontal wheels, such as the need for a fast current or the mill's position atop the water flow.

The horizontal waterwheel is also sometimes called the Norse wheel. It was most likely the first device that allowed for the 'mechanization' of the home. Until the time of the development of the horizontal or Norse wheel, grain had to be ground (to make flour) in hand-operated mills. The women of the household had to do this job, and in large households this could keep the women occupied all day. The water-driven mill for grinding grain changed this. It may seem absurd from our vantage point two millennia later, but the introduction of this labor-saving device, and the freedom from the day-long drudgery of grinding grain by hand, caused some moralists to be concerned about what the women would do with their new-found free time.

The oldest known waterwheel, which was used for grinding corn, was described by Strabo⊗ about 24 B.C. as existing at Cabeira along the southern coast of the Black Sea, having formed part of the property which the last king of that region lost when he was overthrown by Pompey⊗ in 65 B.C. The first literary reference occurs in Greek, about 30 B.C. The mill was praised in the following lines:

> Stop grinding, ye women who toil at the mill,
> Sleep on, though the crowing cocks announce the break of day.
> Demeter has commanded the water nymphs
> to do the work of your hands.
> Jumping onto the wheel, they turn the axle
> Which drives the gears and the heavy millstones.

—Antipater of Thessalonica[2]

⊗ Strabo was probably the world's first geographer. His book *Geography* is especially important because it is the best contemporary record available of the countries and peoples known to the Greeks and Romans around A.D. 20.

⊗ Pompey was a great Roman general and political leader. Along with Julius Caesar, Pompey was one of the three leaders to form the 'triumvirate' that ruled Rome around 60–50 B.C. Pompey was overthrown by Caesar, left Rome for Egypt, and was murdered there in 48 B.C.

In ancient China, waterwheels were connected to bellows that were used to supply a high volume of air to iron-making furnaces. This first happened in the region of modern-day Nanjing, in east-central China, around A.D. 30. This development is credited to Tu Shih, an administrator of the region, who was interested in the use of cast iron for making agricultural implements. In ancient China the waterwheels were also used about this time for operating trip hammers⊗ that would be used to crush ore, or pound iron into shape. Another 1,200 years passed before this technology was adopted in medieval Europe, where it may have been invented independently of the much earlier Chinese development.

⊗ A trip hammer is a type of hammer—usually a very large and heavy one that could not be handled by a person—used in such applications as pounding cloth or leather, crushing ores, or hammering metals into desired shapes. The hammer is raised by a mechanical device like a waterwheel or steam engine (Chapter 10). When the hammer has been raised to a certain point, a 'trip' (a generic term for a device that releases a mechanism) allows the hammer to fall, striking a blow. As the waterwheel continues to operate, the hammer is then raised up again, 'tripped' again, and the cycle repeats to provide a continuous series of hammer blows.

Another type of wheel, also of ancient origin and related to the noria, uses pails hung from, or attached to, the blades of a paddle wheel. This device is also used to draw water from streams for agricultural irrigation or household use. It seems to have been developed in India and the Middle East. Variations of the noria, or the pail-on-the-blade design, are still used in some parts of the world today. They are common, for example, in Egypt and in sub-Saharan Africa.

VITRUVIUS AND THE INVENTION OF GEARS

The first clear description of any type of waterwheel occurs in a book by the Roman engineer, author, and architect Marcus Vitruvius Pollio—the first inventor (or at least chronicler) of an energy-related device that we can actually identify in the historical record. Vitruvius describes the design and use of this innovation in his book *De archi-tectura*, which was first published about 27 B.C. and is still available some 2,000 years later.⚛ Vitruvius claimed that the machines he described were in common use, and he attributes their invention to Greek engineers. (Whether the Greek or Roman origin is truly the case is disputed, and we will look at some counterarguments later.)

⚛ Vitruvius's book on architecture is available as a two-volume bilingual edition from the Loeb Classical Library.

The first form of this wheel was used to drive grinding mills for processing grain into flour around 100 B.C. It was also used for extracting oil from olives. So far as we know, the waterwheel had not yet been adapted for operating any machinery. Vitruvius tells of a vertical wheel that is immersed into a stream, creating an undershot wheel similar to that shown in Figure 8.2. With this type of wheel to turn the millstone, we need to connect the shaft of the wheel to the shaft of the stone via a system of gears (Figure 8.10). When we do

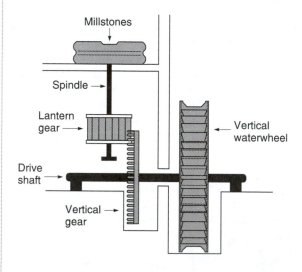

FIGURE 8.10 A water-powered gristmill. The gears in a waterwheel make it possible to transfer the ENERGY into WORK.

this, we are no longer obliged to drive the millstone at the same rate as the waterwheel is turning. By a proper selection of the relative sizes of the gears, we can drive the stone faster (which is usually what's desired) or slower than the wheel.

This development of gearing and its use to change the speed of rotation was the first major achievement in the design of continuously operated machinery. It occurred roughly 2,100 years ago. That seems like a long time ago from our perspective, but, from a different point of view, it means that we humans took about 238,000 years to develop machinery. Even though the undershot wheel and its gear system shown in Figure 8.10 may seem very primitive to us nowadays, we need to appreciate that it's a very recent development in the long span of human history.

WATERWHEELS IN THE MEDIEVAL WORLD

It may be that while we owe Vitruvius the first recorded description, the critical developments were made not in Rome or Greece, but in the north and west of Europe. Part of the reason for believing this is linguistic: most of the words used to describe the technology of water-operated mills derive from Celtic or Germanic languages. There are no corresponding words in the Latin or Greek of the time. The remains of water-operated mills dating from the second and third centuries A.D. have been discovered in the marshes of Denmark. This region is well past the northern limits of the Roman empire. The first mill in Ireland dates from the second half of the third century, near Tara.🔬

🔬 Tara was the site of the court of the ancient Irish kings, until about 600. The name may be more familiar to modern-day readers as Scarlett O'Hara's plantation in the novel and film *Gone with the Wind*.

The overshot wheel was fully developed by about A.D. 400. It may have been used in southern Germany for working a saw used to cut stone, in addition to the usual applications of producing flour and oil. Neither the Romans nor the ancient Chinese adapted the waterwheel for mechanical purposes. During the Middle Ages, however, the use of waterwheels burgeoned. William the Conqueror in 1086 ordered a survey of England, compiled in *The Domesday Book*. The survey showed 5,624 water mills operating in England, one on average for about every 50 families, or one mill per 200–400 people (from which we can infer that the individual mills must likely have been small and of low POWER). Some of these were still in use 700 years later, and a few were modified and were still working into the 20th century! At about the same time, water-powered mills became very important in the Islamic world. By 900 they were supplying about 10 tons of ground grain every day to a city we still hear of in the news—Baghdad.

In the fourth century A.D., the Romans erected a large mill at Barbegal, in what is now modern-day France, in the region of Provence. For its era, this was a prodigious feat of engineering. The installation used eight pairs of overshot wheels. Each individual wheel was about seven feet in diameter and two feet wide. The entire mill was capable of producing an enormous amount of flour per day, variously estimated between 3 and 28 tons. This output was far beyond the needs of the local population; if the high estimate of production is used, enough flour would have been produced for 80,000 people. Most of the flour was shipped to Rome, some was used by the nearby population (estimated at about 10,000), and some went to the Roman troops in southern Gaul (now southern France).

Around 990 a new type of water-operated mill was introduced in the Middle East, in the region of modern-day Iran and Iraq, for processing sugar cane. Such cane-crushing mills are known to have existed at Basra, a city frequently mentioned during the Iran–Iraq and the Persian Gulf wars. Other applications of waterwheels developed in this region during the Islamic period were similar to those being developed in Europe, as, for example, fulling cloth⚛ and preparing pulp for paper.

⚛ 'Fulling' is an operation in the manufacture of cloth, especially woolens. In the early days of clothing manufacture, woven woolen cloth was often quite porous. To make the cloth feel more substantial or 'fuller,' it was pounded with a special type of clay still known as 'fuller's earth' to fill the tiny pores. With modern weaving machinery, fulling is no longer important, but fuller's earth is still used, e.g., as a component of talcum powder.

The importance of water is illustrated, in an odd way perhaps, by the fate of technology when the western half of the Roman empire collapsed around 400. Most of what the Romans achieved, such as their spectacular aqueducts and baths, slowly fell into disuse. In marked contrast, waterwheels were used more widely, as indicated by the *Domesday Book*, and their design continued to be improved.

At about the same time as the Domesday survey, the Chinese were building a large clock. It was operated by a waterwheel about 11 feet in diameter. The wheel revolved at precisely 100 revolutions each day. The mechanism must have been very precise. Before 1300, the Chinese developed the 'great water-driven spinning machine,' for use in making cloth from hemp or ramie.⚛

⚛ Ramie is a tropical Asian plant. The fibers from the stems can be woven to make fabric, cords, or rope.

Early in the medieval period, waterwheels and water-operated mills were an excellent source of wealth for their owners. Typically, the miller was paid for his efforts by being allowed to keep—and then sell or trade—some fraction of the grain being ground. As a result, and as seems to happen at many times in our own society, the wealthy often became the target of jokes, ill feeling, or invective from those who were less well off. This likely arises from jealousy and from the injustice of most such inequities. The medieval millers were no exception to this fate, as is eloquently documented in 'The Miller's Tale' in Geoffrey Chaucer's immortal *Canterbury Tales*:

The miller was a burly fellow—brawn
And muscle, big of bones as well as strong,
As was well seen—he always won the ram
At wrestling matches up and down the land.
He was barrel-chested, rugged and thickset,
And would heave off its hinges any door
Or break it, running at it with his head.
His beard was red as any fox or sow,
And wide at that, as though it were a spade.
And on his nose, right on its tip, he had
A wart, upon which stood a tuft of hairs
Red as the bristles are in a sow's ears.

Black were his nostrils; black and squat and wide.
He bore a sword and buckler by his side.
His big mouth was as big as a furnace.
A loudmouth and a teller of blue stories
(Most of them vicious or scurrilous),
Well versed in stealing corn or trebling dues,
He had a golden thumb—by God he had!

—Chaucer[3]

The kinetic energy of water was the crucial energy source for the society of medieval Europe. Among its applications were crushing wheat for flour, fulling cloth, tanning (crushing the oak bark or 'tan bark'), beer making (crushing the malt and hops), paper mills (shredding and pulping the linen and cotton rags), operating the bellows on iron-making furnaces, hammering iron, operating grindstones for shaping and sharpening tools, and operating sawmills. By the sixth century, it was likely impossible to supply enough flour to feed the population of a large city without using water-driven mills.

When paper-making spread into Europe, the process of pounding the fibers of rags into pulp was carried out mechanically, using a waterwheel to operate trip hammers. This process may have originated in the region around Baghdad. It took about 200 years to pass into northern and western Europe, moving via the region that is modern-day Spain (which at that time was an Islamic region).

By the end of the Middle Ages, waterwheels were also used to raise buckets of ore from mines, to make wire by drawing metal through dies, and in crushing minerals and ores. Perhaps the ultimate application was the use for a water mill for polishing gems (in Paris, in the 1530s). In Peru, water-operated crushing mills were being built in the late 1500s to break up chunks of silver ores. A series of dams and canals was constructed to supply water to the wheels. Unfortunately, a dam broke in 1626, and its failure caused so much damage that the canal/dam/mill system never completely recovered from the resulting devastation.

The use of water in medieval society stands in great distinction with use of water in the ancient world. The medieval societies wholeheartedly adopted water as a source

> WATER WAS AS VITAL AN ENERGY RESOURCE TO MEDIEVAL EUROPE AS PETROLEUM IS TO US TODAY.

of ENERGY for all sorts of mechanical work. However, in the ancient world, little, if any, mechanical WORK was done by water except for the processing of agricultural products (grinding grain and pressing olives). Why this is the case will probably never be known for sure, but two factors are generally agreed upon. First, ancient Greece and Rome were societies that relied on slaves for much of the hard labor. (A seven-foot-diameter wheel is estimated to be able to grind a ton and a half in 10 hours; this work could be achieved by 40 slaves.[4]) The supply of slaves was renewed by captures made during war. The second issue is geography. Many parts of southern Italy and southern Greece do not have reliable supply of fast-moving water that will run the year around. Therefore, water as an energy source, to replace the human muscles of slaves, was not an attractive option in the ancient world.

We owe two other things to the water mills of medieval Europe. First, building a mill with an overshot wheel required a very substantial initial investment, for the mill itself, the wheel, the dam and its sluices. It became difficult for a single individual to have the capital necessary to build a mill. By about 1100 it was common for two to five people to own a mill collectively, taking equal shares in the profits. In 12th century France, a company was established in which there were 'shares' in the mill. The value of each share fluctuated with business conditions. These shares could be bought, sold, traded, or bequeathed. This practice gave rise to the modern stock corporation and the stock exchange. Beginning in the early 13th century, it appears that no millers—the people who actually operated the mill and ground the grain—were themselves among the shared owners (or as we would say nowadays, the stockholders) in the mill. Thus we see the beginnings of a major division in society: that between 'capital,' the people who invest the money and hold the ownership, and 'labor,' those who do the work.

Second, it was quite common for more than one mill to be built along a given stream. This arrangement worked fine provided that nobody later interfered with the water flow. For example, if the upstream mill raised the height of its dam, more water would be impounded in the 'mill pond,' and correspondingly less would be available for the other mills downstream. On June 8, 1278 a law suit was filed by a 'downstream' mill against the 'upstream' mill, which had illegally raised the height of its dam. Suits, counter-suits, and appeals followed, until the case was finally settled when one of the two companies went bankrupt, and there was only one surviving descendant of the original litigants—in 1408, 130 *years* later!

WATERWHEELS IN THE EARLY MODERN WORLD

Water continued to be a major energy source into the modern era. Even though the steam engine was being developed and perfected in the late 18th century and early 19th century, waterwheels could still generate power just as well—and sometimes more cheaply—than steam engines. This was especially the case if a number of factories clustered together around a site of ample, reliable water. Water mills remained the main source of mechanical WORK in England at least into the 1830s.

The pioneer mechanized factory was a mill that produced yarn for making silk stockings, set up at Derby in 1702. Its machines were driven by a waterwheel about 12 feet in diameter. It was taken over by the Lombe brothers, John and Thomas, in 1716. The Lombes' factory in Derby was the prototype of the textile factory whose widespread adoption throughout northern England was a foundation of the Industrial Revolution. It was worked by a single undershot waterwheel, which, by means of gearing, drove the factory machinery, which was in turn tended by some 300 factory workers. For its time, there was no precedent for this scale of mechanization. Although the Lombes' factory was not a commercial success, it did become the model for the use of water to drive spinning machines. It soon became the standard technology in the cotton industry.

The waterwheel was the energy source that allowed for mechanization of the textile industry, which was one of the three major technological developments that constituted the Industrial Revolution in Britain. (The others were the steam engine—to be discussed in Chapter 10—and the emergence of technologies based on iron and coal.) The waterwheel allowed the textile industry to concentrate in large spinning mills located along swiftly moving streams. Those early factories were, by almost all accounts, dreadful places. Child labor was prevalent. The workweek consisted of six 12- or 13-hour days; even in one establishment near Manchester that was considered a 'model mill' the standard workweek for children was 74 hours. The lack of safety guards on machinery caused mutilations such as broken or lost limbs. The damp environment was a breeding ground for tuberculosis.

> *And was Jerusalem builded here*
> *Among these dark Satanic Mills?*

—Blake[5]

A second critical component of the Industrial Revolution was the development of large-scale iron making. This also occurred in Britain, particularly in the region around Coalbrookdale. The crucial technical development was the invention of the blast furnace, that effected the smelting of the iron ore by blowing a rapid stream of high-pressure air (the 'blast') through a bed of iron ore and coke. In the 18th century, the standard blast-furnace technology used a blast supplied by water-operated bellows. Because of the dependence on water for 'blowing' the furnaces, operations sometimes had to stop in dry weather, if there was not sufficient water available to work the bellows. In 1742, a primitive steam engine was installed at Coalbrookdale to pump water needed to keep the water-operated bellows running all the time.

As factories themselves clustered together around good water sources, then the people who worked in them obviously clustered around the factories. This meant a shift of population, with a migration of people into cities that grew up around the factories. In the United States, water power was still the major energy source for factories, textile mills, and metal forges into the 1850s. Industrialization of America began in New England because of the water resources that could be used for water-operated machinery (Figure 8.11). Thousands of such installations were in operation in New England during the early decades of the 19th century.

If we consider the United States as it was in early 1850, with about 30 states mainly on the eastern seaboard, the rivers that were suitable for watermill installations were located primarily in New England and, to a lesser extent, in New York and Pennsylvania. Therefore, most of the factories were in the north and northeast. Because the population concentrates in cities around manufacturing areas, most of the people were in the north and northeast as well. Consequently, when the Civil War broke out, the northern states had an overwhelming advantage in men and material. The southern states had several military advantages, such as being able to fight on their home terrain, having internal lines of communication and transport, and, probably, superior generals. However, once the Civil War became a war of attrition (especially after the battle of Gettysburg) it was inevitable that the north would win, simply on the basis of men and material. This has obviously had very profound implications for our history, and especially for the status of our African-American citizens.

In the late 19th century the waterwheel inspired another invention, which, however, is not particularly related to energy technology. A young man growing up on a ranch in western Nevada watched an undershot wheel, raise buckets of water out of the nearby Carson River. The water was used for watering horses on the ranch. Later he went to college at Rensselaer Polytechnic, in Troy, New York. In those days there was a 60-foot diameter waterwheel operating in Troy, that likely supplied some additional inspiration for the budding young engineer. His name was George Ferris, Jr. Many of us have enjoyed rides on his wheel—the Ferris wheel.

We've seen that waterwheels, operating many kinds of machinery, were the crucial source of WORK in the Middle Ages, and continued to be very important well into the modern era. For the Romans and the medieval Europeans, many of the waterwheel installations were marvels of engineering. Nevertheless, from our vantage point at the dawn of the 21st century, we would find that these massive devices provided remarkably little POWER. A waterwheel might produce about a half horsepower.⚛ By far the largest and most elaborate waterwheel installation of the early modern era was the pumping station at Marly, in France. It was built in the 1680s, involving the labor of nearly 2,000 workers for eight years. Its purpose was to supply water to the fountains of the French court at Versailles. The pumping station itself had 235 pumps. These pumps operated by 13 undershot waterwheels, each about 45 feet in diameter, that pumped

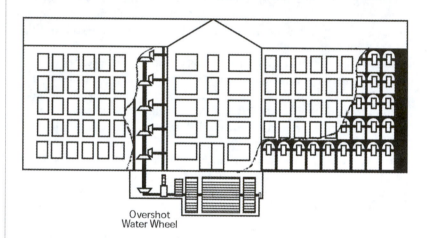

Overshot
Water Wheel

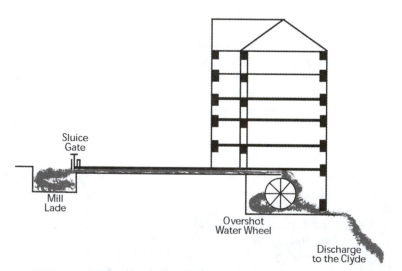

Sluice
Gate

Mill
Lade

Overshot
Water Wheel

Discharge
to the Clyde

FIGURE 8.11 Mills like these used an extremely complex series of gears and pullies to operate vast machinery from the powerful motion of the waterwheel.

water from the river at a rate of about a million gallons per day. The POWER output of this installation was 125 horsepower. Only about 10% of the kinetic energy of the water was converted to WORK. In the early 1800s, this gigantic complex of machinery was replaced by a single 150-horsepower steam engine. Another measure of the relatively low output of POWER and WORK—relative to the size of the wheel, is the estimate that an overshot wheel seven feet in diameter with a water flow of three gallons per *second* would generate a POWER of about 2–2.5 horsepower, equivalent to a small 250 cc motorcycle engine.[6]

⚙ We will discuss the 'horsepower' in more detail in the chapter on steam engines. As the name implies, it is a unit of POWER. A half-horsepower device is not very powerful, having about the same POWER as four 100-watt light bulbs.

Because of its greater efficiency, the overshot wheel was the type most widely used, from about the 18th century to today. However, the critical factor is that the cost of the dam and the necessary millraces and other equipment for handling the flow of water had to be economically justified by the return on the amount of capital invested. Despite its technical disadvantages relative to the overshot wheel, the undershot wheel remained popular, and often proved to be the better choice, at least in certain conditions. For one thing, it is definitely cheaper to build and install than the overshot wheel, especially if it's necessary to build a dam to provide water to the overshot wheel. Therefore, if not much capital is available for investment, the undershot wheel is very attractive. Furthermore, given a fast-moving stream with a reliable flow (i.e, not subject to drastic fluctuations) the POWER of the undershot wheel can be very high, despite its low efficiency. To manage this, the wheel needs to be large. Usually this is not a problem unless the wheel would happen to obstruct navigation in the river. At the end of the 19th century, the largest waterwheel anywhere in the world was a 72-foot-diameter behemoth, an overshot wheel on the Isle of Man. A wheel of this size might produce only 30 horsepower.

Throughout the 18th century, and much of the 19th, waterwheels could provide WORK or generate POWER just about as well as the primitive steam engines of the time. In fact, if there situations where several factories could be built near each other at a site of reliable water flow, it was probably cheaper to rely on water than for each of the factories to have its own steam engine. Well into the middle of the 19th century the kinetic energy of water was still the major source of ENERGY for industry in the United States.

WATER 'POWER' AS A FORM OF SOLAR ENERGY

In the early pages of this chapter, and our discussion of kinetic and potential energies, we've dodged a rather important question: Where does the gravitational potential energy of the water come from in the first place? In a sense, the energy in water is an indirect form of solar energy. The sun is the source of energy for driving the *hydrologic cycle* on Earth (Figure 8.12). Consider water in a lake or in the ocean. As this water is heated by the sun, some evaporates into the atmosphere as water vapor. Since the air over the lake or ocean is also heated, the moisture-laden warm air rises into the atmosphere. Some of this air will be carried over land. The water vapor collects into drops, and collections of drops themselves form clouds. At some point the water will condense, and fall to the ground as rain or snow. The rain will accumulate eventually in streams, which run into larger streams, which run into rivers, which run into lakes or oceans. The cycle can begin all over again. For our purposes, we can tap some of the gravitational potential energy of the water in these streams or rivers as it makes its way back to the oceans. The driving force for the hydrologic cycle is the sun. This is why

WE CAN CONSIDER WATER 'POWER' TO BE AN INDIRECT FORM OF SOLAR ENERGY.

WATER 'POWER' AND THE ENVIRONMENT

The kinetic energy of water is one of the 'cleanest' forms of energy available to us. The water itself is almost pollution free (assuming, of course, that it's not contaminated to begin with). There are no by-products from extracting WORK from the water. There are no emissions to the atmosphere, no contamination of the water, nor is there any radioactivity. As a fluid, water is relatively easy to control. Although most waterwheels are very low POWER, they operate at a high efficiency. Typically 80% to 90% of the kinetic ENERGY of water can be converted to useful WORK. As we will see when we come to examine the concept of efficiency in more detail, this is excellent performance in comparison to efficiencies that are typically less than about 40% for most other energy systems. (As we'll also see in later chapters, efficiency matters greatly when one has to purchase fuel, but an advantage of waterwheels is that running water is free. Furthermore, even a comparatively inefficient undershot wheel requires no elaborate construction work or

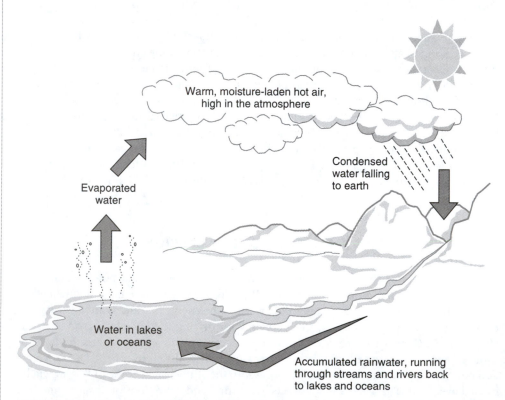

Warm, moisture-laden hot air,
high in the atmosphere

Condensed
water falling
to earth

Evaporated
water

Water in lakes
or oceans

Accumulated rainwater, running
through streams and rivers back
to lakes and oceans

FIGURE 8.12 The Earth's hydrologic cycle is operated by solar energy evaporating water, which later condenses and returns to Earth.

tricky engineering to increase the rate of water flow.) Another advantage is that many times even very small streams can be utilized, with an appropriate-sized wheel, to produce WORK. In fact, waterwheels will operate in streams even that have large fluctuations in the flow rate during the course of a season or year. (However, changing the flow rate of the water will cause a change in the speed of rotation of the wheel.) Of course, the energy source itself—the water—costs nothing.

Although water as a source of ENERGY has numerous advantages, being cheaper than purchased fuels, renewable, and nonpolluting, it is not without its disadvantages as well. Certainly the wheel's operation is halted by freezing, can be 'drowned' in times of flood, or lose water entirely in a sustained drought. Even the dams built to overcome these problems could be made less effective by an accumulation of silt behind the dam. In 19th century America it seemed that getting about 160 days' operation per year was doing well.

CITATIONS

1   de Duve, Christian. *Vital Dust*. Basic Books: New York, 1995; p. 36.
2   Antipater of Thessalonica, in Strandh, Sigvard. *The History of the Machine*. Dorset: New York, 1979.
3   Chaucer, Geoffrey. *The Canterbury Tales*. Oxford University: Oxford, 1998; p. 15. Many other editions of Chaucer are readily available.
4   Gimpel, Jean. *The Medieval Machine*. Penguin, 1976; Chapter 1.
5   William Blake, 'Jerusalem,' in Heaney, S.; Hughes, T. *The Rattle Bag*. Faber and Faber: 1982; p. 221. Blake's poetry is available in many anthologies and editions of Blake's works.
6   Landels, J.G. *Engineering in the Ancient World*. University of California: Berkeley, 1978; Chapter 1.

FOR FURTHER READING

Berry, R. Stephen. *Understanding Energy*. World Scientific: Singapore, 1991. Chapters 2 and 3 of this book provide a discussion to the concepts of kinetic and potential energy and their intercon-version.

Briggs, Asa. *England in the Age of Improvement*; **Porter, Roy.** *England in the Eighteenth Century*. The Folio Society, London, 1998 (Briggs), 1999 (Porter). These two well-written volumes in the Folio Society's *A History of England* series document some of the problems associated with the rise of the factory system in the textile industry.

de Camp, L. Sprague. *The Ancient Engineers*. Barnes and Noble: New York, 1993. This book provides an easily readable history of technology from ancient times through the Renaissance. Chapter 7 discusses Roman waterwheel technology (along with other remarkable inventions, including history's first vending machine—a coin-operated dispenser of holy water).

Derry, T.K.; Williams, Trevor I. *A Short History of Technology*. Dover: New York, 1993. A relatively inexpensive paperback edition of a book that treats the history of technology from ancient times through about 1900. Chapter 8 discusses the use of waterwheels.

Fuller, Edmund. *Tinkers and Genius*. Hastings House: New York, 1955. A history of inventions and technology in New England through the 19th century. Chapter 2 covers the applications of water technology.

Gies, Frances; Gies, Joseph. *Cathedral, Forge, and Waterwheel*. HarperCollins: New York, 1994. A readable and well-illustrated history of technology in the Middle Ages.

Gimpel, Jean. *The Medieval Machine*. Penguin: New York, 1976. This book discusses the development of energy resources in medieval Europe, and how that development triggered an 'industrial revolution' centuries before the better-known one of the 18th century.

Guillerme, André E. *The Age of Water*. Texas A&M: College Station, 1988. A discussion of water systems in northern France over the period 300–1800. It treats not just waterwheels, but many uses of water, including in the military, political, and intellectual communities.

Hawke, David Freeman. *Nuts and Bolts of the Past*. Harper and Row: New York, 1988. This book provides a history of technology in America in the time between the Revolutionary and Civil Wars. Chapter 29 deals with water technology.

Hill, Donald. *A History of Engineering in Classical and Medieval Times*. Barnes and Noble: New York, 1997. Chapters 3, 8, and 9 provide information relating to the use of water as an energy source. The book covers the period from 600 B.C. to about 1450.

Kirby, Richard Shelton; Withington, Sidney; Darling, Arthur Burr; Kilgour, Frederick Gridley. *Engineering in History*. Dover: New York, 1990. This is an inexpensive paperback that provides, in Chapter 5, information on waterwheels.

Macaulay, David. *Mill*. Houghton Mifflin: Boston, 1983. An excellently illustrated description of the building of various water-operated mills in 19th-century New England.

Petroski, Henry. *Remaking the World*. Knopf: New York, 1997. This collection of essays on 'adventures in engineering' (yes, engineers have adventures) includes one on the history of Ferris wheel and its inspiration in an undershot waterwheel used for the watering of horses.

Ramelli, Agostino. *The Various and Ingenious Machines of Agostino Ramelli*. Dover: New York: 1987. This is a relatively inexpensive paperback reprint, with lavish illustrations, of a book first published in 1588. Numerous examples of devices operated by water are provided.

Strandh, Sigvard. *The History of the Machine*. Dorset: New York, 1989. A very well illustrated history of machines, from the use of human muscles through computers. Chapter 4 provides information on water-operated devices.

Usher, Abbott Payson. *A History of Mechanical Inventions*. Dover: New York, 1988. Another in a series of useful, inexpensive paperbacks from this publisher. Chapter VII discusses waterwheel technology through the year 1500.

Vitruvius. *The Ten Books on Architecture*. Dover: New York, 1960. This is the seminal book on architecture, still in print some 2,000 years after it was written. Several editions are available, including a bilingual Latin/English version in the Loeb Classical Library series.

9

WIND ENERGY

To recap briefly from the last chapter, waterwheels were the primary energy source of medieval Europe, as important to the industrial economy of that time as oil is to us today. However, they can only be located near good streams or rivers and become useless if the water freezes. Furthermore, in the Middle Ages, an army attacking a city or castle could starve it by stopping the flow of water to the local mills.

Ancient humans recognized that there is another force in nature that also possesses kinetic energy, but is not subject to the problems associated with water power—wind (Figure 9.1). All of us occasionally experience the force of the wind in various ways—scattering papers that we've dropped, blowing a hat off, turning an umbrella inside out, or even making difficulty in walking into a strong wind.

WINDMILLS IN THE ISLAMIC WORLD

Certainly the ancient Greeks and Romans knew of the energy of the wind, because they had sailing ships. Cloth sails were known to the Sumerians at least by 3500 B.C. The sailing ship represents the first time in history that some source of ENERGY available in nature was successfully harnessed to do a type of WORK. (The use of fire as an ENERGY source predates harnessing the wind, but fire was used to supply heat for warmth and cooking, not for the doing of mechanical WORK.) The Egyptians also had various kinds of sailing ships. As far as we know, the classical Greeks and Romans never developed the sail (from ships) into any rotary device—a windmill—for doing mechanical WORK. Hero of Alexandria (Egypt) describes, around A.D. 50, a wind organ that was supplied air by a piston moved by a wheel equipped with sail-like blades to collect air. Hero's description is about the only mention of windmills in the ancient world. It seems that nobody ever thought that the little toy described by Hero could somehow be scaled up so that the WORK and POWER of wind could be used to relieve humans or draft animals from their chores.

FIGURE 9.1   The stereotypical windmill is a turret mill, originally invented by Leonardo da Vinci.

The first practical development of windmills apparently came in the Islamic countries of the Middle East during the seventh century. The first extensive written description of a windmill is by Abul-Hasan Al-Hasudi, one of the greatest of scientists and historians of the Muslim world, in the 10th century. The Banû Mûsà, 🟠 writing in Baghdad about 850, mentioned a small 'windwheel' that was used to operate a fountain for decorative purposes. Other works of medieval Islamic authors also mention windmills. al-Istakhrî described windmills in the region that is nowadays the western part of Afghanistan. Windmills are described in Syria by the great geographer al-Dimashqî, writing in the latter part of the 13th century. al-Mas'ûdî, writing around the end of the 10th century, tells the story of a Persian inventor who claimed to Caliph 'Umar I that he was able to build a windmill. The caliph (a caliph is a secular and religious leader of a Muslim community or country) called his bluff and made him substantiate his claim by actually doing it. 'Umar I is believed to have ruled in the years 634 to 644, so this story had been around for some 300 years before al-Mas'ûdî recorded it. There is also a legend that a Caliph 'Omar' (possibly the aforementioned 'Umar I) put heavy tax on windmills, making the inventor so outraged that he assassinated the caliph.

🟠 The Banû Mûsà brothers, Muhammad, Ahmad, and al-Hasan, were the sons of noted astronomer, Mûsà Shâkir. When the father died, the three brothers were entrusted to the care of Caliph al-Ma'mûn; they became trusted advisors to this caliph and his successors in Baghdad. They made many contributions to science, especially mathematics and astronomy, and to engineering in waterwheel and windmill technology. Some historians consider that their competence in the technology of devices that operate by differences in wind pressure or water pressure has only been matched in the past century. One of their most important books was the *Book of Ingenious Devices*, published in Baghdad around 850, which contained descriptions of about a hundred mechanical devices.

The stimulus for inventing the windmill in the Islamic world was the need to grind corn in areas lacking steadily flowing streams of water. The Islamic 'vertical mill' looked somewhat like our modern revolving doors. Probably the vertical mill was invented in what is now modern-day Iran in the early seventh century. Some aspects of the design were likely copied, or derived, from sailing ships. One reason for thinking this is that the first 'sails' of windmills were made of cloth, like the sails on ships.

The vertical windmill spread throughout Islam, and eventually to India and then to China. Besides its most important use (grain), it was also used to process sugar cane in medieval Egypt. In China the windmills were used for pumping water and for hauling canal boats. In the Chinese version, the shaft is also vertical, and a bamboo framework carries a circular array of slatted sails, like those of a junk, so pivoted that they catch the wind from any direction.

The mills were supported on foundations built specifically for that purpose, or sometimes either on the towers of castles or on top of a hill. The millstones were housed in an upper chamber or compartment. A lower chamber contained the rotor. The axle was vertical with arms that were covered with a double skin of fabric. The walls of the lower chamber were pierced with funnel-shaped ducts, looking a bit like portholes, with the narrower end pointing inward to increase the speed of the wind when it hit the vanes.

THE MEDIEVAL POST MILL

The design that most of us may mentally associate with windmills apparently was an independent invention in the West, and not a derivative of the Islamic vertical mill. The idea that it would be possible to use the ENERGY of wind (i.e., flowing air), rather than of flowing water, may have come into Europe from the Islamic world. However, the actual design and development may very well derive from the water-operated mills described by Vitruvius. That inference is drawn from the fact that many of the mechanical details of European windmills and the Vitruvius mill are fairly similar.

The idea must have originated at least in the early 12th century. A deed, dated around 1180, recorded a gift of land lying near a windmill to an abbey in Normandy (part of modern-day France). A windmill was built and operating in Yorkshire, England in 1185, based on records of rent payments. Other historical records from the end of the century also mention windmills.

Around the end of the 12th century, Pope Celestine III ruled that windmills should pay tithes to the church. In 1191 a British knight of Bury St. Edmunds refused to pay from his mill. This did not sit well with the local abbot. First, it represented a challenge to, or disrespect for, the authority of the church. Second, the abbot apparently was the owner of the only other mills (operated by animals) in the vicinity, and the knight's refusal may also have represented a business problem for the abbot, since the wind is free, but animals must be fed and sheltered. The knight's argument was based on the proposition that anyone has access to the wind. The abbot replied, 'I thank you as I should thank you if you had cut off both my feet. By God's face, I shall never eat bread till that building be thrown down. You are an old man, and you should know that neither the King nor his Justiciar can change or set up anything within the liberties of this town without the assent of the Abbot and the Convent.'[1] The knight ordered his servants to tear down the mill.

During the 13th century, windmills became common features on the plains of northern and western Europe. As an example, about 120 windmills were built in the vicinity of Ypres (a small city in what is now the southwestern part of Belgium). Illustrations dating from about 1270 show post mills, indicating that, by then, they must have been fairly common.

Wind differs in one crucial aspect from water: water will always flow in a single direction, as determined by the bed of the stream or river, but the direction of wind can shift. Granted, in some regions, including the Islamic regions of ancient Persia, the prevailing wind direction is relatively constant. But in many parts of the world, including much of medieval Europe, the wind direction can change daily, and even sometimes during a single day. Some method has to be found to allow the windmill to take advantage of the energy of the wind, regardless of its direction. Medieval engineers solved this problem by mounting the entire windmill on a large, upright post, so that the whole mechanism is free to turn with the prevailing wind direction.

These devices are called post mills (Figure 9.2). (The post mill design was not necessary in Afghanistan and Iran, because in those regions the wind direction is usually constant.) In post mills the whole structure rotates so that the vanes or sails will be faced to the wind. A few working post mills, which seem to have stepped right out of the illustration of a medieval manuscript, still exist in Britain and Europe. They are not economically useful any longer, but are remarkable testaments to the ingenuity of medieval engineers and to the longevity of their construction practices. Indeed, the post mill is a remarkable engineering challenge. Imagine architects and engineers nowadays proposing to balance an entire structure, complete with its machinery for capturing ENERGY and converting it into useful WORK, on a single post—not only balancing it on the post, but also allowing it to rotate around on that post. Almost certainly we would think them to be incompetent, or possibly completely out of their senses. And yet—based on the surviving examples in Britain and Europe, these devices were not only conceived and built by medieval engineers, they actually worked for about 700 years! Can we expect any factory or power plant built now still to be functioning in the year 2700?

The post mill design spread throughout *northern* Europe. It was used there in the same kinds of applications as were waterwheels. The design was so successful that it was 'exported' to the Middle East by the crusaders (1189–1192). The post mills were probably three to five times as powerful as the Islamic mills. Just as we saw in the last chapter for waterwheels, windmills greatly improved living standards in the medieval world by freeing women from the very time-consuming job of grinding grain by hand.

FIGURE 9.2 A surviving post mill, in which the mill is balanced on a post (inside the building, in this example).

The windmill has several advantages relative to the overshot waterwheel. No capital investment is needed for a dam, sluices, and other water-control equipment. It can operate in arid areas, or in areas where there are no fast moving streams. It will continue to function even in the freezing winter weather of northern Europe. The fact that no one can 'shut off' the wind means that windmills could be used in besieged castles for grinding grain, which is why some medieval castles had windmills built on their walls, for example, a castle built by the Crusaders in 1240 in what is modern-day Syria.

The importance of windmills to the industrial economy of northern Europe was so great that in 1341 the Bishop of Utrecht tried to get a monopoly on the wind blowing in his diocese. The advantages of windmills made them especially important for the cold, windy plains of northern Europe. They were not nearly so important in the warm mountainous regions of southern Europe, with its fast-moving streams that seldom, if ever, froze. Several centuries after the windmills were common in northern Europe, they were still rare in the south, as exemplified by Don Quixote's astonishment at seeing windmills (in Cervantes' novel of about 1605) in Spain. The late appearance of windmills in Spain, which had been an Islamic country for many centuries, suggests that their design migrated into Spain from northern Europe, since windmills with vertical sails were not known to have been used in medieval Islam.

THE TURRET MILL

In time, the post mill became rather impractical because of the difficulty of balancing the large mill on the post, and the problem of turning the entire mill. As happens many times in the history of science and technology, the solution to the problem came from an entirely unexpected source, a person working on a different problem in a seemingly unrelated area.

While the post mill continued to be developed, some visionary individuals were considering an entirely different application of wind energy: the possibility of human flight. From time immemorial our human and hominid ancestors saw birds flying by simply (as it appeared) flapping their wings. It seemed reasonable to expect that some device ought to allow humans to fly too, by using muscles of their bodies. The most notable such attempt may be embodied in the Greek legend of Daedalus and Icarus.

By the 15th century it was obvious, at least to one scientist and inventor living in Milan and, later, Florence, that there were two parts to the problem of flight. One part was simply the propulsion necessary to get off the ground in the first place. The second part was the 'lift' provided by the air flowing past the wings. He managed to find a fairly complete solution to the problem of lift, but never solved the problem of generating enough propulsion energy from human muscles to actually take off. Since many of his notebooks have survived to the present day, we know that eventually his sketches turned from the issue of how to solve the combined propulsion-and-lift problem to new designs for windmills. He is credited with the idea of having *only* the top part of the mill turn with the wind. This design is called a turret mill or tower mill (Figure 9.3). With this new design, builders were no longer limited to using the relatively lightweight material, wood, for the mill. They now could be made larger and built of more permanent materials, such as stone or brick.

The inventor? A person who, very possibly, was the greatest genius who ever lived: Leonardo da Vinci (Figure 9.4).

A left-handed, vegetarian, homosexual bastard, Leonardo da Vinci (1452–1519) contravened most of the accepted norms of his day. Reared by his peasant grandparents in a remote Tuscan village, he had minimal schooling. He was apprenticed as a painter because his illegitimacy debarred him from respectable professions. (Painting in fifteenth-century Tuscany was regarded not as 'creative art' but as a lowly trade, fit for the sons of peasants and artisans.) Lacking literary culture he was scorned in the highbrow Florence of the Medicis. This turned him towards science and observation. 'Anyone who invokes authors in discussion is not using his intelligence but his memory,' he contended.

He was insatiable for newness, both in art and science. His first known drawing was also the first true landscape drawing in western art. He was the first painter to omit haloes from the heads of figures from scripture and show them in ordinary domestic settings, and he was the first to paint portraits that showed the hands as well as the faces of the sitters. His Leda (which does not survive) was the first modern painting inspired

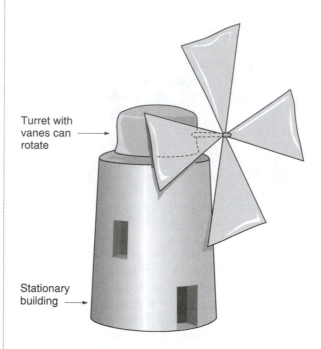

Turret with
vanes can
rotate

Stationary
building

FIGURE 9.3 A turret mill.

RITRATTI DI MEDICI E NATURALISTI ITALIANI DAL SECOLO XV° AL XVIII°

FIGURE 9.4 Leonardo da Vinci.

by pagan myth. His notebooks, of which over 5,000 pages survive, are all written backwards in mirror writing, and are dense with intricate drawings. They record his observations on geology, optics, acoustics, music, botany, mathematics, anatomy, engineering and hydraulics, together with plans for many inventions, including a bicycle, a tank, a machine gun, a folding bed, a diving suit, a parachute, contact lenses, a water-powered alarm clock, and plastics (made of eggs, glue and vegetable dyes).

It is true that Leonardo was not strictly a scientist, not always as original as he seems. His war-machines had already been designed by a German engineer, Konrad Keyser; his 'automobile' by an Italian, Martini. Though he came close to formulating some scientific laws, his insights were sporadic and untested by experiment. He thought of looking at the moon through a telescope a century before Galileo, but he did not construct one. He knew no algebra, and made mistakes in simple arithmetic. His man-powered flying machine, designed to flap its wings like a bird, could never have flown. Apart from anything else it must have weighed about 650 lbs (as against 72 lbs for Daedalus 88, the man-powered aircraft which flew 74 miles over the Aegean in 1988).

Despite these reservations his notebooks give an astonishing preview of the new world science was to open.

—Carey[2]

The turret mill was invented by Leonardo. He designed it around 1500–1502, but it is not certain whether Leonardo's design was ever actually built. The turret mill did not come into general use for about another 200 years. Its practical development was left to the people we stereotypically associate with windmills—the Dutch. The turret mill appears to have originated independently in northwestern Europe in the early 15th century; written descriptions appear in France around 1420. In Holland the most important use of the windmill was for pumping water, to reclaim low-lying wetlands. Other uses included raising materials out of mines, crushing the ores, stretching fibers to make hemp, and operating sawmills. This last application, sawing wood, was especially important for the Dutch navy and its large fleet of commercial merchant ships. Between 1500 and 1650 the amount of land available for farming in the Netherlands increased by about 40%, thanks to the drainage provided by the windmills. In these early Dutch windmills, the sails supplied WORK to move large wooden wheels (designed much like the waterwheels used for irrigation that we discussed in the previous chapter) that removed water from fields in buckets or scoops. This wind-mill-driven 'scoop wheel' may have been invented as early as the 14th century. The use of the windmill for pumping water was common in England by the end of the 17th century.

Despite steady engineering improvements in post mills—such as the invention of the brake, and also a way of adjusting the clearance of the millstones—post mills simply became too unwieldy as they got bigger. This is simply a result of the fact that the entire mill had to be turned, and the bigger they got, the more unmanageable the job of turning became. They were also likely to be wrecked in severe windstorms. In the turret mill, the main portion of the building of the mill is a non-moving solid structure that could be made of brick or stone instead of wood. On top of this solid portion of the building is a revolving turret that holds the sails. With this improvement, mills immediately could be made much larger and more efficient.

The Dutch windmills were one of the first—possibly *the* first—mechanical devices to be the subject of a detailed engineering analysis. This work was done by Simon Stevin. He designed new mills that featured much larger scooping wheels, that revolved more slowly, and geared transmission systems that used more efficient designs of gears. He figured out the design of the gears—that is, how big the wheeled portion should be, and how many teeth each gear should have. He worked out the minimum wind pressure required on a given surface area of each sail or vane that would be needed to lift the water to whatever height would be required for dumping it. On the basis of these calculations, Stevin could also figure out how much water would be raised with each complete revolution of the vanes.

WIND ENERGY IN THE EARLY MODERN AGE

Windmills and waterwheels share the limitation that the work or power has to be used essentially on the spot where it is produced. In fact, it took nearly to the end of the 19th century to figure out how to do work at, or transmit power over, long distances, as we will see later. As late as 1850 most of the corn ground in England and Wales, which in those days were the most populous parts of Europe, was still ground in hundreds of little wind- or watermills scattered all over the countryside.

As a rule, water is a much better source of ENERGY than wind. An important reason for the relatively limited applications of wind is that they are not badly affected by fluctuations in the ENERGY supplied by the wind. This would not be true for, say, operating a factory in cases where a steady, reliable output of mechanical WORK was needed all day long. Even for milling grain, water (if available) was preferred during medieval times. For example, records for the county of Hampshire in England show waterwheels outnumbering windmills by a ratio of nearly 25 to 1 during that time. Why then build a windmill? One obvious reason would be a lack of a water source nearby. Another would be a steady, reliable source of wind at the locality. A third factor might be ease of access. That is, if a farmer had to transport his grain a long distance to get to a watermill and then haul the flour back, it might just be easier to put a windmill on his own property.

Even though the steam engine (Chapter 10) eventually took over as the major source of WORK and POWER in industry, even in the 19th century improvements were still being made to windmills and watermills. By the 19th century, for example, the state-of-the-art windmill design included vanes mounted on springs, a fantail (described in the next paragraph), and a governor to help regulate speed. It's possible that even if the steam engine had never been invented, the 19th century could have achieved significant industrial growth solely on the basis of wind and water.

The fantail was invented in 1745 in England by Edmund Lee. The fantail is mounted at a right angle to the main blades of the windmill (Figure 9.5). As

FIGURE 9.5   A model fantail windmill.

the wind shifts direction, it strikes the vanes of the fantail and causes the body (in the case of a post mill) or turret to turn so that the blades are again moved into the wind. In essence, the fantail is a separate, little windmill mounted on the rear of the turret. If the wind shifts way from the main set of vanes, the fantail will likely catch the wind instead. Then, through a sequence of gears, the wind-operated fantail will turn the turret of the main mill so that it will stay facing the wind. Any slight shift in wind direction will be compensated by the fantail's immediate correcting motion of the turret. This invention, from 19th-century Britain, is one of the first examples of using 'feedback' to establish a self-regulating machine.

Windmills and watermills share another limitation. If we consider a windmill, for example, the air (wind) moving past the blades of the mill still has some kinetic energy. If it didn't, the air would have to be perfectly still just downwind of the mill. The windmill works by having the kinetic energy of the wind do WORK on the sails or blades. Since the air or wind moving past the mill still possesses some kinetic energy (again, we know this is true because if it didn't have any more kinetic energy, the air would be perfectly still or quiet downwind of the mill), the kinetic energy of the wind was not converted completely to WORK. In fact, the same thing applies to waterwheels. Here too we know this because the water leaving the wheel still has some kinetic energy (because it is moving). Thus we have not converted the kinetic energy of the water *completely* to work. (Yet again, if we had done so, the water would have to stop completely.)

This pair of observations leads to an observation that we will explore in more detail later:

> IT APPEARS THAT IT IS IMPOSSIBLE TO CONVERT ENERGY COMPLETELY (I.E., 100%) INTO WORK.

As we will see, this has extremely important implications. In fact, we've already seen something very much like this (in Chapter 4):

> ENERGY CANNOT BE CONVERTED COMPLETELY TO WORK (I.E., WITH 100% EFFICIENCY). SOME ENERGY IS INEVITABLY WASTED.

There seems to be a trend here!

In the 18th and 19th centuries as many as 100,000 windmills may have been operating in Europe. Despite this enormous number of operating windmills, and despite their advantages, it eventually proved that they could not compete with the steam engine. By the middle of the 19th century, milling grain and doing all of the many mechanical jobs of the windmill had largely been taken over by the steam engine. The steam engine didn't have to depend on weather conditions and the prevailing wind. The steam engine—as a locomotive—could also provide cheap transportation of the grain and flour, largely wiping out the issue of distance. Over the last part of the 19th century, and into the 20th, windmills steadily faded as useful devices. The 'last stand' of the traditional windmill was in the American West. The pioneers who settled the west used windmills to pump underground water to the surface for cattle. In a sense, windmills 'won the West,' at least for many ranchers.

WHERE DOES THE WIND COME FROM?

The phenomenon that we recognize as wind is a mass of air that is in motion relative to the Earth. Like any other body or object in motion, the wind possesses kinetic energy. Using a windmill, we can capture some of that kinetic energy and transform it to useful work as, for example, in pumping water. As wind flows, it loses some of its kinetic energy as a result of friction with the Earth's surface and—though not so readily apparent—of friction within the wind itself. If the energy losses due to these kinds of friction were not restored, then eventually so much kinetic energy would be lost that the wind would 'run down' and stop blowing. Of course this does not happen. We all know that wind blows every year, and, in some parts of the world such as the Northern Great Plains or in Patagonia, it seems to blow all the time, day in and day out. From our discussion of the energy balance (Chapter 4), we cannot extract an unlimited amount of ENERGY from a SYSTEM, because at some point all of the stored energy would become depleted and any further removal of ENERGY would be impossible. Somehow the energy that is lost from the wind, or that we extract from it, must be replenished.

The ultimate ENERGY source that causes wind, and that replenishes the energy lost from wind, is the sun. Wind is caused by the unequal heating of various parts of Earth's atmosphere by energy from the sun. Not all locations on Earth's surface are heated to the same extent by the sun. Air that is above the hotter locations expands and, as it does so, rises. As this warm air rises into the atmosphere, air from cooler locations—those that have not received so great a share of heating by the sun—flows toward the area where the warm air is rising. Thus

WINDS ARE A MANIFESTATION OF SOLAR ENERGY.

CITATIONS

1   DeBlieu, Jan. *Wind*. Houghton Mifflin: Boston, 1998; pp. 214–216.
2   Carey, J. *The Faber Book of Science*. Faber and Faber: London, 1995.

FOR FURTHER READING

Brooks, Laura. *Windmills*. MetroBooks: New York, 1999. An extensively illustrated 'coffee table' overview of windmills. Chapter 2 treats their history.

de Camp, L. Sprague. *The Ancient Engineers*. Barnes and Noble: New York, 1993. This book provides an easily readable history of technology from ancient times through the Renaissance. Wind technology is treated in Chapter 9.

Derry, T.K.; Williams, Trevor I. *A Short History of Technology*. Dover: New York, 1993. A relatively inexpensive paperback edition of a book that treats the history of technology from ancient times through about 1900. Chapter 8 discusses wind technology.

Gies, Frances; Gies, Joseph. *Cathedral, Forge, and Waterwheel*. HarperCollins: New York, 1994. A readable and well illustrated history of technology in the Middle Ages, including much information on windmills.

Gimpel, Jean. *The Medieval Machine*. Penguin: New York, 1976. This book discusses the development of energy resources in medieval Europe, and how that development triggered an 'industrial revolution' centuries before the better-known one of the 18th century.

Hill, Donald. *A History of Engineering in Classical and Medieval Times*. Barnes and Noble: New York, 1997. Chapter 9 provides information relating to the use of wind as an energy source, covering the period from 600 B.C. to about 1450.

Hills, Richard L. *Power from Wind*. Cambridge University: Cambridge, 1994. A history of windmill technology, with numerous excellent illustrations.

Ramelli, Agostino. *The Various and Ingenious Machines of Agostino Ramelli*. Dover: New York: 1987. This relatively inexpensive reprint of a book first published in 1588 provides some examples of devices operated by wind. Excellent illustrations clearly indicate how each device functioned.

Strandh, Sigvard. *The History of the Machine*. Dorset: New York, 1989. A very well illustrated history of machines, from the use of human muscles through computers. Chapter 4 provides information on wind-operated devices.

Usher, Abbott Payson. *A History of Mechanical Inventions*. Dover: New York, 1988. Another in a series of useful, inexpensive paperbacks from this publisher. Chapter VII discusses wind technology through the year 1500.

White, Michael. *Leonardo: The First Scientist*. St. Martin's: New York, 2000. This new biography of Leonardo da Vinci is perhaps the first one that provides extensive treatment of his inventions and investigations of science.

10

THE STEAM ENGINE

The highest development of the medieval world was based entirely on three ENERGY sources: draft animals, water mills, and windmills. We've said that the mills were as important to the industrial economy of medieval Europe as oil is to us today. The transition from the medieval to the modern world—in terms of energy and its use—is marked by a shift to reliance on new energy sources, especially coal, and on new devices for doing WORK, of which the first major invention, and one of the most crucial, was the steam engine. As we move from the medieval to the modern world, we will also see that the pace of invention and discovery accelerates rapidly.

The steam engine derives from the confluence of two totally separate issues, which will appear at first to be completely unrelated to each other, and furthermore to be completely unrelated to the steam engine. One was the problem of water in mines. The other, which we'll discuss in more detail in Chapter 11, is the work of the early scientists on studying the properties and behavior of gases.

THE 'PREHISTORY' OF STEAM

Some early probings of the possible uses of steam were made by individuals we've already met. The first demonstrations of the possible motive power of steam are due to Hero of Alexandria (around A.D. 50). Besides the water-powered bellows we've discussed previously, Hero also invented toys that worked by steam. Leonardo da Vinci recorded in his notebooks a steam-operated gun. Denis Papin, a professor of mathematics at the University of Marburg (a city in what is now the west-central part of Germany) designed and built a 'steam digester,' which was the first pressure cooker. He is also credited with inventing the safety valve, which would release steam if the internal pressure got too high. Papin demonstrated his device by cooking some meals with it. Much more importantly, he also conceived and demonstrated how a piston could be moved upward by low-pressure steam and returned downward by atmospheric pressure.

OTTO VON GUERICKE AND THE FORCE OF THE ATMOSPHERE

During the mid-16th century, Girolamo Cardano wrote of steam power and how condensing steam can produce a vacuum. However, the person who truly established the foundation for the eventual development of steam as an energy source was Otto von Guericke (Figure 10.1), the versatile mayor of Magdeburg (a city in what is now western Germany). He was interested in experimenting with vacuums.

von Guericke created an apparatus consisting of two hemispheres that were very carefully fitted together. The apparatus is sometimes named in honor of von Guericke's home town, 'the hemispheres of Magdeburg' (Figure 10.2). Using an air pump, von Guericke created a vacuum inside the hemispheres. Then, in a public exhibition in 1672, von Guericke hitched a team of eight horses to each hemisphere. The two eight-horse teams were unable to pull the hemispheres apart. (Some modern observers believe that, considering the relatively poor status of precision machining and the crude nature of air pumps in von Guericke's time, these must have been very weak horses indeed, but no matter, the essential point was demonstrated.) von Guericke's public exhibition showed that the air itself can exert a tremendous amount of force.

In a related experiment on the same principle, von Guericke showed—in another public demonstration—that when a partial vacuum was created below a large piston working in a cylinder, the combined strengths of 50 men could not prevent atmospheric pressure from driving the piston into the cylinder.

Another, lesser known, experiment provided remarkable evidence of the potential effects of the pressure of the atmosphere. von Guericke connected his air pump to a

FIGURE 10.1 Otto von Guericke, the mayor of Magdeburg.

FIGURE 10.2 A public demonstration of the inability of numerous heavy weights (on the platform) to overcome the force of the atmosphere pushing against the partial vacuum in the hemispheres of Magdeburg.

spherical copper container. Two men worked the pump, to remove as much of the air as possible. The sphere suddenly crumpled, in what was the world's first recorded 'implosion.' Reports of the imploding copper sphere, and the hemispheres of Magdeburg, very likely inspired other scientists to recognize that air pressure, pushing against a vacuum, had considerable force and could be harnessed somehow to do useful WORK.

The two key observations that directed the attention of scientists and engineers to the possibility of using the properties of steam to do WORK were first, the understanding that the atmosphere possessed weight; and second, the discovery that it was possible to create a partial vacuum, either by using an air pump or by condensing steam in an enclosed vessel.

THE ATMOSPHERIC ENGINE

Papin and Savery

von Guericke's demonstration led other scientists to recognize that somehow the air itself could be harnessed to do WORK, if only there were some convenient way to create a vacuum, as von Guericke had done inside the hemispheres.

At the Académie des Sciences in Paris the ideas of the great Dutch scientist Christiaan Huygens encouraged his research assistant, a French scientist, Denis Papin, who sometimes used the first name Dionysius (Figure 10.3). Huygens had been inspired in turn by Leonardo da Vinci's idea: If gunpowder could drive a cannon ball down the bore of a cannon, as it so obviously can, then why couldn't it be used to operate an engine? Evidently Papin was familiar with the experimental devices used in those days to assess the quality of a batch of gunpowder. He then reasoned that if the amount of gunpowder and its explosion within the cylinder could be controlled or regulated, then the movement up and down of the cylinder head (with each succeeding gunpowder explosion) could be used as the driving force for an engine. Perhaps fortunately for posterity, Papin soon realized that a gunpowder engine was rather impractical and turned to steam instead.

> In retrospect, of course, the idea of operating machinery or driving a vehicle using a 'gunpowder engine' may seem more than a little hair-raising. However, we should consider that Huygens, or perhaps more appropriately, Leonardo had hit upon the concept that we now call the internal combustion engine. We will see more about the concept of explosions inside engines when we discuss engine knock and the diesel engine in later chapters.

Around 1690, Papin had the idea that one way, possibly a very useful way, of making use of steam would be to use a piston and cylinder (Figure 10.4). Papin was also the first to suggest the use of steam in this device. He suggested that water could be boiled inside the cylinder. The piston would then be pushed upward, rising on the column of steam made from the boiling water. Then, if the fire is removed, the steam would condense, creating a partial vacuum inside the cylinder. When the steam condensed, the air pressure would force the piston down. If the piston is connected to something, then WORK can be done, e.g., raising a heavy load.

Since it is a property of water that a small quantity of it turned into vapour by heat has an elastic force like that of air, but upon cold supervening is again resolved into water, so that no trace of the said elastic force remains, I concluded that machines could be constructed wherein water, by the help of no very intense heat, and at little cost, could produce that perfect vacuum which could by no means be obtained by gunpowder.

—Papin[1]

FIGURE 10.3 Denis Papin.

In 1690 an engine was built that used steam to raise a piston to the top of a vertical cylinder. At the top of its stroke the piston was held in place by a latch. After the steam had condensed in the cylinder, the latch was released and atmospheric pressure forced the piston down. Historians of technology differ on whether Papin himself built the engine; some accounts suggest that though Papin had conceived the design for what eventually became the steam engine, he never actually built one. Papin died poor and unknown, unrecognized for his work.

The English inventor, Thomas Savery, then made several significant improvements to Papin's original design, around 1698. He introduced the idea of using a separate boiler to create the steam, rather than boiling the water right in the cylinder itself. He also recommended condensing the steam more thoroughly and more quickly by pouring cold water onto the cylinder.

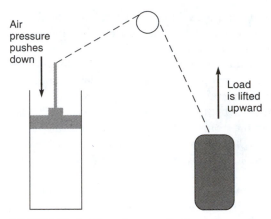

FIGURE 10.4 The force of the atmosphere pushing against a partial vacuum can lift a weight or load.

In the late 17th century in England, coal mines had already been dug so deep into the earth that there were serious problems with water getting into the mines and flooding them. Some of this water was surface water that ran down the mine shaft, but much of it was water from underground streams. Water could be hauled out in buckets, or it could be pumped, over very short vertical distances. The bucket system or the crude pumps could have been worked by draft animals or by the miners themselves.

Savery received a patent for a steam-actuated pump in 1698. In the following year, he demonstrated a working model of an engine to the Royal Society. Savery advertised his machine as the 'Miner's Friend.' It was at best a first, faltering step on the way to modern machinery and engines. It was inefficient, expensive to buy and operate, and, at times, downright dangerous. It operated using high-pressure steam, and unlike Papin's pressure cooker, it didn't have a safety valve. With the comparatively crude metal-working technology of the day, the high temperature of the engine would melt solder and cause the joints to split. Even so, Savery's Miner's Friend deserves recognition as a significant achievement, because, with all its faults, it was still the first real, working device to go beyond the medieval technology of waterwheels and windmills.

> *I am very sensible a great many among you do as yet look on my invention of raising water by the impellent force of fire a useless sort of project that never can answer my designs or pretensions; and that it is altogether impossible that such an engine as this can be wrought underground and succeed in the raising of water, and draining your mines, so as to deserve any encouragement from you. I am not very fond of lying under the scandal of a bare projector, and therefore present you here with a draught of my machine, and lay before you the uses of it, and leave it to your consideration whether it be worth your while to make use of it or no. . . .*
>
> *For draining of mines and coal-pits, the use of the engine will sufficiently recommend itself in raising water so easie and cheap, and I do not doubt but that in a few years it will be a means of making our mining trade, which is no small part of the wealth of this kingdome, double if not treble what it now is. And if such vast quantities of lead, tin and coals are now yearly exported, under the difficulties of such an immense charge and pains as the miners, etc. are now at to discharge their water, how much more may be hereafter exported when the charge will be very much lessen'd by the use of this engine every way fitted for the use of the mines?*
>
> —Savery[2]

Soon it was realized that the energy in steam could be used to obtain the desired reciprocal motion. As steam is forced into the cylinder of the device, a modification of Papin's piston-in-cylinder design, it pushes the piston up, against the weight of the atmosphere. Now, suppose the supply of steam is shut off, and the steam is made to condense back to liquid water. The volume of the liquid water produced will be, roughly, about one-thousandth the volume of the steam. Condensing the steam thus creates a near-vacuum in the cylinder. The weight of the atmosphere on the piston then pushes it down. By connecting this device to a pump, the energy in the steam could be used to operate the pump, and thus to pump water out of the mines. When the piston is pushed up, it pushes—via the connection in the lever—the pump handle down. When the piston falls, it raises the pump handle.

Savery's design was actually used in a number of mines to operate water pumps. It had several defects, however, some of which were mentioned earlier. A major one was that the iron cylinder where the steam was condensed back to liquid water then had to be reheated with every cycle of operation (that is, with every stroke), which is very wasteful of fuel.

Thomas Newcomen

The first truly successful 'steam' engine is attributed to another English inventor, Thomas Newcomen. His design, which dates to about 1712, involves using a very heavy counterweight to raise the piston (Figure 10.5). He also developed a better way of condensing the steam—a spray of cold water *into* the cylinder. (Some historians attribute this invention to Newcomen's observing a Savery engine in which the cylinder had cracked, due to the poor quality of metal in those days, and cold water leaked inside.) Newcomen also realized that it is uneconomic to condense *all* of the steam. It's better, he decided, to condense only part of the steam, keeping the cylinder fairly warm, and letting the engine work more quickly. We must recall that POWER = WORK/time. If we decrease the amount of time required to do a given amount of work, then the engine is more *powerful.*

FIGURE 10.5   An early Newcomen engine.

Thomas Newcomen was a hardware dealer with little formal education. His friend and partner was a plumber, John Calley. Newcomen and Calley tinkered for years with Savery's pump. They developed what is, in essence, a totally different device, but one that derived almost entirely from existing techniques and parts. Their engine worked at low pressure, so it was safer than Savery's, and cheaper to build. It was still very inefficient.

The Newcomen design was made fully usable in the mid-1720s. Newcomen's engine was the first new practical way of doing WORK since the extensive adoption of water-wheels some 700 years earlier. In the course of 40–50 years, Newcomen's engine became very widespread throughout Europe, for operating pumps and other kinds of machinery. Nevertheless, Newcomen's engine had a number of flaws.

The Newcomen engine was cumbersome and very slow. It sometimes required several minutes for the piston to go up and down, and the cylinder had to be heated and cooled on each stroke. Because of the comparatively crude manufacturing methods and metallurgical skills of the time, the engine leaked badly. The ability to do WORK was limited by the fact that the real effort was supplied by the pressure of the atmosphere. The engine was extremely wasteful of energy because, at one point in its operating cycle, the cylinder has to be hot (full of steam) but at another point in the cycle the cylinder has to be cold (to condense the steam). The engines were of enormous size and yet produced about four horsepower. Probably no more than about 1–2% of the chemical potential energy in the coal was eventually transformed into useful WORK. Fortunately, most of these engines were used at coal mines, where the coal was essentially free. And yet despite all these flaws, Newcomen's engine was an immediate commercial success. Forty years later more than a hundred were in use, mostly in pumping water from coal mines. The Newcomen engine was notorious for its high fuel consumption. A contemporary criticism of the Newcomen engine was that it took an iron mine to build one and a coal mine to keep it going. In particularly bad cases, the wastage of fuel was over 99%, meaning that less than 1% of the chemical potential energy in the coal was transformed into useful WORK.

Not only did Newcomen's engine become an immediate commercial success, it was a long-lived one as well. A Newcomen engine used at a coal mine in Derbyshire (England) was in operation from 1791 to 1918. In fact, the last surviving Newcomen engine (in Yorkshire, England) was not scrapped until 1934 (!) and had operated running for more than a hundred years without needing extensive maintenance.

It's important to recognize that in the Newcomen engine the steam actually only functions as an intermediary agent. The steam is used to create the necessary vacuum. It is the pressure of the atmosphere pushing against the vacuum that is the source of WORK. For that reason, the Newcomen engine is more appropriately called an atmospheric steam engine, or an atmospheric engine.

And, despite the notorious inefficiency of the Newcomen engine, it was better than anything else then available. In 1775 a Newcomen engine, built by the English engineer John Smeaton, was sold to Russia for the purpose of pumping out the naval dry docks at Kronstadt. The new engine replaced two windmills, each 100 feet high, that had been installed by Dutch engineers some 60 years earlier. The windmills took a year (!) to pump the water out of the dry dock; the Newcomen engine did it in two weeks.

JAMES WATT AND THE STEAM ENGINE

Despite its inefficiencies, Newcomen's engine gradually gained applications in working pumps and other machinery in England and continental Europe. Even universities began to acquire models of the Newcomen engine to study. One such model engine was at the University of Glasgow. By the 1760s, the university's model had been broken for some time, and finally someone took the broken model engine to the university's scientific instrument maker to have him repair it. On Easter Sunday of 1765 he was thinking about the working of this model engine while walking on Glasgow Green. Suddenly he realized what the major problem was with the Newcomen engine—the one cylinder must alternate between being hot and being cold. This is very inefficient operation and very wasteful of fuel. He recognized that it would be much better to have two chambers—one, the cylinder with piston, that is always hot and one that is always cold (Figure 10.6).

> . . . as steam was an elastic body, it would rush into a vacuum, and if a communication was made between the Cylinder and the exhausted vessel, it would rush into it and might be there condensed, without cooling the Cylinder.
>
> —Watt[3]

This modification incorporated two major advances: By eliminating the temperature cycling (hot-cold-hot-cold, etc.) in the single cylinder, the efficiency of operation is greatly increased. In the Papin, Savery, and Newcomen designs it is really the pressure of the *air* that moves the piston. The function of the steam is simply to provide a convenient way of making a vacuum. The new device actually uses the expansion of the steam to do the work of the engine so, strictly speaking, it is the first true *steam* engine. The instrument maker, the great Scot who came up with these ideas, was James Watt. A steam engine is an apparatus or device for converting heat (in steam) into kinetic energy. A steam engine is one representative of a more general class of devices called *heat engines*. We will examine the concept of heat engines in more detail later.

James Watt was the son of a man who repaired ships and made nautical instruments. Watt was sickly as a boy, so had the great good fortune of not having to go to school. Instead he was schooled at home. His father made his workshop the 'school.' At the comparatively young age of 21, he had become the 'mathematical instrument maker' to the University of Glasgow. In the mid-to-late 18th century, the Scottish universities, Edinburgh and Glasgow, were possibly the best in the world. Watt's employment at Glasgow let him meet some of the world's leading scientists and to learn of their ideas.

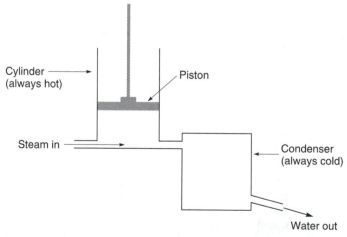

FIGURE 10.6 A schematic of Watt's contribution—the condensing chamber that is always cold, allowing the cylinder always to be hot.

Watt's steam engine rapidly became *the* crucial device for powering all sorts of factory machinery in virtually every industry in Britain, continental Europe, and America. Over a thousand steam engines of Watt's design were installed in factories in the 1790s. This is more than the total number of steam engines of all kinds sold in the period 1700 to 1790. Part of the reason for this is that the Newcomen engine had largely been restricted to use in operating pumps, because they provided WORK only on the downward stroke of the piston. Essentially, this meant that the engine was limited to unidirectional motion. This unidirectional motion for doing WORK meant that there was no easy way to apply the Newcomen engine to operate most kinds of factory machinery. Machinery usually required a continuous rotary motion—in fact, just like that produced by a waterwheel. Watt invented the so-called double-acting cylinder, in which steam that was alternately admitted and released from *both* sides of an enclosed piston provides WORK in both directions.

Watt actually made a series of improvements, both to the earlier engine of Newcomen and to his own original design for the steam engine. As we've seen, one of the crucial improvements was a separate chamber to condense the steam. A second very important contribution of Watt was the invention of a system of so-called sun-and-planet gears⊛ that converted the reciprocating (i.e., up-and-down) motion of the piston into rotary motion of a shaft or axle. Watt worked hard on further improving his invention. He invented a governor that automatically controlled the engine's output of steam, and, later, a gauge for determining pressure in the cylinder and an 'indicator'⊛ for following changes in steam pressure during the operating cycle of the engine.

> *I intend in many cases to employ the expansive force of steam to press on the pistons, or whatever may be used instead of them, in the same manner in which the pressure of the atmosphere is now employed in common [Newcomen] engines.*
>
> —Watt[4]

⊛ The gear systems that we described in Chapter 8, with waterwheel systems, are the simplest application of gears. One gear meshes with the other. One is driven by something (e.g., the waterwheel) and the other is caused to rotate because it meshes directly with the first. Suppose, however, that we had a different system, in which one of the gears was held stationary. The remaining gear would then rotate around it. If the sun-and-planet gears are attached to some sort of casing, the rotation of the planet gear around the sun gear results in a rotary motion. Connecting the casing to the reciprocating piston rod of a steam engine converts the reciprocating motion into rotary motion. This is one method of converting reciprocating motion (as from a steam engine) into rotary motion. Sun-and-planet gear systems of much greater complexity are in everyday use, in, for example, the automatic transmissions of automobiles.

⊛ The 'indicator' of a steam engine, one of the many improvements introduced by Watt, is a device intended to measure the 'indicated work.' One common form of the indicator consists of a very small cylinder with piston that, like the main engine itself, is operated by the pressure of the steam. Through a system of springs and levers, the other end of the piston is connected to a marking pen. The pen rests against a vertical drum, to which is attached a sheet of paper. The drum itself is connected to some part of the engine, and is caused to rotate as the engine operates. The paper moves horizontally as the drum rotates, while the marking pen moves vertically. The extent of horizontal motion of the drum is proportional to the stroke of the engine; the extent of vertical motion of the marker is

> proportional to the steam pressure in the engine. Because both motions occur simultaneously, at the end of test the pen will have traced out a closed, curved shape. This curve is called an indicator diagram. The area enclosed by the closed curve determines the WORK done by the steam in driving the piston. The characteristic shape or form of the closed curve shows how the steam pressure in the cylinder varied from point to point throughout the stroke of the piston, and is useful in checking the operation of various valves in the engine.

The ability of the Watt engine to drive rotating machinery made the steam engine attractive to the British cotton industry, which in those days was the main user of automatic machinery. In turn, this freed industry from a reliance on water. Consequently, manufacturers could build larger factories, and could locate them nearer to sources of raw materials and population, rather than near sources of water.

It was Watt's invention of the governor⚛ that made the easy adaptation of the steam engine to rotary machinery possible. Watt's governor adjusted the amount of steam passing into the cylinders according to how rapidly the machinery was going. This is the first example of real feedback control of a machine. The Greek word for 'governor' is *kybernetes*. It has come into our language as 'cybernetics.' That is, fundamentally, what cybernetics means—governorship. Watt's governor and Papin's safety valve were the two essential devices that made steam engine operation safe, at least in terms of preventing them from running out of control and possibly exploding.

> ⚛ The governor of an engine, another of Watt's inventions, is a device for automatically regulating engine speed. One simple form consists of two weighted balls connected to a hinge at the top of a rotating spindle. The spindle is connected to the engine, and is rotated as the engine operates. As the speed of the engine increases, the balls tend to fly outward; in doing so, they partially close the valve that admits steam to the cylinder. As the engine slows, the balls will drop down toward the spindle; in this case, they will open the steam valve more widely. As the engine speeds up beyond a desired value, the governor restricts steam flow, and the speed drops. If the speed drops below the desired value, the governor opens the valve and speed increases.

By 1790, Watt's engines, steadily improved by his stream of inventions, had almost totally replaced the older Newcomen engines. To improve efficiency even more, Watt installed a steam jacket around the cylinder, to keep it as hot as the steam itself. These improved engines had a fuel consumption one-third that of a typical Newcomen engine for the same amount of WORK. This gain of efficiency, from the 1–2% of the Newcomen engine to 6–7% in the Watt engine, made all the difference in accounting for Watt's engine displacing Newcomen's. As we've seen, the Newcomen engine was so inefficient that it could only be used in coal mines, where the coal cost nothing. By the time of the development of Watt's engine time there were other mines that needed to be pumped, such as the tin and copper mines in Cornwall. The Newcomen engine was not competitive in that application because the coal had to be *purchased* at the Cornish tin and copper mines. The extremely low efficiency of the Newcomen engine made the cost of purchased coal prohibitive.

The engineers of that era spoke of the 'duty' of an engine. The term probably derives from the practice of having humans or animals perform WORK (operating machinery) by serving a turn of duty in a treadmill. The duty was essentially the WORK done multiplied by the time it took to do it. For steam engine operation, the duty of the engine was expressed in terms of the amount of water pumped from a mine for a given quantity of coal that was burned. James Watt measured the amount of work a horse could do by making it pull a known weight, lifted over a pulley, up to a certain height. He conceived the idea of work being the product of force and distance and of power being the rate of doing work. The name of the unit of POWER developed by Watt is, of course, the **horsepower**.

Essentially, James Watt's invention revolutionized industry at the end of the 18th century. The engine quickly became the principal means of doing WORK in factories.

> *Our road turned to the right, and we saw, at the distance of less than a mile, a tall upright building of grey stone, with several men standing upon the roof, as if they were looking out over battlements. It stood beyond the village, upon higher ground, as if presiding over it—a kind of enchanter's castle, which it might have been, a place where Don Quixote would have gloried in. When we drew nearer we saw, coming out of the side of the building, a large machine or lever, in appearance like a great forge-hammer, as we supposed for taking water out of the mines.* It heaved upwards once in half a *minute with a slow motion, and seemed to rest to take a breath at the bottom, its motion being accompanied with a sound between a groan and a 'jike.' There would have been something in this object very striking in any place, as it was impossible not to invest the machine with some faculty of intellect; it seemed to have made the first step from brute matter to life and purpose, showing its progress by great power. William made a remark to this effect, and Coleridge observed that it was like a giant with one idea. At all events, the object produced a striking effect in that place, where everything was in unison with it . . .*
>
> —Wordsworth[5]

This device was the second Watt engine to be built in Scotland (1788).

'William' was the author's brother, William Wordsworth, who is generally considered to have been the greatest of the English Romantic poets. Examples of his work can be found in most anthologies of great poetry, and certainly in anthologies dealing with the poetry of the Romantic era (i.e., the late 18th and early 19th centuries), as well as in collections devoted entirely to his work.

'Coleridge' was Samuel Taylor Coleridge, another of the great English poets of the Romantic era. Perhaps his work best known nowadays is 'The Rime of the Ancient Mariner.' He also had a vigorous and productive career thinking and writing about religion and about politics, as well as being a literary critic.

Humankind was no longer dependent on draft animals, rivers, or the wind. It should not diminish Watt's luster in any way to point out that Watt did not invent the steam engine essentially out of nothing. Watt's tremendous achievement was to improve upon the earlier, existing designs of men like Papin, Savery, and Newcomen to develop the first truly *practical* steam engine. We will meet other cases of how a brilliant inventor, scientist, or engineer built upon earlier, perhaps flawed efforts of others finally to create a practical, useful device. A particular example that we'll see is in the story of the light bulb (Chapter 14).

One of the first important steps in spreading the steam engine away from the mines and into factories was due to the enterprise of Matthew Boulton. He owned the largest button factory in Birmingham, operated by water. Unfortunately, Birmingham is not a hilly area, so there's not much gravitational potential energy, nor is it a very wet place. The millpond used to supply water to the waterwheels dried up in the summer. It was suggested to Watt that his engine could be used to pump the water up from the millpond into the millrace and so work the wheel. In other words, a steam engine would be used to pump water to run a waterwheel, which in turn ran the factory.

The people in London, Manchester and Birmingham are steam mill mad. I don't mean to hurry you, but I think in the course of a month or two, we should determine to take out a patent for certain methods of producing rotative motion from the fire engine.

—Boulton[6]

RICHARD TREVITHICK—STEAM ON WHEELS

The next major development, and another crucial one for society, came about as a result of returning to Savery's 17th-century use of high-pressure steam. We've seen that one of the drawbacks of the Miner's Friend was that it could be dangerous, because of the high-pressure steam. Watt stayed away from high-pressure steam and its potential explosion hazards. Thus Watt never built engines with steam pressures much greater than that of the atmospheric pressure, because he was afraid of boiler explosions and their potentially terrible consequences. (In this regard, Watt's fears were quite justifiable. Consider the single example (from 1850) of a high-pressure steam boiler explosion in a New York City factory that instantaneously demolished the whole factory, with 67 deaths and another 50 injuries.)

In fact, Watt had essentially a patent monopoly on the use of low-pressure steam. Richard Trevithick may have had the insight to realize that high-pressure operation could offer significant advantages (not the least of which is the fact that high-pressure engines would do the same WORK as a low-pressure engine but could be considerably smaller and lighter, and therefore cheaper), or perhaps he was looking for ways to circumvent the patents. If a high-pressure engine could be made smaller—and hence lighter—than a low-pressure engine doing the same amount of WORK, then perhaps it could be made so small, and so light, that it would be portable. (The Watt engines were so heavy that it is quite doubtful that they could have generated enough energy to move themselves, let alone pull a train behind.) Yet another of a long line of exceptionally talented British engineers, Trevithick realized that the moving piston of the steam engine could be made, with appropriate mechanical linkages, to turn wheels. Trevithick built his first steam 'road carriage' in 1801, making the first trial run on Christmas Eve. Two years later he was driving through London at the princely rate of 10 miles per hour. In February of 1804 he mounted a steam carriage on rails, and hauled a load of 10 tons, along with 70 men, on a 10-mile trip in Wales. This was the first successful operation of a locomotive.

By the 1850s, the steam engine had become the dominant source of WORK in most of its possible areas of application. It dominated in the application for which it had first been developed, the pumping of water from mines. It played a major role in the operation of machinery in factories. It was *the* source of locomotion on all the world's railroads. It was beginning to displace sail as the means of propelling ships.

STEAM—ENERGY FOR THE INDUSTRIAL REVOLUTION

Steam was *the* ENERGY source of the Industrial Revolution. Consider: Watt's steam engine operated the pumping machinery to get water out of mines, and operated the hoisting machinery to lift coal and ores from the mines. This made it possible to get coal and ore cheaply from deeper and deeper seams. The steam locomotive made it possible to transport the coal and ore—and other raw materials—cheaply. The steam engine, when applied to the blast furnace, allowed continuous operation and the production of cheap iron and steel. The iron and steel could be used to make a wide variety of industrial machinery that could be operated cheaply and on a large scale by the steam engine. The products of these manufactories could in turn be transported cheaply on land by the steam locomotive, and internationally in steam-worked iron ships.

Steam engines were readily adopted in 19th century Britain. However, in the United States, even up until about the time of the Civil War, waterwheels provided far more WORK than steam engines, and wood supplied much more heat than coal. It was not until about 1875 that steam became dominant in American industry. Nevertheless, steam was instrumental in the growth of industrial cities in the East. Worcester and Fall River, Massachusetts are examples. These cities did not have access to fast-flowing rivers to take advantage of waterwheels. Instead, factories in those cities relied on steam engines, fuelled by coal from Pennsylvania.

> *A water mill is necessarily located in the country afar from the cities, the markets and magazines of labor, upon which it must be dependent. . . . A man sets down his steam engine where he pleases—that is, where it is most to his interest to plant it, in the midst of the industry and markets, both for supply and consumption of a great city—where he is sure of always having hands near him, without loss of time in seeking for them, and where he can buy his raw materials and sell his goods without adding the expense of a double transportation.*
>
> —Scientific American[7]

It's been estimated that a steam engine capable of generating as much POWER as a typical lawnmower engine would fill a two-story building.[1] But harping on the inefficiency of the early steam engine overlooks what was its absolutely critical contribution to society. The steam engine is the first device that provides useful WORK from an ENERGY source (steam) that could be created on demand and as needed by humans. A steam engine could be erected any place that WORK was needed.

The steam engine was the device that freed human society from reliance on ENERGY sources that were supplied by nature (wind, water, or animal muscles). Too, the steam engine was the first step in being liberated from the constraints of geography. That is, it was no longer necessary to build factories on the banks of fast-flowing rivers, or to dam a river for a water supply. It was no longer necessary to find locations that offered a reasonably steady wind.

Another aspect of the dominance of the steam engine in society is the way in which it entered everyday speech. In fact, some of the colloquialisms are still heard once in a while. For example, a busy and vigorous worker was someone who 'got up a head of steam,' meaning that he or she had, in essence, developed a high pressure of steam and was ready to do some job. Many of us are annoyed from time to time by 'high-pressure salesmen.' One reason these people are so obnoxious and unpleasant is that they tend to use 'steamroller tactics' to convince you to buy something. In fact, if you are really annoyed by such a person, it wouldn't be unreasonable if you got all 'steamed up.' When you're feeling angry like that, it's a good idea to find some safe way to relax and 'let off a little steam.' Otherwise, you might just 'blow a gasket,' or, worse, you could actually 'explode.'

The steam engine is only the first step in being freed from the constraints of geography. There is still one constraint in using the steam engine: it provides WORK by moving a piston up and down, or making a shaft or axle turn. This means that the engine itself still must be very close to the device that it will operate. Whatever mechanism is being operated by the steam engine has to be directly, physically connected to it. One consequence of this limitation is that every factory with steam-engine-driven machinery had to have its own engine, in the basement or yard, and its own supply of fuel, usually coal. A little over a century after Watt's great invention, the second liberating step began to emerge from research in a completely different field of science—the study of electricity. This we will explore in future chapters.

CITATIONS

1   Denis Papin, cited in Derry, T.K.; Williams, T.I. *A Short History of Technology*. Dover: New York, 1993; Chapter 11.

2   Thomas Savery, cited in Bernal, J.D. *A History of Classical Physics*. Barnes and Noble: New York, 1972; p. 260.

3   James Watt, cited in Derry, T.K.; Williams, T.I. *A Short History of Technology*. Dover: New York, 1993; Chapter 11.

4   James Watt, cited in Kirby, R.S.; Withington, S.; Darling, A.B.; Kilgour, F.G. *Engineering in History*. Dover: New York, 1990; Chapter 7.

5   Wordsworth, Dorothy. *Recollections of a Tour Made in Scotland*. Yale: New Haven, 1997; p. 50.

6   Matthew Boulton, cited in Pool, Robert. *Beyond Engineering: How Society Shapes Technology*. Oxford: New York, 1997; p. 122.

7   From *Scientific American* magazine, cited in Nye, David E. *Consuming Power*. MIT Press: Cambridge, 1998; p. 71.

FOR FURTHER READING

Bernal, J.D. *A History of Classical Physics*. Barnes and Noble: New York, 1997. A well-written survey of physics from the ancient Greeks through the end of the 19th century, easily readable by the layperson. Chapter 10 discusses the development of the steam engine, and its relationship to the development of theories of heat.

Bernal, J.D. *Science and Industry in the Nineteenth Century*. Indiana University: Bloomington, 1970. The first essay in this book concerns the relationship of the science and technology of the 19th century to various cultural and economic issues. Chapter II of that essay is on the subject of relationships of heat and energy.

Briggs, Asa. *The Power of Steam*. University of Chicago: Chicago, 1982. Chapter 1 covers some of the early history of applications of steam; Chapter 2, the impact of James Watt's engine. This book is exceptionally well illustrated and, in addition, requires little or no previous scientific background.

Bruno, Leonard C. *The Tradition of Technology*. Library of Congress: Washington, 1993. A lavishly illustrated bibliography that discusses the major writings in various fields of technology, including the steam engine.

Buchanan, R.A. *The Power of the Machine*. Penguin: London, 1992. This inexpensive paperback discusses the impact of technology on society since about 1700. Chapter 3 deals with the development of steam engines.

Cardwell, Donald. *The Norton History of Technology*. Norton: New York, 1995. This is a splendid history of the development of technology from ancient times through nuclear energy and computers. Several chapters relate to the development and application of steam engines.

Derry, T.K.; Williams, Trevor I. *A Short History of Technology*. Dover: New York, 1993. A relatively inexpensive paperback edition of a book that treats the history of technology from ancient times through about 1900. Chapter 11 discusses the development of the steam engine and its subsequent evolution.

Hart, Ivor B. *James Watt and the History of Steam Power*. Collier: New York, 1961. A small paperback that delivers exactly what the title indicates.

Hawkins, N. *New Catechism of the Steam Engine*. Lindsay: Bradley, IL. This is a recent reprint, in relatively inexpensive paperback, of a book published in 1904, extensively illustrated with late-19th century diagrams and drawings. This book provides a good introduction to the working of actual engines, and will also be useful to readers interested in the history of steam-engine technology.

Hills, Richard L. *Power from Steam*. Cambridge University: Cambridge, 1989. This book provides a very thorough history of stationary steam engines (i.e., not treating steam locomotives), including uses in pumping water from mines, in the development of the textile industry, and in operating electricity generators.

Kirby, Richard Shelton; Withington, Sidney; Darling, Arthur Burr; Kilgour, Frederick Gridley. *Engineering in History*. Dover: New York, 1990. Chapter 7, provides information on the steam engine and its atmospheric engine forerunners. Available as an inexpensive paperback.

Laidler, Keith J. *To Light Such a Candle*. Oxford University: Oxford, 1998. The focus of this book is the interrelationship between science and technology; how in some cases advances in basic science spark technological advances, and in others a step forward in technology stimulates new work in science. Chapter 2 deals with Watt and the steam engine, and how the steam engine led to developments in the science of thermodynamics.

Nye, David E. *Consuming Power*. MIT: Cambridge, 1998. A social history of the use of energy in the United States. Chapter 3, 'Cities of Steam,' discusses the impact of the steam engine.

Petroski, Henry. *Remaking the World*. Knopf: New York, 1997. This collection of essays on engineering includes one ('Harnessing Steam') on the development of the steam engine.

Rose, Joshua. *Modern Steam Engines*. Lindsay Publications: Bradley, IL, 1993. This is an inexpensive paperback reprint of a book originally published in 1887. Very thoroughly illustrated, and useful for anyone interested in the history of steam engine technology.

Strandh, Sigvard. *The History of the Machine*. Dorset: New York, 1989. A very well illustrated history of machines, from the use of human muscles through computers. Chapter 5 discusses the development and use of steam engines.

11

HEAT AND THERMAL EFFICIENCY

James Watt's invention of the steam engine was probably the most important new energy development in a millennium. Initially applied to manufacturing, it provided a source of energy that could be used to replace human or animal muscles, did not depend on having suitable rivers available, and did not depend on the wind, which might or might not be blowing. Richard Trevithick's extension of Watt's steam engine to the steam locomotive provided an entirely new, cheap, fast, and powerful form of transportation. These two applications of the steam engine—to manufacturing and to transportation—are a sharp dividing line between medieval and modern technologies.

THE NOTION OF EFFICIENCY

There is a second dividing line as well. An advantage of water and wind as energy sources is that they are essentially free. After the initial investment in the mill (and dam, if needed), there is no cost for the water or wind itself. This is not true with the steam engine. The heat needed to generate the steam was obtained by burning a fuel, most commonly coal. This coal had to be purchased. Buying the coal represented a continuing expense for the owner of a steam engine. For this reason, interest quickly developed in determining how much WORK a given engine could do per amount of coal consumed. In modern terminology we would refer to this as the **efficiency** of the engine.

In very broad terms, engineering efficiency (ε) is defined as the ratio of the output to input:

$$\varepsilon = \text{output/input}$$

The ratio may be expressed in terms of any convenient engineering parameter. Energy is of most concern to us, but in principle we could express this formula in terms of anything that we can keep track of: raw materials and products, money flow, the use of time, or even people.

Originally the steam engine, and the earlier atmospheric engines, had been simply a means of raising water from coal mines. Steam engines were being continually improved by hunches, trial and error, and tinkering. The science of steam engines was developed only later. (It comes, as we shall see below, in part from the study of the behavior of gases by the French physicists Charles and Gay-Lussac.) The relationship between science and technological development is a complex one. Sometimes useful, functioning devices are invented and improved with little true understanding of the scientific principles underlying them. It is only later, after the practical success of these devices has been established, that other workers determine the relevant scientific principles.

> In every department of human affairs, Practice long precedes Science: systematic enquiry into the modes of action of the powers of nature, is the tardy product of a long course of efforts to use those powers for practical ends.
>
> —Mill[1]

Or put a bit differently,

> . . . the practical mechanics of the engineer arrives at the concept of work and forces it on the theoreticians.
>
> —Engels[2]

The steam engine is an excellent example of this situation. In other cases the fundamental science is established first, and from that basis in science engineers then design and build devices having practical, commercial applications. An example is provided by the development of the science and technology of electricity, discussed in Chapter 12.

In the early days of the steam engine and Newcomen's 'atmospheric engine,' nothing was known as to what—if anything—limited the amount of WORK that could be done by these engines for a given amount of ENERGY in the form of fuel. In Chapter 10 we introduced the notion of engine efficiency, and saw that the Newcomen atmospheric engine was singularly inefficient at converting the chemical potential energy present in the fuel (usually coal) into useful mechanical WORK. In some cases, less than 1% conversion was achieved. Watt's steam engine was an enormous improvement compared to the Newcomen engine. Nevertheless, by modern standards, it too was rather inefficient. Only some 5–7% of the energy in coal was converted to useful WORK. The necessity of having to buy coal to 'feed' the engine, and the early applications of these engines for pumping water from mines led early engineers to define the 'duty' of a steam engine: the amount of water that could be raised from the mine per bushel of coal burned. Here's an example. In the 1840s, the average duty of a high-pressure Trevithick engine used for pumping was 68 million ft·lb per bushel of coal. That is, the engine could raise 68 million pounds of water one foot, or more reasonably, 680,000 pounds of water (about 85,000 gallons) a hundred feet. For comparison, a good Newcomen engine running in the 1760s had a duty of six million ft·lb per bushel—less than one-tenth the efficiency.

TEMPERATURE AND THERMAL POTENTIAL ENERGY

To develop the concept of efficiency, let's begin by considering some common observations:

- If we have a cup of hot coffee or tea, or a bowl of hot soup, and we let it stand for some time, we find that it is now much cooler. That is, it has become closer to the temperature of the room.
- If we have a dish of ice cream, or a cold beverage, and let it stand for a while, we find that it has warmed up. Again, it has become closer to the temperature of the room.
- No one has *ever* observed the reverse; that is, that a glass of water sitting on a table would spontaneously begin to boil or spontaneously freeze.

The first two of these observations are actually manifestations of the same phenomenon. Heat has moved from one object to another, and has moved in the direction from the hotter object (the cup of coffee in the first case, or the air in the room in the second case) to the colder object (the air in the room in the first case, or the ice cream in the second). What measurement allows us to determine the relative hotness or coldness of an object? The temperature. Thus we can generalize our first two observations as follows:

But in the third case, we see another important point: heat *never* spontaneously flows from low temperature to high temperature. This is now the third time that we've encountered a situation

> HEAT SPONTANEOUSLY FLOWS FROM A HOT OBJECT (ONE AT HIGH TEMPERATURE) TO A COLD OBJECT (ONE AT LOW TEMPERATURE).

in which a spontaneous change in one particular direction never occurs. Recall that water never spontaneously flows uphill, or up a waterfall, and that a mixture of carbon dioxide and water vapor will never convert itself back into methane and oxygen. In the waterfall and the burning methane cases we were able to illustrate the change in the system by means of a simple energy diagram. It seems reasonable to expect, then, that we should be able to make an analogous diagram (Figure 11.1) to explain the flow of heat.

To maintain the analogy with our previous energy diagrams, we must recognize that the heat is flowing spontaneously from high potential energy to low potential energy. To emphasize the point that we are dealing with a SYSTEM in which heat is the kind of energy of interest, we can refer to *thermal potential energy*. To sum up where we've gotten to so far, we can say that heat flows spontaneously from high thermal potential energy to low thermal potential energy. But we've also said, from commonplace observations of the world around us, that heat flows from high temperature to low temperature. What is the difference between these two statements? There is no difference. That is,

Therefore we can write our energy diagram as shown in Figure 11.2.

> TEMPERATURE IS A MEASURE OF THERMAL POTENTIAL ENERGY.

At this point a brief digression on terminology is appropriate. We've just expanded the concept of potential energy by introducing the adjective *thermal* as a

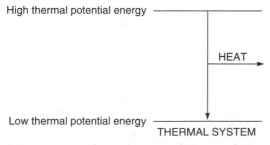

FIGURE 11.1 Thermal potential energy diagram.

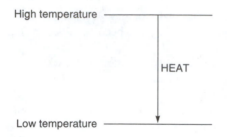

FIGURE 11.2 Thermal energy diagram with the energy correctly labeled as 'temperature.'

reminder that we are dealing with a system in which heat is moving from high to low potential. In a waterfall, the water that is at high potential falls, down the waterfall, under the influence of gravity. In that case, it would be appropriate for us to speak of *gravitational potential energy*. The energy that is 'stored' or 'hidden' in the chemical bonds in molecules of methane and oxygen is a much more subtle concept, but for our purposes we can think of this as a form of *chemical potential energy*.

In all three kinds of system, the key concept is that energy flows spontaneously from high potential (thermal, gravitational, or chemical) to low potential.

There is another very important point of analogy with our waterfall. We saw that ancient humans learned to design waterwheels to convert some of the ENERGY of the moving water into WORK, that is, a device is inserted into the system to extract energy or work. We can do exactly the same thing with a thermal energy system (Figure 11.3). In the thermal system, the kind of device that we use to extract useful WORK from the system is called a **heat engine**. A steam engine is one kind of heat engine.

The point at which technology really took off is with the invention of the heat engine . . .
—Benson[3]

To push the analogy further, we'd seen that the way of obtaining 100% conversion of kinetic energy to work in a windmill or waterwheel would be to have the kinetic energy of the wind or water be zero when it left the mill. A kinetic energy of zero would correspond to the complete conversion of the kinetic energy of the moving wind or water into work. A kinetic energy of zero is also the lowest possible value of kinetic energy, since there is no conceivable way that an object can have a negative kinetic energy. In a thermal system, therefore, we should be able to obtain 100% conversion of thermal energy (that is, heat) into WORK, provided that we could establish a point of zero thermal energy. Because temperature is the measure of thermal energy, this 'zero point' of thermal energy would be, therefore, the lowest temperature that could theoretically be attained. In other words, to find out how to achieve 100% conversion of thermal energy to WORK, we need to find out what the lowest possible temperature is.

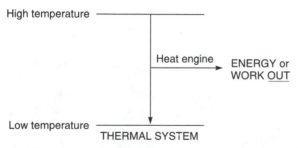

FIGURE 11.3 Energy diagram for a thermal system.

THE LAW OF CHARLES AND GAY-LUSSAC AND THE QUEST FOR ABSOLUTE ZERO

The first detailed study of the effect of temperature changes on gases was done by the French physicist Jacques Alexandre César Charles in 1787, but these studies were carelessly carried out, the quality of the data was poor, and the work was never properly published in the scientific literature. The law governing the thermal expansion and contraction of gases should correctly be attributed to Joseph Louis Gay-Lussac, though it is known in many textbooks as Charles' Law.

> *Charles had remarked on these properties of gases 15 years ago, but not having published the results, I have the great good luck to make them known.*
>
> —Gay-Lussac[4]

Gay-Lussac studied many different gases, but found that, regardless of the specific gas, every gas expands by about 1/300 in volume for each degree Celsius it is heated, and contracts by the same fraction when it is cooled. However, this observation depends critically on two restrictions: The pressure of the gas must remain constant, and the amount of the gas being studied must also remain constant.

To explore the question of finding the lowest possible temperature, we begin with this observation by Charles and Gay-Lussac: If a sample of gas is kept at constant pressure, the volume of the gas is directly proportional to its temperature. This relationship is so important and universal a property of gases that it is now usually known as Charles' Law. Let's look at some data (Table 11.1) that are consistent with this law. (These are not, by the way, original experimental data; these data are an example that is consistent with the law and may make it easier to see the important effect that is occurring as the gas is cooled.) These data show that for every 1°C drop in temperature, a gas contracts by 1/273 of its volume. The extension of this observation suggests that reducing the temperature to −273°C would cause the gas to contract by the fraction 273/273, which is equal to 1. In other words, the gas would lose all its volume and vanish. This is physically impossible, since matter cannot be destroyed. Therefore, −273°C represents the lowest possible temperature. (The argument we are making here ignores the fact that all real gases would actually turn to liquids at some temperature above −273°C. ⚛)

⚛ At any given temperature, the molecules of any gas have a characteristic kinetic energy. The kinetic energy is directly proportional to the temperature, so that if the temperature is reduced, the kinetic energy of the gas molecules is reduced also. In addition, the molecules of any gas exert weak attractive forces among one another. The strength of these intermolecular attractive forces depends on the chemical nature of the gas molecules. At ordinary conditions of temperature and pressure, the kinetic energy of the molecules is far stronger than the weak attractive forces, and the gas molecules whiz around in their container largely independently of each other. As the temperature—and consequently the kinetic energy—is reduced, however, a temperature will be reached at which the intermolecular attractive forces overwhelm the kinetic energy of the molecules. At that point, the gas molecules would condense into the liquid state. Ammonia, a gas that dissolves in water to make a common household cleanser, and neon, a gas used in colored electric lights, illustrate the importance of the intermolecular attractive forces. These forces are exceedingly feeble in neon, which does not liquefy until a temperature of 27 K (−246°C) is reached. A molecule of ammonia, which has about the same mass as an atom of neon, experiences fairly strong attractive forces, so ammonia liquefies at 240 K (−33°C).

TABLE 11.1 EXAMPLE DATA CHART
SHOWING THE RELATIONSHIP BETWEEN
TEMPERATURE AND VOLUME ACCORDING TO
THE GAS LAW OF CHARLES AND GAY-LUSSAC

Temperature	Volume remaining	Volume contracted
0°C	1*	0*
−1°C	272/273	1/273
−2°C	271/273	2/273
−10°C	263/273	10/273
−20°C	253/273	20/273

*These are the values at the beginning of the experiment.

Two temperature scales are in common use, the Fahrenheit scale (in which water freezes at 32°F and boils at 212°F) and the Celsius scale (in which water freezes at 0°C and boils at 100°C). The relationships between these two common temperature scales are

$$°F = (9/5°C) + 32$$
$$°C = 5/9(°F − 32)$$

It would seem, then, that we could define any sort of temperature scale we want—that, in other words, the relationship of temperature to a certain amount of thermal energy could be quite arbitrary. This is not a very satisfactory state of affairs for measuring quantitatively the amounts of thermal energy. It would be helpful if there were some definite, immutable, fixed point on which a temperature scale could be based. We've seen that a temperature of −273°C represents the lowest possible temperature, called **absolute zero**. This allows us to reset the temperature scale so that the zero of temperature is at absolute zero. This temperature scale is called the Kelvin scale. Its units are called kelvins (*not* 'degrees kelvin') and have the symbol K (*not* °K). In comparison with the Fahrenheit and Celsius scales, we find the relationships shown in Table 11.2. Comparing the last two columns of Table 11.2 shows that the temperature in kelvins can be found from

$$K = °C + 273$$

So, on the absolute temperature scale, the zero point is chosen to be the temperature at which the volume of any gas would vanish if it could continue to contract when it is cooled. As we've said, real gases do no such thing, because every real gas will turn to a liquid at some temperature above zero kelvins. The importance of the concept lies in this: Every real gas, regardless of its chemical composition, expands by exactly the same fraction for a given increase in temperature. Because of that constancy, the zero point of the absolute temperature scale is absolutely unambiguous.

To sum up this part of the chapter,

ABSOLUTE ZERO IS THE LOWEST POSSIBLE TEMPERATURE (THE POINT OF ZERO THERMAL ENERGY).

QUANTIFYING EFFICIENCY

The concept of absolute zero should help us now to understand the efficiency of heat engines. Let's again go back to the waterfall (Figure 11.4). We've said that only some fraction of the kinetic energy of the water is converted to WORK, and we know this

TABLE 11.2 COMPARISON OF
BASIC TEMPERATURES ON THE
FAHRENHEIT, CELSIUS, AND
KELVIN SCALES

	°F	°C	K
Water boils	212	100	373
Water freezes	32	0	273
Absolute zero	−459	−273	0

because the water still has some kinetic energy left (i.e., it's still moving) after it's passed the waterwheel. The connection to our present concern lies in the concept, discussed in more detail below, that in a heat engine, only some fraction of the thermal energy (the heat) is converted to WORK.

In the early 19th century, steam engines 'took off.' Their use in operating factory machinery, and the freedom they provided from the geographic constraints of wind and water use, allowed industrial productivity to soar. However, the details of building, operating, and improving these devices were still entirely pragmatic and empirical. There was no theoretical, scientific understanding of the relationship between heat and mechanical WORK, or indeed even much understanding of the nature of heat itself. As noted earlier, the steam engine developed largely as the result of trial-and-error tinkering and experimenting.

The person who established the scientific understanding of the steam engine and who single-handedly laid the foundation of the science of thermodynamics was the French investigator, Nicholas-Leonard Sadi Carnot. He was the first scientist to analyze quantitatively the relationship between heat and WORK and to recognize that an engine functions because heat flows from a hot body to a colder one. In 1824 Carnot wrote a paper on the 'power' of heat. He did not use the term 'power' as we have defined it, but in speaking of the way in which heat causes motion in steam engines. Carnot considered that this 'power' comes from the *one-way-only* transfer of heat, from a hotter to a colder body; during this transfer, heat can be made to do WORK.

For Carnot, heat is *the* 'moving force' of the world:

It causes the agitations of the atmosphere, the ascension of clouds, the fall of rains and of meteors, the currents of water which channel the surface of the globe, and of which man has thus far only employed but a small portion. Even earthquakes and volcanic erup-

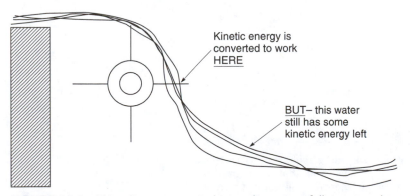

FIGURE 11.4 This diagram reminds us that waterfalls cannot be perfectly efficient.

tions are the result of heat. From this immense reservoir we may draw the moving force necessary for our purposes.

—Carnot[5]

Wherever there is a temperature difference, in any system, there is the possibility of generating useful WORK (or, in Carnot's terminology, 'motive power'). A temperature difference is just like a 'head' or column of water: the greater the temperature difference, the greater the WORK for the same quantity of heat, just as water falling 200 feet will do twice the WORK of the same amount of water falling 100 feet.

Carnot analyzed the workings of a steam engine in a way that is analogous to how we might think about a waterfall. For a given volume of water that flows over the fall, the amount of ENERGY that can be converted to WORK depends on the distance through which the water drops. (In fact, this is true of any form of matter falling in a gravitational potential system. If a bowling ball slips out of your hand and hits you on the foot, it might hurt a bit—to say nothing of being embarrassing. The same bowling ball falling from the roof of a skyscraper and hitting you on the head would probably kill you.) In other words, the critical information is the height of the pool of water at the bottom of the waterfall subtracted from the height of the cliff that is the top of the waterfall—the difference in gravitational potential energies of the two locations.

There are two important points from this little analysis: First, we could, in principle, measure the two heights from any agreed-upon reference point or 'zero point.' For measurements that we make on Earth, the one uniform zero point for height would be distance from the Earth's center. Second, the absolute values of the two heights are not important in deciding the amount of WORK we can extract from the ENERGY of the falling water—what counts is the difference between them.

Now let's consider a heat engine (Figure 11.5). Here, only some fraction of the thermal energy is converted to WORK, because the system still has some thermal energy left (i.e., it's not at absolute zero).

In the waterfall analogy, the water drops to the bottom of the waterfall and forms a pool there. The pool by itself is not capable of turning a waterwheel. However, we need to

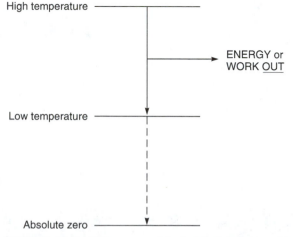

FIGURE 11.5 When we add absolute zero to this thermal energy diagram we see that even at the 'low temperature' point there is still some thermal potential energy remaining that has not been converted to WORK.

draw a distinction between the total energy content of the water and available energy that we can actually use. Yes, the pool at the bottom of the waterfall is not capable of turning a waterwheel, but it still contains much potential energy. Imagine that if someone dug a hole that would let the water in the pool run out. If that person also placed a second waterwheel at the bottom of the hole, an additional amount of the ENERGY in the water could be converted to WORK. Disregarding for a moment the practicality of doing this, imagine that someone could dig a hole all the way to the center of the Earth. Then (and only then!) *all* the gravitational potential energy of the water could be used, conceivably to run a whole series of waterwheels. Of course we do not have the engineering technology to dig a 4,000-mile-deep hole clear to the center of the Earth. Therefore, in practical reality, only the ENERGY of the falling water of the original waterfall is used. That ENERGY is available to us for doing WORK, and it represents only a portion of the total energy of the water. The remaining potential energy of the water, which still 'counts' as part of the total energy of the water, is not available to us by any known technology.

In principle, we can get more WORK out of a waterfall by letting the water fall a greater distance. We ought to be able to do exactly the same thing with a heat engine by allowing the temperature to fall a greater distance, which means that we make the difference between the high temperature and low temperature larger—creating a greater difference in thermal potential energy. So, the bigger the difference between the high and low potential (whether gravitational or thermal), the more the WORK we can extract from the system.

Following exactly the same line of thinking, in a heat engine, it is not the value of either the high or low temperature that dictates the amount of ENERGY that can be converted to WORK, but rather the temperature difference. In the same way, then, the available energy (not the total energy!) can be expressed in terms of the temperature difference within the heat engine. In 1851, a professor of physics at the University of Glasgow, William Thomson, later and better known as Lord Kelvin (Figure 11.6), recognized that there could be a temperature below which no further WORK could be produced. As we'll see later in this chapter, we can express this most conveniently in terms of the concept of the absolute zero of temperature. On the basis of this analogy, available energy can be expressed as $T_h - T_l$. The subscripts h and l stand for *high* and *low*, respectively.

The pool of water at the bottom of the waterfall still had some ENERGY. Analogously, the cold side of the steam engine still contains heat. We found that we could possibly extract more WORK from the water in the pool if we could dig a hole and let the water drop even further. Conceptually, we could do this until we had used up all the available gravitational potential energy of the water—until we had reached the center of the Earth, which is the zero point of gravitational energy. Quite similarly, suppose that the condenser on a steam engine is at, say, 25°C. The water produced in the condenser could, in principle,

FIGURE 11.6 Lord Kelvin (1824–1907). Born in Belfast, Sir William Thomson developed many of his theories while a professor at the University of Glasgow.

be cooled further and further, until we reached a point at which no more work could be extracted from it. This point would be the temperature analog of the center of the Earth, the zero point of potential energy below which it is impossible to go. In a thermal system, this zero point of energy is the temperature called absolute zero. The total energy of the thermal system would be represented by the difference between the temperature of the hot side and absolute zero: $T_h - 0$, or, simply T_h.

The engine efficiency is a measure of the available energy as a fraction of the total energy. This concept can be expressed mathematically in a formula for calculating efficiency:

$$\varepsilon = (T_h - T_l)/T_h$$

where T_h is the thermal potential energy. Using this formula, the efficiency will turn out to be a fraction having a value between 0 and 1.000. Often it is more convenient or easier to express efficiency as a percentage, rather than as a fraction. This is done readily by using the comparable formula:

$$\%\varepsilon = 100[(T_h - T_l)/T_h]$$

In this case, efficiency has some value between 0 and 100%. The formulas are equivalent and provide the same information; the difference is whether the calculated efficiency is expressed as a fraction or as a percentage. However, there is one crucial fact in the use of efficiency calculations: *these formulas only provide the correct answer if the temperature is expressed in kelvins.*

It is important to recognize that this ratio of the available energy to the total energy represents the *maximum* efficiency of a heat engine. This efficiency value, sometimes called the **Carnot efficiency**, is reached only if the engine is mechanically perfect; for example, if there are no losses of energy through friction, or losses of heat to the outside world. In any real heat engine, the efficiency is less—sometimes considerably so—than the predicted maximum (Carnot) efficiency simply because of such problems as friction among the parts and losses of heat to the surroundings.

Carnot readily recognized that as heat 'falls' from a higher to a lower temperature, the 'duty' of the engine will be larger, the larger the difference between the two temperatures. This is exactly analogous to the notion that, in a waterfall, the greater the drop of water (i.e., the greater the change in gravitational potential energy), the more WORK will be done by the waterwheel. Carnot also found that for a given drop in temperature (which we would write as $T_h - T_l$), more WORK is done for lower values of T_h. For example, an engine working between the temperatures of 400 and 300 K will do more work than one working between 1000 and 900 K, even though the temperature drop (100 K) is the same in both cases. Indeed, the efficiency of the first engine is 0.25 and, of the second, only 0.10.

To help visualize this fact, we can again consider the waterfall analogy. Let's suppose we have a waterfall that has a height of 1000 feet. For convenience, we can assume that the base is at sea level, or a nominal elevation of 0, and the top at an elevation of 1000 feet. Now suppose—leaving aside the possibly thorny details of the construction—that a waterwheel was mounted in this waterfall at the 900-foot mark. This wheel would take advantage of a 100-foot drop of water (that is, from 1000 to 900). However, an enormous amount of gravitational potential energy of the water (corresponding to the drop from 900 to 0 feet) is 'thrown away' in this arrangement. Consider now a second waterwheel, some 300 feet high, with a waterwheel mounted at the 200-foot level. The same 100-foot drop of water (now from 300 to 200) occurs for this wheel. But much less potential energy is 'thrown away'—in this case, only the amount corresponding to a drop from 200 to 0. If a heat engine operates between 1000 and 900 K we are, crudely, 'throwing away' all the rest of the thermal potential energy from 900 to 0 K. A comparable heat engine running between 300 and 200 K only discards the thermal energy between 200 and 0. Since the efficiency of the engine depends on the amount of thermal energy that is *not* thrown away, the engine running at lower temperatures will be the more efficient, exactly in accord with our calculation in the previous paragraph.

The value of these formulas (aside from their obvious utility in calculating efficiency) is that, finally, we have arrived at a direct understanding of what it would take to achieve a complete conversion of ENERGY to WORK. This complete conversion would represent an efficiency of 1.000 (or 100%). This value of the efficiency can be achieved for the special case when

$$\varepsilon = (T_h - 0)/T_h = T_h/T_h = 1$$

or

$$\%\varepsilon = 100[(T_h - 0)/T_h] = 100[T_h/T_h] = 100\%$$

Notice how we have achieved $\varepsilon = 1$ (or $\%\varepsilon = 100\%$) in this example: we have made the value of T_l zero. This tells us that to achieve a complete conversion of thermal ENERGY to WORK (an efficiency of 1.00 or 100%) we must let the low temperature be absolute zero (0 K).

Recall from the discussion of Charles' Law that at absolute zero a gas would, in principle, vanish completely; that is, it would have zero volume. Charles' Law suggests that to reach absolute zero we would have to destroy matter. However, an abundance of information collected over centuries in laboratories all over the world shows that matter cannot be destroyed.

Let us explore the consequence of this unfortunate fact. We've discovered, from the mathematical expressions for

IT IS IMPOSSIBLE TO ATTAIN A TEMPERATURE OF ABSOLUTE ZERO (0 K).

efficiency, that to achieve a complete conversion of thermal ENERGY to WORK, we need to have the low temperature 'side' of the thermal energy system be at absolute zero. We have also just concluded that it is impossible to achieve a temperature of absolute zero. Since we need a temperature of absolute zero to have an efficiency of 1, but we can't get to absolute zero, we are brought to the appalling conclusion that

We have already seen some evidence that has hinted at this conclusion in our considerations of other systems. We know that our bodies get

IT IS IMPOSSIBLE TO CONVERT THERMAL ENERGY TO WORK COMPLETELY (WITH 100% EFFICIENCY).

hot when we exercise or work vigorously; some of the chemical potential energy in food is lost as this heat, rather than being converted to WORK. We know that the wind or water leaving a mill still has some kinetic energy left after it's done the work of the mill. Now we have found a system—a thermal system—where we can directly show, and indeed calculate, the fact that the complete conversion of this kind of ENERGY to WORK is impossible.

It is impossible to extract heat from the hot side of a heat engine and convert it entirely into WORK, because it is impossible to attain a temperature of absolute zero. This says, in essence, that some heat must be passed to the cold source. This fundamental limitation means that if we want to achieve a high efficiency from the heat engine, we can't get it by lowering the temperature of the cold side arbitrarily (because we've 'hit a wall' at absolute zero). Therefore, the only practical alternative is for us to raise the temperature of the hot side as much as possible. We are, of course, limited by the high-temperature strength of the materials of the engine, and, in some ultimate sense, by the melting point of the material used to make the engine.

The new scientific understanding of heat engines and thermodynamics was applied to practical aspects of the design and operation of steam engines by the Scottish engineer William Rankine.⊛

⊛ Like so many other of the great scientists and engineers of the 19th century, William Rankine made substantial contributions to many different fields. Early in his career he studied the problem of metal fatigue (a phenomenon that we can easily reproduce by bending a paperclip back and forth until it breaks) and how it applied to the cracking of axles on railway cars. His book *Manual of the Steam Engine and Other Prime Movers* is a seminal text in the application of scientific principles to understanding the steam engine. He also contributed to understanding the principles of construction of dams. The Rankine temperature scale is an alternative approach to absolute temperatures; absolute zero on the Rankine scale (which of course is also absolute zero on the Kelvin scale) is −459°F. Thus

$$°R = °F + 459$$

THE CALORIC THEORY OF HEAT

So far, we have had several indications that there must be some connection between the concept of heat and the concept of energy. We have defined units of measurement of energy in terms relating to heat: the Btu and the calorie. In our discussion of the energy produced by our bodies we indicated that, as we know from our own experience,

$$\text{Energy} \rightarrow \text{Work} + \text{Heat}$$

suggesting some connection between heat and energy (and, indeed, among heat, energy, and work). We introduced two definitions of fuel: something consumed to produce energy, and a material burned to produce heat, suggesting that, since the two definitions apply to the same kind of material (fuel) they must somehow be equivalent.

But then what exactly is *heat*? Intuitively, we recognize from our daily experience that there must be some relationship between our ideas of heat and of temperature. Even without any sort of scientific instrument, we can sense that some objects are hot—relative to ambient temperature—and other objects are cold. In daily conversation we may speak of a hot object as 'having a lot of heat' or of the need to 'heat up' a cold object. We can also make the common observation we've introduced previously, again without needing any special scientific equipment, that heat flows from a hot to a cool object. By adding just a bit of 'equipment' we can make one additional observation: If we are cooking, and leave a utensil in the pot or pan while the food is cooking, we soon observe that the utensil itself gets hot. We observe this also if we leave a spoon in a hot cup of coffee.

In our intuitive concept that if something is hot, it has heat, or if something got cold it lost heat, these simple observations suggest that 'something' must be moving around. When a cup of hot coffee stands until it becomes lukewarm, heat must have gone out of the coffee. When a cold ice cube melts and becomes a puddle of lukewarm water, heat must have gone into the ice. When we put a cold spoon into a cup of hot coffee, and soon find that the handle of the spoon is hot (and the coffee is slightly cooler), heat must have gone from the coffee to the spoon. Again, it is a common observation that one kind of substance that flows easily is a fluid, such as water. Thus the first concept of heat was that heat itself is a fluid—a very special kind of fluid that happened to be invisible and weightless—called **caloric**.

The caloric concept of heat explains many, indeed most, of our common observations about heat. Caloric drains out of a hot object, causing it to cool. Caloric runs into a cold body, causing it to warm up. If hot and cold objects are brought together, caloric flows from the hot one into the cold one. Ice melts when it absorbs enough caloric to become liquid water. Water boils when it absorbs still more caloric and becomes steam. Even more subtle concepts can be explained by the caloric theory. The amount of heat that is necessary to increase a fixed mass a given temperature increment varies from material to another. This is a characteristic property of matter, called the heat capacity. In qualitative observations, we know that some substances heat up easily, and others do not. This might seem quite analogous to the easily made observation that different substances dissolve in water to a greater or lesser extent (compare sand and salt, for example). Therefore, if heat were considered to be a material substance—caloric—the difference in heat capacities between materials could be explained by the analogy that different substances take up different amounts of heat just as different amounts of material are taken up by water.

In science, we accept a theory if it is able to explain some body of observations. However, we try to push further than that. We look around for new observations to test the theory with. If the theory 'works,' then we consider it to be all the stronger and more useful. If the theory is unable to accommodate the new observation, we modify the theory or junk it.

COUNT RUMFORD AND THE END OF THE CALORIC THEORY

The idea that heat is not a fluid but, rather, is a form of motion was occasionally advanced by scientists, many of whom are among the great names of science: Francis Bacon, Robert Boyle, and Isaac Newton are examples.

Heat is a very brisk agitation of the insensible parts of the object (i.e., the atoms), which produces in us that sensation from which we denominate the object hot; so that what in our sensation is heat, in the object is nothing but motion.

—Locke[6]

John Locke is now best known as a philosopher, but he was active as a physician and did a certain amount of work in experimental chemistry.

The person credited with the overthrow of caloric theory was Benjamin Thompson (Count Rumford, Figure 11.7), certainly one of the most colorful characters in the history of science. Thompson was born in Woburn, Massachusetts, and was indentured to a merchant in Salem when he was 13. Thompson learned science on his own, became a schoolmaster, published scientific papers, married a wealthy widow, joined the New Hampshire militia and became a major, and, at the same time, also became a secret agent for the British.

In 1775, Thompson fled Woburn, abandoning his wife and baby daughter, and fled to London when he was increasingly suspected of activities on behalf of the British. Once settled in England, where he was awarded an appointment in the Colonial Office for his service as a secret agent, he contributed to the design of military weapons. This won him a Fellowship of the Royal Society, which to this day is one of the highest

FIGURE 11.7 Benjamin Thompson, Lord Rumford (1753–1814). Born in Massachusetts, this interesting scientist's activities included service in the New Hampshire militia, spying for the British, working for the poor in Bavaria, and discovering that heat is a form of motion.

honors to which a scientist can aspire. While he was working on munitions, he observed that a gun barrel becomes hotter after firing a real bullet than when firing a blank charge. This seemingly small observation suggested to him the idea that heat may be a form of motion. As we will see, experiments he did later in life on a bigger scale (with cannon barrels rather than musket barrels) were crucial in establishing the validity of this idea.

When the Revolutionary War broke out, Thompson returned to America, now as a lieutenant-colonel in the British army. From around New York, he recruited a regiment from persons loyal to Britain and commanded it in several battles. While Thompson's regiment was in winter quarters on Long Island, they behaved in a manner not uncommon among occupation troops, devastating farms and orchards, wrecking churches, and even tipping over gravestones to use as tables.

Despite the way the war turned out and despite the behavior of his own troops, Thompson returned to England as a hero. He was even knighted by King George III. Soon thereafter, he moved to Bavaria, where he served as minister of war and police minister. He reorganized the Bavarian army, and worked to improve living conditions for poor people in Bavaria. He set up schools so that soldiers and their families could receive a basic education. He increased the salaries of enlisted men. In one of the world's first large-scale experiments in 'work-fare,' Thompson had the beggars of Munich rounded up; they were provided with some food and shelter and put to work making equipment for the military. While he was working on designing winter uniforms for soldiers, Thompson discovered that air trapped in fur, feathers, or cloth is what gives these materials their insulating properties. Thompson was the inventor of thermal underwear. Another of his inventions was the forerunner of what we would now recognize as the drip coffee maker. He is also credited with inventing the pencil eraser and improving breeds of horses and cattle. He was awarded the title of Count of the Holy Roman Empire (which, as historians have pointed out, was neither holy nor Roman, and not even much of an empire) by the Bavarian government and chose the name of one of his hometowns, Rumford (now Concord), New Hampshire, to go with the title; therefore Benjamin Thompson is also known as Count Rumford.

In the final twist to a very remarkable life, Rumford moved in 1804 to Paris, where he married the widow of Lavoisier. Nothing in the historical record suggests that this was a happy marriage. Rumford died in 1814.

So far, based on the observations we have been discussing, the caloric theory works just fine. But, now we test it with a new observation. In the late 18th century, it was observed that drilling the 'bore' into cannon barrels makes them very hot. Such an enormous amount of heat is generated by the process that in fact, if the barrel and the drill were covered with water, the water got hot enough to boil. The caloric theory explanation was that as the drill chipped off pieces of metal from the inside of the barrel, the broken surfaces allowed caloric to leak out of the metal into the water. So far, so good. But as the boring was continued, there were two new observations: The water still got hot even when the drill was so dull that no new chips were breaking off. (In this instance, there was no way for the caloric to get out of the metal.) Even if the water boiled away completely, any newly added batch of water would soon boil, and if that second batch of water boiled away and was replaced, this new batch of water would boil also. Yet if caloric is some physically real fluid substance inside the metal, there can't possibly be an inexhaustible supply of it.

According to the caloric theory, caloric was released into the water from the metal being drilled. The caloric ran into the water, heating it. Rumford argued that *if* heat were a fluid that actually was present in the original metal of the cannon barrel, then surely a point would have to be reached sooner or later when all of the caloric originally in the barrel was used up. This should be observable at some late stage of the drilling operation, when just about all the caloric has been used up, by the drilling producing less heat and boiling away less water. Rumford tested this by an experiment in which he submerged a cannon barrel in a tank of water, began boring the cannon, and measured the time it took the water to boil.

> *It would be difficult to describe the surprise and astonishment expressed in the countenances of the bystanders, on seeing so large a quantity of cool water heated, and actually made to boil, without any fire.*
>
> —Rumford[7]

Water would boil for as long as the drill could be turned. There was never a point reached when it seemed as if the caloric was completely consumed. But then where was the heat coming from, if not caloric leaking out of the cannon barrel?

> *Anything which any insulated body, or system of bodies, can continue to furnish without limitation, cannot possibly be a material substance, and it appears to me to be extremely difficult, if not quite impossible, to form any distinct idea of any thing, capable of being excited and communicated in the manner the Heat was excited and communicated in these experiments, except it be Motion.*
>
> —Rumford[7]

It's ironic that Carnot, whose analysis of the efficiency of heat engines is the foundation of the entire science of thermodynamics, started with the accepted concept of caloric, which we now know to be dead wrong. Carnot pictured a heat engine as working by taking caloric from the boiler or source and throwing it out in the condenser or sink, just as a waterwheel uses water. In fact, he considered the analogy of a waterfall causing a waterwheel to turn. We've seen that, in an overshot wheel, the wheel is turned by the impact or force of the water hitting the blades (and by the weight of the water). Since caloric was considered to be a real, physical substance, then, in the same way, the force of 'falling heat' ought to make the piston in an engine move. Furthermore, Carnot regarded the loss of some heat into the environment as inevitable, just like the fact that not all of the energy of falling water is converted to WORK in a waterwheel. From that, Carnot developed the concept that it is impossible to extract WORK from heat without at the same time discarding some heat into the environment; in other words, that it is impossible to convert heat to WORK with 100% efficiency.

THE EQUIVALENCE OF WORK AND HEAT

The only thing going on in Rumford's system (cannon barrel, drill bit, and water) is the turning motion of the drill bit inside the barrel. As a result, this led to the hypothesis that

HEAT IS A FORM OF MOTION.

This hypothesis received substantiation from a simple experiment performed by Humphry Davy.

'Sir Humphry Davy?' said Mr. Brooke, over the soup, in his easy smiling way, taking up Sir James Chettam's remark that he was studying Davy's Agricultural Chemistry. 'Well, now, Sir Humphry Davy: I dined with him years ago at Cartwright's and Wordsworth was there too—the poet Wordsworth, you know.... But Davy was there; he was a poet too.

—Eliot[8]

He went outdoors on a day when the temperature was below freezing and rubbed two ice cubes together. The ice melted. Since the temperature outdoors was too cold to melt ice, the only possible source of heat in this experiment was the motion of the blocks against each other. This very simple, but very profound, experiment gave immense credence to the new concept of heat as a form of motion.

Since heat is a form of motion, we can consider that the heat present in a gas is produced by moving molecules. Then the zero point of heat (absolute zero) should be where all molecular motion in a gas stops. From the reasoning we've developed earlier in this chapter, no more WORK could be produced below that temperature. In fact, it is a convenient and more realistic way of defining absolute zero: the temperature at which substances possess no thermal energy.

We see now that heat is a form of energy. We presented earlier the concept that the energy associated with motion is kinetic energy. Therefore, heat is a special kind of kinetic energy—the kinetic energy of atoms or molecules. Temperature scales actually measure the average thermal kinetic energy per atom in an object. The more kinetic energy each atom has, the more vigorous the thermal motion. We will revisit this concept later in this chapter when we discuss how heat is conducted through an object. The 'microscopic,' indeed atomic scale, work one vibrating atom does on another is what actually passes heat through an object, or from one object to another. The object with more average thermal kinetic energy per atom will pass heat to an object with less. The temperature at which all thermal energy has been removed from an object would be absolute zero.

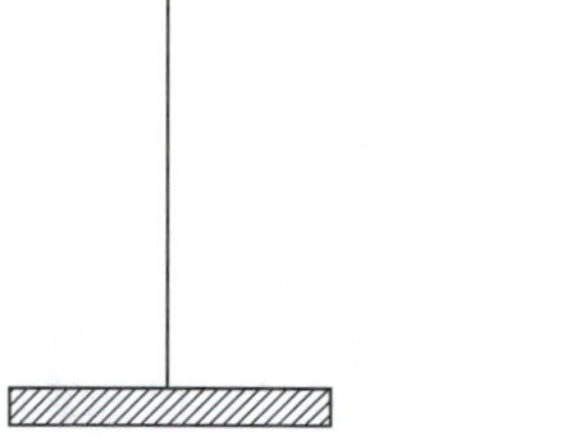

We have defined ENERGY as the capacity for doing work. We have also defined WORK as moving a body against a resistance or a force, one common example being lifting an object against gravity. A simple weight-and-pulley system with the concept for doing WORK is shown in Figure 11.8. If we let the heavy weight fall, it will make the light object rise against gravity, that is, it will do WORK. We've also seen that we can measure the heat released from fuel, or from food, with a calorimeter (as we saw in Chapter 3). James Prescott Joule conducted the key experiment in relating heat to work by connecting these two pieces of equip-

FIGURE 11.8 A simple system with the concept for doing WORK.

ment together. He replaced the burning fuel sample (Figure 11.9a) with the work-producing device, except that instead of raising a weight through the water (which, probably, would only cause a giant splash) a paddle is turned (Figure 11.9b). In this case, releasing the weight causes a vigorous stirring of the water, and the temperature goes up. This experiment established the **mechanical equivalent of heat**: One pound falling 778 feet (or, alternatively, 778 pounds falling one foot) is equivalent to 1 Btu.

In Joule's experiment, the weight of water and the temperature rise of the water could be measured. These data allow determination of how many Btus of thermal energy went into the water. In the very same experiment, we also know how many pounds of weight were allowed to fall, and the distance in feet through which the weight fell. These data allow determination of the mechanical energy in units of foot-pounds. Hence it is immediately possible to determine how many foot-pounds of mechanical energy are equivalent to one Btu of thermal energy. In fact, it turns out, as we've seen above, that

$$778 \text{ft} \cdot \text{lb} = 1 \text{Btu}$$

Joule's experiments killed off the caloric theory for good.

As this simple concept was refined, there emerged some more rigorous definitions: THERMAL ENERGY is a kind of energy due to the random motion of the constituent atoms or molecules of an object (or of a system). HEAT is thermal energy that is in the

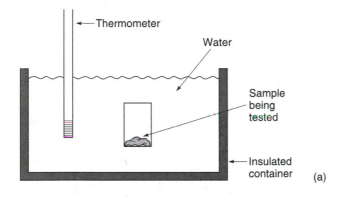

(a)

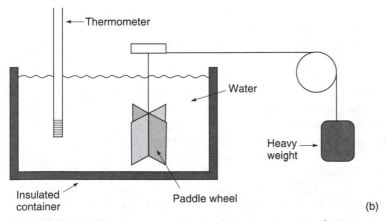

(b)

FIGURE 11.9 The apparatus for Joule's experiment relating heat to work.

process of being transferred from one object to another (or from one system to another). Therefore, HEAT is not a physical fluid, like air or water, but is 'energy in motion' between objects or systems having different amounts of THERMAL ENERGY. How do we measure thermal energy? TEMPERATURE provides the quantitative measure of thermal energy.

If the energy from a mechanical system, such as the one in Joule's experiment, is somehow equivalent to thermal energy (and indeed Joule proved this to be the case), then the conclusion we have reached about the efficiency of a thermal energy system should extend to mechanical systems as well. Indeed it does. So the conclusion we had arrived at regarding efficiencies and conversion of ENERGY to WORK is even worse than it seemed at first. Actually

IT IS IMPOSSIBLE TO CONVERT *ANY KIND OF ENERGY* TO WORK COMPLETELY (WITH 100% EFFICIENCY).

This is a dreadful situation. It means that no matter what we do to convert ENERGY into WORK in any SYSTEM, some of the energy is going to be wasted. Very often (but not always) the wasted energy leaves the system in the form of heat.

Subsequently, other investigators were able to determine, for example, the electrical equivalent of heat, and to amass a growing body of evidence that the various forms of ENERGY can be interconverted. Joule himself measured the heat produced by an electric current, by the friction of water against glass, and by compressing gas. All of the investigators who worked on this problem found that a fixed amount of one kind of energy was converted into a fixed amount of another kind of energy. Furthermore, if energy in all its varieties was considered, then when these transformations take place, no energy was either lost or created. By *very* careful measurements (as, for example, in making sure to account for possible heat losses to the room in the Joule experiment) it was shown that the energy present in one form could be converted completely to another, *but* that no 'extra' ENERGY appeared in the SYSTEM nor disappeared from it.

This growing body of carefully obtained experimental data finally led to an important statement summarizing the results:

THE TOTAL AMOUNT OF ENERGY IN THE UNIVERSE IS CONSTANT. IT CAN BE CONVERTED FROM ONE FORM TO ANOTHER, BUT IS NEITHER CREATED NOR DESTROYED.

This statement was formulated by the German scientist Hermann von Helmholtz (Figure 11.10), at the time he was serving as a doctor in the German army. On his mother's side of the family, Helmholtz is a descendant of William Penn. Helmholtz worked in several scientific fields and surely must be counted as one of the leading scientists of the 19th century. His accomplishments included the ability to speak at least nine languages. Helmholtz's summary statement shows us that the various forms of ENERGY are equivalent and that one form can be converted into another.

Let's think again about waterwheels. We know that not all of the kinetic energy of the water is transferred to the wheel, because the water is still flowing (that is, it still has some kinetic energy) after it leaves the wheel. Therefore, the amount of WORK taken out of this SYSTEM is less than the total ENERGY in the SYSTEM. Some of the ENERGY has been wasted (because it's in the water flowing away from the wheel). Could we get all of the ENERGY transferred into WORK? In principle, yes. All we need to do is stop the water absolutely dead at the instant it hits a blade of the wheel, so that *all* of the kinetic energy of the water is transferred to the wheel. No one has ever succeeded in designing a waterwheel that would function in this way. This suggests that the maximum amount of WORK we can get from a system is limited by the amount of available ENERGY in it.

FIGURE 11.10 Hermann von Helmholtz (1821–1894). Educated at Berlin, Helmholtz made important contributions to mathematics, physics, and physical science.

Consider two different ways that we could position a waterwheel under a waterfall (Figure 11.11). Intuitively, it should seem that the SYSTEM on the left is not very efficient. The wheel is positioned at a point where not very much of the potential energy of the water has been converted to kinetic energy. On the other hand, by putting the wheel as far down the waterfall as we can get it, shown on the right-hand side, we should have a much better SYSTEM, one that takes advantage of maximum conversion of potential energy to kinetic energy.

Since Joule's work showed us that mechanical energy is equivalent to thermal energy, let's now consider a thermal energy system. Figure 11.12 suggests an analogy between the two systems (thanks again to Joule). With the waterwheel we've seen that the lower we position the wheel, the better; and that to get complete conversion of ENERGY to WORK we would have to stop the water dead, so to speak. By analogy, the lower we would make the 'low-temperature side' of the thermal system, the better. So—how low can we go? The answer: absolute zero.

We can sum up by saying that WORK is the result of using ENERGY to cause something to happen in the real world—it is the physical manifestation of ENERGY. In the real world we are never able to convert ENERGY completely into WORK. (If we could, we could make a perpetual motion machine.)

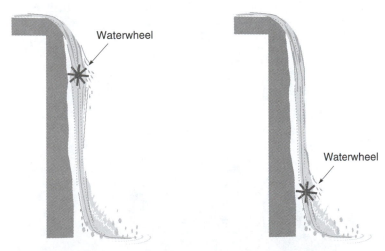

FIGURE 11.11 The picture on the left shows a system that is not very efficient because it has such limited conversion of potential energy to kinetic energy, but the waterfall on the right takes much better advantage of the potential energy.

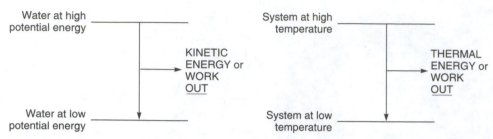

FIGURE 11.12   These two energy diagrams show the similarities between gravitational and thermal energy systems.

HEAT AND HOW IT IS TRANSFERRED

Carnot expressed the essential characteristic that makes any heat engine efficient: There must be a difference in temperature in the engine. The thermal energy in a single object cannot be converted into WORK, but, if we have two objects at different temperatures, that restriction does not apply.

Temperature is the quantity that measures thermal potential energy. Just like gravitational potential energy tells us which way matter will flow (or, put crudely, whether or not it will fall down) in a gravitational system, temperature indicates which way (if any) thermal energy will flow. If no heat flows when two objects touch, then those objects are said to be in thermal equilibrium and, by definition, their temperatures are equal. But heat flows from, say, the first object to the second, then, by definition, the first object is hotter than the second. Another way of saying essentially the same thing is that a hotter object with a temperature will *always* transfer heat (thermal energy) to a colder object. The behavior of the coffee, tea, soup, ice cream, or cold beverage that we've used as examples earlier in this chapter are manifestations of this general rule.

There are three ways that heat can move in a system. They are called conduction, convection, and radiation (Figure 11.13). **Conduction** occurs when heat is transferred directly through the body of an object. An example would be the observation you'd make by holding a metal object in your hand and sticking the other end of it into a flame. All of us realize that sooner or later the end in your hand will get hot enough to

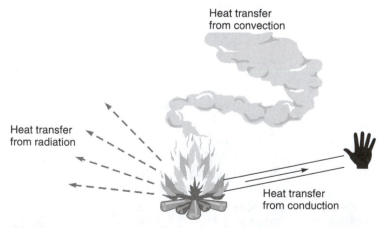

FIGURE 11.13   Heat transfer can occur in three ways: radiation, convection, or conduction.

burn you. In this case the heat of the flame is transferred directly to your hand through the motion of the heated atoms in the metal object. As a rule, conduction is particularly important for heat transfer through solids.

What's happening as heat is conducted through the solid is roughly this: The end of the metal object gets hot, and as it does so, the atoms in that portion of the metal increase in kinetic energy. Assuming for our purposes that the metal object doesn't actually melt, the *average position* of each atom in the solid remains fixed, in some crystalline array that is characteristic of whatever solid we are using. However, at any temperature above absolute zero, each of these atoms that has a fixed average position can, and indeed does, vibrate around that position. As the atoms gain kinetic energy by being heated, these atomic vibrations become more and more rapid, and the movements of the atoms extend further and further from their fixed 'equilibrium' position. The atoms at the end inserted into the flame, which will be the hottest portion, will vibrate the most because they have the most kinetic energy. As they do this 'heat dance,' they will jostle neighboring atoms in the slightly cooler portion. Those atoms in turn will acquire kinetic energy as a result of these impacts. Therefore, they vibrate more energetically themselves, and jostle atoms in an adjoining cooler portion of the object. This goes on and on until the kinetic energy bumps its way, atom by atom, the whole length of the object and we feel the heat in the palm of our hand.

Convection is a process of heat transfer that can occur in gases or liquids. We can see convection, for example, when we're cooking and watch a pot of water come to a 'rolling boil.' We can feel convection, when we're warmed by a hot summer breeze or cooled by a cold draft.

Heat can also be transferred by **electromagnetic radiation**. We can sense the heat from an object that is glowing red hot, even though we are not (let's hope!) in direct contact with it for conduction, nor feeling currents of heated air from convection. Unquestionably, the most vital aspect of heat transfer by radiation is that radiation is the process that brings heat from the sun to the Earth.

Regardless of the method by which the heat is actually transferred—conduction, convection, or radiation—heat will flow from a hot object to a cool object until the temperatures of different portions of the system are equal. Essentially, the purpose of the temperature scale is to classify or 'order' objects according to the direction in which heat (thermal energy) will flow between them. This lets us create a straightforward definition of the concept of *heat*: energy that flows from one object to another because of a difference in their temperatures.

HOW MUCH HEAT FLOWS?

A final question on this topic is to determine the amount of heat that actually flows. To help answer this question, we again rely on some common observations. Some substances, like water, have an ability to remain perceptibly warm for a long period of time. (For example, people who go swimming at night after a very hot day may find the water still quite warm, even though the air has cooled.) Other materials, like silver, lose heat remarkably quickly. We might express this by saying that water has the capacity to store a lot of heat, and silver does not. In fact, we define a fundamental property of materials called the **heat capacity**. The amount of material involved is important. A single drop of boiling water splashed on your skin might cause a momentary irritation or tiny burn, but a bowl full of boiling water could cause a very severe burn. Relative temperatures are also important. A red-hot ingot of iron can be felt to put out a lot of heat, while a block of iron just a few degrees warmer than the room might not even be perceived to be warm. Thus the amount of heat that flows is a function of (1) the amount of material (the mass) in the system, m; (2) the heat capacity of the system, C; and (3) the difference in temperature, ΔT. The amount of heat that flows is given the symbol Q. Then

$$Q = mC\Delta T$$

Normally we can look up values of C for different materials. Then if we know two of the other three terms—Q, m, or ΔT, we can calculate either the heat flow or the temperature change.

As the 19th century drew to a close, engineers felt that the steam engine was reaching perfection. Thanks to the development of the science of thermodynamics by Carnot and his successors, and its practical application to steam by engineers such as Rankine, the efficiency of these engines was continually being improved. Furthermore, there seemed to be no limit to the size of the engines that could be built, nor to the WORK they could do. Clearly steam would transform, and rule, the world. There were only two small clouds on the horizon. In the 1870s, engineers had transformed the laboratory experiments of Michael Faraday into practical, working electric generators and motors. And in the 1880s, the first engine was invented that used a new fuel distilled from petroleum—gasoline.

CITATIONS

1   John Stuart Mill, quoted in Heilbroner, Robert. *Teachings from the Worldly Philosophy*. Norton: New York, 1996; p. 130.
2   Friedrich Engels, *Dialectics of Nature*, London, 1940, p. 75; as cited in Bernal, J.D. *Science and Industry in the Nineteenth Century*. Indiana: Bloomington, 1953.
3   Richard Benson, cited in Brand, Stewart. *The Clock of the Long Now*. Basic Books: New York, 1999; p. 101.
4   Joseph Gay-Lussac, cited in Glashow, S.L. *From Alchemy to Quarks*. Brooks-Cole, Pacific Grove, CA: 1994; p. 177.
5   Sadi Carnot, quoted in von Baeyer, Hans Christian. *Maxwell's Demon*. Random House: New York, 1998; p. 38.
6   John Locke, cited in Laidler, K.J. *To Light Such a Candle*. Oxford: Oxford, 1998.
7   Count Rumford, cited in Glashow, S.L. *From Alchemy to Quarks*. Brooks-Cole, Pacific Grove, CA: 1994, p. 207.
8   Eliot, George. *Middlemarch*. Numerous editions of this classic novel are always in print; the quotation is from p. 27 in the Folio Society, London, edition.

FOR FURTHER READING

Asimov, Isaac. *The History of Physics*. Walker: New York, 1966. A very readable introduction to physics for persons with minimal math and science background. Though some of the later chapters on the most modern discoveries are now badly outdated, much of the book remains very useful. Chapter 14 discusses heat.
Bernal, J.D. *Science and Industry in the Nineteenth Century*. Indiana: Bloomington, 1953. An eloquent pair of essays on the development of scientific concepts and their impact on industry in the 19th century, with much useful information on the history of concepts of heat.

Bloomfield, L.A. *How Things Work*. Wiley: New York, 1997. This introductory physics textbook addresses the subject by analyzing the ways in which common devices that we encounter in everyday life function. Chapters 6 and 7 treat heat and thermodynamics, including analyses of wood stoves, air conditioners, and automobiles.

Cardwell, Donald. *The Norton History of Technology*. Norton: New York, 1995. A superb one-volume survey of all of technology, well written, and available in paperback edition. Chapter 10 includes a discussion of heat engines.

Glashow, S.L. *From Alchemy to Quarks*. Brooks-Cole, Pacific Grove, CA: 1994. This introductory text on physics requires some mathematical sophistication. Chapter 5 covers heat, and its relationship to motion.

Purrington, R.D. *Physics in the Nineteenth Century*. Rutgers: New Brunswick, 1997. More comprehensive than Bernal's book mentioned above, and containing interesting treatment of the history of the concept of heat and its applications.

Strandh, Sigvard. *The History of the Machine*. Dorset: New York, 1979. Chapter 5 treats steam and combustion engines. This is a well-illustrated coverage of all types of machinery, not just limited to energy technology.

Trefil, James S. *The Unexpected Vista*. Macmillan: New York, 1983. Subtitled 'A physicist's view of nature,' this is a well-written account, easily understandable by persons of limited science and math background, of the application of physics to various phenomena in nature. Chapters 7 and 8 are particularly relevant to our discussion of heat and its movement.

von Baeyer, Hans Christian. *Maxwell's Demon*. Random House: New York, 1998. An account, written for readers of limited scientific background, of the behavior of heat. The work of Rumford, Joule, and Carnot, among others is discussed. An excellent introduction to the topic.

12

AN INTRODUCTION TO ELECTRICITY

THE 'PREHISTORY' OF ELECTRICITY

The first recorded observations of electrical phenomena were made by the Greek philosopher and scientist Thales, around 600 B.C. He observed that various naturally occurring substances, such as amber, could be rubbed with a cloth and would then attract light objects, such as pieces of paper. This is a phenomenon that we now call *static electricity* (Figure 12.1). One place we encounter it nowadays is in the 'problem' of 'static cling' when we launder clothes. Like most other ideas and observations of the ancient Greeks, little or nothing was done to develop the practical consequences. The study of electrical phenomena languished for about the next 2,000 years.

THE EARLY 'ELECTRICIANS'

In the beginning of what we might call the modern scientific era—in the 17th century—many investigators in Britain and in Europe continued to add to the fund of experimental observations on static electricity. One of the first significant contributions of the modern era is due to an English physician, William Gilbert. He was the official court physician to Elizabeth I, and, after she died (not, we presume, as a result of Gilbert's ministrations), to James I, who became the new monarch. Gilbert had ample free time to do his own scientific work, and was interested in phenomena related to electricity and magnetism. In 1570 he developed the term 'electrics' to describe substances that would generate these feeble attractive forces when rubbed.

A substance that showed such attractive forces was said to be 'electrified.' An electrified substance was thought to have gained some sort of electrical fluid that—once it was actually in the substance—remained stationary. For that reason, the term static electricity was developed. (About that time, too, those scientists who studied electricity

FIGURE 12.1 Lightning is an electrical phenomenon commonly experienced today. Although lightning was not recognized as electricity until Benjamin Franklin's experiments in the 18th century, Thales theorized about static electricity as early as 600 B.C.

came to be called 'electricians.' Of course the word has a totally different connotation today.)

Around 1660, the famous mayor of Magdeburg, Otto von Guericke, built a machine that consisted of a large ball of sulfur inside a glass globe. The glass globe was rotated at high speeds by means of a crank. It was excited by friction by applying a cloth pad, or even with his hand. Two brushes touching the globe allowed sparks to leap across a narrow air gap between two metal balls. As it turned, he found that it would then attract light objects like feathers and threads. By moistening a string, von Guericke was able to transmit electricity several feet through the string. He rubbed the spinning sulfur globe in the dark and saw it glow. He saw the glow eerily extend from the globe to his hand, some inches away. However, von Guericke never followed up these observations. To him this eerie light was just another aspect of magnetism. von Guericke had no way of realizing it at the time, but he had in fact constructed the first electric generator machine. He not only generated electricity in greater amounts than ever before by means of his machine, but he also demonstrated that it could be *transmitted* over distances.

In 1746 a Dr. Spencer came from Scotland to Boston to demonstrate electrical phenomena. One of Spencer's lectures was attended that summer by a visitor from Philadelphia, Benjamin Franklin (Figure 12.2). He was so intrigued by Spencer's demonstrations that he later purchased various pieces of apparatus and began doing experiments on his own.

Franklin, whom we've already met as the developer of the Franklin stove, and the first American scientist to be recognized as 'world-class,' proposed in the mid-18th century the concept that electricity is a fluid.

. . . the greatest man and ornament of the age and country in which he lived.

—Jefferson[1]

FIGURE 12.2 Benjamin Franklin (1706–1790). This famous American scientist was also renowned as a statesman, philosopher, printer, and author.

We can sometimes demonstrate this for ourselves when our bodies acquire a static charge. (This is especially noticeable in wintertime, when there is little moisture in the air, and we are often shuffling around on carpets in our stocking feet.) Sometimes if we reach out to touch an object we can actually see a spark jump from our fingertip to the other object.

Around 1730, or nearly two centuries after Gilbert, Stephen Gray, an English 'electrician' (who was actually a chemist) learned that rubbing a piece of glass would cause the ends of it to attract light objects. He then probed this phenomenon further by sticking a cork into the open end of a glass tube. After rubbing the glass in the usual manner, he found that now the cork attracted small objects. The next experiment was to stick a nail in the cork. Indeed, now the end of the nail attracted things. He replaced the nail with a metal rod that had a knob on its end—that too attracted objects. After experimenting with the longest metal rod he could find, he next tried thread, stretched out longer and longer until it ran all around the garden of a friend's house—and it still attracted things. In the ultimate 'scale-up' of the experiment, an electrified thread was able to cause a feather to stick to its other end, at a distance of about 800 feet.

Gray realized that he had discovered that electricity was something that would move. He could distinguish between things he called 'electrics,' which could generate electricity, and 'non-electrics,' which electricity would go through. In fact, the electric fluid can move so easily through some bodies that they can't even retain any of the fluid. In our more modern terms, we'd say that they can't retain an electric charge, and we call them conductors. He found that metals were the best conductors, and that water was a conductor. Essentially, Gray recognized that the electric fluid is not necessarily static or stationary, but can actually move through some bodies.

Based on Gray's observations, scientists recognized that substances could be divided into two classes: those that readily pass the electric fluid, which nowadays we call **conductors**; and those that retain the electric fluid, which we now call **insulators**. As a rule, good conductors are metals. Wood, rubber, and plastics are examples of insulators.

Later in his career, Gray suspended an iron rod sharpened at both ends. When he brought an electrified object near the rod in a darkened room, he saw sparks of light shining from it. He also heard crackling noises that accompanied the sparks. The noise and the sparks reminded him of the roar of lightning bolts. That small observation of Gray's was the first realization that lightning was somehow similar to electricity.

Following Gray's lead, in the 1730s the French electrician C.F. Du Fay discovered what seemed to be two 'kinds' of electricity. One is generated by rubbing a piece of glass, and the other from rubbing a piece of resin. He referred to these as vitreous electricity and resinous electricity. He also showed that the same kinds of electricity repelled each other, but opposite kinds attracted. Du Fay also proved that a wet string could conduct electricity better than a dry string, by electrifying a cork ball connected to a glass tube with a string 1,256 feet long.

Franklin realized that the concept of two kinds of electricity—vitreous and resinous—was not necessary. The presence of one ensured the absence of the other. Franklin refined Du Fay's original concept to the idea of positive and negative electricity. He showed that matter under ordinary conditions was not electrified. If some electrical charge were accumulated on an object, it would be said to be positively charged. If charge were removed, then that object would be said to be negatively charged. Furthermore, Franklin had the important realization that electricity was not *created* by rubbing a glass tube; it was only being transferred. In other words, when an 'unelectrified' object was rubbed, it could either gain electric fluid and become positive, or it could lose some electric fluid and become negative.

In Franklin's nomenclature, an excess of electric charge is positive, and deficit is negative. This is analogous to the way we think about money. If we deposit money in the bank, our balance is positive. If we have an overdraft, our balance is negative. As things turned out, Franklin's reasoning was exactly backwards. We know now something that Franklin did not—that the carriers of electric charge are electrons, which themselves have a negative charge. The physical reality of the situation is that a material with an excess of electric charge is negative, and one with a deficit is positive. Franklin's legacy has been an awkward and often confusing system of defining the signs (i.e., positive or negative) of electrical connections and the direction of electricity flow that is different between practical electrical work and physics. This mix-up has persisted for two-and-a-half centuries. However, we must realize that Franklin had no way of knowing that he was making an assignment that was backwards. In Franklin's time the atomic theory had not yet been established, let alone the concept of the components of atoms and the charges of sub-atomic particles. The fact that Franklin happened to get it wrong-way-round should in no way diminish his standing as one of the most eminent scientists of his era.

These emerging concepts were further refined by Franklin. He hypothesized that bodies having an excess of electric fluid have a positive charge, and bodies that have a deficit of electric fluid have a negative charge. When a positively charged body is brought into contact with an uncharged or negatively body, the excess electric fluid that is in the positively charged body should flow into the uncharged or negatively charged one. Similarly, if we were to bring an uncharged object in contact with a negatively charged object (i.e., one that has a deficit of electric fluid), we would expect some fluid to flow from the uncharged to the negatively charged object.

Franklin extended his work by proposing that lightning is electric. He proved it in 1752. His proof came in what must be one of the best-known scientific experiments of history: the kite-in-a-thunderstorm experiment (Figure 12.3). He proved it by hanging a key from the kite, and charged a Leyden jar from a lightning discharge. (A Leyden jar, shown in Figure 12.4, is a device used in experiments on electricity, to accumulate an electric charge.) Remarkably, the experiment did not kill him. The next person to try the experiment, a Russian 'electrician' in St. Petersburg, was struck dead. And despite Franklin's unquestioned brilliance, not much was truly understood about electricity by the early 19th century:

> *I am wonderfully ignorant of the whole subject: is there yet a general theory of electricity—electricity, what it is?*
>
> *Not that I know of. Its effects can be seen and measured, but apart from that and some pretty wild unsubstantiated statements, I do not think we yet know the ABC.*
> —O'Brian[2]

With Franklin's gift for turning scientific observations into practical inventions, he immediately turned his kite experiment into the first practical—indeed, life-saving—application of the new understanding of electricity, the invention of the lightning rod.

Without doubt the lightning rod was an important invention that saved property and lives. In the Middle Ages, for example, there was a custom of ringing church bells during lightning storms, on the ill-founded idea that the noise of the bells would somehow disperse the thunder and the storm. Unfortunately, a bell rope, once it's been wet by the rain, is a fairly good conductor of electricity. As a result, it was not uncommon for the bell-ringers to be electrocuted during these storms. In 1786, the Parlement de Paris passed a law forbidding the practice of bell ringing during lightning storms, because over 100 people had been killed in the previous 30 years at the ends of wet bell ropes.

Franklin reasoned, quite rightly as it turned out, that the lightning rod would work best with a sharp end. However, other scientists, especially in England and Europe, argued for that a rounded end rod would work better. The matter was referred to the Royal Society in Britain to determine which was the correct design. The British settled for the round (i.e., the wrong) design, on the grounds that Franklin was a revolutionary. In fact, the pointed versus round argument never was settled by scientific trial (at the time)—it was settled in favor of the round design by King George III

FIGURE 12.3 Here is a very early image of Franklin completing his experiment with the kite and lightning bolt. The importance of his discovery is demonstrated by the fact that his experiment became a part of American legend almost as soon as he had completed it.

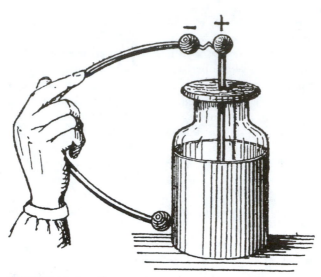

FIGURE 12.4   The Leyden jar: Franklin and his contemporaries thought of electricity as a fluid and called Leyden jars like this one 'condensers.' The scientists recognized that these 'condensers,' invented about 1745, could hold an accumulated electric charge just like glass bottles hold accumulations of condensed water. Most renditions of Franklin's experiment (including Fig. 12.3) do not picture a Leyden jar, but his famous key *was* connected to the metal rod inserted through a Leyden jar's cork top. Therefore, Franklin was able to prove that lightning was electricity by storing its discharge in his 'condenser,' where it remained until it was grounded. The process of instantaneous grounding is shown in the picture.

himself, in a rage against the American revolution. It's said that the president of the Royal Society, Sir John Pringle, took the dangerous course of possibly offending the Society's royal patron, by replying: 'Sire, I cannot reverse the laws and operations of nature.'[3] While this controversy over lightning rod design was raging, Franklin was in France negotiating the alliance between America and France that virtually ensured the success of the Revolutionary War.

> A humorous example is provided by the great satirist Jonathan Swift in his immortal *Gulliver's Travels*. There he described how 'the two great Empires of Lilliput and Blefuscu' actually went to war over the issue of whether it is best to open a boiled egg at the breakfast table at the sharper or at the more rounded end.

> *The succession to Dr. Franklin, at the court of France, was an excellent school of humility. On being presented to any one as the minister of America, the commonplace question was* 'c'est vous, Monsieur, qui remplace le Docteur Franklin?' *'it is you, Sir, who replace Dr. Franklin?' I generally answered, 'no one can replace him, Sir; I am only his successor.'*
> —Jefferson[4]

When he heard of the king's meddling, Franklin expressed the hope that King George III would just dispense with lightning rods altogether—and suffer the obvious result.

This story of King George III and the lightning rod is probably not the first instance of political interference in science, and it surely was not the last. The 20th century has witnessed tragic examples, most especially under Nazi and communist political regimes. Even at the dawn of the 21st century, political issues cloud the pricing and availability of AIDS medication for many millions of sufferers in sub-Saharan Africa.

THE ELECTRIC FLUID AND ITS POTENTIAL

Franklin's hypothesis about the flow of electric fluid immediately suggests the flow of water down hill. We should recognize from the previous discussions of heat and waterfalls that, if something flows, it must do so because of a difference in potential. We should now be able to build immediately on that analogy to describe an electrical SYSTEM (Figure 12.5). To complete the diagram for the electrical system, we need a unit to express electrical potential energy. Electrical potential energy, or electrical potential, is measured in volts, in honor of the Italian 'electrician' (physicist), His Excellency, Count Alessandro Giuseppe Antonio Volta, and is given the symbol V. Thus the **volt** is the measure of electrical potential energy (Figure 12.6). In keeping with the analogy of a waterfall, we sometimes refer to the low-potential side of an electrical system as the 'ground.'

THE ELECTRIC CURRENT

Continuing the analogy, water is moving from high gravitational to low gravitational potential energy. In the more general case of a gravitational SYSTEM, the 'thing' that is moving is *matter*. Water is simply one special kind of matter. Not only is the difference in potential important, but so is the amount or quantity of matter that flows. As an extreme example, a single drop of water falling the height of the world's tallest waterfall (which happens to be the Angel Falls in Venezuela, at 3,280 feet) will be not nearly as effective in doing WORK as thousands of gallons of water flowing down just a few feet in a working water wheel. (The same is true in a thermal system. If our sidewalks are icy, pouring a teacup of boiling water on the ice isn't going to melt much of the ice.)

A difference in electrical potential energy at two different points in space may result in the movement of electrical charge between them. This motion of electrical charge is called an electric **current**. Current is measured by the number of charges going past a selected point in a unit time.

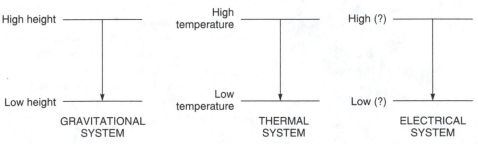

FIGURE 12.5 Here we can see how closely related the gravitational and thermal systems are to an electrical system.

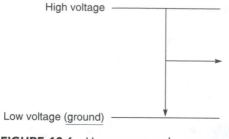

FIGURE 12.6 Here we see the energy diagram for an electrical system, complete with the appropriate unit labels (voltage).

In a gravitational system, what flows is matter, measured in various units of mass. In a thermal system, what flows is heat, measured in units of Btu. In an electrical system the flow is the electric charge. However, it is more common and more convenient to express the flow as an electric current, which is the amount of charge flowing per unit time. That is,

$$\text{Current} = \text{Charge/Time} \qquad (12.1)$$

The units of electric current are named in honor of a French 'electrician' (a physicist), a child prodigy who is said to have mastered *all* of what was then known of mathematics by the time he was 12—André Marie Ampère. Thus the unit of electric current is the **ampère**, or, as we usually call it, the **amp**.

RESISTANCE

We've seen that in the early 18th century the English electrician Stephen Gray made the first observations that some substances (metals, such as copper, aluminum, or silver) conduct electricity easily, and others (e.g., wood, rubber, or plastic) do not. Within these two families of materials, not all are equally good conductors or insulators. The amount of electric current that will flow will depend not only on the potential difference (the voltage) but also on the ease with which the current flows. This latter property is called the **resistance**. About a century after Stephen Gray's work, a professor of mathematics in Cologne, Germany, Georg Simon Ohm (Figure 12.7), worked out the fundamental laws of conductivity.

FIGURE 12.7 George Simon Ohm (1789–1854). Ohm was born in Bavaria, where he conducted his research on resistivity.

Ohm considered the flow of electric current in a wire as analogous to the flow of heat through some material that readily conducts heat. To develop this analogy, he needed a concept that would correspond to the role of temperature in the flow of heat. Ohm conceived of a quantity that he called the electroscopic force (now called **electromotive force**, essentially equivalent to the voltage) which pushed electricity along the wire in the direction of the current. We've seen that the amount of heat that will flow via conduction depends in part on the nature of the material it's flowing through. (Contrast a metal cup with a styrofoam cup, for example.) In the same way, the flow of electric current should vary with the nature of the conductor. From that analogy, Ohm introduced the idea of a resistance to the current. The resistance is property of whatever object the current is flowing through. Using a battery that would allow him to connect different

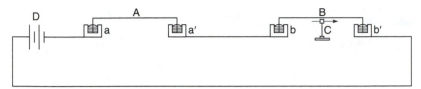

FIGURE 12.8 Ohm's experiment was set up much like this, with battery D providing energy for the entire circuit and compass C measuring the strength of the current. A and B were lengths of various conductors, placed into copper terminal pieces that dip into mercury cups (a, a', b, b'). By changing the conductor used for A and B, Ohm could measure the change in current and determine the proportional resistance.

materials to the terminals, Ohm showed the current obtained was directly proportional to the voltage at the terminals and inversely proportional to the resistance of the conductor (Figure 12.8). This relationship is now known as Ohm's law.

It has become customary in electrical work to consider not the conductivity (that is, the ease of flow of the electrical current) but rather the resistance (the opposition to the current flow). Ohm developed one of the important equations for dealing with electricity:

$$\text{Current} = \text{Potential Difference/Resistance}$$

or, equivalently,

$$\text{Potential} = \text{Current} \times \text{Resistance}$$

Electrical resistance has the units of ohms, given the symbol Ω. Current is given the symbol I. Thus

$$\text{Potential} = \text{Current} \times \text{Resistance}$$
$$\text{Volts} = \text{Amps} \times \text{Ohms}$$
$$V = I\Omega \tag{12.2}$$

There are useful analogies between the flow of electricity and the flow of water, or other fluids. Think of a garden hose used to water your lawn or garden. The pressure of water supplied to the hose is constant. This is comparable to an electrical system with a constant voltage. What happens if someone steps on the hose or drives a car over it? We can immediately see that the water flow diminishes when the hose is squashed. Compressing the hose by standing on it or driving over it increases the resistance to the flow of water. If someone puts a kink in the water hose, the flow will decrease drastically. When the resistance increases and the pressure is held constant, the flow decreases. In exactly the same way, and as can be seen from the equation for Ohm's law, when we increase the resistance in an electrical circuit and keep the 'pressure' (the voltage) constant, the current will drop. Even without kinks and crimps and fools driving their cars over your hose, there is still an inherent resistance to the flow of water. If you could lay the hose out in a perfectly straight line, it would still have some resistance. If we connected a second hose, to double the length, we'd find that the flow would drop. If we got a skinnier hose (with a smaller diameter for the water to flow through), we'd also see that the flow would drop. Electricity flowing through an electrical system will behave in the same way. The relationship between electricity flow and water flow is shown in Figure 12.9.

Indeed, as Gray showed, different materials resist the flow of electrical current by different amounts. Some materials offer little resistance to the movement of electrons. Other materials offer a great deal of resistance. The *resistance*, as it is commonly measured, depends on the amount of the material, as well as on which specific substance it is. It would be useful to have a property that would characterize materials based only on their

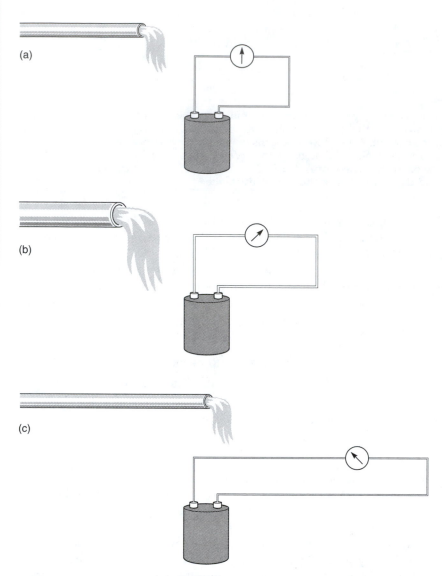

FIGURE 12.9 An illustration of some analogies between water flow and electricity flow. In (a), water flowing through a pipe is comparable to electricity flowing through a circuit. In (b), increasing the diameter of the pipe allows more water to flow; increasing the diameter of the wire reduces resistance and allows more electricity to flow. In (c), lengthening the pipe reduces water flow; similarly, lengthening the wire increases resistance such that less electricity flows.

composition and nature, without having to take into account such factors as amount or size. **Resistivity** characterizes this property of materials in that way, that is, depending only on the nature of the material (Figure 12.10). Thus, for example, all samples of copper have the same resistivity, but a copper wire of a given diameter and, say, 10 feet long has twice the *resistance* of a copper wire of the same diameter but only five feet long. This relationship is expressed by the equation

$$\Omega = (\text{Length} \times \text{Resistivity})/\text{Area} \qquad (12.3)$$

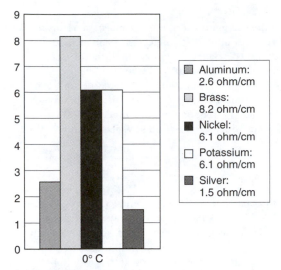

Aluminum:
2.6 ohm/cm

Brass:
8.2 ohm/cm

Nickel:
6.1 ohm/cm

Potassium:
6.1 ohm/cm

Silver:
1.5 ohm/cm

FIGURE 12.10 Conductors have varying resistivities, but their differences seem minimal when compared with insulators like carbon, which at 0°C has a resistivity of 3,500 Ω/cm.

In terms of fundamental properties of the materials, we can say, for example, that iron has a higher resistivity than copper. However, we cannot say that a particular piece of iron wire has a higher resistance than some other piece of copper wire, unless we know the lengths and cross-sectional areas of each, so that we can appropriately calculate the resistances.

When a wire actually breaks, the current in the circuit stops because the air has so high a resistance (or, in other words, is so good an insulator) that the voltage is not sufficient to overcome it. The electricity in our household circuits stays in the wires and does not leak out through the walls because the copper or aluminum in the wires has a much lower resistance (i.e., is a much better conductor) than the plastic or rubber insulation around the wires. Since it is easier for the current to flow through the low-resistance path—the wire—than through some route of higher resistance, the electricity follows the circuits we have created for it.

For all practical purposes, every material exhibits some amount of resistance to current. However, a very few materials, **superconductors**, have no resistance whatsoever. Superconductors began as curiosities in the physics laboratory. They now have some applications, though usually very specialized ones in, for example, scientific apparatus (Figure 12.11). Superconductivity is under intense investigation in many laboratories around the world; it is hoped that future commercial applications might include long-distance transmission of electricity without losses and high-speed magnetically levitated ('maglev') trains to replace conventional railroads. So far, the materials known to be superconductors show this remarkable property only at quite low temperatures (e.g., the temperature of liquid helium, about 4 K or −269°C), and it is this need—to keep them insulated and very cold—that has so far restricted their application from electrical transmission lines, high-speed transport, and ordinary household electrical products. One of the challenges of science in the 21st century is to develop materials that would be superconducting either at ordinary temperatures, or at least at temperatures that could be maintained with common refrigerating equipment. Such a development would represent one of the most revolutionary advances in our use of energy in all of history.

As one example, chemists rely heavily on the technique called nuclear magnetic resonance (NMR) spectroscopy to determine the structures of complex molecules. (The method is akin to magnetic resonance imaging used in medicine for diagnosis of injuries or disease.) NMR is especially powerful for determining the arrangements of carbon and hydrogen atoms in organic chemistry. A scientist skilled in designing NMR experiments and interpreting the data can often unequivocally work out the molecular structure of a newly discovered substance, even from a sample of a tenth of a gram. Most modern NMR instruments rely on superconducting magnets, which the users must vigilantly keep cooled to liquid helium temperatures.

In the late 1980s, physicists discovered materials that display superconducting properties at the temperature of liquid nitrogen, 77 K (−196°C). While to our everyday experience this temperature is still extraordinarily cold (indeed, life as we know it could not exist at such temperatures), the large-scale production of liquid nitrogen is easily within today's technology. Therefore, a device relying on superconductors operating at 77 K would be much easier to build and operate, and far cheaper to maintain at operating temperature, than ones requiring temperatures near that of liquid helium, 4 K. At present, however, virtually all of these so-called high-temperature superconductors (that is, materials that display superconducting properties at or above liquid nitrogen temperatures) are nonmetallic. Fabricating them into wires might be difficult. Further, there seems as yet to be no theoretical understanding of high-temperature superconductivity that is accepted widely in the scientific community. This lack of a generally accepted theory of high-temperature superconductivity may also be impeding progress (though it is useful to recall how the steam engine was developed before there was a satisfactory understanding of heat and efficiency).

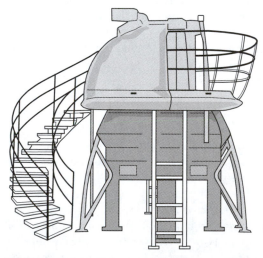

FIGURE 12.11 This spiral staircase wraps around a huge superconductor of the kind used in laboratories for research today.

ELECTRIC POWER

Notice that our definition of electric current has introduced the units of time. We have previously identified POWER as the rate of doing WORK or the rate of using ENERGY. Electrical POWER, P, is determined by

$$\text{Power} = (\text{Charge/Time}) \times \text{Potential}$$

$$\text{Power} = \text{Current} \times \text{Potential}$$

The units of electrical power are **watts**, W. Thus

$$\text{Watts} = \text{Amps} \times \text{Volts}$$

$$P = IV \qquad\qquad (12.4)$$

We should recall our definition that

$$\text{Power} = \text{Energy/Time}$$

Now this equation can be rewritten in a slightly different form,

$$\text{Energy} = \text{Power} \times \text{Time}$$

If you examine your electric bill, you will find that you are being charged a certain amount of money per kilowatt-hour (kWh). Now a kilowatt (1000 watts) is a unit of POWER, and of course an hour is a unit of time. A kilowatt-hour, therefore, is a unit of (POWER × Time), which is energy. When you pay your electric bill, you are *not* purchasing POWER, you are in fact purchasing ENERGY. So, the local 'power company' that produces electricity in 'power plants' is not in the power business, it is producing and selling ENERGY.

In fact, it was the work of these scientists in establishing the exact relationships among the amp, the volt, the watt, and the ohm that made the buying and selling of electrical ENERGY possible. That possibility in turn made it possible for the electrical supply industry of every nation to come into being, a story that we will follow in subsequent chapters.

Most of us encounter the watt when we're purchasing light bulbs. Bulbs come in various ratings, the most commonly used ones being 60, 75, and 100 watts. Suppose that the electricity in your home is 110 volts. Then the 60-watt bulb will have a current of 0.55 amps (60 watts/110 volts), and the 100-watt bulb will have a current of 0.91 amps (100 watts/110 volts). As a rule, the electrical potential difference—the voltage—doesn't change. The power—the wattage—of the bulb is the product of voltage times current. From Ohm's law, the current will depend inversely on the resistance of the filament in the bulb (Figure 12.12). That is, the greater the resistance, the smaller the current. And, the smaller the current, the less power consumed, and the less light will be produced. In other words, we will perceive that a 60-watt bulb is 'dimmer' than a 100-watt bulb.

Generally, most light bulbs use the same material, tungsten, as the material for the filament. If the material is the same from one bulb to another, then so is the *resistivity* of the filament, since all samples of the same material have the same resistivity. What we have to do is change the *resistance* of the filament. Based on Equation (12.3), the easiest way to make the resistance higher is to make the cross-sectional area smaller. In other words, we make the filament skinnier. So, we have then the following relationships: the thinner the filament, the higher the resistance. The higher the resistance, the lower the current. The lower the current, the lower the power (wattage). The lower the wattage, the less light is

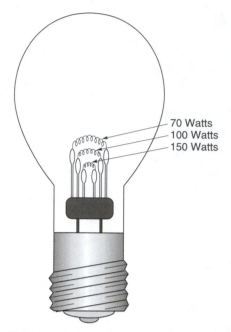

70 Watts
100 Watts
150 Watts

FIGURE 12.12 This cutaway image of a three-way light bulb allows us to see how different wattages are produced. In this case, rather than changing the cross-sectional area of the tungsten, the filaments have varying lengths that affect their resistance to electricity and, therefore, their brightness

produced. Therefore, a 60-watt light bulb has a thinner filament than one rated at, say, 100 watts.

We can imagine, and probably intuitively understand, that a stream of water flowing through a hose at a pressure of 100 pounds per square inch (psi) is more powerful than an equal-sized stream at 25 psi. A high-pressure stream of water can knock a person off his or her feet, or do considerable damage. In the same way, a current of electricity provides more power at the relatively high 'pressure' of 100 volts than the same current would at 25 volts.

Remembering that $P = VI$ and also recalling Ohm's law that $V = I\Omega$, we can combine these two equations to see the relationship among power, current, and resistance:

$$P = \Omega I^2 \qquad (12.5)$$

Regardless of why we need the electricity, or how it is generated, the local utility wants to have smallest possible losses of electricity while it is being distributed from the point of generation to the individual consumers. In other words, the local utility wants to get as much of the electricity generated delivered to you as possible. Even though the wires in electrical distribution system are made of excellent conductors, there is still some resistance to the flow of current. (As mentioned briefly above, the hope for eliminating this problem in the future lies in the development of so-called high-temperature superconductors that could be fabricated into wires.). Sending electricity through wires requires that we do WORK to overcome the natural resistance of the wires. When that WORK is performed, some ENERGY is inevitably wasted. Much of that ENERGY appears as heat, and indeed is sometimes referred to as resistance heating.⚛

⚛ We should recognize that resistance heating is not always a bad thing. Incandescent light bulbs give off light because the resistance heating of the filament is so great that the filament becomes hot enough to glow. The heating coils on an electric stove rely on resistance heating; on many such stoves turning the control to "high" generates enough heat for the coil to become red hot. Many other electrically operated devices that we use to produce useful heat make use of resistance heating.

For a distribution system having a given resistance, the losses are proportional to the current *squared*, as can be seen from Equation (12.5). Significant reductions in losses can be achieved by making the current as small as possible. But a second important relationship is that among power, current, and voltage in Equation (12.4). If the current, I, is made very small to reduce losses during transmission, then the voltage, V, must be made very large to maintain a given value of the power. Consequently, electric companies transmit electricity at extremely high voltages—thousands or even tens of thousands of volts. The higher the voltage for a given amount of power, the lower will be the current, and, therefore, the lower will be the heating losses.

Voltage is changed in an electrical circuit by using a transformer. A transformer can be used to increase or decrease voltage, and would be referred to as a step-up or step-down transformer, respectively. Neglecting the small losses in the transformer itself, the electrical energy going into the transformer must equal the electrical energy coming out. If we check the transformer for a given period of time, then it would also be true that the electrical power must be the same into and out of the transformer. That is,

$$P_{in} = P_{out} \quad \text{or} \quad V_{in}I_{in} = V_{out}I_{out} \tag{12.6}$$

For a 220-volt, 100-amp electrical service to a residence, the same 22,000 watts of power could be transmitted at 20,000 volts and 1.1 amps, and then stepped down to 220 volts in an appropriate transformer.

CITATIONS

1   Jefferson, Thomas. Letter to Samuel Smith, August 22, 1798. (Reprinted in the Library of America volume, *Jefferson*, New York, 1984; pp. 1052–1055.)
2   O'Brian, Patrick. *Blue at the Mizzen*. Norton: New York, 1999; p. 201.
3   Cited in Derry, T.K.; Williams, T.I. *A Short History of Technology*. Dover: New York, 1960; p. 609.
4   Jefferson, Thomas. Letter to Reverend William Smith, February 10, 1791. (Reprinted in the Library of America volume, *Jefferson*, New York, 1984; pp. 973–975.)

FOR FURTHER READING

Asimov, Isaac. *The History of Physics*. Walker: New York, 1984. This book provides an excellent introduction to physics, through about the 1960s, and requires little prior scientific background from the reader. Chapter 25 gives an introduction to electricity.

Aubrecht, G. *Energy*. Merrill: Columbus, 1989. A very useful introductory textbook on energy science and technology, though with a much different organization than is used here. Chapter 4 provides a discussion of electricity.

Bloomfield, L.A. *How Things Work*. Wiley: New York, 1997. This is an introductory college-level textbook on physics, organized around teaching the subject by analyzing the physical principles involved in many common, everyday devices. Chapters 11 and 12 introduce the principles of electricity through such devices as photocopy machines, flashlights, and transformers.

Bronowski, J. *The Ascent of Man*. Little, Brown: Boston, 1973. A well-written and well-illustrated history of science through the 1960s. Chapter 8 discusses (among other topics) Franklin and his lightning rods.

Burke, J. *Connections*. Little, Brown: Boston, 1978. A somewhat quirky history of science focusing on how scientific and technological innovations have spread change through society. Very well illustrated. Chapter 9 includes a discussion of electricity and electrical phenomena.

Cardwell, D. *The Norton History of Technology*. Norton: New York, 1995. An excellent survey of many forms of technology, well beyond just energy-related topics. Chapter 9 discusses electricity.

Glashow, S.L. *From Alchemy to Quarks*. Brooks/Cole: Pacific Grove, CA, 1994. A fine introductory physics text, though requiring some mathematical sophistication. Chapter 7 is an introduction to electricity.

Leon, George deLucenay. *The Story of Electricity*. Dover: New York, 1983. This paperback provides a very useful, and simply explained, introduction to the principles of electricity.

Morgan, Alfred P. *The Boy Electrician*. Lindsay Publications: Bradley, IL, 1995. An inexpensive paperback reprint of a book of experiments and projects in electricity. It is a cornucopia of ideas for those who like to learn by doing, or who simply enjoy tinkering in a home workshop.

Ramage, Janet. *Energy: A Guidebook*. Oxford: Oxford, 1997. A fine introduction to energy technology, available as a reasonably inexpensive paperback. Chapter 7 provides an introduction to the concepts of electricity.

13

HOW ELECTRICITY IS GENERATED

A way in which we can sometimes make our own observation of the flow of electricity is when we have accumulated static electricity on our bodies and can actually see a spark jump between our fingertip and some object we're reaching for. In dim light we can see this momentary flash of electricity as the 'electric fluid' flows from us to the object we're about to touch. But note that the spark happens only once, and it appears to be nearly instantaneous. Similarly, lightning, which is a flow of electricity driven by enormous potential differences, literally occurs in a 'flash.'

LUIGI GALVANI'S FROGS

Up until the end of the 18th century, only these very momentary flows of electricity could be produced. To put electricity to work for us, we need a continuous flow of electricity. But until about 1800, there was no way to make an electric current under controlled conditions. All electrical experimentation related to phenomena of static electricity. The first key observation leading to electric currents and their production was made by an Italian physicist and obstetrician, Luigi Galvani. He was experimenting with frogs' legs. Galvani showed that the leg muscles of frogs would twitch when an electric spark was applied to them, even though the tissue was no longer living (Figure 13.1). Very likely other scientists had also made this observation, but Galvani added two very important new observations. First, he noticed that the 'twitch' was even more pronounced if he happened to be touching the muscle with a metal scalpel at the time. Galvani thought that perhaps the effect—that is, the greater response of the muscle tissue—was due to some sort of induced electric charge in the scalpel.

In the previous chapter, we discussed one of the most famous scientific experiments in history, Benjamin Franklin demonstrating that lightning was a form of electricity, by flying a kite in a thunderstorm. Though we may now mostly remember Franklin for his

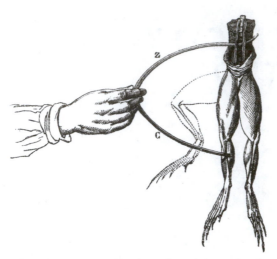

FIGURE 13.1   Luigi Galvani found that when he placed two different metals (shown here as zinc, z, and copper, c) on a frog so that one end touched the nerves and vertebrae while the other touched the leg muscles, a substantial twitch could be observed.

political and diplomatic activities later in his life, in fact Franklin was one of the great scientists of his time. His work was well-known among the European scientific community, very likely including Galvani. Consequently, Galvani reasoned that there ought to be a really considerable effect on the muscle tissue if he exposed samples to a thunderstorm.

This line of reasoning led to the second important observation, which may have been quite accidental. Galvani used brass or silver hooks to attach the leg muscles to an iron railing. Indeed, just as Galvani had hoped, there was a significant effect on the 'twitch' of the leg muscles. Then Galvani discovered that the same effect occurred even if there was no thunderstorm.

ALESSANDRO VOLTA AND THE EARLY 'BATTERY'

For some time scientists thought that the effects observed by Galvani belonged to the study of animal physiology, not the physics of electricity. These observations with frogs' legs seemed not to fit in with the laws of electricity. Du Fay, for example, had shown that a wet string was a better conductor of electricity than a dry one. Put in other terms, the wet string could not maintain a difference in electrical potential energy from one end to the other; instead, electricity was conducted easily along the string. In Galvani's experiments, the tissues of the frogs' legs were wet, with the bodily fluids that permeate the tissues. Therefore, it would have been reasoned, the legs themselves could not sustain an electrical potential difference any more than could a wet string. In the scientific thinking of Galvani's era, this twitching of frogs' legs was a new phenomenon, and one apparently different from Du Fay's demonstration of the easy way in which a wet string conducted electricity. For that reason, Galvani's discovery was at first called 'animal electricity.' Galvani's experiments were taken as proof (quite wrongly, in fact) that there was some kind of 'organic' or animal electricity in the realm of physiology that was different from the 'inorganic' electricity studied by physicists.

FIGURE 13.2 Alessandro Volta (1745–1827). This Italian professor's creation of the Voltaic pile and research were so important that the unit of electrical potential energy, the volt, is named after him.

This infuriated Galvani's countryman, Alessandro Volta, Professor of Physics at Pavia (Figure 13.2). Volta absolutely rejected the idea of animal electricity and recognized that the voltage in Galvani's experiments was due to the different metals used. In fact, Volta had already established that there was a 'contact potential' between two different metals when they were put together. Volta wondered whether this so-called animal electricity was really just the effect of the contact potential from two different metals touching. What had really produced the effect Galvani observed in his experiments was not the thunderstorm, but rather the two different metals (the brass hooks and the iron railing) in contact.

Galvani's discovery remained largely a curiosity until Volta reasoned that two different metals (such as iron and copper) in contact with a solution capable of conducting the electric current (such as the fluids in the frog's leg) would generate an electric current (Figure 13.3). Volta produced devices for generating electric current consisting of two metals, zinc and copper for instance, in contact with some medium that would conduct electricity, such as a piece of blotting paper soaked in a solution of salt. Volta found that it was possible to draw electricity continuously from the ends of this 'stack,' which came to be called a Voltaic pile (Figure 13.4). Volta's demonstration of this device to Napoleon Bonaparte so impressed him that he made Volta Count of Lombardy. (Those who enjoy 'what if' speculations about history might consider how the course of European history would have changed if Volta

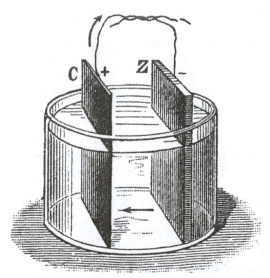

FIGURE 13.3 Some voltaic cells were created with a liquid medium.

FIGURE 13.4 Numerous voltaic cells using blotting paper were stacked by Volta into a Voltaic pile that functioned much like today's batteries. The cells were placed onto the wood base so that copper was on the bottom and zinc on the top, all contained by glass rods.

had overdone it and actually zapped the Emperor.) The Voltaic pile and the remarkable effects of the electricity it generated became a popular scientific demonstration, as in this fictional account of the effect on Jeremiah Dixon (the Dixon of the Mason-Dixon line):

> *So forcefully that his Queue-Tie breaks with a loud Snap, Dixon's Hair springs erect, each Strand a right Line pointing outward along a perfect Radius from the Center of his Head. What might be call'd a Smile, is yet asymmetrick, and a-drool. His Eyeballs, upon inspection, are seen to rotate in opposite Senses, and at differing Speeds.*
>
> —Pynchon[1]

These devices came to be called *voltaic piles.* If we connect the copper and zinc plates, we would discover that an electric current will flow between them. In modern terminology the voltaic pile would be called an electrochemical cell, or perhaps just an electric cell. Two or more cells connected together constitute a 'battery.' However, in our informal usage we nowadays use the term battery to refer even to a single cell. So, in essence, the voltaic pile is the forerunner of our battery.

A pile, in the literal sense, was necessary because Volta's only way of measuring electricity was by getting enough potential to raise a spark. In those days there were no sensitive electrical instruments for measuring voltage or current. Volta had to stack 20–30 of the metal disks on top of one another to get his first 'battery.' But, once he had done this, he realized that the generation of electricity had nothing to do with the frog at all; the frog's body parts were not necessary (a fortunate thing for untold generations of frogs yet to be born).

The voltaic pile made it possible for scientists to experiment with electric currents under reasonably controlled conditions. Humphry Davy is an example. Using the electricity from a voltaic pile, he became the first scientist to isolate several elements—such as sodium, magnesium, and calcium—from their compounds. By stacking up many alternating zinc and copper plates, with salt-soaked blotting paper between, it was possible to produce significant amounts of electric current. Despite the advance that the voltaic pile represented, it was still far from ideal as a steady source of electricity. Just like our modern-day batteries, the pile eventually 'runs down.' It is rather messy to work with. Generating a steady current for a long period of time requires devoting lots of space to setting up the pile.

BATTERIES

Before turning to the series of discoveries that led to the practical method for making large quantities of electricity, let's consider the modern descendant of the Voltaic pile, the battery. How does a battery create the electric potential that is available to us? In any

SYSTEM, providing a high potential energy of electricity requires a 'reverse energy diagram' (Figure 13.5).

> Strictly speaking, *one* of the devices that we commonly refer to as a 'battery' is called a cell. A collection of two or more cells connected together would be a battery. However, there is an almost universal tendency to eschew the pedantic and speak of individual cells as batteries. We will do so here.

A battery 'pumps' electrical charge uphill to establish the high electrical potential energy (high voltage). We know that the spontaneous direction of change is for electricity to flow from the high voltage, or high potential, to the low voltage. Since the battery carries out a non-spontaneous change, WORK or ENERGY is required to effect this change. In exactly the same way, if we want to create a high gravitational potential energy by pumping water up hill, we must supply WORK to the pump to get it to operate. In the case of a water pump, we might supply the WORK manually, or by connecting the pump to a motor of some sort.

> *For every natural event, there is an allowed direction and a forbidden direction. Apples fall; they don't jump up to their branches. Sugar lumps dissolve in your coffee; they don't form by themselves in a cup of sweetened coffee. Hydrogen combines with oxygen to form water; water does not dissociate spontaneously into a mixture of hydrogen and oxygen. All nature's streets are one way. You can take them in the wrong direction, of course, but you have to work for it: lift the apple, extract the sugar, wrench the hydrogen off the water molecules, for example, with electricity.*
>
> *This is the absolutely fundamental way of nature, expressed by what scientists call the second law of thermodynamics. . . . If you have to work for it, you are going in the forbidden direction. In the allowed direction, on the other hand, the phenomenon, properly harnessed, may work for you, though it would never give you as much work as you would have to carry out yourself to reverse it. . . . A convenient image to describe the two directions is downhill for what is allowed and uphill for what is forbidden.*
> —de Duve[2]

How does the battery supply the WORK to create the high electrical potential? There are two types of batteries: **primary cells** and **secondary cells**. Either device uses its chemical potential energy to do this work. The chemical potential energy is stored in the chemical

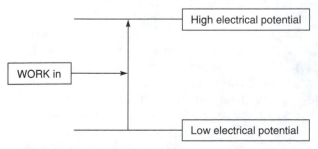

FIGURE 13.5 The simple reverse energy diagram makes it clear that some work must be put into the system for the battery to have high electrical potential that we can use.

components of the battery—in the case of the Voltaic pile, in the copper and zinc and the salt solution. In a primary cell, the chemicals necessary for production of the electricity are sealed inside. Once they have been consumed, the battery is useless. The nearly ubiquitous AA batteries are an example of primary cells. A secondary cell, however, is charged with electricity before it can be used. That initial charging puts the chemicals in the battery into a state from which they can then react to generate electricity for us when we need it. In a sense, we have put electricity into the battery before using it, and withdraw it later as we need it. For that reason, many secondary cells are also referred to as **storage batteries**. The battery in an automobile is an example.

When we use a battery to run a flashlight, or calculator, or other device, the battery has to create the electrical potential energy (voltage) by consuming some of its chemical potential energy to 'pump' the electric charges uphill. Each time it does this, some of the chemical potential energy is consumed. Eventually, not enough of the original chemical components of the battery remain to supply the needed chemical potential. When that happens, we can no longer obtain an electrical potential from the battery. We speak of this effect by saying that the battery has 'run down' or has 'gone dead.' Of course, we usually find that a calculator battery dies in the middle of an examination, or that a car battery is dead on just the day that you need to be at an important meeting or appointment.

Because the secondary cell is charged with electricity to get it to operate in the first place, most such batteries can be recharged. An automobile battery, for example, supplies electricity to start the car, but then is recharged while the motor is running. Cordless power tools often have rechargable batteries; they can be removed from the tool and recharged using household electricity. But, no process of any sort operates with 100% efficiency. Therefore, each recharge cycle does not fully restore the battery to its original condition. After a large number of recharge cycles, even this kind of battery will reach the point at which it no longer functions adequately. In other words, even rechargable batteries will 'go dead' to a point from which they cannot be restored. Modern battery technology is such that good rechargable batteries will last a very long time before they are truly 'dead.'

FIGURE 13.6 Hans Christian Ørsted became a professor at the University of Copenhagen, where he established the connection between electricity and magnetism.

THE GREAT DANE

The next significant observation was made in 1820 by the Danish scientist, Hans Christian Ørsted (Figure 13.6). That was a remarkable year in Ørsted's dual career as a physicist and chemist. Not only did he make a crucial discovery in physics, which we will discuss below, but he also made an important discovery in chemistry. Ørsted discovered the nature of the active poison (a chemical called piperidine) in poison hemlock—the drink that Socrates was required to take to commit suicide in 399 B.C., as described by Plato in his dialog *Phaedo*.

Apparently it is not known what actually motivated Ørsted to try this experiment in physics. Ørsted's apparatus is shown schematically in Figure 13.7. During a lecture demonstration, Ørsted observed that when the switch was closed—that is, the current was flowing—the compass needle responded just as if a magnet were brought near it (Figure 13.8). That is, the needle would swing violently around when the current was flowing. Since it is well known that a compass responds to a magnetic field, Ørsted deduced that

In the previous chapter we met the French physicist André Marie Ampère, of whom it is said that he had mastered all of the known mathematics by the age

A FLOWING ELECTRICAL CURRENT GENERATES A MAGNETIC FIELD.

of 12. Ampère was a self-educated genius whose father had been executed during the French Revolution. Within a week of learning of Ørsted's experiment, Ampère published an extensive theoretical treatment of the subject, showing that an electric current causes a compass needle to deflect. Though there is no question that Ampère was a brilliant scientist, it is likely that he had been thinking along the same lines before Ørsted produced the necessary experimental demonstration.

MICHAEL FARADAY AND THE INVENTION OF THE GENERATOR

Ørsted's experimental observation and his deduction remained mainly a curiosity for about ten years. Then, in 1831, Michael Faraday (Figure 13.9), the great English chemist and physicist, asked an extremely profound question: Will this experiment work backwards? In other words, Faraday's question was: Does a moving magnetic field generate an electric current? A simple schematic diagram of Faraday's experiment is shown in Figure 13.10. By passing the magnet in and out of the loop of wire, Faraday showed that a current is generated as the magnetic field passes through the conductor (the wire). Faraday's very simple experiment answered his original question in the affirmative.

This simple observation is responsible for all electricity generation that we have today, with the exceptions of the small amounts of electricity we obtain from

A MOVING MAGNETIC FIELD GENERATES AN ELECTRIC CURRENT.

batteries and from the direct conversion of solar energy into electricity (photovoltaics, Chapter 36). Faraday showed that mechanical kinetic energy (pushing the magnet back and forth) can be converted into electrical energy. Electric heating and lighting at home,

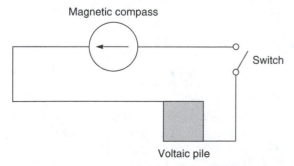

Magnetic compass

Switch

Voltaic pile

FIGURE 13.7 The basic set-up of Ørsted's experiment was very simple: when the switch was closed, creating an electric current, the compass moved because of the magnetic field.

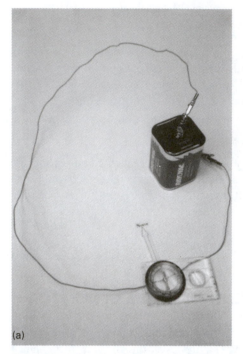 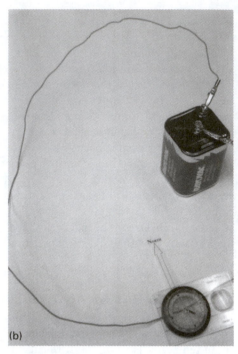

(a) (b)

FIGURE 13.8 Ørsted's experiment can easily be recreated today by using a battery, a simple circuit, and a compass. (a) When the circuit is broken, the compass needle points due North, showing that there is no magnetic field. (b) When both ends of the circuit are attached to the battery so that electricity flows through the wire, the compass needle points South-East, showing that a magnetic field accompanies the electric current.

FIGURE 13.9 Michael Faraday (1791–1867). Responsible for many scientific discoveries, this Englishman's most important contribution was his realization that a moving magnetic field generates an electric field.

all of our electrical appliances, electric motors and other electric devices in manufacturing industries, electrical railway locomotives and streetcars or subways *all* come from this simple little experiment. By showing how mechanical work can be transformed into electricity, Faraday brought electricity into the scope of the Industrial Revolution.

During the decade between Ørsted's observations and Faraday's discovery, other scientists had just placed magnets near the wires carrying currents. They found that nothing happened. The idea that it was necessary to move the magnets, or the wire, was entirely due to Faraday. This discovery was not some serendipitous accident, but rather had been carefully thought through and is recorded in Faraday's laboratory notebooks.

As long as the magnetic field was in motion, an electric current could be generated essentially forever. However, the recipro-

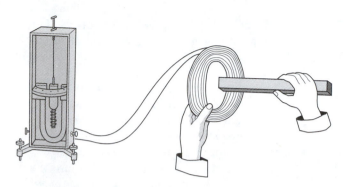

FIGURE 13.10   A schematic diagram of Faraday's demonstration that a moving magnetic field generates an electric current. The device on the left is an early instrument for detecting current flow.

cating (back-and-forth) motion of Faraday's original experiment is not desirable for large-scale generation of electricity. It is not easy to produce smooth reciprocating motion on a large scale. Also, there are always two 'dead spots' when we start and stop the magnet at each end of its motion. Faraday soon refined his original, somewhat primitive experiment with two improvements. First, it's generally much easier to sustain rotary motion than reciprocating motion. Second, he realized that the key is having the magnetic field and electrical conductor moving relative to each other. It doesn't matter, in principle, which one of the two is moving and which is stationary. Moving a magnetic field past a stationary electrical conductor has exactly the same effect as moving the conductor through a stationary magnetic field. Thus Faraday refined his original bar magnet and coil of wire experiment into the concept of an electrical generator or **dynamo** (Figure 13.11).

There are two versions of a story about Faraday and his coil-and-magnet experiment. Perhaps both actually occurred. Some time in 1831 Faraday was demonstrating the experiment and was asked by a woman, 'But, Mr. Faraday, of what use is this?' Supposedly he replied, 'Madam, of what use is a newborn baby?' In the other version, Faraday was asked the same question by a member of Parliament, William Gladstone (the same Gladstone who later served four terms as prime minister of Britain). In this version, Faraday is said to have replied, 'Sir, in twenty years, you will be taxing it.'

We must again emphasize that *all* of the electricity we use, except the little we get from batteries and from direct conversion of solar energy to electricity, comes from this device. Therefore, the essence of electricity generation is to keep the generator turning as cheaply and reliably as possible. An electrical generator is a device that converts kinetic energy (usually rotary) into electrical energy. One cheap, reliable way of spinning the

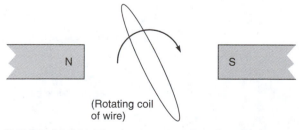

(Rotating coil
of wire)

FIGURE 13.11   The basic principle of a dynamo or generator is a coil of wire rotating around a magnet.

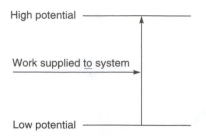

FIGURE 13.12 When a system goes from low potential to high potential due to WORK put into a system, we can represent it with a reverse energy diagram.

generator is to take advantage of the kinetic energy of falling water. As we'll see, all the other approaches make use of steam to turn the generator.

The first full-scale electricity-generating machines were built in the 1840s and 1850s. In those early devices, the magnetic fields were produced by permanent magnets. The use of permanent magnets is a real disadvantage in the construction of these machines, because a very large array of horseshoe magnets was required to provide a sufficient magnetic field. In the 1850s, many designers tried to solve this problem by replacing the permanent magnets by electromagnets. In the ultimate stage of development, a part of the current produced by the generator can be used for the energy needed in the magnetic field windings of the electromagnets. Essentially, this provides a self-generating magnetic field. This specific type of generator is called a dynamo. However, in common usage the terms 'dynamo' and 'generator' are often used interchangeably.

We must realize that in an electrical generator, *we* must supply WORK to the system (the work of spinning the generator). We can illustrate this by means of a 'reverse energy diagram' (Figure 13.12). Why should we do this (i.e., put WORK into spinning the generator)? Because we can take advantage of the generator by physically separating it (even by tens of hundreds of miles) from where we need the electricity, as shown in Figures 13.13 and 13.14. Thus we have a way of taking kinetic energy that might be available to us in one place, and easily and conveniently making it available someplace else.

Finally, we have tried in earlier chapters to show that the energy diagrams for various kinds of SYSTEMS, such as thermal, gravitational, and chemical, are *exactly the same* in concept. Now we have introduced the idea of a 'reverse energy diagram' in which we supply WORK to a SYSTEM to cause a change from low potential to high potential (and note that this would be a non-spontaneous change). Does that mean that there are 'reverse energy diagrams' for all of these systems? Yes. As examples: If we supply WORK, we can cause water to flow uphill. We can do this with a pump. If we supply WORK, we can cause heat to flow from a cold object to a warmer object. One kind of device we can use for this process is a refrigerator. Indeed, if we supply WORK (in the form of chemical

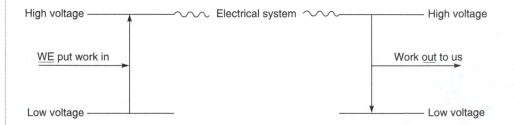

FIGURE 13.13 WORK put into the system results in high electrical potential energy that can be transmitted to a different location before converting back into WORK.

FIGURE 13.14 The principle of the energy diagrams is put into practice through numerous complex transmission lines across the world.

energy) we can even convert carbon dioxide and water back into methane and oxygen. We'll see some of these examples in detail later. The crucial point to remember is that regardless of what kind of SYSTEM we're considering, the principles of the energy diagram and the reverse energy diagram are *exactly the same* (Figure 13.15).

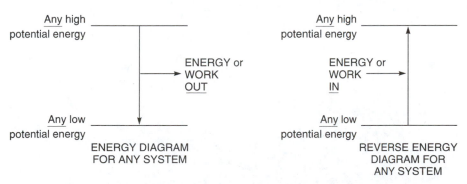

FIGURE 13.15 Any system whatsoever can be described by one of these diagrams: spontaneous flow from high to low potential with energy out (on the left) or nonspontaneous flow from low to high, requiring energy (on the right).

CITATIONS

1   Pynchon, T. *Mason and Dixon*. Holt: 1997, p. 765.
2   de Duve, Christian. *Vital Dust*. Basic Books: New York, 1995; p. 36.

FOR FURTHER READING

Cardwell, D. *The Norton History of Technology*. Norton: New York, 1995. An excellent survey of many forms of technology, not limited to energy-related topics. Chapter 13 discusses the invention of the dynamo and related inventions.

Hawks, Ellison. *Popular Science Mechanical Encyclopedia*. Popular Science: New York, 1941. This is another of the older books that is well worth seeking out in libraries or book sales. The third chapter, 'Electricity: Man's Silent Servant,' contains a well-illustrated discussion of generators and related equipment. The basic principles are still the same, and much can be learned from these older sources.

Kirby, Richard Shelton; Withington, Sidney; Darling, Arthur Burr; Kilgour, Frederick Gridley. *Engineering in History*. Dover: New York, 1990. This is a paperback reprint of a book originally published in 1956, so it provides a survey of engineering and the development of technology through about 1950. Chapter 11 deals with the rise of the electrical generating industry.

Ramage, Janet. *Energy: A Guidebook*. Oxford: Oxford, 1997. An excellent introduction to many aspects of energy technology. Chapter 7, which introduces electricity, discusses generators and how they work.

Strandh, Sigvard. *The History of the Machine*. Dorset: New York, 1979. Chapter 6 of this book is devoted to electricity. Numerous excellent illustrations supplement the text.

14

IMPACTS OF ELECTRICITY ON SOCIETY

In the previous chapter we introduced two fundamental scientific observations. Ørsted showed that a moving electric current will produce a magnetic field. Faraday then turned Ørsted's observation around and showed that a moving magnetic field will produce an electric current. Following Faraday's fundamental discovery in physics came two engineering design modifications of Faraday's original experiment: the use of rotary motion instead of reciprocating motion, and the idea of moving the coil of wire and keeping the magnet stationary. Although many people contributed to perfecting the electrical generator, credit for the development of the truly practical generator is usually awarded to the Belgian 'electrician' Zénobe Gramme. (Notice that there is some similarity between Gramme's achievement and Watt's. Neither can truly claim to have originated or invented the device with which they are associated, but each genuinely deserves credit and praise for bringing the device to a practical state of perfection.) The perfection of the generator by Gramme in 1870 provided for the rapid expansion of the electricity industry. In this chapter we will examine some of the ramifications of this on society.

BREAKING THE GEOGRAPHIC BARRIER

In the early days of electricity, it was assumed that the electricity would be generated on site, that is, generated wherever it was used. One of the first such applications in the United States occurred in 1878, when Wanamaker's department store in Philadelphia was illuminated with arc lights (Figure 14.1), supplied with electricity from a generator in the basement. In the 1870s and 1880s there was a growing realization that the future of electricity was not in local, on-site generation, but rather in distributing electricity produced in large, central power stations. It is important to realize that electricity is different from other kinds of energy because, with electricity, we do not need to have the energy source in the same location (e.g., in the same building) as where we use the energy. Now a

FIGURE 14.1 The 19th century Wanamaker Grand Depot was one of the first department stores to use arc lights, which received power from an on-site generator.

factory or industrial community no longer had to be located near a source of water power. The beauty of electricity as an energy source is that we can operate an electric motor or electric lamp (or any other electrical device, of course) at any place whatsoever, as long as we can run wires to it from some source of electricity.

ENERGY can be transmitted and distributed in its electrical form very cheaply and efficiently. Where the distances between the source of ENERGY and the place where we need to do WORK are considerable, electricity has no peer. If a large amount of ENERGY has to be transmitted to some location many miles away, think of what few choices we have: air under pressure, water under pressure, and electricity. Sending compressed air a long distance would require huge pipes, along with the associated costs of making them, installing them, and maintaining them. Inevitably, there would be some leakage and losses. Hydraulic power, dependent on high-pressure water, would have the same requirements with the additional concern that the pipes would have to be heated, or buried deeply underground, to avoid freezing. For carrying ENERGY as electricity, comparatively small insulated wires will do the job. Freezing is not a problem. The natural obstacles in the landscape, such as rivers or canyons that are tricky to cross with large pipelines, are not a problem either. Figure 14.2 illustrates electric lines carrying electricity to rural customers.

Because of the relatively low cost and high efficiency of transmitting ENERGY over long distances via electricity, it is now possible for us to generate large quantities of ENERGY in central generating stations ('power' plants) and transmit it to locations where it's needed. Though the economics and structuring of the electric utility industry are now undergoing significant change, in most cases it is probably cheaper to have a small number of very large generating stations than a larger number of smaller stations scattered around an area. An additional benefit of electricity as an ENERGY source is that it can be generated either from fuel—coal, natural gas, or oil—that can be transported to the power plant, or from the kinetic energy of falling water, or from nuclear reactions. In the chapters that follow we will examine how these kinds of power plant work, and what some of the advantages and disadvantages are for each.

FIGURE 14.2 The isolation of this Pennyslvania valley, located far from a generating station, poses no problem for residents who receive energy via miles and miles of electric wires strung along the road.

ELECTRIC LIGHTING

As we've seen, Volta developed the voltaic pile around 1800. Within two years, Humphry Davy had shown that a spark could pass between two carbon rods connected to a voltaic pile. Davy's observation was responsible for development of the first practical form of electric lighting, the arc light.

Arc lights first came into practical use in 1844. Initially they had two applications. The first was for street lighting. The lighting of streets certainly improved safety, since a person now could see where he or she was going. It also contributed to a reduction in street crime, by reducing the number of darkened places where criminals could lurk. A second application was for lighting harbors and railway switch yards. This lighting increased the number of hours that they could be in operation, and thus the number of hours during which goods of all sorts could be loaded for shipment or unloaded for distribution. In other words, the shipment and handling of merchandise was increased because of the increased hours during which work could be done.

Because of the inherent glare and brilliance of the light from an arc lamp, a natural application was in lighthouses. In 1858 the South Foreland lighthouse and, in 1862, the Dungeness lighthouse in England were equipped with arc lights. The electricity supply for these lights was supplied by large generators driven by steam engines. The first athletic event played under lights (arc lights) was a soccer match in Sheffield, England, in the late 1870s.

Many inventors began working on lighting systems after the development of a practical electrical generator by the Belgian engineer Zénobe Gramme in 1870, and, close on his heels, the German engineer Werner Siemens. At first most systems of arc lights were arranged in series. This is a significant drawback, because if one lamp failed, the entire system shut down like a chain with a weak link. (We sometimes see this problem with decorative strings of lights put up at the holiday season. Older sets of these lights are wired in series, and, if one bulb burns out, the whole string stops functioning until we have made a frustrating and time-consuming search, bulb by bulb, for the defective one.) The first central electricity-generating station intended to provide electricity for the operation of arc lamps was built in 1879 in San Francisco.

Despite these successes, arc lights have a number of disadvantages, particularly for applications in situations in which a competent, trained electrician may not be available (such as in our homes). Arc lights require dangerously high voltages of electricity. As the light 'burns,' the carbon electrodes are indeed consumed, so the lamp requires occasional adjustment. It was, mechanically, a complicated piece of equipment. Electric relays were needed to keep the tips of the carbon rods positioned at the correct distance for producing an arc. The carbon rods had to be replaced from time to time. In operation, the lamp emitted a hissing noise. All these limitations might be acceptable in a lighthouse, where the keeper could attend to these duties, or in a commercial enterprise, with maintenance workers on the staff, but it would hardly be acceptable in the average home, which is where the potential for a big market lay.

It was soon recognized that several criteria would have to be met for electric lighting to be successful in the home. First, it had to be reasonably safe, operating at potentials of about 100 volts. Second, it had to be simple and easy to use, even for persons with no special training; for example, there had to be an easy way to change bulbs. Third, it needed to produce a 'soft' light.

The concept of an incandescent light originated in 1820. An incandescent light is one in which the electric current heats a filament that glows to give off light. In contrast, an arc lamp functions by the electric current creating a luminous discharge as it crosses the gap between two electrodes.

For several decades many talented scientists in America and Europe experimented with the development of a *practical* incandescent electric light. By 1877 the consensus of the international scientific community was that no practical result would ever come of all these experiments. And then a partially deaf, former paperboy and candy seller came up with two crucial ideas. First, it was important to pump all of the air out of the bulb, to avoid burning up the filament. Second, out of thousands of candidate materials tried, a very thin filament of carbon was superior to all others, including thicker filaments of carbon and special materials like platinum. The first test of an incandescent light bulb that incorporated these new ideas occurred on October 21, 1879. The bulb burned continuously for two days, and was the first truly successful demonstration of an incandescent light.

> *The hunt was a long, tedious one. Many materials which at first seemed promising fell down under later tests and had to be laid aside. Every experiment was recorded methodically in the notebooks. In many there was simply the name of the fiber and after it the initials 'T.A.,' meaning 'Try Again.'*
>
> *Literally hundreds of experiments were made on different sorts of fiber. . . . Threads of cotton, flax, jute silks, cords, manila hemp and even hard woods were tried.*
>
> *Chinese and Italian raw silk both boiled out and otherwise treated were among those used. Others included horsehair, fish line, teak, spruce, boxwood, vulcanized rubber, cork, celluloid, grass fibres from everywhere, linen twine, tar paper, wrapping paper, cardboard, tissue paper, parchment, holly wood, absorbent cotton, rattan, California redwood, raw jute fiber, corn silk, and New Zealand flax.*
>
> *The most interesting material of all . . . was the hair from the luxurious beards of some of the men about the laboratory.*
>
> — Jehl[1]

AND THEN THERE WAS EDISON . . .

Thomas Edison (Figure 14.3) was a deaf, former paperboy and self-taught inventor. Contrary to our popular mythology, Edison did *not* invent the light bulb. Several persons had had the idea of an incandescent light before Edison and had designs that more-or-less worked. What Edison *did do*, and for which he deserves enormous credit, was invent the first incandescent lamp that could be made in large quantities and easily sold.

> At last, on 21 October 1879, Edison made a bulb that did not burn out. Its filament was of carbonized cotton sewing thread. . . . The first commercial bulb, which followed swiftly, had a horseshoe filament of carbonized paper.
>
> Three years later the Pearl Street Central Power Station was completed in New York—the first of the world's great cities to be electrically lit. The coming of electric light was widely seen as banishing the fear and superstition that darkness had bred.
>
> The same point is made, though less poetically, in Conan Doyle's The Hound of the Baskervilles (1902) when the new heir to Baskerville Hall arrives from North America and remarks, on viewing his spooky ancestral home:
>
> It's enough to scare any man. I'll have a row of electric lamps up here inside of six months, and you won't know it again with a thousand-candlepower Swan and Edison right here in front of the hall door.
>
> — Carey[2]

Thomas Alva Edison was born in Milan, Ohio in 1847. His first contact with electrical devices was working, as a young man, as a telegraph operator. He moved to Boston in 1868. During a relatively brief time in Boston, no more than about a year, Edison purchased a set of books describing the work of Michael Faraday. The work of Faraday may have been the inspiration for Edison to begin a serious career as an inventor of electrical equipment. Edison's first triumph as an inventor came in 1870, with the invention of the stock ticker. The stock ticker was a critical device in financial markets for nearly a century, until the coming of the era of instantaneous electronic communications. It actually was a modification of the telegraph, so that it could print, rather than requiring a human operator to listen to the sequence of electrical dots and dashes, and, moreover, could print rapidly, at some 185 characters per minute. Edison sold the invention to the Gold and Silver Telegraph Company for $40,000, an enormous sum in those days. He used the money he got from this invention to set himself up in business in Newark.

In the late 1870s Edison began to think in detail about a *system* of electric lighting. One of Edison's critical contributions—far more important than the mere invention of incandescent light bulb itself—was the recognition that the light bulb is useless as an isolated device by itself. Other inventors perfected the incandescent light at just

FIGURE 14.3 Thomas Alva Edison (1847–1931). Born in Ohio, Edison is famous as the inventor of the incandescent light bulb, but his real genius lay in understanding the need for an electrical system to provide power to the bulb.

about the same time that Edison did. Edison's true brilliance was to recognize a lighting system rather than a new object (the bulb). He recognized that an electric lamp needed to be thought of as part of a system along a central electricity-generating station and all the necessary distribution equipment.

> *It was not only necessary that the lamps should give light and the dynamos generate electric current, but the lamps must be adapted to the current of the dynamos, and the dynamos must be constructed to give the character of current required by the lamps, and likewise all parts of the system must be constructed with reference to all other parts, since, in one sense, all the parts form one machine, and the connections between the parts being electrical instead of mechanical. Like any other machine the failure of one part to cooperate properly with the other part disorganizes the whole and renders it inoperative for the purposes intended.*
>
> *The problem then that I undertook to solve was stated generally, the production of the multifarious apparatus, methods and devices, each adapted for use with every other, and all forming a comprehensive system.*
>
> —Edison[3]

Despite the scattered successes of arc lamps, the major source of illumination through the 1870s was gaslight (Figure 14.4). Most of these lights functioned by burning

FIGURE 14.4 These German stamps celebrate various forms of street lighting in Berlin prior to incandescent lighting. Clockwise from upper left, they show: gas lamp, electric carbon-arc lamp, gas lamps, and chandelier.

a combustible gas made by heating coal under conditions at which it would decompose but not burn. Edison began electrical power generation with the goal of competing with gas lamp illumination in New York City. Edison began by doing an economic analysis to compare the costs of building and operating a central electricity station to the costs associated with gas lighting. Ironically, it's now thought that this original economic analysis was badly flawed, but, despite that, it served to convince Edison and his backers of the commercial possibilities of electric lighting. It was also very helpful that, by this time, Edison had developed a friendship with a prominent attorney, Grosvenor Lowrey. He set up contacts for Edison with Wall Street financiers. He helped get financial backing for Edison from Drexel, Morgan and Company. By one of those interesting confluences of events in history, Edison stood on the brink of international fame for his lighting system just at the same time that Drexel and Morgan came into the hands of the young J.P. Morgan, who became likely the most famous (or notorious) of the Wall Street financiers of the late 19th century.

The development of a successful incandescent light required finding solutions to two problems. The first one was the nature of the material to be used as the filament. It had to be some material that was both an electrical conductor and could be heated to glowing (incandescence) without melting. The second problem was the unfortunate fact that virtually any substance, if heated hot enough, will combine with oxygen—in short, will burn up. So, some way had to be found for removing the air from the bulb. It was not until 1865, when a special mercury pump was invented, that it was possible to obtain a satisfactory vacuum to prevent igniting the filament. Tungsten, the substance now generally used as filaments, came into general use around 1910.

The widespread public acceptance of electric lighting in the home was a 'chicken and egg' story, since of course lighting can't be used without a source of electricity. Edison, in fact, not only developed the first practical light bulb, but he and his associates also actually developed the components of an entire system for electric lighting: improvements in the electricity generator, cables for transmitting electricity to the consumer, schemes for electric wiring in the home, electric meters, fuses and switches, and even the sockets or other fittings for the bulbs.

Edison opened an electricity-generating station on Pearl Street, New York, in 1882. On its first day, Edison had 52 customers. While it was not the first electricity station of any kind, it was the first dedicated especially to supplying electricity for incandescent lighting. The Pearl Street plant was only a tiny cog in the vast system of Edison's enterprises. There were also the Edison Electric Light Company, the Edison Machine Works to build dynamos, the Edison Electric Tube Company to make the underground conductors, and the Edison Lamp Works to manufacture incandescent lamps. The Pearl Street station would have been useless without all of the other components of this system. By about 1890, manufacturing companies and electric utilities owned by Edison and his backers had spread all over the United States and were getting established in Europe.

This single event—the opening of the Pearl Street station—was an absolute disaster for the gas lighting industry, and investment in shares in gaslight companies plummeted.

A sign in hotel rooms in 1892 read, 'This room is equipped with Edison Electric Light. Do not attempt to light with a match. Simply turn key on the wall by the door.'
—Reynolds[4]

Information about Edison's new system spread remarkably quickly. Within a few months people were applying to Edison for licenses to build electricity-generating plants all over the country. Here is where Edison's vision of electricity as an entire system really paid off, because it was the system concept that allowed the very rapid expansion of the electrical network. Within 20 years of Pearl Street there were 2,250 generating stations in the United States, and, by 1920, almost 4,000. By then, about a third of the nation's homes were wired for electricity.

The growth of the electrical industry was so rapid that by the 1890s three large firms had emerged in the United States, the Edison General Electric Company, the Thomson-Houston Company, and the George Westinghouse Company. In 1894 the Edison Company and Thomson-Houston amalgamated to form the General Electric Company. More than a century later, General Electric and Westinghouse remain two of the most important corporations in the United States. The formation and development of these large firms essentially marked the end of an era for electricity. By 1900, electricity had become an advanced technology. It was no longer the domain of the lone inventor, the empirical tinkerer, or the 'electrician' of the days of men like Gray and Franklin. Now electricity was big business, with further research and development in the hands of the electrical engineer.

One of Edison's crucial concepts was the idea of parallel wiring rather than series wiring. If lights are wired in parallel, one bulb burning out does not affect the operation of the others. With series wiring, however, one bulb burning out causes the entire system to shut down. Those who have experienced the slow, trial-and-error hunting of one burned-out bulb in a holiday decoration can imagine the incredible frustration of having an entire house wired in series, and having to search in darkness through the house to find the defective bulb.

Centralized electricity-generating stations were developed around 1880. As we've seen, Edison's first was on Pearl Street, New York, established in 1882. In the early 1880s a pioneering electricity station was also established in Sunbury, Pennsylvania. The first electric rate was 28¢/kWh. Nowadays it is very rare, at least in most parts of the United States, to encounter electric rates higher than about 9¢/kWh.⚛ The difference in the value of money in the 1880s compared to our present time is extraordinary; therefore, the first electric rates were not triple the cost of today's rates, but, in terms of the purchasing power of the dollar then and now, probably more than 30 times higher. It is noteworthy that, when planning his first central generating station, Edison was careful to pick a spot in Manhattan where the transmission lines to customers could be short but the number of customers served could be high.

⚛ In late 2000 and early 2001, severe problems developed in the supply and distribution of electricity in California. In some places electric rates spiked to the unheard-of (at least unheard-of for modern times in America) levels of 48¢/kWh. Compared to some other parts of the world, and compared to what our 19th-century ancestors paid Edison, this is still cheap electricity. However, having one's electric bill suddenly become six times larger is not a pleasant experience; and this was especially troublesome for low-income people or retirees on fixed incomes.

Despite the extreme cost of early electric rates, it is still true that the access to electricity changed living conditions drastically. The first domestic electric appliance was the iron (1889), which began to eliminate some of the laundry drudgery of heating irons on a stove, and periodically reheating them as they cooled. Electric fans followed around 1890; electric refrigerators came into fairly wide use in the 1930s, followed by electric washing machines in the 1940s. Edison himself remarked that he had made electric lighting so cheap that the day would come when only rich people dined by candlelight.

One reason why the early electric rates were so high was that a centralized electricity station represented a major investment of capital. It also used very large equipment (such as steam boilers) that is best utilized if it can operate all the time. The problem with generating electricity only for lighting is that most of us use electric lights only six to ten hours per day, even in the winter time when daylight hours are short. To improve the economics of centralized electricity generation, other uses were needed, especially applications of electricity that could potentially operate round the clock.

ELECTRIC MOTORS

As early as 1837, the inventor Thomas Davenport (of Brandon, Vermont) used primitive motors to do industrial work, such as drilling iron and steel and turning wood in his workshop. These motors were not very satisfactory because they relied on voltaic piles that eventually 'ran down.' In 1851 the English inventor Charles Babbage recognized that there were some applications that cannot be fulfilled by steam engines because such engines are simply too big and too costly. These are applications that require devices that can start and stop at a moment's notice, require no time for maintenance or management (for example, for shoveling coal), have a moderate initial cost, and a low daily operating expense. (Babbage also invented a mechanical device, called the 'analytical engine,' to help him with his calculations. Today we operate analytical engines with electricity—but we call them 'computers.')

In the 1870s the German-born owner of a textile factory in Manchester, England, predicted that if electricity could be transmitted over long distances, it would revolutionize industry. He knew something about revolutionizing—he was Friedrich Engels, the friend, benefactor, and collaborator of Karl Marx. Then in 1873 Zénobe Gramme showed that the electric generator could be operated 'backwards.' That is, if current were supplied *to* the generator, the armature would spin. This was the first practical electric motor.

The first factory to be completely electrified was a cotton mill, built in 1894. Initially, factory designers thought that large steam engines or large waterwheels would be replaced by equally large electric motors. However, it soon became clear that there are more advantages to having a relatively small electric motor on each machine. This eliminates the complicated drive train of belts, pulleys, and gears needed to operate each machine from a centralized source. Having individual motors allows much more precise control of the speeds of individual machines, in turn providing the ability to make new kinds of precision-machined goods. Each machine can be controlled by its individual operator. The independent operation of each machine can also speed up the circulation of raw materials and partially assembled or partially finished goods.

As electric motors replaced steam engines, there were also major shifts in factory design and location. First, it was no longer necessary to build factories that were several stories high to accommodate a huge steam engine and the complex system of shafts and belts used to operate the individual machines. Second, it was no longer necessary to locate them near water sources or coal fields.

Neither Marx nor any other devotee of the economic interpretation of history was able to foresee the coming of electrical power, one of the consequences of which was that factories no longer needed to be built near coalfields; moreover the properties of electrical power are such as to make possible the springing up of numerous small factories containing machine-tools individually powered by their own electrical outlets.

—Medawar[5]

By 1901 almost 400,000 motors had been installed in factories in the United States.

The significant advantages of electric factory machinery combined with electric lighting (which allowed for safe, round-the-clock operation of factories) led to great improvements in industrial output, which in turn helped contribute to a generally rising standard of living.

In 1889, the Deptford Power Station, on the Thames estuary in England, began supplying electricity. The Deptford station is the prototype for the type of large, centralized, coal-fired power plants that we will describe in a later chapter. The location of the plant was not dictated by where the market was, but rather by convenience of the coal supply. (For this plant, coal was brought by ship from the north of England.) The Deptford station supplied alternating current at 10,000 volts to substations in London some six to ten miles away.

The development of the large, centralized power station, as at Deptford, raised speculation as to the possibilities of using electric locomotives on railways (Figure 14.5), operated by electric motors in the locomotives or cars. The beginning of the electric railroad was at Richmond, Virginia in 1881 with a system of streetcars.

By the 1930s, it was apparent that a train hauled by an electric locomotive had certain advantage over a train with a steam locomotive, in particular, a rapid acceleration. In those days, a steam train starting from rest could accelerate at about half a mile per hour every second; but an electric train could accelerate more than a mile per hour per second, and at the end of half a minute would be traveling at 30 to 40 miles per hour. This quick acceleration is exactly what was needed for passenger traffic in urban and suburban areas, where stops are frequent and traffic heavy. The electric locomotive provided a higher average speed, and therefore shorter intervals between successive trains arriving

FIGURE 14.5   A railroad locomotive run by electricity.

and departing a given station. In turn, this meant a great increase in the passenger-carrying capacity of the line.

Suburban electric trains were developed that did not need a separate locomotive. The driving motors were distributed among the coaches. This so-called multiple-unit system allowed any number of coaches to be coupled together without reducing train speed, since each additional coach contributed its own share of the power. This system provided great flexibility. During the course of a day or week, passenger trains could be lengthened or shortened as needed to cope with demand, without any effect on the speed of the trains or the time interval between them.

AC OR DC?

The story was told at about this time [1890] that a Cambridge student, asked in his final viva voce examination what electricity was, hesitated and then stammered that he had known this once but had forgotten. The examiner sighed: 'What a pity, what a pity! Only one man in the world who knew what electricity is, and now he has forgotten!'
—Gratzer[6]

As electric systems began to spread in the 1880s and 1890s, a ferocious controversy erupted regarding the better way of transmitting and using electricity: alternating current or direct current (Figure 14.6). The early generators of Grammé, Siemens, and their contemporaries provided direct current, and this direct current is what Edison built his system around. Notably, Edison's system was based on a whole series of small generating plants located near their customers, rather than a few very large plants serving large numbers of customers over a wide geographical area. In fact, after the success of the Pearl Street station, it was quickly followed by about 60 more small plants just in New York City.

However, electrical engineers began to realize that, for long-distance transmission, alternating current had numerous advantages over direct current. For once in his life, Edison was on what proved to be the wrong side of the argument. Having started with direct current, he resisted alternating current to the last ditch, and had a prominent role in a bitter, acrimonious, and even ignominious dispute on the superiority of alternating versus direct current.

George Westinghouse, whose start in business came from his invention of the air brake for trains, had the vision to take alternating current seriously. The Westinghouse company began, around 1885, to make equipment for high-voltage alternating current. Edison tried to discredit alternating current in the public's mind, but really succeeded in discrediting only himself. Edison and his associates showed public demonstrations of how stray dogs and cats, pushed onto sheets of metal connected to alternating current, were promptly electrocuted by the high voltage. Based on those demonstrations, which were dreadful enough, Edison then proposed using an electric chair to replace the standard (at the time) method of hanging condemned murderers. Of course, for this awful application Edison enthusiastically recommended using alternating current and equipment supplied by Westinghouse. The first such execution by alternating current occurred on August 6, 1890, when the condemned murderer, William Kemmler, was executed at Auburn prison, New York. Indeed, Westinghouse equipment was used. The lowest and ugliest point of this squabble was reached by the suggestion from Edison and his allies that, instead of saying that a person had been electrocuted, we should say that he had been

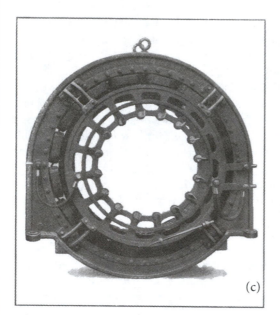

FIGURE 14.6   (a) The field magnet and (b) drum armature of a direct current generator. (c) Armature and (d) field magnet of an alternating current generator.

'Westinghoused.' None of these negative campaign tactics had the slightest effect in counteracting the growing acceptance of alternating current.

The chief source of scientific and engineering expertise for Westinghouse and the commercial development of alternating current was a brilliant, colorful, somewhat bizarre (and possibly crazy) Croatian immigrant, Nikola Tesla. As a student at Graz Polytechnic Institute in Austria, Tesla saw a class demonstration of a generator running 'backwards' as a direct-current motor. Parts of the motor were giving off sparks, and it seemed obvious that it would soon wear out. Five years later, as he was walking with a friend in the Budapest city park, Tesla envisioned a simple way to produce a rotating magnetic field through a particular way of using multiple phase currents. This discovery was the heart of his alternating-current motor.

In 1876, Sir William Siemens had suggested that the energy of the falling water at Niagara Falls could be used to operate generators that would supply electricity economically to New York, Toronto, Philadelphia, and Boston. In the 1890s, Tesla, with Westinghouse's financial backing, developed the key components of an alternating-current electricity system. By 1895, the designs were in place for a complete generating system, and Westinghouse was awarded the contract for the new generating station to be installed at Niagara Falls. (The plant at first transmitted electricity to Buffalo, about 20 miles away.)

Westinghouse, with the funds he had available from the air-brake business, offered to buy Tesla's patents for $60,000 plus a royalty of $2.50 per horsepower. Some years later, when Westinghouse had suffered serious financial setbacks and was in dire straits, Tesla agreed to accept a cash settlement of $216,000 instead of royalties. It has been estimated that this cost (in terms of lost royalties) Tesla 12 million dollars at that time and possibly 50 million dollars over his lifetime. Depending on one's point of view, Tesla's settlement was either an act of extreme loyalty and generosity to the man who helped him get started, or, alternatively, one of the dumbest financial transactions ever.

The use of energy in cities has been marked by a constant displacement of burning fuels by electricity. Gaslights were replaced by electric lamps. Steam-powered factories were modernized with machines having individual electric motors. Steam locomotives were replaced by electric locomotives. Now, as we enter the 21st century, we see the beginning of a trend to replace gasoline-powered autos with electric cars.

CITATIONS

1   Jehl, F. *Menlo Park Reminiscences.* Edison Institution: Dearborn, MI, 1937.
2   Carey, J. *The Faber Book of Science.* Faber and Faber: London, 1995; pp. 169–172.
3   Thomas Edison, cited in Jehl, Francis, *Menlo Park Reminiscences, Volume 2.* Dearborn, MI: 1938; p. 852.
4   Reynolds, Francis D. *Crackpot or Genius?* Chicago Review Press: Chicago, 1993, p. 14.
5   Medawar, Peter. *The Strange Case of the Spotted Mice.* Oxford: Oxford, 1996; p. 184.
6   Gratzer, Walter. *A Bedside Nature.* W.H. Freeman, New York, 1998, p. 91.

FOR FURTHER READING

Adair, Gene. *Thomas Alva Edison.* Oxford, 1996. A brief biography of Edison with some of the technical backgroud to his inventions. Easily accessible to readers with little or no previous science.

Baldwin, Neil. *Edison: Inventing the Century.* Hyperion, 1995. One of two recent full-length biographies of Edison. Easily readable with little or no technical background.

Billington, David P. *The Innovators.* Wiley, 1996. A discussion of the engineers, inventors, and scientists responsible for some of the most important inventions of the period 1776–1883. A useful chapter on Edison and his approach to systems, and the AC/DC wars.

Cardwell, Donald. *The Norton History of Technology.* Norton, 1995. A splendid book that surveys the whole history of technology from the ancient Greeks through the latter part of the 20th century. It encompasses far more than energy technology. Highly recommended to any reader wishing to learn more about the development of technology. The section on the rise of the electrical supply industry is particularly relevant to this chapter.

Cheney, Margaret. *Tesla: Man Out of Time.* Barnes and Noble: New York, 1981. A biography of Nikola Tesla, written with a somewhat melodramatic flair. The book tends to read much like a novel, and requires no scientific background. Often available as a mass-market paperback or on 'remaindered' tables.

Israel, Paul. *Edison: A Life of Invention.* Wiley, 1998. Another recent, full-length biography of the great man. Numerous interesting illustrations from the original laboratory notebooks of Edison and his associates.

Martin, Thomas Commerford. *The Inventions, Researches, and Writings of Nikola Tesla.* Lindsay: Bradley, IL, 1988. This is an inexpensive paperback reprint of a book originally published in 1894. It is, for the most part, an anthology of Tesla's writings. One would need some background in electricity to understand the various articles and speeches in depth. Nevertheless, this book provides an overview of some of the many concepts and devices developed by Tesla, an appreciation of which can be obtained even by glossing over the more technical passages.

Morus, Iwan Rhys. *Frankenstein's Children.* Princeton University Press, 1998. This book discusses some of the 'electricians' who were contemporaries of Faraday, but who nowadays are not nearly so well known. It also discusses some of the many diverse paths by which electricity became part of society, particularly an industrial society reliant on machines and 'driven' in part by consumers.

Platt, Harold L. *The Electric City.* University of Chicago Press, 1991. Using the city of Chicago as an example, the author traces the transition of energy use from wood to coal and gas to electricity. A major focus is the career of Samuel Insull, the president of the Commonwealth Edison Company, who worked both to find ways of providing cheap electricity and to 'sell' people on the idea of electricity when gas already seemed to be the preferred choice. A section of illustrations contains many interesting old advertisements and photographs of the period up to the 1930s.

Rose, Mark H. *Cities of Light and Heat.* The Pennsylvania State University Press, 1995. This book focuses on the experiences in Denver and Kansas City, addressing both the use of electricity and of gas. The effects of geography, politics—particularly city politics—and the choices made by consumers on the development of electric utilities and gas companies are discussed. The book covers the period from about 1860 to 1940. There are numerous contemporary photographs and advertisements.

Routledge, Robert. *Discoveries and Inventions of the 19th Century.* Crescent Books, 1989. This book is a reprint of the 1890 edition. The chapter on electricity, especially the section 'Electric lighting and electric power' is particularly relevant, and provides what is virtually an eye-witness account of the development of the electric industry as it was happening. The entire book, with numerous period illustrations, is a treasure trove for the reader interested in the history of technology.

Verne, Jules. 'In the twenty-ninth century: The day of an American journalist in 2889.' This short story can be found in various anthologies of Verne's work, of science fiction, or of French literature. One useful source is The Folio Society edition of *French Short Stories.* The city in which the journalist works is run totally by electricity; Verne's predictions of the state of technology some 900 years from now are interesting.

15

ELECTRICITY FROM FALLING WATER

THE ROTARY GENERATOR

We've seen that Michael Faraday discovered the concept of the continuous generation of electricity. This discovery was the first crucial step on the way to large-scale availability of electricity. Converting Faraday's discovery into a practical working device invention is credited to several engineers, among them Zénobe Gramme, who developed the so-called armature-wound generator around 1870. The basic principle is that an electrical conductor moving through a magnetic field (or vice versa!) generates an electric current. The generator is, in essence, as shown in Figure 15.1. In this example of a simple generator, the conductor moves through the magnetic field because *we* turn it. In other words, we must do WORK on (or supply ENERGY to) the generator, for which we therefore have the 'reverse energy diagram' shown in Figure 15.2. (Two points to keep in mind regarding these 'reverse energy diagrams' are, first, as the name implies, it is 'backwards' from the energy diagrams we have discussed earlier, and, second, *both* arrows in the diagram have to change direction as we go from one type of energy diagram to the other.) The entire essence of large-scale electric energy generation hinges on finding ways to spin the generator reliably and economically.

THE WATER TURBINE

We have already seen that one source of cheap, reliable rotary motion is the waterwheel. In principle, then, we could use the WORK provided by the waterwheel to operate the generator (Figure 15.3). As shown in the figure, two changes occur in this SYSTEM. In the waterwheel, water flows spontaneously from high to low gravitational potential energy. When this happens, some ENERGY becomes available. The waterwheel itself is the device used to capture or extract a portion of this energy as useful WORK. The generator causes a non-spontaneous movement of electrical charge from low to high electrical potential

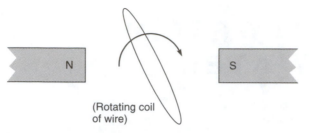

FIGURE 15.1 The essence of the generator: a coil of wire rotating around a magnetic field in order to create an electric current.

energy—from low voltage to high voltage. This non-spontaneous change will happen only if we supply WORK *to the generator*. The WORK being supplied is that provided by the waterwheel.

In our discussion of the use of energy in medieval times, we met two major types of waterwheel: the undershot wheel, operated by a rapidly flowing stream; and the overshot wheel, operated by the weight of falling water. We also saw that these are not very efficient devices for the conversion of kinetic energy into WORK, because the water still possesses significant kinetic energy even after it leaves the wheel. The quest for more efficient waterwheels began around the time when the late Middle Ages was blossoming into the Renaissance. An early design is due to Leonardo da Vinci. Leonardo recognized that water flowing through *curved* blades bears against those blades with a force or impulse. This is illustrated schematically in Figure 15.4. Crudely, this is analogous to the way a stream scours the outer bank of its bed when it bends. The device that Leonardo invented was at first called an 'impulse wheel.' Nowadays it is generally known as a *turbine*. A **turbine** can be defined as a device that converts the kinetic energy of a moving fluid into rotating kinetic energy. The fluid used in the turbine is called the **working fluid**. For the kinds of turbine discussed in this chapter, the working fluid is water. Later we will meet an extremely important family of turbines in which the working fluid is steam.

In a turbine operated by water, the water flows through pipes or ducts over curved vanes. To provide the impulse to the vanes, the turbine has to be enclosed in some sort of casing. The higher the pressure of water supplied to the turbine (essentially, the higher the column of water, or its 'head'), the faster we can get the turbine to spin. The rotational speed of an undershot or overshot waterwheel is limited by the speed of the stream. This is not so with turbines, where it is easy to get rotational speeds as high as 30,000 revolutions per minute (rpm). Turbines can be classified into two general types—impulse and reaction turbines. Some of the features of these designs will be discussed below.

The development of the turbine during the mid-19th century was a major engineering advance in providing a source of WORK for the electrical generator. The combination of a turbine and a generator can be represented as shown in Figure 15.5. This diagram is in fact a generic scheme for generating electricity, regardless of the working fluid in the turbine. In the special case where we take advantage of the kinetic energy of water to drive the turbine/generator set we refer

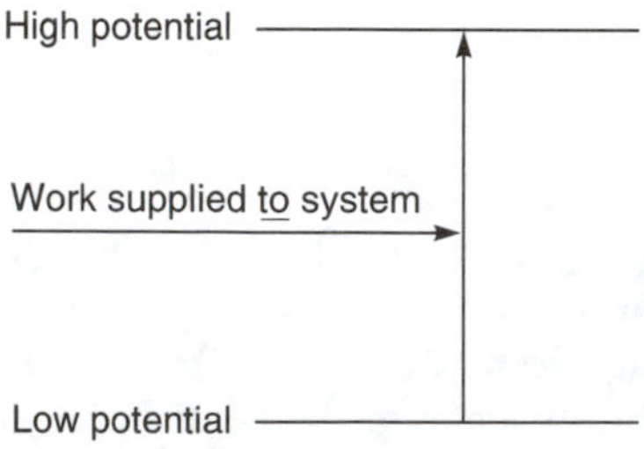

High potential

Work supplied <u>to</u> system

Low potential

FIGURE 15.2 This reverse energy diagram can be applied to all systems.

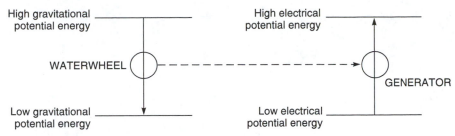

FIGURE 15.3 Hydroelectricity is based on this relationship between gravitational and electrical systems.

to the technology as **hydroelectricity** or hydropower. (The term 'hydroelectricity' does not imply that the electricity is somehow different from electricity generated in other ways. Rather, the prefix 'hydro-' is used simply to remind us that water is the fluid used to operate the turbine.)

It is the flow of the water over curved vanes that is the essential design feature of the turbine that allows it to capture the kinetic energy of the water. The energy is captured by the 'impulses' or 'reactions' between the fast-flowing water and the passages created by the curvature of the vanes, hence the two distinct types, reaction and impulse turbines. These turbine designs will work only if water is confined, and therefore the passageways for the water created by the vanes are enclosed in some sort of external, closely fitting casing.

The principal type of impulse turbine is the Pelton turbine, sometimes called the Pelton wheel. This type of turbine is especially suitable for large heads of water that have a low to moderate flow rate. A virtue of the impulse turbine is that it can be adapted to streams of variable flow, and to streams where a small volume of water is moving over very high falls. A Pelton turbine, and impulse turbines in general, are vertical wheels that are operated by the kinetic energy of a jet of water striking the blades. (Some designs use a series of small buckets to capture the water, rather than only smooth blades.) The high-speed, high-pressure jet is a result of the 'head' of water used to provide the working fluid. The shape of the blades is critical. Ideally, the blades should almost completely reverse the direction of the water; that is, they should turn the flow of water through as near 180° as

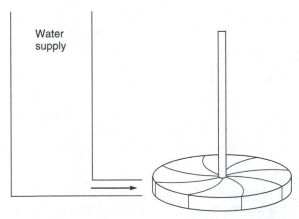

FIGURE 15.4 A basic water-operated turbine has a stream of water directed against curved blades.

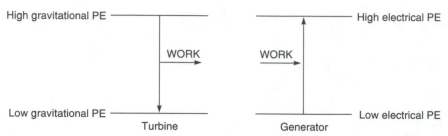

FIGURE 15.5 Here we can see the turbine–generator relationship expressed with energy diagrams.

possible. Doing so provides the best opportunity for capturing as much of the kinetic energy of the water as possible.

This design goes back to the California gold rush in the middle of the 19th century. In those days, one of the most common ways of extracting gold was hydraulic mining. In this process, a jet of water at high pressure was used to flush the sand and gravel away. The water used for the hydraulic mining came from reservoirs up the river and was brought in pipes to the mining location. When the gold finally ran out, all of this system of reservoirs, pipes, hoses, and other paraphernalia was still in place. Rather than have all of the effort in setting up the water systems go to waste, this piped water was used in some places to drive waterwheels. Lester Pelton, a mining engineer who had worked in the gold fields, designed and built a waterwheel with curved blades. One day, the water jet happened—possibly by accident—to strike the outer edges of the blades. So much kinetic energy was transferred to the wheel that its rotational speed increased to the point at which the wheel disintegrated.

Pelton recognized that the kinetic energy of the jet could best be utilized when its direction was almost completely reversed by the blade. To facilitate this, he changed the blade design to one basically made of two buckets, or cups, joined at the edges where they came together to form the middle of the blade. These pairs of cups are distributed around the circumference of the wheel. Because the division between the two cups is very thin, much like a knife blade, this design effectively split the water jet almost in half. At the same time, the shape of the cups caused the jets to reverse direction almost completely. The Pelton turbine, used when the head is very high—that is, is available at very high pressures, spins as the high-speed water jet hits each of the double cups.

The Pelton wheel hydroelectric plant at Fully, Switzerland illustrates some of the remarkable design features of these installations. At the Fully plant, water is led downwards from a lake through a pipe three miles long. In that three-mile distance, the pipe drops in elevation by more than a mile. This very steep gradient results in the water, at the lower end of the pipe, having a pressure of more than 2,000 pounds per square inch—that is, a pressure of a *ton per square inch*. The water leaves the nozzle at a speed of nearly 400 miles per hour. (To indicate some of the extraordinary properties of water under these conditions, it is estimated that a water jet coming out of a three-inch pipe at a pressure of 500 pounds per square inch could not be cut through or deflected even by hitting it with a steel crowbar.) The Pelton wheels are 12 feet in diameter. Water at this pressure and velocity spins the wheels at a rate of 500 revolutions per minute.

Reaction turbines are horizontal wheels. Like the impulse turbines, they move at high speeds. They are made of curved vanes mounted on a central shaft. The pressure of the water flowing through these vanes causes the wheel to rotate, or in somewhat archaic terminology, the wheel is turned by the 'reaction' of the water. A well-designed and operated reaction turbine can convert more than 90% of the kinetic energy of the water into rotary motion, at high rates of revolution.

In the late 1820s Bénoit Fourneyron developed a small turbine capable of producing about six horsepower. Within a few years, he had refined his design to the size of a 50-horsepower wheel. Fourneyron's design featured two concentric horizontal wheels. The fixed inner wheel had curved guide vanes that directed the water against runner blades on the outer wheel, the rotor. The efficiency of this outward-flow radial turbine was as high as 80–85%, but it had a disadvantage in that turbulence occurred at certain speeds when the water left the guide vanes.

An American engineer, James Francis, compared the performance of a Fourneyron turbine with one that he had designed himself. To try to insure that the results obtained from the two turbines could reasonably be compared, both were installed at the same site—a factory in Lowell, Massachusetts. The information obtained on turbine design and performance from this series of head-to-head trials led to the design of what has come to be called the Francis turbine. Subsequently, the design was improved further by James Thomson, Lord Kelvin's brother. Thomson's improvements consisted mainly of adjustable guide vanes and curved blades on the rotor. In the 1860s the Francis turbine began to replace the waterwheel. With the further improvements by Thomson it became, by the early 20th century, the most commonly used turbine in hydroelectric plants.

Nowadays most of the world's large hydroelectric plants still use Francis turbines. The world's largest hydroelectric plant is the Itaipu plant, which is built on the Parana River on the border between Brazil and Paraguay. The Itaipu plant has 18 Francis turbines, *each* having an output of 700 MW. Therefore, the total generating capacity of this plant is about 12 GW (12,000 MW). To put this figure into perspective, the output of the Itaipu plant is some 10–12 times larger than that of a large coal-fired or nuclear electrical plant.

At the other extreme of scale from plants like Fully and Itaipu are the many little 'mini-hydro' plants generating less than a megawatt, indeed sometimes only a few tens of kilowatts. They still use the same basic concepts of turbine design and operation. Small-scale projects, especially ones that don't require constructing dams to form large reservoirs of water, represent one aspect of hydropower that seems to be growing in popularity around the world. (As we will see below, large-scale hydropower projects seem to have lost most of their luster and are actually under attack in many places.) A 200-kilowatt turbine, planned to be installed in the Thames River, could serve as a prototype for similar schemes on many other rivers in Britain. Some Third World countries find the installation of 'mini-hydro' plants to be an attractive alternative, economically, to expanding nation-wide electricity grids to connect a few large, expensive electricity-generating stations.

HYDROPOWER

Hydropower supplies some 10–12% of the electricity demand in the United States. Roughly, this amounts to about 300 billion kWh of energy. This energy production is equivalent to what might otherwise be supplied by 40–50 nuclear or coal-fired plants. Worldwide, hydropower provides about 20% of electricity demand; this is equivalent to about 4% of *total* energy demand. The world electricity production from hydropower plants is about 2400 billion kWh. Worldwide, estimates suggest the potential of producing five times the total output of all present power stations. While this may sound very attractive and promise a great future for hydro, many of the potential sites for new development are places that are inaccessible, far from potential consumers and markets for electricity, and often both. In many of the developed countries, there is growing public concern about the environmental effects of constructing dams to provide hydroelectric sites. In some Third World nations, construction of dams and their large reservoirs could cause many people to have to be displaced from their homes. With these issues taken into consideration, it is likely to be the case that no more than about half of the world's potential hydro capacity will ever really be developed.

Hydropower has a number of advantages. Compared to energy sources such as fossil fuels, hydropower is essentially non-depletable. While we might someday experience shortages of fuels, there is little likelihood that rivers would actually dry up. Hydropower is highly efficient. About 90% of the kinetic energy of water is converted to electrical energy (as compared, for example, to 30–35% when electricity is generated in coal-fired power plants). In addition, hydropower is highly reliable mechanically, since the only moving parts are the turbine and the generator. As a result, maintenance costs are low. Without doubt, hydropower plants are a proven technology. Unlike fossil-fuel or nuclear plants, there are no waste products and virtually no pollution. Since there are no waste products, there are also no recurring costs for waste disposal. Another contrast to fossil-fuel or nuclear plants is that the energy source, the water, is free—there is no recurring cost to buy fuel. If opportunities exist for the development of hydropower, it is still possible for regions or countries that do not have their own indigenous fossil fuel resources to develop heavy industries on a large scale. From the plant operator's perspective, an important advantage of hydropower is that it can easily be switched on and off (by diverting water onto or away from the turbines), making it an excellent choice for providing extra electricity at times of peak demand or in energy emergencies.

Another virtue of hydropower is economic. In most places, hydropower has always been one of the least expensive sources of electricity. For example, some of the large hydropower installations built in the middle of the 20th century today generate electricity at a cost under 1¢/kWh. The initial capital investment required to build a new hydropower installation can be very high. Large plants may cost between $500 and $2,500 per installed kilowatt of capacity; a comparable figure for a coal-fired plant might be about $1,300/kW. However, once the plant is built and running, the recurring costs for operations and for maintenance are usually quite low. An alternative to building an entirely new plant is to augment the capacity of an existing one (sometimes referred to as 'uprating'); where possible, this is a cheaper option than building a new plant.

The first hydropower plant was completed at Niagara Falls in 1878. By 1895 there were three 5,000-horsepower generators set up, producing electricity at 2,200 volts. The local business community and civic boosters made this 'electrification' of the Falls a major theme of the 1901 Pan-American Exposition in Buffalo. By 1905, the two generating stations installed at Niagara Falls produced one-tenth of all the electric energy generated in the United States. Much of this electricity was consumed by the industries in Buffalo, only about 30 miles away. The early days of electricity generation at Niagara Falls are illustrated in Figure 15.6 and a modern rotary generator at Hoover Dam is shown in Figure 15.7.

As a portent of things to come later in the century (and discussed in the next section) not everyone saw the hydro plants at Niagara as an unqualified benefit. The technical success of the plants soon led to plans to install more generating capacity there. Soon some segments of the press were predicting the possibility of the Falls essentially drying up, because of the diversion of water to the hydro plants. Remarkably, by 1900 New York's state government had granted permits to companies that would, in principle at least, use essentially *all* the water flowing over the falls. Some developers went so far as to propose simply converting the entire Falls to

FIGURE 15.6 The power of the early Niagara Falls plant is evident in this long row of generators run by the turbines.

economic uses. The potential of using all the water was a real one. For example, by 1908 one company was pulling 8,000 cubic feet of water *per second* from the river. This enormous diversion of water had the effect of lowering the water level in the river by about six inches. If all of the other companies had actually been able to proceed as planned, the likely effect would have been to reduce the crest of the beautiful Horseshoe Falls from 2,950 feet to 1,600 feet, and that drastic change would have shifted these falls totally to Canada.

The first transmission system intended to carry electricity from a hydroelectric plant was built by the Westinghouse Company in 1891. The Westinghouse system carried electricity from the generating plant at Willamette Falls, Oregon, to the nearby city of Portland. The main transmission line from the plant to the city operated at 3,300 volts. In Portland, transformers reduced the voltage to 1,100 for distribution within the city. In those days, voltage standards had not yet been developed, so, depending on the requirements of the local householders, voltages were reduced further, either to 50 or 100 volts. In the previous chapter, we mentioned the 'battle of the currents' between Edison, favoring DC, and Westinghouse and Tesla, who supported AC. In fact, it was the AC system that made it possible for Westinghouse to construct the transmission line between Willamette Falls and Portland; at the time, there was no economical way to transmit DC that distance (13 miles). It was the victory for Westinghouse and Tesla in this 'battle' that greatly facilitated the growth of hydroelectricity. The AC transmission systems made it possible for electricity to be generated at hydro plants, which might be some distance from large markets for electricity, and yet have the electricity transmitted reasonably economically to those markets.

FIGURE 15.7   The staircases in the picture reveal the immense size of modern rotary generators like these at Hoover Dam.

In the 1910s and 1920s, several electric utilities interconnected their generating plants by means of transmission lines. Of course, the motivation for this was for the companies each to extend their service areas. We now use the terms 'electric grid' or 'power grid' to describe this interconnection of transmission lines. The development of the grid, which by now essentially interconnects the entire nation, provided numerous benefits. First of all, grids provide a way for the most efficient (and therefore most economical) plants to generate most of the electricity. Second, an interconnected grid increases the chances of supplying energy to consumers even if one plant happens to shut down unexpectedly. The grid allows incorporating the electricity-generating capacity of hydro plants, even if they are located far from places of high demand for electricity.

These facts lead to the question: If hydropower is so good, why not convert all of our electricity generating capacity to hydro? One of the main problems is that most of the suitable hydropower sites have already been developed. At first sight, it would be reasonable to think that there is still a lot of potential for further development in the United States. The Federal Energy Regulatory Commission has a listing of over 7,000 prospective hydro sites, which, if developed could support some 147,000 MW of electricity generation. By 1991, only about 2,200 of these sites, providing 73,000 MW, had actually been developed. Other projects, totaling another 3,300 MW, were either in an advanced stage of planning or, at the time, under construction. However, most of this additional 3,300 MW was in the form of expanding or uprating of sites already in operation. On the basis of these figures, it might be expected that the United States could double the present capacity for generating hydroelectricity.

Most of the potential for increasing hydroelectricity generating capacity is in the west. But, many of these opportunities are really for upgrading or expanding some of the existing hydro stations, rather than building entirely new ones. In the 1980s, for example, Grand Coulee Dam (Figure 15.8) was uprated by some 2,000 MW with the addition of three new turbine-generator sets. Other upgrades in the 1980s added another 1,100 MW in hydro capacity.

Despite these optimistic notes, there seems to be a consensus that hydropower is either approaching the limit of its potential in the United States, or has already reached it. The reason for this is largely the growing public concern over the environmental effects of building dams. This issue we will explore in the next section. Because of the public's environmental concerns, it does not seem likely that there will be any major restructuring of regulatory and licensing provisions affecting hydroelectricity. In fact, there is potential for the total hydro capacity to *drop* in the future. There will likely be increasingly fierce competition for use of water, with needs for drinking water and agricultural irrigation contesting with its use in hydro plants. Dams are perceived to have very negative impacts on some species of endangered fish and other wildlife. These same concerns are evident in other nations. In Switzerland, for example, about 15% of per capita energy consumption comes from hydro plants. Environmental considerations are an important factor among the Swiss, who have expressed concern about the environmental impact of dam construction on some of the pristine valleys in the Alps.

Today Bogotá, Colombia, with a population of six million, is one of the largest—possibly *the* largest—cities in the world to derive most of its electrical energy from hydroelectricity. The Magdalena and the Caurca rivers provide the water source. These rivers arise in the Colombian paramos, an unusual tropical, high-altitude ecosystem unique to Colombia, Ecuador, and Venezuela.

FIGURE 15.8 This older picture of Grand Coulee dam shows that it was a very powerful hydroelectric plant even before it was upgraded in the 1980s.

If global climate change (discussed in Chapter 32) causes climate shifts to reduce rainfall, then water resources could become even more scarce. A decrease in hydro capacity due to global climate change would be a sadly ironic situation, because hydroelectricity generation is one of the technologies that does *not* produce emissions thought to contribute to climate change in the first place.

DAMS—PRO AND CON

Not many rivers actually have fast currents or significant waterfalls to provide an adequate head for a turbine. We can solve that problem, though, by creating an artificial head, by building a dam. Dams are by no means new. The documentary history of dams can be traced at least to 1177. Even in the Middle Ages there were arguments and lawsuits over dams, for illegally raising the height of a dam (and thus reducing the water flow downstream) and for interfering with navigation. The building of dams has long been a subject of literature:

> *Daily they would vainly storm,*
> *Pick and shovel, stroke for stroke;*
> *Where the flames would nightly swarm*
> *Was a dam when we awoke.*

—Goethe[1]

Dams provide a number of significant benefits: They create a source for hydropower generation. They control seasonal fluctuations in water flow, helping navigation and downstream flood control. The newly created artificial lake can be used as a source of recreation, or the water in it can be used for agricultural irrigation. However, barring some very major changes in public perception and public policy, it is unlikely that hydropower will expand into a significantly larger share of the United States' energy supply.

In the 1930s and 1940s dam-building in the United States was a highly praised activity, through the Tennessee Valley Authority in the southeast and efforts such as the Grand Coulee Dam and Hoover Dam in the west and northwest, because the dam-building project was seen as a way of bringing electricity to many rural communities and farms, and the availability of cheap electricity would foster the growth of industries and create jobs. This was especially important during the Great Depression of the 1930s. Folk singers of that era, such as Woodie Guthrie and Pete Seeger, who were politically on the left, had a range of songs about—and in favor of—building dams.

Today there has been a swing of opinion. Environmentalists, many of whom are also politically on the left, are now vehemently opposed to dam construction. Many advantages can be listed for dam construction: creation of sites for hydropower generation, controlling seasonal fluctuations in water flow, controlling downstream floods, using the artificially created lake or reservoir for recreation or for crop land irrigation. On the other hand, there are numerous counterarguments: The artificial lake floods valuable farmland, or pristine wilderness. Dams do not last forever. Silt can accumulate behind the dam, eventually filling in the reservoir. If the dam fails, there is great potential for a catastrophic accident with large loss of life and property damage. For example, a dam failure in the Hunan province of China in 1975 cost the lives of 20,000 people.

In the mid-1980s the United States Army Corps of Engineers conducted an inspection of about 4,000 dams. Their inspection showed that 988 were rated 'unsafe' and another 58 'urgently unsafe,' an interesting choice of words that suggests it might not be prudent to live downstream from such a dam. The breaking of a dam is an event that can cause terror to anyone. Consider the quintessential American cowboy story:

> 'What's that?' Hopalong suddenly demanded, drawing rein and listening. A dull roar came from the dam and he instinctively felt that something was radically wrong.
>
> 'Water, of course,' Red replied, impatiently. 'This is a storm,' he explained.
>
> Hopalong rode out along the dam, followed by Red, peering ahead.
>
> 'She's busted! Look there!'
>
> A turbulent flood poured through a cut ten feet wide, and roared down the other side of the embankment, roiled and yellow.
>
> 'God almighty!' Red shouted. 'The dam's broke!'
>
> —Mulford[2]

while half a world away:

> 'A crack in the dam!'
> The lads
>   shudder,
> They jump into cars
> that come flying for them,
> Alarm calls
>   sound at the construction
> Guitars
>   with ribbons
> are flung
>   onto beds,
> Forget dances,
> pictures!
> Everyone
>   like rioters
> runs
>   to the dam!
> 'A crack in the dam!'
> Forgetting the toasts,
> the wedding guests,
> sober up at once.
> The bridegroom runs
>   wearing a butterfly tie,
> cursing
>   his fancy clothes,
> gathering speed
>   as he runs.
> And his welder bride
>   pulling off stiletto heels,
> runs after him barefooted
>   in white
> to the river's edge.
>
> —Yevtushenko[3]

It is important to realize that the turbines and generators themselves have a negligible effect on the environment. That is, the actual generation of the hydroelectricity is a very environmentally friendly technology. However, the damming of a river or stream, which, in most locations, is a necessary prerequisite to provide the needed head of water, has a potentially major impact on the local ecosystem. Dams create an artificial reservoir—a pond or lake—where a flowing stream or river used to exist. Those aquatic plants and animals that are best adapted to free-flowing water may not be able to flourish in the relatively stagnant conditions of the reservoir. Similarly, new species, that *are* adapted to still or sluggish water, may move in. The relatively sluggish or stagnant water in the reservoir can potentially serve as a breeding ground for mosquitoes. Dams can block fish migration and destroy spawning grounds, affecting both commercial and sport fishing. For this reason, in some parts of the United States, such as New England and the Pacific Northwest, there is a growing effort to dismantle small hydroelectric plants to restore trout and salmon to the streams. Not only do the reservoirs potentially change the ecosystem in the river itself, but also, depending on the specific conditions where the dam is built, reservoirs can flood large areas of forest and wildlife habitat (indeed, in some cases they can even flood valuable farmland or towns).

The dams allow for the accumulation of silt in the reservoir. These sediments, which may carry vital nutrients, accumulate in the reservoir instead of being carried downstream. Therefore, there is also a potential impact on ecosystems downstream of the dam, as well as those affected by the upstream reservoir. The downstream situation is further aggravated because the water flowing out of the hydro plant may be warmer, have lower oxygen content, or both, compared to the normal river conditions.

The resulting pond or lake behind a dam also usually raises the water table behind the dam (as a result of seepage) and lowers it below the dam. Evaporation of water from the reservoir increases the amount of dissolved salt and minerals in the water. If the river has a high concentration of dissolved minerals to begin with—an example being the Colorado River—then this can become a serious problem.

As an example of how this can become a serious problem, we have the salt cedar or tamarisk (Figure 15.9). This tree is a native of Asia, but was introduced to the United States in the late 18th century.⊛ It spread only slowly during the 19th century. In the early years of the 20th century, it began to move rapidly through the American southwest. It is possible that dams have given the tamarisk a crucial advantage over competitor trees by suppressing the natural flooding of rivers. The tamarisk produces and releases its seeds on a different cycle than those of the native cottonwoods and willows. Natural flooding favors the seed cycle of the native trees; when this is suppressed, the tamarisk has a chance to take over. What had been brought to the United States as a pleasant tree for shade and for firewood has now become the dominant tree over more than a million acres. As a result, these tamarisks now soak up about twice as much water every year as used by all the cities of southern California. Not only did the tamarisk displace the native cottonwoods and willows, the dominance of the tamarisks created another problem as well. As they drop in the fall, the tamarisk leaves release so much salt that the soil may not support other vegetation.

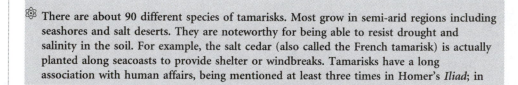

⊛ There are about 90 different species of tamarisks. Most grow in semi-arid regions including seashores and salt deserts. They are noteworthy for being able to resist drought and salinity in the soil. For example, the salt cedar (also called the French tamarisk) is actually planted along seacoasts to provide shelter or windbreaks. Tamarisks have a long association with human affairs, being mentioned at least three times in Homer's *Iliad*; in one case the Trojan warrior Adrestos was captured by Menelaos when the horses pulling his war chariot ran into a tamarisk tree.

Since most—but not all—of the environmental problems associated with dam building derive from the creation of the reservoir, then ideally we should maximize the amount of electricity generated while minimizing the size of the necessary reservoir. One measure of estimating the environmental impact of a hydroelectric plant is the power produced per hectare⊛ of surface covered by the reservoir. The larger this indicator number, the smaller the likelihood of a serious impact on the environment; conversely, the smaller the number, the worse are likely to be the problems. The Balbina hydroelectric plant in Brazil is an example of a major environmental disaster; there, only about 1 kW is generated for each hectare of reservoir surface. To put this number in context, another plant in the same country has an indicator value of 588 kW/hectare, which indicates a substantial generation of electricity for a relatively small area covered by the reservoir.

⊛ A hectare is a metric unit of area, equal to 2.5 acres, or 108,000 square feet.

One of the first battles between environmentalists and developers was the Hetch-Hetchy dam project, near Yosemite. In the previous section we mentioned the controversy over the development of hydroelectric sites at Niagara Falls, where, for a time, the Niagara River was rapidly being diverted to power stations. To some people, the hydroelectric plants at Niagara Falls were popular tourist sites. To others, though, the dams at Hetch-Hetchy and diversion of water from the Niagara River were simply symbols of commercialism run amok. These battles were ignited anew in the 1960s by one of the dumber schemes to come out of Congress—the damming of the Grand Canyon.

But, it is fair to say that in most cases, when the construction of new dams was not perceived to be a threat to popular locations such as Niagara Falls, then there was actually

FIGURE 15.9 Tamarisk (salt cedar) trees are now a common sight in the American southwest.

little opposition. Hundreds of new dams were built in the early part of the 20th century. These dams were, in many cases, symbols of technological progress and of economic prosperity. Particularly during the Great Depression of the 1930s, the federal government worked on many dams in the southeast, via the Tennessee Valley Authority, in the southwest, particularly on the Colorado River, and in the Pacific northwest. Indeed, in that time of severe economic problems, dam construction was widely viewed as a way of stimulating the local economy.

The single project that received most public acclaim was the construction on the Colorado River of the Hoover Dam, which began in 1931 in the depths of the Depression. In late 1930, Elwood Mead, then the Commissioner of Reclamation, noted that the difficult national economic conditions had caused 'pressure for action on this matter, as a means of furnishing employment and encouraging a revival of business.' When the dam was completed in 1936, President Franklin Roosevelt told the nation how badly these projects were needed to 'create a new world of abundance.'

The Hoover Dam was the engineering and construction marvel of the 1930s. Even the construction site became a tourist attraction. The pouring of the concrete began in June 1933, and continued round the clock for two years. In 1934 and 1935 the Hoover Dam construction site was as popular a tourist destination as the Grand Canyon. At the time, the Hoover Dam was the tallest dam anywhere in the world, at 726.4 feet above bedrock. The base of the dam is about 600 feet thick (200 yards, or the length of two football fields), and the top can accommodate three lanes of traffic. The four water intake towers are as tall as 33-story buildings. The finished dam used three million cubic yards of concrete. The new reservoir covered an area of 227 square miles. The turbine and generator installation covers about 10 acres. The 17 turbines generate five billion kilowatt-hours a year. (The indicator value for Hoover Dam is about 10.) The dam was finished in March 1936, two years ahead of schedule. Since then, more than 27 million people have visited the dam. The American Society of Civil Engineers has declared it to be one of the country's Seven Modern Civil Engineering Wonders.

Some observers have written off the construction of large hydroelectric projects as 'yesterday's technology.' This is because of the record of the huge dams for damage to the environment, for massive cost-overruns, failure to meet the promised amounts of electricity and water, and social disruption. Yet, on the other hand, the list of projects in progress is probably at an all-time high. Areas experiencing rapid development of hydroelectricity include the the Mekong River basin in Southeast Asia, Iran, and South Africa, including the kingdom of Lesotho. The construction companies argue that their industry is more socially, environmentally, and financially responsible than indicated by its past record.

On the other hand, the international news is not all positive. Worldwide environmental protests resulted in (or at least contributed to) stopping the Narmada project in India. Similarly, the Cree people in Quebec organized an international campaign to halt a hydroelectric scheme on the James Bay. The Three Gorges dam on the Yangtze River in China, which will probably be the largest engineering project in the history of the world when completed, may already be experiencing financial trouble. Not only that, it is estimated that three million people will be displaced from their homes by the new reservoir. The Ilisu dam is planned to be built in eastern Turkey in the heart of territory whose ownership is disputed by the Kurds. The dam will be built in the headwaters of the Tigris River just before it flows over the Turkish border into Syria and Iraq. This construction is contrary at least to the spirit, if not the letter, of the United Nations Convention on the Non-Navigable Uses of Transboundary Waterways.

As we look to the future in the United States, it does seem that environmental laws will place severe restrictions on new hydropower development. Indeed the environmental effects of very large dams are so severe that there is almost no chance that any more will be built in the United States. First of all, the 1968 National Wild and Scenic Rivers Act simply precludes building facilities on stretches of many virgin rivers. Therefore, this (and related) legislation eliminates at once development of about 40% of the remaining undeveloped hydroelectric resource. In addition, another 19% of potential sites are under a development moratorium until their final status can be decided. It is probable, therefore, that less than half the remaining hydroelectric potential—perhaps 30,000 MW—will actually be available for development under the provisions of current federal legislation. But, out of this, only about 22,000 MW of the undeveloped resource is likely to be economically viable. Of this economically viable fraction, perhaps as little as 8,000 MW will actually wind up being developed, because of the jungle of regulatory complexities and federal, state, and local jurisdictional overlaps in the licensing process. As we have noted before, this situation is rich in irony, because hydroelectric plants produce no air pollution, no radioactive wastes, nor any other harmful emissions to the environment.

CITATIONS

1   Goethe, Johann Wolfgang von. *Faust. Part Two*. Many editions of this work are available; the cited passage is from lines 11123–11126.
2   Mulford, Clarence E. *Hopalong Cassidy*. Tom Doherty Associates: New York; Chapter XX.
3   Yevtushenko, Yevgeny. 'A Crack,' in: *Bratsk Station and Other New Poems*. Doubleday: New York, 1967; p. 135.

FOR FURTHER READING

Brower, Michael. *Cool Energy*. MIT Press: Cambridge, 1994. Chapter 6 discusses issues regarding hydroelectric generation. The entire book surveys various approaches to so-called renewable energy that would have minimal environmental impacts.

Fussell, Paul. *BAD or, the Dumbing of America*. Summit Books: New York, 1991. Those prepared to believe that the country is going to hell in a handbasket will find their darkest suspicions confirmed. The chapter on 'BAD engineering' includes information cited here on unsafe dams as well as many other construction catastrophes seemingly waiting to happen.

Homewood, Brian. 'High and dry in Colombia.' *New Scientist* **1996**, *150*(2036), 34–37; Pearce, Fred. 'And the waters rushed in . . .' *New Scientist* **1999**, *162*(2184), 55. These two relatively recent articles discuss hydroelectric generation. This is an excellent magazine in which most of the articles can be appreciated by persons with modest scientific background.

Macaulay, David. *Building Big*. Houghton Mifflin: Boston, 2000. All of Macaulay's books are treasures, splendidly illustrated. *Building Big* has a section on dams, including Hoover Dam and Itaipu.

Petroski, Henry. *Remaking the World*. Knopf: New York, 1997; pp. 184–193. An excellent source of information on the building of the Hoover Dam. Petroski writes interesting and informative essays on various facets of engineering.

Pohl, Frederick. *Chasing Science*. Tom Doherty Associates: New York, 2000. Pohl is best known as one of the premier science fiction writers of the 20th century. This book, subtitled 'science as a spectator sport' discusses some of Pohl's personal encounters with various aspects of science and technology. Chapter 6 discusses dams.

Ramage, Janet. *Energy: A Guidebook*. Oxford: Oxford, 1997. An excellent introduction to many aspects of energy technology. Chapter 9 discusses hydroelectricity.

Saunders, Robin. 'Water.' In: *Energy Primer*. Portola Institute: Menlo Park, CA, 1974; pp. 52–73. Though parts of this book are now out-of-date, it still provides a great deal of useful information, especially for those who are interested in building or designing energy devices for themselves. This section on water provides useful background on waterwheels and small dams.

Strandh, Sigvard. *The History of the Machine.* Dorset Press: New York, 1979. This is a very well-illustrated treatment of exactly what the title promises. Chapter 4 provides information on the various forms of water turbines and how they are accommodated in dams.

Thomson, J. Arthur. *The Outline of Science.* G.P. Putnam's Sons: New York, 1937. This old book is worth seeking out in libraries or used-book sales. It provides a fairly comprehensive and well-illustrated (for the time) review of science up to the time shortly before the Second World War. Chapter XXIV includes a discussion of 'electricity from waterfalls.'

Usher, Abbott Payson. *A History of Mechanical Inventions.* Dover: New York, 1982. Chapter XV discusses the early development of hydroelectricity. This book is a paperback reprint of a book originally published in 1929; so covers the topic up to the early 1900s.

16

ELECTRICITY FROM STEAM

We saw in the last chapter that the development of the turbine allowed harnessing the kinetic energy of water to operate an electricity generator. Hydroelectricity has many advantages, not the least of which is its being virtually pollution-free. And, we can create the necessary head of water for the turbine by building a dam. Unfortunately, the number of rivers that provide likely sites for hydroelectric generation is limited, and nowadays it is very questionable whether another big hydroelectric project would ever be allowed to proceed in the United States. In particular, the environmental problems associated with the building of dams send the message that we need to find an alternative fluid to use in the turbine, despite the many excellent advantages of hydroelectricity.

Such a 'working fluid' has to satisfy a number of stringent criteria:

- It needs to be cheap and readily available.
- It needs to be easy to produce or generate, and easy to handle on a large scale.
- It should be reasonably safe to work with, and safe in the environment.
- It should have enough mass to provide impulse to the turbine.
- It should be easy to generate a sufficient head to run the turbine.

This is, in fact, a very tough list of criteria to satisfy. In fact, the only obvious candidate for this working fluid is water. While we have ruled out any further significant application of *liquid* water, we need to recognize that there is another useful form of water—steam.

Steam is a form of water that we can get to flow easily, that can be made any place, and satisfies the other criteria we have just listed. The 'other 90%' of our electricity—that is, the electricity not generated by hydropower—all derives from steam-driven turbines.

RECIPROCATING STEAM ENGINES IN ELECTRICITY GENERATION

By the late 19th century, the steam engine had been brought to a high state of development. Therefore, one of the early solutions of the problem of reliable operation of electricity generators was to connect the steam engine to the generator. This is illustrated schematically in Figure 16.1.

We might wonder why we simply don't connect steam engines to generators and continue to produce our electricity that way. Unfortunately, for this type of application, steam engines have several disadvantages: The reciprocating motion of the piston in the cylinder needs to be converted to the rotary motion of the generator. This is inefficient; for one thing, it usually produces significant vibration. Reciprocating steam engines are too slow. The best technology of Edison's day was operation at about 350 revolutions per minute (rpm). This meant that the steam engine had to be 'geared up' to operate a generator at, e.g., the 1,000–1,500 rpm typically required for the generators of that era. Indeed, when the first central generating stations were built, Edison had a very difficult time even finding an engineering firm that could build a reciprocating steam engine capable of running at this upper limit of about 350 rpm. As the demand for electricity continued to grow, bigger and bigger and more and more complex reciprocating steam engines would have been needed to operate the generators.

As the continual improvement of electricity-generating and distribution systems made a public electricity supply increasingly practical, great efforts were made to design high-speed steam engines to drive generators. As the 19th century drew to a close, a major need of the rising electrical industry was for some kind of engine that could both run at very high speeds and provide a smooth, vibration-free operation. A turbine in which steam is the working fluid—the steam turbine—meets both of these requirements. It soon took over the job of driving electricity generators. (At about the same time, the steam turbine found a second important application: propelling high-speed ships.) By about 1910, the reciprocating steam engine had been almost totally displaced by the steam turbine in the electricity-generating application.

THE 'PREHISTORY' OF THE STEAM TURBINE

The first device that might actually be called a steam turbine is ascribed to Hero of Alexandria. In Hero's time (the 1st century A.D.) there was very little demand, and little apparent need, by society for mechanical devices that would provide a steady, reliable source of work. Consequently, Hero's inventions never seemed to pass beyond the stage of

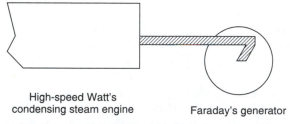

High-speed Watt's
condensing steam engine

Faraday's generator

FIGURE 16.1   Schematic diagram of Faraday's generator run by Watt's steam engine.

FIGURE 16.2 Hero's Sphere of Aeolus.

being regarded as interesting toys. The most notable of these is the so-called 'Sphere of Aeolus' (Figure 16.2). It made direct use of the pressure exerted by a jet of steam in order to spin itself, so technically it can be described as a sort of reaction turbine.

The Sphere of Aeolus was a hollow metal ball that was free to rotate around a horizontal axle above a small boiler. The horizontal axle consisted of a hollow tube that would allow the passage of steam from the boiler into the sphere. Two tubes, bent in opposite directions, protruded from opposite sides of the sphere. Steam from the boiler passed through the stationary tubes forming the axle on which the sphere was pivoted. As the steam escaped from the protruding bent tubes, it provided enough energy to rotate the sphere. Since the motion of the sphere was caused by the reaction of unbalanced forces, this is why we can say that Hero's sphere was a reaction turbine. Nothing in the historical record suggests that it was ever used to do useful mechanical work.

Fifteen hundred years after Hero, societal conditions had changed to the point where now there was a recognized need for, and value of, devices that could provide useful work. In the early 1600s two inventors, Andrea della Porta and Solomon de Caus, proposed methods of harnessing the 'impellent force of fire.' In both cases steam pressure was used to raise water by pushing it up a pipe. About 20 years later, Giovanni Branca invented a simple steam reaction turbine, in which a jet impinged upon blades projecting from a rotating wheel. Although none of these inventions was ever particularly successful, they were the forerunners of the more successful inventions that came by the end of that century. These included a stamp-mill that would be driven by a jet of steam.

In 1769, James Watt experimented with a ring-shaped 'cylinder' in which a piston was to be driven around by steam. At the time, the device could not be made to work, so Watt abandoned the idea. However, in 1784, Watt was suddenly forced to reconsider the idea, since the Hungarian Baron Wolfgang von Kempelen took out a patent on a steam turbine, claiming that his engine was superior to Watt's and would easily push it off the market. von Kempelen patented what was, in his words, a 'reaction machine set in motion by Fire, Air, Water, or any other Fluid.' Despite this grandiose, all-encompassing claim (which is by no means unique in the patent literature), von Kempelen's machine was mainly intended to be driven by 'boiling waters or rather the vapor proceeding therefrom [i.e., steam].' Watt said little about von Kempelen's engine on the presumption that even discussing the device might suggest to von Kempelen some needed improvements. Once Watt set to work analyzing this competitor, and made a thorough examination of the steam consumption, the speed of rotation, and their relationship to the output of work, he concluded that von Kempelen's design was technically impossible. Indeed, von Kempelen is now best remembered for another of his remarkable inventions, the 'chess-playing Turk.'❀

❀ Toward the end of the 1760s, von Kempelen had built an automaton in the form of a male in Turkish dress, which sat beside a large, boxy chest, which had a chessboard on its top. von Kempelen challenged famous chess players to play against the 'Turk,' and for many years toured Europe and the United States with his show. The android could also talk, though with a two-word vocabulary: 'Check!' and 'Gardez!' von Kempelen claimed that in the android he had managed to copy human intelligence and ability to reason. After attending a demonstration of the chess-playing Turk in London, Edgar Allan Poe wrote an essay denouncing any suggestion that the device had an ability to think. The fraud was eventually revealed—von Kempelen had hidden a dwarf, who was a skillful chess player, in the chest, and the android's machinery consisted simply of mechanisms that would allow the hidden dwarf to move the chessmen on the board.

About 30 years after von Kempelen's attempt at a steam turbine, a more serious and successful effort was made by Richard Trevithick. He built a so-called 'whirling-engine' that consisted essentially of a pair of hollow arms mounted on wheel set on a shaft. The wheel was about 15 feet in diameter. Steam at a pressure of 100 pounds per square inch escaped at a tangent from a small hole at the end of each arm, causing them to whirl around. The weakness of Trevithick's design was that the rotational speed of 250–300 rpm, produced only about one-fifth of the work potentially available from the steam.

THE DE LAVAL TURBINE

In a sense, a turbine resembles a fan run backward. While a fan uses rotary motion to push air from low pressure to high pressure, a turbine uses the flow of steam from high pressure to low pressure to create a rotary motion. High-pressure steam exerts pressure on the turbine blades, and those blades rotate away from the steam. The steam does WORK on the turbine needed to spin the generator. A steam turbine is ideally suited to electricity generation because both involve rotary motions. The shaft of the turbine, on which its blades are fastened, accepts WORK from the steam and transfers that WORK to the generator.

The first person to use successfully the energy of steam to produce rotary motion was Carl Gustaf Patrik de Laval, who developed his first invention, the milk separator, in the 1870s. To accomplish the separation of cream from milk, containers of milk had to be rotated at a very high speeds, typically 6,000–10,000 rpm. As we've seen, these rotational speeds cannot be achieved by a reciprocating steam engine, which is limited to about 350 rpm. Though the engine can be 'geared up' to produce higher speeds, a very complex mechanism would be needed for the roughly 20-fold increase in speed for the milk separator. de Laval wanted an engine or device that could be connected directly to the separator without the complexity of an intervening gear mechanism.

Fully convinced that speed is a gift from the gods, I dared, in 1876, to believe in the direct exploitation of steam on a power-generating wheel. It was a bold venture. At that time, only low speeds were employed. The speeds which were considered normal in later separators were never used. In contemporary handbooks they wrote about steam: 'Unfortunately, the density of steam is too low to permit any thought of exploiting it in a power-generating wheel.' In spite of this I succeeded.

—de Laval[1]

de Laval's turbine was perfected in the late 1880s. These turbines were designed to be a single-wheel, single-stage impulse type, in which a jet of steam impinged directly on the wheel blades. This was a simple and efficient device that derived from the design principles of Branca's machine. The wheel diameter ranged from 3 to 30 inches. Because there was only one wheel and one set of nozzles, in which the steam attained high velocity before striking the blades, the speed of the blades at the rim of the wheel was exceedingly high. Some of his machines ran at speeds as high as 40,000 rpm. By 1897 de Laval was using steam at high pressures, about 3,000 pounds per square inch.

In the de Laval turbine the wheel fitted is set in motion by a combination of impulses and reactions, produced by the flowing steam directed against the vanes by one or more nozzles. The channels between the vanes are designed so that no expansion of steam takes place while it is traversing these channels. As a result, there is no change in steam pressure. Expansion of steam takes place in the nozzles, which are designed to allow for this expansion. Perceiving the importance of the expansion of the steam, and then working out the correct design of the nozzles were significant engineering achievements.

The de Laval turbines were built for smaller units, from five to several hundred horsepower. They revolved at very high speeds, from 10,000 to 30,000 rpm for the primary shaft. For connecting the turbine to machinery, these speeds were reduced by reduction gears. Although the de Laval turbines were limited in capacity, compared to the Parsons turbines we will meet in the next section, still a considerable number were placed in commercial service. The de Laval turbine had an extremely high operating efficiency. A comparative test made in 1890 between a six-horsepower steam engine and a five-horsepower de Laval turbine showed that the steam engine consumed 75% more steam. de Laval had many ideas on how his 'steam-motor,' as he called it, could be put to practical use. For a while he considered using it to propel ships. But, in the end, he decided to concentrate on producing units for generating electrical energy.

THE PARSONS TURBINE

The English engineer Charles Parsons (later Sir Charles, Figure 16.3) set about developing a steam turbine specifically for driving an electricity generator. The basic idea for Parsons' turbine was taken from the water turbine designed by Fourneyron. The immediate incentive for Parsons' work was the need for some type of device to drive a generator directly. As we've seen, the required speed was beyond the range of the reciprocating steam engine. Parsons' steam turbine used a sequence of fixed 'guide' blades and moving blades (Figure 16.4). The steam flows between fixed guide vanes, so that it meets the blades of the wheel at right angles. The steam was admitted at the middle of the axle and flowed parallel to it in both directions, while delivering its kinetic energy to the turbine and expanding to atmospheric pressure. The electricity generator was directly connected to the turbine axle.

Parsons' machine was a reaction turbine. As the steam passed through the series of rotors and fixed blades, the size of both the moving and fixed parts increased along the length of the turbine to correspond with the reduction in the pressure of the steam as it expanded. The flow of steam through the turbine was assisted by a condenser which drew off the exhaust steam. In the condenser the steam, after doing work on the turbine (i.e., keeping it spinning), was condensed at atmospheric pressure. In earlier designs the low-pressure, low-temperature 'spent' steam had always been exhausted simply to the outside atmosphere. Now, the condensed steam (which is liquid water) could be pumped away or pumped back to the boiler. This so-called **condensing turbine** proved to have a smaller steam consumption than a conventional steam engine of equal horsepower. It also provided the advantages of space saving, greater reliability, and much lower vibration.

FIGURE 16.3 Charles Parsons (1854–1931). This English engineer invented the first steam turbine that had stationary and moving parts.

In the Parsons turbine the shapes of the blades differ from those used in the impulse turbine. In particular, provision is made for expansion of steam as it flows among the blades. Consequently, a drop in pressure occurs during the passage of the steam through the turbine. A turbine of the Parsons type consists of a long, closed horizontal drum mounted on a shaft in a strong casing, through the ends of which the shaft passes. In a large turbine there may be some hundreds of thousands of blades, of lengths ranging from an inch to over a foot. Steam is admitted at the smaller end of the casing and threads its way through the many rows of guide and moving blades, every one of the moving blades getting a sideways push which is transmitted through drum and shaft to the generator. After expanding greatly, the steam escapes to the condenser. A turbine is so well balanced and revolves so smoothly that a coin stood on edge on the casing is not upset. The ultimate effect is very similar to the way that wind or a person's breath turns a toy pinwheel, except that in the turbine the steam is flowing through a tightly enclosed space so that the amount of kinetic energy transferred from the steam to the turbine is maximized.

The first Parsons turbine/generator set, working at 4,800 rpm, was installed in 1888 at the Forth Banks (England) generating station. It had an initial capacity of 75 kilowatts (kW). Also in 1888, a 32-horsepower (hp) unit was installed at the U.S. Naval Proving Grounds at Newport, Rhode Island. The development of this machine for commercial application in the burgeoning electricity supply industry took place remarkably quickly, so

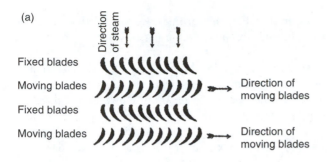

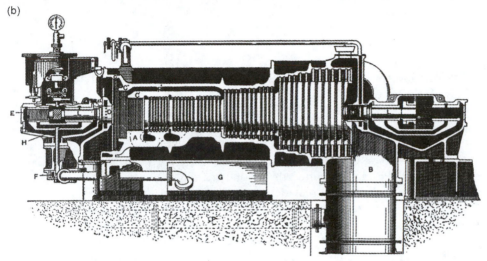

FIGURE 16.4 The elements of the Parsons turbine (a) include fixed guide blades that attach to the casing and rotating blades that give the whole machine (b) its power. Steam enters at E, passes through a series of rotors, and exits the turbine at B.

that steam turbines were coming into widespread use in electricity supply stations by the end of the century. By 1900, Parsons had built two 1,000 kW units for Elberfeld, Germany. In 1901, a 2,000 hp unit was put in operation at Hartford, Connecticut.

Since the 1880s the general style of the Parsons turbine has not been significantly changed, though there are now many variants of this type. One noticeable change that has been made is the remarkable improvement in efficiency. Essentially, turbines have been modified to allow the steam to expand in several stages. In other words, high efficiency, combined with a satisfactory speed of revolution, is obtained by passing the steam through a series of small elemental turbines. By doing this, the total drop of pressure as the steam expands is divided into a number of small stages, which theoretical considerations indicate to be necessary for high efficiency. Thus the first Parsons turbine, built in 1884, drove a generator that produced 7.5 kW (about 10 hp) at 100 volts. Its steam consumption was 130 pounds per kilowatt-hour. Nowadays the steam consumption per kilowatt-hour would be about 8 pounds. In the course of time the small Parsons unit of 1884, generating 10 hp, has evolved into the gigantic units in modern electricity stations that develop 600,000 hp at a fraction of the fuel consumption of earlier steam engines.

Although some of the early Parsons turbines were built as small as 10 hp, large sizes predominated. Thus in general a Parsons turbine is of larger capacity—often much larger—than a de Laval unit. The speeds of Parsons turbines were lower than de Laval turbines, no more than 18,000 rpm. Indeed, some very large Parsons turbines operate at speeds as low as 750 rpm.

Another advantage of the turbine is that it can be built on a larger scale than a reciprocating engine. The limits of the reciprocating engine were not important when one was used simply to meet the needs of a single factory installation. However, the development of large, centralized generating stations in the electrical industry created new demands for work and power that the reciprocating engine could not meet. Between 1890 and 1900 the best marine engine had a maximum capacity of 5,000 to 6,000 kW (about 6,700–8,000 hp), reaching the limit of development of the reciprocating engine. In contrast, steam turbines for luxury ocean liners, discussed in the next section, were already at some 17,500 hp.

A DIGRESSION—STEAM TURBINE APPLICATIONS IN SHIP PROPULSION

Parsons was quick to recognize that, with adequate gearing down, his turbine could be a very effective method of propulsion for ships. In the tradition of bureaucracies the world over, the British Admiralty was reluctant to admit the potential of ship propulsion by steam turbines. Parsons converted them—and marine engineers in other parts of the world—by an unsolicited but sensational demonstration at the Jubilee Naval Review in 1897. Parsons' own experimental turbine-propelled launch, the *Turbinia*, sped past the comparatively slow-moving naval ships at the then-incredible speed of 34 knots.

Thus Parsons crowned his achievement of providing a turbine suitable for the rapidly growing electrical industry by making the first successful application of the steam turbine to ship propulsion. In 1896, he devised a system with three shafts, each of which carried three screws, driven by a connected series of turbines, those for high and intermediate pressure at the sides, with a low-pressure turbine in the center; there was a separate turbine, driving the central shaft, for reversing. The three turbines together developed about 2,000 hp. The turbines ran most efficiently at speeds higher than were suitable for the screws. As the turbine came into wider use for ship propulsion, reduction gearing was developed, even double-reduction gearing for slow merchant ships in which the screws were required to turn at less than 100 rpm. Electric drive was also developed, the constant-speed turbines driving generators which furnished power to the propeller shaft motors, the speed of which can be effectively controlled regardless of the turbine speed.

In the 1890s the steam turbine emerged as a superior form of marine propulsion, especially in larger applications. Within a decade, steam turbines were being extensively adopted for both naval and merchant ships, with the Cunard luxury liner *Mauretania*, launched in 1906, equipped with steam turbines that developed 70,000 hp and a speed of 27 knots. The *Mauretania* and her ill-fated sister ship *Lusitania*,⚛ built in 1907, had turbines directly connected to their screws. There were four turbine units, of about 17,500 hp each, in each vessel. The success of the turbines in these two gigantic, luxurious ships paved the way for the general installation of turbines in other vessels. The *Mauretania* was one of the last large hand-fired ships, requiring a crew of 324 firemen and 'trimmers' to cart and shovel a thousand tons of coal per day into her boilers. The *Lusitania* and the *Mauretania* reduced the transatlantic travel schedules from seven or eight to five days per crossing.

⚛ The *Lusitania*, a British passenger liner, was en route from New York to Liverpool in May, 1915. The ship was carrying a load of munitions for the British war effort in addition to her crew and the passengers. On the night of May 7, she was struck by a torpedo from the German submarine U-20. Although the incident occurred only a few miles off the coast of Ireland, the *Lusitania* sank so quickly that only about 800 people of the nearly 2,000 passengers and crew could be rescued. The casualties included 128 American citizens. The incident led President Woodrow Wilson to insist that Germany should not attack passenger or merchant ships without making some kind of provision for the safety of the people on board. The resumption of unrestricted submarine attacks by Germany in early 1917 ultimately led to American involvement in the First World War. The sinking of the *Lusitania* was the beginning of a two-year-long process that eventually dragged the United States into the war.

In the early years of the 20th century, the navies of the world were involved in intense competition in building new fleets, in preparation for a coming war. Most made a rapid transfer to steam turbines for their capital ships. The naval arms race culminated in the British *Dreadnought* of 1906, a ship of 21,000 tons equipped with ten 12-inch guns that could fire shells eight miles. Its steam-turbine engines were capable of speeds of 21 knots. All other great naval powers immediately followed suit. By 1913, the United States, as a result of the advocacy and prodding of former President Theodore Roosevelt, had built 13 Dreadnought-class battleships. The naval programs in Germany threatened British claims to 'rule the waves.' A similar naval buildup in Japan helped to defeat Russia in 1905. Indeed, the Japanese annihilation of the Russian fleet in the Tsushima Straits is one of the most lopsided naval victories in history. The new technology accelerated the pace of military spending. Even a century ago, cost overruns, deficit spending, and rapid obsolescence of weapons anticipated a pattern repeated in the Cold War and remains familiar as we begin a new century with talk of 'Star Wars' missile defenses. In the three decades between 1884 and the outbreak of the First World War in 1914, British naval budgets increased by a factor of five.

Most large ships for the next generation of ocean liners and naval vessels were equipped with turbines. Steam turbines were eventually displaced by diesel engines for most purposes of marine propulsion, but they remain in use in some larger vessels. However, the steam turbine is still supreme in generating much of the world's electricity. Those turbines use inlet steam at about 600°C and 4,500 pounds per square inch pressure, which expands in 20 or more stages to generate 75,000 hp (nearly 56,000 kW). The efficient, high-speed operation of a steam turbine in any application still requires that we have a way to generate a head of steam. That is the issue we take up next.

ROBERT BOYLE AND THE BEHAVIOR OF GASES

Generating a high-pressure head of water is comparatively easy. We merely need to confine a large vertical column of water. This would be pretty awkward to do with steam. Luckily, there is a solution. We've discussed before that one of the general scientific developments of the 16th and 17th centuries was that scientists began understanding the properties of gases. We've met the French physicists Charles and Gay-Lussac, who showed that, at constant pressure, the volume of a given quantity of gas is directly proportional to its temperature.

In the early 1660s, two English scientists, Henry Power and Richard Towneley, observed that the pressure and volume of a gas are inversely related, provided the temperature is held constant. This relationship can be expressed in mathematical terms as

$$PV = \text{Constant}$$

where P represents pressure and V, volume. Their work was known to several of the eminent scientists of the time, including Robert Boyle, Robert Hooke, and Isaac Newton. In France, Edmé Mariotte published this same relationship without giving credit to Boyle, let alone to poor Power and Towneley, presumably being unaware of Boyle's publications. By a quirk of history, this law of gas behavior is known by Boyle's name in English-speaking countries, and by Mariotte's name in France, but it really should be known as Power and Towneley's law.

Robert Boyle enjoyed a vigorous scientific career during the 17th century, among other things promoting the idea that the scientific study of nature was actually a religious duty. Boyle poured mercury into a long open tube whose end was shaped like the letter J, as illustrated in Figure 16.5. By doing so, the mercury in the tube trapped some of the air in the short closed side. By adding more mercury, he raised the pressure on the trapped air. The trapped gas was always able to support the column of mercury on the long open side once its volume had been reduced by the appropriate amount, so that the pressure it exerted was equal to the pressure exerted upon it. It was relatively easy to measure the pressure increase, by measuring the difference between the lengths of the two columns of mercury in the open and closed sides. He found that doubling the pressure on the trapped gas roughly halved its volume; tripling the pressure reduced the volume to a third, and so on. Furthermore, this effect could be reversed: A gas whose volume has been compressed in increasing the pressure can be made to expand again by lowering the pressure, and vice versa. This is why we can say that, for a given quantity of gas, the pressure is inversely related to the volume, so as one goes up, the other goes down. Therefore, as we have shown above, the product of the two remains constant:

$$PV = \text{Constant}$$

We will retain the custom of most texts, and call this relationship Boyle's law.

However, we must not disregard Mariotte entirely. As we've seen from the work of Charles and Gay-Lussac, a volume of gas (as, for example, one trapped under a column of mercury) will expand if heated and contract if cooled. So, if one were studying the effect of changes in pressure on the volume of gas, one would have to be sure to keep the gas at constant temperature. Otherwise, changes in volume would take place as a result of two effects—one of pressure, which would be what was under investigation, and the other caused inadvertently by changes in temperature. The specific effect that one was trying to isolate for study (in this case, the role of pressure in changing the volume of a gas) would be confounded by a second effect—the effect of temperature. In other words, one would likely measure volume changes for which pressure was not solely

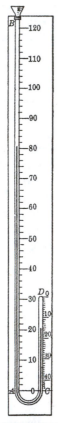

FIGURE 16.5 Boyle's J-tube experiment on gas pressure and volume.

responsible. Boyle did not, apparently, take note of this fact and therefore may not have appreciated its significance. In 1676, Mariotte discovered Boyle's law independently, and he did draw attention to the importance of making the measurements at constant temperature. So, calling the relationship of pressure and volume Mariotte's law rather than Boyle's law is not simply an act of European, or French, chauvinism, but actually there would be some justice in our doing this. But, as for Power and Towneley—well, they're history.

There is an alternative way of stating Boyle's law. Since the product of pressure and volume is a constant (*provided that* neither the temperature nor the quantity of gas is allowed to change), then this constant value must be the same for *any* pair of pressures and volumes. Suppose you have a sample of gas at some particular pressure that we will denote as P_1 and a corresponding volume that we will call V_1. If you then change the pressure to some different value that we will call P_2, the corresponding volume of gas will change to some new value, V_2. Boyle's law tells us that P_1V_1 = constant *and* P_2V_2 = constant. If we have not changed the temperature or the quantity of gas, then the 'constant' must be of exactly the same value in both cases. Therefore, we can also say that

$$P_1V_1 = P_2V_2$$

This equation is also a formulation of Boyle's law.

Why does Boyle's law 'work'? The explanation of Boyle's law is not obvious, and derives from the molecular nature of gases. Think of operating a small hand-pump of the kind used to inflate bicycle tires, sports balls, or air mattresses. If you give the piston a quick shove inward, and let go, you will notice that it jumps back at you. This seemingly odd behavior is a manifestation of what Boyle called 'the spring of air.' In the early days of science, in Boyle's era, no one knew of the existence of atoms or molecules and their behavior. Some of the early scientists thought of the particles of air as having actual little springs on them (illustrated by the diagram in Figure 16.6), and when measuring the spring of air they thought they were measuring the springs that held the particles apart in the air.

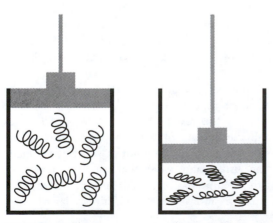

FIGURE 16.6   The conceptualization of air particles as tiny springs can still help us to understand the relationship between pressure and volume for gas molecules.

The first valid explanation was formulated by James Clerk Maxwell, a brilliant theoretical physicist, in the 1870s.⚛ The pressure exerted by a gas confined in some container is actually due to the bouncing of gas particles off the walls of the container. As it collides and bounces, each particle subjected the wall to a tiny force. The total force exerted by all the gas molecules colliding over a unit area is the pressure. That is, what we observe macroscopically as a 'pressure' is the result of an enormous number of separate 'pushes' against the walls of the container by each of the gas molecules that collide with the walls. There are so many of these 'pushes' in an instant of time, and each separate push is so tiny, that the total effect is sensed on a macroscopic scale as a smooth, even pressure. Since the gas molecules are able to move freely and randomly in all directions, the observed pressure is equal in all directions.

> ⚛ It is as difficult to rank scientists as it is to choose the greatest quarterback or left-handed pitcher of all time. But by any reckoning, James Clerk Maxwell is among the elite of the 19th-century physicists. In fact, he is probably the greatest of theoretical physicists in the era after Newton and before Einstein. Among his numerous contributions include Maxwell's equations, a set of four equations that form a complete description of the relationships between electric and magnetic fields; Maxwell's relations, another set of four equations that are central to the science of thermodynamics; and the imaginary creature called Maxwell's demon, which, by segregating fast-moving molecules from slow ones in a container of gas, could bring about a flow of heat.

To begin to see how this explanation of the origin of gas pressure relates to Boyle's law, suppose that the sample of gas is in a container that is enclosed at the top by movable frictionless piston. (For our purposes, we will assume that the piston itself has no weight, and we will neglect any possible friction between the walls of the container and the piston.) The pressure of the gas is caused by the force of the particles bouncing against the walls of the container and the underside of the piston. Assume that the top of the piston has been stacked with just enough weights resting on it to balance exactly the gas pressure. Now suppose we remove one of those weights. If we do that the external force pressing down on the top of the piston is necessarily decreased. Consequently, the upward force of the bouncing gas molecules is greater than the downward force of the piston, and we will observe that the piston moves upward. However, the piston moving upward causes the volume of the container to increase. As the volume of the container increases, each molecule of the gas will have to travel a greater distance to reach the underside of the piston. Because the molecules must travel further to collide with the bottom of the piston, the number of collisions in any given instant of time must necessarily decrease. (This is shown in cartoon fashion in Figure 16.7.)

This decrease in the number of collisions per instant is a direct consequence of the fact that each gas molecule has to spend more time traveling and, therefore, less time colliding. But since the measured pressure is the result of the collisions of the gas molecules against the container walls and the bottom of the piston, we observe that the pressure decreases. In fact, the pressure will drop to the point at which it is exactly balanced by the remaining weights on the piston, and, when that happens, the piston will no longer rise. We will observe that the volume has increased, and the pressure has decreased. This inverse relationship of pressure and volume is exactly as described by Boyle's law.

In the same way, suppose that additional weights had been added to those originally present on this same piston. Now, the extra weights will cause the piston to move downward against the force of the particle collisions. But, as the piston moves

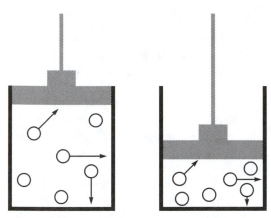

FIGURE 16.7 Individual gas molecules are able to travel further when the volume is increased.

downward, the volume of the container necessarily decreases. In this case, each gas molecule has a smaller distance to travel in order to reach the bottom of the piston. For a gas molecule, less time is spent traveling and more time is spent colliding. In other words, the number of collisions in any given instant rises, and we observe therefore that the pressure increases. Eventually, the pressure will increases to some value at which the additional weight placed on the piston is exactly balanced by the gas pressure. In this example, the gas volume has decreased and pressure has consequently increased. Once again, the inverse relationship of pressure and volume holds, exactly in the manner described by Boyle's law.

One of Boyle's greatest achievements is that he established the first *quantitative* law describing the measurable properties of matter. That is, he gave to science and engineering a mathematical formulation that allows predictions of how the volume of a gas will change to compensate for a change in pressure, or how the pressure will change to compensate for a change in volume. This legacy is remarkably ironic, in that Boyle himself expressed serious reservations about mathematical formulations or expressions of physical phenomena, and his written records of his own experimental work contain few, if any, such mathematical representations—including the 'law' relating the pressure and volume of air for which he is best known. This is a 'law' that Boyle never called a law and to which he never personally gave symbolic mathematical expression.

Nonetheless, the work of Boyle and Mariotte showed that for a fixed quantity of gas at constant temperature,

$$\text{Pressure} \times \text{Volume} = \text{Constant}$$

From the work of Charles and Gay-Lussac, we know also that

$$\text{Volume} \propto \text{Temperature}$$

(where the symbol $\propto$ means 'is proportional to'). If we keep the pressure fixed, an increase in temperature causes the volume to go up. Now, if we combine the two discoveries of Boyle and Charles, we can say that an increase in temperature of a fixed (constant) volume of gas must increase its pressure. Thus Boyle gives us the key to creating a head of steam— we must heat a fixed volume of steam to a high temperature.

TURBINES AS HEAT ENGINES

The high-pressure head of steam entering the turbine therefore will also be at high temperature. The working fluid—the steam—is going to impinge on the turbine blades and get them to spin.

A simple toy pinwheel shows us that a moving stream of gas (in this case, air from our mouths) can spin a rotating device. If we confine some air in our mouths by puffing out our cheeks (the air being at slightly higher pressure than that outside our mouths) and then blow onto the palms of our hands, we can feel the escaping air as cool. 'Expanding' the air from the (relatively) high pressure in our mouths to the lower pressure in the room causes its temperature to drop. (This is a property shared by most, but not all, gases.) To obtain high pressures of steam to operate a turbine, we heat it to high temperatures. In exactly the same way, the high-pressure steam will expand as it goes through the turbine. But since

$$\text{Volume} \quad \times \quad \text{Pressure} \quad = \text{Constant}$$

$$\uparrow \qquad\qquad\qquad \uparrow$$

If this goes up , This has to go down
(working fluid expands)

The essence of a steam turbine, then, is illustrated conceptually in Figure 16.8. However, we have already seen that when a SYSTEM changes temperature, we can—by inserting the right device—extract WORK, as in the energy diagram of Figure 16.9. The generic device for extracting WORK from this spontaneous change in temperature (i.e., thermal potential energy) is a *heat engine*. The steam turbine represents one example of a heat engine.

The *efficiency* of operation of a heat engine is given by the equation

$$\text{Efficiency} = (T_{\text{HIGH}} - T_{\text{LOW}})/T_{\text{HIGH}}$$

where the temperatures are *always* expressed in kelvins. The efficiency calculated in this way will be a number between 0 and 1. Often it is more convenient to express numbers as percentages, rather than as fractions. Thus the alternative equation

$$\text{Efficiency} = 100(T_{\text{HIGH}} - T_{\text{LOW}})/T_{\text{HIGH}}$$

can also be used, *provided that* one is careful to express efficiencies calculated in this way as percentages. For example, we might speak of a device as having an efficiency of 0.45, or 45%. Ideally, of course, we would like the efficiency of any device to be as close to 1 (or 100%) as possible. If we examine either form of the equation, it should be clear that the only way to achieve an efficiency of 1 is to let $T_{\text{LOW}} = 0\,\text{K}$. However, we've already shown that it is impossible to attain a temperature of $0\,\text{K}$ (absolute zero).

In other words,

IT IS IMPOSSIBLE TO ATTAIN AN EFFICIENCY OF 100% FROM ANY HEAT ENGINE.

In addition, it is generally much simpler and less costly to heat something than to cool it below ordinary temperatures. (For a quick example, compare the prices of a stove and a refrigerator of comparable workmanship.) Therefore, the approach to achieving high efficiencies is to make T_{HIGH} as high as possible. In other words, the higher the high-temperature side of the turbine, the greater will be its efficiency.

This concept of efficiency derives from the 'cycle' developed by Sadi Carnot. Efficiencies calculated in this way represent the best efficiencies for a heat engine. That is, Carnot's efficiency represents an ideal efficiency; the efficiency of real-world devices is

(a)

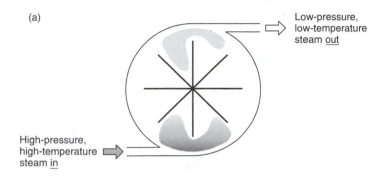

Low-pressure,
low-temperature
steam out

High-pressure,
high-temperature
steam in

(b)

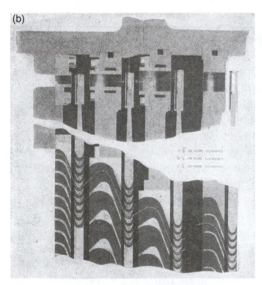

(c)

FIGURE 16.8 (a) As steam passes through a turbine, the pressure and temperature drops, creating an expanded volume that helps the turbine to spin even faster. (b) The layers of steam turbine blades (see Figure 16.4) are designed to allow for increasing volume as the steam passes through the rotor. Each successive row of stationary blades is housed in a larger casing so that the steam can increase its volume and velocity before striking the moving blades. (c) The blades on this rotor augment in size (from left to right) to seize maximum power from the expanding steam. When assembled on a turbine, the rotors' increasing sizes are obvious, as shown in Figure 16.4.

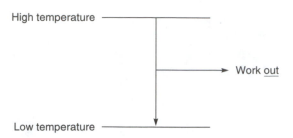

FIGURE 16.9 The familiar energy diagram for a thermal system also applies to a steam turbine.

often less than the Carnot efficiency because of friction, leaks, and other mechanical imperfections. In a modern electricity-generating plant, the high temperature is limited to roughly 800°F, or 1,260 K. The heat eventually winds up in a body of water such as a river. Thus in the summertime the low temperature might be about 70°F or 530 K. The Carnot efficiency would be about 60% for a device operating between 1,260 and 530 K. In other words, if the electricity plant were operating at this efficiency, then 40% of the chemical potential energy in the fuel simply gets dumped into the 'cold side,' the river or an ocean. In reality, efficiencies of most fossil-fuel-fired generating plants are less than the ideal Carnot value, and indeed less than 40%. Consequently, about two-thirds of the chemical potential energy of the fuel simply winds up in the low-temperature reservoir.

Well, why do we put up with this? Why can't we turn all the heat liberated as we burn the fuel straight into electricity? It is because we are stuck. Since the electric current extracts ENERGY from the generator, something has to supply WORK to the generator to keep it turning. The WORK comes from the steam's thermal energy (that is, heat). Since heat can't be converted directly into WORK at 100% efficiency, the steam turbine must and does operate as a heat engine, converting a limited amount of heat into WORK as it flows from a hotter object to a colder object. In the steam turbine, the hotter object is the high-pressure steam, and the colder object is the outside water or air. If we run the plant as a heat engine, we absolutely must have a cold side to the engine (otherwise there is no heat engine). And, since we cannot operate a heat engine at 100% efficiency, the plant must dump heat—a large amount of heat, as it turns out—into the cold reservoir.

Operating electricity-generating plants with steam turbines, and the fact that the steam turbine is a heat engine, leads to two additional consequences. First, because the plant must necessarily dump a great deal of heat into a cold reservoir, many modern electricity plants are built along rivers, on lakes, or on the seashore. For instance, the upper Missouri River in North Dakota was sometimes facetiously referred to as 'power plant alley' because the suitability of this major river made it very convenient to build electricity plants along its banks. Second, dumping so much heat into the cold reservoir of a river or ocean can, over time, actually make that stretch of river or sea coast warmer than it would otherwise have been if there were no generating plant. This localized warming of the water can alter the microclimate of the affected region and possibly even affect the local eco-system. The possible effects of this dumping of heat into the environment are called 'thermal pollution.'

Because turbines avoid many of the mechanical problems (friction and leaks, for example) that reduce the efficiency of reciprocating steam engines, they can also make better use of steam at much higher temperatures and pressures. This lets turbines achieve much higher efficiencies than were ever possible with the reciprocating steam engines. (Turbines are also much quieter because they don't have the severe vibrations generated by the reciprocating motion of the pistons.) Today's nuclear and fossil-fuel-fired electricity-generating stations use turbines that operate at pressures of several thousand pounds per square inch and generate hundreds of thousands of horsepower, as compared to Watt's early, atmospheric-pressure, 50 hp engines. The Newcomen engine consumed 20 pounds of coal per horsepower-hour (lb/hp-hr) when it was first brought out, and eventually was improved sufficiently to reduce this figure to 16 lb/hp-hr. By the 1780s, Watt's engine, which was a significant improvement on the Newcomen engine, consumed about 6 lb/hp-hr. There were no electricity-generating stations in those days, so the engines and their efficiencies were rated on the basis of their 'duty,' their ability to perform mechanical work per amount of coal consumed. In 1900, the Edison electricity-generating station at Chicago used 7 lb of coal per *kilowatt*-hour (which is equivalent to a little more than 5 lb/hp-hr of mechanical work). By 1913, with turbines, coal consumption was less than 3 lb/kW-hr. In 1924, the turbines at the Duquesne Light Company's Colfax station in Pittsburgh consumed only 1.29 lb/kW-hr (equivalent to 0.96 lb/hp-hr). If we disregarded all of the other advantages of the steam turbine relative to the reciprocating engine (its speed of operation, freedom from vibration, quietness, etc.), the steam turbine would still have to be regarded as an invention of absolutely first-class importance, solely on the basis of the extraordinary improvement in efficiency—which translates directly into much less fuel that must be extracted from the earth, shipped, processed, and burned.

THE TURBINE/GENERATOR SET

So why did we get into this discussion in the first place? It is because the steam turbine has become the standard technology for operating the generator in all the world's electricity-generating plants (except hydroelectric plants), and to use the turbine most effectively we need to supply it with a head of steam, as sketched in Figure 16.10. The generation of electricity from steam involves a turbine/generator set in which high-pressure, high-temperature steam is used as the working fluid in the turbine, and electrical energy is provided from the generator. We've discussed previously that the central problem in electricity generation is to find a cheap and reliable way to turn the generator. Now we can take a 'step back' (both figuratively and literally) from the generator and say that the practical problem to be solved is to find a cheap and reliable way to generate high-pressure, high-temperature steam.

Steam under pressure gives an increase in overall efficiency because it raises the temperature and therefore increases the temperature difference between high and low temperature reservoirs. The steam flows through the turbine to a condenser, where what is now a low-pressure, low-temperature steam condenses back to water. This is the lowest temperature point in the cycle. To keep the condenser at ambient temperature,

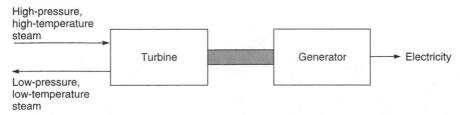

FIGURE 16.10 Most of the electricity we use today comes from generators spun by turbines that move as a result of high-pressure, high-temperature steam.

cooling water from a lake, river, or ocean is employed. This cooling water merely circulates through the system, and is later exhausted at a slightly warmer temperature. The water in the condenser is recycled back to the boiler to close the cycle. To handle such large amounts of steam, many condensers have more than an acre of cooling surfaces.

As we begin the 21st century, the steam turbine coupled with a generator remains the main method of generating electrical energy. Within 25 years after the installation of the last 7,500 hp stationary reciprocating engines, there were steam turbines operating at 240,000 hp. Turbines are now being installed up to 330,000 hp.

CITATION

1   Gustav de Laval, quoted in Strandh, Sigvard. *The History of the Machine*. Dorest Press: New York, 1979; p. 131.

FOR FURTHER READING

Asimov, Isaac. *The History of Physics*. Walker and Co.: New York, 1966. An easily readable introduction to physics intended for readers with little previous science background. Chapter 14 is particularly relevant to the present discussion.

Cardwell, D. *The Norton History of Technology*. Norton: New York, 1995. A fine book on the development of technology. Chapters 13 and 14 relate to the material presented here.

Gebelein, Charles G. *Chemistry and Our World*. Brown: Dubuque, 1997. Most introductory chemistry texts include a discussion of Boyle's law and related aspects of the behavior of gases. This is a well-illustrated book with abundant examples of the relevance of chemistry to daily life. Chapter 2 includes a discussion of Boyle's law and development of the applications of steam.

Hawks, Ellison. *Popular Science Mechanical Encyclopedia*. Popular Science: New York, 1941. An interesting review of many aspects of 'modern' technology up to about 1940. Though some of the material is now out of date, the fundamentals have not changed, and they are presented well in this book. The chapter on 'The Giant Power of Steam' discusses (among other things) steam turbines. Though long out of print, this book is worth seeking out in libraries or used book stores.

Sevenair, John P.; Burkett, Allan R. *Introductory Chemistry*. Brown: Dubuque, 1997. Chapter 11 provides a solid discussion of the development of Boyle's law and related material on relations among pressure, temperature, and volume of gases.

Strandh, Sigvard. *The History of the Machine*. Dorset Press: New York, 1979. A well-illustrated history of machine technology, covering much more than just energy-related devices. Chapter 5 discusses the development of steam turbines.

Usher, Abbott Payson. *A History of Mechanical Inventions*. Dover: New York, 1988. This is an inexpensive paperback reprint of a book originally published in 1929, covering the development of technology through the early decades of the 20th century. Chapter XV provides a good discussion on the development of steam turbines.

17

BOILERS AND HEAT TRANSFER

HOW TO GET UP A HEAD OF STEAM

The sequence of invention and engineering development that we should keep in mind is Ørsted → Faraday → Gramme → Parsons. The invention of the steam turbine and its development, most notably by Charles Parsons in the 1880s, was the key step for the production of large amounts of electrical energy from steam, freeing electricity generation from dependence on the energy of moving or falling water. Indeed, the steam turbine shifts the focus of our concern about electricity generation. Initially we said that our concern centered on finding a way to turn the generator as cheaply and reliably as possible. The 'new' concern—as a result of applying the steam turbine to the technology of electricity generation—is to find a way to generate large amounts of high-pressure, high-temperature steam as cheaply and reliably as possible.

The short answer to the question of how to accomplish this is: boil water! This requires raising its temperature, as shown in Figure 17.1, where the 'energy in' is supplied as heat.

In modern power plants, the 'steam rate,' the amount of steam required to be generated in a given time, is often in excess of a million pounds of steam per hour. (To put this number into some personal context, someone who drives 10,000 miles per year in a car that gets 20 miles per gallon would have to drive the car for *250 years* to consume an amount of gasoline equivalent in volume to the amount of water boiled in *1 hour*.) Our ability to boil water on this prodigious scale is governed by two factors. The first is the rate at which heat is produced in the first place. In steam-turbine plants (except nuclear plants, which we will discuss later) the necessary heat comes from burning the fuel. Therefore the 'heat release rate' is related directly to the rate at which we can burn the fuel. The second factor is the rate at which the heat produced from burning the fuel can be transferred into the water or steam. The 'heat transfer rate' depends on a variety of factors, including the temperature difference between the hot flame and the relatively cool steam, the thickness of the material through which the heat must travel, the ability of that

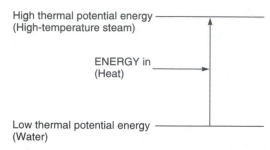

High thermal potential energy
(High-temperature steam)

ENERGY in
(Heat)

Low thermal potential energy
(Water)

FIGURE 17.1 In order to generate electricity, energy in the form of heat must first be put into the system.

material to conduct heat, and the available surface area across which that heat is transferred. Since in most steam-generating equipment the first three factors are, very roughly, the same, surface area is the key to heat transfer.

A steam-generating system consists of two major components: the **furnace**, where heat is liberated by burning the fuel and is transferred to the **boiler**, where water is heated for generating steam. In modern systems, which have evolved through a series of improvements that we will discuss later in the section on boilers, the furnace and the boiler are physically the same piece of equipment. Because of that, we will often use the term 'boiler' to mean both the place where fuel is burned and the place where steam is generated.

Our focus in this and the next chapter will be on power plants in which coal is used as the fuel. There are several reasons for this focus. First, more than 80% of all the coal consumed in the United States is used for generating electricity in power plants. Second, about 60% of the electricity we use is produced in coal-fired power plants. Third, using coal as the fuel provides some challenges for the protecting the environment that are not encountered with oil or natural gas. However, what we encounter in these two chapters (i.e., 17 and 18) will be similar for steam generation from other kinds of fuels.

HEAT TRANSFER

Heat is conducted by three different mechanisms (Figure 17.2). In some SYSTEMS, all three may be operating at once; in others, one may dominate. One mechanism is conduction. When we place a metal object into a source of heat (for example, a metal cooking utensil into a hot pan of food, or a poker into a fire), we notice that the other end—the one we're holding—becomes warm or even uncomfortably hot. This is a result of the conduction of heat through the metal. A second mechanism is convection, the transfer of heat as a result of movement of fluid, often, for example, hot air. A hot stove warms a room chiefly through the convection resulting from movement of heated air from the stove. The third mechanism is radiation. The heat the Earth receives from the sun arrives in the form of radiation. (Since outer space is a near-vacuum, neither conduction nor convection would be effective forms of heat transfer in space.)

In most materials, conduction is a consequence of the motion of the molecules or atoms of the object conducting the heat. (The average positions of the atoms themselves do not change in conduction—only their energy of motion.) We have already seen how the work of Rumford and Davy led to the understanding that heat is a form of motion.

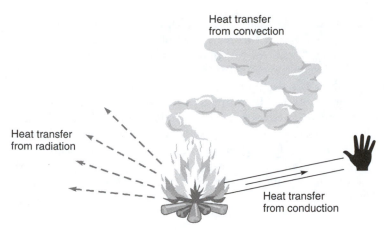

FIGURE 17.2 Heat transfer occurs in three ways: conduction, convection, and radiation.

Atoms at the hot end of a metallic rod—a poker in a fire, for example—vibrate faster and faster as the temperature at that end increases. When these atoms collide with their slower-moving (cooler) neighbors, some of their energy is transferred to the latter. A 'wave' of such atomic collisions allows energy—heat—to travel along the poker. For heat to be conducted through the poker (or any other object), the poker's ends must be at different temperatures. If the entire poker were at the same temperature, all its constituent atoms would have the same average energy, and any one atom would have as much chance of losing energy in a collision with a neighbor as it would of gaining energy from a neighboring atom. Therefore there would be no net flow of energy from a region of rapidly moving atoms (hot end) to an adjacent one of slowly moving atoms (cold end).

There are wide differences in the ability of various substances to conduct heat. Gases are poor conductors, because their molecules are relatively far apart and collisions between them are correspondingly infrequent. The molecules of liquids and nonmetallic solids are closer together, leading to somewhat higher thermal conductivities. Metals have by far the greatest ability to conduct heat. Because of this, it is often the case that metallic objects that must be both heated and handled, such as cooking pans, may have the actual pan made of metal but the handles made of wood or heat-resistant plastic, so that we can pick up the hot pan without being burned.

Convection is much simpler than conduction, since it consists of the actual motion of a volume of hot fluid (which can be either a gas or liquid) from one place to another. Convection may be either natural or forced. In natural convection, the buoyancy of a heated fluid leads to its motion. When a portion of the fluid is heated, it expands, becoming lower in density than the surrounding, cooler fluid, and therefore rises upward. This is the basis of the common expression 'hot air rises.' In forced convection, a blower or pump directs the heated fluid to whatever location we like it to deliver its heat.

When heat is transferred by radiation, the energy is transported by electromagnetic waves. These waves travel at the speed of light, and require no material medium for their passage. An object need not be so hot that it gives off visible light for it to be radiating electromagnetic energy—all objects radiate such energy continuously, whatever their temperature. However, any object will also absorb electromagnetic energy from its surroundings, so that if both the object and its surroundings are at the same temperature, the rates of emission and absorption are the same. When an object is at a higher temperature than its surroundings, it radiates more energy than it absorbs. This excess of radiation is what we detect as heat. The rate of radiation of energy from an object increases rapidly as its temperature increases. A hot body radiates electromagnetic waves of all wavelengths, though the intensities of the different wavelengths may be very different. For example, when we see the glowing embers of a wood fire, or the red-hot coil of an electric stove, enough radiation is emitted as visible light for our eyes to respond. As the temperature of the object increases, the dominant wavelength of the radiation becomes shorter as the temperature of the body increases. A body that glows red is cooler than one that glows bluish-white. Other wavelengths as well are given off, but our eyes are not sensitive to them. Later, when we discuss the problem of global climate change, often called the greenhouse effect, we will come back to the importance of infrared radiation as a mechanism of heat transfer (Chapter 32).

THE BOILER

The burning of the fuel—together with the transfer of its heat—in a modern steam-turbine electricity-generating plant takes place at high temperatures. This is vitally important, because the temperature at which the input heat is supplied has a critical effect on the overall efficiency, as we have seen in the last chapter. As we know, the higher the input temperature, the higher will be the efficiency that can be attained. The highest temperature in the fossil-fired power plant is right in the flame itself, which is around 3,000°F (1,920 K). The hot gaseous products of combustion in the immediate vicinity of the flame are about 2,100 to 2,300°F. These combustion gases actually transfer the heat to the water or steam in the boiler tubes. As this transfer takes place, it is accompanied by a considerable temperature drop. At best, the steam will be only about 950°F (780 K).

The heart of steam generation is the boiler. A *boiler* is, in essence, a device for heating water to convert it to steam. Conceptually the simplest possible coal-fired boiler would look something like that sketched in Figure 17.3. We already know that heat will flow spontaneously from high temperatures (high thermal potential energy) to low temperatures. Thus heat will flow from the hot gases produced by the burning coal—that is, from the fire—to the relatively cool water. The transfer of heat takes place, in this crude cartoon design, across the surface of the water container that is exposed to the heat.

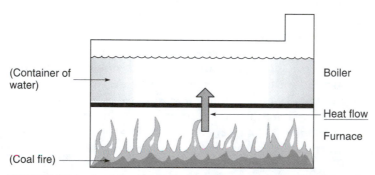

FIGURE 17.3 In this simple boiler, heat is transferred from the furnace across the bottom surface of the water container.

The rate at which we can generate steam is limited by the rate at which heat can be absorbed by the water. The two factors we must consider are the rate at which heat can be produced in the first place, which is determined by the combustion rate of the coal (the heat release rate), and the rate at which heat can be transferred to the water or steam (the heat transfer rate). We will defer the question of heat release rate to the next section. In our simple boiler, steam generation is limited by the heat transfer surface, which is highlighted in Figure 17.4.

This simple design is OK for generating small amounts of steam (such as in a tea kettle) or small amounts of low-pressure, low-temperature steam (e.g., for home heating), but is hopeless for producing the large quantities needed by typical power plants. It is simply impractical for use in power plants because of the immense size of a flat-bottomed boiler that would be needed to produce one million pounds of steam per hour and the great difficulty in distributing the fuel (e.g., coal) evenly in a furnace of such immense size.

The first engineering design improvement in increasing the ability to generate large quantities of steam is to increase the heat transfer surface. We can do this by sending hot combustion gases (using metal pipes or tubes to contain them) right through the water reservoir. This design (illustrated in Figure 17.5) is called a **fire-tube boiler**. In fire-tube boilers, the hot gases pass through tubes surrounded by water. This design represents a substantial increase in the heat transfer surface. For example, if we have a flat-bottom boiler 10 feet on a side, we have $100\,\text{ft}^2$ of heat transfer surface. A 20-foot tall flat-bottom boiler 10 feet on a side with 64 additional two-inch diameter fire tubes provides a heat transfer surface of $670\,\text{ft}^2$—nearly seven times greater. We can do a much better job of heat transfer with a fire-tube boiler. Fire-tube boilers were the kind of boiler used in steam locomotives.

With this design, boilers of a moderate size can be made very much more efficient because of the increased heat transfer surface. Further improvements are obtained by using a large number of small tubes in place of a smaller number of large ones. For example, four tubes that are each two inches in diameter will occupy about the same amount of space in the boiler as one tube of four-inch diameter. However, the four two-inch tubes are more desirable, because their combined surface is twice as large as that of the single four-inch tube. And we can keep going: four times the amount of surface given by the two-inch tubes can be obtained by using 16 one-inch tubes, which thus present a much larger surface. This enables steam to be generated more economically. The use of tubes also enables the heat contained in the hot gases from the incandescent fuel to be

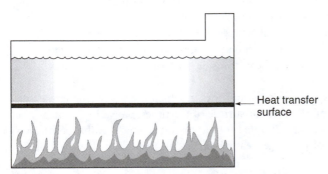

FIGURE 17.4 The heat transfer surface on a basic boiler is very limited.

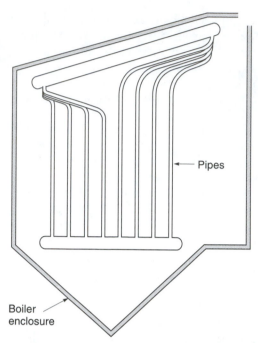

FIGURE 17.5   A fire-tube boiler places the heat in the midst of the water.

better utilized. In the simple flat-bottom boiler design, these gases would otherwise simply go up the chimney, and the full benefit of their heat would be lost.

As the demand for steam (that is, the amount of steam generated per unit time) continued to increase, even the fire-tube boiler could not keep up. There were two limiting factors. The heat transfer rate was still limited by the heat transfer surface, even using multiple fire tubes. The heat release rates required were such that human beings could not shovel coal fast enough. This second factor was addressed by replacing humans with mechanical devices—called stokers—to feed coal to the furnace.

The first issue, the limitation of heat transfer surface, was addressed by, in essence, turning the fire-tube boiler inside out, so that the water or steam is completely surrounded by the high-temperature gases. This design (shown in Figure 17.6) is a **water-tube boiler** or **steam-tube boiler**. The water-tube boiler, with numerous water tubes, is about as good as we can do in increasing the effective heat transfer. We saw how the addition of fire tubes to a hypothetical $10 \times 10 \times 20$ ft boiler could increase the heat transfer surface by a factor of nearly seven, from 100 ft^2 to 670 ft^2. Confining the same volume of water in two-inch-diameter water tubes would provide a heat transfer surface of 4,000 ft^2—a 40-fold increase!

A water-tube boiler consists of a number of tubes in which the water circulates, with the flames and hot gases acting on the exterior surface of the tubes. Nowadays, water-tube boilers have become almost universally adopted for supplying steam to turbines for generating electric power. Because of the large heating surface of the tubes, they are able to raise steam very quickly.

The most economical boilers are compact (thus minimizing heat loss to the surroundings), with a large area of contact between the water and the hot combustion gases.

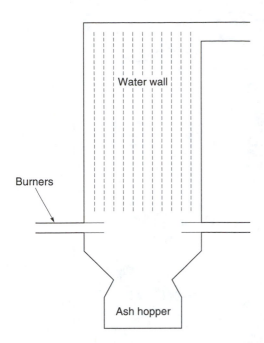

FIGURE 17.6 Water-tube boilers, the opposite of fire-tube boilers, pipe the water through the heat.

The large surface area inherent to the design is further increased by a convoluted system of piping, illustrated in Figure 17.7. The ends of the circulating tubes terminate in so-called 'headers,' which are in turn connected to the steam drum and the water drum. An additional bank of tubes increases the temperature of the steam before it is delivered to the turbines. This additional heating is called 'superheating.' Its use eliminates the moisture present in steam just at its boiling point, and provides significant savings in both fuel and water. Heat is transferred by radiation as well as by conduction and convection. The luminous flame from burning coal emits radiation over a wide range of energy. The combustion gases are highly turbulent. Their turbulence prevents the formation of a stagnant layer of gas, which would not be good for transferring heat, around the water tubes. Radiative transfer is facilitated by a covering of soot or ash on the tubes, but the layer should be kept thin or else it, too, will partially insulate the tubes and decrease the transfer of heat by conduction.

Stoker-fired water-tube boilers worked well in small-sized power plants, and some old ones are still being used today. However, large modern turbines require a great deal of steam at high pressure and high temperature. Even 50 years ago, pressures of 2,000 pounds per square inch and temperatures of 1,000°F (800 K) were common. The large water-tube boilers used to produce steam in electricity-generating stations today are probably the best devices we have for burning fuel and generating steam on the very large scales needed to supply our demand for electricity. What differentiates these boilers from the smaller, stoker-fired units is the method of burning the coal, which will be discussed in the next section. As we will see in the next chapter, the waste products of these huge coal-burning installations can potentially be the source of major pollution problems. Fortunately, there are a variety of technologies that can overcome many of these potential environmental problems.

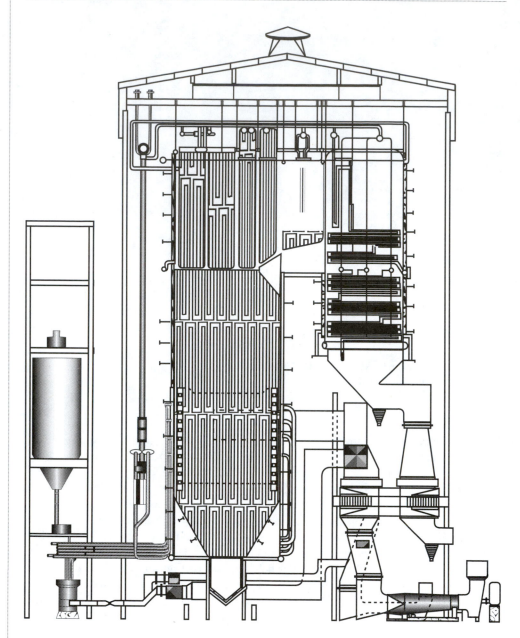

FIGURE 17.7   There is an immense amount of piping in modern water-tube boilers.

HOW TO BURN COAL

As we've seen, the limitations to steam generation are the heat transfer rate (determined by heat transfer surface) and the heat release rate. The water-tube boiler solves the first problem. To address the second problem, we need to think about how coal burns. The burning of the coal is the first step in raising the steam that eventually drives the turbine. Burning releases some of the chemical potential energy that was stored in the chemical bonds in the coal.

He . . . composed in the grate a pyre of crosslaid resintipped sticks and various coloured papers and irregular polygons of best Abram coal at twentyone shillings a ton from the yard of Messrs Flower and M'Donald of 14 D'Olier street, kindled it at three projecting points of paper with one ignited lucifer match, thereby releasing the potential energy contained in the fuel by allowing its carbon and hydrogen elements to enter into free union with the oxygen of the air.

—Joyce[1]

For use in small furnaces, coal from the mine is simply crushed into small lumps. These can then be shoveled into the furnace. For larger furnaces, it is difficult for workers to shovel fast enough or to spread the coal evenly enough. These problems can be overcome by using mechanical 'stokers' to maintain a high combustion rate and an even distribution of heat. Small power plants, or steam plants that supply space heat or process heat for industries, can be well served by a stoker-fired, fire-tube boiler. However, for very large power plants, with steam rates in excess of 100,000 lb/hr, even stoker firing can't keep up. The key to attaining the highest heat release rates is to recognize that coal is a solid, and therefore its combustion is necessarily a chemical reaction at its surface.

For the moment, let's consider that coal is pure carbon. (As we'll see in Chapter 18, the fact that coal really isn't pure carbon causes considerable technical problems in power plant operation.) Using this simplifying assumption for the time being, we can then represent the complete combustion of coal by the equation

$$C + O_2 \rightarrow CO_2$$

Because coal is a solid, the reaction of carbon with oxygen molecules will occur where the oxygen molecules have access to the carbon: on the surface of the coal.

For burning a given amount of coal, the rate of combustion will be limited by the available surface area. Since combustion is a heat-liberating (**exothermic**) reaction, the heat release rate will be governed by the surface area of the coal (via the combustion rate). Therefore, to increase the heat release rate, we must increase the surface area available for reaction. We see examples of the effects of surface area in our daily lives. As examples, we might compare the experience of dissolving crystals of sugar in coffee or tea against the slowness of dissolving a sugar cube, or how we increase the dissolving of a bouillon cube in hot water by mashing it with a spoon. Suppose we had a cube that was one inch on a side. The total surface area of this cube is 6 in^2. Now suppose that we were to divide the cube into smaller cubes each 0.01 inch on a side. The new cubes have a surface area of only 0.0006 in^2. However, there are now one million of these tiny cubes that could be produced from our original one-inch cube, so the *total* surface area in the system is 1,000,000 × 0.0006 in^2 = 600 in^2. In other words, we have increased the surface area by a factor of a hundred! Roughly, therefore, if we did the same thing to a one-inch lump of coal, we would increase the burning rate, and the heat release rate, 100-fold also.

In actual operation, the burning of coal is far more complex than we have suggested it to be by considering coal to be a pure, solid carbon. Furthermore, the burning of coal is vastly more complex than the combustion processes for natural gas or oil. All coal inevitably contains some moisture. When the coal is introduced into the furnace and begins its heat-up to combustion temperature, the water evaporates. Some of the heat energy from the furnace is used up in evaporating this moisture. At still higher temperatures, the molecular structure of the coal begins to break down under the influence of heat. A variety of volatile, gaseous products is produced by this breaking down, and many of these gases are combustible. They will, therefore, ignite and burn in a cloud around the coal particle. The proportion of the coaly material that breaks down to produce these combustible volatile products varies from one kind of coal to another. The solid that remains after the volatiles have been driven out is a char. The char is *mostly* carbon. It will burn according to the equation we've already used:

$$C + O_2 \rightarrow CO_2$$

However, all coals also contain some amount of incombustible minerals. As the coal burns, these minerals are transformed by the heat to the material we recognize as ash. When the volatiles and char have burned away completely, the ash remains. Of course we would like to extract the maximum of the chemical potential energy of the coal as heat. To do this, both the char and the volatiles have to be burned completely, and at about the same rate.

Using finely pulverized coal provides a solution to obtaining the greatly increased heat release rates. Unfortunately, it also introduces a mechanical design problem. In furnaces that burn lumps of coal, the coal is supported on a grate that holds the burning coal off the floor of the furnace and allows air to circulate through the burning coal bed. With pulverized coal, there's no way to support it on a grate. Either the powdered coal would sift through the grate, or it would be so tightly packed on the grate as to limit the access of air and negate the benefits of pulverizing it in the first place. The solution to this problem is to blow the pulverized coal, with air, into the boiler. This approach is called suspension firing or, often, **pulverized coal firing**, illustrated in the sketch in Figure 17.8. Coal is fed into a hopper from a coal bunker, and from the hopper into a pulverizing mill. Most of the moisture is removed from the coal by hot air that enters through a pipe at the lower portion of the pulverizer. An exhaust fan then draws the pulverized coal into a separator. The pulverized coal falls to the bottom of the chamber. The air drawn through the fan is discharged either into the atmosphere or into the furnace. The pulverized coal then passes through an air lock into a storage bunker. The bottom of the storage bunker is connected through a rotary feeder to a duct, at one end of which is another air fan. This fan draws in hot air and drives the powdered fuel into the burner. The burner is cone-shaped and has spiral grooves, much like the rifling of a gun barrel, so that the pulverized coal and air are blown into the boiler in a swirling, turbulent cloud.

An air-blown stream of very fine coal-dust particles (of diameters between 1/500 and 1/1,000 of an inch) enter the hot combustion chamber. As soon as they do, any

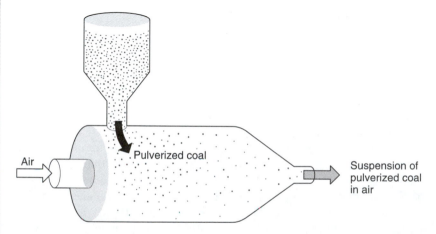

FIGURE 17.8   A conceptual sketch of pulverized coal firing shows how it must be suspended as it enters the combustion chamber.

remaining moisture evaporates, and then, as the particles are heated, the volatile products are released and burned. A second stream of air is often admitted to the boiler to complete the burning of the char. The char and volatiles burn together, so that most of the heat is released in the same region of the furnace. Rather large furnace chambers are needed for complete combustion of the coal. It is not unusual for a boiler in a modern coal-fired plant to be ten or more stories tall.

The first pulverized coal-fired boiler was installed in 1926. This is now the standard practice for coal combustion in large-scale power plants.

Very roughly, a coal-fired steam-turbine generating plant will operate at about 30% efficiency for conversion of the chemical potential energy of the coal into the electrical potential energy (high-voltage electricity) leaving the plant. Certainly the loss of 70% of the heat energy of the coal is high, but it is still a major improvement over early plants. Edison's Pearl Street Station in 1882 is said to have used about 10 pounds of coal per kilowatt-hour of electrical energy generated. By 1900 it was typical that about 7 pounds were required to generate a kilowatt-hour, with this requirement dropping to just over 2 pounds by the 1920s. Nowadays, a kilowatt-hour can be generated with less than a pound of coal. Nevertheless, there is obviously still plenty of room for improvement. One reason for the relatively low efficiencies of approximately 30% is that our present technology takes such a roundabout route to converting the chemical potential energy stored in the coal to electrical energy. Consider this (Figure 17.9): first the chemical potential energy is released as heat; then the heat energy is used to produce high-temperature (high thermal potential energy), high-pressure steam; next the thermal potential energy of the steam is converted into mechanical work in the turbine; finally, the work of the turbine is used in the generator to produce the high electrical potential. With this long sequence of operations, it is inevitable numerous losses and inefficiencies occur.

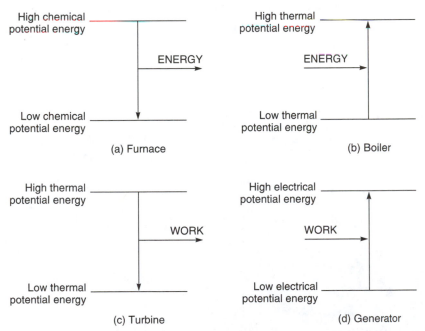

FIGURE 17.9 This quadruple energy diagram shows the conversion of high chemical potential energy to high electrical potential energy.

There ought to be better, more direct ways of producing electricity from a primary energy source; perhaps in this new millennium we will see some of them realized.

CITATION

1   Joyce, James. *Ulysses*. Folio Society: London, 1998; pp. 626–627. Many other editions of this extraordinary novel are available.

FOR FURTHER READING

Babcock and Wilcox Company. *Steam—Its Generation and Use*. Babcock and Wilcox: New York, 1978. A comprehensive book on various types of boilers and the relevant information on heat transfer, principles of combustion, and pulverized coal firing. The book assumes a solid background in science and mathematics.

Cook, A.F. *Raising Steam on the LMS*. Railway Correspondence and Travel Society: Huntingdon, UK, 1999. Likely the best available book on the evolution of design of fire-tube boilers, their operation and use. The main focus is on fire-tube boilers mounted on wheels (i.e., steam locomotives).

de Lorenzi, Otto. *Combustion Engineering*. Combustion Engineering: New York, 1947. A comprehensive book on fuel combustion and boiler design. Though some of the material is now quite out of date, there are many excellent illustrations of different types of boilers.

Hawks, Ellison. *Popular Science Mechanical Encyclopedia*. Popular Science: New York, 1941. The first chapter, 'The Giant Power of Steam,' includes a discussion, with useful illustrations, of various types of boilers.

Singer, Joseph G. (Ed.) *Combustion: Fossil Power Systems*. Combustion Engineering: Winslow, CT, 1981. Like the book from the Babcock and Wilcox Company listed above, this is another comprehensive treatment of combustion processes, boiler design, and raising steam. Intended mainly for readers with a background in principles of science and engineering.

18

ELECTRICITY FROM COAL

We've now seen that the standard approach to the generation of large amounts of high-pressure, high-temperature steam in coal-fired power plants is to use a pulverized coal fired water-tube boiler. It's important to remember that the burning of coal is by no means the only source of heat that can be used to raise a head of steam. Any fuel—natural gas, oil, or wood, for example—can be used. In later chapters we'll see how the heat from nuclear reactions or heat from sunlight can also be used to make steam. Our focus will be on coal for three reasons. First, the majority of electricity produced in the United States, roughly 60%, is generated in coal-fired plants (e.g., Figure 18.1). Second, combustion in electricity plants is by far the dominant use of coal in the United States, accounting for about 80% of annual coal production. Third, this application of coal has in the past caused a variety of environmental problems that have caused us to develop a range of technologies to reduce or prevent their occurrence. Let's now see how we put together a power plant.

A half-century ago, power plants were fairly small—on the order of 5 MW for example—and served a town or city in the immediate vicinity. However, after the Second World War, new plants were constructed to larger and larger sizes. Nowadays, a typical mid-sized unit might generate (in round numbers) 400 MW of electricity and consume 10,000 tons of coal a day. (To put this into perspective, a single railroad car of coal—which typically amounts to a hundred tons of coal—supplies *either* enough coal to heat a modest two-story wood-frame house during the winters for 10 *years*, *or* enough coal to operate a typical modern power plant for 15 *minutes*.) Rather than just providing electricity to people in the region, new plants provide electricity to regional or national power grids, so may serve customers hundreds of miles away.

Along with the growth in capacity of the plant has come a steady increase in requirements for the amount of steam, as well as its temperature and pressure. Some 75 years ago, a boiler that was considered modern by the standards of the time would provide steam at, for example, 650 pounds per square inch (psi) and 725°F. Fifty years ago, boilers were producing steam at 2,000 psi and 1,000°F. Nowadays modern boilers

FIGURE 18.1 The Portland Generating Station in Upper Mount Bethel Township, Northampton County, Pennsylvania is a modern coal-fired power plant.

convert prodigious quantities of water into steam, in some cases several million pounds of water are boiled per hour. The amount of water that's boiled in 1 hour in a boiler with a steam rate of 3,200,000 pounds per hour can be visualized like this: If you have a 20-gallon gasoline tank in your car and used gasoline as fast as that boiler raises steam, you would suck the tank dry six times per *second*!

Today some very large plants have outputs over 1,000 MW. However, in those cases, plants are built with two or more generating units of, say, 500 MW capacity in parallel to generate the total plant output, rather than relying on a single gigantic turbine and generator.

THE BOILER

The crucial point is that any steam turbine needs a head of steam, and the essence of electric power generation is to find a way of raising that head of steam as reliably and economically as possible. We've seen in the last chapter how boilers have evolved from the simple concept of a tea kettle and work on the same principles. Even though a modern power-plant boiler is vastly larger than a simple tea kettle, the same fundamental issues apply to each. There has to be a source of heat, as hot as possible (think of how when we're in a hurry to get hot water, we all have the tendency to turn the stove setting to 'high'). This is the high heat-release rate. Further, there needs to be excellent contact between the heat source and the water or steam, to assure a high heat-transfer rate.

It's important not to waste the heat generated inside the boiler (but as we'll see later in this chapter, a lot of it is lost anyway, even in the best of designs). For this reason, boilers are configured to enclose as large a volume as possible while presenting a relatively small surface to the outside world. This helps minimize the heat loss from the boiler to its surroundings. The large heat-transfer surface is achieved by the water-tube design. We've seen in the previous chapter that a water-tube boiler uses a number of tubes in which the water or steam circulates. The hot combustion gases act on the exterior surfaces of the tubes. One of the advantages of this design (aside from the fact that the large heat-transfer surface of the tubes allows the boiler to raise steam very quickly) is that, owing to their particular construction, they are relatively compact and hence minimize heat loss. Transfer of the heat from the coal flame to the water or steam is achieved by radiation and by convection.

The boiler design that seems to do the best job at accommodating all these factors is a big rectangular box. Depending on the size of the plant (the generating capacity) and the calorific value of the coal used, it might be 10–20 stories tall. At least one entire wall of the boiler will be made of water tubes welded together with a steel 'membrane,' as sketched in Figure 18.2. This is called the **water wall**.

The boiler will be equipped with numerous burners, usually four or more. They are mounted in the walls so as to produce a flame pattern that is very turbulent, to assure good mixing of the coal particles and air. The flame temperature might be around 3,000°F (2,000 K). In many designs, the air that carries the coal into the burners (called the primary air) may not be sufficient to assure complete combustion of the coal. Additional air (called secondary air) may be supplied a little higher in the boiler. Inside the 'box' most of the heat is transferred by radiation.

Coal is pulverized and driven into a burner and so into the boiler, as described in Chapter 17. The pulverized coal is ignited by the hot 'fireball' of burning coal already in the boiler. (The fire is started in the first place by blowing in the pulverized coal and using a kerosene torch, or even rags soaked in kerosene, for ignition.)

All coals will contain some amount of incombustible materials that remain behind as an ash. Some provision is made to catch and collect the ash residue from the coal in an ash hopper. The basic components are laid out as sketched in Figure 18.3.

The combustion gases leaving the furnace chamber and passing into the flue are still very hot (e.g., 1,000–1,200°F). To capture some of this heat, additional steam tubes are mounted in the flue. Here, heat is transferred mainly by convection. Thus the full system looks like Figure 18.4.

Water tube

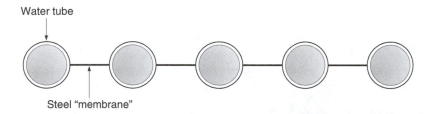

Steel "membrane"

FIGURE 18.2 Cut-away view of water from above, showing the tubes and steel membrane.

THE BY-PRODUCTS OF COAL COMBUSTION—INTRODUCING ENVIRONMENTAL ISSUES

Up to now, we've assumed for simplicity's sake that coal is pure carbon. Unfortunately, that's an oversimplification that we can no longer rely on. As we take coal from the ground, it consists of (1) a carbonaceous portion, which may contain 70–95% carbon, 2–6% hydrogen, 2–20% oxygen, and some nitrogen and sulfur; (2) a noncombustible inorganic portion, consisting of minerals that accumulated with the carbonaceous portion, and any rocks or minerals that might accidentally have been mixed with the coal during mining, and (3) some amount of moisture, much of which is removed by drying during the pulverizing process. One way of remembering the constituents of coal is by the mnemonic NO CASH, in which the letters stand for nitrogen, oxygen, carbon, ash (which is, strictly speaking, not a constituent, but rather a product of heating the inorganic portion of coal), sulfur, and hydrogen.

The molecular composition and structure of coal has been a subject of intense research and considerable debate. The molecular structure varies from one kind of coal to another, and likely even among coals of the same kind. Luckily, we need not worry about the molecular structure of coal, because we can treat the combustion of coal as if it were a simple mixture of the elements. Thus, for complete combustion we can write

$$C + O_2 \rightarrow CO_2$$
$$4H + O_2 \rightarrow 2H_2O$$
$$S + O_2 \rightarrow SO_2$$
$$2S + 3O_2 \rightarrow 2SO_3$$
$$2N + O_2 \rightarrow 2NO$$
$$N + O_2 \rightarrow NO_2$$
$$\text{Minerals} \rightarrow \text{Ash}$$

Both sulfur and nitrogen form more than one oxide. Often, we can simplify this situation by not worrying about which specific oxide has formed. Rather, we lump all the oxides together as SO_x (pronounced 'socks') and NO_x (pronounced 'knocks'). The noncombustible minerals in the coal undergo a variety of chemical and physical transformations to form the ash residue. The ash is sometimes humorously referred to as RO_x ('rocks').

If these products of coal combustion could simply be released into the air and thoroughly diluted into the total volume of the atmosphere, the potential harm they cause would not be very apparent. (This gives rise to the somewhat facetious—and false—statement, 'The solution to pollution is dilution.') But being able to 'dilute' pollutants to levels at which they appear to be harmless is seldom pos-

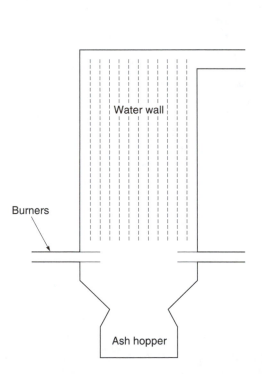

FIGURE 18.3 Cut-away frontal view of boiler, showing the main three components: water wall (here the tubes are represented by dashes), burners (through which the pulverized coal is fired), and the ash hopper.

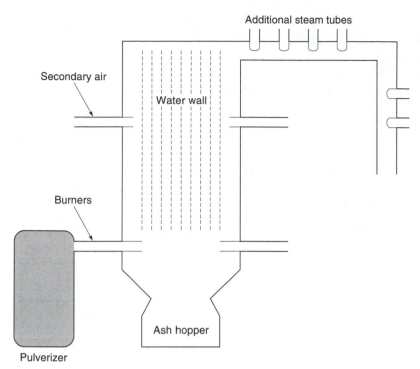

FIGURE 18.4 The boiler's full system.

sible. In many cases, these combustion products are generated and released in or near large centers of population. In many such cases, the local geography and the prevailing weather patterns make it difficult to achieve an extensive dilution of emissions. Consequently, cities or regions may suffer severely from various kinds of atmospheric pollution. In British and Australian slang, large cities such as London used to be called 'the big smoke.'❀ This term derives from the very high concentration of coal-burning equipment, especially household stoves and fireplaces, in the comparatively small area of a city. Many of these devices did not burn coal very efficiently or effectively, producing a smoky fire; when thousands upon thousands of these smoke-producers were concentrated in a city, the total smoke emission was enormous.

> ❀ As early as 1860, British tramps used the term 'the smoke' in their argot to refer to London. By 1900, 'the great smoke' or the 'the big smoke' had become colloquial in Britain. For example, one might speak of having to 'go up the smoke' as meaning to have to go to London. Meanwhile, Australians adopted 'the big smoke' as a slang term for any city during the mid-19th century; by about 1920, it was usually taken to refer to Sydney.

Concerns about the impact of coal combustion on the environment have a remarkably long history. Seven centuries ago, in 1306, the English king, Edward I, banned the burning of 'sea coles' while Parliament was in session. The penalty for breaking this law was to ensure that the offender would never do it again, by the simple, but effective, process of cutting off his head. Throughout much of the intervening 700 years, histories, memoirs, and novels, especially in Britain, have made occasional reference to the smoke, stench, and dirt that were associated with coal.

Unfortunately, there seems to be a tendency not to take action about environmental problems until it is we ourselves who are directly threatened by the problem. Two air pollution disasters about a half-century ago helped focus public concern on the possible consequences of emissions from coal combustion. The first occurred in October of 1948, in Donora, Pennsylvania. Highly polluted air was trapped over the town for days, and contributed to about 20 deaths. Far worse was the so-called 'Black Fog' disaster in London in December of 1952, which killed about 4,000 people in the course of several days.

London has a long history of notoriety for its 'pea soup' fogs. This type of air pollution problem results when smoke (that contains tiny particles of soot and ash) and SO_x from burning high-sulfur coal are emitted into a damp, cold atmosphere. The relatively cool water vapor condenses around the particles. This condensation results in a suspension of water droplets in air, a condition that is a fog. Both oxides of sulfur readily dissolve into the water droplets. Sulfur dioxide dissolves to form the relatively weak acid, sulfurous acid,

$$SO_2 + H_2O \rightarrow H_2SO_3$$

Sulfur trioxide, on the other hand, dissolves to form sulfuric acid, which is a quite strong acid,

$$SO_3 + H_2O \rightarrow H_2SO_4$$

The situation is made all the worse by the interaction of sulfur dioxide or sulfurous acid with oxygen in the air,

$$H_2SO_3 + \tfrac{1}{2}O_2 \rightarrow H_2SO_4$$

The conversion of the dioxide to the trioxide, or sulfurous to sulfuric acid, seems to be facilitated by the components of the ash particle around which the droplet formed. The net effect is to produce a thick fog, not of water, but of dilute sulfuric acid. When these droplets of dilute sulfuric and sulfurous acids enter the human respiratory tract, as would unavoidably happen to anyone outdoors in such a situation, this exposure can be extremely harmful. The sulfur oxides or the corresponding acids irritate and attack the mucous membranes and sensitive tissues that line the respiratory tract. The effects of SO_x exposure include rhinorrhea, a form of excessive mucous discharge from the nose; bronchoconstriction, a decrease of the diameter of the pulmonary air passages; and destruction of the mucous membranes. Three classes of people are especially vulnerable to air pollution: those whose respiratory tracts have not yet fully developed, meaning young children; those whose respiratory tracts are beginning to wear out, meaning the elderly; and those whose respiratory tracts aren't working very well anyway, meaning those with chronic respiratory problems.

This type of air pollution problem was so severe that a special word was invented to describe it. The seriousness comes from the fact that the smoke particles cause the formation of the fog and help convert SO_2 and H_2SO_3 to the more formidable H_2SO_4. That is, it's the smoke plus fog that makes things so nasty, and gives us

$$SMoke + fOG \rightarrow SMOG$$

These pea soup smogs have been described at least since the time of the first Queen Elizabeth, at the end of the 16th century.

After the 1952 tragedy in London, the British Parliament began to take action on regulating emissions from coal combustion. The original legislation was followed by further clean air acts in 1956 and 1968. The United States was not so quick, but did enact clean air legislation in 1970, followed by amendments in 1977 and in the 1990s. The tragic smogs that affected Donora and London *were not* the result of coal burning for electricity generation. In London particularly, part of the problem was the result of millions of homes, each with coal stoves, fireplaces, and furnaces, burning relatively high-sulfur coal and pumping out both smoke and SO_x. However, these air quality laws have had a direct impact on the emissions from coal burning in power plants and were significant in the impetus to develop some of the technologies we will discuss in the following sections.

There are no pea soup smogs any more. Requirements for using smokeless fuels and increasing the efficiency of coal combustion to reduce smoke formation has taken away one-half of the 'equation' for smog formation. Putting an end to the dangerous threat of pea soup smogs is one example of how environmental regulations or legislation actually can be effective.

SULFUR OXIDE EMISSIONS

SO_x and NO_x are of concern because of their role in the formation of acid rain (Chapter 30). Rain that is highly acidic has been falling for decades in the industrial regions of the United States, particularly in the northeast; in the neighboring parts of Canada, where the smelting of sulfide ores may be an important contributor to the problem; in England, especially in the industrialized north; and in heavily industrialized parts of Europe. It is not uncommon for this rain to be 10 times more acidic than ordinary rainfall in unpolluted areas. If the pollution is particularly severe, the rain may be 1,000 times more acidic than normal rain. Acid rain harms crops and forests, rivers and lakes. As it accumulates in lakes and causes them to become increasingly acidic, it severely disrupts aquatic ecosystems. It damages buildings, corrodes statues and monuments, and attacks metal.

It is important to know that coal combustion—indeed the combustion of fuels of any kind—is by no means the only source of SO_x emissions. Sulfur oxide emissions caused by human beings, called **anthropogenic** SO_x, arises from two main sources. One is indeed the burning of a sulfur-containing fuel, usually coal or a fuel oil. The other occurs in the metallurgical industry, when sulfide ores of metals are 'roasted' to produce the oxide, for example,

$$NiS + \tfrac{3}{2}O_2 \rightarrow NiO + SO_2$$

There are also natural sources of SO_x, such as volcanoes. The amount of sulfur released into the world's atmosphere annually, from all sources, is about 182 million metric tons. Anthropogenic sulfur emissions are 25% of the total worldwide, but in densely populated urban areas can account for up to 90% of the total sulfur emission. With regard to the anthropogenic sulfur emissions, the United States and China are the main 'bad guys,' accounting together for 75% of the total.

There are several approaches to mitigating SO_x emissions. One is **coal cleaning**, which removes some (but usually not all) of the sulfur from the coal. A second is to switch to burning a coal that has less sulfur than the one presently being used in the plant. This approach may bring with it some possible problems, because some boilers have been built to fire coals with a narrow range of specifications. A third approach is to switch the plant to burn an entirely different kind of fuel of lower sulfur content, such as natural gas. This option can be extremely expensive, since it may involve rebuilding a considerable section of the boiler. All three of these are precombustion strategies, since they involve doing something before the coal is burned to reduce SO_x formation. The alternative is to deal with the SO_x after it has formed, the approach of postcombustion clean-up.

The first approach to dealing with the problem was one mentioned above—the unfortunate idea that 'the solution to pollution is dilution.' In practice, this meant building very tall smokestacks on power plants. If the SO_x and other combustion products were lifted high up into the air and then caught by prevailing winds, with a bit of luck (luck for the firm causing the problem, that is), they might not come back down for many hundreds of miles. The unlucky recipients of the pollution falling back to earth would have a nearly impossible job in tracing the problem unequivocally back to its source. The acid rain can cause damage in locations that had nothing to do with its generation in the first place. Furthermore, the 500 or more miles that the pollution travels is often enough to carry it across national boundaries, sometimes exacerbating tensions between countries.

The present clean air standards require that sulfur dioxide emissions be reduced by 90%. The best available technology to achieve that is **flue-gas desulfurization**, often called simply FGD. The hardware or device in which FGD is accomplished is called a **scrubber** (Figure 18.5). Scrubbers were the FGD technology specified in the 1977 Clean Air Act

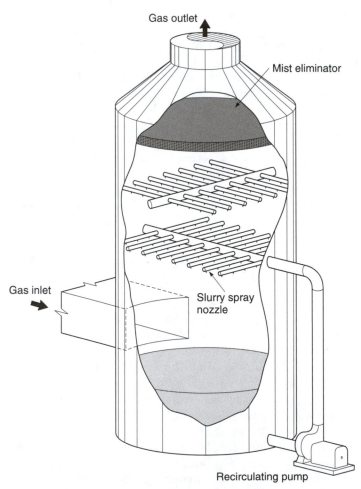

FIGURE 18.5 Scrubbers designed like this can significantly reduce SO_x emissions released into the environment.

Amendments. In the scrubber, the combustion gases pass through a spray of wet alkaline solution or slurry, usually of lime or limestone. The scrubber takes advantage of the fact that SO_x dissolves in water to produce acids, sulfurous and sulfuric. Because lime and limestone are alkaline, they will react with the acid. Lime and limestone offer two other advantages. First, they are about the cheapest alkaline substances available. Second, the reaction products, calcium sulfite and calcium sulfate, are not soluble in water, and so form a precipitate sometimes called 'scrubber sludge.' The reactions can be thought of as the following:

$$CaO + SO_2 \rightarrow CaSO_3$$
$$CaO + SO_3 \rightarrow CaSO_4$$

The most important message about scrubbers is that they work—they work very well indeed. In the last quarter of the 20th century, when the use of coal in the United States roughly doubled (so it might not be unreasonable to expect that sulfur emissions would also double) sulfur emissions actually *dropped* by 30%. But, the other important message about scrubbers is that they do not *destroy* pollution. The sulfur isn't gone. It has only been converted from a very dilute form hard to handle—SO_x in the combustion gases—to a much more concentrated form that is easier to handle—scrubber sludge. In fact, scrubbers produce a lot of sludge. A 1,000 MW plant, with, say, two 500 MW boilers side-by-side, each with a scrubber, could produce enough scrubber sludge to coat a square mile of land a foot deep. That results in a serious problem for getting rid of the sludge. Furthermore, scrubbers use a lot of water. Using this same 1,000 MW hypothetical plant as an example, the water consumption in the scrubbers would be about a thousand gallons per *minute*. This will be of increasing concern in areas facing water supply problems.

There is a partial solution to the problem of what to do with scrubber sludge. The calcium sulfate that forms is usually associated with two molecules of water ($CaSO_4 \cdot 2H_2O$) as the material commonly known as gypsum. One use of this material is in the manufacture of sheet rock or wallboard, and indeed scrubber sludge has been used in this application. Gypsum also has other uses, in cement manufacture and as a soil treatment, for example. The income from sale of some of the by-product gypsum could offset some of the costs of operating the scrubber. However, doing so means that the power plant must take on some aspects of being a chemical manufacturer, something they may be reluctant or unable to do.

Some utilities have elected to rely on low-sulfur coal. Some coals, called 'compliance coals' have low enough sulfur content that their use can reduce significantly the need to install scrubbers. As a rough rule-of-thumb, low-sulfur coals in the United States are located west of the Mississippi River, and high-sulfur coals are in the east. Therefore, a utility in the east would face significant transportation costs to buy the low-sulfur coal in the first place. This recurring cost for low-sulfur fuel has to be balanced against the cost of a scrubber and the lime or limestone needed in it. In some cases, the savings on capital equipment may more than balance the extra transportation costs for the low-sulfur coal; in other cases, perhaps it would be better to buy the scrubber.

Alternate fuels are sometimes considered to replace coal, a strategy called fuel switching. Both natural gas and petroleum products are more expensive fuels than coal, but using them would reduce SO_x emissions because they usually have little or no sulfur content. Ethanol, methanol, wood, or biomass—fuels that we will consider later, in Chapter 34—could also provide substantial reductions in SO_x emissions, but they too are more expensive than coal, and in some cases yield substantially less heat per pound of fuel burned. One option that is emission-free and close to coal in cost is nuclear energy.

NITROGEN OXIDE EMISSIONS

NO_x is more difficult to deal with than SO_x. First, there are actually two sources of NO_x in a combustion system. **Fuel NO_x** is NO_x generated from the reaction of nitrogen atoms chemically incorporated in the fuel molecule; **thermal NO_x** is generated from the air used for combustion when molecules of nitrogen and oxygen react at the high temperatures of the combustion system:

$$N_2 + 2O_2 \rightarrow 2NO$$

Virtually any time a high-temperature combustion process occurs, some thermal NO_x formation will take place.

Air is about 80% nitrogen. Whenever the temperature of air is raised above about 900°F, the nitrogen in the air burns to form nitrogen oxide (NO), as shown above. In the presence of more air, NO is quickly converted into nitrogen dioxide:

$$NO + \tfrac{1}{2}O_2 \rightarrow NO_2$$

Nitrogen dioxide is toxic. In the presence of water droplets, it dissolves to produce nitric acid, the other component of acid rain. Nitrogen dioxide is brown; its presence in air that is badly polluted gives the sky a sickly brownish color. Because of the formation of thermal NO_x in this way, even if it were possible to reduce the nitrogen content of a fuel to 0.000% there would still be NO_x emissions from a combustion system. (In contrast, if the sulfur content of a fuel were reduced to 0.000%, there would be no SO_x emissions at all.)

There are several reasons why NO_x is hard to deal with. First, there is no easy, cheap way to remove nitrogen from coal. Much of the sulfur in coal is present as the mineral pyrite. Pyrite is more dense than the carbonaceous portion of the coal; when coal is pulverized the pyrite is relatively easy to separate by taking advantage of this density difference. Nitrogen, on the other hand, is chemically incorporated in the molecular structure of the coal. Second, we've seen that scrubbers work well in part because the sulfur is tied up as calcium sulfate, which is not soluble in water and forms a precipitate or 'sludge.' An analogous system for 'flue gas denitrogenation' would not work this way, because all nitrate salts are soluble in water. Third, thermal NO_x formation is almost inevitable, and the higher the combustion temperature (which we want, both to drive heat transfer and to increase the steam temperature to the turbine), the more thermal NO_x is formed.

Today, there is no standard commercially available NO_x reduction or removal method for coal-fired power plants. However, the development of approaches to NO_x reduction or removal is a vigorous subject of research and development, for example, in the U.S. Department of Energy's Clean Coal Technology Program. In Europe and Japan, many boilers are equipped with a selective catalytic nitrogen oxide reduction system, where ammonia is mixed with the nitrogen oxides in the stack. This process produces harmless nitrogen and water vapor:

$$2NH_3 + NO + NO_2 \rightarrow 3H_2O + 2N_2$$

This process is gaining ground in the United States. American efforts to reduce NO_x have also focused on the combustion process itself. New, so-called low-NO_x burners are being designed in which the fuel and air mixture and the temperatures attained are more care-fully controlled to lower NO_x production.

ASH AND PARTICULATE EMISSIONS

In a pulverized-coal-fired boiler it is hoped that the ash residue from the coal would fall to the bottom of the boiler, where it could be removed via the ash hopper. Some 80% or more of the ash does that. However, the extremely turbulent fireball results in some ash particles being carried out of the boiler with the flue gas. These ash particles are called **fly ash**. If they get out into the environment they are also termed **particulate emissions**. Particulate emissions, which can include soot (from incomplete combustion) and coal dust as well, are undesirable for several reasons, including the aesthetics of having fine ash particles deposit on objects outdoors, their potential of exacerbating respiratory problems such as emphysema, and the possibility that the tiny ash particles could also carry car-cinogens into the lungs.

Particles in the stack gases of coal combustion vary in size from 0.01 microns (μm) to 10 μm in diameter. (A micron is one-millionth of a meter and is about 0.00004 inch.) The smallest particles, in the range 0.01 to 1 μm, are the most dangerous to health. If they are inhaled, they can be trapped in fine passages of the respiratory system and add to pollution-related respiratory distress. First of all, particulates lodged in the lungs can irritate and damage the lungs. Second, the particulates have large surface areas with many tiny cracks and crevices and therefore can absorb other pollutants on their surface or inside these crevices. When these particles enter the lungs, the pollutants are released at a higher concentration to the surrounding tissue than if they were simply inhaled directly from the atmosphere. This could exacerbate a health problem in the sick individual and, in extreme cases, could even be fatal.

Particulates are released when coal burns, unless they are trapped or removed from the effluent smokestack gases. The basic approaches to reducing particulate emissions are a baghouse, which uses a fine fabric filter to trap the ash particles (crudely, working like a gigantic vacuum cleaner) and an electrostatic precipitator (ESP), that uses a high-voltage electric field to give an electric charge to the fly ash particles, trapping them on high-voltage plates. A third alternative is the cyclone collector, which whirls the combustion gas around like a miniature cyclone or tornado, forcing the entrained particles to the walls by centrifugal force.

Electrostatic precipitators have long been used to remove suspended particulates from coal smoke. The particles become ionized by a strong electric field between two electrodes of an electrostatic precipitator. Once ionized, the particles are drawn to and captured on the electrodes, as shown schematically in Figure 18.6. Periodically, the par-ticles are removed from the electrodes and disposed of.

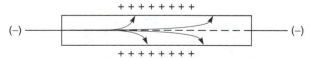

FIGURE 18.6 Section of an electrostatic precipitator, showing how the ionized coal smoke particles are drawn toward the charged electrodes.

Every coal-burning plant produces ash as a waste product (both fly ash from the flue and 'bottom ash' from the ash hopper on the boiler). Somehow this ash has to be disposed of. Even though ash is only a small percentage of the coal burned, the absolute amounts of ash produced are huge because of the massive quantities of coal used. If a plant burns 10,000 tons of coal a day, which is not an unreasonable figure for a large plant, even if the ash yield from the coal is 10%, then 1,000 tons of ash have to be disposed of every day. One obvious solution is to send the ash to a landfill, but this requires a significant amount of space, either adjacent to the plant or at least somewhere within range of low-cost transportation for the waste. Landfills are becoming increasingly difficult to find, especially when a plant is located in a densely populated region. Some ashes can be returned to the mine for use in mined-land reclamation. Fly ash can also be used to make concrete, and scrubber sludge can be used to make gypsum wallboard. Periodically, interest surfaces in extracting valuable metals from the ash, for instance, gallium and germanium, which could be used in the solid-state electronics industries, as well as cobalt and chromium. Another concept is to refine the scrubber sludge, rather than disposing of it, to recover elemental sulfur or sulfuric acid. If appropriate processes could be developed that would allow utilities to sell the ash and sludge as by-products and make money doing so, rather than paying a cost for disposal of these wastes, the net financial gain could perhaps reduce environmental control costs overall, or even pay the costs of environmental compliance.

THE OVERALL PLANT LAYOUT

The overall layout of a coal-fired power plant might look like Figure 18.7 (neglecting the turbine and generator portion of the plant). The combination of scrubber and ESP or

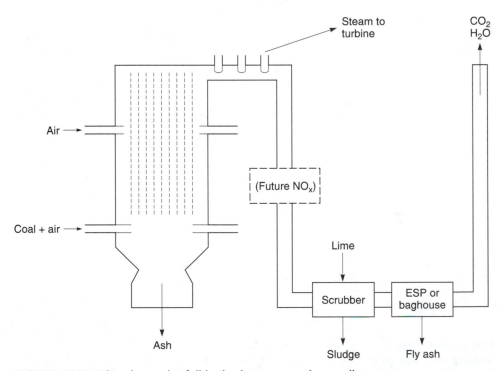

FIGURE 18.7 This shows the full boiler layout to reduce pollutants.

baghouse, together an appropriate NO_x control strategy, would make it seem that we have removed or reduced all of the harmful pollutants. Only carbon dioxide and water go 'up the stack.' In recent years, however, there has been growing concern about carbon dioxide emissions and their role in the environmental problem called the greenhouse effect. We will return to this issue later (Chapter 32).

After doing its WORK, the steam leaving the turbine passes through large condensers using cooling water, and the resultant condensate (condensed steam) is reused as feedwater in the boiler. This system is sketched in Figure 18.8. The low-pressure, low-temperature steam that is leaving the turbine must be condensed back to water, using a device called a **condenser**. To handle the large amounts of steam used in today's plants, many a condenser has more than an acre of cooling surfaces. Cold water from a natural source, a river, a lake, or the ocean, is brought in to extract heat from the steam. The loss of heat from the steam causes the cold condenser water to warm up, or even get hot because the cooling water that passes through the condenser absorbs the heat given off as the steam condenses back to water.

The hot water from the condenser cannot be discharged back to the environment as-is. Doing so could, potentially, upset the local environment by causing an unnatural growth of aquatic plants or some kinds of aquatic animals. The hot water from the condenser has to be cooled back to ambient temperature before it can be discharged. This cooling is done in a **cooling tower**. Thus the full system looks like that in Figure 18.9. Nowadays power plants are generally built close to rivers, lakes, and oceans, which provide a ready source of water for cooling.

The efficiency of conversion of chemical potential energy in the coal to electrical potential energy leaving the plant is about 35%. This figure depends of course on the specific plant, and may get close to 40% in a very modern and well-run plant and dip into the 20% range in old plants.

Given an overall efficiency of about 35%, where does the rest of the ENERGY go? If the generator is running well, the energy loss there is only about 1%. The turbine may have an efficiency of some 45%. About 10% of the energy in the coal goes straight up the

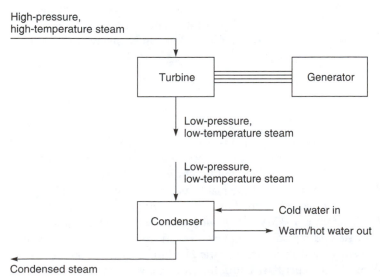

FIGURE 18.8 When the steam leaves the boiler, it travels through the turbine to the condenser, where it returns to liquid form.

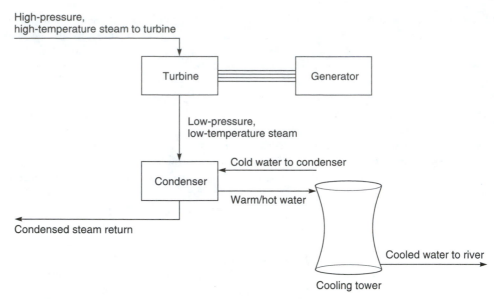

High-pressure,
high-temperature steam to turbine

Turbine

Generator

Low-pressure,
low-temperature steam

Cold water to condenser

Condenser

Warm/hot water

Condensed steam return

Cooled water to river

Cooling tower

FIGURE 18.9 Water from the condenser is cooled before being returned to the environment. Fresh, cold water is sent to the condenser to begin the process again.

stack in the form of hot gases and particulates. Some heat is lost from the steam just in moving it around the plant, from the boiler to the turbine and from the turbine to the condenser. More heat is wasted in warming the cooling water in the condensers. Fans are needed to provide the primary air; pumps are needed to circulate water through the condensers and cooling towers; scrubbers must be operated. All of these consume more ENERGY. A plant that is highly efficient by today's standards will throw away the equivalent of 2 MW in wasted energy for every 1 MW of electrical energy put into the distribution system. (Another way of putting it is that two-thirds of all the energy extracted from the ground by coalminers is thrown away as hot air and hot water.) By the time the electricity actually reaches the consumer, more losses have been encountered in the wires and transformers. Consequently, only about 30% of the chemical potential energy in the coal is available to the consumer in the form of electrical potential energy at the outlet in the wall.

THE COSTS

This preliminary introduction of environmental issues, to which we will return in detail in later chapters, helps address this question: If all we need to do is boil water to raise steam and then use the steam to turn the turbine/generator, why isn't electricity free? First, there is a very considerable expense just to build the plant. As a rough estimate, a new power plant with a full complement of environmental protection devices on its 'back end' might cost $1,250 per kW. Therefore a 400 MW plant would cost a half-*billion* dollars! This money has to be borrowed, and repaid *with interest*. Second, coal is not free. A relatively low grade of coal might still cost $15 per ton. A power plant that burns 10,000 tons of coal per day has a fuel bill of $150,000 per *day*. Third, the entire power plant is an assemblage of complex, precision machinery. There is a significant, continuing cost for maintenance. Fourth, the electric company needs to make some kind of financial return to its investors—after paying the cost of the plant, the cost of the coal, the cost of maintenance and upkeep, there needs to be some money returned to investors. Where does all this money come from? It comes from us, once a month, when we write a check or stop at the electric company office to pay our electric bills.

To add a scrubber to an old plant can cost as much as $300 per kilowatt of capacity. That amount of money is more than an entire new coal-fired generating plant would have cost in the 1960s. On an equivalent cost-of-coal basis, the present scrubber technology adds as much as $31 per ton and particulate control adds about $3 per ton. Costs of some of the new NO_x control strategies are discussed in Chapter 29.

Environmental control is an expense that the utility may pay initially, but consumers of the electricity will inevitably pay in the end. In other words, if we want a cleaner environment, *we* will pay for it. The costs of the controls could increase the effective cost of coal by anywhere in the range of $25–50 per ton for underground-mined coal and about $18 per ton for surface-mined coal. Even so, with all of the environmental controls added, coal is by no means uncompetitive with other fuels such as oil or natural gas. Furthermore, coal-burning electricity-generating plants, even with emissions controls, are much less expensive to build and operate than nuclear plants; consequently, coal maintains a cost advantage there as well. Though many of the public have concerns about potential environmental problems associated with coal-fired electricity plants, often these concerns are much less than those for nuclear plants.

FOR FURTHER READING

Fowler, John M. *Energy and the Environment*. McGraw-Hill: New York, 1984. Though some of the numerical data in this book are now out of date, it otherwise remains a fine introduction to the subject for readers with minimal science or mathematics background. Chapter 8, on air pollution associated with energy use, is particularly useful for this present discussion.

Gebelein, Charles G. *Chemistry and Our World*. Brown: Dubuque, 1997. A well-illustrated introductory text that relates chemistry to many of the phenomena in our everyday world. Chapter 18, 'Air Pollution, Energy, and Fuels,' relates to the present discussion.

Graedel, Thomas E.; Crutzen, Paul J. 'The Changing Atmosphere.' In: *Managing Planet Earth*. Freeman: New York, 1990; Chapter 2. This book is a collection of articles that were originally published in *Scientific American* magazine; its focus is on environmental problems.

Kraushaar, Jack J.; Ristinen, Robert A. *Energy and Problems of a Technical Society*. Wiley: New York, 1988. Chapter 13 of this book, 'Pollution and the Atmosphere,' is relevant to the present chapter. This book assumes little prior science or mathematics background on the part of the reader.

McNeill, J.R. *Something New Under the Sun*. Norton: New York, 2000. This remarkable book, subtitled 'an environmental history of the twentieth-century world,' surveys the history of the last century's technological developments, politics, and international relations in terms of humankind's dramatic impact on the environment. Chapters 3 and 4 discuss the atmosphere and what we have done, and may still be doing, to it.

Ristinen, Robert A.; Kraushaar, Jack J. *Energy and the Environment*. Wiley: New York, 1999. An outgrowth and revision of a different book by these authors, listed above. Chapter 9 discusses air pollution. Again, minimal science or mathematics background is needed.

Schwartz, A. Truman; Bunce, Diane M.; Silberman, Robert G.; Stanitski, Conrad L; Stratton, Wilmer J.; Zipp, Arden P. *Chemistry in Context*. McGraw-Hill: New York, 1997. An introductory text on chemistry with emphasis on relating chemistry to issues that affect us all, such as the environment, nutrition, and consumer goods. Chapter 6 deals with acid rain and ways of combating it.

19

ENERGY FOR TRANSPORTATION

For most of us, our direct, personal utilization of energy is mainly for two reasons. The first is the ubiquitous use of electricity for lighting, cooking, and much of our entertainment (such as stereos, televisions, and radios). Those of us who live in 'all electric' houses or apartments also depend on electricity for heat and, for many of us, for air conditioning. The second is the use of energy to transport ourselves, and to ship our goods, from place to place. A good transportation system is vital in a modern industrialized society, because it lets us move ourselves and our belongings and lets others move to us the food and consumer goods we need, easily, at reasonable cost, providing us with remarkable convenience.

HUMANS AND OTHER ANIMALS

For almost all of human history, transportation of ourselves, our belongings, or other cargo has depended either on the use of human muscles or on the muscles of certain animals (horses, camels, oxen, and occasionally more exotic animals like elephants). From the dawn of early hominids several million years ago, the obvious form of 'transportation' was to use the energy of our own bodies for walking or running. We can also carry or pull things. Photographs of some of the primitive peoples of the world show them carrying animal carcasses or other burdens slung from a pole carried along by two or more people.

Animals were being used for transportation by about 3000 B.C. Over the years an extraordinary variety of animals have been pressed into service as draft animals: horses, oxen, asses, camels, reindeer, llamas, elephants, water buffalo, and yaks.

Humans in northern Europe had been pulling loads on sledges since about 5000 B.C. No doubt the idea spread that one could just as easily hitch an animal to the sledge rather than packing the load onto the animal itself. Then some unknown inventor mounted a sledge on wheels. Drawings of these first vehicles have been found on tablets excavated in Mesopotamia, dating from about 3500 B.C. (Figure 19.1). Actual models of vehicles, and

FIGURE 19.1 This is one artist's representation of the pictographs still visible in Mesopotamia today. The 'first vehicles' represented were a sled and a wagon (with wheels).

sometimes even the vehicles themselves, have been found in Middle Eastern tombs that date from 3000–2000 B.C.

Nowadays probably none of us can appreciate, or even imagine, what life must have been like before the development of modern transportation systems. Almost everyone lived their lives in the immediate vicinity of the village or farm where they had been born. There might be occasional visits to the local market or fair, or, rarely, to the regional capital for some unusual event. For a few privileged people, there might have been a once-in-a-lifetime religious pilgrimage. But for almost all the people almost all the time, life was confined to a region limited by the distance that could be traversed on foot or on horseback. Even just 200 years ago, people tended to remain settled and live out their lives close to where they were born. While there may have been other sociological or psychological factors involved, a major reason for this state of affairs was the sheer physical difficulty of moving oneself and one's belongings in the absence of reliable means of transport.

The problem was particularly severe for people who did not have access to large bodies of water, such as the sea or large, navigable rivers. Inland, transport depended on two factors: the availability of draft animals and a good system of roads. In fact, the latter was much more the severe problem. In western Europe, the road system had been neglected ever since the collapse of the western half of Roman Empire in the early 400s. In many places roads consisted only of tracks that became rutted by the continued passage of vehicles and animals and, in bad weather, turned into an impassable mudbath. Any vehicle that could traverse these roads had, of necessity, to be large and heavy, simply to survive the jolting and jarring of the alleged roads. But a large, heavy vehicle is also a slow one, especially if it has to be pulled by animals. This situation set up a vicious cycle: to avoid being rattled to pieces on the poor roads, a vehicle had to be large, heavy, and slow; but, a large, heavy, slow vehicle was superbly designed to wear the road surfaces even more. It was probably better to ride *on* the horse than *in* a horse-drawn vehicle.

In the early 1800s, a Scot, working as a surveyor of road systems, thought up a better way of making roads than the rutted, eroded, muddy travesties that then existed. The idea was to begin with a bottom layer of stone or gravel to allow rain water to drain away and to put on top of the gravel layer some impervious material that would not turn to mud but would have some 'give' (i.e., elasticity) to it so that the road would not wear so fast. The Scottish surveyor was John Loudon McAdam. His better way of building roads was

FIGURE 19.2 The speed and regularity of stagecoach travel depended on improved road quality.

quickly recognized, and his expertise soon became in demand throughout Britain. By the 1830s, he had been responsible for improving hundreds of miles of road in the British Isles. We remember him today when we speak of a 'macadam' road.

Once a good road system had been developed based on McAdam's ideas, it became possible to establish regular passenger transportation via fast stagecoach services (Figure 19.2). Improvements in the design of the coaches themselves, and setting up a system to allow for regular changes of horse teams, allowed the stagecoach companies to maintain unprecedented speeds over long distances. The first horse-drawn coaches, or omnibuses, began public service on regular schedules in Paris in 1662. The first commercially successful service began in 1827, also in Paris; two years later the omnibus appeared in London.

It was not long before inventors got the idea of designing a 'horseless' carriage, propelled by the major energy source of the time—steam. The steam carriages actually developed a modest amount of popularity, despite their quirk of blowing up once in a while.

SAILING SHIPS—MOVING WITH THE WIND

Early humans may have gotten the idea of building rafts by seeing logs floating down rivers. Boats certainly emerged very early as a technological innovation among people who lived along rivers. They provided an effective means of transport for people and their goods. Primitive rafts were simple bundles of logs or reeds, but as early as 4000 B.C. some Egyptian rafts had an early form of sail. These very early designs were improved by the Cretans and Phoenicians into two kinds of boats: rowing galleys, used for warfare and powered by humans, and sailing ships, used for commerce and powered by the wind. The invention of improved sails, probably by Arabs, may have occurred around A.D. 200. The Islamic conquests, followed by extensive Chinese voyages of exploration, made the sailing vessel dominant by the 1400s.

In the West, the steerable rudder was invented by the Dutch, in about A.D. 1200. (The Chinese had already known about it for a thousand years.) Magnetic compasses were first used around A.D. 1000. With perfected sails, rudder, and compass, it was possible to navigate and sail accurately and to develop genuine sea routes, first in the Mediterranean Sea, and then the great European voyages of discovery in the 16th century. Sea-going boats tended to be larger than those developed for river transport. The improvements in ship construction and the increasing mastery of sailing techniques and navigation, which collectively took centuries to occur, gave Western societies a form of transportation that required no fuel and that became capable of carrying a substantial cargo over the seas of the world. Sailing ships provided western Europe, and, later, America, with much of the wealth without which the subsequent processes of industrialization could likely never have started.

By 1800, then, there were three sources of transportation energy: human muscles, the muscles of draft animals, and the wind. A dramatic change occurred with James Watt's perfection of the steam engine in the late 18th century.

STEAM FOR TRANSPORTATION

Locomotives

We've already seen how Richard Trevithick used the steam engine to turn wheels, creating the first steam locomotive. Trevithick's designs actually did work. But, somewhat analogous to Newcomen's building of a workable engine that was later perfected by Watt, Trevithick's locomotive designs had plenty of room for improvement. Many engineers worked on the problem, but the credit for the development of the first practical steam locomotive is generally awarded to George Stephenson.

An important distinction is that the locomotive steam engine emerged as an application of *high-pressure* steam. Watt's low-pressure, stationary, engine had dominated the steam-engine market until 1800, and had totally swept aside the earlier Newcomen engine. For all its fine qualities, the Watt engine was too big and too heavy to be adapted easily for locomotive purposes, although William Murdoch succeeded in making a small steam engine drive a three-wheeled carriage. It is questionable whether many of the low-pressure Watt engines of the time could have produced sufficient WORK to move themselves, let alone to haul a train of cars. Trevithick demonstrated his high-pressure machine, smaller and more compact than a Watt engine (and therefore also lighter), at the beginning of the 19th century and applied it to a locomotive on a colliery tramway in Penydarren (Wales). This successful demonstration led people to realize that the steam locomotive could be a viable idea. George Stephenson's role was to introduce further improvements. He put his engine, *Locomotion Number 1*, into commission on the Stockton and Darlington Railway in 1825. Four years later, the success of Stephenson and his son Robert at the Rainhill Trials of 1829 established their locomotive *Rocket* as the prototype for the Liverpool and Manchester Railway. When this railway opened in 1830, it was the first railway in the world in the sense that we would recognize now, i.e., with passenger and freight trains operating on a regular timetable.

In 1829, Stephenson's locomotive *Rocket* demonstrated the ability to haul a 13-ton train 30 miles (from Liverpool to Manchester)—at an average speed of 14 miles per hour. Stephenson's demonstration set the stage for an enormous expansion of railways throughout the industrialized world. Throughout the 19th century there was an immense, sustained activity in building railways worldwide. The dominant fuel in steam locomotives has been coal. Thus the sequence Newcomen → Watt → Trevithick → Stephenson added a new energy source for transportation for the first time since the invention of the sail.

Over the next century and a half, steam locomotives increased massively in size and power. In most countries, they were replaced by diesel or electric traction in the last half of the 20th century, except for China, one of the last bastions of steam power on railways. It is noteworthy that the locomotives produced and used there are still of the same basic type, and work on the same basic design, as those built by the Stephensons in the 1830s.

It is remarkable, perhaps, that in the lengthy evolution from Stephenson's $7\frac{1}{4}$-ton *Rocket* to the most recent steam locomotives of several hundred tons there were no revolutionary changes in fundamental design. The line of development of steam locomotives was a steady increase in efficiency and in tractive effort (the capacity for hauling heavy trains). Since condensing of steam after it leaves the cylinder is not practical on a locomotive, it is exhausted directly into the atmosphere still at a relatively high temperature. Therefore the efficiencies of steam locomotives are no better than about 8%, and frequently less than 5%. Furthermore, there are practical upper limits to the height and width of locomotives, because they have to be able to pass through tunnels and fit between station platforms.

Throughout almost the entire steam locomotive era, coal was the fuel of choice for generating the steam. Wood was used on early locomotives in the United States, and stayed in use in some countries that had ample wood supplies but few resources of coal or petroleum. One of the first successful oil-fired locomotives was the *Petrolea*, built for England's Great Eastern Railway in 1886. By 1900, several American railroads were using oil as fuel for steam locomotives, but there were never more than 15% of steam locomotives fired with oil in the United States.

The railways provided a new and extraordinary facility for personal transportation that was quickly adopted not only in the advanced industrial countries, but also in places like India, China, and South America. The development of the locomotives and tracks themselves in turn created a need for 'infrastructure,' such as station buildings. By the end of the 19th century the railways had entered into the imaginative experience of more people in the world more generally and in a shorter space of time than any previous technological innovation in the entire history of the world.

The use of railways for transporting passengers and freight spread very rapidly throughout the 19th century. Even so, it was not until 1869 that there was a transcontinental railway link across the United States—just a little more than 125 years ago. Though the rail system became very widespread in most industrialized nations by the end of the 19th century, both personal travel and shipping of goods were still limited to places linked by rails, and then only at the time scheduled and the rates established by the railway companies. As late as the turn of the 20th century, most of the people in modern industrialized nations still depended on themselves or horses for personal transportation.

Ships

We have previously met Denis Papin, the French scientist who had the concept for a movable piston inside a cylinder as a way to do WORK. In 1685 Papin suggested the possibility to using steam propulsion as a way of propelling ships. About a hundred years later (in 1783) a ship called the *Pyroscaphe* was propelled by steam on the Saóne River in France—for about 15 minutes. The first practical steamship was the *Charlotte Dundas*, built in 1802 to tow canal barges on the Forth and Clyde Canal in Scotland. It was soon retired from service because the waves created by its paddle wheel blades caused severe erosion to the canal banks. In the United States, credit for the first practical steamship is given to Robert Fulton for *The Claremont*, launched in 1807.

At the beginning of the 19th century, the finest development of the sailing ship was represented by the East India merchant ships and the principal ships of battle of the navies of the world. Those ships represented the ultimate in the technological development of wind as the energy source and in the building of ships from wood. Using steam to propel ships and converting their construction to iron or steel caused a profound transformation of maritime technology, in both its civilian and naval applications. Although Robert Fulton did not invent the steamboat, for all practical purposes his was the first viable one. It was operated by a Boulton and Watt engine imported from England. Some five years after Fulton, Henry Bell began to operate a successful steamship service on the River Clyde (Scotland) with his paddle steamer *Comet*. All the early steamships were propelled by paddle wheels, which were adapted from waterwheel technology, quite familiar to engineers of the day. We've seen in previous chapters that the steam engine is not a very efficient device for converting ENERGY to WORK. Consequently, a large ship was likely to require an immense fuel supply, which in turn posed formidable problems of how to store it on board the vessel. For this reason, early steamships were not thought to be practical except for estuaries, lakes, and rivers.

By virtue of geography, England and France approached steam propulsion on water from a different perspective than did America. Both England and France already had efficient, well-run systems of inland waterways on rivers and canals and so thought of steam primarily in terms of ocean-going naval or merchant vessels. This meant large vessels and long voyages. In early America, however, the interest was in building smaller boats to navigate the rivers, especially the large rivers like the Ohio, the Mississippi, and the Missouri. These trips would be short, quick, and cheap.

As early as 1787, the *John Fitch* (modestly named for its builder, John Fitch) was navigating the Delaware River at Philadelphia. It was a 50-foot boat with a paddle wheel in the stern, and was operated by a steam engine that was built by Fitch's friend Henry Voight, a local clockmaker. In 1790, Fitch operated a summertime steamboat service between Pennsylvania and New Jersey. During its relatively brief summer service, the steamboat logged some two to three thousand miles and achieved speeds as high as 7–8 miles per hour. However, it was not until the efforts of Fulton that it was possible to produce an efficient, reliable, steamboat that—most importantly—would actually turn a profit. His contribution, much like that of James Watt, was for building, logically and systematically, on the work of others, for understanding the principles involved, and learning from mistakes—his and others. By early 1807, Fulton's 150-foot vessel, *Steamboat*, was able to make the 150-mile run from New York to Albany in 32 hours, compared to four days by sail. Commercial trips began in September of that year, and in the next year the famous *Claremont* was launched. Fulton's first steamboats traveled 3–5 miles per hour. As steamboats were improved, they became an extraordinarily successful business. Fulton's first New Orleans boat resulted in a net profit in its first year of $20,000 on an initial investment of $40,000.

By 1838, the Great Western Steamship Company began to operate a trans-Atlantic service with the first large steamship, designed by the archetype of 19th century engineers, Isambard Kingdom Brunel. The ship was the *Great Western*. It was a wooden ship and paddle-wheel propelled, but Brunel had correctly calculated that the volume of space required to carry fuel, expressed as a fraction of the total volume of the ship, would decrease as the size of the ship increased. The *Great Western* arrived in New York 15 days out of Bristol with plenty of coal remaining. Brunel's innovations were the first key steps that enabled steamships to take over from sailing ships most of the traffic in goods and passengers on the lucrative North Atlantic crossing.

When profits like those earned by Fulton are possible, competition develops quickly. One of the factors that provides an advantage in this competition was speed. The way to get higher speed is to use higher-pressure engines, since the higher the pressure, the smaller the pistons and other engine parts could be for the same amount of power. The higher-pressure engine is therefore lighter, and the resulting lower weight would yield better ship performance. However, owners and operators were tempted to operate even these engines at pressures above design levels and to lower costs by compromising on maintenance and safety margins. The result was a rash of boiler explosions. A 1-ton boiler linked to a 10 hp engine contains enough energy to throw the boiler nearly four miles high with an initial velocity of projection of 1,100 feet per second. When the steamboat *Oronoko* exploded in 1838, steam swept through the whole length of the boat as if it were hit by a tornado, killing more than a hundred passengers in seconds.

Therefore, the steamboat was also the subject of legislation that first established the authority of the federal government to regulate an industry in the interest of public safety. In the 1840s some 70 explosions killed 625 people. In December 1848, the commission of patents estimated that over the years 1816–1848, 233 steamboat explosions had killed 2,563 people. In 1850 alone, 277 people were killed in boiler explosions, and a year later the number rose to 407. The steadily escalating death toll resulted in a public outcry. In 1852, Congress created the Joint Regulatory Agency of the Federal Government. This agency actually was successful in lowering fatalities by about a third up to 1860. Eventually, uniform codes and regulations for the construction and operation of high-pressure boilers were adopted, and boiler explosions on boats became a thing of the past.

Despite the success, tremendous controversy was associated with the creation of this agency, and bitter opposition came from those in the steamboat business. (Anyone following current events will realize that some things never change!) To free-market conservatives, government regulation of business was economically and philosophically distasteful. A Congressional inquiry had actually started in 1824, after an accident that May killed 13 people. Six years later, another inquiry looked into an accident near Memphis that killed 50–60 people. (In the meantime, about 270 people had been wiped out in less spectacular blasts.) This 1830 inquiry led to legislation requiring inspection of steamboat boilers every three months, but it failed to win passage. Among the arguments raised against it was the idea that the Constitution did not include 'insuring the public safety' among the powers reserved to Congress. Seven years later, President Van Buren urged passage in his 1837 State of the Union address. Congress finally acted the following year, after a truly spectacular explosion near Charleston took the lives of 140 people.

This law, which had caused such furor and took so long to win passage, provided that each federal judge would appoint a boiler inspector, who was to examine every steamboat boiler in his jurisdiction twice a year. For this service, the owner of the boiler was to pay the inspector $5, in return for which his license to navigate would be certified. The law also provided that, in suits against boiler owners for damage to persons or property, the fact that a boiler had burst was to be considered evidence of negligence. Congress strengthened the law in 1852. Until the problem of so many tragic deaths resulting from bursting steamboat boilers, most Americans had believed that the federal government could not, indeed ought not, interfere with the rights of personal property. The several hundred boiler-related deaths finally convinced many Americans, and most congressmen, that there had to be a point at which property rights of some individuals had to give way to the civil rights of others. This boiler legislation established the precedent on which all succeeding federal safety-regulating legislation would be based.

Through most of the 19th century, sail competed with steam, but as the century progressed, steam became increasingly important for ocean-going cargo, passenger, and war ships. By the end of the 19th century, steam—generated most often by burning coal—was dominant in both land and water transportation. By the mid-20th century, there were four important general types of ship propulsion in service: (1) reciprocating steam engines directly connected to the propeller shafts, (2) steam turbines driving directly or through gears (recall Charles Parsons), (3) diesel engines (discussed below, and in Chapter 24) directly connected to the shaft, and (4) steam turbines or diesel engines driving electricity generators that operated motors turning the shafts. Early steamships had reciprocating engines, as indeed did all commercial steamships in the 19th century. The reciprocating steam engine was been very popular, and it was not until the middle of the 20th century that the combined tonnage of turbine- and diesel-operated ships exceeded the tonnage of ships with reciprocating steam engines. Steam turbines were eventually displaced by diesel engines for most purposes of marine propulsion, but they still remain in use in some larger vessels.

Cars, Trucks, and Buses

The dominant forms of land transportation of the 19th century had some significant drawbacks. Railroads could certainly haul large amounts of freight, or large numbers of passengers, swiftly, but only where there were railway tracks. Horses required constant food and care (regardless of whether they were being used or not) and room for stables. There is also the common observation that what goes into the front end of a horse sooner or later emerges from the back end as a generally unpleasant 'by-product' that has to be collected and disposed of. During the 19th century, and even before, engineers dreamed of developing the 'horseless carriage.' All sorts of propulsion schemes were tried. Steam might be an obvious choice, because of its success with locomotives.

We've seen that Trevithick's idea was to rely on high-pressure steam to build a compact, and therefore light, engine that could be movable, and propel a vehicle. On Christmas Eve, 1801, he was ready to test his steam-driven carriage (Figure 19.3). The first test was not successful, because the boiler could not produce enough steam. But, three days later he did successfully drive the vehicle up a hill.

Soon after Trevithick's successful trial of a steam carriage, the Philadelphia Board of Health asked the American inventor Oliver Evans to construct a steam dredge to deepen Philadelphia's harbor. (Although he is virtually unknown today, Oliver Evans was one of the first real American technological geniuses. He invented a fully automated production line for a flour mill by the time he was 30 years old.) Evans seized the opportunity to advance his ideas for steam-propelled vehicles. Since the dredge would have to be moved to the harbor in any case, Evans placed it on a wheeled frame. A belt ran from the crankshaft of the dredge's engine to the rear axle of the frame. Evans, in doing so, produced America's first self-propelled road vehicle, which he named the *Orukter Amphibolos* (the 'amphibious digger,' Figure 19.4). This dredge/boat/car used a steam engine to drive wheels on land and a paddle wheel when it was in the water, as well as a bucket chain dredge. On August 12, 1805, Evans drove it around the streets of

FIGURE 19.3 Trevithick's steam carriage, designed in 1801, was one of the earliest attempts at a steam-propelled vehicle.

Philadelphia at 4 miles an hour. This, of course, is just about the same speed that can be attained in any large, congested city today with the most modern automotive technology of the 21st century. Philadelphia got the last laugh—the rough cobblestone streets broke the wheels of this remarkable contraption. Several months later, the Pennsylvania legislature banned 'steam wagons' from turnpikes, partly because of the experiences with Evans' vehicle.

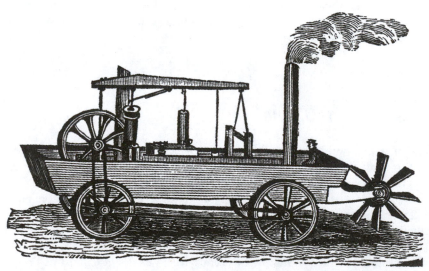

FIGURE 19.4 Evans' amphibious digger not only traveled on land and in water, but it also functioned very well as a mechanical shovel driven by steam.

After Stephenson developed the essentials of the modern rail system in the early 19th century, further experimentation and tinkering with steam-propelled automobiles was mostly, though not completely, abandoned for further improvements in locomotive design. It was felt that steam automobiles would always encounter some fears from the public, as a result of the spectacular and disastrous explosions of high-pressure engines on steamboats. The public fears translated into attempts to regulate them, even on private property, and virtually to prohibit them on public streets. In the decades before 1890, the early steam-propelled cars failed mostly because of governmental regulation and prohibitions, and not because of perceived limitations in their mechanical efficiency. Compared to a horse-drawn vehicle of comparable capacity, a steam vehicle provided greater speed, lower operating costs (you only need to buy fuel when the vehicle is operating, but you have to feed a horse all the time, whether it is working or not), and less pollution than horses (again, horses are rather efficient converters of food into a waste 'by-product'). None of those considerations mattered—the public was still afraid of the steam vehicles. The government regulators simply reflected public opinion when they banned steamers. The regulators cited the issues of their dizzying speed (maybe 10 miles per hour), the inevitable smoke and steam exhaust, and their potential for blowing themselves up (along with the driver and any hapless passengers).

Steam engines are necessarily slow to start up, since it takes a while to heat the water in the boiler. Starting in less than five minutes became possible only after the invention of so-called flash boilers after 1900. Imagine getting in your car today, turning the key, and then waiting five minutes until you could drive off! High rates of water consumption limited early steamers to a 6–8 mile operating radius, possibly acceptable for commuting to work or doing the shopping, but scarcely useful for long trips. Eventually the range of steam automobiles was extended considerably by adding condensers to convert the steam exhaust back to water and recirculate it to the boiler.

One day in 1876, a young man in rural Michigan watched a steam traction engine at work. The device was rather interesting, and the experience led him to believe that he might make some improvements and developments in the field of 'road locomotion.' His name was Henry Ford.

Steam as an energy source for transportation has not yet totally disappeared, and of course the use of steam turbines for generating electricity is of immense importance to our society. But, in most of its direct applications to transport, steam has been almost entirely replaced by internal combustion engines and electricity.

THE COMING OF PETROLEUM

One of the first steps toward a practical new form of personal land transport occurred around 1850 in Scotland. James Young, a chemist, was experimenting with an oil that seeped out of the rocks in a coal mine in England. His experiments at first showed that the oil could be separated into a paraffin wax, an oil that could be used for lubrication, and a liquid like kerosene. Later, Young found that by heating the oil he could drive off—and subsequently collect—a liquid that we would recognize today as gasoline.

In 1855, Benjamin Silliman Jr., a professor of chemistry at Yale, was asked to analyze some samples provided to him by the Pennsylvania Rock Oil Company. One of the tests he used was distillation, a method of analysis that separates the components of a mixture based on differences in the temperatures at which they boil. Silliman found that about 50% of the rock oil could be distilled to kerosene and a light lubricating oil, called paraffin oil. The kerosene was especially useful as a fuel for lamps, replacing whale oil. Silliman also noted that another 40% of the rock oil could be distilled into other products. On the basis of Silliman's one laboratory report to them, the Pennsylvania Rock Oil Company decided to commercialize their product. Rock oil is much better known to us in the Latinized form of its name, petroleum (*petro*, rock, and *oleum*, oil). The Pennsylvania Rock Oil Company drilled the first successful oil well, the Drake Well, near Oil Creek, Pennsylvania, in 1859.

Kerosene and lubricating oils were valuable products from petroleum. Some of the paraffin oil could also be used as a fuel. One by-product of distilling petroleum was a useless nuisance—gasoline. The temperature at which gasoline boils is very low, and therefore, even though it had no particular use, it was the first component to distill when the petroleum was treated. In other words, while distilling petroleum to make kerosene and paraffin oil, gasoline was an inevitable by-product.

As the 19th century drew to a close, small engines operating on the gas (not gasoline) that can be made by heating coal were able to compete successfully with steam engines for many small and medium duties. This competition was possible because of the widespread availability of coal gas at the time. Gas engines also provided the practical advantage of not needing a boiler to raise steam. However, there seemed to be absolutely no way that a gas engine would compete with a steam engine for transportation, because the engine had to be connected to the gas supply through pipes. But, when it was recognized that an alternative was available—the easy-to-vaporize gasoline from petroleum, this situation was revolutionized. Gasoline (or, for that matter, kerosene or diesel oil) could be easily carried in a tank to supply the engine.

At the same time (i.e., the end of the 19th century), liquid fuels derived from petroleum began to supersede gas derived from coal. There are two principal reasons. First, a hundred years ago there were no long-distance gas pipelines. Gas was available only relatively close to the place at which it was manufactured. In other words, gas generally served domestic and industrial customers in a single urban area. (In fact, one form of gas that can be made from coal is called town gas.) Although the energy requirements of nonurbanized areas in many parts of the world were growing, only the biggest installations would justify the establishment of gas manufacturing alone. In comparison, liquid fuels offered several advantages. For one thing, liquid fuels are easy to transport and store. Second, they are generally easy to feed into the engine, sometimes just by gravity flow alone. Also, when compared on an equal weight basis, the liquid fuels derived from petroleum provide more heat per unit weight than does coal. At the time that the petroleum industry was developing as a supplier of illuminating fuel (kerosene for oil lamps), it was, as we've seen, also producing quantities of the relatively useless by-product, gasoline. Therefore, as interest developed in using a liquid fuel for vehicle engines, the petroleum industry just happened to be in position to offer such a fuel at an attractive price.

In the steam engine, as developed by Newcomen and later perfected by Watt, the fuel (quite often coal) is burned outside the cylinders and is used to 'raise' steam. The steam pushes the piston. This type of device is called an external combustion engine. Adapting an external combustion engine for a horseless carriage would require a firebox to burn the fuel and a separate boiler to raise steam and deliver it to the engine. The alternative to external combustion is the *internal combustion engine*. In this case, the fuel is burned *inside* the cylinder. The heat liberated by the burning fuel is transformed directly to mechanical energy, eliminating entirely the steam-generating step (and also eliminating the hardware needed to produce the steam).

OTTO AND THE AUTO

The Prehistory of the Automobile

The internal combustion engine had a long development period before it became an effective successor to the reciprocating steam engine at the end of the 19th century. The very earliest ideas on internal combustion engines may have developed from early firearms and cannons. In a sense, a gun is an internal combustion engine, since the fuel (gunpowder) is burned inside the cylinder (the barrel of the gun), and the chemical potential energy liberated in the burning is transformed into mechanical energy on the piston (the bullet or cannonball). Of course in a real gun or cannon there was only one 'cycle' of fuel burning and the 'piston' receiving the resulting energy. Then, the gun had to be reloaded with gunpowder fuel and the bullet or cannonball. The nonrepeatability of the action prevented the development of a genuine gunpowder engine, but the conceptual possibility of such an 'internal combustion' engine was certainly established, and inventors were alert to the potential of such an engine once an appropriate fuel became available.

The idea of obtaining work from combustion or explosion in a confined space—the gunpowder engine—dates from the 17th century, when several inventors, including Denis Papin, independently attempted to make gunpowder-fueled pumps. In 1684, a French monk, Jean de Hautefeuille, fared no better with suggestions for a gunpowder-driven fountain (but it is interesting to imagine what such a thing might have been like in operation!). Starting from about the 1790s, many people worked on various designs of internal combustion engines, using coal gas, wood gas, or volatile hydrocarbons as fuel. It was not until the late 1860s, when the French inventor Joseph Etienne Lenoir produced small, quiet gas engines, that the internal combustion engine became a commercial success.

The fuel that enabled the internal combustion engine to become a practical proposition was coal gas. This gaseous fuel had been first produced from coal by William Murdoch, an agent of Boulton and Watt, in the 1790s. Its first use was in gas lighting, initially at the Boulton and Watt factory in 1798. After that, its production and use for lighting spread rapidly, so that by the middle of the 19th century virtually every town of any size in Britain, and many in Europe and America, were equipped with gas lights and a place to make a supply of coal gas. In 1794, an Englishman, Robert Street, tried to build an engine in which the piston was to be driven by the combustion of gasified tar or turpentine oil. In 1801, the Frenchman Philippe Lebon came up with a similar idea and tried out a kind of double-acting engine driven by the internal combustion of a mixture of coal gas and air, ignited electrically. In 1809, George Cayley described how a gas engine ought to work, but he did not attempt any experiments.

In 1859, the French engineer Etienne Lenoir succeeded in making an engine that ignited coal gas with an electric spark in a horizontal cylinder, thus driving a piston and using a flywheel to return the piston to the firing end of the cylinder. The machine was noisy and cumbersome, but it worked, and with sufficient development it could be transformed into a viable engine. Smooth action of the Lenoir engine was achieved by the adoption of the 'Otto cycle,' which we will examine in detail later.

Nikolaus Otto and His Contemporaries

The German merchant Nikolaus August Otto became so interested in the Lenoir gas engine that he had a couple made under license by a workshop in Cologne. They did not work very well, so Otto, who was convinced of the engine's great potential, sat down and redesigned the whole thing. In this endeavor, Otto teamed up with an engineer, Eugen Langen. The operating efficiency of the Otto–Langen engine was much better than that of Lenoir's gas engine. Otto and Langen carried on experimenting, and the ultimate outcome of their work, in 1876, was a great breakthrough, the first four-stroke engine.

In Otto's engine (Figure 19.5) the fuel is burned inside the cylinder. No steam is generated. The high-temperature, high-pressure gases from the combustion of the fuel do the WORK on the piston. This approach represents the *internal combustion engine*. Otto did not invent the concept of the internal combustion engine. However, he did develop the concept into a practical device. We'll see how Otto's engine works when we talk about gasoline (Chapter 21).

The successful operation of Otto's internal combustion engine design required a very fast combustion inside the cylinder. The fuel needed to mix almost instantaneously with air, and that condition required that it be a material that would vaporize very easily. A cheap material that fitted this criterion very nicely was the useless by-product of distilling petroleum—gasoline.

It was still necessary to find a way of getting gasoline into the cylinders quickly and efficiently. This problem was solved in 1893 by a German engineer, Wilhelm Maybach. The inspiration for his device was a very popular fashion accessory of that era, the perfume atomizer. Maybach and his colleague, Gottlieb Daimler, adapted the concept of the perfume atomizer to inject or spray gasoline into the cylinders of Otto's internal combustion engine. They had invented the carburetor (Figure 19.6). Otto's engine fixed up with the carburetor was a new contraption that Maybach and Daimler decided to name in honor of the boss's daughter, a young woman named Mercedes.

Gottlieb Daimler realized the development potential of the Otto engine. He planned to use volatile petroleum derivatives as fuel. His third experimental model was used to drive a bicycle by *Benzin* (gasoline)—an almost useless liquid that, in those days, could be bought over the counter in drugstores.

About 10 years after Otto developed the internal combustion engine, Karl Friedrich Benz used it to build the first practical automobile, a three-wheeled vehicle. It was Benz

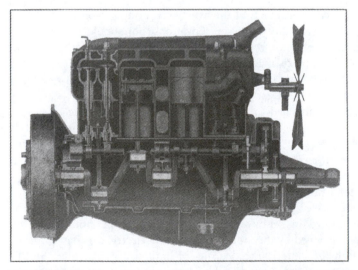

FIGURE 19.5 The interior of an early four-stroke Otto engine.

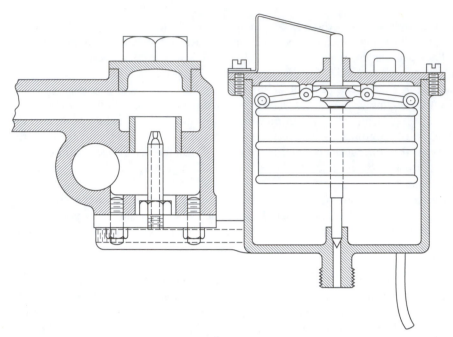

FIGURE 19.6 Although somewhat more complicated than a perfume atomizer, Maybach's carburetor worked in a similar way as it sprayed gasoline into the engine cylinders.

who created the 'horseless carriage,' the first genuine motor car. In 1901, Benz eliminated the last vestiges of carriage design by placing the engine under a hood in front of its car. Daimler built a motor truck in 1891 and Benz, the first motorbus in 1895. Though Benz developed the first automobile around 1886, it was not until the early decades of the 20th century that automobiles became cheap enough for most individuals to be able to obtain one. Henry Ford built his first motor car in 1896. In 1913, Ford dropped the price of his Model T to $500. This action meant that the automobile was no longer mainly a plaything for the rich, but was within economic reach of the average working person.

It does not decrease the originality of these innovations, nor should it in any way diminish the accomplishments of those who made them, to realize that they are, in essence, inspired assemblies of preexisting parts. First, the internal combustion engine had been developing steadily since Lenoir's gas engine of 1859. We've seen that as long as internal combustion depended on coal gas for its fuel, it was useful only for stationary engines, because it was directly attached to its source of supply—usually the town gasworks. The advent of petroleum-derived liquid fuels opened the way to a genuinely mobile internal combustion engine, carrying its supply in a tank on the vehicle. The gasoline engine, with a carburetor to atomize the fuel before injection into the cylinders, held out the promise of being a versatile, lightweight engine suitable for horseless carriages.

By the end of the 19th century there were three choices for powering the transportation device that the French called an 'automobile': steam, gasoline, and electricity. The very first automobile speed record was set by an electric car, 24.5 miles per hour. However, those early electric cars were plagued by the same problems that electric cars have today: a heavy weight, because of all the batteries needed, and a limited driving range. Steam-powered cars were somewhat successful, notably the Stanley 'Steamer,' which was probably faster than most other cars of its time, but had the disadvantage that the driver had to wait for the car to develop a head of steam before driving off. By about 1910, the gasoline-fueled internal combustion engine was dominant, just in time for Henry Ford and his 'Model T.' The great advantage of the internal combustion engine was its use of energy derived directly from burning fuel, i.e., gasoline. All the apparatus connected with the use of steam could be eliminated. Both the engine and the fuel supply in the automobile were relatively compact, so that it could be economical to build a vehicle that would carry only a few people at a time—that is, the personal passenger car as we would recognize it today.

The engine was almost invariably a four-stroke gasoline engine operating on the Otto cycle, though two-stroke engines were introduced and became standard for motorcycle use, and eventually diesel engines were manufactured small enough to be competitive in automobiles. Otto's engine was not only a major stride in the development of land transportation as we know it today, but also the parent of powered aviation.

The 1890s gasoline-powered, internal combustion engine inventors chose that technology for more than just its greater potential. They were concerned about a possible revival of both legislative bans on, and consumer resistance to, steam, even though steam worked better than internal combustion engines for the automobiles of the 1890s. The switch of Locomobile (the leading American manufacturer at the time) and Bollée (a leading French manufacturer) from steam to internal combustion just after the turn of the century established the triumph of internal combustion. Nevertheless, in 1906 a steamer still held the land speed record, and as late as 1917 some steamers lingered on for heavy trucks. Another advantage for the internal combustion engine was that the driving rules of that era typically required that steam-car drivers hold a boiler operator's license, at a time when drivers of gasoline autos needed no license at all.

The Impact of the Early Automobile

A major effect of the automobile was that it revived road transport systems. We have seen that they had been in dreadful shape in Europe for some 1,500 years, ever since the fall of the western portion of the Roman Empire. In the 19th century, roads languished in competition with the railways. The revival and rebuilding of roads promoted a massive increase in facilities for personal transport and movement of freight to suit the precise location and convenience of the customer.

Oddly, one of the major forces pushing for the rebuilding of the road system was the reversion to the use of human energy in the form of the bicycle. The bicycling craze in the late years of the 19th century helped people, especially young people and women, to travel more widely than ever before, thereby generating a new demand for even more freedom of movement in the future. The bicycle not only created a market of potential customers, but also was a fertile source of engineering expertise. Many of the techniques devised for bicycle construction were used in car-making too.

THE WRIGHT STUFF

James Watt did not invent the steam engine, Robert Fulton did not invent the steamboat, and Henry Ford did not invent the automobile. Not only that, the Wright Brothers did not invent the airplane. George Cayley, in 1799, outlined the principles of aircraft design (Figure 19.7), with stiff wings and a body, and a stabilizing tail unit with control surfaces. These principles are in fact the ones on which aircraft design is based today, but Cayley's contemporaries did not even take him seriously. After all, Cayley needed a prime

FIGURE 19.7   This aircraft designed by George Cayley in the 18th century shows that, more than a hundred years before the Wright brothers' flight, scientists were approaching plausible ideas of air transportation.

mover—a source of motive power for the airplane. A steam engine was obviously out of the question, being far too bulky and heavy. He even touched on the possibility of using a 'gunpowder engine,' but went no further than to suggest that a heat engine with air as its medium must be the best means of powering an aircraft.

In 1842, John Pennington of Baltimore applied for a patent for a gas-filled, dolphin-shaped flying machine to be driven by a small steam engine and steered by means of a rudder. For the sake of light weight, the engine was to be made of steel, which seems a curious choice with the hindsight provided by today's high-strength, light-weight materials. The boiler was to be filled with alcohol, which has a lower boiling point than water, so less heat would be required to generate the alcohol vapor. (Strictly speaking, the engine was not a 'steam' engine, since it was designed to operate on the vapors of alcohol instead of water.)

In the latter years of the 19th century other inventors took up the challenge of powered flight. The early years of the 20th century saw the gasoline engine used to drive propellers in prototype aircraft. In this application, there was no real contest among different types of engines. The gasoline-fueled internal combustion engine was the only choice for a lightweight power source. (In 1894, Hiram Maxim proposed a high-pressure steam engine for *airplanes*, with the steam being generated on the ground. Though this concept was clearly a dud, one of Maxim's other inventions likely changed the course of human history in 1914–1918: the machine gun.)

Around 1896, Professor Samuel Pierpont Langley, of the Smithsonian Institution, was experimenting on the Potomac below Washington with model airplanes. Some of these models, which had wingspreads of up to 14 feet, were powered with small steam engines. Then, in 1903, he built a full-size one-man airplane and provided it with a remarkable gasoline motor designed by Charles Manly. This engine was a five-cylinder water-cooled gasoline engine of 52 horsepower that weighed only 125 pounds ($2\frac{1}{2}$ pounds per horsepower). Unfortunately, Langley's plane twice collapsed as it was being launched from the roof of his houseboat, nearly drowning Manly, the pilot. Langley became discouraged and refused to experiment further. Manly could not have been too happy either.

The Wright brothers' gasoline motor, which they made themselves, was a four-cylinder, four-inch bore, and four-inch stroke, and developed 12 horsepower. The motor weighed 179 pounds (15.9 pounds per horsepower—compare this with Manly's engine). The whole airplane weighed 750 pounds.

The Wright brothers, Wilbur (1867–1912) and Orville (1871–1948), sons of a bishop of the United Brethren Church, had a keen interest in mechanical inventions from boyhood. Their thoughts turned towards flying machines in June 1878 when their father gave them a toy helicopter designed by the Frenchman Alphonse Penaud, who first used rubber bands to power model aircraft. As young men, they experimented with kites and gliders, while running a business repairing and building bicycles. It was Orville who had the idea of constructing an aircraft wing with movable sections (ailerons), so that the pilot could vary their inclination. This was the original Wright Brothers patent. Their first powered machine was a 40 ft. wingspan biplane with a 16 h.p. four-cylinder motor. To reduce the risk of its falling on the pilot, the motor was mounted on the lower wing right of centre, and the pilot lay flat, left of centre, to balance it. It had two propellors, and sledge-like runners instead of wheels. For take-off it was put on a truck with wheels that fitted into the groove of a monorail track. It first flew on 17 December 1903 at Kill Devil Hill, a few miles south of the remote coastal hamlet of Kitty Hawk, North Carolina, watched by five locals from the Kill Devil Life Saving Station, two small boys and a dog. The brothers tossed a coin to decide who should be pilot first, and Wilbur won.

—Carey[1]

The Wright brothers first flew on December 17, 1903 (Figure 19.8).

Success four flights thursday morning all against twenty one mile wind started from Level with engine power alone average speed through air thirty one miles longest 57 seconds inform Press home Christmas.

—O. Wright[2]

This flight lasted only 12 seconds, but it was nevertheless the first in the history of the world in which a machine carrying a man had raised itself by its own power into the air in full flight, had sailed forward without reduction of speed, and had finally landed at a point as high as that from which it had started.

—O. Wright[3]

By 1905, they were carrying passengers.

Well, sir, we pulled that fool thing around over the ground of Huffman Prairie about thirty or forty times, hoisting it up on the derrick so it would get a good start, and we were all hot and sweaty and about played out. What was the use of our wasting our

FIGURE 19.8   The Wright brothers' first flight, December 17, 1903.

time over such a ridiculous thing any longer? But once more we pulled her up again and let her go. The old engine seemed to be working a little better than normal. Orville stuck his head and nodded to Wilbur and Wilbur turned her loose. <u>And by God the damn thing flew</u>.

—Webbert[4]

THE DIESEL

Internal combustion engines had one additional, very significant difference from external combustion steam-engine technology. With steam engines, the technology developed first, and the scientific understanding of the engine followed later (e.g., Watt preceded Carnot). But in the case of the internal combustion engines, a much higher level of scientific understanding was required of both inventors and manufacturers virtually from the very beginning. The steam engine had been a product of the empirical tradition of practical millwrights. Speculation about the fundamental behavior of the engine (the way in which it transformed heat into WORK) and the practical concern to measure the efficiency of the engine prompted the development of thermodynamics by Carnot and his successors. The new science then was available for, and became important to, the development of the internal combustion engine. In particular, the high-compression engine developed by Rudolf Diesel in 1892 was inspired by the scientific principle that it should be possible to induce self-ignition in the fuel by intense compression.

At this time there was interest in developing internal combustion engines that would operate not on gasoline, but on fuels of poor quality, such as pulverized coal or the high-boiling components of petroleum. The fuel that Diesel first intended to use in his engine was finely powdered coal, a fuel that many designers were then thinking of using for various kinds of engines. Diesel soon abandoned that idea and instead started to experiment with crude oil and fuels from heavier petroleum fractions. Since we have said that gasoline became a favored fuel for internal combustion engines, since it vaporizes so easily, a line of development using heavy, hard-to-vaporize materials might sound unpromising, since these substances not only vaporize less readily than the lighter ones, but also ignite less readily if sparked. In 1892, Diesel patented an engine in which a mixture of air and heavy oil vapor could be ignited by compression alone.

In the 1920s, oil became the dominant fuel for ships, thanks in large part to a decision made in 1911 by Winston Churchill, then serving as First Lord of the Admiralty, to convert the British navy from coal to oil. Oil firing for steamships is now used in 79% of steam tonnage and has greatly reduced costs. In 1903, five years after the Diesel engine was first produced commercially, this type of engine was installed in two Russian tankers for service in the Caspian Sea. The first important ocean-going vessel to be diesel powered was the *Selandia*, built in 1912. Marine diesels, like locomotive diesels, have a higher thermal efficiency than steam engines. Diesel had developed one form of his early high-compression oil-burning engine specifically for use in submarines, but it was quickly recognized that it provided an excellent engine for any small or medium-sized ship, and it was eventually developed to challenge the steam turbine in even the largest marine applications. Virtually all large ships are now diesel powered.

Although Diesel had to overcome formidable operating difficulties before his engine could become a commercial success, by the second decade of the 20th century it had been adopted for a wide variety of heavy-duty engines, in ships, tractors, and buses, and continues to enjoy enormous success today in virtually every country of the world. Railway locomotives with Diesel engines were first used in the 1920s. They caught on slowly, because of the economic crisis of the Great Depression in the 1930s, followed by the tremendous burdens on manufacturing and logistics of the Second World War. American railroads put the first high-speed passenger-service diesel–electric locomotives into operation about 1935. Freight-service diesels appeared in 1937, and the dual-purpose road-switcher type came into use in 1940. The diesel–electric locomotive began to replace the steam locomotive for mainline operation in the United States during the 1940s. During the 1950s, diesel locomotives totally displaced the coal-burning steam locomotive from railways in the United States.

The diesel locomotive has important advantages relative to the steam locomotive. The useful power of a diesel locomotive may be greater than that of a steam locomotive of the same nominal rated capacity because the diesel power is available over a wider range of speeds. The engine terminal expenses are generally a good bit less, because there is less work to be done at the terminal. Steam locomotives require a supply of water; in turn, this requirement creates a need for water tanks and water purification facilities. With a diesel locomotive, there are no complications of supplying pure water for the boiler, because no water is used. Diesel fuel is easy to handle and store at the terminals. Virtually all diesel locomotives used in the United States are actually diesel–electric locomotives. The diesel engine operates an electricity generator, which supplies electricity to the traction motors that actually propel the locomotive. The diesel–electric locomotive has an efficiency of about 30%, which is about five times that of the steam locomotive. Diesel locomotives can be assembled easily in multiple units under the control of one crew, and these multiple units may produce tractive effort and power concentrations far greater than a steam locomotive.

THE JET

In 1929, the Swedish company Bofors decided to develop a gas turbine engine for airplanes. They had in mind an 'aeroplane without a propellor,' where the propellor's powerful air stream would be replaced by the gas jet from a steam turbine. Bofors tried to interest the Royal Swedish Air Force in the project, but this happened at a time when the defense budget had been greatly cut, so the scheme was abandoned in 1935.

In the 1930s, engineers, most notably the English engineer Frank Whittle, began serious work on the jet engine. By this time, electric lighting had become widespread in homes, drastically reducing the demand for kerosene as a fuel for lamps. But at the same time, gasoline was becoming in shorter supply due in part to building up stocks of fuel for the impending war. As a result, Whittle reworked his design of an engine to use kerosene as the fuel. Today's jet engines use a slightly modified and refined form of kerosene as jet fuel.

Many designers in the German aviation industry had been trying since about 1930 to produce a gas turbine which could be used in airplanes. A team led by Ernst Heinkel was the first to reach a solution. A few days before the outbreak of the Second World War in 1939, the world's first jet plane, the Heinkel He 178, flew for the first time, at Warnemünde. The Junkers firm started to develop another gas turbine engine around the New Year of 1940, and the first flights took place two years later with the Jumo 004 engine. This later became the first jet engine to be mass produced. A slightly modified version was installed in the German twin-engine fighter-bomber aircraft, Messerschmidt Me 262, which began to be produced in 1943.

These startling breakthroughs of gas turbine (jet) engines, associated mainly with aviation in the Second World War, opened car manufacturers' eyes to this new power source. In 1947, Rover began to work intensely on development. In the early 1950s, Jet 1, the turbine-driven Rover, became one of the world's most talked about cars, but never went into mass production. In the late 1950s, Renault put a great deal of effort into gas turbine propulsion, but that project too was abandoned after a few years. Chrysler's first gas-turbine-driven car was completed in 1950, and in 1963 the company took its Chrysler turbine on a European tour that aroused a great deal of interest. Once again, this vehicle was never put in production. A major reason for the failure to commercialize these successful engineering prototypes is that the turbine motor was too expensive to operate and more expensive to build than a conventional car motor.

ELECTRICITY FOR TRANSPORTATION

Through the second half of the 19th century the steam locomotive was virtually the universal source of power on railways. However, it was recognized that, for certain purposes, electric locomotives had enormous advantages over conventional smoke-generating steam locomotives, and were adopted in London on the new Central Line of the Underground Railway in the 1890s. These advantages are especially important in metropolitan regions where the smoke from the steam locomotives would be a major pollution problem.

Where oil fuels were cheap, as in the United States, some of the leading mainline services were converted to diesel traction, which emerged as a serious competitor to steam after the Second World War, as we've seen. In Europe, when the national railway enterprises were almost totally rebuilding their systems after the war, the favorite choice of traction power was electricity, supplied at high voltage as alternating current through overhead cables. In the United States, only 2% of the total mainline mileage is electrified, but abroad where fuel is scarce, and especially in sections where water power is plentiful (i.e., for hydroelectricity), railroad electrification has developed more extensively.

The electric street railway has had a brief but hectic history in the United States. From the late 1880s to 1950 it furnished an enormous amount of urban transportation before being largely displaced by a motorized version of one of its predecessors, the omnibus. Ernst Werner von Siemens built the world's first public-service electric street railway at Lichterfelde, a suburb of Berlin, in 1881. Charles C. Henry of Indiana introduced interurban electric lines into the United States in 1894 as extensions of street railways. The interurbans, which carried freight as well as passengers, reached their peak operation about 1910, but had almost entirely disappeared in the United States by 1930. The rapid increase in the number of private automobiles in the United States brought about the decline of the electric street and interurban railway.

THE ENERGY DEMANDS OF TRANSPORTATION

In the United States, the transportation sector accounts for about 27% of our total energy consumption. This fact, though, doesn't reflect our fixation on a single kind of energy source, petroleum, that exists in transportation. There are ongoing efforts to examine alternatives to petroleum for transportation energy. In some parts of the country, railway locomotives are electric, and the electricity can of course be generated from many different sources. (In many areas of Europe and Japan, the great majority of railway locomotives are already electric.) Natural gas is a possible replacement for diesel engines in city buses. Methanol, which could be made from natural gas or from coal, is a possible automotive fuel. Ethanol, which can be made from corn or other agricultural products, is also advocated as an alternative automotive fuel.

Despite these energy alternatives, it is still true that a single source—petroleum—accounts for about 99% of all our transportation energy. Looked at from a different perspective, nearly two-thirds (actually, 62%) of all the petroleum we use is consumed by the transportation sector. If we round the numbers off a bit, we can say that (a) a quarter of all the energy we use is used for transportation; (b) all of that energy derives from petroleum; and (c) this use accounts for two-thirds of all the petroleum we consume.

These facts are also impacted by one other key point: about 50% of the petroleum we use is imported. Because, in a sense, the United States is 'hooked on oil' and because half or more of it is imported, we have become very vulnerable to the availability of imported oil and its cost. There have been three major shifts in oil pricing and availability during the past two decades: the oil price 'shock' of 1973; the oil embargo and consequent price rise in 1979 and the early 1980s; and the oil price 'spike' during the Persian Gulf war of the early 1990s.

When these changes occur, they cause a number of dislocations in society. A rise in oil prices usually kicks off a spurt of general inflation in the economy. When oil is expensive or is in short supply (or both), we get serious about such matters as energy conservation, including buying fuel-efficient cars, increasing domestic oil exploration and production, and seeking alternatives, such as making liquid fuels from coal. We have fought one war (the Persian Gulf war) over oil (despite the official rhetoric about defending democracy in Kuwait). Some cynics observe that the best thing the people of war-torn

Bosnia and Kosovo could have done to secure Western intervention and aid was to strike oil.

It's important to realize that hominids that we would recognize as our human-like ancestors existed about 5,000,000 years ago. Modern humans (*Homo sapiens*) first evolved some 240,000 years ago. Yet there are still a few people alive today who were born before airplanes existed or before automobiles were readily available. The alternative to human or animal muscle for convenient personal transportation has existed for less than 90 years (if we take the availability of the Model T Ford as the start of the 'boom' in personal transportation). Nevertheless, virtually all of us own our own car, or have ready access to a car through families or friends. For many students who don't already own one, a car very often becomes one of the first major purchases made after graduation. So, at the beginning of the 21st century, petroleum is the dominant transportation energy source: gasoline for cars, light trucks, and piston-engine planes, kerosene (jet fuel) for jet aircraft, diesel fuel for heavy trucks, railway locomotives, and small ships, and fuel oil for large ships all derive from it.

Usually when oil is readily available and relatively cheap, we promptly forget about all of these concerns. It happens that nowadays the price of oil seems to be slowly creeping up again, although we are still able to import as much as we want. Is the next 'oil crisis' coming?

A serious rise in the price of oil fuels could encourage the development of alternatives to the internal combustion engine as the form of propulsion, with various electric motors being the most obvious candidates. Both electricity and steam provided strong competition to the gasoline engine at the beginning of the century, but the need to carry bulky batteries or steam-generating equipment weakened their challenge. Changing circumstances, however, and especially new environmental concerns about the emission of polluting gases, could provide new opportunities. Spark-ignition engines can be made to run well on fuels such as ethanol, liquefied petroleum gases (LPG), and compressed natural gas. Compared with gasoline, these fuels generate less energy per pound of fuel carried. In some countries, such as the Netherlands, which already have an extensive LPG network, cars are adapted to run on either gasoline or LPG and have one tank for each fuel. In those parts of the world where petroleum products are not readily available, diesel engines could be run with fuels of similar molecular weight, such as oils from coconut, soya, and sunflower. But in considering any of these alternatives we must take into account the fact that the cost and upheaval needed to extend or change any country's fuel distribution system would be enormous.

CITATIONS

1   Carey, J. *The Faber Book of Science*. Faber and Faber: London, 1995; 236–240.
2   Wright, O. Text of telegram sent to his father, December 17, 1903.
3   Wright, O. *Flying* [magazine] December 1913.
4   Webbert, C. In: *Grand Eccentrics*. Bernstein, M. (Ed.). Orange Frazer: Wilmington, OH, 1996; 91.

FOR FURTHER READING

Chapelon, André. *La Locomotive a Vapeur*. Camden Miniature Steam Services: Bath, UK, 2000. One of the greatest books ever written on steam locomotives, now available in English translation.

Chapelon ranks among the finest of mechanical engineers involved in design and construction of steam locomotives.

Cross, Gary; Szostak, Rick. *Technology and American Society*. Prentice Hall: Englewood Cliffs, 1995. This book discusses the interplay between inventions and technological development on the one hand with social, economic, and cultural change on the other. Chapters 6, 15, and 18 are particularly focused on the development of transportation technology.

Ethell, Jeffrey L. *Frontiers of Flight*. Smithsonian Institution: Washington, 1992. A well-illustrated history of the development of aviation from the earliest powered flight to space exploration.

Jakab, Peter L. *Visions of a Flying Machine*. Smithsonian Institution: Washington, 1990. This book is in essence a 'prehistory' of the Wright brothers' first successful flight. It describes the experiments and developmental work that led to their success.

Kirby, Richard Shelton; Withington, Sidney; Darling, Arthur Burr; Kilgour, Frederick Gridley. *Engineering in History*. Dover: New York, 1990. This is an inexpensive reprint of a book originally published in 1956. Chapter 12 provides a history of transportation from early 19th century to the middle of the 20th century; and Chapter 8, a history of the development of roads, canals, and bridges.

McShane, Clay. *Down the Asphalt Path*. Columbia University: New York, 1994. A history of the interactions between the development of the automobile and increasing urbanization.

Newcomb, T.P.; Spurr, R.T. *A Technical History of the Motor Car*. Adam Hilger: Bristol, 1989. This is a history of the engineering development of automobiles and their components, such as tires, steering systems, and brakes, through the early 1980s.

O'Brian, Patrick. *Master and Commander*. Norton: New York, 1970. There is no better way to learn the intricacies of the design, nomenclature, and operation of sailing ships than by following the adventures of Jack Aubrey of the Royal Navy, and his friend Dr. Stephen Maturin. *Master and Commander* is the first of a 20-volume series, all still in print in relatively inexpensive paperback editions.

Riley, C.J. *The Encyclopedia of Trains and Locomotives*. MetroBooks: New York, 2000. A nicely illustrated overview of steam, diesel, and electric locomotives and the history of their development.

Stoff, Joshua. *Picture History of Early Aviation, 1903–1913*. Dover: New York, 1996. This book delivers exactly what the title indicates, with over 300 photographs from the period.

Strandh, Sigvard. *The History of the Machine*. Dorest: New York, 1979. A very-well-illustrated history of machines of all sorts. Chapter 4 includes information on the use of human and animal muscle power; Chapter 5 is devoted to various kinds of engines.

Suzuki, Takashi. *The Romance of Engines*. Society of Automotive Engineers: Warrendale, PA, 1997. This book provides an idiosyncratic, but enjoyable, history of the development of engines, with the major focus on internal combustion engines. An extraordinary mélange of engines, vehicles, and aircraft is discussed.

Wescott, Lynanne; Degen, Paula. *Wind and Sand*. Eastern Acorn Press: Washington, 1996. Subtitled 'The Story of the Wright Brothers at Kitty Hawk,' this book details the development of aviation by the Wrights through 1911. It relies heavily on the brothers' own writings and photographs.

20

PETROLEUM AND ITS PRODUCTS

FOSSIL FUELS AND THE GLOBAL CARBON CYCLE

We should first begin by inquiring what the term *fossil fuels* actually means. A *fossil* is a remnant of formerly living organisms preserved in the Earth's crust. A *fuel* is a substance that we use as a source of ENERGY, usually by burning it to liberate heat. So, fossil fuels are energy sources originating from the remains of once-living organisms that have accumulated inside the Earth. The principal fossil fuels are coal, petroleum, and natural gas. Their importance to us lies in the fact that about 95% of all the energy used in the United States derives from fossil fuels. Collectively, they represent the most important source of ENERGY that we have, for electric power production, transportation, manufacturing, and use in the home.

At first sight, fossil fuels are very diverse materials: a colorless gas, a liquid that ranges in color from a pale yellow to black, and a brown or black solid. Despite these great differences fossil fuels as a group have two things in common: First, they are, by definition, fossils—remains of animals and plants. Second, their major chemical component is carbon. To begin the story of the origin of fossil fuels, we consider how carbon is dispersed throughout the world. This can be represented by the **global carbon cycle** (Figure 20.1). The key step that starts the cycle going is photosynthesis. Photosynthesis occurs within plants to convert carbon dioxide from the atmosphere and water into sugars, which the growing plants then use as an energy source to help produce all of the other chemicals needed for their life cycles. In other words, solar-powered life processes convert inorganic materials—carbon dioxide and water—into the organic compounds needed for life and growth. These compounds include cellulose, starches, and sugars; proteins and amino acids; and fats and oils. These are the compounds that will eventually transform, through a series of biochemical and geochemical processes, into the fossil fuels. It is important to notice

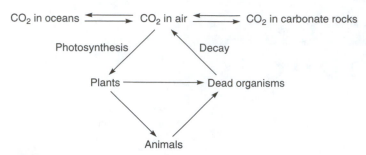

FIGURE 20.1 Carbon moves through our entire world as it passes through different stages of the global carbon cycle.

that the energy source that drives the global carbon cycle is sunlight (solar energy). Essentially,

THE FOSSIL FUELS REPRESENT A STORED-UP SUPPLY OF SOLAR ENERGY.

In principle, this cycle accounts for the fate of all the carbon in the world. Then where do fossil fuels come from? The preservation of plant and animal remains—the word *fossil* in the name fossil fuels—suggests that it is the decay process that must be interrupted.

Needing these vital materials, we of course would like natural processes to maximize the amount of fossil fuels that are produced. If we consider the formation of fossil fuels as a detour in the operation of the global carbon cycle, as can be seen in Figure 20.2, then two options present themselves: The amount of fuel that can be produced can be affected by increasing preservation at the expense of the decay process or by increasing the production and input of the dead organisms. In fact, both happen in nature.

Maximizing the production of organisms requires abundant light, to drive the photosynthesis process important in the growth of plants, ample warmth (there are few living things in the Arctic, for example), and ample amounts of moisture (similarly, there are few living things in deserts). These criteria are satisfied by tropical or subtropical environments such as swamps, river deltas, lakes, lagoons, and shallow seas. The watery environment is important for two reasons. First, the abundance of water stimulates growth of organisms. Second, the water will eventually help in the preservation of these organisms. In fact, virtually all petroleum is found in what are clearly marine sediments collected beneath the seas and oceans.

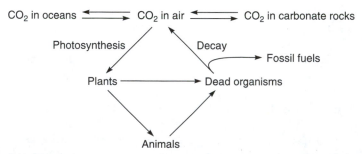

FIGURE 20.2 When some dead organisms are trapped in the Earth, their carbon becomes a part of fossil fuels rather than returning to the air.

An observation so obvious that we probably never give it much thought is that there is usually not an extensive accumulation of dead plants on land. A day's outing in a forest doesn't require that we clamber over the remains of trees that died and toppled decades or centuries ago. On land, exposure of dead organisms to the oxygen in the air causes virtually complete decay to carbon dioxide and water. Decay is prevented by excluding the principal agent of decay, which is oxygen in the air. This can be accomplished by covering the dead organisms with some sediment (i.e., mud) or having them accumulate in stagnant or slow-moving bodies of water, where the amount of oxygen dissolved in the water is not large. In a watery environment, particularly if the water is stagnant and contains little dissolved oxygen, the water helps to protect accumulated dead organisms from the ravages of oxygen from the air. Here, the main process in the decomposition of dead organisms is effected by anaerobic bacteria.

The kinds of organisms that accumulate and gradually transform to fossil fuels include plankton (single-celled animals or plants that live in aquatic environments), algae, and the higher plants—grasses, shrubs, and trees. Chemically, these organisms are composed of carbohydrates (sugars, starches, and cellulose); fats, oils, and waxes; proteins; and—in the special case of higher plants—lignin, the special material that provides the mechanical structure and rigidity for these plants. As these various chemical components are decomposed, they gradually transform into a family of new substances called **kerogens**. These are brown or black solids that do not dissolve in ordinary chemical solvents and have large, complicated molecular structures. Kerogen is rich in carbon and hydrogen, and contains some nitrogen, sulfur, and oxygen. A familiar form of kerogen is peat moss.

COOKING KEROGEN

Kerogen can be thought of as being the 'halfway point' to petroleum. Eventually (on a long geological time scale) the accumulation of dead organisms and their conversion to kerogen stops. If the environment dries out, the accumulated kerogen could be exposed to the atmosphere and destroyed. Alternatively, the accumulated kerogen could be covered by sediments, such as mud, sand, or silt. Now a new process can begin. As the kerogen is buried deeper and deeper inside the Earth, it becomes exposed to high temperatures, and sometimes high pressures for thousands, hundreds of thousands, millions, or even hundreds of millions of years. The temperature increase inside the Earth is due to a natural phenomenon called the geothermal gradient, an increase of temperature with depth. One of the phenomena that cause the geothermal gradient is the decay of naturally occurring radioactive materials (Chapter 26) in the Earth's crust. The long, slow 'cooking' inside the Earth opens a new chapter in the formation of fossil fuels.

Normally as kerogen accumulates it is mixed with the surrounding inorganic sediments; typically, a so-called oil source rock will contain about 1% kerogen and 99% inorganic rocks.

When kerogens have formed from algae, or plankton and algae, most of their chemical components have derived from fats, oils, and waxes. The molecules are characterized by long, continuous chains of carbon atoms. As the kerogen 'cooks,' some of the chemical bonds between these carbon atoms break apart. The effect is just like breaking any other sort of chain—smaller pieces of chain are produced. Ultimately, the fragments of chain become small enough that they are actually in the liquid state. This is the material that we would recognize as petroleum or crude oil.

There is a rough relationship between molecular size and the physical state in which the molecule exists at ordinary temperatures and pressures. (This rule tends to hold as long as we confine the comparison to molecules within a specific family of chemical compounds, as is the case when we consider the molecules in petroleum.) Large molecules, with more than about 18 carbon atoms, tend to be solids. Medium-sized molecules, with 5–17 carbon atoms, are usually liquids. Small molecules, with four or fewer carbon atoms, are gases at ordinary conditions. As the kerogen is broken apart by heat, eventually some molecules are produced that exist as liquid. This liquid is petroleum or crude oil. (We will use these terms interchangeably.)

The amount of crude oil that forms from kerogen and the sizes of molecules in the oil depend on how much the kerogen has been 'cooked' inside the Earth. This in turn depends on the depth to which the kerogen has been buried and the accumulated time during which it has been buried. As the conversion of kerogen to petroleum occurs, it may be that some of the new molecules are small enough to be gases. This is the material that we recognize as natural gas. Many deposits of petroleum around the world are accompanied by some natural gas. Thus the continued breaking down of the chains of carbon atoms, with higher and higher temperatures, longer and longer times, or both, eventually leads to chain fragments so short that they contain four or fewer carbon atoms. These molecules are gaseous at ordinary conditions. Thus the product of extended 'cooking' of kerogens from algae or plankton is natural gas.

If the kerogen contains abundant lignin, contributed by the higher plants, the chemical story becomes much different. Lignin is characterized by many rings of carbon atoms, interconnected by short chains of carbon atoms. The rings resist being broken apart chemically. As the kerogen undergoes transformation, the structure actually becomes more and more rigid, as rings unite to form larger, rather than smaller, molecular fragments. This lignin-rich kerogen transforms into a hard, carbon-rich solid, coal.

As the hydrocarbons develop from the kerogen, molecule by molecule, they migrate with the water out of the mud and shale strata into the more open strata. Usually this oil will migrate through the interior of the Earth, moving away from the source rocks through other porous rocks in the Earth. Three things might happen to the oil. It could migrate to the surface of the Earth, where it may possibly be destroyed by contact with the air. It is very likely that far more petroleum has disappeared in this manner than has been captured in the Earth. It could migrate away through the porous rocks until it is completely dissipated. Or, it could migrate until it encounters a formation of an impervious layer of shale, salt, or some other nonporous rock, which stops the migration and allows the oil to collect. This third option is the one that is desired. The oil accumulates in the pores of the porous rock, but it cannot migrate further because it's stopped by the nonporous rock structure. This trapped deposit of oil is called a **reservoir**. The oil reservoir contains an accumulation of oil held in place by nonporous rocks. This concept is shown schematically in Figure 20.3. If the accumulation is large enough—and if we can find it—it is worth drilling.

Three factors make an oil reservoir valuable: its location, its size, and the quality of the oil. Factors affecting the location of the reservoir include its depth and the hardness of the rock intervening between us (on the surface) and the reservoir. These factors indicate how difficult it will be to drill to the reservoir. The size of the reservoir indicates something about how much money might be made by extracting and selling the oil. The quality of the crude oil determines, first of all, how easy it is to get out of the ground. The ability of the oil to flow, or to be pumped, improves as the size of the molecules decreases. The resistance to flow, the *viscosity*, is the property that is usually measured. A second concern about the quality of the oil is its chemical composition, one point in particular being its sulfur content. A third property of crude oil of interest is the range of boiling points of its

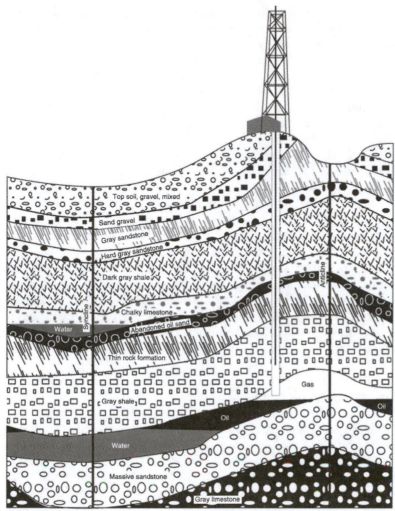

FIGURE 20.3 Oil reservoirs can be difficult to find because they are formed in pockets where kerogen 'cooked' deep beneath many layers of rock.

components. The boiling temperatures and viscosities are functions of the size of the molecules. As useful rules-of-thumb, the larger a molecule, the higher will be its boiling point and the higher will be its viscosity. (Again, these rules work only as long as we confine comparisons to molecules within similar chemical families.) The sulfur content of a crude oil depends on the extent to which the chemical bonds holding the sulfur atoms in the molecular components of the oil have been broken.

NATURAL GAS

Because natural gas is formed under conditions similar to those that formed crude oil, and usually from the same kinds of kerogen, gas and oil often occur together. At high pressures inside the Earth, the gas can even dissolve in the oil (and is referred to as **dissolved gas**), and it is released when the oil is brought to the surface. A rough analogy is the fizzing when a soft drink bottle is opened, caused by the release of dissolved carbon dioxide that had been held under mild pressure. Alternatively, the gas can exist as a separate material, usually over the pool of oil (a case known as **gas cap gas**).

In addition, the gas can also migrate away from the oil. The only thing that can keep the gas inside the Earth—that is, keep it from being lost to the atmosphere—is a layer of impervious, nonporous rock that serves as a 'cap' to prevent the further migration and escape of the gas. The gas itself is held in some sort of porous rock, such as sandstone.

The principal chemical component of most natural gas deposits is methane, CH_4. Many gas deposits will contain the related compounds ethane, C_2H_6, propane, C_3H_8, and butane, C_4H_{10}. Some gas deposits contain compounds that would be liquids at ordinary temperatures, but are present as vapors in the gas because of the higher temperature inside the Earth. These compounds include pentane, C_5H_{12}, and hexane, C_6H_{14}. When such gases are brought up to the surface, the pentane and hexane will condense back to the liquid state, forming the useful product known as **natural gasoline**.

If the natural gas is cooled below ambient temperatures, the butane and propane can be condensed to liquids. The condensed butane and propane are sometimes referred to as **liquefied petroleum gases**, more familiarly known by the initials LPG. In areas that have no easy access to natural gas distribution systems, LPG makes a very convenient fuel gas for home heating and cooking. Butane and propane can also be used by the chemical industry for conversion into many useful materials, such as polypropylene or synthetic rubber.

Some deposits of natural gas also contain hydrogen sulfide, H_2S. Gases that contain hydrogen sulfide are called **sour** gas. Hydrogen sulfide is a very undesirable component of natural gas, for several reasons. It has a vile odor—in fact, it is the compound that gives rotten eggs their characteristic odor—and is poisonous. Hydrogen sulfide dissolves in water to form a mild acid that could corrode pipes, valves, and other components, in systems for gas distribution and use. If the gas were burned, the sulfur in the H_2S would be converted to the oxides of sulfur. These gases are health hazards and contribute to the serious environmental problem of acid rain (Chapter 30). Since hydrogen sulfide is a mild acid, it can be removed from natural gas by absorbing it in a chemical that is a mild base. This process is called **sweetening**.

After natural gas has been sweetened, and the butane and propane have been removed for LPG or for the chemical industry, the gas that remains is mostly methane, also containing some ethane. This is the product that actually comes to us as consumers of natural gas. Most commercial natural gas also contains a very small amount of sulfur compounds that have been added to the gas deliberately to serve as odorants. These odorants function as a warning signal to our noses to indicate a gas leak. It may seem odd, even self-defeating, that we go to the trouble to sweeten natural gas and then add back a sulfur compound. However, the sulfur compounds that are used as odorants have such extraordinarily horrendous odors that we can detect even one part of odorant mixed with a million parts of natural gas. Thus the possible corrosion, or the contribution to acid rain, is miniscule, and is certainly outweighed by their value as a warning that a gas leak is occurring.

Especially for domestic use, natural gas is truly a premium fuel. It has the highest calorific value of all the fossil fuels, about 24,000 Btu/lb or 1,000 Btu/ft^3. Natural gas heating systems are easy to control from a thermostat. These systems can essentially be turned on instantaneously (as, for example, compared to the time it takes to get a wood fire or coal fire burning well). Natural gas is a 'clean' heat—it leaves no ashes and produces no mess. No space in the home is required for storage of gas, in contrast to space needed for oil tanks, a coal bin, or a wood pile. There are no delivery problems—the furnace draws gas from the distribution system as needed, and there is no necessity to call a dealer to arrange for delivery.

PETROLEUM AND ITS AGE–DEPTH CLASSIFICATION

Petroleum varies somewhat in chemical composition, but as a rule it contains 82–85% carbon and 12–14% hydrogen. Notice that these numbers do not sum exactly to 100%; the balance of the composition is accounted for by small amounts of oxygen, nitrogen, and sulfur. The range and complexity of naturally occurring hydrocarbons is extremely large, and the variation in composition from one deposit to another is enormous. All crude oils differ in the fractions of the various hydrocarbons (described below) they contain. The specific chemical compounds can and do have molecules of different shapes and vary in size or weight from those based on the single carbon atom to those based on large and complex molecules of 80 or more carbon atoms. Physically, crudes may be black, heavy, and thick, like tar, or brown, green, to nearly clear straw color, with low viscosity. Many crudes are less dense than water, running from 79 to 95% of the weight of an equivalent volume of water. But some are heavier than water.

The temperatures inside the Earth facilitate the breakdown of the long chains of carbon atoms in the kerogen into shorter chains. Most of the compounds that are produced belong to a family of compounds called alkanes, or often, in the terminology of the petroleum industry, paraffins. The physical state in which a particular paraffin exists at a given set of temperatures and pressures depends on the number of carbon atoms. When the carbon atoms are connected in a linear chain, that relationship is as summarized in Table 20.1.

A liquid oil is likely to contain a mixture of paraffins having five or more carbon atoms. (In fact, many crude oils also contain some dissolved gases.) That alone would make oils chemically complex substances. However, the situation is made even more complicated by the many ways in which carbon atoms can join together. For example, the structure of the compound with four carbon atoms, butane, could be drawn as shown in Figure 20.4. However, there is another way in which four carbon atoms can be joined together (Figure 20.5). As the number of carbon atoms in the molecule increases, so does the number of possible ways for linking them together. The five-carbon-atom compound, pentane, can have three possible arrangements (Figure 20.6). Notice that in the first structure, the carbon atoms are linked together in a chain that is more-or-less 'straight'. In the two others, there is not a single chain containing all five carbon atoms. Rather, these structures can be visualized as consisting of chains having branches. The number of possible molecular structures rises drastically as the number of carbon atoms increases, as shown in Table 20.2. To distinguish between the two broad categories of linking carbon atoms, a straight chain or a chain with branches, we use the terms straight-chain paraffins

TABLE 20.1 RELATIONSHIP OF THE NUMBER OF CARBON ATOMS TO PHYSICAL STATE OF OIL

Number of carbon atoms in the chain	Physical state at room temperature and pressure
<5	Gaseous
5–15	Liquid
>15	Highly viscous liquids to waxy solids

FIGURE 20.4 The butane molecular structure is one way that four carbon atoms can be arranged.

FIGURE 20.5 Isobutane's structure shows a second possible arrangement of four carbon atoms.

and branched-chain paraffins. Both kinds can occur in petroleum, though usually the straight-chain form dominates.

> The apparent straightness of a chain of carbon atoms is an illusion caused by the fact that we are forced to represent a three-dimensional molecule on a two-dimensional sheet of paper. In the paraffins, each carbon atom forms chemical bonds with four other atoms (hydrogen or carbon). These four chemical bonds are at angles of 109.5° to each other, directed toward the corners of a tetrahedron. Therefore there is no such thing as a truly straight chain of carbon atoms in these compounds. Rather, the chain of carbon atoms proceeds in a somewhat zig-zag fashion (Figure 20.7). Chemists have developed conventions for representing specific three-dimensional arrangements on paper for those situations where it is very important to attend to, and keep track of, the exact three-dimensional structure of a molecule. For our purposes, it is not necessary that we do so; consequently we can write molecular structures *as if* the four bonds that a carbon atom forms with its neighbors were at 90° angles to each other.

A further possible structural arrangement of carbon atoms is a ring (e.g., Figure 20.8). This family is called cycloalkanes (the prefix 'cyclo-' indicating the ring arrangement), cyclic paraffins, or naphthenes. The naphthenes constitute a distinct family from the paraffins, because there are slightly fewer hydrogen atoms in the naphthenes. As an example, the paraffin with six carbon atoms (hexane) contains 14 hydrogen atoms. The naphthene with six carbon atoms (cyclohexane) contains only 12 hydrogen atoms.

Finally, a very special family of cyclic compounds is the aromatic hydrocarbons. The parent compound in this family is benzene (Figure 20.9; notice that benzene has only six hydrogen atoms, compared to 12 in cyclohexane). In principle, the naphthenes can have any number of carbon atoms greater than three in a ring. The aromatics are built only of hexagonal rings. Aromatics as a family have some very special chemical properties that make them quite distinct from the paraffins and naphthenes, and this distinction has important consequences in organic chemistry and the chemical industry. We, however, will gloss over almost all of these, and only touch as occasion arises on some of the special properties that apply particularly to aromatics when they are present in petroleum or its products.

> Benzene was discovered in 1825 by Michael Faraday. It is the parent of the whole family of aromatic compounds, all of which are built of joined hexagonal rings. The characteristic properties of these compounds derive from the nature of the bonds in the molecules. In any molecule, a carbon atom always contributes four electrons. In the paraffins and

naphthenes, these four electrons are used to form bonds with four other carbon or hydrogen atoms. In aromatic compounds, a carbon atom forms bonds with only three other atoms. The fourth, 'extra' electron from each carbon atom contributes to the formation of a type of bond that is spread over the entire molecule, rather than being tightly held (localized) between two adjacent atoms. These so-called 'delocalized' electrons make aromatic molecules particularly stable in many chemical reactions. For instance, comparing the molecular formulas might suggest that the conversion of benzene, C_6H_6, to the corresponding naphthene, C_6H_{12}, should be easy. It actually is fairly difficult. Similarly, the stability of aromatic molecules results in less heat being evolved when they burn than would be predicted from the simple molecular composition.

Aromatics are usually not desirable components of petroleum products. In jet fuel and diesel fuel (Chapters 23 and 24), they can lead to formation of soot as the fuel burns. Soot contributes to air pollution, as has been experienced by anyone exposed to the billowing cloud of soot in the exhaust of a bus or large truck with a diesel engine. Many of the components of soot are large aromatic molecules, which are also suspect carcinogens. Steady exposure to soot could possibly lead to various forms of cancer. Small aromatic molecules, such as benzene, used to be added to gasoline to enhance the octane number (Chapter 21). However, they too have come to be recognized as suspect carcinogens. Chronic exposure to benzene vapor may lead to leukemia. Since most crude oils, other than Pennsylvania-crude-quality oils, contain some aromatic compounds, refiners are faced with the challenge of removing them to meet concerns for air quality and human health.

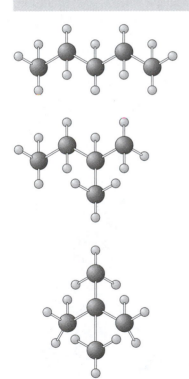

FIGURE 20.6 Pentane (a five-carbon molecule) has three different isomers, all with different possible arrangements.

To sum up, petroleum contains four kinds of hydrocarbons: straight- and branched-chain paraffins, naphthenes, and aromatics. Because of the many possible molecular structural arrangements that can exist for compounds with, say, eight or more carbon atoms, any sample of petroleum will contain hundreds of individual specific compounds.

In addition to the hydrocarbons, which, strictly speaking, are compounds that contain only carbon and hydrogen, there can be many molecules that contain one or more atoms of nitrogen, sulfur, or oxygen. These are sometimes grouped together in the jargon of the petroleum business as NSOs. For the most part, we will not be too concerned about the NSOs. The sulfur-containing compounds are an exception. Burning them, as we will see (Chapter 30), contributes to the environmental problem of acid rain. In addition, many sulfur compounds have stupendously awful odors. We have mentioned hydrogen sulfide as an impurity in natural gas. The liquid sulfur-containing compounds are even worse. Reading their descriptions uncovers a litany of comments such as 'odor of skunk,' 'rotten onions or leeks,' 'repulsive garlic,' and even just plain 'repulsive.' Petroleum containing sulfur compounds is called a high-sulfur crude, or sour crude. This sulfur will need to be removed at some point in the refining process.

TABLE 20.2 RELATIONSHIP OF NUMBER OF CARBON ATOMS TO POSSIBLE PARAFFIN STRUCTURES

Number of carbon atoms	Number of possible structures
1–3	1
4	2
5	3
7	9
10	75
20	366,319

The properties of a given sample of crude oil are determined by its constituent molecules. These in turn are determined by the kinds of organisms that originally accumulated, the chemical changes experienced during the formation of the original kerogen, and the depth and duration of burial of the kerogen. Thus various crude oils from around the world may vary greatly in their molecular compositions. The chemical complexity of one sample of oil is illustrated in Figure 20.10.

Both the molecular size and the sulfur content will depend on the extent to which the kerogen has been 'cooked' inside the Earth. The extent of cooking is determined by two factors: how much the kerogen has been heated, which is determined by its depth of burial inside the Earth and how long it has been heated, which is determined by its geological age. In other words, the properties of a particular crude oil are determined by its age and its depth. This concept provides a way of classifying or categorizing crude oils, and relating these categories to oil properties.

Let's consider a petroleum that is just at the beginning of the oil formation process. The long chains of carbon atoms have not been broken down very much, for two reasons: First, the kerogen has likely not been buried very deeply and, as a consequence, has not been exposed to high temperatures inside the earth. Second, the kerogen has likely not been

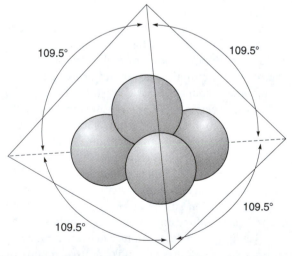

FIGURE 20.7 Although hydrocarbon structures are commonly drawn as if they were two dimensional, paraffinic hydrocarbons have tetrahedral geometry around each carbon atom, as illustrated here for methane, CH_4.

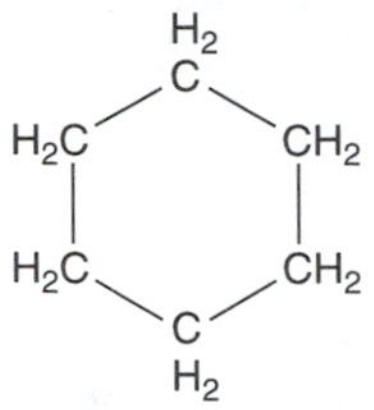

FIGURE 20.8 The name cyclic paraffins hints at the ringed arrangement of this paraffin family, which is also called cyclo-alkanes or naphthenes.

buried for a long time. Thus the chains of carbon atoms are still fairly long. The dominance of long-chain compounds gives this crude oil some distinctive properties: Comparatively, the oil is likely to be fairly dense. The long chains of carbon atoms can tangle with each other as the oil tries to flow, so the oil is also likely very viscous. It also has a high percentage of high-boiling components. In addition, there has been little chance for any sulfur compounds to break down. Thus the oil likely has a high sulfur content. Because of the short time (geologically) and shallow depth of burial, these oils are classified as *young–shallow* crudes. Young–shallow crudes are usually heavy, viscous, and sour.

Examples of young–shallow crudes are some of the oils in the Ventura, California area in the United States, and some Iranian crudes. This kind of crude oil will require a considerable amount of refining before use, and therefore is not a very desirable material for using in an oil refinery.

With a deeper burial comes exposure to higher temperatures; in turn, the higher temperatures facilitate a further breaking down of the carbon chains. In addition, at least some of the chemical bonds retaining sulfur atoms in the molecules are broken. This oil would be classified as a *young–deep* crude. Compared to other types of oils, we would expect that the young–deep crudes would be of moderate density and viscosity and moderate sulfur content. Sometimes in life, especially in business transactions, we sometimes encounter the expression 'time is money.' But to a geologist, time is temperature. The same changes that were effected by a deep burial and exposure to high temperatures could still occur with shallow burial (hence low temperatures) but over very long periods of time. Cooking provides a rough analogy: If we need something to cook more quickly, we turn up the heat. On the other hand, if not everything is yet ready, or if the meal needs to be delayed a while, we can let the cooking proceed for a longer period of time by turning the heat down. A young–deep crude developed its characteristic properties because it was 'cooked' for relatively short times at high temperatures. We ought to be able to arrive at almost the same point by doing the geological cooking for a long time at low temperatures. Such oils would be called *old–shallow* crudes, and indeed these oils often show similarities with the young–deep crudes. Oil from the region around Oxnard, California is young–deep oil, as are Libyan crudes. The Scarborough, Texas area is an example of a source of old–shallow oil.

The final case is represented by very long periods of burial that are also very deep inside the Earth. We should infer that an extensive breakdown of the molecules has occurred and hence expect an oil that is rich in low-boiling components, has low viscosity,

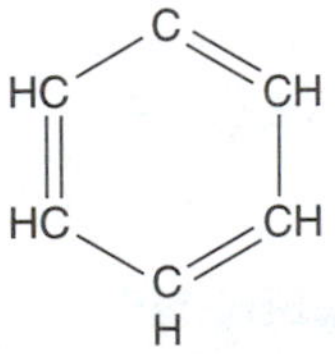

FIGURE 20.9 Aromatic hydrocarbons like this benzene molecule are only composed in hexagonal rings.

and has a low sulfur content (i.e., is sweet). The long-time exposure to high temperatures provides the impetus for breaking the carbon chains extensively, to relatively small molecules; and, most of the sulfur has been driven out by this geological cooking process. The product, *old–deep* crudes, have a low density, low viscosity (i.e., they have little resistance to flow), and very low sulfur content. These low-sulfur oils are called **sweet** crudes. Old–deep crudes were found in Pennsylvania when the American oil industry was in its infancy. The region around Bradford, Pennsylvania produces old–deep crudes. The term 'Pennsylvania crude' is sometimes used as a standard of comparison to denote oil of this high quality, regardless of its source. For example,

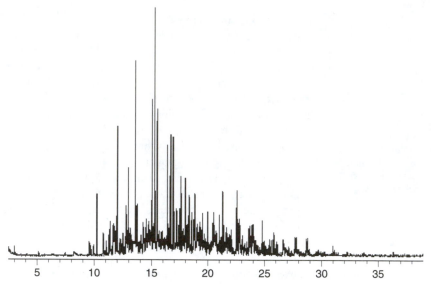

FIGURE 20.10 An analysis of a crude oil sample by gas chromatography. Each 'spike' in the graph represents a different chemical component of the oil.

Moroccan oil is often old–deep oil. Certainly most refineries, if given a choice, would prefer to operate on Pennsylvania-crude-quality oil. It would provide a high yield of desirable low-boiling-range products, be easy to pump and handle, and require little or no sulfur removal. Unfortunately, only about 2% of the world's remaining oil reserves is of Pennsylvania crude quality. (That means that oil refiners have to work harder and harder to produce the products that we want, and that satisfy increasingly strict environmental regulations, from crudes of lesser quality than old–deep crudes.). Old–deep crudes are highly desirable feedstocks for refineries because they provide high yields of valuable products. Refiners prefer light, paraffinic, sweet crudes, since they contain a large proportion of gasoline components. These crudes are less corrosive, burn cleaner, and require less processing to yield valuable products. They command a premium price.

We've seen that there are four broad categories of crude oils, based on their geological age and depth of burial. If we could examine samples of crude oils from all over the world, we would find that they range from lightly colored free-flowing liquids with a mild but not unpleasant odor to thick dark liquids with the consistency of road tar and that, frankly, stink. This is the stuff that goes into the refinery. At the other end of the refinery, we as customers expect and demand products that are of consistent quality. That demand arises because all of the items that we have built to use petroleum products—automobile engines, for example—are designed to handle a specific fuel of reasonably consistent properties and behavior. So the first challenge that refiners face is how to take this highly variable mixture of 'stuff' and convert it into products of consistent quality.

PRODUCTS FROM PETROLEUM—INTRODUCTION TO REFINING

As we've seen, natural gas, coming from the ground, has six or seven hydrocarbon constituents. Treating gas involves sweetening, if necessary, condensing the liquid natural gasoline, and removing the propane and butane for LPG. The resulting product, as sold to us, has very uniform properties and consists almost entirely of a single component.

Now consider crude oils. If we were to analyze crude oils, we would find that they contain literally hundreds of individual chemical compounds. In principle, we could separate the crude oil into each of these components, which would certainly guarantee products of consistent quality. However, there are two pretty serious problems with this idea. First, separating crude oil into its individual molecular components would be so difficult that the products would be extremely expensive, perhaps $8–10 per gallon for gasoline, for example. Second, the percentages of all the components must, by definition, add up to 100 (since nothing can have more than 100% of ingredients); since crude oil has several hundred components, most of them are, necessarily, present at less than 1% each. Therefore the yields of these very expensive components would be way too small to meet market demands. The compromise is to separate the oils not into pure components, but into groups of components of which each group has properties that are relatively consistent, and that vary only over narrow ranges. This compromise approach of separating crude oils of variable quality into affordable products of consistent properties is the essence of petroleum refining.

The technology for making this separation is distillation. This is a separation process that relies on differences in boiling behavior of the components of a mixture. In an oil refinery, distillation is done in a large, vertical cylindrical vessel called a distillation column or distillation tower (Figure 20.11). Distillation is the heart of an oil refinery and is the single most important process in refining crude oil into useful products. In the laboratory, we use distillation to separate individual components of a mixture at specific temperatures, their boiling points. In a refinery, a given distillation product is produced over a range of temperatures, the boiling range.

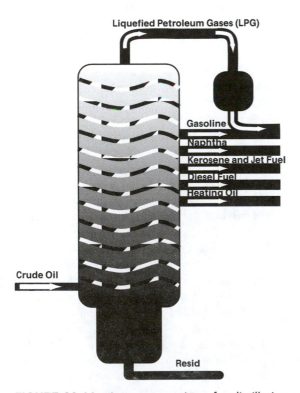

FIGURE 20.11 A representation of a distillation column in an oil refinery and some of the principal distillation products.

The first product to come out of the crude oil in a distillation tower is gases that were dissolved in the oil. These are collected and liquefied by mild pressure and refrigeration. LPG can be used as a convenient fuel gas, especially in places where natural gas is not available, or can be converted to useful chemical products. Then, in increasing order of boiling range, the next products are gasoline, naphtha (which can be converted to gasoline, or used as a solvent), kerosene and jet fuel, diesel fuel, and heating oil or fuel oil. Some material will not distill at all. It is called the residuum, more commonly known as *resid*. Resids can be treated to make lubricating oils, asphalts, and paraffin wax.

Distillation separates components of a mixture based on differences in their boiling temperatures or, put in other terms, on differences in volatility. For any liquid substance, at any temperature, some of the molecules of this substance will exist in the vapor (gas) phase. The contribution of these molecules to the pressure of the atmosphere above the liquid is called the **vapor pressure** of that liquid. As the temperature of the liquid rises, more of the molecules enter the vapor phase; that is, the vapor pressure increases.

If we heat the liquid hot enough, some temperature will be reached at which the vapor pressure becomes equal to the prevailing pressure of the atmosphere. When that point is reached, the liquid boils. The temperature at which this happens is called the boiling point of that liquid substance.

At any given temperature, the proportion of molecules in the vapor phase (the vapor pressure) is a characteristic property of a material. If we were to tabulate the vapor pressures of different liquids at a particular temperature, we would find they vary among the liquids in the same way that the boiling temperatures do. As examples, ethyl ether, which used to be the common 'ether' used as an anesthetic for surgery, has a boiling point of 35°C. Water has a boiling point of 100°C. When compared at a constant temperature—say, ordinary room temperature, about 22°C—the vapor pressure of ethyl ether would be much higher than that of water. Saying that ethyl ether is more volatile than water expresses the same idea in slightly different terminology.

Now consider a mixture of two liquids; call them A and B. It doesn't matter what, specifically, the liquids are, because the concepts apply to any mixture. Suppose that at the beginning the SYSTEM consists of a 50:50 mixture of the two liquids and that B is much more volatile than A; in other words B has a much higher vapor pressure than A. If we examined the vapor, we would expect that there would be a much higher proportion of B molecules than A.

What if we remove this vapor from the SYSTEM, and, in a separate container, condensed it back to a liquid? We would discover that the liquid left behind is now much richer in A than it was originally, and that the 'new liquid' made by condensing the vapor is much richer in B than was the original liquid.

If we were to continue this through several steps, we could eventually arrive at a situa-

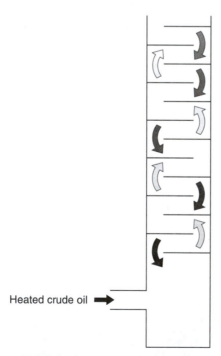

Heated crude oil ➡

FIGURE 20.12 As crude oil is heated in a distillation column, vapors of low-boiling materials ascend the column, while vapors of higher-boiling materials condense and flow back down the column.

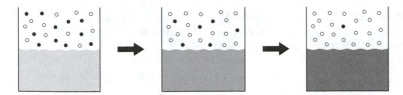

FIGURE 20.13 The continuous fractionation of a mixture of low-boiling material (open circles) and high-boiling material (dark circles) results in progressive enrichment of vapor with low-boiling material and enrichment of liquid with the high-boiling material.

tion in which the original 'left-behind' liquid is essentially 100% A, and the liquid condensed from it would be essentially 100% B. In other words, we have effected a separation of the original 50:50 mixture of A and B into its two components. Note that the more volatile of the components, B, is the one that came away with the vapor.

The description we have just gone through is actually a perfectly generic one. The same concept applies to mixtures that are other than 50:50; and it applies to mixtures of more than two components. In the distillation tower in a refinery, vapors of the volatile components of petroleum rise and recondense continuously as they ascend the column, as is shown in Figure 20.12.

Internally, this has the equivalent effect of an enormous series of the open beakers used in Figure 20.13. The more volatile materials (that is, those of higher vapor pressure or lower boiling point)

> DISTILLATION IS THE MOST IMPORTANT STEP IN PETROLEUM REFINING.

become enriched at the top of the column. At any given location along the column a mixture of the vapors of the petroleum components is condensing to a liquid of particular composition. Removing and condensing those vapors gives the liquid products that we as consumers recognize: gasoline, kerosene, diesel fuel, and heating oil.

Nothing much happens in a refinery until the crude is first distilled. Many other processing steps are then applied to convert the distillation 'cuts' into salable products. Some of these operations will be discussed in the coming chapters. In the next few chapters we will focus on certain of these products, particularly those that are relevant to transportation.

FOR FURTHER READING

Anderson, Robert O. *Fundamentals of the Petroleum Industry*. University of Oklahoma: Norman, 1984. Though many of the statistics are now out of date, this well-illustrated book still provides a good overview of the industry.

Berger, Bill D.; Anderson, Kenneth E. *Modern Petroleum: A Basic Primer of the Industry*. PennWell: Tulsa, 1978. An overview of the petroleum industry for readers with little technical background.

Conaway, Charles F. *The Petroleum Industry: A Nontechnical Guide*. PennWell: Tulsa, 1999. A book designed for nonspecialists. It covers virtually every aspect of the petroleum industry, beginning with how the Earth was formed, through the marketing of petroleum products.

Sinclair, Upton. *Oil!* University of California: Berkeley, 1954. Sinclair is now largely forgotten, but in the early decades of the 20th century he was a widely known author of several muckraking novels exposing social or political problems. This book is a fictionalized account of the development of the oil industry in Southern California and the corruption in the administration of President Harding, set in the early 1920s.

Yergin, Daniel. *The Prize*. Simon and Schuster: New York, 1992. A superb book on the rise of the oil industry and its impact on national and international politics, economics, and society.

21

GASOLINE

Except for electricity, gasoline is probably the energy source that all of us deal with the most in our daily lives. Gasoline is unquestionably the most important product from petroleum. The market demand for gasoline in the United States is such that oil refiners need to produce 45–50 barrels of gasoline from every 100 barrels of crude oil refined.

Historically, this has not always been the case. Originally the most important product from petroleum was kerosene. In the mid-19th century many homes were illuminated by oil lamps. The most popular fuel for these lamps was whale oil, produced from the carcasses of whales.⊛ As whalers slaughtered more and more whales and whale oil became less available, kerosene supplanted whale oil as a fuel for lamps.

⊛ One of the best ways to learn more about the capturing of whales and the production of whale oil is from one of the greatest of American novels, Herman Melville's *Moby Dick*. Many editions of this book are available, including inexpensive paperbacks.

As the 20th century dawned, Henry Ford began to realize his dream of building a cheap, reliable automobile that everyone could afford. Not only did the coming of the automobile revolutionize American society, but it was fueled by gasoline, the waste product of the oil industry. Kerosene remained the most important petroleum product until about 1920. Then two separate factors converged. First, the increasing availability of low-cost automobiles made it easy for more people to own cars. This led to more driving, which in turn increased the demand for gasoline. Second, the steady expansion of electric power networks made electric lighting increasingly available, and this in turn reduced the demand for kerosene.

Therefore the roles reversed. Gasoline became the dominant petroleum product with kerosene as a 'throw-away.' Kerosene has made a comeback since the 1960s, due to the wide acceptance of jet aircraft for civilian air transport (jet fuel is basically a refined form of kerosene), but gasoline is still the main petroleum product by far.

Two issues dominate the production and use of gasoline: its performance in the engine, and the yield (that is, the amount produced) per barrel of crude oil. We will consider engine combustion performance first, since that issue has a bearing on strategies used for producing gasoline.

OTTO-CYCLE ENGINES AND THEIR PERFORMANCE

Virtually all automobile engines work on the four-stroke principle invented by Nikolaus Otto over a hundred years ago. Although engines come in various sizes, the combustion process is easiest to visualize if we consider a four-cylinder engine. It's also easiest if we assume two valves per cylinder. Sketches showing the operation of a simple engine of this kind are shown in Figure 21.1.

The ratio between the volume of the cylinder when the piston is at its lowest point and the volume when it's at its highest point is called the **compression ratio**. The compression ratio provides an approximate indication of the power of the engine, the acceleration available, and the maximum speed. (It also indicates the maximum pressure in the cylinder, since the higher the compression ratio, the more the air/fuel mixture is compressed—raised to a high pressure—during operation.) As examples, the engine in a Model T Ford from the 1920s had a compression ratio of about 4:1. Typical mid-sized cars on the market today have engine compression ratios around 8:1. High-performance cars may have compression ratios of 10:1 or higher. The compression ratio has an important effect on the combustion of the fuel, as we will see.

Our concern focuses on what happens during the power stroke. Ignition of the gasoline/air mixture by the spark plug causes a flame to progress through the mixture as combustion proceeds. The ignition causes an immediate rise in the temperature (and therefore the pressure too, thanks to Robert Boyle) of the mixture in the cylinder. As the hot mixture expands, it pushes down on the piston, doing the WORK necessary to propel

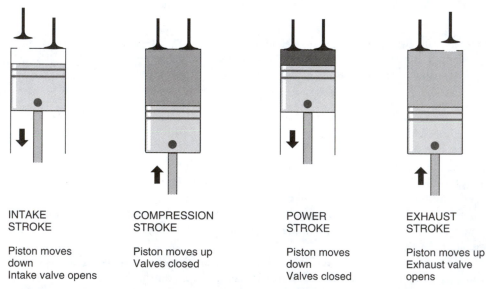

INTAKE STROKE	COMPRESSION STROKE	POWER STROKE	EXHAUST STROKE
Piston moves down	Piston moves up	Piston moves down	Piston moves up
Intake valve opens	Valves closed	Valves closed	Exhaust valve opens

FIGURE 21.1 The Otto engine operates with four strokes: intake, compression, power, and exhaust.

the car. (Note that, as the mixture of gasoline, air, and combustion products expands when the piston moves, its temperature and pressure will necessarily drop. Therefore our automobile engines are another type of heat engine.) In normal operation, the hot, burning gasoline/air mixture expands smoothly and delivers its chemical energy to the piston as WORK.

Under certain conditions of engine operation, the pressure and temperature of the fuel can get so high that the remaining unburned air/fuel mixture explodes, rather than continuing to burn smoothly (Figure 21.2). When the explosion of the unburned air/fuel mixture occurs in the cylinder, it is so violent that we can hear it inside the car. (On an equal weight basis, a gasoline–air mixture is a more devastating explosive than dynamite.) This phenomenon is called *engine knock*.

Engine knock is undesirable for a number of reasons. It reduces gasoline mileage; that is, it wastes gasoline. It reduces the apparent power and acceleration of the car. It leads to considerable wear and tear on engine parts and increased maintenance. Engine knock became more and more of a problem beginning in the 1930s, as engine designs evolved to higher compression ratios. Knock is affected by two parameters: engine design and fuel quality.

The key feature of engine design is compression ratio. The higher the compression ratio, the higher will be the pressure inside the cylinder at the moment of ignition. But, the higher the initial pressure inside the cylinder at the instant the fuel ignites, the easier and more likely it will be for the pressure to get high enough to cause the engine to knock. In other words,

THE HIGHER THE COMPRESSION RATIO, THE MORE LIKELY THE ENGINE IS TO KNOCK (WITH A GIVEN TYPE OF FUEL).

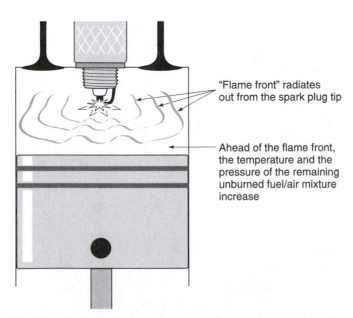

"Flame front" radiates out from the spark plug tip

Ahead of the flame front, the temperature and the pressure of the remaining unburned fuel/air mixture increase

FIGURE 21.2 In normal combustion, the gasoline/air mixture is ignited by the spark plug, and burns completely and evenly through the cylinder.

OCTANE NUMBER

An important property of hydrocarbon liquids is their **autoignition temperature**, the temperature at which the liquid will ignite and burn without a source of ignition. The autoignition temperature of a hydrocarbon is related to its molecular composition and structure. Large, straight-chain hydrocarbon molecules have much lower autoignition temperatures than do branched-chain and smaller molecules. Because it consists mainly of smaller molecules, gasoline has a fairly high autoignition temperature. That's why a gasoline engine normally requires a source of ignition—a spark from the spark plug. However, inside modern gasoline engines, temperatures and pressures get high enough for autoignition to occur. When premature ignition occurs, the engine 'pings' and, under severe conditions, a 'knock' occurs.

The octane number of gasoline is a measure of its ability to burn smoothly without engine knocking. Straight-chain alkanes are less thermally stable and burn less smoothly than branched-chain alkanes. The 'straight-run' gasoline fraction obtained directly from petroleum is a poor motor fuel and needs additional refinement because it contains primarily straight-chain hydrocarbons that preignite too easily to be suitable for use as a fuel in internal combustion engines.

To investigate the relationship between gasoline composition and engine knock, automotive engineers studied a variety of pure chemical compounds and their behavior in engines under standardized test conditions. The essence of their findings is summarized in Figure 21.3. To describe the knocking behavior quantitatively, two standards of comparison were established. The linear paraffin heptane was assigned a value of 0. The highly branched paraffin called 2,2,4-trimethylpentane (and informally, but quite incorrectly, called 'iso-octane' or even 'octane') was assigned a value of 100. The *octane number* of gasoline is equal to the percentage of 'octane' (2,2,4-trimethylpentane) in a blend of heptane + 'octane' that has the same knocking characteristics as does the gasoline being rated, when both are compared in a standardized engine test.

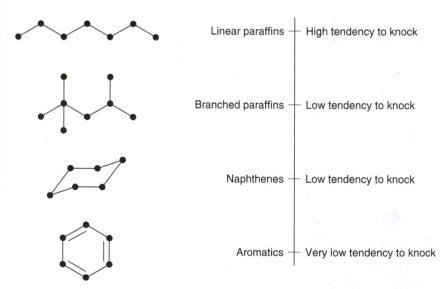

Linear paraffins	High tendency to knock
Branched paraffins	Low tendency to knock
Naphthenes	Low tendency to knock
Aromatics	Very low tendency to knock

FIGURE 21.3 The different families of components of gasoline have characteristic knocking behavior. The molecules shown in this chart are representative of each family; only the carbon atoms are shown.

The higher the octane number of a gasoline, the lower is its tendency to knock. Therefore, increasing the octane number of the gasoline counteracts the effects of increasing the compression ratio of the engine.

> THE HIGHER THE COMPRESSION RATIO OF THE ENGINE, THE HIGHER THE OCTANE NUMBER OF THE FUEL NEEDED TO AVOID KNOCKING.

Nowadays there are three grades of gasoline sold at most service stations: 86–87 octane is intended for most small or mid-sized cars; 89 octane is used in some mid-size or large cars; 92–94 octane is used in cars with large engines or in high-performance cars.

Based on market demand, and on the need to meet engine performance requirements, there are two key points regarding gasoline: We want to have gasoline available in the 86–94 octane range, and we want to produce 45–50 barrels of gasoline for every 100 barrels of crude oil refined. The octane number of a gasoline can be increased either by increasing the percentage of branched-chain and aromatic hydrocarbons or by adding so-called octane enhancers (or a combination of both).

PRODUCING GASOLINE IN THE REFINERY

Straight-Run Gasoline from Distillation

One way to think about, and classify, refinery processes is to consider all the kinds of things we can do with molecules. We can separate them without any chemical change at all (that is, these could be called physical processes, because they take advantage of physical properties of molecules, and not their chemical properties). We can take small molecules and put them together to make bigger ones, so-called buildup processes. We can take big molecules and break them apart to make smaller ones in breakdown processes. And, finally, we can change the molecular structure without changing its size (sometimes called change processes).

Distillation is an example of a physical process. The gasoline that is produced from the distillation of crude oil is called **straight-run gasoline**. There are two problems with straight-run gasoline: First, even with the best Pennsylvania-crude-quality oils we only get 20 barrels of gasoline per 100 barrels of crude; with poor-quality crudes, we may get less than 10 barrels of straight-run gasoline. Second, the octane number of straight-run gasoline is about 35. As a result, we need to enhance both the yield of gasoline and its octane number to meet present-day market demand and performance requirements.

The key to enhancing yield comes from the approximate relationship between boiling point and size of molecules. As long as we compare substances that are fairly similar chemically, we find that the higher the boiling point, the larger the size of the molecules, and vice versa. We can illustrate this with Table 21.1, which is arranged in order of increasing boiling range *and* increasing molecular size (i.e., number of carbon atoms). So, to increase the yield of gasoline, two strategies can be used. We can combine molecules that have fewer than five carbon atoms to build up molecules in the gasoline-size range. Or, we can also break apart molecules that have more than nine carbon atoms to make smaller molecules that are in the gasoline-size range. Normally, many refineries do both.

TABLE 21.1   THE LINEAR PARAFFIN MOLECULES IN
VARIOUS DISTILLATION PRODUCTS OF PETROLEUM
STEADILY INCREASE IN SIZE AS THE BOILING RANGE
INCREASES. MOST OF THE PARAFFINS IN GASOLINE
HAVE FIVE TO NINE CARBON ATOMS

Product	Number of carbon atoms	Boiling range of paraffins
LPG	3–4	<−1°C
Gasoline	5–9	36–150°C
Kerosene	10–14	174–250°C
Diesel fuel	12–18	217–317°C
Fuel oil	12–20	217–350°C
Resid	(very large)	> 350°C

Alkylation

Alkylation is the building up of small molecules of three or four carbon atoms to products of six to eight carbon atoms. Buildup processes are used to form larger molecules of high octane number and boiling in the gasoline range by combining smaller molecules. Cracking processes (discussed below) can produce as by-products molecules that are too small to be in the gasoline range. Unless they can be sold to chemical manufacturers, these small hydrocarbon molecules are of relatively low value. Alkylation uses these small molecules to form compounds in the gasoline range with very high octane numbers. The product is called **motor fuel alkylate**. Usually the molecules combine in such a way that branched paraffins are formed. The octane number of motor fuel alkylate is about 100. Alkylates are important constituents of high-octane gasoline or aviation gasolines.

In 1938, it was discovered that gasoline of a much higher octane than typical straight-run gasolines could be produced by treating small hydrocarbon molecules with sulfuric acid. This process, called sulfuric acid alkylation, provided most of the high-octane aviation fuel used during the Second World War and became, for a time, one of the most important processes in the refinery.

Thermal Cracking

Breakdown processes represent a third route to gasoline, i.e., in addition to physical processes and alkylation processes that build up molecules. Breakdown encompass all types of 'cracking' processes that use heat, pressure, and catalysis. The oldest of these operations is thermal cracking, now generally used in the United States only on distillation residua (resids) to make a solid coke and to increase the yield of light products.

Cracking is the breaking down of molecules of more than nine carbon atoms to the range of five to nine carbon atoms. To do this, a refinery will select a material of high boiling range for which little demand or market exists. One possibility is to use some of the grades of fuel oil. We know that heat, from the geothermal gradient in the Earth, can break molecules apart. The simplest approach to cracking is to expose the material to high temperatures, a process called **thermal cracking**. The high temperatures break apart carbon–carbon bonds in molecules, and form molecules of smaller size. Unfortunately, the products of thermal cracking tend to be in the 55–75 octane number range, not good enough for today's engines.

As the 1930s drew to a close, it was apparent that the automobile engines were becoming more powerful, of higher compression ratio, and demanded gasoline of higher octane rating. The initial goal was to increase the yield of gasoline from a barrel of crude, which was about 20% for straight-run gasoline from a good-quality crude and had an octane rating around 50. By 1935, the octane rating had been boosted to 71 with corresponding increases in the gasoline-to-crude ratio. One of the thermal cracking processes important in the 1920s and 1930s was invented by a man with the extraordinary name of Carbon Petroleum Dubbs, who, perhaps understandably, went by the initials 'C.P.' Despite its oddity, his name was very apt, because one of his most important inventions was a technique to reduce the formation of undesirable solid carbon (or coke) when petroleum was processed to enhance the yield of gasoline.

> Though he may have been with stuck with a bizarre name, Dubbs was both an accomplished research scientist and a very fine businessman. Proof of the latter fact is that C.P. Dubbs managed to become a millionaire during the Great Depression. There is no evidence that he ever murdered his parents for giving him such an unusual name.

Catalytic Cracking

The most widely used breakdown process today is catalytic cracking, which efficiently converts gas oils into high-octane gasoline and other lighter products. (Gas oils are petroleum fractions that boil in the range of about 450–800°F, such as diesel fuel, heating oils, and fuel oils.) Around the time of the Second World War, Eugene Houdry discovered that if cracking was done in the presence of clay minerals, not only did the large molecules break apart, but also the products were converted to branched paraffins, naphthenes, or aromatics. That is, both the yield and the octane number go up. The clay does not enter into the cracking reaction; it serves only to speed up the reaction and to enhance the octane number of the product. Substances that affect the outcome of a chemical reaction without themselves participating as reactants or being consumed are called **catalysts**. Therefore this kind of process is called **catalytic cracking**.

Houdry was the rich, successful heir to his father's French steel firm. He was a First World War hero, a good athlete, and interested in auto racing. The former Vacuum Oil Co. invited him to New Jersey and offered to purchase his laboratory. Their joint efforts succeeded, but a sudden drop in the price of crude oil caused a loss of interest. A third partner joined the effort, and in 1936 the first commercial plant went into operation using bentonite clay as the catalyst. By the time the Second World War broke out there were 12 plants in the United States, providing 132,000 barrels per day of high-octane gasoline. By the end of the war, some 34 plants were in operation with a capacity of 500,000 barrels per day.

> An enormous family of clay minerals exists throughout the world. The chemical structures of clays derive from the oxides of silicon and aluminum, usually with small amounts of other elements and some molecules of water. The chemical composition and structures of the clays vary widely across the whole family. One major type of clays was first discovered in 1874 near the town of Montmorillon in France; these clays have since been called montmorillonites. Those montmorillonites that have significant commerical application are called bentonites. Large deposits of bentonites occur in, as examples, Wyoming and Alberta.

Catalytic cracking enables the production of 40–45 barrels of gasoline per 100 barrels of crude, with octane numbers over 90. Nowadays many of the catalysts used are zeolites.❀ The discovery of catalytic cracking revolutionized the refinery industry. Catalytic cracking (Figure 21.4) is the second most important process in a refinery. Some historians of technology consider the development of catalytic cracking to be the greatest triumph of chemical engineering.

❀ Zeolites are another large family of naturally occurring minerals based on a framework of aluminum and silicon oxides along with atoms of such elements as sodium, potassium, calcium, or magnesium and some molecules of water. The zeolites were first characterized in the 1750s by the Swedish mineralogist Axel Cronstedt. One of the first commercial uses of zeolites was in water softeners. Today, many of the zeolite catalysts used in petroleum refining are synthetic, rather than natural materials. The synthesis of zeolites allows petroleum chemists to control precisely the activity of the catalyst and the size of the molecules that will react in its presence.

The next innovation in gasoline production technology was hydrocracking, which utilizes both a catalyst and hydrogen to process resids or product in the middle-boiling range to high-octane gasoline, jet fuel, and high-grade fuel oil. In this process, hydrogen, in the presence of a catalyst containing platinum, reacts with and breaks apart heavy, high-boiling oils to gasoline, jet, and diesel fuels. An odd measure of the importance of hydrocracking is that it is the petroleum industry, not the jewelry trade, that is our largest user of platinum. Hydrocracking was developed at almost the same time as kerosene reemerged as a demand product. The jet engine (Chapter 23) had been introduced during the Second World War and, in the following two decades, most piston-engined military and commercial aircraft had been replaced by jets.

Catalytic Reforming

Change processes are a fourth route (in addition, that is, to physical, buildup, and break-down processes) to gasoline. After the cracking processes have arranged the product distribution (i.e., the amounts of gasoline, jet fuel, diesel fuel, and heating oils) to match the market demand, some materials remain whose molecules are the right size to be in the boiling range of a particular product, but are of the wrong molecular configuration to have the properties desired in the product. For example, straight-run gasolines or naphthas have desirable boiling temperatures, but octane numbers that are too low.

In other words, the trouble with straight-run gasoline molecules is that they're the wrong shape—they're mostly low-octane-number linear paraffins. To enhance the octane number of straight-run gasoline, we don't need alkylation or cracking processes, all we need is to change the shape—to 're-form,' so to speak—the molecules we already have. The process of 're-forming' molecules also uses special catalysts (though different from the ones we use for cracking). Thus this process is called **catalytic reforming**. In the presence of certain catalysts, such as finely divided platinum placed onto a support of aluminum oxide, straight-chain hydrocarbons with low octane numbers can be re-formed into their branched-chain isomers. This important gasoline-upgrading process produces gasoline of about 95 octane. Catalytic reforming can also be applied to straight-run naphtha.

Another change process is isomerization, which converts linear, straight-chain paraffins to branched-chain compounds. This also results in an increased octane number.

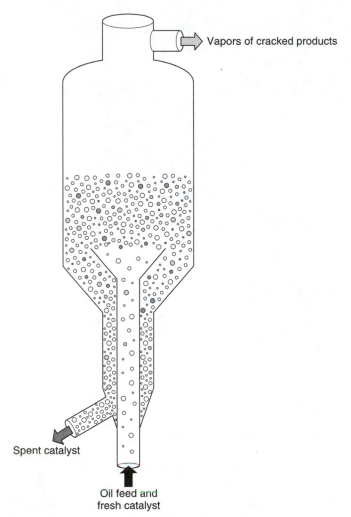

Vapors of cracked products

Spent catalyst

Oil feed and
fresh catalyst

FIGURE 21.4 A schematic diagram of a catalytic cracking unit. The material to be cracked and fresh catalyst are fed into the bottom. The cracked products are withdrawn from the top. Used catalyst is removed from the bottom, to be regenerated and recycled to the unit.

These processes are summarized in the diagram of Figure 21.5.

Octane Enhancers

The octane number of gasoline can also be increased by adding special chemicals called antiknock agents or octane enhancers. In 1921, Thomas Midgley discovered that tetraethyllead was an outstanding antiknock agent for low-grade fuels. It became the most widely used antiknock agent in the United States, up to 1975. The addition of three grams of tetraethyllead per gallon of gasoline increased the octane number by 10–15. From 1925 to 1975, both regular and premium grades of gasoline contained an average of three grams of tetraethyllead per gallon. This product was often called 'leaded gasoline,' although it

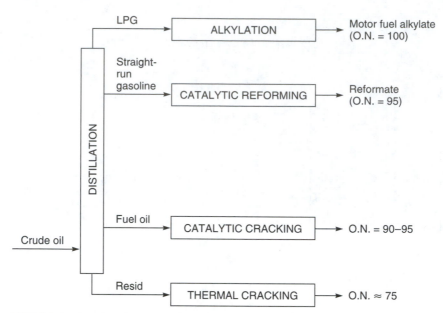

FIGURE 21.5 Gasoline is produced from many processes in an oil refinery, including distillation, alkylation, reforming, catalytic cracking, and thermal cracking. O.N. = octane number.

most certainly did not contain metallic lead, as the name might imply. Sometimes it was also called 'ethyl' gasoline. Eventually, lead's toxicity encouraged fuel producers to remove tetraethyllead from gasolines; the driving force behind this decision was the worry that lead could 'poison' the catalysts used in the catalytic converters installed on cars to reduce air pollution (Chapter 31). Without lead, less expensive gasoline blends with low octane ratings did not burn efficiently. This problem was partially alleviated by engine redesign and partially by blending higher octane components (such as benzene and toluene) into the fuel, but both approaches increased gasoline prices significantly.

The exhaust emissions of automobile engines contain CO, oxides of nitrogen, and unburned hydrocarbons, all of which contribute to air pollution. (We'll look at this problem in detail in Chapter 30, on acid rain.) As urban air pollution worsened in the 1950s and 1960s, Congress eventually passed the Clean Air Act of 1970, which, among many other things, required that 1975-model-year cars emit no more than 10% of the carbon monoxide and hydrocarbons emitted by 1970 models. The solution to lowering these emissions was a platinum-based catalytic converter. The only problem was that it required lead-free gasolines, since lead deactivates the platinum catalyst by coating its surface. As a result, automobiles manufactured since 1975 have been required to use lead-free gasoline to protect the catalytic converter. Even now there are a few pre-1975 cars on the road, some of which are rolling rust buckets and others of which are vintage autos restored and cared for by auto buffs. However, it is now virtually impossible to purchase leaded gasoline in the United States.

Because tetraethyllead can no longer be used in the United States and a few other countries, other octane enhancers are now added to gasoline. These include toluene, tertiary butyl alcohol, methyl tertiary butyl ether (MTBE), methanol, and ethanol. The most popular octane enhancer has been MTBE. Unfortunately, MTBE is not only a good octane enhancer, it is also fairly soluble in water. In recent years, there has been rapidly growing concern about gasoline containing MTBE leaking out of underground storage tanks (at service stations, for example) and eventually mixing with ground water. Although gasoline doesn't dissolve in water, the MTBE will. This leads to contamination of ground water, and, potentially, of drinking water. A move is afoot in many locations to ban the use of MTBE in gasolines.

Blending

In modern refineries, depending on the refinery configuration (that is, the specific set of these operations that are available in the refinery), the refinery can vary its mix of products to meet market demands, within rather wide limits, regardless of the composition of the crude coming into the refinery. The products we've illustrated in this chapter are called **blend stocks**. The refiner can blend various portions of these materials to obtain the 87, 89, and 93 octane gasolines that are actually available in service stations. This approach meets both the market demand (for about 50 barrels of gasoline per 100 barrels of crude processed in the refinery) and performance requirements.

The gasoline must also be stabilized. This involves a closely controlled distillation that removes enough lighter hydrocarbons to give the fuel the desired volatility. Too much volatility produces **vapor lock** in an engine; in cold weather, too little volatility leads to difficulty in starting an engine. (Vapor lock is a condition in which the gasoline turns to vapor in the fuel line before getting to the engine, instead of vaporizing when injected into the cylinder. This can occur, for example, on exceptionally hot days. Vapor lock will cause the car to stop because it interferes with the normal flow of liquid gasoline to the engine. Though it is often maddening and frustrating, there is often little one can do other than coast to the side of the road and wait for the engine and fuel lines to cool off.) Refinery products must be blended to yield a final product of the desired viscosity, volatility, and octane number.

As we've seen, certain additives promote burning and reduce engine knocking. Other additives improve combustion by cleaning the engine of carbon deposits. Poor combustion causes carbon deposits to build up inside the automobile engine; chemicals are sometimes blended into gasoline to help prevent this. Most engine-cleaning products are hydrocarbons. Toluene is used in solvent systems to clean carburetors.

Refineries have always made gasoline. We've seen that there was a time when gasoline was a nuisance, a dangerous by-product with little practical use. Of course, since the advent of the automobile, and particularly once Ford dropped the price into the range of affordability by the ordinary working person, gasoline has become the premier petroleum product. A problem for refiners has been what to do with what is left over from the petroleum after the gasoline has been produced. There is, after all a market for only so much kerosene, diesel fuel, heating oil, solvents, lubricating oil, and waxes. Satisfying the primary gasoline demand could potentially cause surpluses of other products. The solution has been and continues to be the development of new processes that allow the refiner to convert the residue from one process into the feed for another.

Costs

To build a modern 200,000 barrel-per-day refinery today from the ground up would cost well over a billion dollars. This 200,000 barrel-per-day plant would produce about 100,000 barrels of gasoline, 5,200 barrels of liquefied petroleum gas, 12,600 barrels of jet fuel, 7,000 barrels of kerosene, nearly 42,000 barrels of heating oil and diesel fuel, and coke, asphalt, lubricating oils stocks, and even 170 tons of sulfur.

A Preview of Coming Attractions

A problem on the horizon for gasoline is the mandated removal of aromatic compounds. Aromatics are desirable in gasoline because of their very high octane numbers. However, aromatics also contribute to air pollution, as the precursors to smoke and soot formation, and some aromatics (such as benzene) are suspected carcinogens. Thus refiners now are in the process of changing some of the strategies to formulating gasolines.

All gasolines of course are highly volatile. Because gasolines are so volatile and because the vapors are so easy to ignite, you can start your car even in the coldest of weather. However, this same high volatility means that some hydrocarbons get into the atmosphere as a result of accidental spills and evaporation during normal filling operations at the service station. Gasoline vapors also evaporate from storage tanks. Hydrocarbons in the atmosphere play an important role in a series of reactions that contribute to urban air pollution. Reformulated gasolines are oxygenated gasolines that contain a lower percentage of aromatic hydrocarbons and have a lower volatility than ordinary gasoline. Nine cities with the most serious ozone pollution were required by the 1990 regulations to use reformulated gasolines starting in 1995, and another 87 cities that are not meeting ozone air quality standards can choose to use them. In late 1996 more than half of the over 40 areas that did not meet ozone standards in 1990 and 28 of the 42 areas that did not meet the carbon monoxide standards in 1990 now do so.

The 1990 Clean Air Act Amendments require cities with excessive carbon monoxide pollution to use oxygenated gasolines during the winter. Oxygenated gasolines are blends of gasoline with organic compounds that contain oxygen, such as MTBE, methanol, and ethanol. Oxygenated gasolines burn more completely than nonoxygenated gasoline and can reduce carbon monoxide emissions in urban areas by up to 17%. The 1990 regulations also require oxygenated gasolines to contain 2.7% oxygen by weight. The use of oxygenated gasoline is currently required in about 40 cities in the United States.

Oxygenated chemicals are now added to gasoline to promote cleaner air, smoother combustion, and higher octane numbers at low cost. Ethanol is the main alcohol used for the purpose. A blend called 'gasohol' introduced during the petroleum shortages of the 1970s contained about 10% ethanol. Ethanol's main drawback is that it absorbs water from the atmosphere, which causes corrosion to fuel system components and can also harm the combustion performance. Ethanol has a high octane number (108). Methanol is often used as a gas line antifreeze. Methanol's high octane number (107) also reduces knocking, but it too absorbs water and causes severe corrosion. The less soluble t-butyl alcohol avoids this problem, since it does not absorb water well; in fact, t-butyl alcohol reduces the tendency of gasoline blends to pick up water. A typical blend may contain 2.5% methanol, 2.5% t-butyl alcohol, and 95% gasoline. As we've seen, MTBE is added to gasoline to promote smoother burning and to promote hydrocarbon burning. Unresolved questions continue to arise about the toxicity of MTBE if it eventually gets into the drinking water supply. Because of this, MTBE is now quite controversial, and it seems likely that its use will be phased out.

Although these oxygenated compounds have high octane numbers, a high octane number relates only to a tendency for engine knocking. An octane number does not itself relate directly to the energy content—in Btu per pound, or Btu per gallon—of a fuel. In general, all oxygenated additives contain less energy than gasoline because they are already partially oxidized. In a practical sense, this means that you get less knocking in the engine, but you also get less 'mileage'—the number of miles you can drive per gallon of fuel. Despite that, because ethanol and methanol can be made from biomass or other renewable sources, they are considered to be good liquid fuel options by advocates of reducing our dependence on fossil fuels.

Although ethanol and methanol add oxygen to gasoline, they are not totally non-polluting fuels. Both can produce toxic aldehydes⚛ on incomplete oxidation and, like all carbon-containing fuels, they produce CO_2. Consumers sometimes express one other complaint: Additives, such as ethanol, methanol, and MTBE increase the cost of gasoline without increasing the energy it provides (indeed, the oxygenates actually decrease the energy provided per gallon).

⚛ Aldehydes are a type of organic compound containing, in addition to carbon and hydrogen, an oxygen atom. The oxygen atom is connected by a double chemical bond to the end of the chain of carbon atoms, as in acetaldehyde, $CH_3CH=O$, a material sometimes used as a flavoring agent in beverages and some foods. The simplest aldehyde is formaldehyde, $H_2C=O$, the water solution of which (formalin) is used as a preservative and embalming agent. Exposure to aldehydes can irritate the eyes, nose, and throat.

FOR FURTHER READING

Anderson, Robert O. *Fundamentals of the Petroleum Industry*. University of Oklahoma: Norman, 1984. Chapter 21 is a good introduction to refinery operations for readers with little prior technical background.

Berger, Bill D.; Anderson, Kenneth E. *Modern Petroleum*. PennWell: Tulsa, 1992. This book is a useful primer on all aspects of the petroleum industry. Chapter 11 gives an overview of refinery technology.

Bleviss, Deborah L.; Walzer, Peter. 'Energy for motor vehicles.' In: *Energy for Planet Earth*. Freeman: New York, 1991; Chapter 5. This book is a collection of articles that originally appeared in *Scientific American* magazine. This chapter discusses developments in fuel efficiency, alternatives to gasoline, and alternatives to cars.

McNeill, J.R. *Something New Under the Sun*. Norton: New York, 2000. This book includes the story of Thomas Midgeley, inventor of leaded gasoline, an excellent scientist and fine person, and who, according to the author, did more to wreck the environment of our planet than any other person who has ever lived.

22

CARS AND THEIR IMPACT

Let's consider how the application of steam and petroleum represented such a radical historical change in transportation. In 219–217 B.C. the Carthaginian general Hannibal assembled what was likely the greatest invasion army of the classical world. Hannibal first crossed the Pyrenees, and then the Alps, to invade Italy. Two thousand years later, in 1812, Napoleon assembled the greatest invasion army known prior to the 20th century, for his invasion of Russia. *Both armies traveled at exactly the same speed.* In 45 B.C., Julius Caesar was ruler of the greatest empire the world had yet known. In 1895, Queen Victoria was ruler of the greatest empire in history. Nearly 2,000 years after Caesar, Victoria moved around her capital, London, in the same way, and probably at the same speed, as Caesar had moved around his capital of Rome.

We've already seen how James Watt perfected Thomas Newcomen's engine and in doing so developed the first practical steam engine in the last years of the 18th century. Early in the 19th century, Richard Trevithick developed the steam locomotive, and, by 1829, George Stephenson perfected the design. After several early European attempts, Robert Fulton developed the steamboat in the early years of the 19th century. So, in a relatively short time, Watt's engine was applied to railroads and ships. Yet it took almost another century before the practical development of the automobile. Without doubt, the automobile was the outstanding feature of land-transport technology in the 20th century.

At the beginning of that century, the coming of the automobile offered a promise of personal freedom and affluence. But just like every other invention or technological development that was thought finally to lead humankind into the promised land of freedom, wealth, and ease, the automobile turned out not to be the economic and social panacea that many expected. Nonetheless, over the course of the 20th century, the automobile probably changed the lives of most people, both in good ways and in bad ones, more than any other invention. Figure 22.1 provides a time line of the developments in transportation.

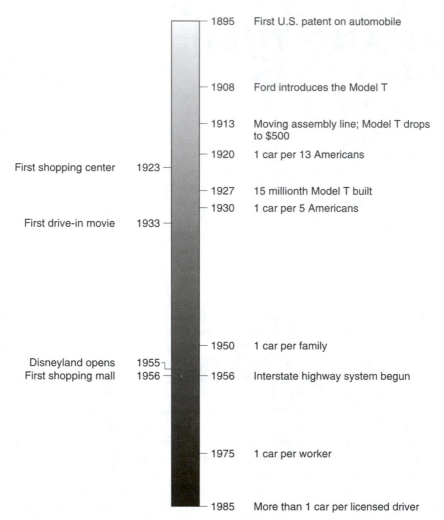

	1895	First U.S. patent on automobile
	1908	Ford introduces the Model T
	1913	Moving assembly line; Model T drops to $500
	1920	1 car per 13 Americans
First shopping center 1923		
	1927	15 millionth Model T built
	1930	1 car per 5 Americans
First drive-in movie 1933		
	1950	1 car per family
Disneyland opens 1955		
First shopping mall 1956	1956	Interstate highway system begun
	1975	1 car per worker
	1985	More than 1 car per licensed driver

FIGURE 22.1   A time line of the major developments in the history of the automobile and its impact on society.

⚛ Television is perhaps the runner-up to this claim. However, television only began to penetrate most households in the 1950s, whereas car ownership became common 30–40 years earlier.

THE CYCLING CRAZE

We have seen how, in principle, a steam engine could be used to propel a 'horseless carriage.' In fact, there were a number of early trials of this concept—as early as the first attempts with locomotives and boats (Chapter 19). Furthermore, improved engineering did eventually lead to some practical steam automobiles with good performance, the Stanley Steamer being a notable example. Other steam-propelled vehicles, such as steam tractors were also developed. So why didn't equally ingenious individuals—the counterparts of Trevithick, Stephenson, and Fulton—develop steam-powered automobiles in the early 19th century? *No roads* (or at least no good roads).

Roads in the early 19th century needed only to accommodate foot traffic or slow-moving horse-drawn vehicles. The roads were not paved, were full of potholes, were not lighted nor marked, and in many cases simply meandered through the countryside. What caused the change to create roads that allowed the widespread use of the automobile? *The bicycle.*

Bicycles began to be popular in the 1870s and 1880s. In the 1890s cycling developed into an extremely popular craze (Figure 22.2). The bicycling craze of the 1890s brought with it numerous benefits. It demonstrated the possibility of long-distance travel over ordinary highways.⚛ It led to the renovation of existing roads, including erection of signposts and the printing of road maps. It stimulated the construction of new roads. And, it led to the establishment of roadside repair facilities, which were the forerunners of the service station. Some of these benefits derived from pressure being brought on legislators by organized groups of cycling enthusiasts, for example, the League of American Wheelmen.

⚛ Long-distance cycling involved trips of many tens of miles; in some cases, more than 100. The first recorded bicycle race took place in 1868 and covered a distance of 83 miles. Perhaps more remarkable than the distance ridden is the fact that the winning cyclist, James Moore, is said to have ridden a bicycle weighing *160 pounds*!

The revival of interest in mechanical means of road transport came with a surprising reversion to the use of human ENERGY for transportation, in the form of the bicycle. To establish a viable industry for a new form of transportation—the automobile—the early innovators and entrepreneurs first had to perceive a market for personal transportation that did not rely on the horse. For most people of the 19th century, personal transporta-

FIGURE 22.2 The popularity of bicycling—the so-called cycling craze—in the latter decades of the 19th century generated interest in personal transport and in improved roads that subsequently helped the automobile gain popularity.

tion meant being pulled by, or riding on, a horse. There can be little doubt that the bicycle itself, and its extraordinary popularity in the 1890s, helped people, especially young people and women, to travel more widely than ever before. This in turn helped create a demand for even more freedom of movement in the future. Thus the existence of the new market was established by the bicycle, which became an item of mass consumption in the 1880s and wildly popular in the 1890s. By 1915 there would be six million bicycles in the United States, five million in Britain, and four million in France. Not only did the bicycle help create a market of potential customers for the automobile, but also many of the techniques devised for bicycle construction were used later in car making. Light tubular frames, ball bearings, chain drive, pneumatic tires, gears, wire wheels, and brake cables were adapted from the bicycle by early car manufacturers. In some cases, the bicycle firms themselves switched from making bicycles to building cars.

HOW RAILROADS PUT PEOPLE IN CARS

A second factor that set the scene for the rise of the automobile was, curiously, the steady expansion of the railroads. This occurred in two ways.

First, as railroads expanded and carried more and more passengers, there were a great number of highly publicized train wrecks. These caused considerable public indignation. Somehow, we seem to be willing to accept a steady diet of small crashes and disasters, whereas the large, spectacular ones attract our attention and outrage. Few people can escape a horrible feeling about a major airline crash (the TWA Flight 800 crash in the summer of 1996, or the Concorde crash four years later are just two examples), yet in just about two days we had managed to slaughter in automobile accidents the same number of people as died in those plane crashes. Except for the immediate families and friends of the victims of those car crashes, probably no one paid any attention. Public concern about railway safety, because of the generally spectacular nature of passenger train wrecks, led to interest in alternative, and hopefully safer, modes of transportation.

Second, the success of railroads meant that more and more people and goods were piling into the cities each day. This caused increased congestion on the streets (Figure 22.3), and especially problems with the horses used to haul buses, carriages, and freight wagons. One particular horse stable was a block long and seven stories high. New York City's horses produced 300 million pounds of manure each year. In addition, horses can be very difficult to handle. When they are enraged or terrified, as in an accident, they can kick, bite, and bolt. In recent years the number of traffic deaths in the United States has been about 160 per million people. A century ago, a time that many people think was a kinder, gentler, slower paced era, the rate was still 110 per million. So, it was clear for various reasons that the horse had to fade from the scene as the dominant source of energy for urban transportation. And then along came Ford.

AMERICAN DOMINANCE IN AUTOMOBILE MANUFACTURING

We've seen (Chapter 19) that the major inventions that made the gasoline-powered automobile possible all came out of Germany, due to such inventors as Nikolaus Otto, Wilhelm Maybach, Gottfried Daimler, and Karl Benz. After a time, the leadership in automobile technology passed to France. Such terms as *automobile, chauffeur, garage,* and *chassis* are all French in origin. The first patent on an automobile in the United

FIGURE 22.3 The problem of congested streets has been with us for a long time. The automobile was not the sole culprit in causing the problem.

States was issued to a lawyer, George Selden. The patent was originally filed in 1879 and finally issued in 1895.

> *In his original patent Selden claimed, 'The combination in a road vehicle equipped with an appropriate transmission, driving wheels, and steering, of an internal combustion engine of one or more cylinders, a fuel tank, a transmission shaft designed to run at a speed higher than that of the driving wheels, a clutch, and coachwork adapted for the transport of persons or goods.' Cars today still fall under that wonderfully broad claim.*
> —Reynolds[1]

By the second decade of the 20th century, automotive leadership had passed to the United States. There are several reasons for this: a generally high standard of living, a large population living in cities separated by long distances, abundant supplies of petroleum, and a vigorous innovative and entrepreneurial spirit.

At first, automobiles were toys for the rich. Henry Ford recognized that for the automobile to be truly successful, it had to be priced within the reach of the ordinary working person of modest financial means. In his continuing effort to reduce the price of the Model T (first introduced in 1908), Ford developed in 1913 a revolutionary industrial idea—the moving assembly line (Figure 22.4). Some earlier assembly lines had been set up, as for example in the manufacture of military firearms, but normally the workers collected

FIGURE 22.4 Henry Ford's moving assembly line had a major role in reducing the price of automobiles, making them available to families of modest financial means.

parts and brought them to one spot where the article was being assembled. On Ford's moving assembly line, the worker stayed in one place, and the car moved along the line. The success of the moving assembly line depends on two key factors: each worker repetitively carrying out one very simplified job, and the parts being highly standardized and completely interchangeable. Though many people were involved in the development of the automobile industry in the United States early in the 20th century, it is usually Henry Ford who is given credit for putting the automobile within reach financially of persons of average means. Ford had little education and many character flaws—including racial bigotry and a domineering personality that drove away many talented associates. Indeed, Ford was about as ornery and reprehensible a person as most of us would care to know. Still, he is perhaps the most important industrialist of that century.

By 1913, due in large part to assembly-line production, Ford was able to sell the Model T for $500.⚛ Before production of the Model T was curtailed in 1927, Ford had sold 15 million of them. Whereas there had been one car for every 265 Americans as late as 1910, there was one car for every five at the end of the 1920s. Though assembly-line production was a major factor in this extraordinary growth, other factors had a role as well. For instance, a car purchase was made easier with installment plans; by 1923 over three-quarters of cars were purchased on credit.

⚛ In the early years of the 20th century, automobiles were toys for the well-to-do. Many cost well over a thousand dollars. In fact, in 1908, the Model T was selling at $850. By 1916, further improvements in manufacturing and the use of the assembly line allowed Ford to drop the price to $400, making the Model T the lowest-priced car in America. To put these prices in context, this was an era in which an unskilled laborer might be paid $1.00, and a highly skilled craftsman $2.50, per *day*. (When Henry Ford offered to pay workers $5.00 per day, so many people showed up to apply for the jobs that a near-riot ensued.) Assuming a six-day workweek with no vacations, a skilled worker would bring home an annual income of $780. Today the median household income in the United States is about $47,000 per year. Very roughly, an automobile purchased in 1920 might cost the equivalent of half a year's salary. Eighty years later, approximately the same ratio still holds.

Early in the 1920s, executives of car companies were already beginning to worry about growth of car sales and impending saturation of the market. They noticed, however, that many people did not want to possess exactly the same car as their neighbors. Instead, consumers seemed willing to accept some increase in cost to obtain variety in styles, colors, or other attributes that would make their car somehow 'different.' It began to be appreciated that car owners could get the idea of trading in their existing cars for a new one that would indeed be different from the one parked at the neighbor's house. This growing appreciation of consumer behavior led the car companies to produce a variety of different models and to change models on a regular basis, often annually, to encourage trade-ins.

A decline in the rate of technical improvements to the basic components of the car also made it worthwhile for the car companies to call attention to changes in style and the diversity of models offered, rather than using small, incremental technical changes as a selling point. By the start of the 1920s, most of the important elements of the modern car had essentially been developed; brakes, tires, valves, air cleaners, oil purifiers, and shock absorbers are some of the examples. Although assembly-line production lowered the cost of cars, and made car ownership possible for most people, it brought with it a new problem. The capital investment that would be required to build an assembly-line facility, or to make radical changes in an existing facility, was so high that it discouraged product innovation, because any major changes in the car would require expensive retooling of the assembly line.

By 1920, there was already one registered automobile for every 13 Americans. A decade later, the figure had risen to one for every five. Not long after the Second World War, there was one car per family in America and, in the 1970s, one car per worker. By 1970, there were more cars than there were households in the United States; in Los Angeles, there were more cars than there were people. By 1985, the number of cars in the United States actually exceeded the number of licensed drivers, and nowadays there are about 20% more cars than there are drivers. While it may seem bizarre that there could possibly be more cars than there are people to drive them, we need to realize that today many cars are owned by corporations and rental companies. And, many families now own recreational vehicles and campers, which might be used only occasionally, in addition to their other vehicles that are used for everyday commuting to work or shopping.

THE EFFECTS OF GROWING AUTOMOBILE USE

What hath Ford wrought? Perhaps the most important single contribution is that the car gave the common person a feeling of independence, particularly independence from the rich and independence from transportation companies that provided travel only to fixed destinations on fixed schedules. Now, a person of average financial means could get around without being dependent on the railroad companies or streetcar companies. Further, by being able to live outside of the center city, the average person was also no longer dependent on big-city landlords or rental companies.

Cars and Industrial Growth

The car actually benefited a large cross-section of industry. Of course, the obvious ben-eficiaries were the car companies and the producers and refiners of petroleum products. But, besides stimulating the development of the oil industry, the automobile also stimu-lated highway and bridge construction, and led to the development of the 'management science' (at first in the petroleum and construction industries) of how to run large engineering companies. In addition, all sorts of small, local businesses benefited: trucking companies, suburb developers, construction contractors, auto parts dealers and suppliers, auto mechanics, and service stations.

In the period from 1920 to 1940, tires and inner tubes accounted for 85% of the sales of the American rubber industry. The automobile also stimulated the application of continuous processing to making plate glass for windshields and windows. The Ford Company itself pioneered this method shortly after the end of the First World War. Plate glass output tripled in the 1920s.

As we've seen in Chapter 21, after the First World War, gasoline consumption by automobiles became the dominant market for petroleum (replacing lubricants and illu-minants). The enormous increase in gasoline consumption would not have been possible without significant advances in the technology of oil refining.

Suburbs

The car made possible a major decentralization of cities. Actually, it was not the car that was first responsible for the flight to the suburbs. As early as 1814, people working in Manhattan began taking a steam-driven ferry to live in a quiet, rural, backwater town called Brooklyn. With the advent of railroads and streetcar lines, suburbs could continue to develop, but only wherever the tracks went. The pattern of development often looked like fingers on a hand (Figure 22.5).

An example is Philadelphia's 'Main Line' suburb. The streetcar companies had helped begin the exodus to the suburbs by extending their lines from the central cities into previously undeveloped districts; but in 'steetcar suburbs,' the only useful residen-

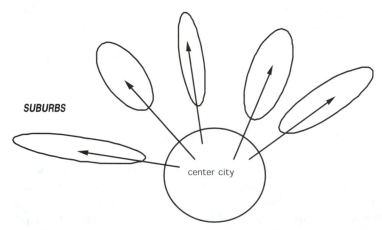

FIGURE 22.5 The development of suburbs initially proceeded like the growth of fingers on a hand, often following railroad and trolley lines.

tial property was on or within walking distance of the streetcar line. The automobile would make it possible to develop suburban property that had previously been inaccessible; that is, to fill in the spaces between the spokes of the streetcar lines, as if filling the spaces between the fingers of the streetcar 'hand.' Furthermore, light trucks made it possible to transport raw materials and finished goods out to the suburbs, so that warehouses and factories could also be moved to the suburbs. These effects led to the phenomenon (or problem, depending on one's perspective) of suburban sprawl (Figure 22.6).

Probably the most important effect of the automobile was how it encouraged suburban sprawl, especially in the last half of the 20th century. The interstate highway system (begun in 1956) facilitated a rush to the suburbs. The suburb became accessible to less affluent people, thanks in large part to the car and the bus. For many people, the suburb offers the potential of owning a private home on one's own lot, instead of an apartment. As land and housing costs go up and up in cities, affordable housing may be restricted to apartments or condominiums. In the suburbs, there may still be the potential to be able to afford a private, single-family house.

The developers of suburbs assumed that families living in those new homes would have a car (or, in most cases, more than one car). Based on this assumption, schools, hospitals, and shopping districts in the suburbs were dispersed across the landscape, rather than being clustered together near a streetcar or rail line, on the presumption that residents would be able to access these facilities with their cars.

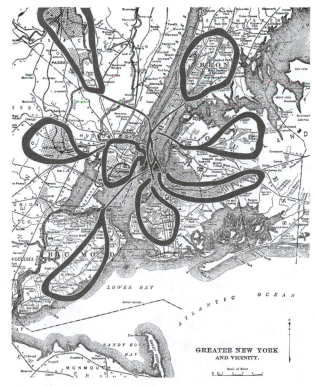

FIGURE 22.6 The phenomenon of suburban sprawl was an outgrowth of the increased ownership and use of automobiles.

There's a stereotyped view of commuting that presumes workers are traveling from the suburbs into the center cities and then back out when the workday is done. In fact, more than a third of metropolitan-area commuters really go from one suburb to another, and about 5% actually go from the center city out to the suburbs to work. Only about a fifth of metropolitan commuters travel from the suburbs to center city, even though it is that direction of travel that represents the stereotype of commuting to work. Suburban living also gives rise to 'two commuter' families, in which two wage earners in a family commute to jobs in different locations—even in diametrically opposite directions—from the family home, and gives rise to running multiple errands in cars, especially on weekends. The two-wage-earner family, with both adults away from the house on weekdays, has caused home errands and recreation to be compressed into the weekend days. As a result, Saturday afternoons are now the peak time for automobile use in the United States. In the 1990s, the phenomenon of the 'soccer mom' driving children to and from school and to numerous extracurricular activities, became a feature of the American cultural landscape.

There is no doubt that the automobile provided ordinary people with greater mobility than they had ever enjoyed previously. This gain, though, had to be balanced against several negative aspects, particularly the ways in which interpersonal life changed. People with cars, and thus the mobility provided by cars, were no longer dependent only on the facilities and social life in their immediate neighborhoods. Consequently, most of the lifestyle associated with the city street neighborhood disappeared: street vendors, delivery boys from local stores, persons simply going for a walk, and, eventually, even the local businesses themselves. Those persons who chose to remain in the center city, or those who unfortunately were economically trapped there, found the street radically changed. A place that had once been the location for social encounters with the neighbors degenerated into simply a route for transportation. As recently as 1910, three-quarters of the streets, even in major cities, were either unpaved or gravel. The principal highways and thoroughfares in cities were surfaced with cobblestones, brick, concrete, or asphalt. In a few cases, even wooden blocks were used for paving. These kinds of road surfaces do a superb job of discouraging rapid driving. With slow-moving traffic, much of it horse-drawn, streets were places where children could play. This kind of traffic could make its way around or through other activities going on in the street, and with comparative safety to those using the streets for play or social activities. The rise of the automobile, along with the electric trolley, made travel, not neighborhood activities, the function of the street.

> To leave our house meant that, once we had crossed our threshhold, we were in danger of being killed by the passing cars. I think back twenty years [to 1904], to my youth as a student: the road belonged to us then, we sang in it, we argued in it, while the horse-bus flowed softly by.
>
> —Le Corbusier[2]

Improved paving provided streets with smoother surfaces that allowed traffic to move more quickly, and made it unsafe to conduct games or social functions in the street.

Though we have been considering mainly automobiles, we have to remember that the internal combustion engine can drive trucks just as well as automobiles. This means that, like automobiles, trucks can 'fill in the blanks' between rail lines in the urban growth ring. Consequently, almost as soon as people started migrating to the suburbs, factories and warehouses started migrating with them. From the very beginnings of urban expansion, there were jobs available in the suburbs.

Road Building

The automobile revived the road transport systems that, in the 19th century, had languished in competition with the railways. By doing so, the automobile indirectly promoted a massive increase in facilities for personal transportation, such as gas stations and roadside restaurants. At the same time, the truck made it possible to move freight to the precise location required by the consumer, not to the local railroad depot, and often did so at a time suiting the convenience of the customer, not at the schedule established by the railroad. Cars and trucks required good roads that were surfaced for smooth, rapid travel.

Car and truck drivers quickly discovered, just as the bicycle riders of the late 19th century had done, that America's roads were not equipped to handle this new traffic. Local automobile clubs, usually voluntary associations of car owners, began forming early in the 20th century. One of their objectives was to apply political pressure to governments at all levels—city, county, state, and federal—to improve the roads. The improvements they wanted included things we take for granted today: widening streets to accommodate more traffic; smoothing the original cobblestone or brick streets with paving materials like asphalt; building wide, high-speed, limited-access highways; and paving dirt or gravel roads in the countryside. There actually was a national network of highways in the 1920s, but these roads were often poorly constructed and maintained. One of the major lobbying organizations of that era was The Auto Club, which used the motto, 'A Paved United States in Our Day.' The system of paved highways in the United States did double in length in the 1920s. Ironically though, not much progress was made in addressing the needs of an affluent, car-owning population until the Great Depression of the 1930s cut sharply into that affluence. Then, the federal government began massive public works projects, and began providing funding to states for such projects, to create jobs of people and to stimulate a devastated economy. Those projects included designing and constructing new roads, specifically to accommodate automobile traffic. As a result, the 1930s saw a second doubling in length of the paved road system in the United States.

The internal combustion engine made possible the cars and trucks that stimulated the demand for new and improved roads and wider city streets. At the same time, it also had a major role in reducing the costs of building the new roads. By about 1910, the horse-drawn road graders began to be replaced by power shovels and motorized graders. Similarly, horse-drawn wagons were replaced by trucks for the delivery of construction materials and for the removal of dirt and construction debris.

Before the Second World War, most Americans used their cars principally for local driving, that is, to get to work, to run errands, to make visits, all within a relatively short distance of home. High-speed driving over long distances on interstate highways or turnpikes did not arrive on the American scene until after the end of the war. But even before the war, stores and production facilities were already relocating to suburbia. The decision by the Eisenhower administration to build a federally funded interstate highway system was the decisive blow to the health of urban communities. The interstate highway system contained a remarkable contradiction, which was perhaps not recognized, or even thought of, at the time the system was proposed. At first, it was thought that the interstate highway system simply would provide a means of alleviating downtown traffic congestion. But, as things turned out, the possibility of high-speed travel between cities meant that the interstate routes often bypassed downtown areas or cut straight through city neighborhoods that had limited access to the highway itself. As a result, businesses began moving out of the city centers. In some places this outward movement was led by the white-collar firms. By 1960, more than half of all industrial jobs could be found outside the core of the city.

Transforming the American Scene

The secret of success of Ford and of the other manufacturers who followed his example was cheapness. The only way that mass production of automobiles (as opposed to building them by hand) could be justified was to tap into the mass market. This mass market consisted of blue-collar workers and farmers. That is, Ford recognized that the automobile had to be made at a cost that was within the reach of the people who made the car in the first place. The successful mass production of automobiles had dramatic effects on workers' wages and on labor relationships. Mass marketing of cars led to the ways for providing credit to buyers, pioneered by the General Motors Acceptance Corporation. Mass marketing also led to ingenious ways of advertising by rival manufacturers. The automobile became a ubiquitous feature of 20th century society. Now, in the early years of the 21st century, there is still no alternative to the private automobile as a convenient means of personal transportation. A serious rise in the price of gasoline, a rise that most people would have to believe likely to be permanent, might encourage the development of alternatives to the internal combustion engine. Environmental concerns about air pollution from vehicles (Chapter 31) could give the electric car or hybrid vehicles significant market opportunities. In the 1950s, there was even a flicker of interest in the nuclear car, an idea that now seems extinct. Some experimenters have tinkered with solar cars. But whatever the source of energy for the vehicle, it will still be recognizable to us as being a *car*. Furthermore, for these or other alternatives to achieve significant market share (at the expense of the gasoline-engine car), the rise of gasoline prices would have to be very dramatic.

It might be said of the nuclear car that not only is it an idea whose time has passed, it's an idea whose time should never have come in the first place.

The freedom of movement provided by the automobile resonated with longstanding American traits of individualism and a desire for geographical freedom. In addition, the automobile helped achieve the middle-class American dream of private home space by encouraging a migration from the center city to the suburb. (Actually, this process began in the railroad era of the latter part of the 19th century, but it was greatly accelerated by the advent of the automobile.) Though the car revolutionized travel by liberating the individual from the timetables and routes of the streetcar and train, it also forced people to join the car culture, whether they wanted to or not. Mass ownership and use of automobiles led to the steep decline of public transportation. In the 1920s, car ownership forced working families to make choices about how best to use their discretionary income. For families of limited means, buying a car replaced the purchase of other 'luxuries.' A common rural saying was, 'you can't go to town in a bathtub,' implying that it was preferable to buy a car rather than indoor plumbing fixtures. By the 1930s the car had often become a necessity for getting to work, thanks to the dispersion of people into suburbs and the decline of public mass transit. There is perhaps no story of greater power and impact than John Steinbeck's *The Grapes of Wrath* to illustrate the extraordinary dependence of impoverished people on their automobile.

The extension of the paved highway system and improvements in road surfaces made during the 1920s helped stimulate the new recreational activity of 'auto touring' (Figure 22.7). Families could pile in the car and head off for the mountains—Adirondacks, Poconos, or New England—or to the shore—from Massachusetts to Florida. Clever entrepreneurs lost little time in recognizing the potential of making money from these people on the move. As a result, this new recreation soon gave rise to the whole array of

FIGURE 22.7 Auto touring—vacationing by car—was another activity that became increasingly popular with the continued growth of automobile ownership.

roadside tourist attractions, from fine museums and nature sites to the tawdry reptile zoos and amusement arcades. Some of these new enterprises clearly embodied the old adage that you can never go broke by underestimating the taste of the American public. Auto touring also stimulated family trips to national parks. With the spread of the interstate highway system, beginning in the late 1950s, chains of motels clustered around the highway exits replaced the small-time 'motor courts' that could be found on the older roads.

The first drive-in movie appeared in 1933. By the early 1950s, there were 4,000 of them all across America. The drive-in restaurant—Royce Hailey's Pig Stand—first appeared in Dallas. In the 1940s, these drive-ins (often providing 'car-hop' waitresses) became the haunts of millions of adolescents and their cars (Figure 22.8). In 1955, the automobile helped make possible the opening of Disneyland, located along the Santa Monica freeway in Anaheim. Kansas City's Country Club Plaza, built in 1923, was probably the first shopping center of the auto age. After 1960, the shopping mall would replace 'main street' to a large extent.

All of these developments resulted in changing the environment through which the automobile moves, and the changes were done to fit the automobile. As much as half of total land area of a city is now dedicated to roads, driveways, parking lots, service stations. Things that were not friendly or useful to the automobile, such as narrow streets and low clearances under bridges, were eliminated, right along with most of the alternative forms of transportation. There were no bicycle lanes. Streetcar tracks were torn up. Some urban streets nowadays even lack sidewalks.

The changes that the automobile caused in patterns of social life also came rapidly. They affected many aspects of life, including activities such as dating, union membership, religion, leisure and recreation, and women's role in the home. 'Lovers' lanes' and 'parking' emerged as institutions among young people. Cars allowed them to cruise into the night, and well out of the city. Social conservatives fulminated that the car was simply a

FIGURE 22.8   The drive-in restaurant was another social phenomenon made possible by the growth of automobile ownership.

'bedroom on wheels.' In 1890, union meetings were a focus of working-class life; by the 1920s they were poorly attended, partly because car ownership gave people a much greater range of choices of after-work activities. Similarly, by the middle of the 1920s, civic holidays were celebrated with much less vigor than they had been just a few decades earlier. Nowadays few people seem to pay attention to what the Fourth of July, Memorial Day, and Labor Day are about; they're just taken to be a chance for a three-day weekend, sales at the nearby mall, and an opportunity to go on a short trip.

The automobile, along with the refrigerator, totally redefined shopping. The car had a crucial effect in changing the role of housewives from women who produced the family's food from scratch and sewed their clothing to consumers who shopped for (and bought) national-brand canned goods, prepared foods, and ready-made clothes. As women drove around to shop or to taxi their children around, the automobile became, in a sense, an extension of the home, used for shopping trips and for driving the children to school and many after-school activities.

By the 1970s the center of consumption had moved to the shopping malls. A forerunner was the suburban shopping district built by a Kansas City developer in the 1920s. At the end of the Second World War there were fewer than a dozen such shopping centers, and even those were just rows of stores. By 1960 there were more than 3,800 shopping centers, and the new ones were being built at exits from interstate highways or, better yet, at places where two interstates crossed. Enclosed centers (i.e., shopping malls), the first example of which appeared outside Minneapolis in 1956, grew into vast windowless sheds surrounded by parking lots. Without traffic or weather, they replaced the range of stores and leisure activities once located on Main Street.

At the end of the 1970s the average car was actually being driven 12% less and 6 miles per hour slower than a decade earlier, and getting 15 miles per gallon instead of 13. But the number of cars had shot up during those 10 years, and both the consumption of gasoline and the total number of miles driven had each increased by about 20%. Between

1969 and 1983, the number of miles driven by the average American household rose nearly 30%. There were about 40% more shopping trips, and the distance traveled on these trips increased by 20%. People were driving further to shop at enormous enclosed malls, while the downtown store areas withered. People were also driving more because new houses were being built further out in the country.

It's obvious to all of us that cars give us a phenomenal amount of personal freedom. Not only are we not restricted to going only where the railroad tracks run, at the scheduled train times, but we seem to have developed the idea of going anywhere we want, whenever we want to go, any day of the year. Of course, cars have brought with them the problems of traffic accidents and rather considerable pollution. We'll discuss the problems of air pollution by vehicle exhaust in more detail later (Chapter 31).

CITATIONS

1   Reynolds, Francis D. *Crackpot or Genius?* Chicago Review Press: Chicago, 1993; p. 30.
2   Le Corbusier. *The City of To-morrow and Its Planning.* Dover: Mineola, NY, 1987.

FOR FURTHER READING

Baritz, Loren. *The Good Life.* Knopf: New York, 1988. This book is a history of the middle class in America, from the founding up to the early 1980s. Several chapters include discussions on the impact of automobile ownership on the middle class, including such topics as Henry Ford and his cars in the early decades of the 20th century, and the rise of suburbia in the 1950s.

Berman, Marshall. *All that Is Solid Melts into Air.* Penguin: New York, 1988. This book discusses the evolution of the concept, and the experience, of modernity, from the 19th century to the 1970s. Though not a history of technology per se, much of the discussion is devoted to the changing role of streets and the effect of building or rebuilding streets and highways, from the Haussmann's boulevards in Paris to the megalomania of Robert Moses, who boasted of driving parkways through New York City neighborhoods 'with a meat ax.'

23

JET ENGINES AND JET FUEL

KEROSENE AS A REFINERY PRODUCT

The story of jet fuel starts with 'Kier's Rock Oil,' a patent-medicine cure-all that was on the market in the early 1850s. This boon to humankind, peddled by Samuel M. Kier, was in fact nothing but good old Pennsylvania crude. It's perhaps not surprising that Kier had far more oil than he had suckers to buy it, this being before the days of telemarketing and infomercials. He decided to see if the oil could be processed into kerosene. He sought the aid of a Philadelphia chemist, J.C. Booth, who converted an old iron kettle into a primitive distillation unit. The vapors boiling off the five barrels of Pennsylvania crude yielded a few gallons of product per day. Kier originally planned to market this product as medicine too, but the design of a new lamp, not using whale oil as the fuel, created a ready market for his product. Eventually he began buying Pennsylvania crude from the Drake well. By 1859, there were some 80 'coal-oil' (as kerosene was sometimes called) plants in the United States.

In 1854, the developers of the Pennsylvania Rock Oil Company sent a barrel of their crude to Prof. Benjamin Silliman Jr. of Yale to have it analyzed. Silliman believed that he could separate the oil into various components by distilling at different temperatures. Distillation is a time-honored process in the chemical laboratory for studying mixtures of substances, so this was not some special procedure designed specifically for the study of oil. Silliman's first experiment produced a thin, clear liquid with a strong odor, a product that we now recognize as gasoline. He heated the remainder to a higher temperature and this time condensed a straw-colored liquid, kerosene.

Edison's development of a practical electric light put an end to the need for kerosene for lighting, although it took a long time to do so. Not until after the Second World War did the Rural Electrification Program finally make electricity available to all parts of the country. Not only did the electric light reduce demand for kerosene, but at the same time the use of kerosene for cooking dwindled as housewives turned to electricity or natural gas.

From the origin of the modern petroleum industry in 1859🜨 to the decade of the 1910s, kerosene was the most important product from petroleum, because of its use for domestic lighting and for space heating. Gasoline was a nuisance. After 1920, kerosene demand dropped, and gasoline became the dominant petroleum product, thanks to the enormous expansion of the private use of automobiles, which increased the demand for gasoline, and the rapid expansion of electric power networks, which simultaneously decreased the demand for kerosene in household applications. Gasoline still remains the dominant product, accounting for some 45–50% of refinery output. However, the development of the jet engine and its subsequent adoption for most military and commercial applications has led to a modest comeback for kerosene as a petroleum product. Jet fuel is essentially kerosene that has been carefully distilled and purified.

> 🜨 The birth of the American oil industry occurred on August 27, 1859. On that day, Colonel Edwin Drake struck oil in the first well drilled in America, near Titusville, Pennsylvania. Drake assembled an outfit based largely on the technology of drilling for salt, and managed to hit oil at a depth of 69 feet. He sold the oil, to producers of illuminating oils for lamps, at $20 per barrel, an astronomical price at the time. In 1859, the median family income in the United States was probably a few hundred dollars per *year*. If the ratio of the cost of a barrel of oil to median income had remained constant over the last 140 years, petroleum today would be selling at about $2,000 per barrel!

THE JET ENGINE

The Early History of the Jet Engine

The successful development of a jet-propelled aircraft emphasizes how few are the really new ideas and how long it takes to realize some of the old ones.

Hero of Alexandria described about AD 50 a machine demonstrating the principle of propulsion by the reaction of jets (of steam) and the Abbé Miolan attempted in 1784 to apply it to the navigation of a hot-air balloon. The project was dismissed as having 'no aviating merit,' but this did not deter J.W. Butler and E. Edwards from patenting in 1867 (36 years before the Wright brothers flew) a design for an aeroplane which envisaged propulsion by the reaction of jets of compressed air or gas, or by 'the explosion of a mixture of inflammable gas or air' emitted through jets.

—Davy[1]

The earliest powered aircraft, for example, the Wright brothers' plane of 1903, used a gasoline-powered reciprocating piston engine. In 1929, the Swedish company, Bofors, decided to develop a gas turbine engine for airplanes, to create, as they put it, an 'airplane without a propeller.' In such a plane the powerful air stream normally produced by the propellor would be replaced by a gas jet.🜨 In 1934, two different versions of the 'pro-peller-jet turbine' (the term 'turboprop,' discussed later in this chapter, had not yet been coined) were developed. The Bofors project never got off the ground—both literally and figuratively speaking—because of deep cuts in the Royal Swedish Air Force budget in the mid-1930s.

The word 'jet' has several meanings. One, as used in this context, is a stream of liquid or gas that is shot out at high pressure from a small opening. The gases produced by combustion of the fuel produce the jet. A jet of water is produced by, e.g., a fire hose or a water pistol. Another meaning is a fast and narrow current in the ocean or atmosphere, as in the 'jet stream' we hear about in weather reports. 'Jet' also refers to a jet engine or to an airplane equipped with jet engines.

At about the same time as the Swedish efforts, the German aviation industry was trying to produce a gas turbine for use in airplanes. The aeronautical engineer Hans von Ohain patented a design for a jet engine in 1935, and began developing it in the following year for aircraft manufacturer Ernst Heinkel. On August 27, 1939, a Heinkel He 178 became the world's first jet-propelled aircraft to fly, at Warnemünde.

The hideous wail of the engine was music to our ears.

—Heinkel[2]

The Junkers firm started to develop another gas turbine engine in early 1940, and the first flights took place two years later with the Jumo 004 engine. This later became the first jet engine to be mass produced. For several years thereafter the German authorities did not show an interest in the jet engine. For one thing, Hitler did not expect a long war, and, for another, rockets and guided missiles were accorded a higher priority. Eventually, a slightly modified version of the Jumo 004 engine was installed in the German twin-engine fighter-bomber aircraft, the Messerschmidt 262 'Stormbird' (Figure 23.1), which began to be produced in 1943.

The course of the development of the jet engine in Britain began with Frank Whittle. He entered the Royal Air Force as an apprentice and was, in due course, selected for training that would lead to a commission. During his training at the RAF College, he was required to write a number of essays; in his fourth term he chose, as a subject for an essay, the 'Future Development in Aircraft Design.' Whittle, who was 22 at the time, argued that high-speed long-distance flights would have to be at high altitudes, where air resistance was much less. Piston engines would not be suitable and he therefore considered the possibility of the turbine, which would perform far more satisfactorily at high altitudes. As we will see shortly, there was nothing new or novel about the idea of a gas turbine. At that time, Whittle thought of the turbine as driving a propeller (again leading to the 'propeller-jet turbine' design—the turboprop— which we will discuss later). He imagined the first use for his turbine engine would be to propel a small, high-speed mail plane. Whittle patented a gas turbine in 1930. His superiors in the RAF showed little sympathy for his ideas at first, but in March 1936, he was given permission to devote part of his time to experimental work. He founded the company Power Jets Ltd. to develop his invention. The following year Whittle began to test his gas turbine. At the outbreak of the Second World War, Whittle received an order for *one* engine—called the W1—while, at the same time, the Gloster Aircraft Co. was given the job of building an experimental airplane to fit it. This plane, the Gloster E28/39, made its first flight in May 1941.

'Frank, it flies!'
'That was what it was bloody well designed to do, wasn't it?'

—Whittle[3]

FIGURE 23.1   The Messerschmidt Me 262 Stormbird was the first jet fighter to enter combat service.

Further development of Whittle's engine was carried out mainly by British Thomson-Houston Co. and by Rolls-Royce. Whittle was not, at the time he made his invention and during the early period of its development, connected in any way with the airplane-engine manufacturing industry. Like James Watt a century and a half before, he was an outsider to the engine-building business.

By the end of the war, both Germany and Britain had put their first jet-engine aircraft into operational combat service. These planes came too late to have much effect on the war, but subsequently jet engines were adopted for most military uses because of the greatly improved performance that they gave in speed and power. From the point of view of the airplane, the great advantage of the jet engine is its high ratio of power to size, which more than compensates for its noise and high fuel consumption. Another advantage is that it enables airplanes to travel faster than the speed of sound, and this capability is now available in many military aircraft. So far, the Anglo-French Concorde is the only civilian passenger airplane with supersonic capability. Nevertheless, all large passenger planes now use jet engines, even though the emphasis in passenger aircraft development has been on jumbo jets capable of carrying several hundred passengers on long-haul flights, rather than on the attainment of exotic speeds.

It is worth recalling the relationship of James Watt's steam engine and Charles Parsons' steam turbine. Watt's engine, which was a reciprocating design, was unquestionably a successful development. Nonetheless, these engines tended to be very large, very heavy, 'rough' in operation due to the vibration of the reciprocating motion, slow, and have a fairly low ratio of power output to weight of engine. The Parsons steam turbine could operate at much higher speeds, was smaller and lighter, and offered a higher power-to-weight ratio. In both, the steam needed for operation was generated in a separate boiler, so these engines are thought of as external combustion engines. Nikolaus Otto's four-stroke reciprocating internal combustion engine was also a successful development. The question remaining is whether an analogous internal combustion turbine engine could be developed. That is, what is the missing component of Table 23.1? In fact, the first suitable engine design was produced in 1791, by the English inventor John Barber. Demonstration of a working turbine engine that could be used in aviation was accomplished by the German engineer Hans-Joachim von Ohain in 1937.

In its design, the gas turbine is the simplest of the combustion engines. In 1897, Nils Gustaf Dalén, a Swedish engineer who later won a Nobel Prize in physics, tried to persuade a friend to become a partner in the development of a gas turbine engine. He described its principle as follows:

> Imagine a stove! If you blow through the grate, the hot smoke goes out through the chimney. But the clever thing is, if you blow in one cubic meter, two or three will go out. A pump up there, in other words, could drive another pump down there to suck in the air. Now, if we substitute these pumps with turbines . . .
>
> —Dalén[4]

From the time of Barber at the end of the 18th century, many inventors had devoted themselves to finding a practical application for this simple principle. The first recorded effort is actually by Leonardo da Vinci, who invented a roasting spit driven by the smoke and combustion gases of the cooking fire. The device actually worked, but would not be called a gas turbine in our modern usage of the term.

There are many problems in constructing a successful gas turbine, and they're not easy to solve. First, the speed of rotation must be very high to give adequate efficiency, a point that Watt noted when von Kempelen claimed, in 1784, that he had invented a steam turbine. (von Kempelen's other bogus invention, the chess-playing Turk, is mentioned in Chapter 16.) Second, if this speed could be achieved, the temperature of the turbine blades would be extremely high, so that a heat-resistant material must be used. Many of these materials are expensive and difficult to shape or machine. Third, a gas turbine needs to operate at high efficiency, since two-thirds of the energy of the turbine is consumed by the compressor that forces air into the combustion chamber (described in the next section). The total efficiency must exceed 54% just to keep it going by itself. Not until the end of the 19th century did anyone achieve an efficiency exceeding this figure. One of the first was Egidius Eling, a Norwegian engineer. He had patented his

TABLE 23.1 PRIOR TO WHITTLE'S PIONEERING WORK ON THE AVIATION GAS TURBINE (JET ENGINE), THE DOMINANT FORMS OF ENGINES WERE THE WATT RECIPROCATING STEAM ENGINE, PARSONS' STEAM TURBINE, AND THE OTTO AND DIESEL RECIPROCATING INTERNAL COMBUSTION ENGINES. THE DEVELOPMENT OF THE JET ENGINE ESSENTIALLY 'FILLS IN THE BLANK' FOR THE POSSIBLE TYPES OF ENGINES

Type of combustion	Reciprocating engine	Rotary engine
External	Watt steam engine	Parsons steam turbine
Internal	Otto four-stroke engine	???

first gas turbine in 1884, which coincidentally was the year in which Charles Parsons was granted a patent for his steam turbine. Eling needed another 20 years to modify his original design sufficiently to make it efficient.

At the end of the 1890s, various inventors were attempting to build flying machines. Dalén and his partner aimed at producing a light and compact engine to propel these early aircraft. Their first experimental engine gave promising results to begin with. Its efficiency was 50%—close to the crucial 54% needed—and its turbine achieved 25,000 rpm. However, the turbine lasted only a short time, being unable to take the extremely high temperatures of the products of combustion. They tested another engine with a redesigned turbine wheel in January 1899, and achieved an efficiency of 58%. Even this new turbine could not withstand the high temperatures of the combustion products. Dalén continued his development work for another couple of years but never did succeed in solving the problem of finding materials with adequate strength at high temperatures. Two French engineers, Marcel Armengaud and Charles Lemale, experimented with gas turbines in the early 20th century. Their first engine, which they made by rebuilding an old de Laval turbine, was completed in 1905. It burned petroleum distillates injected into the combustion chamber. Water was also injected and vaporized in the combustion chamber, the idea being to cool the combustion gases before they reached the turbine, and thus protect the turbine blades from extreme temperatures. Its output was only a few percent above the 54% efficiency needed just to keep the engine going by itself, but it proved the basic principle to be correct.

When Germany's Messerschmitt Me 262, powered by two of von Ohain's engines, flew in 1942, it became the world's first jet fighter, with a top speed of 540 miles an hour. The British Gloster Meteor, with a top speed of 480 miles an hour, followed in 1943. After some modifications, the Meteor became the first jet fighter to enter military service the next year. The faster and superior German Me 262 first entered operational combat service in 1944, and was the first jet fighter to engage an enemy plane. More Me 262s were produced during the war than any other jet fighter, but fortunately for all of us it was too late to help Hitler's cause. Using Whittle's engine design, Lockheed developed the F-80 Shooting Star jet fighter in 1943. Although it was initially plagued with problems, the Shooting Star was significantly improved by 1947, and set the world speed record of 623 miles per hour. Civilian jet aircraft became operational in the mid-1950s. The British DeHaviland 'Comet' was the first commercial jet airliner. It was not particularly successful, mainly because of an unfortunate tendency for the wings to fall off due to 'metal fatigue,' the problem of breakage of a metal.⚛ The airplane that really established the jet age in passenger flying was the Boeing 707.

⚛ Metal fatigue can easily be illustrated by straightening out a metal paper clip, and bending it back and forth until it breaks. Of course, paper clips are neither designed nor meant to withstand metal fatigue; still, this breakage occurs after a surprisingly short number of bendings.

Compared to piston engines, all jet engines have the disadvantage of a greater fuel consumption per unit of energy output. However, jet engines have several advantages: They provide much greater speeds, especially at high altitudes; they are structurally and mechanically simpler; and they are of much lighter weight, about half the weight of a piston engine of comparable energy output.

The Turbojet Engine

We will first look at how jet engines work, and then discuss some of the issues relating to fuel used in these engines. To begin with, consider what happens when you blow up a balloon and let go of it. Contrary to what might be assumed, the balloon is moved by the pressure of the air inside, not by the air escaping. Air escaping from the balloon reduces the interior pressure on the 'nozzle end' of the balloon. The high pressure continues to push out on the other end of the balloon. This unequal pressure propels the balloon in the direction of the greater pressure. In the same way, the jet engine moves in the direction of the greater pressure inside of it. The propulsion is possible because of Newton's third law of motion. According to this law, which is seldom known by its name, but is familiar when cited, 'for every action there is an equal and opposite reaction.' In the balloon, the *action* is the air rushing out of the neck. In the jet engine, the *action* is the hot gases rushing out the rear of the engine at a high pressure. The *reaction* in both cases is the unequal pressure inside.

Most jet engines nowadays are so-called turbojets. They are, in principle, very simple machines; simpler than the Watt or Otto engines, for example. A turbojet engine consists of a metal tube, open at both ends. At the leading (front) end is a rotary air compressor that draws in and compresses air. This compressed air is then passed to combustion chambers, sometimes called burner cans (shown in the diagram in Figure 23.2). There, the combustion process involves injecting the fuel into the stream of compressed air. Injection takes place through burner nozzles arrayed around the burned can. Depending on the size of the engine, there may be 8, 16, or 32 nozzles per burner can. The burning fuel maintains the air at a very high temperature and pressure. The high-temperature, high-pressure combustion gases are directed against the blades of a turbine. We should recall that a turbine is a device that converts the kinetic energy of a moving fluid—in this case, the hot, gaseous products of combustion—into rotary motion. Why do we need rotating motion in a jet engine? To run the air compressor. The combustion gases drive a turbine that is directly coupled to the compressor by a shaft. After leaving the turbine, the hot compressed air enters the tail pipe that ends in a nozzle. The function of this nozzle is to convert the high ENERGY of the air (pressure and heat) into energy of motion, or kinetic energy. The 'reaction' of the mass of ejected exhaust gases pushes the plane forward. It is this basic scheme that gives rise to the *turbojet* engine, sketched in Figure 23.3.

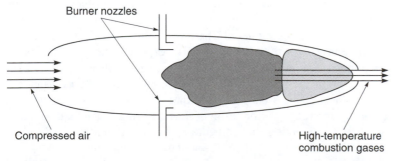

FIGURE 23.2 Schematic diagram of a burner can in a jet engine. The fuel is burned in a steam of compressed air, forming high-temperature, high-pressure combustion gases.

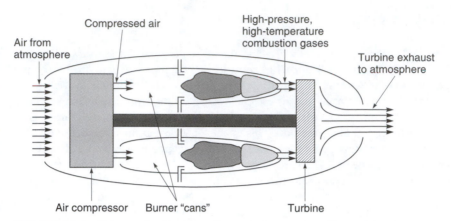

FIGURE 23.3 Schematic diagram of a turbojet engine. The turbine is operated by the combustion gases from the burner cans; the air compressor is operated by the turbine.

During flight, air comes into the engine's inlet whatever the speed of the plane is, which could be in well excess of 500 miles per hour. Inside the inlet, the air slows down and its pressure increases, but its total ENERGY is unchanged. The air then passes through a series of fanlike compressor blades that push it forward, doing WORK on it and increasing both its pressure and its total ENERGY. By the time the air arrives at the combustion chamber, its pressure is far above atmospheric. Since a given mass of hot air occupies a greater volume than cold air (Charles' law), the hot exhaust gas takes up more volume than it did before the combustion took place. The hot exhaust gas exits the combustion chamber traveling faster than when it entered. Its pressure is still very high and it accelerates out of the back of the engine, exchanging its pressure for speed. On its route out of the engine, the exhaust gas does WORK on this turbine and gives up a little of its ENERGY in the process. The turbine drives the compressor for the incoming air. Overall, the engine has slowed the air down, then added ENERGY to it, and finally accelerated it back to high speed. Because the engine has added ENERGY to the air, the air leaves the engine traveling faster than when it arrived. The jet engine has exerted a force on the air to accelerate it rearward; the air has pushed back and produced a thrust that propels the airplane forward.

In the turbojet engine, the working fluid in the turbine is the hot air and combustion products. This hot gas mixture has two roles: It spins the turbine, but also provides some jet thrust. Just as in the steam turbine, the high-pressure, high-temperature gas stream expands going through the turbine, and expands again as it exits the nozzle at the back of the engine. Recall from the studies of Charles and Boyle that we should expect the temperature to drop, especially as the combustion gases expand as they exit the nozzle. Therefore, the jet engine is a type of heat engine. The thrust from the back end of the engine can be achieved only if the speed of the gases exiting the back of the engine is greater than the speed of the airplane itself. To ensure that this condition will prevail, the ENERGY supplied to the compressor by the turbine is used to raise the pressure, and of course the combustion of the fuel raises the temperature. Furthermore, the nozzle at the back end of the engine helps to accelerate the gases as they leave the engine.

Over the years, there has been a steady increase in horsepower ratings of gas turbine engines. In the 50 years following the Wright's first flight, the maximum horsepower of piston engines increased from 12 to 3,500, but in only 15 years the horsepower rating of jet engines climbed to 25,000, more than seven times the maximum horsepower of a piston engine. Although jet engines are heat engines, just as Watt's steam engine and Parsons' steam turbine, we should appreciate that the modern jet engine depends on materials that must be able to withstand temperatures and mechanical stresses that are far greater than the steam-operated devices experience. Additionally, the jet engine components have to be machined to finer tolerances and greater accuracies than are required in a steam turbine and far beyond anything James Watt could ever have hoped to be produced in the machine shops of his day.

Fan Jet or Turbofan Engines

Two more recent engine developments include the bypass engine, the so-called *fan jet*, and the 'prop jet' or *turboprop*. We'll discuss the fan jet first, and the turboprop in the next subsection. Figure 23.4 is a sketch of the fan jet engine. If the highest possible speed is not needed (in other words, neglecting military aircraft), a fan jet engine is more efficient than a turbojet. Fan jet engines are now common on civilian aircraft. They're noticeable because of the conspicuously wider diameter of the engine (to accommodate the fan) at the front end.

The thrust exerted by a jet engine must, assuming the plane is flying level and at a uniform speed, exactly equal the air resistance the plane encounters. For a large commercial plane, flying at a subsonic speed (i.e., below the speed of sound), the goal of the engine designers is to achieve the optimum thrust from the engine for the minimum fuel consumption. Fuel consumption is an issue for two reasons. First, of course, the airlines must buy the fuel, and therefore reduced fuel consumption translates directly into cost savings. Second, any given airplane is limited in the amount of fuel it can carry. The lower the fuel consumption, the greater the flying range of the plane for a fixed fuel load. In the fan jet engine, only a fraction of the air pulled into the engine is sent to the burner cans. The rest of the air is blown out immediately, adding to the jet of air expelled from the engine. This means that the front section of the compressor must be of wider diameter than the rest, which is why the fan jet engines of commercial jet planes have such conspicuously large 'bulbs' in front.

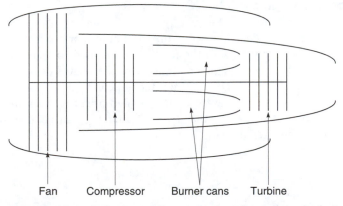

Fan   Compressor   Burner cans   Turbine

FIGURE 23.4 Schematic diagram of a fan jet (turbofan) engine. The large fan at the front of the engine helps some of the air to bypass around the compressor and burner cans.

The turbojet engine moves too little air and gives the air too much kinetic energy. In doing so, it incurs an efficiency penalty. To make a jet engine more efficient, it should move lots of air. That way, the air will leave the engine traveling relatively slowly and will thus take away relatively little energy. The thrust of an engine is proportional to the mass of air expelled per second and the increase in velocity of the air as it passes through the engine. We can write this as

$$T \propto (m/\sec) \cdot \Delta v$$

where T is thrust, m the mass of air expelled, and Δv the increase in velocity. Fuel consumption is proportional to mass of air expelled per second and the square of the velocity increase,

$$C \propto (m/\sec) \cdot (\Delta v)^2$$

where C is the fuel comsumption. (The ENERGY expended has to come from somewhere, and it is provided by the ENERGY released as the fuel burns—the spontaneous change from high chemical potential energy to low chemical potential energy in a combustion process.) Fuel consumption is, therefore, proportional to thrust times velocity. Since we can rewrite the previous relationship as

$$C \propto (m/\sec) \cdot \Delta v \cdot \Delta v$$

then

$$C \propto T \cdot \Delta v$$

So, if the total mass of air expelled per second can be increased for the same velocity, a greater thrust can be obtained at a given fuel consumption. As in the turbojet, the engine has slowed the air down, added ENERGY to it, and returned it to high speed. Again, the air leaves the engine traveling faster than when it arrived. But, the fan jet engine moves more air than a simple turbojet engine does, giving that air less ENERGY and, therefore, consuming less fuel in adding ENERGY to the air.

This is why a fan jet is more economical than a turbojet, but why the turbojet gives higher speed. The fan jet, however, is not as economical as the turboprop. In the turboprop engine, little energy is left for the jet. Most of it is used to drive the propeller so that as much air as possible 'bypasses' the combustion chamber and turbine.

Turboprop and Turboshaft Engines

The turboprop engine (Figure 23.5) represents, in a sense, the fruition of Frank Whittle's original idea. In a turboprop engine, about 80% of the total energy is used to spin the propeller, and only about 20% provides thrust via the nozzle at the rear. A related design is the turboshaft engine. In this engine, all of the ENERGY from the hot combustion gases is used to operate the turbine, and none is available for jet thrust. Turboshaft engines are used in helicopters. They are also used on some other types of aircraft as an auxiliary source of energy needed on board the aircraft.

Jet Cars?

The use of the jet engine for airplanes led many observers to anticipate a similarly bright future for gas turbine engines in road vehicles (Figure 23.6). These hopes have never been realized. One reason is the high fuel consumption of the gas turbine engine. Another

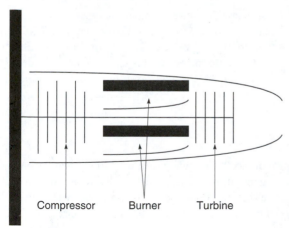

Propeller

Compressor   Burner   Turbine

FIGURE 23.5 Schematic diagram of a prop jet (turboprop) engine. In such an engine the turbine operates a propeller that provides most of the propulsion for the airplane.

major concern of the production process for gas turbines is its required manufacturing accuracy, as mentioned above. The gas turbine must be machined to tolerances 10 times finer than the usual automobile engine. Thus there is a high manufacturing cost combined with higher operating costs (because of the high fuel consumption).

In diesel engines (discussed in Chapter 24), the maximum gas temperature in the cylinder rises above 2,300 K, but the cylinder must endure that temperature only periodically, at the peak of combustion. In the gas turbine engine, though, combustion is occurring continuously. As a result, the temperature of the combustion gases entering the turbine is limited to about 1,300 K even when the turbine is made of special (and hence expensive) heat-resisting alloys. If it were possible to develop and mass produce turbines produced

FIGURE 23.6 The gas turbine (jet) automobile has fascinated car designers for decades, but has never managed to be a commercial success.

from ceramic materials that would resist heat even better than specialty alloys, the turbine inlet temperature could be allowed reach about 1,650 K. The Carnot equation for heat engine efficiency (Chapter 11) shows that, for a fixed value of T_{low}, an increase in efficiency can be achieved by increasing T_{high}. Even if no other design changes were made, this increased temperature alone can result in a 20–30% reduction in fuel consumption.

Electricity from Jet Engines

In Chapter 15 we saw that the essence of electricity generation is to find a way to provide WORK to the generator, to turn the generator as cheaply and reliably as possible. The solution is to connect the generator to a turbine. We've seen the use of water (Chapter 15) and steam (Chapter 16) as the working fluids in turbines used in electricity generation. Now we have met the gas turbine, and it would be reasonable to wonder if a gas turbine could also be used to operate a generator. The answer is yes.

Gas turbines can be used to provide electricity generation. These turbines are similar to those used as aircraft engines (and are sometimes called 'aero-derivative' turbines) but, in electricity applications, are often fired with natural gas. Though these are often in installations of much smaller generating capacity than the hydro- or fossil-fuel plants we have discussed earlier, they offer many advantages. First, they can be purchased and installed relatively quickly, much faster than dams or large fossil-fuel plants. Therefore they offer a rapid way of expanding the total generating capacity of a utility system. Second, a gas turbine can be started and brought up to speed very rapidly, far faster than, say, a pulverized-coal-fired boiler. Gas turbines make excellent reserve units that can be quickly brought on line during periods of the day when demand for electricity peaks, or during unusual episodes of demand (for example, on an extremely hot and humid summer day, when consumers are switching on their air conditioners and causing electricity demand to skyrocket). Third, the fuel, natural gas, is relatively clean burning, so these plants do not need the scrubbers and particulate contol equipment associated with a coal-fired plant.

JET FUEL

While it's possible in principle to design a jet engine that would operate on gasoline, the tremendous demand for gasoline for military purposes made it seem sensible to use a fuel other than gasoline. By the late 1930s and early 1940s the market demand for kerosene had dropped greatly. The civilian demand for gasoline was already fairly high, and with the outbreak of war the added military demand pushed gasoline 'off the chart.' Thus jet engine development centered around kerosene as the fuel.

While jet engine fuel has about the same boiling range as ordinary kerosene, it is usually more highly refined than the kerosene we would buy for space heating or kerosene lamps. The issues of concern in jet fuel performance include low-temperature behavior and smoke formation.

At high altitudes the atmosphere is very cold, well below 0°F. In fact, at high altitudes, in transcontinental or intercontinental flights, for example, the outside air temperature can be around −75°F. Any liquid is more difficult to pump or to get to flow at low temperatures. (This behavior is reflected in the folk saying that someone, or something, is 'slower than molasses in January.') That is, the resistance to flow, or the viscosity, of the fuel increases as the temperature decreases. At the low temperatures of high-altitude flight the fuel still has to be able move through the lines from the fuel tank to the engine, and consequently the physical behavior of the fuel at very low temperatures is vitally important. As a rule, the larger the molecules, the more difficult it is for them to flow, and this relationship becomes especially pronounced at low temperatures. Thus the composition of jet fuel is adjusted to ensure that it will flow reliably at low temperatures. In addition, at very low temperatures some of the very large, waxlike molecules that might be dissolved in the fuel will precipitate as a solid. Thus kerosene is also 'dewaxed' to make jet fuel.

Two properties used to measure the low-temperature behavior of fuels are the cloud point and pour point. The cloud point is the temperature at which the first crystals of wax (or other solid) begin to form as the fuel is cooled. This initial formation of small wax crystals gives the liquid a translucent, cloudy appearance. The pour point, at even lower temperatures, is that temperature at which the liquid will not flow at all, even if one were to attempt to pour it out of an open container. While these may seem like rather 'low-tech' measurements, the determination of the cloud point and pour point is carried out under very carefully standardized conditions, and the values of these properties are used to establish specifications for the fuel.

Smoke is a nuisance and a pollutant from civilian aircraft. In addition, it could possibly be used to detect military aircraft, so smoke formation is also a concern to the military. As a rule, aromatic compounds are the 'bad guys' for smoke formation. Thus smoke formation can be adjusted by the removal of aromatics during refining. This property of the fuel is determined by the smoke point. Like the cloud and pour points, the smoke point also might appear to be a rather 'low-tech' test, but it too is carried out under very rigidly standardized conditions. The smoke point determination involves adjusting the height of a wick used in burning the fuel being tested, to find the maximum height at which the flame can be sustained without producing visible smoke or soot.

Sulfur is also removed from jet fuel during refining, both to reduce SO_x emissions and because sulfur compounds in the fuel are mildly corrosive to the metal parts in fuel storage and handling systems. The SO_x emissions are a concern because they can contribute to serious air pollution problems, notably acid rain (Chapter 30). As a secondary issue, many sulfur compounds have absolutely dreadful odors.

Jet fuels are also treated with a variety of specialized chemicals, added in very small quantities, to adjust or tailor the properties of the fuel. These so-called additives do a variety of jobs. They improve the lubricity of the fuel (its ability to lubricate) to help operation of fuel pumps. They prevent the buildup of static electricity, reducing the danger of an accidental fire caused by an electric spark. They help prevent the slow degradation of the fuel by oxygen during the time that it is in storage. They combat the formation of ice. Even after the dewaxing, desulfurization, and other refining steps, a fuel will then be treated with a specified 'additive package' before it is actually sold and used.

The production of jet fuel, expressed as a fraction of the total amount of liquid fuels produced from distillation of petroleum, rose sharply in the 1950s and 1960s. In the mid-1950s, when most of the operational jet airplanes were used by the military, jet fuel accounted for about 4% of the total amount of petroleum distillate fuels. By the mid-1960s, as jet airplanes steadily became more important in civilian air transportation (time line, Figure 23.7), jet fuel production grew to about 10% of total distillate fuels. In the subsequent 30 years, there was continued, but very small, growth of the share of distillate fuel production represented by jet fuel, so that today it is about 12%. (It's important to remember that this figure is a percentage, applied to the total amount of petroleum consumed. Even though the *percentage* of total distillate fuel production represented by

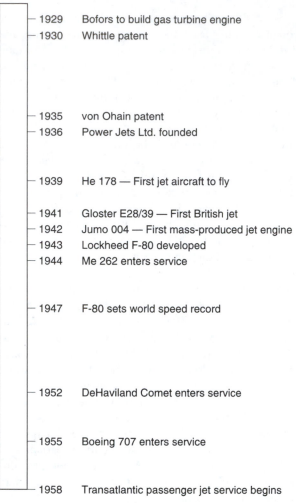

1929	Bofors to build gas turbine engine
1930	Whittle patent
1935	von Ohain patent
1936	Power Jets Ltd. founded
1939	He 178 — First jet aircraft to fly
1941	Gloster E28/39 — First British jet
1942	Jumo 004 — First mass-produced jet engine
1943	Lockheed F-80 developed
1944	Me 262 enters service
1947	F-80 sets world speed record
1952	DeHaviland Comet enters service
1955	Boeing 707 enters service
1958	Transatlantic passenger jet service begins

FIGURE 23.7   A time line of some of the major events in the development of jet aircraft.

jet fuel has held nearly constant at 10–12% over the past three decades, the actual *amount* of jet fuel produced has increased because of the increase of total petroleum consumption.) Not all of the products from petroleum are distillate fuels. If we were to express jet fuel production in terms of its proportion of *all* petroleum products, jet fuel represents about 10%. In other words, for every 100 barrels of petroleum fed to refineries, 10 barrels of jet fuel are produced.

CITATIONS

1   Davy, M.J.B., quoted in Gregory, Kenneth. *The Next to Last Cuckoo*. Akadine Press: Pleasantville, NY, 1997; pp. 67–68.

2   Heinkel, Ernst, quoted in Chase, Alex. *Technology in the 20th Century*. Bluewood Books: San Mateo, CA, 1997; p. 100.

3   Whittle, Frank, quoted in Chase, Alex. *Technology in the 20th Century*. Bluewood Books: San Mateo, CA, 1997; p. 102.

4   Dalén, Nils, quoted in Strandh, Sigvard. *The History of the Machine*. Dorset: New York, 1989; p. 150.

FOR FURTHER READING

Bentele, Max. *Engine Revolutions*. Society of Automotive Engineers: Warrendale, PA, 1991. The autobiography of a German engineer whose long career involved him in developments of a variety of engines. Many of the chapters detail the development of jet engines for aircraft and attempts to develop the gas turbine automobile.

Cardwell, Donald. *The Norton History of Technology*. Norton: New York, 1995. Chapter 16 provides a useful history of the development of jet engines, and information on the various types, such as the fan jet and turboprop.

Ethell, Jeffrey L. *Frontiers of Flight*. Smithsonian Books: Washington, 1992. This is a well-illustrated book on the history of aviation. Chapter 8 discusses the development of the jet engine, and Chapter 10, the history of jet engines for civilian passenger airplanes.

Guibert, J.C. *Fuels and Engines*. Editions Technip: Paris, 1999. Chapter 7 discusses engine operation, characteristics of jet fuel, and how fuels are formulated. Intended mainly for readers with technical background.

Gunston, Bill. *Jet and Turbine Aero Engines*. Patrick Stephens: Sparkford, UK, 1997. A history of the development and applications of jet engines, written mainly for readers with little science or mathematics background.

Hünecke, Klaus. *Modern Combat Aircraft Design*. Naval Institute Press: Annapolis, 1987. Though this book does not have the most recent details on the subject matter, Chapter 6 provides an excellent introduction—for the reader with some technical background—to jet engines and their operation.

Hünecke, Klaus. *Jet Engines*. Motorbooks International: Osceola, WI, 1997. Subtitled 'Fundamentals of theory, design and operation,' this book provides a fine, detailed discussion of most aspects of jet engines and how they work.

Morgan, Hugh. *Me262: Stormbird Rising*. Osprey Aerospace: London, 1994. This book provides a history of the development and operation of the world's first successful turbojet military airplane.

Suzuki, Takashi. *The Romance of Engines*. Society of Automotive Engineers: Warrendale, PA, 1997. Chapter 40 discusses many of the attempts to adapt the gas turbine engine to cars, trucks, and buses.

24

DIESEL ENGINES AND DIESEL FUEL

LIFE WITHOUT MATCHES

The story of diesel engines actually begins with the question of how one actually lights a fire. The common, ordinary match (Chapter 7) is only about 150 years old. Lighters—the Zippo, Bic, and similar—are more recent. Before the development of the match, there were various cumbersome ways of lighting fires, such as flint-and-steel sets to strike sparks, or devices for rapidly rotating wooden rods and using the friction heat to ignite tinder. (Tinder is some readily combustible material such as dried cloth, wood shavings, or small twigs.) Indeed the easiest way to start a fire was to 'borrow fire' from the neighbors, by obtaining an already glowing piece of coal or charcoal from them. However useful borrowing fire from the neighbors may be, there was still a need to have a portable and convenient way of starting a fire. One device for doing that, which might be considered an early forerunner of the lighter, was the fire piston.

The fire piston may actually have originated in primitive southeast Asia, but was reinvented in France and Britain early in the 19th century. In 1804, a French inventor, Joseph Mollet, produced a model that represents the European reinvention of the 'fire piston' (Figure 24.1). When you pump air into a bicycle tire, the nozzle of the pump gets warm, and if you compress air suddenly to one-fifth of its original volume, the high temperature produced is enough to ignite a little tinder on the underside of the piston. Similarly, a sharp blow on the end of the piston compressed the air and heated it so hot (remember Boyle) that the hot air ignited the tinder. The burning or glowing tinder could then be used to light a fireplace fire or whatever else it was desired to burn. Fire pistons were manufactured commercially in France, but were made obsolete by the development of the match in the mid-19th century. By the late 19th century, fire pistons were museum pieces. In Munich, about 1888, a lecturer on heat transfer lit his cigar by this method; in the audience was a young man named Rudolf Christian Karl Diesel, who later said that the demonstration was a major inspiration for his new engine, in which ignition is achieved by compression alone.

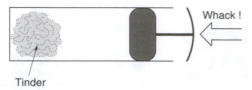

Whack !

Tinder

FIGURE 24.1 The fire piston was a portable fire-starter in the days before matches and pocket lighters. A sharp blow on the end of the piston raised the air temperature inside high enough to ignite the tinder.

Actually, there are several stories as to the source of the inspiration that triggered Diesel's invention of his engine. It may have been a fire piston in a Paris museum that inspired him during a time when he was living in Paris and studying various mechanical devices. Diesel attended the Technische Hochschule (Polytechnic) in Munich. In 1878, he heard a lecture there on Carnot's theorem concerning the ideal conditions for expansion of gases in an engine's cylinder. In the lecture that inspired Diesel, his professor pointed out that steam engines transform only 6–10% of the chemical energy of their fuel into useful WORK (i.e., their efficiencies are in the range of 6–10%). Diesel was convinced that he could develop an engine that would do better. Throughout the rest of his studies and into his early career, which involved research work on refrigerators, Diesel never let the idea drop. Eventually, in 1893, he came up with a scheme for an engine that would indeed operate much more closely to the ideal (Carnot) efficiency for a heat engine. The first diesel engine was built soon after.

THE DIESEL ENGINE AND HOW IT WORKS

The Early Development Work

Diesel designed his engine on the general pattern of the gas engine perfected by Otto in 1876. He set about developing an engine that would be more efficient than existing steam or gasoline engines, and that could, in principle, be operated on a whole variety of fuels: gases, powered coal, coal tar (a liquid by-product of converting coal into the coke used in metallurgical furnaces), or higher-boiling petroleum products. (This is not to imply that all these fuels could be used in the same engine, but rather that variants of Diesel's fundamental design could be adapted to these different kinds of fuels.) The use of pulverized solid fuels quickly seemed impractical.⚛ The vast majority of diesel engines nowadays operate on the petroleum fraction that we commonly call diesel fuel.

⚛ Several reasons contribute to making the idea of a solid-fuel diesel (or other internal combustion) engine impractical. First, solids are harder to handle, to get to flow, and to meter or regulate than fluids. In particular, it is comparatively easy to inject a fluid into a chamber (such as an engine cylinder) that is at high pressure, whereas this can be difficult to do with solids. Second, solids, such as many coals, can be abrasive, and steadily wear away the internal parts of the fuel-handling and injection system. Third, most prospective solid fuels—particularly coal—leave an ash when they burn. This ash cannot be allowed to accumulate in the cylinder. Therefore, it has to be swept out of the cylinder during each cycle of the engine. The ashes of some coals contain quartz (the principal mineral component of sand), which is a very hard and abrasive substance. Running the engine would result in, essentially, continuously sand-blasting the exhaust valves and other parts of the exhaust system. Also, the ash has to go someplace—either into the air or into some kind of special collection hopper on board the vehicle. Stimulated in part by the oil shortages of the 1970s, the U.S. Department of Energy has shown occasional interest in operating a diesel engine on a suspension of finely pulverized coal in oil or in water. Doing so would eliminate the problem of handling solid fuel, and would probably reduce substantially the abrasion problem in fuel injection. However, it's hard to beat the ash problem.

Diesel first intended to use finely powdered coal, a fuel that many designers were, at the time, thinking of using for various kinds of engines. Diesel soon abandoned that idea. Instead, he began to experiment with crude oil. Diesel's first engine, a single cylinder machine using oil as fuel, was built in 1893. In 1897, after a period of hard work on further development of the engine, which took a toll on his health, Diesel built an engine with an efficiency clearly superior to that of any other combustion engines. The 1897 version of his engine showed a mechanical efficiency of 34%, a great advance over all other engines.

In its original form, the diesel engine was designed to operate at low speeds. It was expected to serve as an alternative to reciprocating steam engines for use both in ships and in stationary installations. These engines usually ran at speeds from 150 to 250 rpm and weighed between 150 and 350 pounds per horsepower. By 1909, Diesel had developed a small four-cylinder engine of 30 hp rated at 600 rpm. It was intended for use as a heavy-duty automobile engine, but was not pursued at the time.

How the Diesel Engine Functions

Diesel's engine can be considered as a variant of Otto's four-stroke cycle. The key components are the cylinders and pistons, intake and exhaust valves, and a fuel injector. The intake and compression strokes deal only with air. The first, intake stroke (intake valve open) draws air into the cylinder (Figure 24.2). The second stroke (compression) compresses the air in the cylinder such that it is above the temperature needed to ignite the fuel (Figure 24.3). Note that both valves are closed: The compression ratios, which are usually in the range 14:1 to 20:1, are appreciably higher than those in conventional gasoline engines.

At this point, it is worth noting the similarity of operation of the diesel engine and the fire piston. We need once again to recall Charles,

$$V \propto T$$

and Boyle,

$$PV = \text{Constant}$$

If we combine the work of these two scientists into a single relationship, we can say that, for a given quantity of gas,

$$PV \propto T$$

So, if we rapidly increase the pressure, the temperature also has to go up. We can see this phenomenon for ourselves in tire pumps or the small air pumps used to inflate various kinds of sports balls. The 'barrel' of the pump can get quite hot after vigorous pumping (of course, some of this heat simply comes from the friction of the pump's piston against the cylinder walls).

Let's look at this in another way. In the diesel engine, we want to get the air hot enough to ignite the fuel. *But*—increasing

FIGURE 24.2 The intake stroke of a diesel engine draws air into the cylinder.

FIGURE 24.3 In a compression stroke of a diesel engine, only air is compressed by the piston as it rises. The compression ratio of a diesel is so large that the air temperature at the end of the compression stroke will be high enough to ignite the fuel.

temperature (Figure 24.4) is an unspontaneous change and we should realize that such a change can occur only if we do WORK on the SYSTEM (Figure 24.5). In this specific case, we sometimes refer to this WORK as *compression work*.

In the diesel engine the first stroke of the piston draws air into the cylinder and the second stroke, the return, compresses the air to such an extent that the temperature rises far above the autoignition point of the fuel. In the third stroke (the power or combustion stroke) the fuel injector squirts fuel into the cylinder. The fuel is injected at high pressure as a fine spray into the compressed air, which is swirled to obtain turbulent mixing. This injection continues for a substantial part of the power stroke. Because the air is above the temperature at which the fuel will ignite, it immediately begins to burn. The technical term for the diesel engine is the 'compression ignition engine' and it is so called because the compression of the charge is sufficient to cause the fuel to ignite. Note here that both valves are closed (Figure 24.6). The piston begins moving downward. We should already know from our discussion on turbines that the consequent expansion of the gas in the cylinder should cause its temperature to drop. But, in this case, the temperature of the gas does not drop, *because heat is being released from the burning fuel*. Initially, the heat ENERGY of the burning fuel is converted directly to useful WORK pushing down the piston in the cylinder.

THE KEY DIFFERENCE BETWEEN DIESEL'S ENGINE AND ALL OTHER INTERNAL COMBUSTION ENGINES IS THAT HEAT FROM THE BURNING FUEL IS NOT 'WASTED' TO HEAT UP THE AIR IN THE CYLINDER.

The air is already hot from the compression work we supplied to it.

As the piston moves downwards the air would normally be cooled, but the heat released by the burning fuel prevents this. Instead, the temperature of the air does not change significantly and the heat ENERGY of the burning fuel is converted into useful WORK driving the piston down the cylinder. This accords with Carnot's concept of maximizing efficiency, that there must be no useless flow of heat (ENERGY) from the hot body (the burning fuel) to a cold body (the air). Compared to other engines, the operation of the diesel engine affords production of the highest temperature of the cycle not by combustion, but rather before the combustion has occurred, entirely by compression of the air in the cylinder.

Once the air has expanded without losing heat, the fuel injector shuts off, and the air can then expand further with the accompanying drop in temperature and pressure that we would expect. So, while fuel is being injected into the cylinder, the air is expanding without a change in temperature (a process called **isothermal expansion**). After the fuel supply is turned off, and the expansion continues with a drop in temperature and pressure (a process called **adiabatic expansion**). The temperature and pressure fall because the heat energy in the air is now converted into

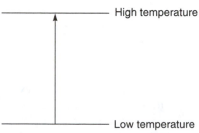

FIGURE 24.4 The compression stroke in a diesel engine raises the air in the cylinder from low to high temperature—an unspontaneous change.

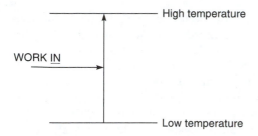

FIGURE 24.5 Because the temperature change in the compression stroke is unspontaneous, we must supply ENERGY or WORK to make it happen. In the diesel engine, the necessary WORK is supplied by the engine during the compression.

useful mechanical work. When the temperature and pressure have fallen—ideally to those of the atmosphere—the exhaust valve opens and the piston expels the air and burned fuel in the exhaust stroke. As in the Otto engine, the piston then returns, drawing in the next charge of air in the intake stroke. Although the ideal is to allow temperature and pressure to fall to the level of the atmosphere in order to maximize efficiency, this concept was not practical, because an enormously long cylinder would have been required.

The fourth stroke (exhaust) occurs as the piston sweeps the combustion gases out of the cylinder (Figure 24.7). In this case, the exhaust valve is open.

Various methods of starting the engine in the first place had to be developed to overcome the high compression ratios. One method uses a heater that is installed in the cylinder head. On modern engines, this heater resembles a spark plug, but instead of a spark gap, a heating coil is provided. The plug is connected to the battery, and when the current is switched on, the coil is heated to the temperature necessary to ignite the fuel.

Owing to the high pressure generated in the compression ignition engine (about 500 lb/in.²) the various parts have to be very strongly made. The necessary strength adds considerable weight, and although this may not be of great importance in stationary engines, it is a serious concern when the engine is intended for transport purposes.

The best diesels have efficiencies of up to 42%, and performances above 30% are routinely achievable with any reasonably well-maintained engine. Although the ideal Carnot efficiency of the Otto cycle is 60%, the best practical performance of gasoline engines is only around 32%.

Usually, diesel engines have a weight per horsepower about 1.5 to 3 times greater than gasoline engines. The reason for this increased weight is the diesel's reliance on self-ignition, which in turn requires the high-temperature compressed air resulting from the high compression ratio. As a result, the pressure in the cylinders of a diesel engine is much higher than a gasoline engine. To withstand the high pressure, a sturdy, rugged (hence heavy) cylinder is required. When the volume of the cylinders is the same (or, more specifically, the volume swept out by the piston through its stroke) is the same, the diesel engine produces only about two-thirds the horsepower of the gasoline engine.

FIGURE 24.6 The diesel engine has no spark plugs. In the combustion stroke, fuel injected into the cylinder ignites because of the high air temperature.

As we've seen, in the diesel engine, the fuel is injected near the end of the compression stroke and ignites spontaneously. As mixing between fuel and air occurs, burning continues. This combustion process is very heterogeneous, that is, it involves a reaction of liquid fuel droplets with the gaseous oxygen in the air. Because of the heterogeneous combustion, some of this fuel burns when oxygen is insufficient for complete combustion. As

FIGURE 24.7 In the exhaust stroke of a diesel engine, the gaseous products of combustion are swept out of the cylinder as the piston rises.

fuel continues to be injected during the power stroke, oxygen is being consumed and the potential exists for more and more soot to be produced. To avoid an excessive sooting problem, less fuel is present in the diesel engine cylinder than in the cylinder of the gasoline engine, and diesel engine power is therefore reduced. But, on the other hand, diesel engines use only about 70–80% of the fuel used by a gasoline engine with the same volume of cylinder swept by the pistons.

Under partial load, the fuel consumption of the diesel engine can be as little as 60% of that required by a gasoline engine. It is worth repeating that the essential difference between Diesel's engine and all other internal combustion engines is that, in the diesel, the burning fuel does *not* heat up the air—compression has already heated it up to the temperature of the burning fuel. The reason that a diesel engine has such good fuel consumption is its high compression ratio required for self-ignition. The higher the compression ratio, the better the thermal efficiency.

Another advantage of a diesel engine is that it has a higher durability, or a longer service life, as a result of the robust, heavy structure needed to sustain the high combustion pressures. Because of this long service life and excellent fuel economy, diesel engines are used extensively in commercial vehicles, such as long-haul trucks and many buses.

COMMERCIAL APPLICATIONS OF THE DIESEL ENGINE

Rudolf Diesel confidently predicted that his engine could have many possible applications. It could replace the steam locomotive for railway traction. It could be useful in road vehicles, such as trucks. It could be used for propulsion of ships and boats, and even as a stationary power source in various kinds of devices. In 1910, Diesel's assertions were dismissed as being ludicrous. The steam locomotive reigned supreme on the world's railways. The gasoline engine was steadily displacing electric motors and steam engines to become dominant in the automotive market. Small ships were powered by reciprocating steam engines, and large ones by steam turbines. Steam engines of various sorts almost totally monopolized the applications for stationary power sources. It was quite obvious to everyone that there was no market for the diesel engine.

But sometimes the 'obvious' can turn out to be dreadfully wrong. Diesels made their first major inroads in marine propulsion, both in ships and submarines, during the First World War, and their dominance in this market (except for nuclear propulsion in the largest submarines) was complete by 1960. On land they began to be used first in heavy earth-moving and farming machines. As they became lighter, they displaced steam locomotives for railway use, and even lighter engines took over most of the truck and bus transport market virtually worldwide.

STATIONARY ENGINES

In the United States the Diesel Motor Co. of America was founded in St. Louis in 1896, at the instigation of the master brewer Adolphus Busch. The first diesel engine made in the United States was delivered in 1898 and generated 60 hp. It was used to power an electricity generator. Nowadays the diesel engine totally dominates the market for stationary power sources, an example being its application to operate auxiliary or emergency electricity generators. Such generators may be either standbys in central electricity-generating stations or used in remote locations where a transmission grid from a central station would be too expensive.

Marine Applications

When the First World War broke out in August, 1914, the diesel engine found its first market niche: *das untersee Boot*, better known in the English-speaking world as the U-boat. Diesel engines offer numerous opportunities for submarine propulsion. One is their relatively compact size, much smaller than a steam engine of comparable power. A second is their good fuel economy, which extends the submarine's cruising range between refueling. A third is that the fuel itself is not nearly so flammable or explosive as gasoline, an important consideration in the confined space of a submarine.

The diesel engine has totally replaced the reciprocating steam engine in small and medium-sized ships (though the steam turbine continues to dominate in large vessels). Virtually all large ships are now diesel powered. For such applications the diesel engine has many advantages when compared with steam engines, especially those depending on coal-fired boilers. The diesel provides not only a substantial saving in the number of crew needed to operate the engine, but also considerable saving in the space required for carrying fuel. In 1910 the first sea-going motor ship, the *Vulcanus*, was built in Holland and fitted with a 650 hp engine. One of the largest diesel-engine vessels was the Cunard-White Star liner *Britannic*, which had two 10,000 hp engines.

Railways

In many nations, including the United States, the diesel engine dominates as the 'prime mover' in railway locomotives. In railway transportation the diesel engine supplies the power for generating electric energy in the diesel-electric locomotive. These locomotives can be adapted to a wider range of service (i.e., passenger, freight, or switching) than any single type of steam locomotive and provide significant economies in fuel consumption and in servicing relative to a steam locomotive.

The diesel-electric locomotive began to replace the steam locomotive for mainline operation in the United States during the 1940s. The principal components in the diesel-electric locomotive are the diesel engine, the electric generator, and the traction motors. Like the automobile engine, the diesel locomotive engine requires some type of variable drive mechanism between it and the driving wheels to take care of varying speed and power demands in starting and in climbing hills, while the engine itself works at generally constant speed. Neither mechanical gears nor a fluid (hydraulic) transmission, such as are used in automobiles, is adequate for the high power required in a locomotive, though some railway systems have used diesel-hydraulic locomotives. The variable drive in the diesel-electric locomotive is obtained using an electricity generator connected directly to the engine shaft. The generator supplies the traction motors with electricity at varying voltages. The function of traction motors can be electrically reversed to generate electricity as the engine is going downhill. This energy is absorbed and dissipated by converting it to heat through electrical resistance. This arrangement acts as a brake on the train going downhill, since the mechanical energy of the train is changed to heat. This so-called dynamic braking saves wear and tear on the brake shoes and wheels.

Compared with a steam locomotive of equal nominal horsepower, the diesel locomotive has important advantages. The useful power of a diesel locomotive may be greater than that of a steam locomotive because the diesel power is available over a wider range of speeds. The engine terminal expenses are less with the diesel locomotive because routine maintenance and refueling require less work. There is no need to provide water for a boiler. The diesel-electric locomotive has an efficiency of about 30%, about five times that of the steam locomotive. Moreover, diesel locomotives can be coupled in multiple units under the control of one crew, and these multiple units may produce tractive effort and power concentrations far greater than a single steam locomotive. Finally, many diesel locomotives can operate readily in either direction, eliminating the need for a turntable or other way of turning the locomotive around.

Road Transportation

Almost all of today's large trucks and buses use a diesel engine as their power source. The first successful application (1924) of a diesel engine was in the Benz 50 hp, 5-ton vehicle. The market position of the diesel was profoundly changed by the development of the high-speed engines designed to compete with gasoline motors. The speed of the engine was increased to 1,000 rpm, and by the late 1930s maximum speeds of 3,000 rpm were achieved in commercial types. The weight of the engine was greatly reduced; standard models were produced ranging in weight between 9 and 15 pounds per horsepower. The diesel engine enjoys worldwide dominance in heavy trucks, is widely used in buses, and has some applications in cars and light trucks.

The Diesel Airplane

The only internal-combustion-engine market in which diesels could not compete, let alone dominate, was in airplanes. The best-ever diesel airplane engine, the Jumo 207 developed in Germany for the Second World War, was still too heavy to compete with spark-ignition gasoline engines. But this was not for want of trying, nor for lack of some early, promising successes.

In September 1928, the Packard Company succeeded in the first flight with a diesel engine mounted on a Stinson-Detroiter plane. The engine was designed to compete in power and weight with the highly successful Wright 'Whirlwind' spark-ignition engine. In May 1929, the plane crossed the United States in only 6.5 hours (a good time even by today's standards for a coast-to-coast flight), demonstrating that the fuel cost for this flight was only $4.68, about one-fifth that for a gasoline engine. The next year, Packard began commercial production. In 1931, a Bellanca airplane with the Packard diesel engine established a new world endurance flight record, 84 hours and 33 minutes. A year later, this engine helped the Goodyear blimp (a type of airship) establish a world altitude record with use of a diesel engine. These successes attracted the attention of the aviation community around the world.

In 1932, President Herbert Hoover awarded the Collier Trophy, for outstanding performance in the aeronautical field, to the Packard Motor Car Company. Unfortunately, only a year later this project was abandoned, closing the factory and putting the workers out of jobs. What happened? Certainly a major factor was that, by 1933, the world was in the depths of the Great Depression; economic activity and employment were substantially curtailed throughout the entire industrialized world. The Packard diesel engine cost about 35% more than the competing Wright Whirlwind engine. Also, the Packard engine was notorious for vibration, exhaust odor, smoke, and soot, characteristics that made it unpopular with pilots and passengers.

DIESEL FUEL

Diesel fuel (occasionally also called diesel oil) boils in the range 175–345°C. This is higher than kerosene or jet fuel, and much higher than gasoline. The boiling range of diesel overlaps somewhat that of kerosene because commercial diesel fuel is often made as a blend of several refinery streams.

Let's consider the issues of fuel quality. First, recall that the fuel is ignited as a result of the very high air temperature in the cylinder, which itself is a result of the compression work. A diesel engine actually operates by self-ignition of the fuel in the cylinder, which, in a very crude sense, would let us say that the engine works by knocking. Thus the fuel characteristics we want are the inverse of gasoline's, for which self-ignition leads to knock.

Because diesel fuel has a much higher boiling range than gasoline, its molecules are larger, on average, so we can't apply the octane scale for measuring fuel quality. Instead, as the standard we select the cetane molecule (more formally known as hexadecane), a linear paraffin molecule that should knock very well, as the standard, and assign it a value of 100 (Figure 24.8). Recall that aromatic hydrocarbons are good antiknock components of

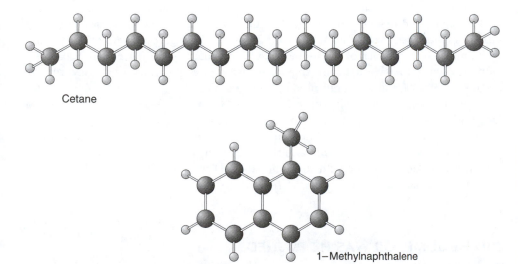

Cetane

1–Methylnaphthalene

FIGURE 24.8   The cetane number standard of diesel fuel performance is determined by comparison with a blend of cetane (hexadecane), a linear paraffin of 16 carbon atoms, and 1-methylnaphthalene, an aromatic compound. Cetane is assigned a value of 100 and 1-methylnaphthalene, 0.

gasolines. For diesel fuel we take the aromatic compound 1-methylnaphthalene to have a value of 0. Thus we can assign a **cetane number** of diesel fuel as the percentage of cetane in a test blend of cetane and 1-methylnaphthalene that gives the same engine performance as the fuel being evaluated. Good-quality diesel fuels have cetane numbers around 50–55.

The other concerns of fuel quality are much the same as for jet fuel. Aromatics should be low, because of their role in smoke or soot production. This is an especially noticeable problem when a vehicle is accelerating, and so the combustion conditions are fuel-rich. This can be seen, for example, in city buses accelerating away from a stop, or on large trucks on the highway, especially accelerating uphill. As we've seen with jet fuel, aromatic compounds are the 'bad guys' in smoke and soot formation. Sulfur content should be low, to reduce air pollution from SO_x. Viscosity must be controlled to provide good flow properties at low winter temperatures. We've seen that cold-weather performance is an issue with jet engines, because the molecules in jet fuel, being larger than those in gasoline, flow less readily at low temperatures. Since the molecules in diesel fuel might be bigger yet, cold-weather performance is even more of a concern, in two regards. Cold-weather starts must overcome the problem of getting the fuel to vaporize when it is injected into the cylinder. The flow of fuel in the fuel lines is a problem at extremely low temperatures. In very cold weather, such as on Alaska's North Slope, it's not uncommon to allow diesel engines in trucks and machinery to run round-the-clock to keep them warm and avoid starting problems.

The diesel engine offers several advantages relative to other internal combustion engines. A prime advantage is its good fuel economy (Rudolf Diesel was right—he did develop a more efficient engine). The complicated electrical ignition system is completely eliminated—note that there are no spark plugs in a diesel engine. The fuel is much less susceptible to accidental ignition and fires than is gasoline. On the other hand, diesels have some disadvantages. For one thing, they are quite heavy, and they are expensive to manufacture. Diesel engines have much higher compression ratios than gasoline engines. Also, they have to be much sturdier than gasoline engines, simply to withstand the mode of operation. (Recall that the higher the compression ratio, the higher the tendency of an engine to knock. Since, in a sense, we want a diesel engine to knock, diesel engines are built with very high compression ratios, 13:1 or higher.) A second disadvantage is that diesel engines tend to be noisy. The fact that the engine has to be very heavy and sturdily built does, however, provide an additional advantage, its longevity. There's a rough rule-of-thumb that the owner of a gasoline engine is doing well to drive it 100,000 miles without having to have a major engine overhaul. Diesel engines can easily run for 300,000 to 350,000 miles without needing to be rebuilt.

Many diesel engines will run on cruder fuel than is needed for spark ignition. Obviously no additives are needed to suppress knocking, and in fact 'pro-knocking' compounds can be added as ignition promoters. In those parts of the world where petroleum products are not readily available, diesel engines could be run using fuels of similar molecular weight and with abundant straight carbon chains, such as the oils from coconut, soy, and sunflower.

DID HE JUMP, OR WAS HE PUSHED?

Diesel foresaw that his engine would replace steam locomotives on railroads as well as be used for streetcars and other road vehicles; he supposed that it would be applied to propulsion of all sizes of ships and boats and that it would meet requirements for large and small stationary power sources on land. His engine came at a time when the electric motor was beginning to show great promise for the future, when the gas engine was proving itself economical, convenient, and reliable for a multitude of purposes, when oil and gasoline engines with flame or electric ignition were available, and when the marine steam engine was well established. The steam turbine, with steam now generated in oil-fired boilers, was a near-ideal power source for large, fast ships, while the simple, sturdy, and cheap reciprocating engine was well suited for smaller ships down to fishing boats and tugs. Yet his prophecies have proved, in retrospect, remarkably accurate. From its first commercial applications early in the 20th century, the diesel engine slowly blossomed and virtually fulfilled all of Rudolf Diesel's predictions that were once dismissed as absurd. In fact, the diesel engine has become so successful that it is the only engine—and diesel fuel is the only petroleum product—that is commonly associated with the name of its inventor. We never hear of Watt engines or Otto fuel, for example.

In 1913, Rudolf Diesel was taking an overnight ferry from The Hook in Holland to Harwich, England. Although he seemed to be in good health and in good spirits, after saying goodnight to the friend with whom he was traveling, he disappeared from the ship while it was en route across the North Sea. Some believe that Diesel committed suicide by throwing himself into the sea, despondent over the apparent lack of enthusiasm for his engine and, possibly, despondent over the likelihood that someday soon a major war would erupt in Europe. A more romantic version of the story suggests that the British secret service thought that Diesel was coming to England as a German spy, so they 'helped him' throw himself overboard.

> *It is with great regret that we record the disappearance of Dr. Rudolf Diesel from the G.E.R. steamer* Dresden *on the voyage from Antwerp to Harwich on the night of September 29; the circumstances are such as to leave no hope of his being alive. Dr. Diesel will be remembered as the inventor of the oil engine which bears his name. Born in Paris in 1858, of German parentage, his training included courses at the Augsburg technical schools and at the Munich Technical College. His first published description of the Diesel engine appeared in 1893; aided financially by Messrs. Krupp and others, the next few years were spent in arduous efforts to realize the principle of his engine in a commercially successful machine. The difficulties to be overcome were very great. In the earliest attempt, compression of the air was effected in the motor cylinder and the fuel injected direct. This engine exploded with its first charge and nearly killed the inventor. The modern Diesel engine compresses the air in the motor cylinder to a pressure above 400 lb. per square inch, during which process the air becomes hot enough to ignite the fuel. At the end of compression, the fuel is injected by means of a separate air supply at a pressure higher than that in the cylinder. Nothing of the nature of an explosion occurs in the cylinder; the oil burns as it is injected, and, as the piston is moving outwards at the same time, the pressure does not rise to any extent. The fuel consumption of these engines is remarkable, being roughly one-half of any other type of oil motor. Engines both of a two-stroke cycle and of a four-stroke cycle are now being developed by many firms both on the Continent and in Britain. In Dr. Diesel's opinion the two-stroke engine would probably be the standard type for marine purposes. Marine Diesel engines of very large power have not yet been constructed, but many important experiments in this direction are being made. Dr. Diesel's loss will be regretted by men of science on account of his efforts to interpret practically the Carnot ideal cycle, and by engineers on account of the immense strides which his untiring energy and indomitable pluck have made possible.*
>
> —Nature[1]

CITATION

1   Gratzer, Walter. *A Bedside Nature*. Freeman: New York, 1998; p. 140 (*Nature* **1913**, *92*, 173).

FOR FURTHER READING

Cummins, Lyle. *Internal Fire*. Society of Automotive Engineers: Warrendale, PA, 1989. This book, well illustrated with period photographs and drawings, traces the history of the internal combustion engine from 1673 up to 1900. Chapter 14 discusses Diesel and his engine.

Cummins, Lyle. *Diesel's Engine*. Carnot Press: Wilsonville, OR, 1993. This book, the first volume of a projected multi-volume series, treats the development of the diesel engine from Diesel's earliest ideas through 1918. An enormous amount of detail, including, e.g., original patent drawings, is presented.

Guibet, J.C. *Fuels and Engines*. Editions Technip, Paris, 1997. Chapter 4 provides a thorough discussion of diesel fuel, and an introduction to diesel engines. This book presumes that the reader has a background in science and mathematics.

Hawks, Ellison. *Popular Science Mechanical Encyclopedia*. Popular Science: New York, 1941. The second chapter, on 'bottled energy' is a well-illustrated discussion of diesel (as well as gasoline and gas) engines. Though long out of print, this book is worth tracking down in libraries or used-book stores, particularly by anyone interested in the technology of the first half of the 20th century.

Norman, Andrew; Corinchock, John; Scharff, Robert. *Diesel Technology*. Goodheart-Willcox: Tinley Park, IL, 1998. Intended as a textbook for students in vocational-technical programs in diesel engines. This book provides a wealth of detail on engine components and operations.

Suzuki, Takashi. *The Romance of Engines*. Society of Automotive Engineers: Warrendale, PA, 1997. This book treats the history of the development of internal combustion engines. It contains considerable information on diesel engines. In particular, Chapters 32 to 34 discuss the ill-fated Packard diesel intended for airplanes.

25

ATOMS AND 'ATOMIC' ENERGY

Think of the world a century ago. Edison had perfected the incandescent light, along with all of the components of the electrical distribution system needed to get electricity into homes. Otto had developed the four-stroke gasoline engine, which Ford would soon use to make the automobile a reality for persons of average means. Steam locomotives—the evolution of the work of Watt, Trevithick, and Stephenson—ruled the rails and moved passengers at speeds up to 100 miles per hour. Charles Parsons' steam turbine allowed ship crossings of the Atlantic in a week or less, similar to times required today. The Wright brothers demonstrated the possibility of powered flight. Electrically powered subways and trolley cars were helping to move people in and between cities. Niagara Falls had been tapped as a source of hydroelectric power. The diesel engine was beginning to find applications in small ships and as a stationary power source. Many other developments not directly related to energy—photography, skyscrapers, phonographs, and radio, as examples—came also at about the same time.

In essence, a very large number of the things we associate with modern-day life were in existence in some form a hundred years ago. Scientists and engineers of the time seemed to feel that they understood the basic underlying principles of these devices. Because of that, some people associated with the scientific community had come to believe that science was 'over'; that everything that there was to discover had already been discovered. All that remained to be done was to improve the precision of measurements, and, by incremental, evolutionary design changes, to improve the speed and efficiency of devices already in existence. (Ironically, we are going through the same exercise now, at the beginning of the 21st century, believing that this time we *really* have reached the end of science.) But, a few surprises were lurking.

THE DISCOVERY OF X-RAYS

People will say that Röntgen has probably gone crazy.

—Röntgen[1]

In 1895, a German physicist, Wilhelm Konrad Röntgen, discovered a new form of electromagnetic radiation that we now call X-rays. Röntgen was interested in the behavior of streams of electrons emitted from negatively charged electrodes, the so-called cathode rays. Specifically, he wanted to investigate the luminescence⚛ caused when cathode rays impinged on certain chemicals. The apparatus Röntgen needed for his work was fairly simple. It consisted of a source of high-voltage electricity and a cathode-ray tube to discharge it through (Figure 25.1). The cathode-ray tube was a bulb of glass, evacuated to a fairly low pressure and equipped with metal electrodes to allow passage of the electric discharge.

⚛ The phenomenon of luminescence is the production of light by means other than heat. Several kinds of luminescence are recognized. A fluorescent object, such as a TV screen, produces light while it is exposed to a source of radiation. A phosphorescent object continues to emit light even after the source of radiation is removed. A firefly is also luminescent because it produces light, but not heat, in this case, from chemical reactions in its body (chemiluminescence).

To observe any faint luminescence, Röntgen darkened his laboratory room. In November 1895, he observed a flash of light coming some distance from the tube. A sheet of paper coated with a chemical compound known to be fluorescent was the source of the light.⚛ Because the cathode rays themselves were shielded such that they could not escape from the apparatus, the energy source causing the fluorescence had to be something other than cathode rays. Röntgen recognized that some sort of radiation was emerging from the cathode-ray tube, presumably produced by the impact of the cathode rays on the solid material with which they collided. That is, the kinetic energy lost by the cathode rays as they were stopped was converted into a different form of energy that could be observed as radiation. Besides finding out where the new radiation was produced, Röntgen also found that it traveled in straight lines, it would penetrate some materials but be stopped by others, and that it would 'expose' a photographic plate, just as ordinary light would.⚛ However, the most remarkable feature of this newly discovered radiation was that it could pass through considerable thicknesses of paper and even through thin layers of metals or other seemingly solid materials. At first, this new form of radiation was sometimes called 'Röntgen rays' in honor of the discoverer, but Röntgen himself used the universal math-

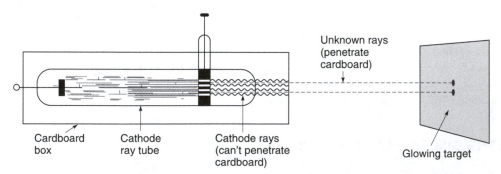

FIGURE 25.1 An example of a cathode-ray tube, used in many physics laboratories around the turn of the last century, and instrumental in Röntgen's discovery of X-rays.

ematical symbol of the unknown, calling the radiation 'X-rays.' (We still use the name even though the nature and origin of X-rays is now well understood.)

Specifically, the compound used in this experiment was barium cyanoplatinate, $Ba[Pt(CN)_6]$.

What we recognize as ordinary, or visible, light is only a small portion of the entire electromagnetic spectrum. The electromagnetic spectrum comprises many kinds of radiation, all of which were shown to be fundamentally similar in the brilliant work of the English physicist James Clerk Maxwell. All electromagnetic radiation is propagated by waves; the distinction among the various forms rests in the wavelength—the distance from the crest of one wave to the crest of the next. The electromagnetic spectrum includes radio waves (with very long wavelengths, typically 1–2,000 m); microwaves, used in the nearly ubiquitous kitchen appliance (0.1 m to about 0.5 cm); infrared radiation (0.5 to about 7.5 $\times$ 10^{-5} cm); visible light (4×10^{-5} to 7.5×10^{-5} cm); ultraviolet radiation, responsible for suntans (4×10^{-5} to 5×10^{-7} cm); and, at the shortest wavelengths, X-rays (5×10^{-7} to 1×10^{-9} cm) and γ-rays (less than 1×10^{-9} cm).

Because metal and bone block the passage of X-rays, whereas thin wood and human flesh do not, Röntgen was able to produce pictures of a set of weights sealed in a small box and the bones of his wife's hand. They were the first X-ray photographs.

Röntgen's first scientific paper, presented in December 1895, began:

> If the discharge of a fairly large Rühmkorff induction coil is allowed to pass through a vacuum tube, and if one covers the tube with a fairly close-fitting mantle of thin black cardboard, one observes in the completely darkened room that a paper screen painted with barium platinocyanide placed near the apparatus glows brightly, regardless of whether the coated surface or the other side is turned toward the discharge tube. The fluorescence is still visible at a distance of two meters from the apparatus.
>
> —Röntgen[1]

On New Year's Day, 1896, Röntgen mailed copies of this first publication on X-rays to the leading physicists of Europe, along with the X-ray pictures he had taken. X-rays immediately caught the fancy of both the scientific community and the general public. Much of the fascination lay in the ability of X-rays to penetrate some kinds matter, but not others. Particularly amazing was the effect of X-rays on humans. As most of us have seen for ourselves, so-called hard tissue, such as bones and teeth, show up easily on an X-ray photograph, because they stop the X-rays, while soft tissue, such as flesh and blood, does not.

> Prof. Röntgen, of Wurzburg, at the end of last year published an account of a discovery which has excited an interest unparalleled in the history of physical science. In his paper read before the Wurzburg Physical Society, he announced the existence of an agent which is able to affect a photographic plate placed behind substances, such as wood or aluminum, which are opaque to ordinary light. This agent, though able to pass with considerable freedom through light substances, such as wood or flesh, is stopped to a much greater extent by heavy ones, such as the heavy metals and the bones; hence, if the hand, or a wooden box containing metal objects, is placed between the source of the Röntgen rays and a photographic plate, photographs ... are obtained. This discovery

appeals strongly to one of the most powerful passions of human nature, curiosity, and it is not surprising that it attracted an amount of attention quite disproportionate to that usually given to questions of physical science. Though appearing at a time of great political excitement, the account of it occupied the most prominent parts of the newspapers. . . . The interest this discovery aroused in men of science was equal to that shown by the general public. Reports of experiments on the Röntgen rays have poured in from almost every country in the world, and quite a voluminous literature on the subject has already sprung up.

—Thomson[2]

By early 1896, experimenters in many countries began to publish pictures of bones. The usefulness of this work to medicine was obvious, and X-rays were immediately applied in medicine and in dentistry. Röntgen became one of the most celebrated scientists of the time. Unfortunately, the danger lurking in X-rays—that extensive exposure of a person to X-rays can cause cancer—was not appreciated for a number of years.

HENRI BECQUEREL'S EXPERIMENT

In January 1896, members of the French *Académie des Sciences* were shown the first French X-rays of the bones of a hand. In attendance at that meeting was Henri Becquerel, the third of four generations of an illustrious scientific family and, like his father and grandfather had been, Professor of Physics at the Museum of Natural History. Becquerel had devoted himself to the study of luminescence. Therefore, what most interested him about X-rays was a report that they came from the fluorescent spot on the wall of the cathode ray tube. Röntgen himself had specifically stated that where the cathode-ray tube fluoresced the strongest was the center from which X-rays radiated. Becquerel wondered whether there could be some sort of connection between X-rays and fluorescence. That is, if the fluorescence of the cathode rays contained X-rays, then it seemed reasonable to suspect that X-rays could be produced in other forms of the fluorescence. In other words, does a fluorescent material also produce X-rays?

I thought immediately of investigating whether the X-rays could not be a manifestation of the vibratory movements which gave rise to the fluorescence and whether all phosphorescent bodies could not emit similar rays. The very next day I began a series of experiments along this line of thought.

—Becquerel[3]

Seeing fluorescence in the visible region is easy, because our eyes respond to light in that region of the electromagnetic spectrum. We can't see X-rays. Becquerel needed some way other than relying on vision to test for the production of X-rays during fluorescence. X-rays do 'expose' some kinds of photographic films or plates. (In fact, this is how physicians or dentists are able to make permanent records of our X-ray examinations—our 'X-rays' are photographs in which the film has been exposed by X-rays rather than by visible light.) Becquerel's experiments involved wrapping a photographic plate in dark paper to prevent visible light from exposing the plate. He put the wrapped plate onto a window sill, where it would be in the sunlight, but, because of the wrapping, the sunlight itself could not affect the plate. Then he put specimens of the minerals he was testing on top of the plate. Because of the wrapping, any visible light that was emitted by the mineral as it fluoresced would not affect the plate any more than would the sunlight. X-rays, though, could easily penetrate the paper wrapping. Any blackening of the photographic plate would mean that X-rays had been given off by the mineral specimen.

Becquerel's father had worked with the fluorescent compound potassium uranyl sulfate, $K_2UO_2(SO_4)_2$. Since samples of this compound were on hand in the laboratory, Becquerel used it in his experiments. He quickly discovered that after exposure to the sun, the fluorescent radiation from the compound would penetrate black paper (which is opaque to ordinary light) and darken a photographic plate on the other side. In other words, this test gave Becquerel a positive result—the photographic plate was blackened, suggesting (but not proving!) that the uranium compound was giving off X-rays when it fluoresced. Becquerel had designed and performed this experiment to test his hypothesis that fluorescent objects also emit X-rays, and it had given the results as predicted. So far, so good—except that, if Becquerel had stopped his work at that point, the first, and apparently positive, results would have been quite misleading.

To prove that the rays causing the plate to blacken really were X-rays, Becquerel had to demonstrate that the rays behaved in *all* respects like X-rays, not just in their ability to penetrate paper. (More generally, the fact some material or phenomenon A behaves in one respect, or has one characteristic, similar to B does not itself prove that A *is* B; rather, that proof requires that they be alike in *all* possible respects or characteristics.) Röntgen had already shown that a characteristic property of X-rays was their tendency to travel in straight lines, and that the penetrating power of X-rays is different for different materials (the reason that we see teeth and bones in X-ray photographs, but not skin or flesh). Continuing with his work, Becquerel tried an experiment in which he would put a small copper cross between the uranium sample and the photographic plate. Figure 25.2 presents a schematic diagram of Becquerel's apparatus.

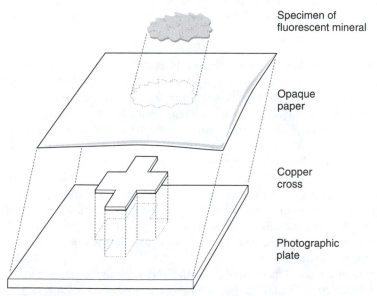

Specimen of
fluorescent mineral

Opaque
paper

Copper
cross

Photographic
plate

FIGURE 25.2 A conceptual diagram of Becquerel's apparatus. It consisted of a fluorescent mineral specimen, a copper cross (opaque to both light and X-rays), and a photographic plate wrapped in heavy paper (opaque to light, but not to X-rays).

If the rays from the uranium sample traveled in a straight line, and penetrated materials to different extents (specifically, passing through paper easily, but not through the copper), just as X-rays were already known to do, then an outline of the copper cross should appear on the photographic plate. All that remained actually to perform the experiment was to put the cross between the mineral and the carefully wrapped plate, set this assembly in the sunlight so that the uranium mineral would fluoresce, and then develop the plate.

> Henri Becquerel (1852–1908), Professor of Physics at the Ecole Polytechnique in Paris, read about X-rays soon after their discovery. He thought that similar penetrating rays might be emitted by phosphorescent substances when exposed to sunlight. So he took some phosphorescent crystals of a uranium compound, in the form of a thin crust, and placed them on a photographic plate which he had previously wrapped in thick black paper to keep the light out. Then he exposed the whole thing to the sun for a few hours. When he developed the photographic plate he found a silhouette of the crystals in black on the negative—which seemed to confirm his idea that sunlight made the crystals emit radiation.
>
> His discovery that they emitted radiation even in the dark was a matter of chance. The sun did not shine in Paris for several days, but, as he had set up his apparatus, he decided to develop the plate nevertheless ...
>
> —Carey[4]

Then an unexpected complication came up—bad weather. On the day that Becquerel wanted to try his experiment, the sky was completely overcast. Rather than take the whole uranium mineral/copper cross/wrapped photographic plate assembly apart, Becquerel decided to store it, intact, inside a drawer in one of the laboratory cabinets. The inside of the drawer provided a dark place that would ensure that the mineral couldn't accidentally be exposed to light (and, presumably, fluoresce) until he was ready to try the experiment again. Five days went by without his being able to try exposing the assembly to sunlight. Finally, he took the apparatus out of the drawer and developed the photographic plate anyway.

> ... experiments had been made ready on Wednesday the 26th and Thursday the 27th of February and as on those days the sun only showed itself intermittently I kept my arrangements all prepared and put back the holders in the dark in the drawer of the case, and left in place the crusts of uranium salt. Since the sun did not show itself again for several days I developed the photographic plates on the 1st of March, expecting to find the images very feeble. The silhouettes appeared on the contrary with great intensity. I at once thought that the action might be able to go on in the dark ...
>
> It is important to notice that this phenomenon seems not to be attributable to luminous radiation emitted by phosphorescence. The radiations of uranium salts are emitted not only when the substances are exposed to light but when they are kept in the dark, and for more than two months the same pieces of different salts, kept protected from all known exciting radiations, continued to emit, almost without perceptible enfeeblement, the new radiations.
>
> All the salts of uranium that I have studied, whether they become phosphorescent or not in the light, whether crystallized, cast, or in solution, have given me similar results. I have thus been led to think that the effect is a consequence of the presence of the element uranium in those salts ...
>
> —Magie (quoting Becquerel)[3]

This is the curious aspect of Becquerel's work. There really was no reason for him to have developed the plate, because he certainly knew that it had not been exposed to sunlight the way he had planned. That is, he had not actually done the experiment the way he had planned it. But, when he did develop the plate anyway, he discovered that it had the image of the copper cross. Since the assembly—uranium mineral, copper cross, and photographic plate—had not ever been placed in sunlight, Becquerel's original idea, that exposure to the sunlight caused the mineral to fluoresce, and the light given off as a result of the fluorescence exposed the photographic plate, could not be correct. The only explanation for the source of the image of the cross on the photographic plate was that some kind of radiation had to be coming out of the uranium mineral itself. Since the mineral had not been exposed to sunlight, the radiation was *not* the result of fluorescence. Furthermore, Becquerel hadn't heated the mineral, exposed it to other chemicals, or indeed done anything else other than simply store it in a drawer in his laboratory for those five days. So, the radiation that produced the image of the copper cross on the photographic plate could not have been caused by anything that Becquerel did *to* the mineral specimen. Something *in* the mineral was giving off radiation.

In fact, Becquerel found that the uranium compound constantly and ceaselessly emitted a strong, penetrating radiation. This property of constantly emitting penetrating radiation was termed **radioactivity** by the Polish–French physicist Marie Sklodowska Curie in 1898. She went on to show that all uranium compounds that she tested were radioactive, and that the intensity of radioactivity was directly proportional to the uranium contents of the various compounds tested. The uranium therefore seemed to be the source of the radioactivity.

What was puzzling about all this was the ENERGY involved. It took ENERGY to expose the photographic plates. This ENERGY was apparently some that the crystals had somehow stored away. We can think of this in terms of the energy balance equation:

$$E_{IN} - E_{OUT} = E_{STORED}$$

In the SYSTEM Becquerel was studying, E_{OUT} is the ENERGY that was put out by the uranium-containing compounds and that darkened the photographic plates. Because the uranium compounds had not been exposed to light, heated, or treated with other chemicals, for Becquerel's experiments E_{IN} is zero. Therefore, there can only be energy coming out of the crystals (E_{OUT}) if there is a substantial amount of energy stored in them somehow (E_{STORED}). As Becquerel continued his studies, he observed another strange aspect of this radiation. We all know from common experience that a hot object gradually cools until it is the same temperature as its surroundings. The amount of heat given off— from a hot cup of coffee, for example—steadily becomes less and less until no more is available. The same is true of fluorescence—the light being emitted weakens and 'goes out.' But this wasn't the case for the radiation coming from the mineral. It went on steadily, at apparently the same intensity, for months.

The one constant factor in all of Becquerel's and Curie's experiments was the presence of uranium in the compounds they were testing. So long as a test material contained uranium it did not matter whether it was fluorescent or not, whether it was exposed to light or not, whether it had been heated or not, or even whether it was solid or in solution. At that time, elemental uranium had not yet been isolated, but, coincidentally, Henri Moissan, at the School of Pharmacy in Paris, was working on a process for producing pure uranium. When Becquerel finally had the opportunity to test a sample of pure uranium metal, it was more intensely radioactive than any other substance yet tested. This was exactly in keeping with Marie Curie's observation that the intensity of radioactivity was proportional to the amount of uranium in the sample; the most intensely radioactive specimen of all was, at least nominally, 100% uranium.

 Ferdinand Frédéric Henri Moissan had quite a career, in which his work on isolating uranium was almost a sidelight. His most enduring scientific accomplishment was the isolation of fluorine in its elemental form. Fluorine is perhaps the most voraciously reactive of the elements. Moissan's work on isolating fluorine and studying its properties brought him the Nobel Prize in chemistry (1906), but also ruined his health. Moissan invented the electric arc furnace, which enabled scientists to achieve extremely high temperatures (approaching 4,000°F). The arc furnace also led to Moissan's most enduring embarrassment. He seemed to have produced diamonds in his furnace, and claimed to be the discoverer of synthetic diamond. Sadly, that work is now widely believed to be false. Moissan was apparently the unwitting victim of overzealous assistants, who 'salted' the experiments with real diamonds in a genuine, but grossly misguided, effort to be helpful.

RADIATION

Very early in the study of radioactivity, it appeared that the radiation given off by uranium (and by thorium, which Marie Curie had also discovered to be radioactive) was not homogeneous in its properties. This observation suggested that more than one kind of radiation was given off. For example, in a magnetic field, part of the radiation was deflected in one direction by a very slight amount, part was deflected in the opposite direction by a considerable amount, and some wasn't affected at all. Since the nature of the radiation was, at that time, not understood, the various kinds of radiation were simply named using the first three letters of the Greek alphabet: α-rays, β-rays, and γ-rays, respectively. The γ-rays displayed an ability to penetrate through materials much like X-rays. In contrast, β-rays were much less penetrating, and α-rays were scarcely penetrating at all. From the direction and extent of the β-ray deflection, Becquerel recognized that it must contain negatively charged particles like those in cathode rays. β-rays proved to be streams of rapidly moving electrons.

THE STRUCTURES OF ATOMS

To develop an understanding of what is happening in radioactivity, we begin by considering the structure of an atom. For our purposes, we consider the so-called solar system atom (Figure 25.3).

> *Have you ever* seen *one?*
>
> —Mach[6]

An atom consists of a **nucleus** that has a positive electrical charge and is surrounded by negatively charged **electrons**. In the solar-system model of the atom, the nucleus is analogous to the sun, and the electrons, to planets revolving around the sun in fixed orbits. On an atomic size scale, the atom is mostly empty space. The atomic nucleus is about 10^{-13} to 10^{-12} cm in diameter, only 1/100,000 to 1/10,000 the diameter of the entire atom. Yet virtually all the mass of the atom (99.9%) is concentrated in that tiny nucleus.

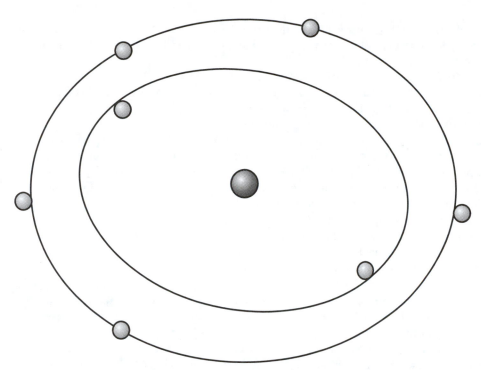

FIGURE 25.3 Atoms consist of a small, massive nucleus with one or more electrons orbiting around it. This diagram depicts the seven electrons of a nitrogen atom around its nucleus.

The ratio of the volumes of the nucleus and atom relates to the cubes of their diameters. Thus the volume of the nucleus is less than one-trillionth (1/1,000,000,000,000) the volume of the atom. If the nucleus were a pea sitting on the 50-yard line of a football stadium, the electron would be a gnat flying around the upper row of seats in the stands. If the nucleus were a 3-foot diameter globe in a classroom, the electron would be about 60 miles away. If an atom could be visualized expanded to the size of the Earth, the nucleus would be a sphere at the center, only about 700 feet in diameter.

The many discoveries in the science of quantum mechanics during the first few decades of the 20th century have shown the solar-system model of atomic structure to be a grossly simplified version of what we now understand about the stuctures of atoms and how the electrons are arrayed around the nucleus. However, it remains a very useful way of thinking pictorially about nuclei and the electrons around them, and about the sharing or exhange of electrons in chemical reactions. The solar system atom is an excellent example of the 'modelers' creed,' which asserts that, 'all models are wrong—some models are useful.' The texts by Sheldon Glashow and Paul Hewitt, listed in the Further Reading section, provide good introductions to the quantum mechanical approach to atomic structure.

The nucleus consists of two kinds of particles: **protons**, which have a positive charge, and **neutrons**, which are electrically neutral. (Actually, there are many other kinds of subatomic or nuclear species shown or postulated to exist, but for our purposes we need only concern ourselves with protons and neutrons.) Since the mass of a proton or neutron is about 1,800 times greater than the mass of an electron, the nucleus contains almost the entire mass of the atom. Even the lightest nucleus, that of the hydrogen atom, is 1,836 times the mass of an electron.

In ordinary chemical reactions, the reactions depend on the donation, transfer, or receipt of electrons. (In other words, chemistry happens to electrons.) The chemical behavior and identity of an element will therefore be determined by the number of its electrons; since in a neutral atom the number of electrons and protons are equal, the chemical identity of an element is also determined by the number of protons in its nucleus. But since the number of electrons might be changed by the atom's gaining or losing some in a chemical reaction, the number of protons in the nucleus is a more reliable indicator of the chemical identity of an element. The number of protons in the nucleus is called the **atomic number** of the element.

In other words, the atomic number (nuclear charge) is a more fundamental property of atoms than is the electron number. Electrons can be removed from atoms, or added to them by heat, by light, or by any chemical processes. The electrically charged particles that result are **ions**. While these ions have properties that differ greatly from those of the neutral atom, they do not represent a new element. For example, the sodium ion, which is one of the constituents of ordinary table salt, is very different from the sodium atom. One can be changed into the other by chemical or physical procedures that were already well known to chemists of the 19th century. Different though they may be, both the neutral sodium atom and the sodium ion are representatives of the same chemical element and have the same atomic number. However, neither the sodium ion nor atom can be changed into an ion or atom of, say, potassium by any of those known 19th-century chemical or physical processes. Therefore, changes in the electron number in an atom are not necessarily crucial to determining its chemical identity, and, consequently, it is not by means of the number of electrons within an atom that elements are best distinguished. On the other hand, the nuclear charge (atomic number) most certainly could not be altered by any method known to the chemists and physicists of 1900. Furthermore, no alteration of the number of electrons associated with an atom, either increasing or decreasing their number, would alter the atomic number of the nucleus. It is, therefore, the atomic number (or the size of the nuclear charge) that best characterizes the different varieties of atoms and best discriminates among the chemical elements.

For nuclear reactions we disregard the electrons, and instead focus on the behavior of the nucleus. We define a second characteristic property, the **mass number**,⚛ as being the sum of the numbers of protons plus neutrons in the nucleus. (Note that the number of neutrons in the nucleus is the difference between the mass number and the atomic number.) In cases in which it is convenient to lump the protons and neutrons together, as in determining the mass number, we refer to them by the generic term **nucleons**. Thus the mass number is equal to the number of nucleons.

⚛ The mass number is a convenient whole number that allows us to characterize a particular nucleus. It is close to, but not identical with, the atomic weight. The accepted convention for atomic weights now in use was established 40 years ago. Atomic weights are measured in atomic mass units (amu), defined to be 1/12 the mass of a single atom of a specific isotope of carbon, C^{12}. The value of the atomic mass unit is known to an accuracy of better than one part per million: 1 amu = 1.660540×10^{-27} kg. The atomic weight of $_6C^{12}$, in this convention, is exactly 12.00000. However, the atomic weight of $_1H^1$ is not 1.00000 but 1.007825. In terms of amu, the atomic weight of any pure isotope is approximately equal to its mass number.

TABLE 25.1 NUCLEI OF THE THREE ISOTOPES OF HYDROGEN HAVE THE SAME ATOMIC NUMBER (BECAUSE THEY ARE THE SAME CHEMICAL ELEMENT) BUT DIFFERENT MASS NUMBERS

Number of protons	Number of neutrons	Atomic number	Mass number
1	0	1	1
1	1	1	2
1	2	1	3

To have an electrically neutral atom, the number of protons must equal the number of electrons. Since neutrons are electrically neutral themselves, it is not necessary to restrict the number of neutrons to achieve electrical neutrality. Thus it is entirely possible that nuclei of the same chemical element could have different numbers of neutrons. Consider the nuclei listed in Table 25.1. The atomic number (the number of protons) is the same in all three cases. Therefore, all of these nuclei are atomic nuclei of the same chemical element, which, in this example, is hydrogen, H. Atoms or nuclei having the same atomic number by different mass numbers are called **isotopes**.

When we write nuclear reactions, it will be helpful to be able to keep track of the fate of the protons and neutrons. We will use a notation scheme that shows the chemical symbol of the element, the atomic number as a subscript, and the mass number as a superscript. The three isotopes of hydrogen shown above are therefore written as $_1H^1$, $_1H^2$, and $_1H^3$, respectively.

CITATIONS

1   Röntgen, W., quoted in Glashow, Sheldon L. *From Alchemy to Quarks*. Brooks-Cole: Pacific Grove, 1994; pp. 417–423.
2   Thomson, J.J., quoted in Gratzer, Walter. *A Bedside Nature*. Freeman: New York, 1998; p. 100.
3   Becquerel, H., quoted in Glashow, Sheldon L. *From Alchemy to Quarks*. Brooks-Cole: Pacific Grove, 1994; pp. 417–423.
4   Carey, J. *The Faber Book of Science*. Faber and Faber: London, 1995; pp. 188–192.
5   Magie, W.F. *A Source Book in Physics*. McGraw-Hill: New York, 1935.
6   Mach, Ernst, quoted in Holton, Gerald. *Einstein, History, and Other Passions*. American Institute of Physics: Woodbury, NY, 1995; p. 161.

FOR FURTHER READING

Asimov, Isaac. *The History of Physics*. Walker: New York, 1966. An introductory survey of all of physics, through the 1960s, intended for the reader with little previous science background. Chapters 36 though 40 cover some of the topics introduced here, though in somewhat different order.

Glashow, Sheldon L. *From Alchemy to Quarks*. Brooks-Cole: Pacific Grove, CA, 1994. An excellent introductory physics text. Chapter 10 discusses the discoveries of X-rays and radioactivity, as well as the discovery of the nucleus itself.

Hazen, Robert M.; Trefil, James. *Science Matters*. Doubleday: New York, 1991. An inexpensive paperback that seeks to provide scientific literacy for readers with little or no prior scientific background. Chapter 4 discusses the atom.

Hewitt, Paul G. *Conceptual Physics*. Addison-Wesley: Reading, MA, 1998. Another splendid introductory physics text, with very informative illustrations. Chapter 32 discusses the atomic nucleus and the phenomenon of radioactivity.

Mann, Thomas. *The Magic Mountain*. (Numerous editions of this classic are available, including relatively inexpensive paperbacks.) This novel is frequently cited as one of the greatest works of literature of the 20th century. For our purposes, it is noteworthy in containing perhaps the first account in fiction of the use of X-rays in medicine—in this case, in a tuberculosis sanatorium.

Pullman, Bernard. *The Atom in the History of Human Thought*. Oxford University Press: New York, 1998. A survey of concepts of atoms and our knowledge about atoms through almost the entire span of human history, from the ancient Greeks to the 1990s.

Romer, Alfred. *The Restless Atom*. Doubleday: Garden City, NY, 1960. A history of the discoveries of the early years of atomic structure and radioactivity, including the work of Röntgen, Becquerel, and Curie. Written for the reader with little scientific background.

26

RADIOACTIVITY, FISSION, AND CHAIN REACTIONS

RADIOACTIVITY IS A NUCLEAR PROCESS

As scientists began to investigate radioactivity, it became clear that it was not like any known chemical process. Chemists had learned a number of ways to influence the rate or course of a chemical process: as examples, by changing the temperature or pressure, by adding a catalyst, or by exposing the system to light. None of these has the slightest effect on the rate of radioactivity. Therefore, if radioactivity is not a chemical process, it must be a physical process. Furthermore, if it is not a chemical process, it must not be occurring in the swarm of electrons, because chemical processes involve gaining, losing, or sharing of electrons. If radioactivity is not occurring among the electrons, the only other place it can be happening is in the nucleus.

Marie Curie defined radioactivity as the *spontaneous* emission of radiation. She also showed that radioactive materials produce a prodigious amount of energy—millions of times more than any chemical process. (On an equal mass basis, nuclear reactions release about 10 million times as much energy as chemical reactions.) This energy keeps pouring out in apparently endless quantity. We have seen that a spontaneous change is one that involves a transition from a state of high potential energy to one of low potential energy. Thus we can immediately draw an energy diagram (Figure 26.1). To begin to understand how to tap into this enormous and seemingly limitless supply of energy, we must begin by inquiring into the nature of the nuclear potential energy and how to measure it.

NUCLEAR BINDING ENERGY

Since the nuclei of all atoms (other than hydrogen) have more than one proton and since electrical charges of the same sign repel each other, we might ask why then do nuclei stay together. In other words, why do any atoms larger than hydrogen even exist? Certainly if we consider an atom the size of uranium, with 92 protons, there would be a tremendous

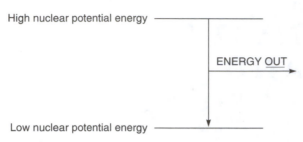

FIGURE 26.1   Radioactivity is a spontaneous change from high to low nuclear potential energy, which means that ENERGY is released from the SYSTEM.

release of energy (from repulsion of particles of the same electric charge) if the nucleus could just fly apart. Some force stronger than the tendency of like charges to repel each other must be holding the nucleus together—as if we had somehow wrapped a tiny, nucleus-sized rubber band around the nucleus to hold all the pieces in place.

Inside the nucleus, nucleons experience two competing forces. In addition to the repulsion between protons, a second force, one of attraction instead of repulsion, holds the nucleus together. This second force is called the nuclear binding energy and is strong enough to overwhelm the weaker force of electrical repulsion.⚛ However, the nuclear binding energy acts only over very small distances, indeed, only when the nucleons are in contact.

⚛ During the 1930s, atomic physicists proposed the concept of the 'strong nuclear force' to explain how nuclei are held together. To break nuclei apart, the strong nuclear force must be overcome, that is, ENERGY or WORK must be supplied to the nucleus to overcome this force. The amount of WORK needed to separate a nucleus into its individual neutrons and protons is the binding energy of that nucleus. The greater the binding energy of a nucleus, the more stable it is, because the more WORK that would have to be done to break the nucleus apart. The origin of the strong nuclear force lies in the fact that the protons and neutrons themselves are made up of smaller and more fundamental particles called quarks. The force that links quarks together to make protons and neutrons also links the protons and neutrons together to make nuclei, and is responsible for the strong nuclear force.

Since nuclear binding energy acts only over the extremely short distances between nucleons in a nucleus (about 10^{-15} m, or 0.00000000000004 inches!), it represents a short-range force. The binding energy drops essentially to zero at distances much greater than the contact distance between nucleons. In contrast, the repulsion between protons weakens only with the square of this distance; so, comparatively speaking, it is a relatively long-range force. When protons are close together, as they are in most small nuclei, the strong but short-range nuclear binding energy easily overcomes the long-range but weaker electrical repulsion force. But, when two protons are relatively far from one another, as they might be if they were on opposite edges of a large nucleus for example, these two protons would experience only very small attractive nuclear binding energy but a repulsive force that is still appreciable. The large nucleus is not as stable as a smaller one, because the repulsive forces are 'catching up' with the binding energy.

The nuclear binding energy keeps the nucleus from coming apart. It is, in effect, the 'rubber band' that prevents the nucleons from flying apart. For most nuclei, some external force that adds ENERGY to the nucleus to help overcome the nuclear binding energy is necessary to break the nucleus apart; otherwise, it will stay together forever. While the nucleons normally stay in contact with one another for an extremely long time, a very large nucleus always has a small chance that nucleons will find themselves separated by a distance beyond the reach of the binding energy, but still within the realm of the repulsive force. If that happens, the nucleons will then suddenly be free of one another; the electric repulsion of similar charges 'wins' and pushes the nucleons apart, overcoming the binding energy. This is the process that we recognize as radioactive decay.

The more protons in a nucleus (that is, the higher the atomic number), the more the repulsion between one another and the more likely is the nucleus to experience radioactive decay. Having additional neutrons in the nucleus reduces the proton-to-proton repulsion by increasing the size of the nucleus without adding to its positive charge.⚛ (Adding too many neutrons can also destabilize the nucleus. So, a stable nucleus results from a finely tuned balancing act among the neutrons and protons.) The neutrons play a large role in nuclear stability. A proton and a neutron can be bound together a little more tightly, on average, than two protons or two neutrons can be.

⚛ One conceptual model of the atomic nucleus envisions protons and neutrons arranged in shells, much as the electrons are arranged in shells around the nucleus itself. Certain numbers of electrons—2, 10, 18, 36, 54, and 86—confer exceptional stability toward *chemical* reactions. The atoms having these numbers of electrons are often said to be inert, though in fact they can take part in a few reactions under very severe conditions. Similarly, certain numbers of protons or neutrons—2, 8, 20, 28, 50, 82, and 126—confer exceptional stability on the nucleus. In the jargon of nuclear physics, these are referred to as 'magic numbers.' *All* radioactive processes known to occur in nature involve a sequence of transmutations until they finally stop with lead (with the magic number 82 protons) or with bismuth (magic number 126 neutrons). More and more neutrons added to a nucleus will take it further and further from being at or near a stable magic number composition, and thus destabilize the nucleus.

Many of the elements with atomic numbers up to 20 have equal numbers of protons and neutrons. The situation is different for elements with larger nuclei, because protons repel each other electrically and neutrons do not. A nucleus with 30 protons and 30 neutrons, for example, could be made more stable by replacing two of the protons with neutrons. In other words, we would have a nucleus of 28 protons and 32 neutrons, the number of protons and neutrons no longer being equal. This inequality of numbers of neutrons and protons that begins to develop above atomic numbers of about 20 becomes more pronounced for even heavier elements, as indicated in Figure 26.2. All nuclei having more than 82 protons are unstable. For example, in $_{92}U^{238}$, which has 92 protons, there are 146 neutrons. If the uranium nucleus were to have equal numbers of protons and neutrons, it would fly apart at once because of the electrical repulsion forces. The extra 54 neutrons (extra, that is, in the sense of making up the difference between the 92 to equal the number of protons and the 146 that actually exist in this nucleus) are needed for relative stability. Even so, the $_{92}U^{238}$ nucleus is still unstable. That is because there is an electrical repulsion between *every* pair of protons in the nucleus, but there is not a substantial nuclear binding force between every pair. Every proton in the uranium nucleus exerts a repulsion on each of the other 91 protons. However, each proton (and neutron) exerts an appreciable nuclear attraction only on those nucleons that happen to be near it.

The relationship between the number of nucleons and the average binding energy per nucleon is called the **curve of binding energy** (Figure 26.3). The highest binding

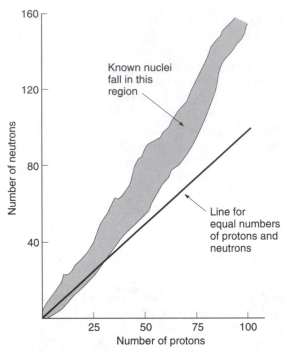

FIGURE 26.2 In small nuclei, the numbers of protons and neutrons are about equal, but in larger nuclei, neutrons dominate.

energy occurs at a mass number of 56, which happens to be an isotope of iron. This is the most stable nucleus. At low mass numbers (<56) the curve rises very steeply. At high mass numbers (>56) the curve drops, but slowly.

Stars in the normal state, whether big or small, burned the lightest element, hydrogen, and formed from it helium, the next heaviest element. The process gave off copious energy. In very massive stars, or in less massive stars going through a phase of internal collapse, the temperature might climb high enough for helium to burn. It changed into carbon and oxygen, with a further release of energy. Then the carbon and oxygen could burn, too, to form still heavier elements.

The escalation through the table of elements became progressively more difficult. The heavier the element, the more protons it had in each nucleus, and the more powerful was the electric repulsion between two nuclei, preventing them from fusing together. By the time you wanted oxygen to burn to make sulphur and silicon, or silicon to burn to make iron, you needed temperatures of billions of degrees so that the nuclei were colliding with sufficient frenzy to crash through the electric barrier. Iron-making marked the limit to nuclear burning in stars, and there was known to be a great deal of iron about. The Earth inherited a huge core of molten iron and meteorites often contained iron, too, all of it forged in stars. If the nuclear forces had their way, the whole universe would consist of iron.

After iron, the making of heavier elements in stars began to consume energy rather than releasing it. No star could earn a steady living that way. But in the explosion of a big star some of the enormous energy released went into building up dozens of chemical elements heavier than iron: gold, lead, all the way through the table of elements to

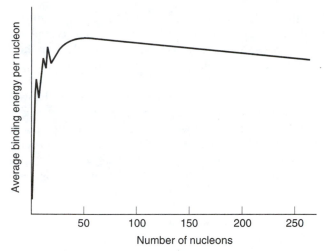

FIGURE 26.3 The curve of nuclear binding energy, which plots the average binding energy per nucleon against the number of nucleons.

uranium and beyond. Even so, heavy elements remained far less abundant in the cosmos than the lighter elements.

Many of the atoms so formed, and later incorporated into the Earth, were radioactive. Their nuclei were overcharged with energy and unable to survive indefinitely. But 'not indefinitely' could mean billions of years. From uranium, thorium, potassium and other radioactive elements, energy stored during the explosions of the ancestral stars slowly trickled out into the rocks of the Earth. It generated the heat that fired volcanoes, shifted the continents and built mountains. The great creakings called earthquakes, which accompanied these processes, were thus direct consequences—albeit greatly delayed and translocated—of those stellar explosions that made the stuff of the Earth available.

—Calder[1]

There is one point that we need to interject in regard to the curve of binding energy. *Relative to all of the energy diagrams we've seen previously, the curve of binding energy is upside down. In the curve of binding energy, stability increases as we go up the curve.* Therefore,

The binding energy per nucleon is a measure of the stability of a nucleus. The higher the binding energy, the more stable the nucleus. Nuclear processes release far more energy per reaction than do chemical processes because the 'glue' holding nuclei together is much more powerful than is the electric force that binds electrons to their atoms. The binding energy of a nucleus is the energy needed to take the nucleus apart into its constituent nucleons.

> A HIGH NUCLEAR POTENTIAL ENERGY REPRESENTS A LOW BINDING ENERGY, AND A LOW NUCLEAR POTENTIAL ENERGY IS A HIGH BINDING ENERGY.

In small nuclei, those with low atomic numbers, only a few protons, the nuclear binding energy overwhelms the repulsive force and the nucleons are tightly bound together. The rising slope of the binding energy curve for small nuclei tells us that the average binding energy of the nucleons would increase even if these nuclei had more protons and neutrons than they already do. At the exact crest of the curve of nuclear binding energy, for nuclei with about 26 protons, the nuclear binding energy and the electrical repulsive force are well balanced. Such nuclei are extremely stable. For these nuclei, there is no increase in the average binding energy of their nucleons by adding or subtracting nucleons.

In contrast to the situation for small nuclei, the situation is reversed for nuclei with many protons: electrical repulsion now overwhelms binding energy. For these large nuclei, the nuclear binding energy cannot hold the nucleus together. Rather, these large nuclei break apart in the process of radioactive decay. The decreasing slope of the binding energy curve in the region of large nuclei signals us that average binding energy of these large nuclei would increase—they would become more stable if they had *fewer* nucleons, not more. The greater stability afforded by a smaller number of nucleons is achieved by breaking apart.

To understand why the various kinds of radioactive decays (that is, α, β, and γ) take place, we consider the plot showing the number of neutrons versus the number of protons in stable nuclei (Figure 26.2). As we've seen, for small, stable nuclei the number of neutrons and protons are approximately the same, while for heavier nuclei slightly more neutrons than protons are required for stability. An element of a given number of protons (i.e., given atomic number) has only a very narrow range of possible numbers of neutrons if it is to be stable.

There can be two reasons why an unstable nucleus is unstable: Either it has too many neutrons relative to the number of protons, or else it has too few. Consider the first case. If one of the 'extra' neutrons could change into a proton, its doing so would simultaneously reduce the number of neutrons while increasing the number of protons. This represents a big step toward getting into the stable balance between numbers of neutrons and protons. But, in this process a neutron of zero electrical charge has (at least conceptually) changed into a proton of +1 charge. To make sure that the net electric charge in the SYSTEM does not change, this neutron-to-proton transformation requires the emission of a negatively charged electron, as

$$_0n^1 \rightarrow {}_1H^1 + {}_{-1}\beta^0$$

The electron formed in this process leaves the nucleus, and is detectable as β radiation. It's important to note that the electron emitted as β radiation in this process was not one of the electrons normally circling around outside the nucleus; rather, it was *created inside* the nucleus as a consequence of the neutron-to-proton transformation. (Notice also that, in this equation, the sums of the mass numbers on both sides of the equation are exactly equal, and the sums of the atomic numbers are also exactly equal. We will explore this point further in the next section.) The nucleus may be left with some extra energy because of its shifted binding energy; and this is given off in the form of γ-rays. Sometimes a series of β decays is required for a particular unstable nucleus to reach a stable configuration. Another way of altering nuclear structure to achieve stability involves α decay, which decreases both the number of neutrons and the number of protons by two. The α-particles are nuclei of helium, $_2He^4$. In some cases a succession of α and β decays, with accompanying γ decays to carry off excess energy, is required before a nucleus reaches stability.

Nuclei that contain more than about 210 nucleons are so large that the short-range forces holding them together are barely able to counterbalance the long-range electrostatic repulsive forces of their protons. Such a nucleus can reduce its bulk and, in doing so, achieve greater stability by emitting an α-particle.

TRANSMUTATION OF ELEMENTS

As scientists followed up Becquerel's discovery of radioactivity (Chapter 25) they found that the isotope $_{92}U^{238}$ emits α radiation. We can begin to write the process of uranium 'decay' as

$$_{92}U^{238} \rightarrow \,_2He^4 + \ldots$$

Examining this partially completed equation shows that we have removed two protons (because $_2He^4$ has an atomic number of 2) and two neutrons (the difference between the mass number and atomic number of $_2He^4$) from $_{92}U^{238}$. Therefore, we must have another product that has 90 (92 − 2) protons and 144 (146 − 2) neutrons. Since this product has a different atomic number (90), it must be a different chemical element. It is in fact the element thorium, Th. Thus the complete nuclear reaction can be written as

$$_{92}U^{238} \rightarrow \,_2He^4 + \,_{90}Th^{234}$$

The essential feature of writing nuclear reactions such as this is that the sums of the atomic numbers and the sums of the mass numbers must be equal on each side of the equation. (Notice in this example that 92 = 90 + 2, and 238 = 4 + 234.)

A consequence of this radioactive process is that one chemical element, uranium, has been turned into a new one, thorium. This change of one element into another is called **transmutation**. Transmutation was a goal of early scientists, the alchemists, for centuries. Their objective was to turn inexpensive metals such as lead or iron into precious metals such as gold. With the hindsight of knowledge of atomic structure, we know that transmutation of elements through chemical reactions is impossible, because chemical reactions involve only electrons, and not nuclei. To transmute one element to another requires, we now know, a change in atomic number—that is, a change in the number of protons in the nucleus. No chemical reaction can do this, because chemical reactions occur only among the electrons.

The alchemists had no way of knowing it, but processes of transmutation were constantly going on all the time, all around them. (In fact, they are going on all the time, all around us, too.) Radioactive decay of minerals in rocks has been occurring since the Earth was formed. But alchemists had no knowledge of radioactivity, nor of how to measure or detect it.

As a consequence of changing uranium into thorium, we have moved slightly along the curve of binding energy, as shown in Figure 26.4. Because a nucleus of mass number 234 has a slightly higher binding energy than one of mass number 238, it is slightly more stable. When a SYSTEM spontaneously changes from being less stable to being more stable, some ENERGY is given off. We've seen this several times in the past, one example being the change of gravitational potential energy in a waterfall. Therefore, we can apply our standard energy diagram (Figure 26.5) to this nuclear process.

Where does the energy come from? If we made a very scrupulous account of all the mass on both sides of the equation, we would discover that some very tiny amount of mass appears to be 'missing' after the reaction. This very tiny amount of 'missing mass' has been converted to energy. The relationship between missing mass and energy is given by Einstein's (Figure 26.6) equation

$$E = mc^2$$

which is probably the most famous equation in all of science. Here E is the energy liberated in the nuclear process, m is the amount of mass that is lost (the missing mass), and c is the speed of light. Although m may be a very tiny number, c is very

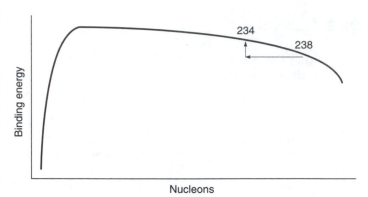

FIGURE 26.4 The transformation from uranium to thorium is illustrated on the curve of binding energy. For illustration purposes, the curve has been 'smoothed' and the horizontal scale is exaggerated.

large (and consequently c^2 is extremely large); thus the amount of energy liberated, E, can be enormous.

When it became appreciated that natural radioactivity involved transmutation, nuclear physicists tried to see if transmutation could be induced in the laboratory, by deliberately causing nuclear reactions to occur.

> From 1900 onwards the Curies had been in correspondence with scientists from all over the world, responding to requests for information. Research workers from other countries joined the search for unknown radioactive elements. In 1903, two English scientists, Ramsay and Soddy, demonstrated that radium continually disengaged a small quantity of gas, helium. This was the first known example of the transformation of atoms. A little later Rutherford and Soddy, taking up a hypothesis considered by Marie Curie as early as 1900, published their Theory of Radioactive Transformation, affirming that radioactive elements, even when they seemed unchangeable, were in a state of spontaneous evolution. Of this Pierre Curie wrote: 'Here we have a veritable theory of the transformation of elements, but not as the alchemists understood it. Inorganic matter must have evolved through the ages, following immutable laws.'
>
> — Carey[2]

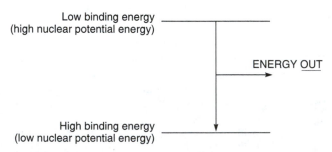

FIGURE 26.5 Because a high binding energy represents a stable nucleus, a spontaneous change in a nuclear SYSTEM proceeds from *low to high* nuclear binding energy.

FIGURE 26.6 Albert Einstein.

The first artificial transmutation was performed by Ernest Rutherford in Cambridge, England.

> *Ernest Rutherford (1871–1937) was a New Zealander, the son of an odd-job man, who won a scholarship to university and came to England to work under J.J. Thomson at Cambridge.*
>
> —Carey[3]

He bombarded nitrogen with α-particles and observed the reaction

$$_7N^{14} + {_2}He^4 \rightarrow {_8}O^{17} + {_1}H^1$$

(Note that the sums of mass numbers are equal on both sides of the equation ($14 + 4 = 17 + 1$), as are the sums of the atomic numbers ($7 + 2 = 8 + 1$).)

> *The Curies had shown that radium emits various kinds of 'radiation', and one of these was now known to consist of a stream of electrically charged particles. These 'alpha particles' were identical to helium atoms with their electrons removed; but they originated not from helium gas but sprang spontaneously from the radium atoms as they disintegrated.*
>
> *Even though atomic disintegration was still little understood, Rutherford saw these high-speed alpha particles as useful projectiles. He intercepted them with a thin sheet of gold foil, to see what happened as they passed through. If atoms were diffuse spheres of electrical charge, as Thomson had imagined, then most of the alpha particles should have gone straight through; a few should have been deflected slightly. But some of the alpha*

particles bounced straight back again. It was like firing artillery shells at a piece of tissue paper, and getting some of them returning in the direction of the gun.

Rutherford could only explain this by postulating that these alpha particles were hitting small, massive concentrations within the atoms. He thus concluded that most of an atom's mass resided in a minute, positively charged nucleus at the centre, while the electrons went around the outside—very much like the planets orbiting the massive sun. Most of the atom was just empty space. If an atom were expanded to the size of the dome of St. Paul's Cathedral, virtually all its mass would lie within a central nucleus no larger than an orange.

In 1919, he started firing alpha particles at nitrogen atoms. Nothing much should have happened. A great deal did.

... the alpha particles came from radium. This time he was directing them down a tube filled with nitrogen gas. At the far end, he found he was detecting not just alpha particles, but also particles with all the properties of hydrogen nuclei. There was, however, no hydrogen in the tube. With his high-speed alpha-particle projectiles, Rutherford had actually broken them off the nuclei of the nitrogen atoms.

The discovery of radioactivity had earlier shown that certain, rare types of atom could spontaneously disintegrate. Now Rutherford had shown that ordinary atoms were not indestructible. By knocking out a hydrogen nucleus (later called a proton) from the nucleus of nitrogen he had converted it into another element, oxygen. Rutherford had, to a limited extent, achieved the dream of alchemists and changed one element to another.

— Snow[4]

THE DISCOVERY OF NUCLEAR FISSION

Rutherford's successful artificial transmutation of nitrogen to oxygen was an important step forward in nuclear physics, but had little practical application. The alchemists wanted to turn inexpensive lead into gold. No one is likely to get rich making oxygen out of nitrogen. Oxygen constitutes about 21% of the atmosphere, 88% of the oceans, and is the most abundant element in the Earth's crust.

In 1934, the brilliant Italian physicist Enrico Fermi and his colleagues were studying a process called β decay, the emission of β radiation during radioactivity. They were adding neutrons to the nuclei of many different atoms to see what nuclear processes would stimulate β decay.

He was compact, small, powerful, penetrating, very sporty, and always with the direction in which he was going as clear in his mind as if he could see to the very bottom of things.

—Bronowski[5]

Nuclear physicists had recognized that bombarding a nucleus with neutrons can stimulate the emission of β-rays. As we've seen, β-rays, which are high-energy electrons created in the nucleus, have an 'atomic number' of −1. Therefore, the atomic number of the product will be one unit *higher* than the nucleus that emitted the β. (Why? Because the atomic number changes by −(−1), which is mathematically equivalent to +1.)

In the mid-1930s, Fermi and his research team showed that dozens of new radio-active materials are produced as a result of neutron bombardment. In most cases, neutron absorption is followed by β decay, yielding an element one higher in the periodic table. When they added neutrons to uranium, Fermi and his colleagues observed the production of some very short-lived radioactive systems. They thought that they had formed ultra-heavy nuclei of atomic numbers greater than 92. Fermi felt that $_{92}U^{239}$ might be formed from $_{92}U^{238}$. By emitting β particles this would become an isotope of the element of atomic number 93, and then possibly, by a second β emission, of element 94.

Fermi indeed discovered the presence of several new radioactive species in neutron-activated uranium. Their chemical properties did not match those of uranium, nor of any of the elements in the range of atomic numbers from 82 to 92, which would be the logical products of known (at the time) nuclear processes involving uranium. Fermi thought at first that he had indeed demonstrated the transmutation to an element of atomic number 93, which he referred to as 'uranium X.' However, the German chemist Ida Noddack found a flaw in Fermi's reasoning. In 1934, she wrote,

> It is conceivable that the nucleus breaks up into several large fragments which would, of course, be isotopes of known elements but would not be neighbors of uranium.
> —Noddack[6]

Though she did not use the term in any of her formal scientific publications, she was the first scientist to call attention to possible occurrence of nuclear fission.

The radioactive by-products produced from neutron bombardment of uranium have a variety of chemical properties. Otto Hahn, Lise Meitner, and Fritz Strassman, working in Germany, attempted to identify these materials. All nuclear reactions known up to that time (the mid-1930s), whether natural or artificially produced, had involved the emission of small particles, no more massive than an α. It was not unreasonable, then, that they and other nuclear scientists tried to associate the various types of radioactivity in the bombarded uranium with atoms only slightly smaller than uranium. Then, in 1938, Hahn and his coworkers found that when barium compounds were added to the bombarded uranium, one form of the radioactivity they were measuring followed the barium through all of the subsequent chemical manipulations they carried out. Since barium is chemically similar to radium, and since radium (atomic number 88) has a nucleus only slightly smaller than that of uranium, Hahn and his colleagues supposed, not unreasonably, that they were dealing with a radium isotope. However, no chemical process known would separate the added barium carrier from the radium that they supposed was accompanying it. When separation after separation failed, and it became quite clear that the presumed 'radium' could not be separated from the barium, the only remaining conclusion was that it was not a radium isotope formed in the process, but rather a radioactive barium isotope.

> It is an old maxim of mine that when you have excluded the impossible, whatever remains, however improbable, must be the truth.
> —Sherlock Holmes[7]

Barium isotopes have an atomic number of 56, which is 36 less than that of uranium. To form a barium nucleus, a uranium atom would have to lose 36 protons, which, by the processes known to nuclear scientists at the time, could be done only by emitting 18 α-particles. Such a vast production of radiation would be a veritable flood of α-particles; nothing like this was ever detected in the neutron bombardment of uranium.

Lise Meitner, Hahn's long-term collaborator on the research program that had led to the discovery, was Austrian. She also happened to be Jewish. She had worked with Hahn in Berlin at the Kaiser Wilhelm Institute for Chemistry until the intensifying persecution of the Jews by the Nazis, coupled with some political vulnerability as an Austrian citizen after the annexation of Austria, forced her to flee (with Hahn's help) in July 1938. With hindsight, it seems likely that her departure may have saved her from the Holocaust. A few months later, Hahn wrote to Meitner of the remarkable conclusion to their research.

Perhaps you can suggest some fantastic explanation. We understand that it really can't break up into barium. So try to think of some other explanation.

—Hahn[8]

Like Hahn, Meitner at first found the production of barium puzzling and difficult to explain. Over the Christmas holidays, she discussed it with her nephew Otto Frisch, also a German refugee, visiting her from Denmark where he worked with Niels Bohr. Together they came up with the explanation.

But how can one get a nucleus of barium from one of uranium? We walked up and down in the snow trying to think of some explanation. Could it be that the nucleus got cleaved right across with a chisel? It seemed impossible that a neutron could act like a chisel, and anyhow, the idea of a nucleus as a solid object that could be cleaved was all wrong; a nucleus was much more like a liquid drop. Here we stopped and looked at each other.

—Frisch[9]

The uranium nucleus, activated by neutron bombardment, had split roughly in half. To Frisch, this splitting process was so similar to the division of a biological cell that he suggested the name '**fission**' for the new phenomenon.

By any reckoning, Niels Bohr was one of the greatest of 20th century physicists, perhaps second only to Einstein for his impact on the science. Bohr had an active career that lasted nearly half a century. Among his numerous contributions to science, perhaps the two best known are the liquid drop model of the atomic nucleus and his model of atomic structure. Bohr's model of atomic structure, proposed in 1913, explained almost perfectly the patterns of light emitted from atoms (especially from hydrogen) and introduced the concept of specific 'energy levels' of the electrons orbiting the nucleus. Though nearly a century old, Bohr's model remains an excellent conceptual tool for thinking about atomic structure. The liquid drop model suggests that the nucleus behaves like a drop of liquid— if it is hit very hard by a small object (i.e., a neutron) it can split into two smaller droplets of roughly equal size.

A few years earlier, Bohr and his former student, the American physicist John Wheeler, had argued that a large nucleus should behave like a droplet of liquid. When struck by a particle, it could become lopsided. If it were struck hard enough, it could split up into two smaller droplets. Frisch and Meitner pointed out to Bohr that his theory could explain nuclear fission.

I had hardly begun to tell him about Hahn's experiments and the conclusions that Lise Meitner and I had come to when he struck his forehead with his hand and exclaimed,

'Oh, what idiots we have been. We could have foreseen it all! This is just as it must be!'
And yet even Bohr, perhaps the greatest physicist of his time, had not foreseen it.

—Frisch[9]

John Archibald Wheeler, an American physicist who had studied with Bohr and collaborated in the development of the liquid drop model, has made numerous contributions to science on his own, including such fields as cosmology and nuclear weapons control. Wheeler's memoir of his life in science, *Geons, Black Holes, and Quantum Foam* (Norton, 1998), is an engaging and readable discussion of aspects of his career and developments in physics and cosmology over the last half-century.

Bohr and Wheeler postulated that fission was more likely to occur in the light isotope of uranium ($_{92}U^{235}$) than in the regular or natural uranium ($_{92}U^{238}$):

$$_{92}U^{235} + _0n^1 \rightarrow _{56}Ba^{144} + _{36}Kr^{89}$$

(Notice that this equation is not balanced. We will see below how to complete the equation to balance it.) The products of this reaction are, however, roughly half the size of the original $_{92}U^{235}$ nucleus. Although the atomic numbers sum on each side of the equation, we are 'short' three mass numbers on the right-hand side. The missing product must have an atomic number of 0. There is no nuclear particle with atomic number of 0 and mass number of 3 that would allow us to complete this equation. However, we have seen that the neutron is the species with atomic number 0, and since it has a mass number of 1, the equation can be balanced by having three neutrons (giving a total mass number of 3) as products. Thus

$$_{92}U^{235} + _0n^1 \rightarrow _{56}Ba^{144} + _{36}Kr^{89} + _0n^1 + _0n^1 + _0n^1$$

When a uranium nucleus breaks in two in this fashion, it does not always divide in exactly the same way. The nucleus may well break at one point in one case and at a slightly different point in another. For this reason, a great variety of radioisotopes are produced, depending on just how the division takes place. They are lumped together as **fission products**. The possibilities are highest that the division will be slightly unequal, with a more massive half in the mass number region of 135 to 145 and a less massive half in the region of 90 to 100.

As we have seen (Figure 26.2), the ratio of neutrons to protons decreases as the atomic number decreases, so that, when fission into two pieces occurs, there are extra neutrons in the product nuclei. These are rapidly emitted to allow the product nuclei to become stable. Most fission reactions end up producing two or three 'extra' neutrons. Usually the fission products are still somewhat unstable, and may undergo β decays (accompanied by γ decays) to achieve the appropriate neutron-to-proton ratios for stability. Because of this, many fission products are themselves radioactive, a problem that we will discuss in Chapter 28.

If we accept the picture of a nucleus as a liquid drop, we can suppose that the absorption of a neutron by a heavy nucleus is enough to set it vibrating. The difference between an ordinary liquid drop and a nucleus is that, when the nucleus is distorted from a spherical shape, the short-range nuclear binding energy holding it together loses much of its ability to do so, but the repulsive electrical forces among the protons are scarcely affected. Nuclear fission essentially involves disrupting the delicate balance within the nucleus between the attraction of the nuclear binding energy and the electrical repulsion between protons. As long as the attractive forces predominate, the additional energy provided to the nucleus when it is struck by a neutron will eventually be lost: the neutron will be captured by the nucleus and a γ-ray will be emitted. But, if the repulsive forces get to predominate, the distortion of the nucleus passes a point beyond which it keeps getting larger and larger until it splits roughly in half. The absorption of a neutron by a $_{92}U^{235}$ nucleus supplies enough energy to distort the nucleus past this critical point beyond which it becomes more and more distorted until it splits. Even though the nucleus breaks apart in fissioning, it does not break apart in exactly the same way every time. Instead, the fissioning of the nucleus can produce many different combinations of smaller nuclei. The barium and krypton we have used in the equations above represent illustrative examples; they are not the sole products of fission.

We saw how the transmutation of uranium to thorium resulted in a move a small distance along the curve of binding energy, and released a small amount of energy. Now, however, we have a different situation, indicated in Figure 26.7.

Uranium fission will yield some four times as much energy, gram for gram, as ordinary radioactivity will. So, why doesn't uranium spontaneously undergo fission rather than radioactive decay? Before fission can occur, the nucleus must absorb the small quantity of energy needed to distort it sufficiently to overcome the short-range nuclear binding energy. Essentially, the neutron absorbed into the nucleus is analogous to the small input of energy from a match that initiates the much larger release of energy from a fire. Many substances—gasoline, for example—are very flammable yet never ignite spontaneously; some small amount of energy must be provided first from a match or spark plug. Similarly, even though a nucleus can release large amounts of energy from fission, it will not do so spontaneously.

The energy release from fission is prodigious. Consider: On an equal weight basis, the fission of $_{92}U^{235}$ releases two million times as much energy as is obtained from burning coal. For example, fission of a pound of $_{92}U^{235}$ would release the energy equivalent to about 1,500 tons of coal. A 600 MW electricity-generating station could run for a year on the $_{92}U^{235}$ contained in 100 tons of naturally occurring uranium (which, as we will see in the next chapter, contains only a small fraction of $_{92}U^{235}$). The energy equivalent of a full 20-gallon tank of gasoline is a tiny sphere of $_{92}U^{235}$ about a tenth of an inch in diameter.

What is the source of all this ENERGY? One answer is much like the answer to the similar question about the ENERGY released in combustion. There (Chapter 5), we saw

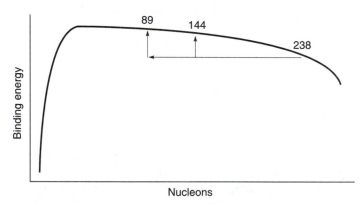

FIGURE 26.7 Nuclear fission is illustrated on the curve of binding energy. As in Figure 26.5, the curve has been 'smoothed' and the horizontal scale is exaggerated for illustration.

that ENERGY is needed to break the chemical bonds in a molecule of fuel, such as methane. In the same way, but at the nuclear scale, ENERGY would be needed to break apart a nucleus completely into its constituent nucleons. ENERGY is released in a chemical combustion process when the constituent oxygen, carbon, and hydrogen atoms are reassembled into the product carbon dioxide and water molecules. Similarly, ENERGY, in this case the nuclear binding energy, would be released when a product nucleus is assembled from its constituent nucleons. Since the binding energy of the original uranium nucleus is smaller than the combined binding energies of the product nuclei, there will be ENERGY left over—and released from the SYSTEM—when fission occurs. Regardless of whether we deal with chemical or nuclear reactions, it takes ENERGY to undo binding—chemical bonds or nuclear binding energy. Conversely, ENERGY is released whenever a system becomes more bound, in the formation of product molecules or product nuclei.

Consider how the energy balance equation describes this situation. As before,

$$E_{in} - E_{out} = E_{stored}$$

In this case, E_{in} is the ENERGY we must supply to the SYSTEM to, conceptually, decompose the uranium nucleus into its constituent nucleons. E_{out} would be the ENERGY released when the nuclei of the fission products are formed from those nucleons. Since the fission products are more stable than the original uranium, E_{out} is greater than E_{in}. In that case, E_{stored}, the ENERGY stored in the SYSTEM, is negative, which tells us that some ENERGY must be released from (or withdrawn from) the SYSTEM. That ENERGY is liberated to us.

CHAIN REACTIONS

The practical utilization of the abundant store of energy locked up in every atom of matter is a problem which only the future can answer. Remember, at the dawn of electricity, it was looked on as a mere toy.

—Becquerel[10]

Because each fission event liberates two or three neutrons, while only one neutron is required to initiate fission, a rapidly multiplying sequence of fission events can occur. In principle, then, we ought to be able to take one of the neutrons that is a product of the reaction and use it to start a second reaction that will generate three more neutrons. If we use one of the product neutrons from that second reaction we can start a third reaction, and so on. A process in which the product of one step is the reactant in, or initiator of, a subsequent step is called a **chain reaction**.

An uncontrolled chain reaction releases an immense amount of energy in a short time. If we assume that two neutrons emitted in each fission are able to induce further fissions and that a microsecond (i.e., one millionth of a second) elapses between the emission of a neutron and its subsequent absorption by another nucleus, a chain reaction starting with a single fission will release nearly 20 *billion* Btus of energy in less than a thousandth of a second. This is the basis of the atomic bomb. (For comparison, if we took a coal that would release 10,000 Btus per pound, that energy release from the atomic bomb is the same as would be achieved by burning one ton of that coal in one millionth of a second.) But, a properly controlled chain reaction (in which only one neutron per fission causes another fission) can be made to release ENERGY at a constant output.

... the nuclei of uranium atoms. A certain number of them explode when neutrons collide with them. Neutrons are among the so-called elementary particles—that is to say, particles which have not yet been broken up, such as electrons, protons, and perhaps a few others. This does not mean that they will never be broken up.

Ordinary atoms hold together when they collide at a speed of about a mile a second, as they do in air. When the temperature is raised and the speed of collisions goes up to ten miles or so a second, they cannot hold together, but electrons ... are torn off them. That is why a flame conducts electricity.

But at moderate speeds—say, a few thousand miles per second—collisions only break up the atoms temporarily. They soon pick up their lost electrons. When the speed rises to tens or hundreds of thousands of miles per second, the nuclei ... are sometimes broken up.

When a current is passed through ... heavy ... hydrogen at a voltage of half a million or so, the atomic nuclei become formidable projectiles, and if they hit ... lithium they break up its atomic nuclei and let neutrons loose. Neutrons can penetrate the nuclei of many atoms even when moving slowly and cause still further changes.

Generally they only chip a piece off. But when they attack uranium ... the uranium nuclei split up. The new fact, first discovered by Joliot and his colleagues in Paris, is that when the uranium nucleus splits, it produces neutrons also....

So most of the neutrons, which can penetrate even metals for some distance, get out. But if the neutrons are liberated in the middle of a sufficiently large lump of uranium, they will cause further nuclei to break up, and the process will spread. The principle involved is quite simple. A single stick burns with difficulty, because most of the heat gets away. But a large pile of sticks will blaze, even if most of them are damp.

If the experiment succeeds, several things may happen. The change may take place slowly, the metal gradually warming up. It may occur fairly quickly, in which case there will be a mild explosion, and the lump will fly apart into vapour before one atom in a million has been affected. Or there may be a really big explosion. For if about one four-hundredth of the mass of the exploding uranium is converted into energy, as seems to be probable, an ounce would produce enough heat to boil about 1,000 tons of water. So 1 oz. of uranium, if it exploded suddenly, would be equivalent to over 100,000 tons of high explosive.

... Most probably, however, nothing much will happen. It may be, for example, that the majority of uranium atoms are stable, and only one of the several isotopes ... is explosive. If so it will take several years to separate the isotopes.

Nevertheless, the next few months may see the problem solved in principle. If so, power will be available in vast quantities. There will be a colossal economic crisis in capitalist countries. There is plenty of uranium in different parts of the world, notably northern Canada, the Belgian Congo, Czechoslovakia and in several parts of the Soviet Union.

So the owners of uranium ores will make vast fortunes and millions of coal miners will be thrown out of work. The Soviet Union will adopt the new energy source on a vast scale, but the rest of the world will have a much tougher job to do so. Fortunately, uranium bombs cannot at once be adapted for war, as the apparatus needed is very heavy and also very delicate, so it cannot at present be dropped from an aeroplane. But doubtless uranium will be used for killing in some way.

An intelligent reader may well ask why, if uranium is so explosive, under certain conditions, explosions do not occur in Nature. The answer is that uranium does not occur in Nature in a pure state. It is generally found combined with oxygen, and neutrons would be stopped by the oxygen atoms to such an extent that an explosion could not possibly spread.

. . . this article is highly speculative. I am prepared to bet against immediate 'success' in these experiments. Nevertheless, some of the world's ablest physicians [sic] are hard at work on the problem. And the time has gone past when the ordinary man and woman can neglect what they are doing.

—Haldane[11]

If we could figure out how to control a chain reaction of nuclear fission, then we should be able to obtain a sustained release of ENERGY that we can use for some useful purpose, i.e., WORK.

Why doesn't a chain reaction start in naturally occurring uranium deposits? The reason lies in the fact that natural uranium contains only about 0.7% of the fissionable isotope $_{92}U^{235}$. The dominant isotope, $_{92}U^{238}$, absorbs neutrons but does so without undergoing fission.

Geologic evidence suggests that a fission chain reaction occurred at least once in nature. Uranium ore mined near Oklo, in Gabon, has much less than the normal 0.7% $_{92}U^{235}$. In addition, the proportions of the isotopes of the element neodymium are not typical of neodymium found elsewhere on Earth, but are enriched in those isotopes that would form as fission products. The loss of fissionable $_{92}U^{235}$ and the appearance of isotopes of neodymium (and other elements) that would be fission products together suggest that fission must have occurred in this ore deposit. The Oklo deposit formed about two billion years ago, when the concentration of $_{92}U^{235}$ in natural uranium would have been about 3%. Likely, solid, relatively insoluble UO_2 was oxidized to soluble UO_2^{+2}. The soluble uranium was concentrated by transport in water solution; at some later time it was reduced back to insoluble UO_2 under anaerobic conditions. The water remaining in the system would serve as a moderator. It is believed that such an event cannot happen now, because two billion years of further radioactive decay of $_{92}U^{235}$ (half-life of 710 million years) have reduced its concentration to the 0.7% that is too low to sustain a chain reaction.

We know by now that, whenever we have an energy transformation that involves a spontaneous change from high potential to low potential, if we are clever we can insert some kind of device into the SYSTEM to capture some of that ENERGY as useful WORK for us. By 1939, physicists recognized that although the nucleus represented tremendous quantities of potential energy, there was no practical way of converting it into useful work. Some eminent scientists believed that no practical method ever could be found.

In fact, in 1933, Ernest Rutherford declared that the development of a practical source of large-scale nuclear energy was an idle dream.

These transformations of the atom are of extraordinary interest to scientists but we cannot control atomic energy to an extent would be of any value commercially, and I believe we are not likely ever to be able to do so. A lot of nonsense has been talked about transmutation.

—Rutherford[12]

Leo Szilard was the sort of person of the 'never say never' philosophy. In a flash of inspiration, said to have happened while waiting for a traffic light to change as he was walking to work, he realized that if an atom requires one neutron for fission but releases two or three, then it is possible to establish a chain reaction by capturing one of the product neutrons to initiate the next step. He patented the idea, in a 1934 patent application that contains the term 'chain reaction.' Szilard wanted to keep the patent secret to prevent the fundamental scientific ideas from being misused. He assigned the patent to the British Admiralty, so that it was not published until after the Second World War. Szilard's basic idea for the chain reaction process leads to some important questions about putting it into practice. First, of course, is the question of whether it is actually possible to create and control the chain reaction process in a practical device. Assuming that were possible, then it becomes important to know how much energy is liberated when the fission chain reaction takes place. It is also crucial to know how to control the process while it is going on, and how to stop it when desired. The translation of Szilard's fundamental scientific idea into engineering practice is the topic of the next chapter.

 Leo Szilard was born in Hungary and began a university career in Germany. In 1929, he had published a seminal paper in the science that is nowadays called information theory. By the early 1930s, Szilard became certain that Hitler would eventually come to power, and that another European war was inevitable. He kept two bags packed in his room, to be ready for precipitous flight, and by 1933 he managed to move to England.

THE KEY TO RELEASE OF USEFUL ENERGY FROM NUCLEAR FISSION IS TO ESTABLISH A SUSTAINED, CONTROLLED, CHAIN REACTION.

CITATIONS

1   Calder, N. *The Key to the Universe*. British Broadcasting Company: London, 1977.
2   Carey, J. *The Faber Book of Science*. Faber and Faber: London, 1995; pp. 260–261.
3   Carey, J. *The Faber Book of Science*. Faber and Faber: London, 1995; p. 198.
4   Snow, C.P. *The Physicists*. Macmillan: London, 1981.
5   Bronowski, Jacob. *The Ascent of Man*. Little, Brown: Boston, 1973; p. 343.
6   Noddack, I.; quoted in Glashow, Sheldon L. *From Alchemy to Quarks*. Brooks/Cole: Pacific Grove, CA, 1994; p. 570.
7   Doyle, A.C. 'The Adventure of the Beryl Coronet.' Available in any of the numerous editions of the cases of Sherlock Holmes that are always in print.
8   Hahn, O., quoted in Glashow, Sheldon L. *From Alchemy to Quarks*. Brooks/Cole: Pacific Grove, CA, 1994; p. 570.
9   Frisch, O., quoted in Glashow, Sheldon L. *From Alchemy to Quarks*. Brooks/Cole: Pacific Grove, CA, 1994; p. 570.
10   Becquerel, Henri, quoted in Glashow, Sheldon L. *From Alchemy to Quarks*. Brooks/Cole: Pacific Grove, CA, 1994; p. 535.
11   Haldane, J.B.S. *Science in Peace and War*. Lawrence and Wishart: London, 1940.
12   Rutherford, Ernest, quoted in Boorstin, Daniel J. *Cleopatra's Nose*. Random House: New York, 1994; p. 149.

FOR FURTHER READING

Asimov, Isaac. *The History of Physics*. Walker: New York, 1966. An introductory survey of all of physics, through the 1960s, intended for the reader with little previous science background.

Chapters 36 through 40 cover some of the topics introduced here, though in somewhat different order.

Aubrecht, Gordon. *Energy.* Merrill: Columbus, OH, 1989. Chapter 6 of this useful introductory text on energy discusses the atomic nucleus and nuclear fission.

Bodanis, David. $E = mc^2$. Walker: New York, 2000. Subtitled 'A biography of the world's most famous equation,' this book provides an excellent survey of the scientific discoveries that led up to the development of this equation, from Faraday through Rutherford, Fermi, Meitner, and Einstein himself. Requires little previous scientific background.

Glashow, Sheldon. *From Alchemy to Quarks.* Brooks/Cole: Pacific Grove, CA, 1994. An excellent introductory physics text primarily for liberal arts students, though requiring some mathematics background; or, as the author says, 'physics for poets who can count.' Radioactivity, the structure of nuclei, and the discovery of fission are treated in Chapter 14.

Glasstone, Samuel. *Sourcebook on Atomic Energy.* Van Nostrand: New York, 1950. Though long out of print, this book is perhaps the seminal text on radioactivity, fission, and other nuclear processes. Its publication revealed much that had been learned during the effort to develop the atomic bomb during the Second World War. Worth searching out in libraries or used book stores.

Hewitt, Paul G. *Conceptual Physics.* Addison-Wesley: Reading, MA, 1998. Another splendid introductory physics text, with very informative illustrations. Chapter 32 discusses the atomic nucleus and the phenomenon of radioactivity; Chapter 33, nuclear fission.

Pasachoff, Naomi. *Marie Curie and the Science of Radioactivity.* Oxford University: New York, 1996. A brief biography of Madame Curie, including her scientific contributions and subsequent life as a celebrity. Requires no prior scientific background.

Romer, Alfred (Ed.) *The Discovery of Radioactivity and Transmutation.* Dover: New York, 1964. This book is a collection of original scientific papers written by some of the early investigators, including Becquerel, Rutherford, and the Curies. The original papers were written for professional journals, so necessarily assume some background in chemistry and physics on the part of the reader.

Rutherford, Ernest; Chadwick, James; Ellis, C.D. *Radiations from Radioactive Substances.* Cambridge University: Cambridge, 1930. This monograph, one of the classic professional science books of the 20th century, summarizes the work of Rutherford and his colleagues in radioactivity and transmutation of elements.

Sime, Ruth Lewis. *Lise Meitner.* University of California: Berkeley, 1996. A biography of Meitner, her role in the discovery of nuclear fission, and an argument that she was cheated out of a share in the Nobel Prize that she deserved for her work on fission.

27

NUCLEAR POWER PLANTS

THE COMPONENTS OF A NUCLEAR REACTOR

Soon after scientists realized that bombardment of uranium with neutrons led to nuclear fission, Niels Bohr and John Wheeler determined theoretically that $_{92}U^{235}$ was much more likely to undergo fission than was $_{92}U^{238}$. Their theoretical prediction was soon confirmed by experiments. Naturally occurring uranium is about 99.3% $_{92}U^{238}$, and only 0.7% $_{92}U^{235}$. Consequently, natural uranium, even if it were produced in very high chemical purity, offers essentially no chance for a sustained nuclear chain reaction, because 993 atoms out of every 1,000 would be the $_{92}U^{238}$ isotope, which absorbs neutrons without undergoing fission. To sustain a nuclear chain reaction, the uranium fuel has to be prepared such that $_{92}U^{235}$ is present in greater concentration than in natural uranium. The process for doing this is called **uranium enrichment** (shown conceptually in Figure 27.1). It requires effecting a separation of isotopes, which is not an easy task. In large part this is because isotopes are, after all, the same chemical element, and therefore can't be separated by processes that rely on differences in chemical behavior. The only distinction is the slight difference in mass number—a difference of about 1.3% in the case of the two uranium isotopes. Therefore, separation processes must rely upon small differences in those physical properties that depend upon differences in atomic or molecular weights.

One approach to uranium enrichment involves the formation of uranium hexa-fluoride (UF_6), which, despite its very high molecular mass, can be vaporized at relatively mild temperatures. (This compound, with its very massive molecules, boils at about 57°C, a temperature well below the boiling point of water.) UF_6 molecules containing U^{235} will be slightly lighter than those containing U^{238}, and will diffuse very slightly faster. As the molecules diffuse over very long distances, a portion of the gas will contain a higher proportion of the lighter $U^{235}F_6$ molecules, while the gas lagging a bit behind will have a higher proportion of the heavier $U^{238}F_6$.

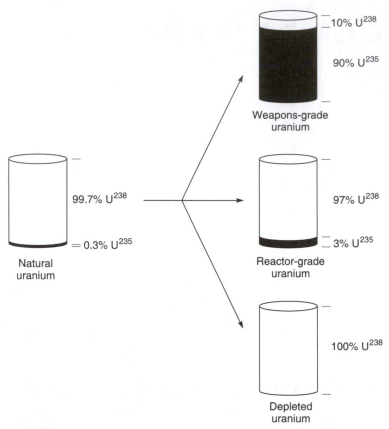

FIGURE 27.1 Uranium enrichment separates the isotopes in uranium ore to produce weapons-grade uranium, 90% U^{235}; reactor-grade uranium, 3% U^{235}; and leave behind depleted uranium, essentially 100% U^{238}.

To make a suitable fuel for the types of reactors discussed in this chapter, the uranium must be enriched to about 3–4% U^{235}. This form of enriched uranium is sometimes called 'reactor-grade' or 'power-grade' uranium. A second form of enriched uranium, containing about 90% U^{235}, has been used for the construction of nuclear weapons, and is sometimes referred to as 'weapons-grade' uranium. The difference in U^{235} concentration between reactor-grade and weapons-grade uranium removes the possibility of a nuclear reactor exploding like an atomic bomb, because the concentration of the fissionable U^{235} isotope in the reactor is too low for a nuclear explosion.

The enriched UF_6 is converted to solid uranium dioxide (UO_2), which is shaped into pellets typically an inch in diameter and about two-thirds of an inch tall, with each pellet being encased in a ceramic shell. These are packed into long, thin zirconium alloy tubes (nominally about an inch in diameter and 12 feet long) to make **fuel rods**. A group of fuel rods packaged together makes up a **fuel assembly** (Figure 27.2).

As the reactor operates, the concentration of fissionable U^{235} drops steadily, since it is being consumed to release energy, and fission products, some of which are good neutron absorbers, increase. Eventually a time comes when the combined effects of losing U^{235} and building up neutron-absorbing fission products leaves a fuel rod unable to contribute further to the energy production in the reactor. At that point the fuel rod is said to be **spent**, but it still contains some 1–2% U^{235}.

To control the nuclear chain reaction, some material is required that will readily absorb the neutrons without reemitting them. The chain reaction is controlled with neutron-absorbing rods, **control rods**, usually containing cadmium or boron.

Control rods can speed up, slow down, or shut down the fission process altogether. The rate of the chain reaction depends on the number of control rods inserted into the reactor, and the extent of their insertion into the reactor (Figure 27.3). The control rods can be moved into and out of the reactor to control the flux of neutrons. Control rods are somewhat analogous to the accelerator in a car, in providing the operator a way of speeding up or slowing down. When the control rods are fully inserted into the reactor, they stop the chain reaction. In a typical American power reactor design, 47 fuel rods and 12 control rods make up a single fuel assembly. Several hundred such fuel assemblies would be used in a single reactor.

When the fission process occurs, the disintegrating $_{92}U^{235}$ nuclei emit fast-moving neutrons of high kinetic energy. Atoms of $_{92}U^{238}$, which still amounts to about 97% of the uranium atoms in the fuel, readily capture the fast neutrons released in fission. $_{92}U^{238}$ does not undergo fission. Slower-moving neutrons are far more likely to be captured by $_{92}U^{235}$

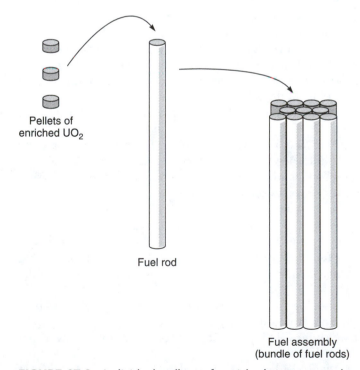

Pellets of
enriched UO_2

Fuel rod

Fuel assembly
(bundle of fuel rods)

FIGURE 27.2 Individual pellets of enriched, reactor-grade uranium are loaded into metal tubes to make fuel rods. A group of fuel rods together make up a fuel assembly.

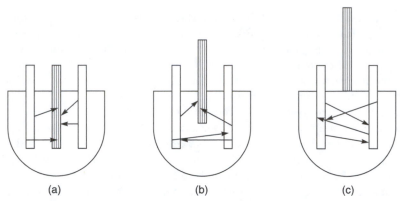

(a)   (b)   (c)

FIGURE 27.3 This diagram of a reactor with two fuel rods and one control rod shows the role of the control rod. (a) The control rod is fully inserted into the core and essentially absorbs all the neutrons, shutting down the reactor. (b) The control rod is partly withdrawn and allows some neutrons to enter fuel rods and initiate fission. (c) The rod is fully withdrawn, permitting the maximum amount of fission and allowing the reactor to run at full output.

than by $_{92}U^{238}$. These slower neutrons would be able to contribute to sustaining the chain reaction when absorbed by the fissionable U^{235}.

A **moderator** slows the neutrons so that U^{235} nuclei will absorb them and continue the fission process. The ideal moderator should reduce neutron energies as efficiently as possible, quickly and with minimum neutron loss. The best are nuclei of low mass, that rarely or never absorb neutrons, and that don't undergo fission themselves. Hydrogen, deuterium, helium, and carbon are all good moderators. The importance of light nuclei in a good moderating medium is so that each collision of a neutron with an atom in the moderator causes only a partial loss of the velocity of the neutron.

When a fission neutron leaves a good moderator, it has only thermal energy left. When a fast-moving neutron hits the nucleus of one of these atoms, the collision resembles that between a golf ball and a baseball. If a golf ball smacks into a baseball, it will bounce back, but at much lower velocity than it had before the collision. Similarly, when the neutron rebounds from a relatively light nucleus, it loses considerable velocity, and, consequently, loses kinetic energy. In comparison, when a golf ball whacks into a solid wall, it flies back with considerable velocity, as empirically demonstrated by anyone who has ever tried to play golf indoors. In the same way, if a neutron bounces back from a heavy nucleus, it loses hardly any of its velocity or kinetic energy. As an additional requirement for a good moderator, it should not contain nuclei that would absorb the neutrons upon collision, because that would cause their loss from the chain reaction.

Hydrogen exists as three isotopes, $_1H^1$, the most common form and called ordinary or light hydrogen; $_1H^2$, called deuterium or heavy hydrogen; and the radioactive $_1H^3$, tritium. Water molecules in which the light isotope $_1H^1$ is the dominant form of hydrogen is called light water.

In principle, hydrogen should be best as a moderator because it has the lightest atoms, which means that a neutron gives up its energy in the fewest collisions. Every molecule of ordinary water contains two atoms of hydrogen, is cheap, and has the further advantage that it can also act as a coolant (described below). Unfortunately, however, light hydrogen is also a good absorber of neutrons. The absorption of neutrons by water is so great that natural uranium, with only 0.7% $_{92}U^{235}$, is unable to sustain a chain reaction. Because of this, reactors using light water must be fueled with enriched uranium.

Heavy water slows the neutrons quickly without absorbing them at all. However, heavy water is expensive because only 0.015% (or one out of 7,000) of hydrogen atoms is deuterium, and separating heavy water from light water—as is the case with any isotope separation—is difficult. Because the neutrons need more collisions with deuterium atoms than with light hydrogen, heavy water reactors tend to use large quantities of heavy water—roughly 500 tons in a large reactor. Some Canadian reactors use heavy water as moderator.

Graphite moderators were used in many early reactors because graphite is relatively inexpensive and easy to work with. However, graphite is a less efficient moderator than heavy water, so graphite reactors had to be big. Furthermore, graphite, when exposed to air at high temperatures, can ignite and burn (as we will see in the next chapter). Graphite fires were partly responsible for two of the world's three major reactor accidents. Although the carbon atoms are relatively heavy, compared to light hydrogen or deuterium, the neutron loss by absorption is only a quarter of that in light water. However, because graphite is a solid, it cannot be pumped or made to circulate through the reactor. Consequently, unlike light or heavy water, graphite cannot do double duty as both moderator and coolant (discussed below). Some reactors in Britain and in countries of the former Soviet Union, including the notorious Chernobyl-type reactors, are graphite moderated.

As fission occurs inside the core, large amounts of heat will be released. The heat provided by a fission reactor comes from the kinetic energy of the fragments of the broken-up nuclei. About 85% of the liberated energy goes into the nuclear fragments, which rapidly collide with other atoms within the fuel rod. When these nuclear fragments collide within the fuel rod, they give up their kinetic energy in the form of heat, thereby heating the rod. The temperature in a light-water reactor (the most common type of reactor) typically reaches 600°F, while in the fuel rods it may reach 4,000°F. A **coolant** serves two purposes: It keeps the core's temperature under control, and it moves the heat generated in the core to some device that is able to make use of it. Light or heavy water can be used as the coolant as well as being the moderator. In a graphite-moderated reactor, gases, such as carbon dioxide, can be used as coolant.

The heart of any nuclear plant is the reactor **core**, which typically contains the fuel assemblies (including control rods), moderator, and coolant.

THE ESSENTIAL COMPONENTS OF A NUCLEAR REACTOR CORE ARE THE FUEL, CONTROL RODS, MODERATOR, AND COOLANT.

The core is enclosed in a very thick, very strong steel shell called a **pressure vessel.** In most reactors the pressure vessel is enclosed in a **containment structure** (sometimes called containment building), a steel-reinforced concrete structure typically 3 feet thick that serves as the outermost barrier between the plant and the environment, preventing most radiation from escaping the plant. As its name implies, the function of the containment building is to contain the reactor contents in the event of a leak or an accident. This structure is used in virtually all reactors in modern industrialized nations. As we will see (Chapter 28), one reason why the Chernobyl reactor accident was so severe is that power reactors of that type built in the former Soviet Union do not have containment structures.

PROCESSES IN THE REACTOR'S CORE

We've seen that the concept of a sustained nuclear chain reaction depends on using a neutron produced in a fission event to start the next fission. Chain reactions would not be achievable if, on average, fewer than one neutron was produced per fission. There would not be enough neutrons available to allow one new fission to take place for every one that has just occurred. Even if *exactly* one neutron was produced per fission event, that would not be sufficient to sustain a chain reaction. The reason is that inevitably some of the neutrons would escape from the core, or would be absorbed by nonfissioning nuclei, and thus these few neutrons would not interact with further U^{235} nuclei. If, on average, one of the fission neutrons produces a further fission the reaction is self-sustaining. In this case we say that the reaction is critical, and we would observe that the fission rate does not change over a period of time. If more than one fission neutron produces a further fission, we say that the reaction is supercritical, in which case the number of fissions increases with time. One other possibility is that less than one fission neutron produces a further fission; in such a case the reaction is subcritical, and the number of fissions gradually decreases with time.

To produce a working, controllable nuclear reactor it must be possible to start the chain reaction, allow the number of nuclei involved to increase gradually to a fission rate sufficient to generate the required energy, and then to maintain this rate. Of course, it must also be possible to decrease the reaction rate in a controlled manner to shut down the reactor. The control of the chain reaction is achieved by controlling the number of neutrons present in the core. Whether the reactor is supercritical or subcritical can be established by manipulating the control rods that are inserted into the core. Pulling control rods out of the core increases the chance that each neutron will induce a new fission, rather than being absorbed in a control rod, and so moves the core toward supercriticality. Inserting control rods into the core increases the chance that each neutron will be absorbed before it can induce a new fission, and moves the core toward subcriticality.

The three possibilities are defined by the reaction multiplication constant, sometimes also called the **multiplication factor**. In the core a certain number of uranium atoms (call this number n) will undergo fission in a unit of time. These n uranium atoms will produce some number, x, of neutrons as they undergo fission. But, of these total x neutrons, some other number, y, will not encounter a U^{235} nucleus. (The main reasons for this are that they may be absorbed by nonfissionable nuclei or escape out of the reactor.) This means that only $x - y$ neutrons actually strike a U^{235} atom and cause a new fission event. The ratio $(x - y)/n$ is the multiplication factor, k.

If the multiplication factor is less than 1, then at each succeeding step in the chain reaction, fewer atoms will undergo fission and, hence, fewer neutrons will be produced. The nuclear chain reaction rapidly comes to a stop. On the other hand, if the multiplication factor is greater than 1, then at each step in the chain reaction process a larger number of uranium atoms undergo fission, and a greater number of neutrons are therefore produced. Right at critical mass, where $k = 1$ and the average fission induces just one subsequent fission, the fission rate remains essentially constant. The fissionable fuel steadily releases ENERGY that can be used to operate an electricity generator.

When U^{235} nuclei undergo fission, they release an average of 2.47 neutrons, which induce other fissions within a thousandth of a second. Some of the neutrons released in fission emerge immediately (these are called **prompt neutrons**). But some of the fission fragments are unstable nuclei that decay and release neutrons long after the original fission. These neutrons are called **delayed neutrons**. On average, each U^{235} fission eventually produces 0.0075 of these delayed neutrons (or 0.75%), which can still go on to induce other fissions. Their time delays range from about one second to about one minute, averaging about 14 seconds. The delayed neutrons allow greater operational safety in reactors by giving operators some time to start shutting the reactor down if any indication of serious problems or danger is detected.

Delayed neutrons keep the reactor from going supercritical. A balance of prompt and delayed allows a total flux of prompt neutrons that is just subcritical. The delayed neutrons allow the reactor to go critical. The delayed neutrons determine the rate of increase of the fission process when the control rods, are withdrawn or the rate of its slowing down when the rods are inserted. Reactors don't respond quickly to movements of the control rods, because, as we've seen, the final release of delayed neutrons following a fission can take a minute or so, slowing the response of the reactor. The fission rate can't increase quickly because it takes time for the delayed neutrons to build up. Similarly, the fission rate can't decrease quickly because it takes some time until the final emission of delayed neutrons is over. The consequence of these long average delay times is a much slower growth rate of fission reactions than those that would result from prompt neutrons alone. A light-water-moderated reactor requires several minutes to go from low to high levels of chain reaction (and power). This allows the operator to monitor the increase of fission reactions, and, if need be, to manually control the reactor.

To summarize the processes that take place in the reactor's core: The fission reactions in the fuel rods result in the highly energetic neutrons and fission fragments moving rapidly away from the site of the reaction. These fission fragments quickly collide with other nuclei in the same fuel rod and, in doing so, give up their kinetic energy as heat. Most of the high-energy neutrons escape from the fuel rod and enter the moderator. In the moderator, these neutrons experience multiple collisions with hydrogen (or carbon) nuclei, each time losing a little of their kinetic energy to the nuclei of the moderator. The moderator slows the neutrons so that their capture by another fissionable U^{235} nucleus is more likely. Some neutrons may encounter a control rod and be absorbed, and are no longer able to initiate another fission reaction. The coolant flowing among the rods carries away the heat generated.

THE FIRST REACTOR

Enrico Fermi led the construction of the first nuclear reactor in a squash court underneath the stands of the University of Chicago's football stadium, Stagg Field. Fermi had emigrated from Italy to the United States in 1938, but was not yet an American citizen and, since the United States was now at war with Italy, was technically, and somewhat ironically considering his contributions to the development of the atomic bomb, an 'enemy alien.'

In 1942, some pure uranium was available as both the metal and its oxide. It was not enriched, so the amount needed for a reactor to become critical was extremely high. A very large atomic pile had to be built. (It was called a 'pile' because it was, literally, a pile of bricks of uranium, uranium oxide, and graphite.) As we have seen, three fates are possible for a neutron in a fission reactor. It can cause fission of another U^{235} atom, can escape into nonfissionable materials, or can be absorbed by U^{238} without causing fission. To make the first fate more probable, the uranium was divided into small pieces and buried at regular intervals in nearly 400 tons of graphite. When this first nuclear reactor was completed, it was 30 feet wide, 32 feet long, and $21\frac{1}{2}$ feet high. It weighed 1,400 tons, of which 52 tons was uranium. The uranium, uranium oxide, and graphite were arranged in alternating layers with holes into which cadmium control rods could be inserted.

On December 2, 1942, at 3:45 P.M., the control rods were pulled out just enough to produce a self-sustaining chain reaction. On average, slightly more than one neutron was produced per fission that was not absorbed by either U^{238}, or by some impurity, or lost to the pile. This meant that, for the first time, a chain reaction was in process. That day and minute are taken to mark the beginning of the atomic age. Because of wartime security considerations, the news of this success was relayed to Washington by a telegram reading, 'The Italian navigator has entered the new world.' There came a questioning wire in return, 'How were the natives?' The immediate reply was: 'Very friendly.' By December 12, the pile was releasing enough heat to be the equivalent of the energy needed to operate two light bulbs (i.e., 200 watts). Fermi's success showed that both nuclear energy and the atomic bomb were practical possibilities.

FIRST APPLICATIONS OF NUCLEAR FISSION

In 1954, the first nuclear submarine, the U.S.S. *Nautilus*, was launched in the United States. Its energy was obtained entirely from a nuclear reactor, and, unlike conventional submarines, it did not have to rise to the surface at short intervals to recharge a set of batteries. Since it could remain underwater for extended periods of time, it was that much safer from enemy detection and attack. The first American atomic surface ship was the N.S. *Savannah*, launched in 1959.

In the mid-1950s, nuclear power stations were designed for the production of electricity for civilian use. The Soviet Union built a small station of this sort in 1954, with a capacity of 50 MW. The British soon followed with one of 92 MW capacity at Calder Hall. The first American nuclear reactor for civilian purposes began operations at Shippingport, Pennsylvania in 1958. These early commercial reactors were designed for what we would nowadays consider a modest generating capacity of 100 MW. Later reactors were much larger; by the late 1970s, plants in the range of 1,000 MW were being built. (A 1,000 MW generating capacity provides approximately enough electricity for about a million homes.)

THE BOILING WATER REACTOR

All commercial nuclear power plants in the United States and more than three-quarters of those worldwide are light-water-moderated reactors of two types: boiling water reactors, and pressurized water reactors. Both types use ordinary, or light, water as both coolant and moderator and require enriched uranium fuel.

It is important to recognize that nuclear power plants differ from conventional ones only by using the heat from a nuclear reactor to raise steam for their steam turbines. The steam turbine–electricity generator sets are identical with those of conventional fossil-fuel-fired stations. The temperatures and pressures of the steam fed to turbines in nuclear plants are not as high as those of fossil-fuel-fired systems. This is due partly to the more severe conditions that the materials used in a nuclear plant must withstand. Because the steam temperatures (T_{high} in Carnot's efficiency equation) and pressures are lower, the resulting thermal efficiencies are also lower (a 34% turbine efficiency is considered good). Since the conversion of ENERGY to WORK is lower, the waste energy—in this case, waste

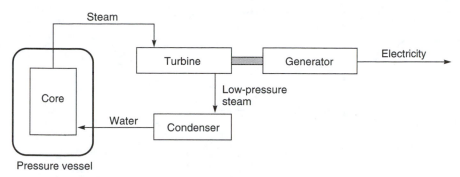

FIGURE 27.4   In a boiling water reactor, water in the reactor's core is allowed to boil to produce the steam needed to operate the turbine.

heat output—is necessarily higher. In principle, heat losses should be least if the coolant itself becomes the steam, and this is the method adopted in boiling water reactors. Since water is used as the coolant, we can generate steam directly from the water. This leads to the simplest design of a power reactor system, a **boiling water reactor**, or BWR (Figure 27.4).

Fuel assemblies in boiling water reactors are surrounded by cooling water. As fission occurs, fission fragments move through the fuel elements and the water; their kinetic energy becomes transformed, via many collisions with other atoms, to thermal energy. This thermal energy—heat—boils the water in the core, bringing the steam to 545°F under a pressure of 1,000 psi. The steam produced is taken from the core to the turbines. The steam drives the turbine, after which it is condensed to water and returned to the reactor, very much like in a fossil-fuel plant (Chapter 18). The maximum Carnot efficiency for this process is about 45% and the actual efficiency is about 30%, which is less than can be achieved in the newest fossil-fuel boilers.

The BWR works perfectly well for electric power generation. However, some of the radioactive fission products can diffuse through the metal cladding of the fuel rods. Also, in the United States, reactors are allowed to stay in operation with up to 1% of the fuel rods actually leaking. As water circulates through the core, it dissolves some of the fission products, some of which are highly radioactive. Thus the steam leaving the containment building and the water returning to it are radioactive. If an accident occurred that ruptured the steam or water lines outside the containment building, then radioactivity could be released to the environment. (This *does not* mean that radioactive material is released during normal operation of the reactor.)

PRESSURIZED WATER REACTOR

The more commonly used **pressurized water reactor** (PWR) addresses this potential emission of radiation into the environment by sealing the cooling water in a closed loop and adding a heat exchanger (Figure 27.5). Water in the reactor core is kept under high pressure (1,500–2,250 psi, much higher than in a BWR) so, even though the water gets very hot, above 300°C, it does not boil to steam. The hot water is piped through a heat exchanger that serves as the steam generator, where the extremely hot,

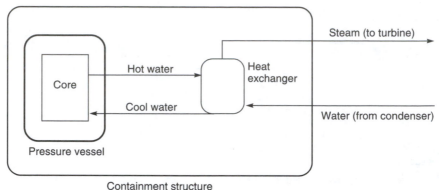

Containment structure

FIGURE 27.5 In a pressurized water reactor, hot, high-pressure water from the core is circulated through a heat exchanger to generate the steam used in the turbine. The steam fed to the turbine is not radioactive.

high-pressure water converts a secondary water supply into steam to operate the turbine. The water in the heat exchanger vessel boils as a result of the heat provided by the reactor (via the hot, high-pressure water), and this steam runs through a turbine just as in a conventional fossil fuel plant. The ideal Carnot efficiency for this process is nearly 50%, but the actual efficiency is about 30%.

With a PWR, there is no release of radiation if pipes outside the containment building are ruptured. Because heat from the high-pressure water inside the core is used to raise steam in the heat exchanger, no radioactive material is ever outside the containment building during normal operation. There is no radioactive steam going to the turbine. This of course does not mean that there can never be a problem with operating such a reactor, nor that there are no potential environmental problems. (These issues will be discussed in the next chapter.) However, the likelihood of an accidental release of radiation to the environment is much, much less from a PWR than from a BWR.

Most of the nuclear power stations worldwide use the PWR. A typical PWR design uses uranium oxide fuel, enriched to around 3% U^{235}, encased in zirconium alloy fuel rods. Water is used as a moderator and a coolant. In 1957, a PWR was installed at Shippingport, Pennsylvania for the Duquesne Light Company, and generated 60 MW (later increased 150 MW). The first truly commercial design of the PWR was the Yankee plant at Rowe, Massachusetts, in 1958. Twenty more PWRs were installed in the 1960s.

Operators start the reactor by retracting the control rods. As rods are removed, the flux of neutrons grows. Soon, the chain reaction begins. As more control rods are retracted, the reactor passes from subcritical through the critical stage, and temporarily into the supercritical range to sustain growth. As the number of fission reactions increases, the thermal power rate grows as well. This growth continues until the full power level is attained, at which time control rods are adjusted to keep the reaction rate steady in time. Under normal conditions, the reactor is operated in the critical range. Thermal energy is carried by the coolant. The water travels in a loop from the reactor to the steam generator and back. The process is self-contained and discharges no effluent to the atmosphere other than steam exhaust and waste heat. There are no piles of fuel or waste products outside the plant, nor any fuel deliveries via railroad cars or barges around the plant.

PRESSURE TUBE REACTORS

The graphite–water reactors used in the former Soviet Union are known as **pressure-tube reactors** (known by the initials RBMK). They operate on very slightly enriched uranium fuel, and differ from BWRs and PWRs in that it is possible to refuel the reactor as it operates, without shutting down. RBMK reactors use graphite moderators. A 100 MW reactor design incorporates about 2,500 columns of graphite blocks, over 200 boron control rods, and over 1,600 zirconium alloy fuel elements. The water system for this reactor has been characterized as a 'plumber's nightmare.' This type of reactor, the RBMK-1000, was involved in the 1986 disaster at Chernobyl in the Ukraine.

THE STEAM CYCLE

The steam cycle in a nuclear fission plant is very similar to that discussed in Chapters 17 and 18 for fossil-fuel-fired plants. Thus a very similar set of energy diagrams (Figure 27.6) apply. A high-temperature, high-pressure head of steam is fed to the turbines. There, the thermal ENERGY of the steam is converted to mechanical WORK, which is used to turn the generators. The steam leaving the turbines is at lower pressure and temperature. That steam is led to a condenser, where it is condensed back to liquid water, which is returned

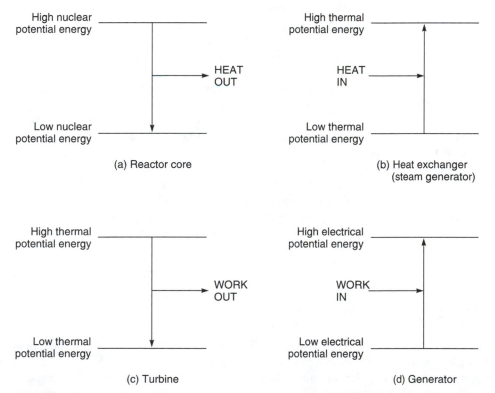

FIGURE 27.6 The sequence of energy transformations in a nuclear plant is almost identical to that of a fossil-fuel-fired power plant, except that the heat necessary to raise the steam comes from a nuclear transformation, rather than from burning a fuel.

FIGURE 27.7   The cooling tower helps to reduce the temperature of the condenser cooling water before it is returned to the environment. Cooling towers have come to symbolize nuclear power, even though *no* nuclear processes occur there.

to the steam generator. The condensers require a source of cool water; often this is obtained from some natural source such as a river. The cooling water that leaves the condenser is considerably warmer, because it has absorbed heat from the steam as the steam condenses. To avoid upsetting the local ecosystem by discharging large quantities of warm water from the condensers, this water is usually passed through a cooling tower (Figure 27.7) before its ultimate return to the environment. Because most cooling towers have a distinctive shape, and often dwarf the reactor building, cooling towers have sometimes been used in the news media as a symbol of nuclear power. This unfortunate choice has led some to believe that the cooling tower *is* the reactor, whereas in fact it is probably the most innocuous part of the plant.

The fossil-fuel-fired boilers, such as the pulverized-coal-fired water-tube boiler discussed in Chapter 17, and the nuclear reactors discussed here, are simply different ways of boiling water to feed a head of steam to the turbines. All of the public concern about the environmental consequences of burning fossil fuels, and all of the concern, or occasional uproar, over nuclear reactor safety and nuclear waste, are both ultimately focused on the question of how to boil water.

REACTOR SAFETY SYSTEMS

Of course, it is always appropriate—indeed, important—to ask what happens if something does go haywire in the reactor. The first line of defense, so to speak, is the *s*elf-controlled *r*emote *a*utomatic insertion of *c*ontrol rods, known from the first letters of its name as **scramming**. This procedure is the emergency insertion of *all* the control rods into the core. With all control rods inserted, this should shut down the reactor. The second safety feature is the *emergency core cooling system*—the **ECCS**—that floods the core with large volumes of cold water. This flooding should reduce the temperature in the core substantially. In addition, the water acts as a moderator to help slow down neutrons.

RADIATION FROM NUCLEAR PLANTS

In ordinary operation, the radiation from a nuclear power plant is virtually nil. By one scale of radiation dosage (the millirem, discussed in Chapter 28) the background radiation that we are all exposed to in our environment, from cosmic rays and naturally occurring radioactive materials, amounts to about 100 millirem per year. The *additional* radiation experienced by people working directly in the vicinity of a reactor is about one millirem per year. The additional radiation to which the general public is exposed amounts only to 0.05 millirem per year. Virtually all radioactive materials are contained within the reactor system, and small amounts only of short-lived radioactive elements are allowed to vent into the atmosphere. Shielding of the reactor is required by various laws and regulations to protect the health and safety of workers inside the plant. This shielding, along with the limitations on radioactive discharges, is also designed to meet public health standards for the population outside the plant.

THE END OF THE STORY?

The history of civil nuclear energy has been a checkered one indeed. After the Second World War, when the general public first became aware of the potential of releasing enormous quantities of energy via nuclear fission, there were widespread hopes of extremely cheap energy. The expression 'too cheap to meter' came into currency, meaning that electricity could be produced in such vast quantities, and so incredibly cheaply, in fission reactors, that it would not be cost-effective to install electricity meters at consumers' sites, nor to bother mailing out bills for the electricity. We could use all the electricity we wanted, free. In 1956, the first nuclear power station at Calder Hall, part of the Windscale complex in Britain, began to feed electricity into the public electricity supply. At the time, Britain was dependent on increasingly expensive, locally mined coal.

Unfortunately, it soon became clear that electricity from nuclear plants was not, and would likely never be, 'too cheap to meter.' This realization was followed by the first of the three major, well-publicized reactor accidents. This one, in October 1957, involved the Windscale reactor and resulted in a fire that polluted an area of farmland in the northwest of England. In 1979, the Three Mile Island accident, followed in 1986 by the major disaster of Chernobyl, have resulted in significant public disillusionment with, indeed even outright hostility for, nuclear energy. (A time line of nuclear events is given in Figure 27.8.) In addition, there remain lingering concerns, even serious feelings of guilt on the part of some individuals, over the atomic bombing of Hiroshima and Nagasaki. Nuclear energy is now regarded as, at best, a dubious blessing. The inability of engineers to give a cast-iron guarantee that there can be no future accidents, and the inability of most of the public to understand that *no* technology can expect to have such a guarantee, increases the negative attitude toward nuclear energy. The situation is complicated even further by the unsolved problem of the disposal of nuclear waste—where are we going to put the stuff to store it safely?

The collective impact of concerns about economics (i.e., it's not 'too cheap to meter'), nuclear accidents, and what to do with the radioactive waste have, not surpris-

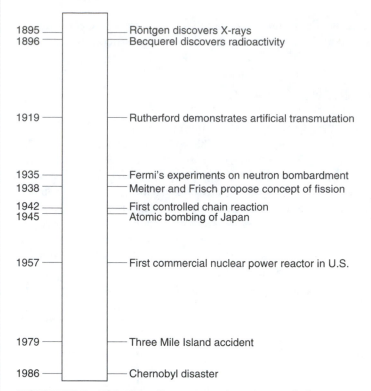

1895	Röntgen discovers X-rays
1896	Becquerel discovers radioactivity
1919	Rutherford demonstrates artificial transmutation
1935	Fermi's experiments on neutron bombardment
1938	Meitner and Frisch propose concept of fission
1942	First controlled chain reaction
1945	Atomic bombing of Japan
1957	First commercial nuclear power reactor in U.S.
1979	Three Mile Island accident
1986	Chernobyl disaster

FIGURE 27.8   This time line summarizes some of the major events in the history of nuclear energy.

ingly, led to enormous problems for the nuclear power industry in the United States. The uncertainties of governmental regulations, combined with what seems at times to be a never-ending, but seemingly always-escalating, sequence of lawsuits, appeals, and counter-suits, have ended the construction of new nuclear reactors in the United States. One example of the impact of delays caused by regulation and litigation is the dramatic increase in construction costs. In 1973, the Shoreham, Long Island nuclear power station was projected to cost $300 million to build. By the time it was completed (in 1984), its cost had escalated to $5.5 billion. And, it has never been put into operation because of public opposition.

More than a hundred nuclear plants (most of which are under 1,000 MW capacity) are operating in the United States and over 350 more in the rest of the world. Nuclear generation accounts for about 15% of electric energy generated in the United States and nearly 20% worldwide. As we begin the new century, the world has about 450 nuclear power stations in 30 different countries. They represent about a sixth of the world's total generating capacity and produce a little over a ninth of the total annual output of electricity.

While many observers expect, or, in some cases hope for, the demise of the nuclear industry, it may be premature to consign nuclear energy to the scrap heap. Concerns are increasing about the impact of human sources of carbon dioxide—mainly from the burning of fossil fuels—on changing the global climate (Chapter 32). Proponents of the large-scale use of nuclear energy have sometimes argued that nuclear energy is the only 'emission-free' source of electricity. Though one might argue about this choice of words, certainly in the event that major reductions in CO_2 emissions are mandated, nuclear fission is the only existing, demonstrated technology for large-scale electricity generation that does not release CO_2. There may be some life in the old corpse yet.

FOR FURTHER READING

Asimov, Isaac. *The History of Physics*. Walker: New York, 1966. An introductory survey of all of physics, through the 1960s, intended for the reader with little previous science background. Chapter 41 discusses nuclear reactors.

Aubrecht, Gordon. *Energy*. Merrill: Columbus, OH, 1989. A solidly useful introductory text on energy, though with considerably different organization than the present book. Chapter 14 discusses the technology for obtaining energy from nuclear processes.

Cardwell, Donald. *The Norton History of Technology*. Norton: New York, 1995. Chapter 17 is a discussion of the history of nuclear energy development.

Cassedy, Edward S.; Grossman, Peter Z. *Introduction to Energy*. Cambridge University: Cambridge, 1998. An excellent introductory book on the benefits and problems of many forms of energy technology. Chapter 7 is a good overview of nuclear reactors.

Cohn, Steven Mark. *Too Cheap to Meter*. State University of New York: Albany, 1997. The title derives from the dream of the early days of the nuclear industry: that electricity could be produced so cheaply there would be no reason to keep track of its use nor send out bills. The book treats the development of the nuclear industry in a societal context.

Collier, John G.; Hewitt, Geoffrey F. *Introduction to Nuclear Power*. Hemisphere: Washington, 1987. A fairly thorough overview of most aspects of reactor design, operation, and safety. This book was intended for readers who are not professional nuclear engineers, but who are interested in learning about nuclear energy.

Galperin, Anne L. *Nuclear Energy/Nuclear Waste*. Chelsea House: New York, 1992. The first half of this book, written for persons having little scientific background, deals with nuclear reactors and their operation.

Hazen, Robert M.; Trefil, James. *Science Matters*. Doubleday: New York, 1991. Subtitled 'achieving scientific literacy,' this book is aimed at readers with little or no prior scientific background. Chapter 8 discusses nuclear energy processes.

Josephson, Paul R. *Red Atom*. Freeman: New York, 2000. A history of the nuclear energy program in the former Soviet Union. This book traces the evolution of the early hope of the Communist party that nuclear energy would provide massive amounts of cheap energy to the tragic reactor accident at Chernobyl. As the dust jacket indicates, it is a story of 'big science run amok.'

Kraushaar, Jack J.; Ristinen, Robert A. *Energy and Problems of a Technical Society*. Wiley: New York, 1988. Chapter 4 is a useful discussion of nuclear energy and various types of reactors.

National Research Council. *Nuclear Power*. National Academy Press: Washington, 1992. This book provides an evaluation of possible near-term and long-term technologies for use of nuclear reactors in the United States. It is not a discussion of the desirability of using nuclear energy in the future, but rather focuses on the question of what technological options would be most desirable if nuclear energy were to continue to be used.

Peierls, Rudolf E. *Atomic Histories*. American Institute of Physics: Woodbury, 1997. This book is a collection of short essays, most of which are accessible to persons with nontechnical backgrounds, on a variety of topics including many on the issues of nuclear weapons control and nuclear energy. Peierls made many contributions to the early development of nuclear physics.

Rose, Paul Lawrence. *Heisenberg and the Nazi Atomic Bomb Project*. University of California: Berkeley, 1998. This book centers on the brilliant physicist Werner Heisenberg, who, unlike many of his colleagues and contemporaries, chose to remain in Germany during the Nazi years. His role, or lack thereof, in a possible Nazi atomic bomb project has been the subject of controversy for many years. This book also provides some of the scientific background on the early days of nuclear science and technology in Germany.

Seaborg, Glenn T. *A Chemist in the White House*. American Chemical Society: Washington, 1998. Glenn Seaborg was a Nobel Prize-winning nuclear chemist who was involved in the discovery of some of the 'transuranic' elements, those with atomic number greater than 92. One of these elements is now named seaborgium in his honor. This book is a memoir of his public service as related to nuclear weapons and the nuclear industry. Seaborg served every president from Roosevelt to Clinton in some capacity.

Shepherd, W.; Shepherd, D.W. *Energy Studies*. Imperial College: London, 1998. Chapter 8 covers most aspects of nuclear energy, from introductory concepts of nuclear reactions to issues of nuclear waste disposal.

Smith, Howard Bud. *Energy*. Goodheart-Willcox: Tinley Park, IL, 1993. An elementary text surveying most aspects of energy technology. Chapter 12 deals with nuclear energy.

28

THE NUCLEAR CONTROVERSY

Construction and operation of nuclear power plants is a very controversial issue. In the United States, there has been for many years a strong antinuclear sentiment. In this chapter we will discuss some of the issues regarding radiation safety, environmental concerns, nuclear weapons proliferation, and reactor accidents. Those are the issues that shape public perception, good or bad, of nuclear energy. At the heart of many of these concerns is the nature of radiation and its effects on human health.

One of the sources of our concerns or fears about radiation is that we cannot see it, or detect it with our other senses, or even feel it affecting us. Most other energy sources and forms of fuels are materials that we can see (such as coal or gasoline) or at least feel (such as heat or sunlight). There are two main factors that determine the effect of radiation on our bodies or on the environment: the type of radiation, and the intensity of the radiation (that is, the rate at which it is being produced).

HEALTH EFFECTS OF RADIATION

Not long after X-rays were discovered, the scientists working with them learned that overexposure to X-rays caused skin inflammations and burns that healed very slowly. The same proved to be true of the radiations from radioactive substances. The acute effects of exposure to radiation are fairly well known from studies of the victims of the bombs dropped on Hiroshima and Nagasaki, Japan in the Second World War, and from accidents in the early days of the atomic weapons program and the nuclear electricity industry. For the most part, all of this knowledge has come the 'hard way,' that is, from studying the effects of radiation on those who have, whether inadvertently or deliberately, been exposed to very large doses.

As we have seen (Chapter 25), there are three major types of radiation. Alpha (α) particles are identical with the helium nucleus ($_2He^4$). They are the least able to penetrate materials; α-particles travel only a few inches in air and can be stopped by a thin shield of material like cloth or cardboard, or even tissue paper. Although α-particles are easily stopped, if they do penetrate tissue they are over 10 times as dangerous to health as γ-rays, but about the only way they can cause such harm is if the radioactive source is eaten or inhaled into the body. Beta (β) particles are identical with electrons. They are given the symbol $_{-1}\beta^0$, to indicate that their mass number is effectively zero, in comparison with the proton and neutron, and that their negative charge can be considered equivalent of an atomic number of -1. They are intermediate in penetrating ability and health effects. β-particles can travel up to a few meters in air and can be stopped by varying thicknesses of sheets of such metals as aluminum, iron, or lead. Gamma (γ) rays are essentially identical to high-energy X-rays. They are very dangerous because of their ability to penetrate through thick layers of materials and because of their health effects. γ-rays are stopped only by substantial thicknesses of lead or thick concrete walls. These relationships are shown schematically in Figure 28.1.

The second issue is the rate at which the radiation is being emitted. In other words, how radioactive is the substance in question? One measure is the rate at which an isotope is decomposing, which is given by the half-life. The **half-life** of an isotope is the time required for half the amount on hand to decay. For example, suppose we have 100 grams of an isotope that has a half-life of 400 days. If we measured the amount of isotope on hand, we would find the relationship shown in Table 28.1. Although the half-lives of radioactive substances vary enormously, from fractions of a second to billions of years, a plot of the amount remaining as a function of time (Figure 28.2) would always have the same shape. In a sense, the half-life issue is a 'good news–bad news' situation. A material with a short half-life will be intensely radioactive, but won't be around long. A material with a long half-life has a low level of radioactivity, but is going to be around for a long time. The amount of radioactive nuclei present in a material can be characterized by the number of nuclear disintegrations per unit time. The **curie** (Ci) is defined as 3.7×10^{10} disintegrations per second. The curie represents the activity of one gram of radium. An alternative unit is the **becquerel**, defined as one disintegration per second.

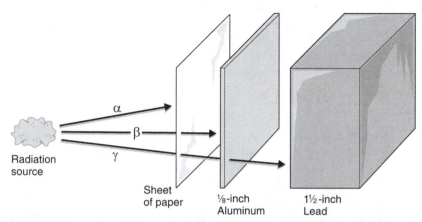

FIGURE 28.1 The different kinds of radiation have markedly different abilities to penetrate matter. α radiation can be stopped by relatively thin shields, such as cardboard. β radiation is more penetrating, and may require several inches of material as a shield. γ radiation is the most dangerous, and can only be stopped by thick concrete or metal (e.g., lead) shields.

TABLE 28.1 HYPOTHETICAL EXAMPLE OF THE DECAY OF 100 GRAMS OF A RADIOACTIVE SUBSTANCE WITH A HALF-LIFE OF 400 DAYS

Time elapsed (days)	Number of half-lives elapsed	Amount remaining (g)
0	0	100
400	1	50
800	2	25
1,200	3	12.5
1,600	4	6.25

We must combine the half-life and the type of radiation emitted to assess the prospective danger of a material. An α-emitter with a very long half-life (e.g., the common isotope of uranium, $_{92}U^{238}$) is relatively safe. A γ-emitter with a short half-life is very dangerous.

Radiation affects us through its action on cells. The energies of X-rays, γ-rays, or high-speed α- or β-particles can be high enough to break chemical bonds in molecules. When that happens, very reactive molecular fragments, **free radicals** or ions⊛ are formed. These very reactive species can, in turn, react with other molecules. Also, an α- or β-particle that is absorbed by an atom will alter its nature (that is, the reaction of an atomic nucleus with either an α or β will change the atomic number, and hence chemical identity, of that nucleus, transmuting one element into another) and, consequently, will alter the nature of the molecule of which that atom is a part. If the newly formed atom is radioactive and decays, that process will also break apart the molecule even if it had survived intact to that point. The biological damage caused depends therefore not only on the type of radiation and its energy, but also on the chemical composition and types of atoms in the body tissue absorbing the particles.

> ⊛ Free radicals are created when a chemical bond consisting of a pair of electrons between two atoms breaks in such a way that one of the electrons goes with each of the fragments. Using ethane as an example, the breaking of the bond between the carbon atoms could be represented as
>
> $$H_3C : CH_3 \rightarrow H_3C \cdot + \cdot CH_3$$
>
> Unlike ions, free radicals have no electric charge. They are, in most instances, extremely reactive species and readily interact with other nearby molecules.

The chemical changes caused by radiation can disrupt the many complex biochemical processes inside a cell. When that happens, these changes also upset the sequences of reactions that control how a particular cell functions and upset the various biochemical interactions among cells. In some cases, it is possible that biological changes may be induced that will allow the unrestrained growth of certain kinds of cells at the expense of their neighbors. This is the pathological condition that we recognize as cancer. Those body parts that are the most vulnerable to the effects of radiation include the skin, which of course is the first part of the body experiencing the radiation exposure, and those parts of the body that produce blood cells, such as the lymphoid tissue⊛ and bone marrow. Leukemia, the unrestrained production of white blood cells (a condition that is slowly, but invariably, fatal) is one of the more likely results of excessive exposure to radiation. Both Marie Curie and her daughter, Irène Joliot-Curie,⊛ died of leukemia, which may have resulted from the long exposure to radiation that both of them experienced during their scientific careers. When radiation exposure is particularly severe, very extensive destruction of tissues can completely break down the functioning of cells. This extreme condition can cause death in a period from months to as little as a few days. Such radiation sickness was observed on a large scale among the survivors of the atomic bombings of Hiroshima and Nagasaki.

The particular importance of the lymphoid tissues derives from the fact that they constitute the body's immune system. For example, the lymph nodes, the spleen, bone marrow, and the thymus are all lymphoid tissue. The lymph nodes help to overwhelm bacteria. The spleen filters bacteria and other microorganisms from the blood. The thymus shrinks and apparently loses its functioning around the onset of puberty, but if it is damaged in infants, severe problems with the infant's immune system can occur. Injury or damage to the bone marrow sharply reduces the production of blood cells.

Irène Curie, daughter of Marie Curie, married the physicist Frédéric Joliot, with whom she collaborated in numerous scientific endeavors relating to radioactivity, fission, and the development of nuclear energy. Irène and Frédéric published their work using their hyphenated surnames, Joliot-Curie. Among the many contributions of this brilliant couple were the discovery of artificial radioactivity and studies that contributed to the eventual discovery of the neutron by James Chadwick. The Joliot-Curies had in fact observed neutrons first, but did not want to suggest the existence of another subatomic particle to explain their observations, so discovery of the neutron is credited to Chadwick. Nevertheless, the Joliot-Curies shared the 1935 Nobel Prize in Chemistry for their splendid contributions to science.

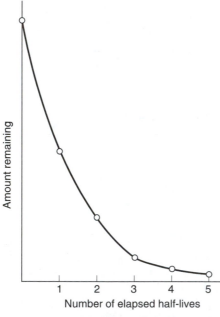

FIGURE 28.2 This generic graph illustrates the decay of any radioactive substance through 5 half-lives.

Concern about the use of nuclear energy for electricity generation arises from the knowledge that nuclear radiation can harm living tissue. The effects can be considered in two separate categories, called somatic and genetic effects. Somatic effects are those that occur in the person who was exposed to the radiation; genetic effects, which might be even more frightening, appear in the children of such a person.

Though the death of an individual from radiation exposure is a sad and unfortunate event, there is also danger—perhaps even worse than the death of the person affected—in that long-term genetic effects may continue over many generations. Sometimes the damaging or destruction of a molecule by radiation may not affect the individual very seriously, especially if the damage is confined to only a few cells. But, the molecules damaged by radiation may be in the genes and chromosomes of the individual who had been exposed. Then this radiation damage could be transmitted to a child of that person, and, if that happens, the child might have the damaged molecule in every cell of its body. The child might suffer far more than the parent, who was actually the person exposed to the radiation.

During the natural progression of generations, mutations in genetic material can occur as a result of random chemical imperfections that become incorporated in key molecules as they are reproduced. Mutations can also arise as the result of natural background radiation (discussed below), such as from radioactive substances in the soil, or cosmic rays from outer space. All of us have imperfect genetic material to some extent, and it is passed on to the next generation. However, the rate of generic mutation in a population will increase as the total amount of radiation in the environment—from natural background sources plus that introduced from human activities—increases. Commonly, genetic mutations are generally for the worse, in that the affected individual may be less strong, less healthy, or deformed as a result of the mutation. (We must recognize, though, that mutated genetic material that is injurious to some people could be advantageous to others in different circumstances. This helps give humans—and other organisms—the genetic ability to adjust to change and to evolve.) Some mutations may be fatal, condemning the afflicted individual to an early death. The nonlethal mutations are the greatest long-term worry for society, however, because the person having a genetic mutation may have children of his or her own. If those offspring survive, they in turn can pass on the damaged genetic material to yet another generation, spreading the mutation further and further into the population.

UNITS OF RADIATION

There are several ways of measuring the radiation dose received by an individual. The **rad** (the name being short for 'radiation') is a direct measure of energy. One rad is equivalent to the absorption of enough radiation in any form to liberate 100 ergs of energy per gram of absorbing material. (One hundred ergs is an extremely tiny quantity of energy, amounting to 0.000002 calories.) The rad is approximately equal to the röntgen, a unit that is discussed below.

Doses larger than 10,000 rads damage the central nervous system. Death occurs relatively quickly, often within one or two days. A dose of one thousand to 10,000 rads has a major effect on the gastrointestinal tract, causing vomiting, fever, and dehydration. Death is likely within one to two weeks. In the dose range 300–1,000 rads, blood cells are the worst affected. Hemorrhaging and diarrhea can lead to death in three to six weeks. Doses below 300 rads, and certainly below 50 rads, seldom lead to death, even within 30 days. There may be loss of hair, loss of appetite, and diarrhea. There is also an increased risk of leukemia and other cancers for the survivors, often developing years after the actual radiation exposure. The biological effects of radiation that cause, or at least are capable of causing, mutations are proportional to the radiation dose rate down to about 100 rads. A rule of thumb suggests that about 100 additional cancers per million people will result from a one-rad dose of radiation. It is not certain whether this rule-of-thumb relationship holds for even lower doses.

The **röntgen** is formally defined as the quantity of X-rays or γ-rays required to produce a number of ions equivalent to one electrostatic unit of charge in a cubic centimeter of dry air at $0°C$ and one atmosphere pressure. This electric charge is equivalent to the formation of about two billion ions. (Two billion ions may seem to be an enormous quantity, but, to put this into perspective, one cubic centimeter of air at these conditions will contain about 25 quintillion—that is, 25 billion billion—molecules.)

> The electrostatic system of units derives from the gram–centimeter–second system of measurement. When two equal electrical charges are separated by a distance of 1 cm and exert a force of 1 g·cm/s^2 (i.e., one dyne) on each other, the magnitude charges is one electrostatic unit (abbreviated *esu*).

A 'röntgen equivalent physical,' shortened to rep, is the amount of radiation that, when it is absorbed by living tissue, produces the same effect as absorption of one rad of X-rays or γ-rays. However, a given quantity of radiation does not produce the same effect on all living species. To take this difference among species into account, the 'roentgen equivalent man,' or **rem** for short, is defined to be that quantity of radiation which, on absorption by the tissues of a living human, produces the same effect as the absorption of one röntgen of X-rays or γ-rays. As a rule of thumb, exposure to 0.5 rem, or 500 millirems (mrem) per year is tolerable.

α-Particles are particularly dangerous. One rad of X-rays, γ-rays, or β-particles is nearly equivalent to one rem. However 1 röntgen of α-particles is equivalent to a dose of 10–20 rem. That is, the absorption of α-particles is at least 10 times as dangerous as the absorption of the same amount of β radiation. In that respect, it's fortunate for us that α-particles are the least penetrating form of radiation and the easiest to protect ourselves against. Exposure to protons and neutrons is also quite dangerous; 1 rad of either is equivalent to about 10 rem.

Many of the fission products produced inside a reactor are radioactive. Although in previous chapters we have illustrated the fission of $_{92}U^{235}$ as its transformation to barium and krypton, in fact, a variety of possible fission products form, whose mass numbers are roughly half that of uranium. Further, not only are some of the original fission products themselves radioactive, they may decay into other, secondary products that are also radioactive.

Among the more dangerous fission fragments are radioactive isotopes of strontium and cesium. The isotope $_{38}Sr^{90}$, a β-emitter, has a half-life of 28 years. Even after a century, $_{38}Sr^{90}$ will have undergone less than four half-lives, so a material containing this isotope is still dangerously radioactive. The problem with $_{38}Sr^{90}$ is that strontium is in the same chemical family as calcium. Therefore, $_{38}Sr^{90}$ in the environment can become concentrated in the milk of cows that might happen to graze on $_{38}Sr^{90}$-contaminated grass. Most of us will remember being admonished as children to drink more milk, because we need it (specifically, the calcium in it) for our bones. Unfortunately, children who drink milk contaminated with radioactive strontium will concentrate the $_{38}Sr^{90}$ in their bones. Calcium compounds are the major inorganic constituents of bone tissue, and strontium atoms, in the same chemical family as calcium, are able to replace calcium atoms in bones. Once bone tissue has formed, the atoms in the bones are replaced only very slowly. Consequently the $_{38}Sr^{90}$, once assimilated, will remain in the bones for a long time and stay in dangerously close contact with the blood-cell forming tissues. $_{55}Cs^{137}$, with a half-life of 30 years, remains in the soft tissues. It is exchanged with other atoms more quickly than is $_{38}Sr^{90}$ in the bones, so stays in the body for a shorter time. However, $_{55}Cs^{137}$ emits energetic γ-rays and, for that time it remains in the body, can do significant damage.

BACKGROUND RADIATION

While the focus of public attention is usually on radiation that might be produced from nuclear waste, atomic weapons testing, or reactor accidents, it is important to recognize that we are all constantly bathed in natural radiation. Anyone, anywhere in the world, will be exposed to radiation. This so-called background radiation is unavoidable. It comes in the form of cosmic rays from outer space, from naturally occurring radioactive substances in the environment, and from medical procedures. On average, we receive 50 mrem per year from outer space and another 50 mrem per year from the natural radioactivity of the soil. For example, the radioactive isotopes $_{90}Th^{232}$ and $_{92}U^{238}$ occur in various kinds of rocks and soil. Further, when these elements happen to be present in clays used to make bricks, the bricks will then have a low level of natural radioactivity. The estimate of radiation exposure for medical uses (such as diagnostic X-rays) in the United States is about 75 mrem per year. In fact, we irradiate ourselves to a small extent (about 25 mrem per year) from radioactive potassium and carbon isotopes, $_{19}K^{40}$ and $_6C^{14}$, that become incorporated in our body tissues. The total background radiation for most people is in the range of some 125–350 mrem per year, well below the 500 mrem federal background radiation standard for the general public. Despite what many people may believe, *less than 1%* of our background radiation comes from sources such as nuclear electricity-generating plants and their wastes, and nuclear fuel reprocessing.

> We are 'carbon-based life forms.' All of our soft tissues are made of compounds of carbon; carbon atoms are the stuff we are made of. The isotope $_6C^{14}$ occurs in trace quantities in our bodies. Potassium is essential to maintaining the proper balance of fluids inside and outside the cells of our bodies. (The fluid balance is determined by the Na^+/K^+ ratio.) About 12 potassium atoms out of every 100,000 (0.012%) are the radioactive isotope $_{19}K^{40}$.

In recent years there has been a growing awareness of the possible background radiation from radon. The isotope of radon $_{86}Rn^{222}$ is an α-emitter that occurs naturally in the environment as a result of the decay of $_{92}U^{238}$. Radon is a chemically unreactive gas in the same chemical family as helium and neon. Because radon comes from natural uranium deposits, the amount present in the environment depends highly on the local geology, that is, on the amount of uranium present in the rocks and soil of a particular locality. However, in some extreme cases radon can amount to about 55% of natural background radiation. Because radon is inert chemically, when it is produced it will not be trapped by any chemical processes in the surrounding soil or water. That trapping in soil or water, could it have taken place, would help us, because it would immobilize the radon. Instead, radon seeps from the ground into basements or lower levels of buildings through cracks or pores in concrete walls or floors.

When air containing radon is breathed, the radon can decay inside the lungs to give polonium, a radioactive element that is neither a gas (so it is not likely to be removed in normal exhalation) nor chemically inert (so it is, unfortunately, likely to interact with surrounding tissue). The isotope $_{84}Po^{218}$ is also an α-emitter:

$$_{86}Rn^{222} \rightarrow {}_2He^4 + {}_{84}Po^{218}$$
$$_{84}Po^{218} \rightarrow {}_2He^4 + {}_{82}Pb^{214}$$

The lead isotope $_{82}Pb^{214}$ produced in this decay is itself radioactive, a β-emitter. So, as the radon initially inhaled decays, it produces other radioactive materials that are more likely to stay in the body. As we've seen, the range of an α-particle in air is quite small, and it is relatively easy to shield against external α-particles. But in this case, the α-emitters are already inside the body, and α-radiation can cause serious biological damage. In the lungs, the range of penetration of the α-particles is about equal to the thickness of the cells that line the respiratory tract passages. Radiation damage to these cells can induce lung cancer.

Many homes in the United States may have some level of radon in the indoor atmosphere. However, there is some controversy regarding the level of radon that can be considered to be safe. In some parts of the country one can buy radon detection kits to use in the home, or hire someone to come in and test the home for high levels of radon.

IS THERE A THRESHOLD RADIATION EXPOSURE?

The concern over whether there is some minimum level of radon that can be considered to be safe is just one facet of the more general argument about the level of radiation exposure of any sort that can be considered reasonably harmless. Over the years, attempts have been made to determine the amount of radiation that can reasonably be borne by individuals (and by mankind in general) without representing a serious danger.

In addition to considering a division of radiation hazards into somatic or genetic effects, as discussed earlier, it can also be useful to divide the effect into acute, which would be due to large amounts of radiation but received in a single dose, and chronic, in this case due to small amounts of radiation but repeated exposures over some period of time. The acute effects, though they can be horrific in the worst cases, are at least straightforward, and there seems to be consensus about what those effects will be. In contrast, the effects of long-term exposure to low levels of radiation are not well known. In fact, it is not even clear whether there *are* such effects, let alone what they might be if they do exist, and whether they can be distinguished statistically among the potentially large number of similar effects that happen to be due to other causes.

It is often possible to establish some sort of relationship between the health effects and radiation exposure in one person, and then, from that, to estimate the incidence of that same effect in the population as a whole. By making that extension to the total population, it is possible to predict that, if the population were exposed to some particular radiation dose, there would be a corresponding incidence of that related health effect. There are two problems with this approach. First, individuals vary widely in how their bodies function. We all see, for instance, among our family and friends that some individuals gain or lose weight much more easily than others. Second, there is the severe complication that many kinds of observable health effects likely have multiple causes. Because of these problems, it becomes very difficult to detect effects that are specifically due to the one event of radiation exposure within a background of similar effects that originated not from radiation exposure, but from other causes. For example, in the United States about 550,000 people die each year of cancer. If the population happened to be exposed to a radiation dose double the background level, there would, in principle, be an additional 2,000 cancer deaths. However, it would be extremely difficult to determine unequivocally that a cancer death rate of 552,000 instead of 550,000 was due to that event of radiation exposure and not simply to random fluctuations in the population. When it comes to humankind's worst known radiation disaster, the reactor accident at Chernobyl (discussed later in this chapter), the estimated effect is 800 additional deaths on top of the 30,000 normally expected for those unlucky people who lived within 20 miles of the reactor. A change of 800 out of 30,000 would be noticeable. The Chernobyl accident is estimated to add possibly additional 40,000 deaths to some 500 million for the entire world's population. That effect is too small to be detected with certainty.

It has often been argued that a threshold for radiation damage exists. Some experiments have compared the effect of delivering the same total dose of radiation either in a single large dose (that is, equivalent to an acute exposure) or by the accumulation of many separate, smaller doses (simulating chronic exposure). A single larger dose can be fatal in cases where the several smaller doses have little apparent effect. These observations suggest that there is a 'threshold' dose below which there is no radiation damage. They also suggest that some sort of repair mechanism allows the body to recover from small amounts of radiation damage. However, there also seems to be a reduced life expectancy for those receiving the several smaller doses. That is, the body's repair mechanism may not provide permanent repairs. Though there are vigorous proponents on both sides, it seems on balance that it is not yet possible to show unequivocally that a safe threshold exists for chronic exposure. Also, it's not yet possible to sort out clearly the effects of radiation exposure in a population from other natural causes or just from random, year-to-year statistical fluctuations.

NUCLEAR REACTOR SAFETY

In terms of the safety of nuclear reactors, several concerns are usually raised. One is the issue of whether a nuclear reactor can blow up just like an atomic bomb. This is impossible. We've seen previously that the amount of fissionable $_{92}U^{235}$ in a reactor is about 3% of the total amount of uranium. In an atomic weapon, the proportion of $_{92}U^{235}$ would be about 90%.

A second issue is whether a nuclear reactor emits radiation to the environment during normal, routine operation. To begin to address this issue, we need to consider again the design of the PWR, which now accounts for about two-thirds of the 'power reactors' now in operation. The heat exchanger system ensures that all radioactive water is confined inside the containment building. The pressure vessel around the core acts as a radiation shield. The containment structure around the pressure vessel and heat exchanger also serves as a radiation shield. During normal operation the radiation level outside the plant is so low that it poses no concern to human health or safety. The containment structure is usually designed to withstand more than the most severe weather or other natural conditions (such as floods or earthquakes) ever anticipated in the region. Indeed, some people have suggested quite seriously that in the event of a weather emergency, people should go *to* the reactor, since it is probably by far the safest building around.

However, some low-level radioactive material is periodically released to the environment, for example through vents or in cooling water, during routine operation. In the United States, the Nuclear Regulatory Commission monitors such radiation releases and sets limits both on worker exposure inside the plants and on releases to the environment. From the design stage onward, nuclear plants must prove that they are operating within these boundaries. Nuclear plant worker exposures and radiation releases are guided by the ALARA principle, which stands for '*as low as reasonably achievable*.' Plants are nominally limited to no more than five curies of radioactivity in vented liquids and gases each year. Individuals outside the plant may be exposed to three to five millirems of radioactivity annually. This dosage for persons outside the plant represents no more than 1% of the 500 mrem allowable background radiation.

Of greatest concern for most people is the prospect that a nuclear reactor could release substantial amounts of radiation to the environment in the event of a serious accident. This is amply demonstrated by the two most famous (or notorious) nuclear reactor accidents, Three Mile Island in the United States, and Chernobyl in the Ukraine (in the former Soviet Union). In both cases, human error played a significant role, either in triggering the incident or in making it worse once it had begun. (In a sense, it's unfortunate that nuclear plants have to be run and operated by human beings.) We'll examine these accidents in some detail later in this chapter.

Unquestionably, the most feared nuclear accident is a **meltdown**. In such an accident, the core overheats, most likely because of a loss of the coolant, causing the fuel to melt. Molten fuel could conceivably burn its way through the steel pressure vessel and then through the containment structure, contaminating the surrounding environment. In the early days of the nuclear industry, it was thought that the molten fuel could continue to burn its way deep into the Earth, possibly even coming out the other side. That quite erroneous belief led to the term **China syndrome** to describe this fearsome event. (An old, and quite incorrect, notion is that a hole dug straight through the ground in the United States would eventually come out in China.) Currently it is thought that the molten fuel would only (!) penetrate about 10 feet into the ground below the plant. In the event of a meltdown that breached both the pressure vessel and the containment structure, the molten core would continue to melt its way downward until the rate at which heat released by the core equals to the rate at which the heat can be conducted away from the molten rock around the core. It turns out that these rates become equal at a depth of about 10 feet. By that time, a volume of soil would have melted and congealed into a glass that might, it would be hoped, contain the remains of the fuel. However, the radiation that escaped during this event could certainly contaminate a huge area around the plant, especially if it happened to get into the surrounding groundwater supplies.

Several small, though by no means inconsequential, reactor accidents occurred prior to the now-notorious events at Three Mile Island and Chernobyl. There was a release in 1957 from the Windscale plant (now called Sellafield) north of Liverpool, England. Windscale was a graphite-moderated reactor used to produce plutonium for nuclear weapons. Because of human error in a routine maintenance procedure, the graphite moderator overheated and caught fire. Although the reactor was flooded with water and the fire was eventually contained—after four days—a large amount of radiation escaped, contaminating 200 square miles of the surrounding countryside. More than a half-million gallons of milk contaminated with radioactive iodine were thrown away (into nearby rivers or the Atlantic Ocean) in the weeks following the accident. The Windscale accident released less than 1% (actually, about 1/150th) of the amount of radiation released at Chernobyl, yet the incidence of leukemia in the area is now approximately 10 times the national average.

Some time in late 1957 or early 1958 an accident occurred in the former Soviet Union, at the Chelyabinsk-40 facility in the Ural Mountains. A tank holding radioactive gases exploded, contaminating thousands of square miles around the facility. Little is known in the West about this accident, which is believed to have taken place at a plutonium processing plant. Since the thugs who ruled the Soviet Union did not exactly encourage the open exchange of information, it was not until some 30 years later, in 1988, that Soviet officials finally acknowledged that an accident had even occurred. The region around Chelyabinsk is now sealed off, and the names of about 30 towns in the area simply disappeared from Soviet-era maps. In the bizarre fantasyland of Soviet officialdom, nothing could have happened to the people in those towns, because there never were such towns.

On March 22, 1975, at the Brown's Ferry nuclear plant near Decatur, Alabama, a maintenance worker used a candle—against regulations—to test for air leaks around electrical cables. Yet another example of human stupidity started a fire that spread to the reactor. Several of the core cooling systems were shut down by this accident. By great good fortune, a supplemental pump provided just enough coolant to the core to prevent a meltdown from happening.

A final concern about reactor safety is the production of radioactive waste. This is a significant issue that we will now consider in detail, before moving on to an autopsy of Three Mile Island and Chernobyl.

RADIOACTIVE WASTE

As the nuclear processes in a reactor generate ENERGY, a wide range of radioactive nuclei are formed. The radioactive fission products formed during operation of the reactor fall into three general categories. The first category is fission products which themselves are radioactive. Most of these unstable products emit β or γ radiation; collectively they account for about three-fourths of the radioactivity of the spent fuel. Fortunately, the half-lives of most of these products are relatively short, usually no more than a few years. A second category of product is the so-called heavy isotopes, meaning elements of atomic numbers above that of uranium (i.e., >92). These products belong to the family of elements called **actinides**. They are produced from nuclear processes, but not fission reactions. Most of these heavy isotopes emit α or γ radiation. Their half-lives tend to be longer than those of the fission products, ranging from some tens of years to hundreds of thousands of years. They account for the remainder of the radioactivity (roughly one-fourth) in the spent fuel rods. These two categories of products remain inside the fuel rods. Finally, as the reactor operates, there is an intense flux of neutrons in the core. Not all of these neutrons interact with the fuel. Neutrons also bombard the materials in the fuel rod cladding and the internal structures in the pressure vessel. Those nuclear reactions also produce new isotopes, some of which are radioactive.

During the course of operation of a nuclear reactor, the amount of fissionable $_{92}U^{235}$ in a fuel rod gradually decreases, because it is being consumed in the fission reactions to liberate energy. In BWRs and PWRs used in commercial power plants, fuel rods have a life expectancy of three to four years. Eventually a point is reached at which the amount remaining will not be sufficient to sustain a chain reaction. When reactor fuel is depleted, it must be removed and replaced. Then the fuel rod is withdrawn from the reactor and replaced with a fresh one. Routine plant maintenance includes shutting down the reactor, removing old fuel assemblies, and replacing them with fresh fuel. Each year, one-third to one-fourth of the fuel rods are removed and replaced.

Some fissionable $_{92}U^{235}$ still remains in the rods, and it could be recovered and recycled to the reactor. Many of the other products in the categories mentioned above cause significant concern. One of the actinides, plutonium ($_{94}Pu^{239}$), is fissionable and can be used to make atomic weapons. It is made inside the reactor by the process

$$_{92}U^{238} + _{0}n^{1} \rightarrow {}_{92}U^{239} \rightarrow {}_{93}Np^{239} + {}_{-1}\beta^{0} \rightarrow {}_{94}Pu^{239} + {}_{-1}\beta^{0}$$

A variety of the other radioactive isotopes are useful in technology or in medicine, an example being the radioactive cobalt that is used in radiation therapy for cancer. And, some radioactive material is produced for which there is no use.

Like $_{92}U^{235}$, the isotope $_{94}Pu^{239}$ is fissionable. Therefore, it can be used as an ingredient in atomic weapons. It could, of course, also be used as the fuel in a nuclear fission reactor, but nowadays no serious consideration seems to be given to that use, due to the potential of the plutonium falling into the hands of criminal or terrorist organizations. Plutonium was the active ingredient in the atomic bomb dropped on Nagasaki, Japan, in August, 1945. Plutonium is so dangerous to human health that the estimated amount that can be tolerated in the body without harm is about 0.1 µg, or 0.000000004 oz.

The 'spent' fuel rod is intensely radioactive. When it is removed from a reactor core, the spent rod still emits millions of rems of radiation. Normally these rods are stored at the reactor site until the most intense radiation has decayed. Nuclear power plants generally store their spent fuel rods on-site in lead-lined concrete pools of water. These pools keep the rods relatively cool and contain the γ radiation. An average commercial power plant would put about 60 spent fuel assemblies into temporary storage each year. In fact, until all the issues surrounding permanent disposal of radioactive waste are resolved, the wastes are being temporarily stored until some decision is eventually reached on permanent disposal.

Nuclear waste is categorized as being high-level or low-level material. Both kinds must eventually be dealt with. The high-level waste consists of spent fuel rods from both commercial power plants and military weapons production. High-level waste derives its name from the fact that it can emit large amounts of radiation for hundreds of thousands of years. The commercial nuclear power plants in the United States produce about 3,000 tons of high-level waste each year.

Low-level wastes come from a much wider range of sources. This type of waste includes the radioactive materials produced in a power plant other than the spent fuel rods themselves; as an example, the radioactive water is classified as a low-level waste. These low-level wastes also include radioactive isotopes used in a range of medical, industrial, and scientific research processes. Mill tailings (discussed below), a waste from the processing of uranium ore, are also classified as low-level waste. Wastes in this category usually release smaller amounts of radiation for a shorter amount of time than do the high-level wastes. However, the term 'low-level' does *not* mean that these wastes are not potentially dangerous. Though the radioactivity is less than that of high-level waste, low-level waste can still be radioactive—and dangerous—for tens of thousands of years.

Slightly more than half (about 54%) of low-level wastes come from nuclear reactors. This low-level waste can be further divided into fuel and nonfuel categories. The fuel wastes are products of fission reactions that leak out of fuel rods and into the surrounding cooling water. The nonfuel wastes are produced as a result of interactions of neutrons with anything in the core other than the fuel itself. The rest of the material, comprising low-level wastes, comes from medical, industrial, or research facilities. These wastes are often kept on-site for the short time (usually a few days or weeks) needed for them to decay to safe levels. In some cases, they are then sent to landfills for disposal.

Mill tailings are the material that is left over when uranium ore is processed. Only about 1% of uranium ore is actually uranium. After processing, the remaining 99% is left on-site as a residue, the so-called tailings. These tailings are left outdoors in piles. The action of wind and rainwater blows radioactive thorium, radium, and radon into the surrounding air and leaches the thorium and radium into water. Ten years ago there were over 140 million tons of mill tailings that had accumulated in the United States. Ten to fifteen million tons more are added each year. When measured by volume, mill tailings constitute largest source of any form of radioactive waste. Mill tailings were sometimes used as foundation and building materials, especially in western states, until the time came when someone realized that this material is radioactive, and therefore that exposure to it brings a risk.

In the early years of the nuclear power industry, it was believed that spent fuel could be reprocessed to reenrich the uranium content and extract the plutonium. As we've seen, the spent fuel is still highly radioactive. It presents a serious disposal problem. The spent fuel rods can be subjected to a process called spent fuel reprocessing. This process extracts the plutonium, the fissionable U^{235}, and useful isotopes (such as the radioactive cobalt). There are several serious concerns about nuclear fuel reprocessing. An accident during transportation of the spent fuel to the reprocessing plant could release radioactivity to the environment. Criminals or terrorists could steal plutonium or U^{235}. These stolen materials could be used to fabricate crude nuclear weapons. Consequently, reprocessing requires extreme security measures to prevent the theft of weapons-grade plutonium. There is still some unusable radioactive waste that requires some kind of disposal. Nevertheless, the remaining waste would be both smaller in volume and dangerously radioactive for a much shorter time period—10,000 rather than 240,000 years—relative to the untreated spent fuel rods.

The question of nuclear weapons proliferation centers around the production of $_{94}Pu^{239}$ and the handling of 'enriched' $_{92}U^{235}$. Both of these isotopes are fissionable. Therefore, they are, in principle, capable of being used in atomic weapons. *Any* nation that has a nuclear energy program based on uranium-fueled fission reactors inevitably produces these isotopes. So far, outside of the former Communist nations, four countries have nuclear fuel reprocessing capabilities—France, Britain, India, and Japan. (Although the United States has the technology, it does not presently reprocess nuclear fuel.) An argument raised by other countries is that their energy and economic development is restricted by the near-monopoly on nuclear fuel reprocessing held by these four nations. This is perceived to be unfair, especially by Third World nations who seek to enhance their gross domestic product and overall standard of living by relying on energy from nuclear reactors. The counterargument is that the more reprocessing plants there are—especially if they are built in countries that are politically unstable—the greater is the risk of despotic governments or terrorists obtaining weapons-grade material.

The concern about terrorists relates in part to the fact that plutonium is, on a weight basis, perhaps the most poisonous and deadly substance known. A nuclear bomb operates by having a charge of conventional explosives blast two portions of the fissionable material together to create a supercritical mass. In a crude device made by terrorists, even if the plutonium does not become supercritical and set off a nuclear explosion, the deadly, poisonous plutonium would be blown over a very large area by the conventional explosive, causing health effects from the radiation and making buildings and land uninhabitable. Further, it takes only a small amount of plutonium to make an atomic bomb, much less than the amount of U^{235} needed for a bomb based on uranium.

In the early days of the nuclear industry, the difficulty of waste disposal was not considered a deterrent to commercial electricity generation. It was assumed that somehow, someone would come up with a good idea to allow waste to be reprocessed or buried. From the beginning of the nuclear era in 1940s, and continuing through the 1960s, barrels of radioactive waste (Figure 28.3) were frequently dumped into the oceans.

The practice was halted in 1970; by 1980 at least one-fourth of these barrels were leaking, and therefore slowly discharging radioactive material into the oceans. The disposal of waste on land is extremely controversial, and many concerns have been raised about this idea. Transportation of the waste from the reprocessing plant to any burial site is an issue, because of the potential for accidents that would release radioactivity.

FIGURE 28.3 Some kinds of radioactive waste are simply held in metal drums until some kind of treatment or storage becomes available.

Unfortunately, the deceptively simple problem of finding safe ways to store high-level radioactive wastes has proved more difficult than anticipated. In 1983, the National Academy of Sciences estimated that it will take three million years for radioactive waste to decay to background levels.

> *We nuclear people have made a Faustian bargain with society. On the one hand, we offer ... an inexhaustible source of energy. ... But the price we demand of society for this magical energy source is both a vigilance and a longevity of social institutions that we are quite unaccustomed to.*
>
> —Weinberg[1]

Currently the basic concept for handling the radioactive waste is to bury it deeply inside the earth, in very deep geological formations that should be stable for at least 10,000 years. One technique for packaging high-level wastes involves melting them with glass and pouring the molten material into impermeable containers. For safe, secure, long-term storage of high-level waste these containers must be waterproof and leak proof. The kind of packaging used has to be tailored to the volume of the waste, the isotopes it contains, how radioactive it is, its isotopes' half-lives, and how much heat it generates. These containers then must be put into a repository that is geologically stable; an earthquake could split open the cavern in which the waste is stored, and release radioactivity to the environment.

Groundwater naturally seeping through the earth could dissolve some radioactive substances, if the containers split open or if the groundwater eventually corrodes its way into the interior from the outside. Dissolved radioactive substances could potentially get into public water supplies. In that case, the radioactive species could then be absorbed by vegetation or ingested by marine and animal life. Eventually humans could ingest these radioactive materials with drinking water and food.

In the United States, the 1982 Nuclear Waste Policy Act established a plan for the first permanent high-level commercial nuclear waste storage repository. Five years later, a site was chosen at Yucca Mountain, Nevada, about 100 miles northwest of Las Vegas. It is not yet in use, as a result of numerous political, regulatory, and legal hassles. While many people involved with the project believe the site to be dry, so that groundwater poses no risk, others disagree. The waste disposal issue remains controversial, due in part to the NIMBY syndrome. (NIMBY stands for *not in my back yard!*). Because of the political, social, and technical problems associated with radioactive waste disposal, no one wants it stored near them.

As one example of the knotty problems associated with long-term storage of nuclear waste, consider the issue of how to create warning signs that people will still be able to read several thousand years from now. Think of the problem this way: Readers will find some unfamiliar words or passages in the novels of Dickens, written about 150 years ago. Many words or phrases in Shakespeare's plays, about 400 years old, sound strange to us now. Chaucer's works, written some 700 years ago, are extremely difficult to read in Chaucer's original English. *Beowulf*, about 1,200 years old, is unreadable in the original by anyone other than specialists knowledgeable in medieval English. How much will our early 21st-century version of English have evolved by the year 3200?

The creation of huge quantities of long-lived radioactive waste is one of the most formidable problems facing the nuclear power industry today. The other is the public perception of nuclear power plants as being very unsafe, due to the possibility of serious accidents releasing radiation into the environment. Public concern was first stimulated by an accident at the Three Mile Island plant near Harrisburg, Pennsylvania. This concern was heightened greatly by the tragedy at the Chernobyl plant, in the former Soviet Union (now the Ukraine). We will now examine what went wrong in these two events.

THE THREE MILE ISLAND ACCIDENT

One of the two most famous (but by no means only!) nuclear reactor accidents in history occurred at the Three Mile Island (TMI) reactor near Harrisburg, Pennsylvania. The incident began on March 28, 1979. Certainly the TMI incident was the worst commercial nuclear reactor accident in the United States. The clean-up, waste storage, entombment, and decommissioning is expected to cost at least two billion dollars. It was not until 1990, 11 years after the accident, that most of this work had been completed.

The Three Mile Island reactors were water-moderated PWRs. Unit 2 had been in operation for only about four months when the accident occurred. The problem began when a cooling system water pump failed. Good engineering design principles always involve redundancy—having extra or back-up units of critical components of a system. There were three back-up pumps to feed water to the reactor if the main pump stopped working. Normally these pumps would swing into operation and easily remedy the problem. But, two weeks before the accident, someone had shut off the water supply to the back-up pumps. Two valves, shutting off water to these pumps, were mistakenly left closed after a routine maintenance check. With no cooling water flowing, the control systems in the plant automatically stopped the turbine, which, in turn, halted the generation of electricity. However, the reactor itself was still at its full level of energy production. Temperatures in the cooling water circuit quickly rose.

Since heat removal was inadequate with the cooling water flow, the temperature rose, increasing the pressure inside the reactor vessel. An automatic reactor shutdown, scramming, occurred because of the high pressure. To this point, the various control systems were doing their jobs appropriately, trying to effect an orderly shutdown of the turbine and reactor.

With water no longer coming into the core, the temperature in the core went up. As temperature goes up, so does the pressure. But, the operators misread the pressure gauges. They were not aware that the pressure was excessively high. As the pressure continued to increase, eventually something had to 'give.' When the pressure rose, unnoticed by the operators, past 2,250 psi, an automatic pressure-relief valve opened, exactly as it was designed to do, to reduce the pressure. Indeed the pressure fell back to the value designed for normal operation of the reactor. But now the pressure-relief valve failed to close as the pressure dropped past the normal operating pressure (2,200 psi). Because of that failure, the pressure inside the reactor continued to drop, well below design level. When the pressure hit 1,600 psi, the high-pressure coolant injection system, part of the emergency core cooling system (ECCS), automatically turned on.

The plant operators were relying on a back-up cooling system to keep the reactor core covered with water. Assuming that the back-up system was adequate, they turned off most of the emergency coolant pumps, in part because the water level gauges incorrectly indicated that the water level in the reactor was OK. What the operators did not know was that their backup system was turned off as well.

The operators' action of turning off the ECCS was done because a water indicator gave a faulty reading that the water level was high enough. This act, which was based on erroneous information, guaranteed that the core would be uncovered. At this point, there was no water being supplied to the core from any source. What little coolant was left in the system turned into steam, partially exposing the reactor core. Since the water level in the reactor was not adequate to cover the core completely, the now-exposed core began to heat up.

As the temperature continued to rise in the core, the zirconium metal cladding of the fuel rods began to rupture. So much heat was generated that the zirconium, a metal deliberately chosen to withstand extremely high temperatures, began to burn. Molten, burning zirconium began to run down the fuel assemblies. Molten zirconium oxide, ZrO_2, and UO_2 from the fuel pellets began flowing to the bottom of the reactor.

At the worst point of the accident, 15–30% of the core was uncovered and apparently remained so for about 14 hours. Temperatures within the reactor core shot up; fuel rods ruptured, and meltdown began. An estimated 45% of the core melted and as much as 70% was damaged. The fuel assemblies had burst open, and, along with a few hundred tons of radioactive rubble, the vessel also contained 20 tons—more than half the core—of melted uranium fuel, which had cooled and hardened. For uranium fuel to melt, temperatures of at least 5,100°F must be reached.

Remarkably, the pressure vessel managed to contain the molten material, although theoretically it should not have been able to do so. Indeed, it was found later that the molten fuel had almost completely burned through the eight-inch-thick steel reactor vessel. With erroneous information displayed on the control system and reactor safety systems unwittingly disabled, the plant operators had no idea of what was really happening. Six hours after the accident started, a representative of the Nuclear Regulatory Commission became convinced that the reactor core had begun meltdown. At about the same time, plant engineers also deduced that the core was uncovered. The engineers convinced operators to increase coolant flow. A total meltdown was prevented only by a last-minute rush of cooling water.

Eventually the operators got the water to the core turned back on. This sudden injection of cold water caused some of the extremely hot fuel assemblies to fragment, just as we sometimes accidentally shatter a glass that's hot by putting ice or a cold beverage into it. Also, extremely hot zirconium will react chemically with water:

$$Zr + 2H_2O \rightarrow ZrO_2 + 2H_2$$

This reaction produced a bubble of hydrogen inside the reactor. But when mixed with air (specifically, the oxygen in air) hydrogen can, under the 'wrong' circumstances, blow up:

$$2H_2 + O_2 \rightarrow 2H_2O$$

As workers strove to bring the situation under control, there were fears that the hydrogen bubble could wreak havoc by exploding. (It didn't.)

Recall that one of the contributing factors to the Three Mile Island accident was that the pressure-relief valve failed to close properly after fulfilling its original design function of reducing the too-high pressure in the reactor. The water flooding into the reactor was able to exit via this valve that had stuck open. So much water came pouring out that it ruptured a special holding tank that had been designed to contain it. So, at the end of the incident, the core had been totally ruined and there were 400,000 gallons of radioactive water sitting on the floor of the containment structure. Radioactive gas had also flooded a water supply tank. In the process of transferring this gas to another tank, some of it was released through venting to the air and was carried toward nearby towns by the wind.

The governor of Pennsylvania at the time, Richard Thornburgh, decided to evacuate young children and pregnant women from a five-mile radius surrounding the plant. It is estimated that more than 150,000 residents voluntarily fled the area.

THE CHERNOBYL ACCIDENT

Many of us occasionally use the expression, 'Well, it seemed like a good idea at the time.' That common expression summarizes the events that initiated the Chernobyl accident. Unlike the case at TMI, where there were mechanical problems and failures contributing to that accident, the Chernobyl accident occurred mainly because of human error. The operators committed at least six serious violations of the operations protocol, including, as we will see, the disabling all the reactor's safety systems.

All power plants—nuclear, fossil fuel, or hydro—require some electricity for their operation and safety. For example, the pumps, control valves, and other equipment usually operate with electricity. Electricity is also needed for operating the various computer-based control and data acquisition systems. The usual way of ensuring that the plant will continue to operate safely is to bring the electricity needed for its operation from some other power plant. That way, if there is a problem at Plant #1, say, the electricity needed to operate control systems, pumps, and other equipment is still available, because it's coming from Plant #2. What if the power from Plant #2 should fail? The conditions that led to the accident arose because of the operators' concern about what would happen if there were a failure in the electrical supply coming into the plant from elsewhere. To be prepared for that situation, most power plants have back-up generators, usually operated by diesel engines, to supply emergency power.

Diesel-engine generators are reliable, rugged machines. Their only fault is that it takes 15–60 seconds to reach full output. The question then becomes how to obtain the necessary back-up electricity in that tiny 15–60 second window of time, between the instant of the power failure and the diesel coming up to full output. A possible solution comes from watching what happens when we turn off an electric fan. When we turn the switch to 'off,' the fan does not stop dead; rather it will continue to turn for a short time as it coasts to a stop. In the same way, when a power plant shuts down, the turbine and generator do not stop instantly, but they, like the fan, coast to a stop. This suggested that, for those critical seconds before the diesel generator is up to full capacity, it might be possible to continue to draw some power from the turbine and generator as they 'wind down.' The engineers at the Chernobyl plant designed a test to see how far the plant could be turned down, while retaining the ability to restart the reactor. This is what 'seemed like a good idea at the time.' Unfortunately, the way the test was carried out was an absolute disaster, stemming from one bad decision after another. Were it not for the appalling scale of human tragedy that ensued, the Chernobyl incident could almost have been the basis for comedy.

The Chernobyl reactor complex is located on the Pripet River about 80 miles from Kiev, which was, at the time, a city in the Soviet Union and is now the capital of Ukraine. The accident began late on the night of April 25, 1986; a time line is given in Figure 28.4. The operators undertook a test of one of the turbines connected to Unit 4, a boiling-water, graphite-moderated reactor. It is ironic that this reactor, now infamous for being the site of the worst known civilian nuclear disaster, had actually performed flawlessly since going on-line in 1983. Furthermore, the method of using the residual kinetic energy of the turbines to supply electricity to the vital pumps, computers, and other components had already been successfully tested at various power stations, including earlier tests at Chernobyl.

Initially, the plant was brought down to half power (that is, to a point of generating only half the normal electrical output) of 1600 MW over a 12-hour period. This took until 1 o'clock the following afternoon. At 1:05, one of the turbines was switched off. At this point the shutdown was stopped because the electricity being produced from the reactor, even at half power, was needed in the Soviet distribution system. Stopping a shutdown in this manner is a violation of experimental and operating protocol. At 11:10 P.M. the shutdown resumed. When the test was resumed, the operators shut off the local automatic control system. This was done because the reactor, at such low power output, would be close to the level at which the automatic control system would shut the reactor down. If this automatic shutdown occurred, it would abort the test. A year's delay would occur before the test could be tried again.

One of the serious problems that contributed to the accident was this temporary stoppage of the experimental test to continue to produce electricity. Doing that had allowed xenon (a fission product) to build up inside the reactor. This build-up of xenon caused a problem because xenon is an excellent absorber of neutrons, so the presence of xenon in the reactor core was as if there were extra control rods in place. Normally the Chernobyl plant operated with a generating capacity of 3,200 MW. We've seen that the operators had, at the start of the test, brought the plant down to 1,600 MW, half power. The plans for the test called for an attempt to restart the reactor when the power had been brought down to 700–1,000 MW. But, with the excellent neutron absorber xenon abundantly present in the core, when the operators shut off the local automatic control system the actual power output dropped to 30 MW.

At this point, being desperate to restart the reactor, the operators pulled almost all of the manually operated control rods out of the reactor. In fact, they were pulled out beyond the physical limits that would allow for their easy reinsertion. Six of the reactor's eight water pumps were running to feed water into the core. Because only six of the eight pumps

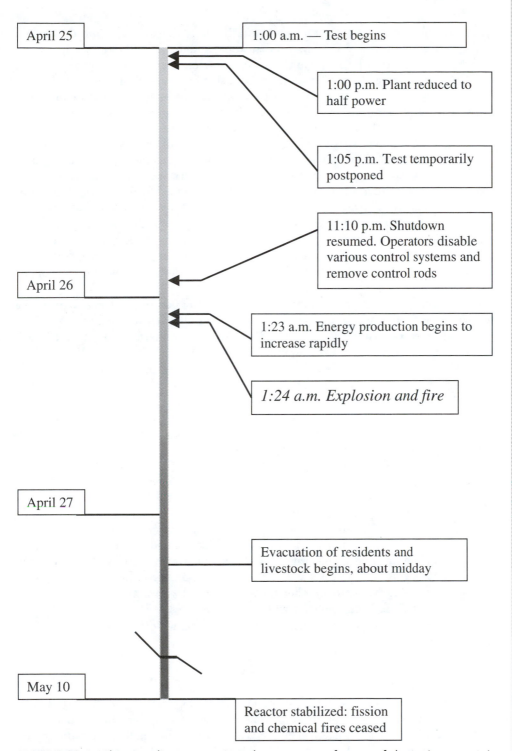

FIGURE 28.4 This time line summarizes the sequence of some of the major events in the Chernobyl reactor accident.

were running, this caused some problems with the stable flow of water into the reactor. The operators first turned off the system that would automatically shut down the reactor in case of some upset or interruption in the flow of water. Then, because there were some problems maintaining a stable flow of water with six of a total eight pumps running, the operators started up the remaining two pumps. The two additional pumps that were started caused the flow rate to jump and the reactor steam pressure to drop toward the emergency trip level. This was a violation of operating instructions.

Eight pumps were now supplying water to the reactor. In normal operation, the usual number is four. The additional supply of water meant that the core was getting cold, and steam generation had almost stopped. Because the reactor's general control system recognized the abnormally low steam generation, it responded by withdrawing the automatic control rods. At 1:23 A.M. the operators also turned off the system that closes the control valves in an emergency. This was yet another violation of all operating procedures. They also ignored the computer command for total shutdown of the reactor. All of these deliberate disablings of control systems were intended to prevent an automatic shutdown of the reactor that would abort the test, forcing a delay of about a year before it could be tried again.

In an effort to let the core temperature rise, the water flow, which had been twice the normal amount (that is, eight pumps running instead of four), was reduced to 75% of normal (three pumps running). As it should, the temperature started rising, boiling began again in the core, and the power output rose to 530 MW.

The All-Soviet regulations for operation of a nuclear reactor required a minimum of 30 control rods to be in the core under any and all circumstances. In principle, even if Mikhail Gorbachev, leader of the Soviet Union at the time, had been standing in the control room himself, he did not have the authority to order or allow removal of more control rods. The operators, however, left only six to eight control rods in the core, much fewer than the legally permissible 30.

By now all control rods were out of the core, and all the automatic circuits that would 'trip' and shut down the reactor if some design parameter were exceeded had been shut off. In another 2.5 seconds, the power output surged to 3,840 MW. (Recall that the plant was designed for a maximum production of 3,200 MW.) In only 1.5 seconds more, the power skyrocketed to an incredible 384,000 MW. The rate at which heat was being generated was much, much faster than the rate at which it could be transferred out of the core. Fuel rods disintegrated and evaporated the cooling water. All of the liquid water in the reactor was converted to gaseous steam virtually instantaneously. This is a phenomenon called **flashing**. The extremely hot zirconium reacted with water to produce gaseous hydrogen, as at Three Mile Island. Recall that the Chernobyl reactor used graphite as the moderator. Graphite (which is pure carbon) can also react with steam to make hydrogen and gaseous carbon monoxide:

$$C + H_2O \rightarrow CO + H_2$$

The combined result of these three factors was that the pressure inside the reactor went sky-high. Presumably, the high pressure broke the seals around some of the many pipes that penetrated the pressure vessel, allowing oxygen inside. This led to immediate ignition of the hydrogen at the high temperature inside the core. The lid of the pressure vessel, which weighed 2,200 *tons*, was blown off. The hydrogen ignited and exploded. The remaining graphite ignited to cause a chemical (rather than nuclear) fire:

$$C + O_2 \rightarrow CO_2$$

and caused flames to carry the radioactive reactor contents up and out of the pressure vessel. Since there was no containment building, they were immediately released to the atmosphere. Flaming debris from the reactor showered over the complex's buildings, setting them on fire. As cooling water reacted with the graphite core, hydrogen gas built up, causing a second explosion that shot radioactive material one mile up into the sky.

At 1:24 A.M. there was a loud explosion, followed seconds later by a fireball and two more explosions. The explosions pulverized fuel rods, and the rising plume from the blasts carried the debris upwards at least 1,500 feet. This tremendous plume was probably responsible for protecting the region right around the reactor from immediate contamination. Evidence of severe irradiation of firefighters and workers did not appear until the next day in most cases. Flames rose at least a hundred feet into the air when the graphite moderator caught fire.

The resulting fires raged throughout the whole reactor complex. The high flames carried most radiation upward, so that relatively few deaths occurred (all 31 of the dead were plant workers or firefighters). The fire itself was put out by about five o'clock that morning, although it took 10 days to control the smoldering core. Several thousand tons of radiation-absorbing boron and lead were dropped into the exposed reactor from helicopters. Nitrogen was pumped under the exposed reactor vessel to cool it, and a huge concrete slab was placed below the reactor to keep molten fuel from burning through. By the end of 1986, the plant was completely entombed in a steel-and-concrete reinforced radiation containment structure.

Local residents were not evacuated until over 36 hours after the start of the accident. Within two days of the accident, 50,000 people had been evacuated from the town of Pripet. Eventually more than 100,000 people and nearly as many cattle were removed from nearby villages and towns and permanently resettled elsewhere.

So great was the release of radiation from Chernobyl that the first inkling Western nations had that something was wrong was when radiation detectors in Sweden began responding, three days after the initial accident. The level of radiation detected at the Forsmark nuclear reactor outside Stockholm was so high that the initial thought was that there must have been a release of radiation from a Swedish reactor. Other Scandinavian countries were showing similarly high readings. Winds were coming from the direction of the Soviet Union, so there was some suspicion that a disaster may have occurred there, but no information was available from the Soviet government. Later that day, an official announcement of the accident was broadcast on Soviet television.

Debris from the accident was carried so far because it was carried upward by the hot gas plume from the graphite fire. The accident released about 2.5% of the radioactivity in the core into the environment. That makes it the largest nonweapon release of radiation publicly known. For at least a month after the accident, all the buildings in Kiev were hosed down daily to remove radioactive particles. Approximately 1,000 square miles of land, the equivalent of the area of Rhode Island, around the reactor complex are permanently contaminated. Financial losses have been estimated at 10 billion dollars, but may eventually prove to be much higher.

Half of the radioactive fallout dropped within 30 miles of the reactor complex. The remainder traveled in a radioactive cloud carried by prevailing winds over parts of the Soviet Union, Scandinavia, continental Europe, Great Britain, and Ireland. Smaller amounts were measured as far away as the United States and the Middle East.

The Chernobyl accident is the most terrible nuclear event (that is, not counting the atomic bombs dropped on Hiroshima and Nagasaki in August 1945) of which we have detailed records. The human suffering caused by radiation exposure continues to this day.

A study conducted by the United States Department of Energy predicted that Chernobyl will ultimately cause 40,000 cases of cancer worldwide, approximately one-third of them in the area of the former Soviet Union, mostly in Ukraine. There are persistent rumors that even more horrific nuclear accidents occurred in the former Soviet Union or in China in the 1950s and 60s, but, if such accidents really occurred, the totalitarian regimes in those countries suppressed dissemination of information about them. Details will not be forthcoming until (or if) their archives are opened to the public. We may never learn the truth.

IS THERE A FUTURE FOR NUCLEAR ENERGY?

Despite some very notorious accidents, it does appear that nuclear power plants—at least PWRs with containment structures—can be operated with good safety records. The Three Mile Island accident, *in which no one was killed or seriously injured*, effectively halted the development of the nuclear power industry in the United States. The Chernobyl accident, with dreadful human suffering and contamination of prime agricultural land, virtually killed any likelihood of a comeback of nuclear power in the United States in the near future. These accidents have certainly fueled mounting public resistance to nuclear energy. Will nuclear energy ever come back? Of course, no one knows for sure. The one factor that is likely to lead to the resurrection of the nuclear power industry is the greenhouse effect (Chapter 32), global climate change that may be due, in part, to emissions of carbon dioxide from the combustion of fossil fuels. Nuclear energy, for all its faults—perceived and real—has the undeniable advantage of producing no carbon dioxide. Therefore, it represents a real option if there is concern for significant cuts in carbon dioxide emissions while maintaining a high level of energy production and consumption.

For that reason, interest is again growing in nuclear energy. New reactor designs are being developed that are 'inherently safe.' Some designs are also based on using thorium as the fuel, to avoid making and handling weapons-grade uranium and plutonium. Though many wrote the nuclear industry off as being dead, especially after the tragic Chernobyl accident, like Mark Twain's comment on reading his own obituary.

Just say the report of my death has been grossly exaggerated.

—Twain[2]

Indeed, the significant problems facing the nuclear energy industry may be more social or political than technical.

The 1950s goal of nuclear-energy-generated electricity, 'too cheap to meter,' has not been achieved. Newer U.S. nuclear plants generate electricity at an average cost of more than 13 cents per kilowatt-hour, more than twice that of coal-burning plants equipped with the latest air pollution control equipment. (This price does not include most of the expenses of radioactive waste disposal and plant decommissioning.) Many feel that the greatest hazard of nuclear power is its contributions to nuclear weapons proliferation. Even commercial power reactors provide the raw materials—plutonium and some uranium isotopes—for nuclear weapons. Perhaps it is this concern, in a politically unstable world that never seems truly to settle down, that may ultimately have a greater effect on the nuclear industry than issues of reactor safety and waste disposal.

At the end of the useful working life of a nuclear plant, it must be decommissioned, or taken out of service. The plant contains some radioactive material in the fuel rods and spent rods. Some radioactive material has likely contaminated the pressure vessel, piping, pumps, or other components because it has leaked out of fuel rods into the reactor during operation. In addition, the high flux of neutrons inside the reactor during the regular operation may have made some parts of the pressure vessel or other internal components radioactive via nuclear reactions that the neutrons induced. As a result, the decommissioning of a nuclear reactor, along with the allied process of decontamination, can be a difficult, expensive, and possibly hazardous operation.

CITATIONS

1   Weinberg, Alvin. 'Social institutions and nuclear energy.' *Science* **1972**, *175*, 27–34.
2   Twain, Mark. 1897. Slight variations have been attributed to Twain by various biographers, including Bernard deVoto and Albert Bigelow Paine. The remark was made in response to a query from a London reporter who told Twain that his death had been reported in New York, and wondered what news he should send in reply.

FOR FURTHER READING

Cartledge, Bryan (Ed.). *Energy and the Environment*. Oxford University: Oxford, 1993. Two chapters are particularly relevant here: 'The legacy of Chernobyl: the prospects for energy in the former Soviet Union,' by Zhores Medvedev; and 'Nuclear power and the environment,' by Walter Marshall.

Cassedy, Edward S.; Grossman, Peter Z. *Introduction to Energy*. Cambridge University: Cambridge, 1998. Chapter 8, on the nuclear fuel cycle, discusses issues relating to nuclear waste and nuclear weapons proliferation.

Fowler, John M. *Energy and the Environment*. McGraw-Hill: New York, 1984. Chapter 11, 'The mixed blessings of nuclear energy,' relates to such issues as environmental effects of nuclear energy, human health and safety, and nuclear weapons proliferation.

Galperin, Anne L. *Nuclear Energy/Nuclear Waste*. Chelsea House: New York, 1992. The last three chapters of this book discuss the issues of nuclear waste, reactor accidents, and the possible future of nuclear energy.

Häfele, Wolf. 'Energy from nuclear power.' In: *Energy for Planet Earth*. Freeman: New York, 1991; Chapter 9. A discussion of safety, nuclear waste disposal, and nuclear weapons proliferation. The author recommends a worldwide organization to address these issues.

Hill, Robert; O'Keefe, Phil; Snape, Colin. *The Future of Energy Use*. Earthscan: London, 1995; Chapter 6. This book, available as a relatively inexpensive paperback, focuses on fossil fuels, nuclear, and renewables. It discusses how these energy sources are used, environmental and social effects, and economic issues. Little prior technical background is needed.

Hohenemser, Christoph; Goble, Robert L.; Slovic, Paul. 'Nuclear power.' In: *The Energy–Environment Connection*. Hollander, Jack M. (Ed.). Island Press: Washington, 1992. This article treats such topics as reactor safety, risk management, and the accidents at Three Mile Island and Chernobyl.

Josephson, Paul R. *Red Atom*. Freeman: New York, 2000. This book provides a history of the nuclear energy and nuclear weapons programs in the former Soviet Union and present-day Russia. It blends discussions of nuclear physics, science policy, and political decisions, all of which culminated in the tragedy at Chernobyl.

Kalfus, Ken. *Pu-239*. Milkweed: Minneapolis, 1999. The title story in this collection of short stories about modern Russia is a harrowing account of a worker from a Russian nuclear plant attempting to sell stolen plutonium on the black market.

Kraushaar, Jack J.; Ristinen, Robert A. *Energy and Problems of a Technical Society*. Wiley: New York, 1988. Chapters 5 and 11 provide information on nuclear reactor safety, environmental issues relating to nuclear energy, and radiation.

League of Women Voters. *The Nuclear Waste Primer*. Lyons and Burford: New York, 1993. This useful paperback book provides an introduction to, and an overview of, many of the issues surrounding the nuclear waste controversy—environmental, health, political, and social issues. Although some of the numerical data may now be out of date, and some of the political and social issues overtaken by more recent events, this book is still a very useful discussion of the issues. It is easily readable, even for those with little or no scientific background. A good glossary is also included.

Ramage, Janet. *Energy: A Guidebook*. Oxford: Oxford, 1997; Chapter 8. This book, available as a relatively inexpensive paperback, covers many energy sources, including fossil fuels, nuclear, solar, and wind, and describes both present applications and possibilities for the future. Readily accessible to the reader with little or no technical background.

Seaborg, Glenn T. *A Chemist in the White House*. American Chemical Society: Washington, 1998. Glenn Seaborg was involved with most aspects of the development of nuclear science and technology, beginning with his work on the Manhattan Project and his discovery of several new elements (for which he received the Nobel Prize), through involvement with the newly formed Atomic Energy Commission and its successor agencies. Seaborg provided scientific advice to every president from Franklin Roosevelt to Bill Clinton. This book is based on Seaborg's diary and provides an insider's look at the making of nuclear policy.

Wendt, Gerald. *The Atomic Age Opens*. Pocket Books: New York, 1945. This is one of the first books published on the promise of 'atomic energy' when the wartime secrecy was lifted. It is of most interest now as an historical document, indicating what people perceived at the beginning of the nuclear era. Written for readers with little science background.

29

ENERGY AND THE ENVIRONMENT

With this chapter we begin considering the implications of energy use for the environment. Because fossil fuels dominate in our energy economy, we will focus heavily on the environmental consequences of fossil fuel utilization. Virtually all of our transportation energy, about 70% of our electricity, and most of our industrial and domestic heating derive either directly from burning of fossil fuels, or indirectly via electricity produced in fossil fuel power plants.

Broadly, the use of fossil fuels involves three main steps. **Extraction** actually gets the fuel out of the ground. **Beneficiation** or refining (Chapter 21) encompasses the various steps in improving the quality of the fuel before it's used. Finally, **utilization** involves the actual burning of the fuel to liberate useful energy. Each of these steps can impact the environment in various ways.

EXTRACTION OF FUELS—MINES AND WELLS

Those who live in coal-producing states have probably seen the tremendous environmental devastation caused by strip mining. Strip mining involves removal of the topsoil, and any plants that happen to be growing in it, and then removal of subsoil to expose the coal seam. Topsoil and subsoil are heaped into 'spoil' piles that continue to grow as mining progresses. Rain can wash some of the spoil away, polluting whatever bodies of water it runs into.

Inevitably a strip mine site is an eyesore during mining. The temporary disruption of the environment is something we must tolerate while the mine is in active operation. The crucial issue, affecting permanent disruption of the environment, is what happens after the coal has been removed, or the mine ceases to operate. Many states have passed **mined land reclamation** laws that require the mining company to restore the mining area to the condition in which it was before mining began.

Mined land reclamation laws have two major components, recontouring and ferti-lizing and seeding (or replanting). The objective is to restore the land to approximately the same topography it had before mining, and to restore the plant community that existed before mining. Mined land reclamation laws sensibly applied do provide excellent restora-tion of the environment. A superb example is in the region near Cologne, Germany, which is home to one of the world's largest coal strip mines. A visitor could scarcely guess that some of the land is in fact reclaimed mines. Some have argued that, in a sense, the requirement of restoring the land to *exactly* the same condition it was prior to mining is perhaps too limiting. For example, if the mined land had no particular use prior to mining, wouldn't it be a benefit to society to restore the land not to the way it was, but, say, into productive agricultural land?

That issue aside, the real problems occur if there are no mined land reclamation laws or if the mining company somehow avoids compliance. (This can be done via so-called grandfather clauses, so that the mine is exempt from reclamation if it was in operation prior to passage of the law, or by the simple expedient of the mining company's declaring bankruptcy and vanishing.) Unreclaimed strip mines are serious problems for several reasons. The obvious one is that they are ugly eyesores. They continue to pollute the local environment by the washing away of the spoil banks. They are also dangerous. Some partially fill with water and are used as unsupervised swimming holes, occasionally result-ing in accidental drownings. Some fools who erroneously believe that a four-wheel-drive vehicle can be driven literally anywhere like the challenge of off-road driving in abandoned strip mines, only to discover that '4 × 4s' or sport utility vehicles can tip over or go out of control and result in injury or death.

Usually, underground coal mines are not so disruptive of a wide area as are strip mines. Today, underground mining is a remarkably safe occupation, with an annual death rate (about four workers in 10,000) that is actually lower than that for general manufac-turing industries. In recent years, underground coal mining has ranked 25th among industries in terms of number of fatalities per thousand workers. Such was not always the case, however, and the early days of coal mining in the United States—and else-where—are a history written in blood. In the anthracite-mining region of Pennsylvania, the annual death toll of miners in the years around the turn of the 20th century was about 3%—that is, three out of every hundred miners were killed every year. Even today there are potential long-term consequences for miners, most notably black lung disease, which is a result of chronic inhalation of coal dust.⚛

⚛ Black lung disease, also called anthracosis, is a form of pneumoconiosis, which, in general, is a disease caused by the inhalation of irritants; the severity depends on what kind of irritant is inhaled, and how much. Coal miners, particularly those working in underground mines, inhale fine particles of coal dust. These very small particles can penetrate into the air sacs in the lung. There, the surrounding lung tissue is converted into scar tissue in the body's effort to encase these foreign, irritating particles. Over a period of time, the victim's breathing becomes more difficult, and the ability of the lungs to exchange waste carbon dioxide for fresh oxygen is reduced. The victim also has a much higher likelihood of developing chronic bronchitis or emphysema. Those who have spent much of their working lives in underground coal mines seem particularly at risk. The last few, unpleasant years of the victim's life become a struggle to breathe.

Underground mining has other hazards as well. Water percolating through mines can react with sulfur compounds in the coal to produce, through a sequence of reactions, what is essentially a solution of sulfuric acid. This so-called acid mine drainage can be so acidic as to kill all aquatic life forms in a stream. Some streams are so acidic they have been used by nearby residents as a folk medicine to remove warts! Abandoned mines can cave in, causing subsidence to the ground above and seriously damaging buildings.

In contrast, the extraction of petroleum and natural gas usually does not do so much harm to the environment, except for localized disturbances around the well sites. However, dreadful conditions can arise in those cases when companies are not committed—or required—to protect the environment. Consider this from the late days of the Russian Empire:

The oil fields left in my memory a brilliant picture of dark hell. I do not joke. The impression was stunning. … Amidst the chaos of derricks, long low workers' barracks, built of rust-colored and gray stones and resembling very much the dwellings of pre-historic peoples, pressed against the ground. I never saw so much dirt and trash around human habitation, so much broken glass in windows and such miserable poverty in the rooms, which looked like caves. Not a single light in the windows, and around not a piece of earth covered in grass, not a tree, not a shrub.

—Gorky[1]

Fortunately, conditions like that would not be tolerated in the United States or most other industrialized nations nowadays!

Many of the spectacular environmental problems usually blamed on the oil industry are actually associated with transportation of crude oil to refineries. Empty tankers returning to their home port for another load often carry water as a ballast. Some tankers, when switching their cargo from one petroleum product to another, used to flush their tanks with water between loads. In either case, this oily water used to be discharged into the ocean before the ship entered port to be loaded. Fortunately, this practice has now been discontinued. When an oil tanker or supertanker breaks up or starts leaking, the oil is spilled and can cause severe environmental damage. A particularly heinous example was the *Exxon Valdez* running aground in Prince William Sound, Alaska. As much as a century can be required for the local area to recover from such an ecological disaster.

There are numerous negative consequences of oil spills. If crude oil is ingested, it can potentially kill or injure fish and other aquatic animals and aquatic plants. They kill waterfowl, otters, and other animals that are not aquatic but that do periodically enter the water. They contaminate local beaches. Perhaps worse, oil that is transported long distances by ocean currents can contribute to pollution in areas far away from the original spill. Fortunately, however, some experience suggests that the damage, however unpleasant it may be for a short time in a local area, can be healed over the long term. A specific example comes from experience in the Second World War. Submarine warfare resulted in the sinking of many oil tankers, with the spillage of millions of tons of oil into the oceans around the globe. There seems to be no permanent environmental damage attributable to those dreadful times. (However, the typical tanker launched in the 1930s or 1940s carried substantially less oil per ship than today's supertankers.) Indeed, the oceans are far more polluted by plastic packaging wastes or discarded plastic items tossed from ships than from oil spills.

Unfortunately, oil spills sometimes bring with them attempts at clean-up that verge on the bizarre. In 1967, the supertanker *Torrey Canyon* broke up off the coast of England. In an attempt to prevent the oil spill from coming ashore, the British bombed it with napalm, hoping to set it afire and have it burn away. When that didn't work, chemicals were added to the oil to try to disperse it into tiny particles that would mix with the seawater and sink. Unfortunately, these chemicals were lethal to some of the aquatic organisms that had managed, up till then, to survive the oil spill itself. A different tack was taken in response to the *Exxon Valdez* disaster. Very hot water was squirted onto the contaminated shore through high-velocity pumps (Figure 29.1). This approach, which cost about two billion dollars, again killed organisms that had survived the spill itself, and the high-velocity water blasted barnacles and limpets loose from their abodes.⚛

> ⚛ Barnacles are crustaceans, the class of aquatic animals that includes lobsters, crabs, and shrimps. Barnacles cement themselves to all sorts of things in the water—rocks, ships' hulls, driftwood, and even to whales. Barnacles on a ship's hull can be a particular nuisance. The natural cement secreted by barnacles is so strong, and resists water so well, that it has been studied to learn how to make superior dental cements. Limpets are types of marine-living snails; most cling to rocks near the shore.

In the spring of 1997, a bizarre situation developed in which an oil spill nearly caused a nuclear accident. A Russian oil tanker sailing near the coast of Japan spilled oil that the Japanese feared would clog the water intake system in a nuclear power plant built on the coast. Luckily, most of the serious problems were averted.

On land, oil is usually transported in pipelines. The first oil pipeline in the United States was built in 1865, to deliver oil to a railroad loading facility in Oil City, Pennsylvania. That same year also saw the invention of the railroad tank car for transporting oil. For many years, most of the long-distance transportation of oil was done by rail. One of the first long pipelines was built in 1918 from Kansas City to Chicago, some 670 miles. The laborers connected the eight-inch diameter pipes *by hand*. The Second World War spurred the building of pipelines, especially when German submarines sank numer-

FIGURE 29.1 Cleaning up from an oil spill is a hard and difficult job, sometimes itself disruptive to the environment.

ous oil tankers transporting oil from the Gulf of Mexico up the East Coast. There are two concerns with pipelines. One is that they will, necessarily, disrupt some bit of the environment as they are built through agricultural land, forests, or wilderness. Once the pipeline is in operation, the potential exists for oil spills from corrosion of the pipes, leaking joints or valves, or catastrophic accidents that rupture the pipe. The most controversial pipeline project has been the Alaskan pipeline, which transports oil from the North Slope of Alaska to a tanker loading facility in Valdez. It went into service in 1977; proponents argue that the environmental impact has been minor.

REFINING AND BENEFICIATION

The potential environmental problems associated with oil refining include leaks from process equipment or pipes, production of waste materials (such as sulfur compounds removed from sour crudes), and spills or leaks of some of the chemicals used in the refining process. In addition the heat or steam needed for operating refinery processes has to be generated in the refinery by burning some sort of fuel. In some cases, the burning of the fuel to obtain heat for the processes causes more air pollution than do the refinery processes themselves. In the United States, oil refineries operate under very stringent environmental regulations. They are so stringent, in fact, that some people believe that there will never be another oil refinery built in the United States.

The waste materials that can be produced in refineries include such noxious gases as ammonia and hydrogen sulfide, as well as poisonous materials such as cyanides. Usually these materials are produced in small quantities that are contaminants of other process streams. The contaminated streams are collected and purified; the contaminants are removed and destroyed.

The first antipollution legislation dealing with the oil industry in the United States was enacted in 1863, in Pennsylvania. It addressed problems of the wastes from crude oil distillation being dumped into streams. One approach to preventing such pollution was the digging of pits that would impound oil-contaminated water in artificial ponds. Since oil floats on water, devices called skimmers could be used to remove the oil layer from the impounded water.

Coal is not ordinarily 'refined' in the same sense that oil is. However, many coals are 'cleaned' prior to use. All coals contain some amount of incombustible material that got incorporated in the coal as it formed in nature. As coals are mined, they may also become adulterated with some of the surrounding rock material that is unavoidably removed with the coal. These incombustible minerals and rocks transform to ash when the coal is burned. Ash has no calorific value, so its presence in the coal is useless in terms of the production of energy. Worse, the ash needs to be collected and disposed of.

Coal cleaning, or coal beneficiation, reduces the amount of ash-forming material. Because much of the sulfur in many coals is associated with the mineral pyrite, coal cleaning also reduces the sulfur content. A very simple strategy for coal cleaning relies on the difference in density between the coal itself and the rocks or minerals. If crushed coal is placed into a liquid that has a density intermediate between that of the coal and that of the minerals, the coal will float and the minerals will sink (Figure 29.2). The separation is never perfect. The cleaned coal still contains some minerals, and the refuse contains some coal.

The refuse from coal cleaning has many names that often vary from one coal mining region to another: culm, slag, and bony coal are examples. For many years the refuse was

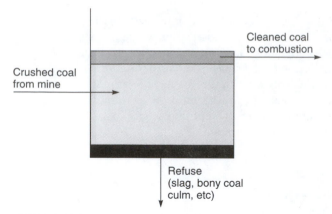

FIGURE 29.2 Most processes for cleaning coal take advantage of differences in physical properties between the coal itself and the rocks or minerals associated with it. In this example, coal from the mine is separated in a liquid medium that is dense enough to allow the coal to float but the minerals to sink.

dumped on the ground, creating ugly miniature mountains that supported very little in the way of plant life. In recent years it has been recognized that these culm banks or slag heaps are potential energy sources, since about 30% of the material is actually coal. Combustion equipment is now coming into use to utilize this unusual form of fuel, generating useful energy and cleaning up the environment at the same time.

The principal technology is fluidized-bed combustion. Air is blown upward through a bed of limestone particles, at a velocity sufficient to suspend the particles in the air stream. This process is called fluidization, because the bed of suspended particles looks and acts much like a fluid. The behavior of the particles in the air stream is roughly similar to demonstrations of vacuum cleaners in stores, in which a ping-pong ball is suspended in mid-air by a stream of air. Coal, culm, slag, and bony coal (or most any other solid fuel) can be burned in a fluidized-bed combustor. The use of limestone as a bed material captures SO_x as it forms. Ash is retained in the bed. The temperature of combustion is much lower than that of a pc-fired boiler, substantially reducing thermal NO_x formation.

FUEL UTILIZATION

Eventually, of course, the fuel is burned. All fossil fuels contain carbon and hydrogen. Petroleum and coal contain some amounts of sulfur, nitrogen, and oxygen. Coal contains some incombustible materials. For our purposes we need not worry about the specific chemical compounds that are present in the fuels. Instead, we can treat the fuel combustion as if the fuel were simply a mixture of the individual chemical elements.

Depending on the combustion conditions, particularly the amount of oxygen available, the carbon in fuel can produce carbon dioxide, carbon monoxide, or soot (Chapter 7). Sulfur is capable of forming more than one oxide (Chapter 18), conveniently lumped together with the symbol SO_x. Similarly, nitrogen can produce more than one oxide; they are lumped together as NO_x. The situation with NO_x is further complicated by the fact that there are actually two sources of nitrogen in a combustion system, fuel NO_x and thermal NO_x (Chapter 18). Sometimes the generic symbol R is used to represent the metallic elements (such as silicon, aluminum, and iron) present in the various constituents of ash. Then the notation RO_x can be taken as a symbol for ash. It's then possible to refer to the potential pollutants as socks, knocks, and rocks.

Ultimately, *all* processes of fuel utilization involve the combustion of the fuel to release chemical potential energy in the form of heat, as shown in the energy diagram of Figure 29.3. We can then use the heat directly for warmth, cooking, or carrying out various manufacturing processes. We can use the heat in a heat engine, or use the heat to raise steam, which we then use in a different kind of heat engine. In all these cases, the objective is to do WORK for ourselves.

Every combustion process is necessarily accompanied by the formation of chemical products of combustion. From the preceding chapters, we now have an extensive catalog of what those products are: CO_2, CO, soot or unburned hydrocarbons, H_2O, SO_x, NO_x, and 'RO_x' (ash). All of these are products of a process (combustion) that we carry out to liberate ENERGY that allows us to do useful WORK (Figure 29.4). For many centuries, the combustion products were either ignored or considered simply to be an inevitable nuisance consequence of fuel utilization. (There are a few notable examples to the contrary, such as complaints about the pollution from coal fires in England that date at least from the 14th century. King Edward I issued a proclamation in 1306 banning the burning of coal under certain circumstances; in a sense, this is the first recorded clean air act, more than 650 years prior to the Clean Air Act in the United States.)

Today there are parts of the world with tremendous pollution problems resulting largely, but not entirely, from energy use. Most of these places are the developing nations or Communist or formerly Communist countries. As examples, the concentrations of SO_2 in the air in Beijing and Shanghai are double the standard established by the World Health Organization. Mexico City suffers from very bad air quality from emissions of soot, lead (from using 'leaded' gasoline), and various unburned hydrocarbon compounds that have evaporated into the air.

No process that produces ENERGY, consumes ENERGY, or transforms ENERGY from one form to another does so with 100% efficiency. Some amount of ENERGY is wasted, and that wasted energy is dumped into the environment, commonly in the form of heat. For example, in a coal-fired electricity generating plant, only about one-third of the chemical potential energy in the coal is ultimately transformed to electrical energy leaving

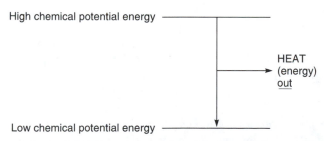

High chemical potential energy

HEAT
(energy)
<u>out</u>

Low chemical potential energy

FIGURE 29.3 The whole reason for using fuel is to take advantage of the spontaneous change from high to low chemical potential energy, to liberate ENERGY in the form of heat.

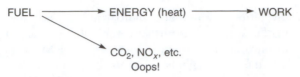

FIGURE 29.4 When fuel is burned to produce heat, some by-products are produced as well. These include carbon dioxide, as well as nitrogen and sulfur oxides.

the plant to the consumers. The other two-thirds is wasted, much of it in the form of heat that leaves the condenser or the cooling tower.

If waste heat is transferred directly to aquatic systems—for example, if the heated water leaving a condenser is put directly back into a nearby lake or river—several serious problems can occur. The ability of gases to dissolve into liquids decreases as the temperature of the liquid increases. (This is why, for example, a warm carbonated beverage is much more likely to foam when it's opened than is a cold one; the carbon dioxide responsible for the carbonation is less soluble in the warm beverage, and comes out of solution when the can or bottle is opened, generating the foam.) Warm water can hold less dissolved oxygen than can colder water. However, the oxygen dissolved in water is essential for reacting with, and helping to destroy or neutralize, any biological or chemical pollutants in the water. If water is warmed by waste heat, it contains less dissolved oxygen and is less able to cleanse itself of other pollutants.

Warm water can also sustain higher rates of aquatic plant growth. The warming of water by waste heat could lead to unwanted amounts of aquatic plants, such as algae. When these algae eventually die, their decomposition uses up even more dissolved oxygen. Many kinds of fish are sensitive to the surrounding water temperature; some prefer relatively cool water, while others are tolerant of warmer water. Heat stress on fishes can affect their resistance to disease and their reproductive cycles.

Even cooling towers, which are intended to dissipate waste heat before it can enter a river or lake, can have impacts on the local environment. Evaporation of water from cooling towers can increase the incidence of fog in the area. It can also change local patterns and frequency of rain or snow.

It has only been in the last half of the 20th century—indeed, in about the last quarter of that century—that we have seriously considered the environmental issues that are associated with those combustion products, most notably acid rain (Chapter 30), vehicular smog (Chapter 31), and the greenhouse effect (Chapter 32). We know, of course that we need ENERGY because we want to do useful WORK. But now we clearly recognize that the production of energy brings with it the inadvertent production of substances that are harmful to the environment, and, because we are inescapably part of the environment, that are also harmful to us. But, we cannot lose sight of the fact that environmental problems associated with energy use are *not* inevitable, and, should they occur, *are* able to be corrected.

We wring our hands over the miscarriages of technology and take its benefactions for granted. We are dismayed by air pollution but not proportionately cheered up by, say, the virtual abolition of poliomyelitis. (Nearly 5,000 cases of poliomyelitis were recorded in England and Wales in 1957. In 1967 there were less than thirty.) There is a tendency, even a perverse willingness, to suppose that the despoliation sometimes produced by technology is an inevitable and irremediable process, a trampling down of nature by the big machine. Of course it is nothing of the kind. The deterioration of the environment produced by technology is a technological problem for which technology has found, is finding, and will continue to find solutions.

—Medawar[2]

ADDRESSING THE ENVIRONMENTAL CHALLENGE

Broadly, there are two strategies we can adopt. The first is to reduce substantially our energy consumption. But since there appears to be some connection between energy consumption and standard of living (Chapter 1), this strategy does not seem likely to be willingly adopted. The second strategy is to develop technological 'fixes' to reduce or remove these potential pollutants before they are released to the environment. The second strategy is the one we have been following. As we will see, the operation of this strategy is governed by one of life's most important rules:

In other words, anything we do has some consequence for the environment, and those consequences have costs associated with them. The costs may be those for the installation and operation of control equipment to prevent pollutants from escaping into the environment; costs for the refining or beneficiation of fuel to prevent pollutants from forming in the first place; or the costs of clean-up and remediation of damage caused by pollutants in the environment. Sometimes those costs are fairly easy to calculate. For example, an expert in coal beneficiation can calculate the additional cost per ton of coal resulting from reducing the amounts of sulfur and ash-forming minerals in the coal. Sometimes, though, the costs are much more difficult to assess. Many health-related issues fall into this category, because it is often hard to attribute a health problem unequivocally to a single factor in the environment, a problem we've already touched on while discussing effects of radiation in the previous chapter. (We should recognize that there is, in a sense, a third strategy as well—simply to do nothing. For a long time in our history this was of course the operant strategy, but it appears now, at least in the United States and many of the other industrialized nations, that this third strategy is no longer socially or politically acceptable.)

THERE'S NO SUCH THING AS A FREE LUNCH.

Virtually all energy technologies produce some amount of pollution. Regardless of what specific device or system is considered—for example, whether a nuclear power plant, an automobile, or an oil refinery—it serves as a source of pollution. That pollution is measured in terms of its emissions, as for example, a number of pounds of SO_x emitted per million Btus of energy generated. Those emissions can ultimately be quantified as a burden on the environment, that might be expressed as parts per million of SO_x in the local air. In turn, the burden has identifiable impacts. Continuing this example, increased concentrations of SO_x in air may lead to human health effects, corrosion of marble and limestone buildings, or crop damage. It should, in principle, be possible to put a cost on each of those impacts. From that information one could assess the cost of environmental impacts of pollution from a particular source.

Unfortunately, we are a long way from being able to do that with much degree of certainty, except in the most egregious cases of emissions causing immediate, local damage. In working backwards, from impacts to burdens to emissions to pollution, there are uncertainties in every link of this chain. First, it is not always clear that a particular impact can be the unequivocal result of one specific environmental burden. Suppose, for example, a person has respiratory tract problems and has been living for a time in a region of relatively high airborne SO_x concentration. Suppose further that that individual is also a heavy smoker. Can it be said with certainty that his or her problems are due to the SO_x? The situation is further complicated by the fact that some impacts might not be immediate; that is, it might be years before an organism, or a whole ecosystem, shows the effects of pollution. A second uncertainty lies in relating the burden to the emissions. Weather patterns may transport pollutants long distances from their sources. Pollutants first emitted to the environment, called primary pollutants, might undergo further chemical reactions in the atmosphere, water, or soil to produce new materials, secondary pollutants. Formation of secondary pollutants in the environment complicates even further the job of tracing the path of pollution back to its origin. Issues like these make it very difficult both to draw a direct line from environmental impact to a specific pollution source, and to quantify and place a value on the costs of that environmental impact.

The reduction of SO_x emissions from stationary sources, mainly power plants, has been mandated by the 1990 Clean Air Act Amendments. Three strategies are mentioned in that legislation: switching to low-sulfur coal (unfortunately, there is no such thing as a truly sulfur-free coal); removing or, more realistically, reducing the sulfur content of the fuel before combustion, or capturing the SO_x after the fuel has burned but before it can be released to the environment. All of these options have real financial costs associated with them. Companies that own deposits of low-sulfur coals are not fools; they know they have a premium commodity and can charge accordingly. Coal cleaning to reduce the sulfur content could add, say, two to three dollars per ton to the cost of the coal. Recalling our example of a power plant that burns 10,000 tons of coal per day, the cost of coal cleaning would represent an extra cost of 20,000 dollars per *day* to the utility, and, ultimately, to us in our electric bills. Scrubbers (Chapter 18) now do an excellent job of SO_x capture. However, they can represent a capital investment roughly equivalent to a third of the cost of the entire power plant. Once purchased and installed, scrubbers represent a continuous cost for lime or limestone, maintenance, and the workers who would operate or maintain them. These costs have to be recouped some-how, and they are often recouped through the medium of the inevitable monthly electric bill.

While there are no coals that are truly sulfur-free, there are other fuels that are either completely sulfur-free or have so low a sulfur content that SO_x emissions from the use of these fuels would be negligible. Examples of such fuels include natural gas, wood, ethanol (which can be produced from, e.g., corn starch), and methanol (currently made from natural gas). In all cases these fuels are more expensive than coal, particularly when compared on a dollars-per-Btu basis. Furthermore, since natural gas is, of course, a gas, and ethanol and methanol are liquids, significant modifications to the combustion equip-ment itself might be required to handle these fuels instead of coal. This retrofitting would also represent a cost to be considered.

The options to sulfur control using coal cleaning or SO_x capture also have 'energy costs' associated with them. Running a coal cleaning plant requires ENERGY that must be purchased from someplace. This ENERGY is then no longer available for other applica-tions, specifically, for the WORK we might wish to do. Similarly, running a scrubber means that we must also use energy from someplace. Most commonly, this energy would come from the power plant itself. Using some of the plant's energy in this way decreases the net efficiency of converting the chemical potential energy in the coal to electrical potential energy leaving the plant to the consumers. Roughly, the energy cost of a flue gas desulfurization (FGD) system reduces the so-called 'coal pile to busbar' efficiency, meaning the conversion of chemical potential energy in the fuel to electrical potential energy leaving the generator, from about 34% to about 30%.

The main fix for SO_x emissions from mobile sources (vehicles) is desulfurization of the fuel before use. This is done in oil refineries by treating the fuel with hydrogen, a process called hydrodesulfurization. For producing clean diesel fuel, hydrodesulfurization adds about three to five cents to the cost of a gallon of fuel.

The problem of dealing with NO_x is compounded by the fact that much of the NO_x produced is actually thermal NO_x, not fuel NO_x. In stationary sources, there are three approaches to NO_x control. The first is postcombustion NO_x capture. This approach is in the development stage. When these devices are perfected for commercial use on power plants, then, just like FGD systems, they will represent an initial capital cost and a continuing cost for operation. The second approach is the so-called low NO_x burner. This, too, is in a promising stage of development. The third, and perhaps most straightforward approach, is simply to reduce the combustion temperature. Unfortunately, this approach can seriously impact the efficiency of a process.

NO_x emissions from mobile sources are primarily addressed by the catalytic converter on the vehicle (Chapter 31). This NO_x is almost entirely thermal NO_x. Nitrogen incorporated in the fuel is removed during refining and, like sulfur removal, by treatment with hydrogen. A catalytic converter adds several hundred dollars to the cost of a car. The second approach is a switch to reformulated gasoline. One of the environmental benefits of reformulated gasoline is that it 'burns cooler' and thereby reduces formation of thermal NO_x. Reformulated gasoline is approximately 10 cents per gallon more expensive than conventional gasoline.

Carbon monoxide and unburned hydrocarbons are generally not problems in stationary sources. The emissions of these substances from mobile sources are generally handled via the catalytic converter, and secondarily through the use of reformulated gasoline. Thus the same issues apply as those that we have just discussed for NO_x. Soot emissions from diesel-powered vehicles are addressed by reducing the concentration of aromatics in the fuel during refining. This, like sulfur and nitrogen removal, is achieved by reacting the fuel with hydrogen, a cost that is about three to five cents per gallon. However, because aromatics are undesirable for good cetane rating of diesel fuel (Chapter 24), the reduction of the aromatic content to avoid sooting gives us, at the same time, the bonus of improved cetane number.

Whether we like it or not, there is a direct connection between air pollution and energy utilization. Very significant strides have been made in the United States and other industrialized nations at reducing this pollution. Progress continues to be made in those countries. The problem for the future may be in the developing nations. These countries represent nearly two-thirds of the total population of the world. Many are in the stage of economic development in which there is a strong relationship between per capita energy consumption and gross domestic product (Chapter 1). For these nations, the key to increasing their standard of living is to increase energy use. In many of these countries, the financial resources or the legislative mandates (or both) are lacking to install and use pollution control devices such as scrubbers. Thus it is very likely that the world is in for a continued period of increasing air pollution as these developing nations continue to industrialize, and continue to increase their energy use.

We must also recognize that some air pollution comes from natural, rather than human, sources. Gaseous sulfur compounds are emitted into the environment from volcanoes. Natural forest fires produce smoke and soot. Lightning flashes generate NO_x.

Other nitrogen-containing gases, such as ammonia, are produced in the decomposition of animal wastes. Natural oil seeps contribute about 10% of the oil in the oceans. This is by no means a complete catalog of natural processes that result in materials that we humans classify as pollutants. Rather, these examples should remind us that the most comprehensive, indeed draconian, worldwide clean air act we can possibly imagine will never reduce air pollution to zero. Mother Nature wins every time.

It will likely prove that the most difficult problem of all is CO_2 emissions. How do we handle this? CO_2 emissions are an inevitable consequence of the combustion of fossil fuels. *All* of the other emissions we have been discussing are not inevitable since they can, at least in principle, be addressed by precombustion treatment of the fuel, postcombustion capture of the product, or both. But not CO_2!

Many ideas, some sound, some zany, have been proposed for postcombustion CO_2 capture. A significant part of the problem is the immense amount of CO_2 to be dealt with. Burning a ton of coal that is about 70% carbon will produce 2.5 tons of CO_2. A 'typical' 10,000 tons of coal per day power plant produces 25,000 tons of CO_2 per day.

The most realistic short-term solution to the problem would seem to be simply to produce less CO_2. There are two ways to do that: reduce the amount of useful energy we produce, at the present levels of efficiency; or, maintain about the same amount of energy consumption, but produce it at much greater efficiency. Since it is unlikely that many people would voluntarily select the first option, the key seems to lie in enhanced energy efficiency.

If we use the same amount of energy, but can produce that energy with greater efficiency, then we can burn less fuel. For example, if we doubled the fuel economy of a vehicle, we could drive exactly the same amount of miles, but consume only half the amount of gasoline (and hence produce half the amount of CO_2 as before). We need to recollect that

$$\text{Efficiency} = (T_{high} - T_{low})/T_{high}$$

and, as we've seen, the key to getting high efficiency is to make T_{high} as large as possible, because the ideal solution, $T_{low} = 0K$, is not possible. Unfortunately, the very high combustion temperatures that provide a large T_{high} lead to greater thermal NO_x production.

If it becomes agreed that CO_2 emissions and the greenhouse effect are a serious problem, this would be the major issue that moves us as a society away from our present heavy reliance on fossil fuels. At present, the only large-scale technology available for generation of electricity is nuclear. Though nuclear energy unquestionably has its own set of problems (Chapter 28), it is a technology that can produce tens or hundreds of gigawatts of electricity with no CO_2, no SO_x, no NO_x, and no RO_x emissions. Other energy sources, notably wind (Chapter 35) and solar (Chapter 36), can also make the claim of being emissions-free, but have yet to be developed to the scale of nuclear energy.

In the following several chapters, we will examine some of the environmental issues associated with these products of fossil fuel combustion.

The environment is not an 'other' to us. It is not a collection of things that we encounter. Rather, it is part of our being. It is the locus of our existence and identity. We cannot and do not exist apart from it.

—Lakoff and Johnson[3]

CITATIONS

1   Maxim Gorky, quoted in Knight, Amy. *Who Killed Kirov?* Hill and Wang: New York, 1999; p. 96.

2   Medawar, Peter. *The Strange Case of the Spotted Mice.* Oxford: Oxford, 1996; p. 117.

3   Lakoff, George; Johnson, Mark. *Philosophy in the Flesh.* Basic Books: New York, 1999; p. 566.

FOR FURTHER READING

Anderson, Robert O. *Fundamentals of the Petroleum Industry.* University of Oklahoma: Norman, 1984. This is a well-illustrated book that discusses most aspects of the oil industry, and can easily be understood by the reader with little scientific background. Chapter 24 discusses some of the environmental issues associated with petroleum extraction, transportation, and refining.

Aubrecht, Gordon. *Energy.* Merrill: Columbus, OH, 1989. Chapter 22, on pollution from fossil fuels, provides additional information on many of the topics treated in this chapter.

Berger, Bill D.; Anderson, Kenneth E. *Modern Petroleum.* PennWell: Tulsa, 1992. A useful introductory book on all aspects of the petroleum industry, for readers with little previous technical background. Chapter 18 discusses the environmental issues associated with the petroleum industry.

Edelson, Edward. *Clean Air.* Chelsea House: New York, 1992. Though some of the statistics are now out of date, this book serves as a good general introduction to problems of air pollution for readers with little previous scientific background.

Fowler, John M. *Energy and the Environment.* McGraw-Hill: New York, 1984. Aside from the fact that many of the statistics in this book are now dated, it remains a useful and well-written introductory text. Chapters 8 and 9 relate to environmental issues associated with fossil fuel use.

Gebelein, Charles G. *Chemistry and Our World.* Brown: Dubuque, IA, 1997. Chapter 18 of this introductory chemistry text discusses air pollution associated with energy production.

Goldemberg, José. *Energy, Environment, and Development.* Earthscan: London, 1996. This slim volume, packed with data tables and graphics, discusses the relationships among energy, the environment, and economic or industrial development. Chapters 4 and 5 focus heavily on the energy and environment relationship.

Hill, Robert; O'Keefe, Phil; Snape, Colin. *The Future of Energy Use.* Earthscan: London, 1995. A useful feature of this book is the attempt to relate economic costs to environmental and social impacts of energy use. Chapter 2, on the cost of energy (including its impact on the environment) is particularly relevant to the discussion in this chapter.

Hollander, Jack M. (Ed.) *The Energy–Environment Connection.* Island Press: Washington, 1992. Part I of this book consists of seven chapters, by various experts in respective fields, on environmental impacts of various energy sources or energy technologies.

Margolis, Jonathan. *A Brief History of Tomorrow.* Bloomsbury: New York, 2000. Chapter 3, 'Global Warning,' focuses on predictions of the future as they apply to the environment. The entire book is an interesting look at the discipline of 'futurology.'

McNeill, J.R. *Something New Under the Sun.* Norton: New York, 2000. Subtitled 'An Environmental History of the Twentieth-Century World,' this book reports on the global-scale, and essentially uncontrolled, experiment that we are performing on ourselves in altering the environment. Though not all of the discussion relates to energy use, this is an important and thought-provoking book.

National Research Council. *Surface Mining: Soil, Coal, and Society.* National Academy Press: Washington, 1981. This is a report of a task force that examined possible impacts of increased strip mining on available agricultural land. Though it is now two decades old, much of the material still makes a useful background to the topic.

Ristinen, Robert A.; Kraushaar, Jack J. *Energy and the Environment*. Wiley: New York, 1999. This is quite useful resource for learning more about the relationships between energy and the environment. Chapter 9 discusses air pollution related to energy use.

Tenner, Edward. *Why Things Bite Back*. Knopf: New York, 1996. This book is a fascinating chronicle of the 'unintended consequences' of technology, that is, how various technological developments intended to make life simpler or easier have taken their revenge on us by causing entirely new problems that no one had anticipated. Chapter 4 discusses both natural and man-made environmental disasters.

Verburg, Carol J. (Ed.) *The Environmental Predicament*. St. Martin's Press: Boston, 1995. This book is a collection of articles from various sources, grouped around four major themes. The fourth unit, 'How Can We Solve Our Environmental Predicament?' is relevant to the present chapter.

30

ACID RAIN

In our discussion of electricity generation in coal-fired power plants (Chapter 18), and again in the previous chapter, we saw that the combustion of *any* fossil fuel converts its constituent chemical elements into their respective oxides. Some petroleum products and virtually all coals contain nitrogen and sulfur. These two elements result in combustion products that are of great concern because of their effects on the environment: SO_x and NO_x. (We should also recall that there are two sources of NO_x in combustion systems: fuel NO_x from nitrogen atoms chemically bonded to molecules of the fuel, and thermal NO_x from the high-temperature reaction of nitrogen molecules in the air.) A major reason why these substances are of concern is their participation in the environmental problem of **acid rain**. Before we discuss the origins and environmental effects of acid rain, we will first digress to the subject of acids, specifically, what an acid is and how we measure acidity.

THE pH SCALE

In aqueous solutions (i.e., solutions in water), acids are conveniently defined as substances that release hydrogen ions. For a simple compound such as hydrogen chloride (HCl), we might write the equation

$$HCl \rightarrow H^+ + Cl^-$$

In aqueous solution, hydrogen chloride produces hydrogen ions, H^+, and chloride ions, Cl^-. The resulting solution is acidic, and for this reason we also call HCl (especially when it is understood to be present in solution) hydrochloric acid. Though the conceptual scheme represented by the equation above provides us an easy way to think of what an acid is, or how it behaves in water, there is a complication with this definition. Since an atom of hydrogen has only one electron, and the nucleus of the common isotope of hydrogen is a single proton, the H^+ ion, a hydrogen atom with the electron missing, is the same thing as a proton. Protons will not exist by themselves in solutions. Rather, when a substance like hydrogen chloride dissolves in water the H^+ ions that would be released associate with one or more water molecules, forming the H_3O^+ ion (called the hydronium ion). Thus it is more accurate to write

$$HCl + H_2O \rightarrow H_3O^+ + Cl^-$$

The resulting solution of hydrochloric acid has the characteristics that we normally associate with an acid in water (such as a sour taste, or the ability to corrode chemically active metals) because of the presence of H_3O^+ ions. It is usually convenient to write the reactions in aqueous solutions as if the active species were H^+ but it should be understood that this is done for convenience or simplicity and that the actual acidic species is H_3O^+.⚛

> ⚛ It's perhaps ironic that water, a simple three-atom molecule and one of the most abundant substances on the planet, is also an extraordinarily complex material. The H_3O^+ ion, which is the H^+ ion hydrated by a water molecule, itself can associate with more molecules of water. Chemists have characterized such ions as $H_5O_2^+$, $H_7O_3^+$, and $H_9O_4^+$. That's one reason why it's convenient to write simply H^+ when considering the reactions of acids in solution, even though the H^+ ion never exists by itself in solids or liquids.

For any solution in water, we can describe its acidity (or basicity) in terms of the **pH** scale (Figure 30.1). There are several important facts to keep in mind about the pH scale. Pure water is chemically neutral, and has a pH of 7.00. pH values less than 7 represent acidic solutions; the smaller the number, the greater the acidity, and the higher the concentration of H^+ in solution. pH values greater than 7 represent basic (alkaline) solutions; the larger the number, the greater the basicity. The pH scale is logarithmic (pH is the negative of the logarithm of the H^+ ion concentration in units of moles per liter). In fact, the term 'pH' means the 'power of hydrogen.' Because of the logarithmic nature of pH, every unit in the pH scale represents a change by a factor of 10 from the previous unit. For example, a solution of pH 2 is not two times as acidic as one of pH 4 (that is, a change of $4 - 2 = 2$), it is 100 times (i.e., 10^2) more acidic. Similarly, a pH of 3 is a solution that is one million times (i.e., 10^6) more acidic than one of pH 9, not six times more acidic.

Substances that we encounter in everyday life have a wide range of pH values. Bases include household ammonia, 11; baking soda, 9; and blood, 7.5 (which is nearly, but not quite, neutral). Acids include milk, 6.5; black coffee, 5.0; tomatoes, 4; lemon juice, 2; stomach acid, 1.5; and the acid in automobile batteries, 0. These relationships are illustrated in Figure 30.2. Notice that because the pH scale is logarithmic, our stomach fluid is not roughly six times more acidic ($7.5 - 1.3 = 6.2$) than blood, it's 1.5 *million* ($10^{6.2}$) times more acidic. The tastes of some foods are due in part to the acids that occur naturally in them, or that are added to produce them. Lemons (2.2) contain citric acid; vinegar is basically a solution of acetic acid (2.5); and apples contain malic acid (3.0). A carbonated water, such as club soda, has a pH of about 4.8 because it is made by dissolving carbon dioxide (forming carbonic acid in solution) into the water in the carbonation

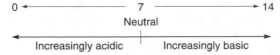

FIGURE 30.1 The key relationships on the pH scale.

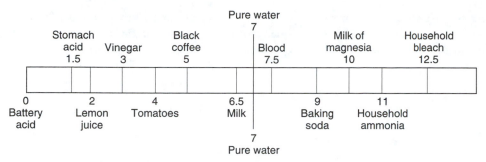

FIGURE 30.2   The pH values of a variety of common solutions.

process. Many colas have pH values close to 3, because they contain phosphoric acid in addition to the carbonic acid from the carbonation process.

'NATURAL' ACID RAIN

If we examined rain falling in some pristine, unpolluted environment (assuming such a place still exists somewhere), we might expect it to have a pH of 7, since pure water is chemically neutral. However, carbon dioxide, a naturally occurring constituent of the environment, is slightly soluble in water. As rain falls through the air, some carbon dioxide dissolves to form the weakly acidic solution of carbonic acid

$$H_2O + CO_2 \rightarrow H_2CO_3$$

This mildly acidic solution of carbon dioxide in rainwater has a pH of 5.6. In other words, even rain falling in a completely nonpolluted environment will still have a pH of 5.6 and be mildly acidic. Therefore, only when the pH of rain is below this value can we suspect the presence of pollutants.

Small amounts of other natural acids, including formic acid and acetic acid, are almost always present in rain and contribute slightly to its acidity.

> ACID RAIN IS RAINFALL HAVING A pH LESS THAN 5.6 (I.E., IS MORE ACIDIC THAN RAIN IN AN UNPOLLUTED AREA).

Although it is common to speak of this problem as being acid 'rain,' in principle, acidic pollution can be deposited not just as rain but as snow, hail, sleet, fog, mist, or dew as well. In fact, sometimes acids can become attached to dry particles that simply drop out of the sky without any form of water present. A more encompassing term, therefore, is acid precipitation, or acid deposition.

Even in the absence of pollution sources, rain or other precipitation can temporarily become more even acidic (that is, of pH less than 5.6) because of natural events such as volcanic eruptions, which send large amounts of sulfur oxides into the atmosphere, and forest fires and lightning flashes, the high temperatures of which promote the formation of thermal NO_x. Decaying organic matter can also contribute nitrogen and sulfur compounds to the atmosphere. These natural phenomena can all contribute to acidic precipitation. Therefore, it is important to remember that an acidic rainfall or other precipitation is not necessarily evidence of pollution caused by human activities.

The worst examples of acid precipitation can be very acidic indeed. In 1982, fog in Pasadena, California was found to have a pH of 2.5. Being outdoors in this fog would be roughly equivalent to breathing a fine mist of vinegar. A pH of 2.5 is roughly a *thousand times* more acidic than normal, unpolluted rain (pH 5.6) and 10 times more acidic than would kill all fish in most lakes. In a small city on the Tibetan Plateau of China, which one might naively assume to be as far away from human influence as one can get, rain of pH 2.25 has been observed—like strolling in a shower of lemon juice. Fogs at Corona del Mar, south of Los Angeles, have been observed that are 10 times more acidic than even the one at Pasadena, with a pH of 1.5.

SO$_x$ AND NO$_x$ AS CAUSES OF ACID RAIN

In 1852, a British chemist, Angus Smith, was studying the problems of the rusting of metals and fading of dyed materials in the heavily industrialized city of Manchester. He recognized that sulfuric acid in the air was the likely source of these problems. About 20 years after his original insight, he published a book, *Air and Rain*, which qualifies Smith to be considered the 'father' of acid rain chemistry. Unfortunately, the book rather promptly vanished into obscurity, taking along with it the concept of acid rain.

> *... truth prematurely uttered is of scarcely greater value than error.*
>
> —Kennan[1]

Then, in the 1950s, the effects of acid rain were rediscovered. Scientists working in the northeastern United States, in Scandinavia, and in the Lake District of England noticed the acidification of lakes and its resulting environmental effects. Their investigations began a new era in the studies of acid rain.

During the 1950s, the Canadian scientist Eville Gorham recognized that acid rain could overwhelm the buffering, or neutralizing, capacity of lakes, soils, and bedrock. Many natural soils or rocks contain basic compounds; an example is limestone, which is essentially calcium carbonate, $CaCO_3$. Basic substances in rocks and soils, or dissolved in natural water, can react with acids—as in acid rain—to neutralize them. However, continual exposure to acids over a long period of time could eventually exhaust this natural neutralization. Once that point had been reached, then the effect of further acidic precipitation would be to make soils and water increasingly acidic. Gorham discovered a number of acidified lakes in England's Lake District. He suggested that the acidification of these small mountain lakes (called tarns) was caused by air pollution.

A decade later, Svante Odén, a Swedish scientist, suggested that the increase in acidity in the lakes of his home country was a result of pollution blowing into Sweden from other countries. Odén found that both rain and surface waters throughout Europe were becoming more acidic, and that winds were responsible for spreading airborne sulfur and nitrogen compounds across the continent. His work also demonstrated how acid rain could cause declines in fish populations, reduction of forest growth, and corrosive disintegration of materials.

It is now well established that the dominant contributors to acid rain are SO$_x$ and NO$_x$. For example, in virtually all parts of the world, increases in acid deposition correlate very well with local increases in SO$_x$ emissions, and the highest concentrations of acid rain are found where SO$_x$ levels are highest. SO$_x$ and NO$_x$ are not, strictly speaking, acids themselves in the way that we have defined the term.⚛ Rather, they would be called acid anhydrides, or 'acids without water.' They dissolve in water—including rain— to form sulfurous and sulfuric acids, and nitrous and nitric acids, respectively. The two major acids in acid rain are sulfuric and nitric acids. In their concentrated forms, sulfuric and nitric acids are extremely powerful acids.

> Chemists recognize more than one classification of acids (and bases). Those substances that behave as acids in water by forming H$^+$ ions are called Brønsted, or Brønsted–Lowry acids. For our purposes in considering acid rain and its effects, this is only type of acid with which we need be concerned. Another generally important category of acids is the Lewis acids. In Lewis's classification, an acid is a substance capable of accepting a pair of electrons from another species. Lewis's classification is much broader than Brønsted's, because it incorporates species that do not even contain hydrogen. For example, SO$_2$ cannot possibly act as a Brønsted acid until it interacts with water to produce H$^+$ from the water molecules, but SO$_2$ can function as a Lewis acid in some reactions.

The sequence of chemical reactions that leads to the production of sulfuric acid from SO$_x$ seems not yet completely understood. The conversion of SO$_2$ to SO$_3$ is fairly slow, but it is catalyzed by the presence of finely divided solid particles. Fly ash, which goes up the stack of a coal-fired power plant along with the SO$_x$ (Chapter 18) provides such catalytic particles. Once SO$_3$ has been formed, it reacts rapidly with water droplets in the atmosphere to form sulfuric acid. Some naturally occurring compounds also facilitate the conversion of SO$_2$ into sulfuric acid. These include ozone, hydrogen peroxide, and the hydroxyl free radical (·OH, which is formed from ozone and water in the presence of sunlight), all of which occur as trace constituents in the atmosphere. The hydroxyl radical and hydrogen peroxide may in fact have major roles in this process. For example, the reaction of SO$_2$ with ·OH can account for about one-fifth to one-fourth of the sulfuric acid in the atmosphere. The reaction is also speeded up by sunlight, so is more important in the summertime and, in any season, during the middle of the day.

Coals from various parts of the world differ considerably in their sulfur content, but the combustion of any coals will inevitably produce some SO$_x$. In coal-fired electricity-generating stations and industrial plants that burn coal for process heat, the SO$_x$ would be emitted to the environment (unless control measures—Chapter 18—are used) along with the carbon dioxide, water vapor, and ash. Electric power plants produce by far the largest amount of sulfur pollution in the United States. Coal and oil-fired power plants generated about two-thirds of the SO$_x$ emissions. Industrial operations such as boilers, smelters, and refineries contributed about 30%; transportation and commercial or home heating added about 3% each. Usually the most acidic rain falls in the eastern third of the United States, with the lowest pH region being in the heavily industrialized Ohio River Valley. Rainfall over the eastern part of North America now often has a pH of about 4.5 (in terms of acidity, this is like having it rain tomato juice). A map of the United States showing pH values of precipitation is given in Figure 30.3.

It is very important to recognize that fossil fuel combustion is *not* the only human source of SO$_x$ and NO$_x$ emissions. The production of metals such as copper and nickel generates huge quantities of SO$_x$. The most common ores of these metals are sulfides, which are compounds of the metal with sulfur. The conversion of sulfide ores in the metallurgical industry involves heating them with oxygen (a process called 'roasting') and is responsible for the SO$_x$ emissions. The world's largest smelter, in Sudbury, Ontario, produces nickel from nickel sulfide. The environment in the immediate vicinity of the plant has been ruined by the uncontrolled release of SO$_x$ in years past. Some who have visited the region suggest that it is about as close to a barren lunar landscape as we might see on Earth. Even now, with government-mandated emission controls, the Sudbury smelter puts out about 2,000 tons of SO$_x$ per day from the world's tallest stack, one

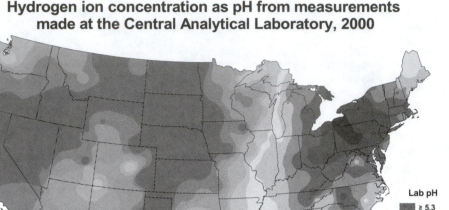

FIGURE 30.3 This map shows the distribution of pH values of rain falling across the United States.

that is equal in height (1,250 feet) to the Empire State Building. This remarkable facility by itself accounts for about 20% of all SO_x emissions in Canada and about 1% of the total sulfur emissions from all human activities worldwide. The extremely tall stack on the smelter provides a superb way of ensuring that the SO_x emissions are distributed over a wide geographical region. A tall stack on any sort of combustion or smelting facility is used to get the emissions as high up into the air as possible, so that they will be swept out of the immediate vicinity of the plant and both diluted and dispersed by winds. This practice led to the facetious comment that, 'The solution to pollution is dilution.' The contribution of the metal industry was dramatically highlighted by the effect of a labor strike at copper smelters in the southwestern United States in 1980. During the time the workers were out on strike, SO_x emissions in Arizona and New Mexico fell by about 90%.

Worldwide, the atmosphere receives about as much sulfur from human activities as it does from natural sources, 75 to 100 million tons per year from each. Sulfur is released naturally by plankton in the oceans, decaying vegetation in swamps, and from volcanoes. The eruption of Mount Pinatubo in the Philippines, in June 1991, injected between 15 and 30 million tons of SO_x into the stratosphere. The emitted SO_x reacted relatively quickly to form small droplets of sulfuric acid. Those small droplets remained suspended in the atmosphere for more than two years, where they reflected and absorbed sunlight. In late 1991, after the eruption, and continuing through 1992, there was a decrease in the average global temperature. This temperature drop was contrary to an otherwise steady increase in average global temperature (Chapter 32), and has been attributed to the effects of the tiny droplets or frozen crystals of sulfuric acid absorbing or reflecting sunlight.

Air is always a source of nitrogen for the production of thermal NO$_x$. The energy necessary for the reaction of the relatively unreactive nitrogen molecules in the air with oxygen can come from the high temperatures in combustion systems, such as in the burning of the gasoline in automobile engines. Thermal NO$_x$ is by no means limited to the automobile engine. There is also concern over high-altitude NO$_x$ emissions from jet aircraft engines, and NO$_x$ production from diesel engines. The same reaction of nitrogen molecules in the air occurs when air is heated to a very high temperature in the furnace section of a coal-fired electricity generating plant. (So, these plants can potentially contribute both sulfur and nitrogen oxide emissions to the formation of acid precipitation.) Nationwide, **stationary sources**, such as power plants, release more NO$_x$ than **mobile sources**, but in urban locations, cars and trucks account for most of the NO$_x$ emissions. In the United States, slightly less than half (about 45%) of NO$_x$ emissions come from transportation sources. The rest came from power plants (about 30%), industrial sources (about one-fourth), with a small amount from factories and other commercial sources, institutions, and homes. Lumped together, electricity generation and industrial processes account for more than half of the NO$_x$ emissions in the United States.

As we will see in Chapter 31, nitrogen oxides are fairly reactive compounds, and are of concern not just because of their contribution to acid precipitation. At high altitudes NO$_x$ can react with upper-atmosphere ozone, contributing to the destruction of the ozone layer.⊛ Closer to the surface of the Earth, NO$_x$ reacts with other components of vehicle exhaust to create the pollution problem known as photochemical smog.

⊛ At altitudes of some 6–18 miles above Earth's surface, the atmosphere contains increased concentrations of ozone, the form of oxygen with three oxygen atoms per molecule (O$_3$). The thickness of the ozone layer varies from one season to another, and also varies with latitude. The ozone layer is *not* some sort of spherical shell of pure ozone that surrounds our planet. Under the best of circumstances the ozone concentration in the 'ozone layer' is relatively small. However, the ozone layer is crucial for human health. High-altitude ozone intercepts some of the ultraviolet light that comes to Earth as part of the incoming solar radiation. Problems attributed to ultraviolet exposure include an increased susceptibility to skin cancer and cataracts, and, possibly, reduced ability of the immune system to ward off disease. For every 1% reduction in ozone concentration, the amount of ultraviolet reaching the surface increases by about 2%. NO$_x$ reacts with ozone; an increase in high-altitude NO$_x$ concentrations could therefore contribute to reduced ozone concentrations, which in turn lead to more exposure to ultraviolet for us.

In the United States, the annual anthropogenic emissions of SO$_x$ and NO$_x$ are roughly equal. However, on a worldwide scale, human activities release almost twice as much SO$_x$ as NO$_x$. If we were to restrict the comparison only to SO$_x$ and NO$_x$ from fossil fuel combustion, then the annual NO$_x$ emission is less than 40% of the SO$_x$ emissions. During the 1980s, world SO$_x$ emissions from the burning of fossil fuels increased by about 10% while NO$_x$ emissions increased by 20%. The effect that is masked by a simple overall global comparison is the major decrease in SO$_x$ emissions from the industrialized nations. (Scrubbers work!) Unfortunately, the drop in SO$_x$ emissions from industrialized nations was more than offset by a huge increase in SO$_x$ emissions from the developing countries. A comparison of SO$_x$ emissions from the United States and China serves to illustrate this. In 1970, the United States produced substantially more SO$_x$, roughly 30 million tons vs. China's 10 million tons. Twenty years later, the SO$_x$ emissions from both countries were essentially equal, at about 22 million tons each. Even though energy consumption in the United States increased steadily in that time period, SO$_x$ emissions were actually reduced, thanks in part to various emission control strategies (Chapter 18). However, China more than doubled its SO_x emissions. Many developing nations do not have the financial ability to afford emission-control technologies or low-sulfur fuels that have been adopted by the industrialized nations. Some of the developing nations have not enacted environmental control legislation or regulations. Though this example has focused on SO_x, there is great concern that NO_x emissions might be the more serious long-range problem, because they are more difficult to control than SO_x.

Just as with SO_x emissions, human activities are not the only sources of NO_x. Plants, soil, and lightning are major natural sources of NO_x. A lightning bolt, for example, is a splendid way to generate thermal NO_x. In a year that is not marked by some dramatic event like the Mount Pinatubo eruption, natural emissions would account for about 20% of SO_x and about 40% of the NO_x released to the atmosphere.

ENVIRONMENTAL CONSEQUENCES OF ACID RAIN

Acid precipitation is an environmental problem for many reasons. It corrodes the limestone, marble, or metals used in buildings. Some ancient monuments and statues that have survived unscathed for millennia, since the time of ancient Egypt, for example, are now being rapidly destroyed by acid rain. It accumulates in, and acidifies lakes and streams and contributes to the destruction of aquatic life. It destroys terrestrial plant life, probably the most dramatic example being in the Black Forest of Germany, where more than two-thirds of the trees are now dead. In the United States, acid rain appears to be affecting forests in northern New England and in the southern Appalachians. Inhalation of acid rain mist irritates or exacerbates human respiratory problems.

Effects on Buildings and Statues

Many priceless and irreplaceable marble or limestone statues and buildings are being attacked by acid precipitation. In many cases these buildings or monuments are irreplaceable artifacts of human civilization. The beautiful sculptures adorning the exteriors of cathedrals, churches, or other historic buildings throughout Europe are eroding. Buildings famous around the world, such as the Parthenon in Greece, the Taj Mahal in India, and United States Capitol show signs of corrosion that is likely due to acid precipitation. Indeed, the ancient monuments of classical Greek civilization, including the Parthenon, have deteriorated more in the past two or three decades than they had in the previous two millennia. In the United States, icons of American history, including the monuments at the Gettysburg battlefield, the Washington monument, the Statue of Liberty, and the Independence Hall in Philadelphia, have all suffered from acid rain. Many old gravestones, especially in the eastern United States, are no longer legible.

The eastern half of the United States, especially in the summertime, experiences a haze that clouds the landscape. It has become steadily worse over the past 20–30 years. Coal-fired electricity generating plants in the Ohio Valley have been implicated as a major cause of the haze. Sometimes it is thick enough to be easily observed from the air, if one is traveling to the East Coast from the west. The haze is an aerosol of particles containing, among other things, sulfuric acid. It is most pronounced in summer because then there is more sunlight to enhance the reactions that produce the sulfuric acid. The Grand Canyon is also experiencing reduced visibility. A possible source of this problem might be the SO_x emissions from the Four Corners electricity plant, located in the northeast corner of Arizona, near the point where Arizona, New Mexico, Colorado, and Utah all meet.

 An aerosol is any suspension of very fine particles of solid or very fine droplets of liquid in a gas. As the name implies, the gas is usually (but does not necessarily have to be) air. The haze discussed here is a suspension of particles of sulfuric acid, ammonium sulfate, and ammonium hydrogen sulfate in the air.

Acidification of Natural Waters

Whenever acidified water enters a lake, whether from acid precipitation, acidic streams flowing into the lake, or runoff from melting acidified ice and snow, the pH of the lake will necessarily drop unless the lake itself, or something in the surrounding soils, contains basic compounds that can neutralize the acid. The capacity of a lake to resist change in pH when acids are added is called its **acid neutralizing capacity**. In those parts of the world, including the American Midwest, where the surface rock is mostly limestone, the limestone itself can neutralize acids. Unfortunately, many lakes in New England and upper New York, and in Scandinavia, are surrounded by granite rather than limestone. Granite will not react with acids nearly so readily as limestone. As a rule, there will be very little acid neutralizing capacity in such lakes, and they will necessarily show a steady, though usually gradual acidification.

Two other mechanisms also affect the ability of a body of water to withstand acidification. First, some lakes or streams may possess a buffering capacity. Over many years, the natural weathering of surrounding rocks may have resulted in the presence of bicarbonate ions in water. The bicarbonate ions can act as natural buffers that prevent a substantial drop in pH when acid precipitation occurs. Second, the H^+ ions added to the water by the acid precipitation can exchange with other positively charged ions (such as Ca^{2+} or K^+) that are naturally present in soils or in sediments on the lake bottom. This process is called ion exchange. It helps counteract the effects of acid addition because the calcium or potassium ions that exchanged for the H^+ ions are not acidic. But, just as in the case of the limestone lakes, the continual addition of acid can eventually get to a point at which the natural buffering capacity of the water, or the ion exchange capacity of soils and sediments, can be overwhelmed.

Buffers are substances that resist changes in pH, even when strong acids or bases are added to the system. Most common buffers are aqueous solutions of a weak acid (that is, one that does not produce much H^+ in solution) and one of its salts. An example would be a solution of acetic acid and an acetate salt. When a strong acid is added, the H^+ ions that it provides are consumed in converting acetate back to acetic acid. A strong base added to the system converts acetic acid to acetate. In either case, little pH change occurs. An application in daily life is the use of buffers as food additives, to prevent pH changes during storage. Potassium hydrogen tartarate is an example of a buffer used in this way.

Some substances, such as the zeolites we met in Chapter 21, contain ions that are relatively loosely held to the molecular structure. When such substances are exposed to a solution that contains ions of a different kind, some of the loosely held ions can enter the solution, and some of the ions originally in solution can be taken up instead. An everyday example is the home water softener. Most such units use a synthetic polymer containing sodium ions. In the presence of so-called hard water, which contains calcium and magnesium ions, an exchange occurs such that sodium ions from the polymer enter the water, and calcium and magnesium are taken up by the polymer. This ion exchange does away with the behavior—forming scums with soap and leaving residues in teakettles, for example—that we associate with hard water.

At least 10% of the lakes and streams in the eastern United States have been adversely affected by acid precipitation. The decreasing pH is often accompanied by a decline in the population of fish and other aquatic species. By the time the pH is reduced to 4, a lake is essentially dead. The ecosystems that are especially at risk are lakes located in areas with high levels of SO_x and NO_x emissions coupled with thin soils, low buffering capacity, or granite rather than limestone. In the United States, such highly susceptible areas (many of which are already acidified) are in the southern Adirondacks, New England, the mid-Atlantic highlands, and the eastern portion of the Midwest. (For example, in the past half-century the average acidity of Adirondack lakes has increased by a factor of 40.) The contiguous regions of eastern Canada are similarly affected; in southeastern Ontario, the average pH of lakes is now about 5.0. In many of these places the water contains increased concentrations of sulfate ions, consistent with sulfuric acid being the acidifying agent.

Effects on Aquatic Ecosystems

In many cases, pollution has a visual impact, in addition to its chemical or biological effects. As examples, we see sooty clouds of smoke, the discoloration of water by something dumped into it, or accumulated dust and dirt. The acidification of natural waters by acid precipitation is unusual in this respect. A lake or stream that has been damaged by acid precipitation does not usually have the appearance of being polluted. The water seems to be pristine and sparkling clear. It would be easy to misinterpret this appearance as signifying clean healthy water, whereas the truth is just the opposite—the water is biologically dead. Let's see what happens with continued acidification and decreasing pH (Figure 30.4).

Any organism will have a particular range—sometimes a rather narrow range—of pH at which it thrives best. One of the first warning signs that acid was a serious problem, noticed in the late 1960s, was a decline in fish populations in Scandinavian lakes. At about the same time, similar observations were made in New Hampshire, correlated with an increase in the acidity of lakes and streams in New England.

The effect on fish is often the first noticeable impact, because jobs and revenue are lost when commercial fisheries become less productive, and recreational opportunities are lost for sport fishing. However, long before fish are lost, other, more humble life forms that have critical positions in the aquatic food chain of the ecosystem are lost.

When the pH of a lake drops below 6, many kinds of small organisms begin to die off. For example, phytoplankton, the tiny, free-floating aquatic plants that serve as the base of the aquatic food chain, have low reproduction and high mortality at pHs below 6. Similarly, many species of free-floating aquatic animals (zooplankton) are adversely affected. The loss of these organisms begins the disruption of the entire food chain, thus threatening the viability of many other species.

Irreversible changes will begin to occur in lakes when the pH drops to about 5.8. At this pH the tiny crustaceans that are another important link in the aquatic food chain

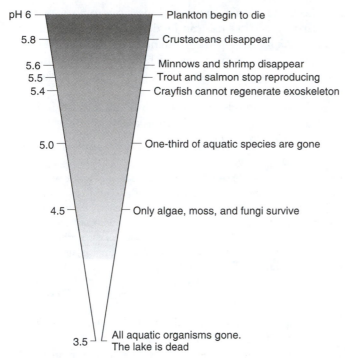

pH 6 — — Plankton begin to die
5.8 — — Crustaceans disappear
5.6 — — Minnows and shrimp disappear
5.5 — — Trout and salmon stop reproducing
5.4 — — Crayfish cannot regenerate exoskeleton
5.0 — — One-third of aquatic species are gone
4.5 — — Only algae, moss, and fungi survive
3.5 — — All aquatic organisms gone.
         The lake is dead

FIGURE 30.4   This section of the pH scale summarizes some of the effects on aquatic ecosystems as they become increasingly acidic.

(copepods), disappear.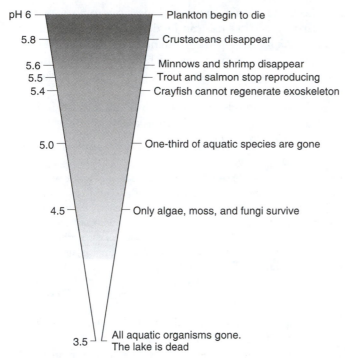 As the pH drops further, to 5.6, the populations of minnows, sculpins, and at least one type of shrimp begin to decline. The shrimp are a major food source for lake trout.

> Copepods, though perhaps known only to specialists in aquatic ecology, are of enormous ecological importance because they provide a food source for many species of fish. Most copepods are very tiny, usually no more than one-tenth of an inch long. They eat phyto- and zooplankton and each other.

> Sculpins are small fish characterized by wide and heavy heads but long, tapered bodies. They occur in both fresh and salt waters of the northern hemisphere. Sculpins stay close to the bottom of the water and usually are not very active. They are of no direct use to humans (in contrast to, e.g., trout or salmon) but of course are still a vital link in the total ecosystem.

Acidity then begins to affect the reproduction of fish. Many fish, particularly those of the family that includes trout and salmon, will not reproduce successfully at pHs below 5.5. As the pH drops below this value, most species of fish are endangered. Numerous problems beset the fish at this acidity: growth is retarded, the salt balance in the blood is upset, calcium is depleted from the skeleton (leading to malformation), aluminum clogs their gills, and mercury may be absorbed. Fish are not the only organisms to suffer under these conditions. Amphibians are also hard hit by the effects of an acidic environment.

The leaching of elements like cadmium, aluminum, and mercury from surrounding soils or rock can poison fish. (Toxic effects on fish can be detected at aluminum concentrations as low as 100 parts per *billion*.) Also, fish contaminated with potentially toxic elements, such as mercury, pose a threat to humans as well. Such elements leached into the water supply can also affect humans directly if these elements make their way into aquifers and from there into drinking water.

At pH 5.4, additional species of copepods vanish. Crayfish suffer extensively, by being unable to regenerate their hard and protective exoskeletons after molting. Fish are unable to reproduce at all at pH values below 5.4.

> Crayfish are crustaceans, the class of animals that includes lobsters, shrimp, and crabs. Unlike mammals, birds, and fish, whose skeletons are inside the body, crustaceans have an exoskeleton; that is, their body parts are supported by a skeleton or shell on the outside of the body. The only way such organisms can grow is by the periodic shedding of the exoskeleton—molting. Molting occurs at longer and longer frequencies as the animal reaches adulthood. The process is usually quick, but nevertheless the animal is almost totally helpless, especially vulnerable both to enemies and to pollution, until a new exoskeleton has grown.

By the time the pH has dropped to 5, about a third of all aquatic species are eliminated from the ecosystem. Most adult salmon or trout die because of the environmental stress resulting from life in water with pH below 5. As an example, many of the rivers in Nova Scotia that provide locations for salmon fishing have become so acidic that they are unable to support fish life. At least eight such rivers have pH below 4.7, with another 20 having pHs in the range 4.7 to 5.4. A century's worth of sport fishing records show that these rivers provided regular good catches until the 1950s, after which they began steadily to decline. Twenty years later essentially no fish were being caught there, and the salmon spawning runs have ceased.

Below a pH of 4.5, most life is gone except for some very hardy aquatic plants, such as algae, moss, and fungi. At pH 3.5, virtually all aquatic organisms die off. At that point the stream or lake is 'dead.' Many lakes in Norway and Sweden are effectively dead in this sense.

Remediation of Aquatic Systems

Although the acidification of natural waters can have these devastating effects on ecosystems, there is hope that, with proper treatment, many acidified lakes can be reclaimed. Since the problem is caused by acids, the short-term remediation relies on neutralizing the acid. This requires adding some basic compound to the water. The base selected for this application must be inexpensive, because large quantities will be needed, and should not be poisonous or otherwise harmful in itself. The best candidate for this application is lime (calcium hydroxide). Liming, the large-scale addition of lime to acidified waters, has been tried in various locations in the United States and Europe. Liming can raise the pH of the water to levels that allow various aquatic species to live successfully. The affected lake can be restored within a few years. However, over a period of time the added lime could itself be overwhelmed by further acid precipitation events. The only long-term solution is to reduce acid precipitation by reducing the amounts of SO_x and NO_x emitted to the environment.

Destruction of Terrestrial Plants

Acid precipitation by itself—that is, in the absence of any other factors—does not appear to kill plants. However, acid precipitation in conjunction with other environmental stresses can have significant impacts on plants (Figure 30.5). The acid precipitation may weaken the plant to the point where other environmental factors, which might not kill a healthier plant, would kill or seriously harm the acid-weakened plant.

In some cases, weather-related stresses likely have significant responsibility for the damage. Acid precipitation slowly weakens trees, reducing their ability to withstand extremes of cold or heat. Acid rain can make conifers more susceptible to cold. In the fall, conifers normally prepare for the freezing temperatures of the coming winter by withdrawing water from their needles. This is part of a natural process in the life cycle of the tree, and is called cold hardening. The process is started by the tree roots, which slow down the release of nitrogen-bearing nutrients into the soil, storing them up in the tree instead. But, nitrogen-containing acids soaking into the needles might override that signal from the roots, because now the tree could mistakenly sense an abundance of nitrogen compounds. That 'wrong signal' caused by acid rain soaking the needles could delay the cold-hardening process and leave the tree vulnerable to freezing. In addition to conifers, the maple trees of New England and Canada may be damaged by weather-related stress that has been exacerbated by acid precipitation.

Acid precipitation weakens the ability of trees to resist attacks by insects or by viruses. Gypsy moth damage is more severe if the tree has also been exposed to acid rain. Similarly, blight-causing viruses cause more damage to trees weakened by acid rain. The destruction of fir trees in the southern Appalachians is a manifestation of this type of problem.

The effects of ozone or other air pollutants in damaging trees can also be made worse if the trees have also been exposed to acid precipitation. Ozone and nitrogen oxides attack the waxy coating on leaves. Once this coating has been breached, the H^+ ions can deplete vital nutrients, such as magnesium, calcium, and potassium. The waxy coating also retards water loss from the tree. In addition, ozone can damage the chlorophyll that facilitates photosynthesis. If that happens, photosynthesis, and, consequently, tree growth, are slowed. Damage to the pine trees in the San Bernardino Mountains of California is an example of the combined effects of acid precipitation and ozone attack.

We've seen how acidification can leach elements from the soil, sediments, or rocks in lakes, with some of these elements then causing toxic effects on the aquatic life. In a similar manner, acidification of soil beneath the trees mobilizes toxic metals, such as aluminum, that attack the tree roots and prevent the absorption of nutrients and water. But, there is at the same time

FIGURE 30.5 This photograph illustrates the effects of acid rain on a forest.

a second damaging effect, in that elements such as potassium, calcium and magnesium, essential for plant growth, are also leached out of the soil. Thus potentially toxic elements are mobilized and attack the tree while at the same time nutrient elements are also mobilized and lost. If the toxic metals were originally chemically combined in soil particles, they may have been biologically unavailable to the plant, and essentially harmless. Only when acidification of the soil, caused by acid precipitation, makes them soluble do they become accessible for uptake by the plant. The essential elements were originally present in the soil in chemical forms such that their uptake could proceed at a rate consistent with the normal biological needs of the plant. When these nutrients become soluble as a result of leaching by acid precipitation, they could flow (in the water) away from the root zone and no longer be available to meet the requirements of the plant. Some soils are better able to withstand acidic precipitation because they contain more alkaline substances, such as limestone, that can neutralize or buffer acids. But, as in the case of lakes and streams, the neutralizing or buffering capacity declines with continual, long-term exposure to acid precipitation.

Acid precipitation may also interfere with the various mechanisms by which trees obtain essential nutrients such as nitrogen and phosphorus. Many plants obtain essential nitrogen by 'fixing' (i.e., capturing and converting into a chemically useful form) nitrogen from the air. This so-called nitrogen fixation process is facilitated by microorganisms that live in soil and are often symbiotically associated with the root structure of the plants. Acid rain can inhibit nitrogen fixation by the microorganisms, by killing them.

In summary, there is strong circumstantial evidence that acidic precipitation is at least a contributing factor to the declining health of trees. These effects leave the trees susceptible to destruction by such other factors as disease, insects, or weather extremes. The damage is enormous on a worldwide scale.

In North America, forest damage has been most dramatic in portions of the Adirondack and Appalachian Mountains and southeastern Canada. Acid precipitation has contributed to the decline of red spruce in the northern Appalachians by reducing the cold tolerance of those trees. About 70% of all British beech trees and 80% of all yew trees show signs of injury associated with acid precipitation. In western Europe, tree loss spread during the 1980s. Damaged forests were reported in Italy, France, the Netherlands, Sweden, Norway, and Germany. Almost one in four European trees has lost more than a quarter of its leaves or needles. Trees in northern Germany have been stripped of many of their leaves. About 125 million acres of European forests have been damaged by acid precipitation, about one-third of Europe's total forested land. In the eastern European countries, former satellites of the Soviet Union, or regions that were part of the former Soviet Union, air pollution controls were essentially nonexistent. Reliable information on environmental damage in this region is just becoming available, but the emerging story is not good. Over 80% of the forests in eastern Germany, Poland, and the Czech Republic may be damaged. In some regions of these countries, forests have been totally destroyed by air pollution. In northern Siberia, reindeer have changed their habitat because lichen, their principal food, has suffered severely from acid precipitation.⚛ In this heavily polluted region, lichen growth has dropped to just 1–2% of normal levels, and the number of lichen species has dropped from about 50 to three or four.

⚛ Lichens are not one plant, but two, growing together in symbiosis—a relationship that benefits each. Lichens consist of algae growing together with fungi. The alga performs photosynthesis and manufactures food for both itself and the fungus. For its part, the fungus shelters the alga and keeps it moist, allowing the alga to grow in harsh environments where it could not survive on its own. Lichens are very abundant in the treeless regions of the Arctic. The plant (or, strictly speaking, plant community) known as reindeer moss, that is a valuable food item for reindeer, is a lichen.

Trees on mountainsides or tops of mountains face a more serious risk of damage from acid precipitation. Because they are at higher elevations, they are the first trees to be struck by the acid precipitation. In some regions, these trees can often be shrouded for long periods in clouds and mist, which can be just as acidic as the precipitation itself. The problem is compounded because these are the trees more likely to be stressed by cold.

Human Health Effects

Acid precipitation affects human life in many ways. The damage or destruction caused to ecosystems can deprive us of various kinds of outdoor recreation or just the aesthetic pleasure of enjoying the natural landscape. The damage to ecosystems more directly affects the benefits that we derive from them, such as clean drinking water and food in the form of fish or agricultural crops. But, perhaps most importantly, acid precipitation can directly lead to severe effects on human health.

When a person is outdoors during a rainstorm, or even in an acidic fog or mist, tiny droplets of acid precipitation are deposited directly in the sensitive tissues of the nose, bronchial tract, and lungs (Figure 30.6). Three groups in the population are especially at risk. One group is the elderly, those whose respiratory systems are starting to wear out or shut down. A second group is the very young, whose respiratory systems may not yet be

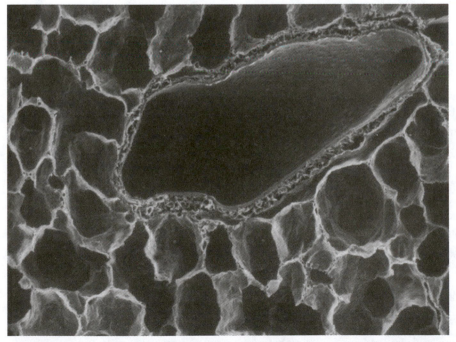

FIGURE 30.6 This is a magnified view of human lung tissue. Each of the passageways is lined with sensitive tissue that can be harmed by chronic exposure to acid rain.

fully developed. The third group is people of any age who already have other respiratory problems such as asthma, chronic bronchitis, or emphysema.

One of the worst recorded instances of pollution-related respiratory illness occurred in London in December 1952 (Chapter 18). This particular pollution incident lasted for five days and claimed about 4,000 lives. A similar incident in Donora, Pennsylvania in 1948 resulted in 20 deaths and caused illness in about 40% of the residents.

Many potentially toxic elements, such as lead, cadmium, and mercury, are normally chemically combined with minerals present in rocks and soil. When these minerals are continually soaked by acid precipitation, some of these elements are leached from the mineral into the acidic water. Once that happens, these elements can eventually wind up in the public water supply, where they can lead to chronic toxicity or serious health threats.

Aluminum, for example, is one of the most abundant elements in the Earth's crust. Aluminum appears to be harmless to most animals as long as acceptable levels of calcium are maintained in the body. If calcium becomes in short supply, aluminum may replace it in tissues or organs, causing problems. Aluminum can be toxic in humans, and does not appear to be an essential nutrient. No disease or condition resulting from aluminum deficiency has ever been identified in living organisms. An excess of aluminum has been found in damaged brain tissue of victims of Alzheimer's disease and of amyotrophic lateral sclerosis (often known as Lou Gehrig's disease). However, it is *not* proven that an excess of aluminum in the body *causes* Alzheimer's or Lou Gehrig's disease; and, even if such connection were to be established unequivocally, it is *not* proven that the aluminum was mobilized as a result of acid precipitation. Much remains to be learned about how acid precipitation or other forms of pollution result in the liberation of potentially harmful elements and the impact of those elements on human health.

ACID RAIN ON THE MOVE

Let's consider two statements:

(1) What goes up must sooner or later come down.
(2) The solution to pollution is dilution.

The implication of the first statement is that acid rain is often regional in character, in that what goes up (SO_x and NO_x) comes down (as acid precipitation) in the same general locality. However, the second statement has been put into practice by the building of very tall stacks on such emission sources as electricity-generating plants and smelters. Doing so provides a 'technological fix' for the problem of a utility or industry polluting its own neighborhood. But, by trying to use a technological fix for one problem we inadvertently create a new one. Tall stacks were built to eject pollutants high into the atmosphere where they could become diluted, thus improving local air quality, and ultimately come down somewhere else, sometimes hundreds of miles away. As a result, local pollution problems are converted into regional ones.

We might expect that the geographic regions with the most acidic precipitation should also be the regions of highest emissions of SO_x and NO_x. This relationship is generally confirmed. For example, SO_x emissions tend to be highest in regions where there are many coal-fired electric power plants, steel mills, and other heavy industries that rely on coal. Although power plants also generate nitrogen oxides, the highest NO_x emissions are usually found in states with large urban areas and heavy population density, factors that translate directly into much automobile traffic. As we will see in Chapter 31, cars and trucks can be significant sources of NO$_x$.

However, acid precipitation does not recognize state or national boundaries, thanks in part to the use of tall stacks. This leads to intense political controversies and accusations. SO$_x$ and NO$_x$ emissions from the Midwest are carried to the northeast by prevailing winds and contribute to acid precipitation in New York, New England, and eastern Canada. These emissions are also carried to the southeast, producing acid precipitation in Tennessee and North Carolina. Similarly, SO$_x$ and NO$_x$ emissions generated in Germany, Poland, and the United Kingdom are carried into Norway and Sweden.

Over a century ago, the Norwegian playwright Henrik Ibsen✵ complained about airborne pollution from Britain crossing the North Sea to Scandinavia. But it was not until the 1960s that Scandinavian scientists showed that acid precipitation, derived in part from burning of high-sulfur coals in Britain, had severely affected rivers and lakes of southern Sweden and Norway. (Norway and Sweden receive 90–95% of their SO$_x$-related pollution from other countries.) Subsequent meteorological studies showed enormous international flows of air pollution, generally moving from west to east with the prevailing winds across Europe. By 1994, satellite monitoring had shown a region of regular, highly acidic precipitation stretching from Britain to all the way to Central Asia.

✵ The Norwegian writer Henrik Ibsen is generally regarded as one of the great playwrights of the late 19th and early 20th centuries, perhaps one of the best ever. Many of his plays are noted for a realistic portrayal of the tragic aspects of society. Some of his better-known works include *Peer Gynt*, *Hedda Gabler*, and *A Doll's House*. Ibsen's major works are always in print, including inexpensive paperback editions.

DEALING WITH SO$_x$ AND NO$_x$ EMISSIONS

The 1990 Clean Air Act has made the reduction of SO$_x$ and NO$_x$ emissions a major priority for the United States. This legislation seeks to reduce annual emissions of SO$_x$ from the 20 million tons produced in 1980 to 10 million tons. NO$_x$ emissions will be cut by two to four million tons annually. Most of the acid rain provisions of the 1990 Clean Air Act are aimed primarily at power plants, which are responsible for SO$_x$ emissions put into the air yearly by combustion. There will certainly be an economic impact of this legislation. Across the country, the average utility bill will go up by at least 2.5%, and possibly over 5%. However, in some locations the increase could be in the range 10–15%.

Most of the industrialized nations are making, and indeed have made, significant strides in controlling SO$_x$ and NO$_x$ emissions. In our new century it will probably be the developing or underdeveloped nations that will face the most significant threats from acid precipitation and that will become significant emitters of SO$_x$ and NO$_x$. One reason for this is that many of these countries lack the legislative or regulatory structures needed to control emissions. Second, impoverished countries lack the financial resources to install modern, but expensive, emission control equipment. A third reason is that these countries have rapidly growing populations; when there are more people, there is more energy consumption, and that increased energy consumption brings with it increased emissions to the environment.

Sulfur Oxides

The fact that most of the anthropogenic SO_x emissions come from a limited number of stationary sources makes the problem of reducing or controlling these emissions easier to attack. We have introduced (Chapter 18) some of the approaches to doing this. They include the precombustion strategies of coal cleaning, switching to a fuel of lower sulfur content, or switching to a different kind of fuel, as well as the postcombustion strategy of flue gas desulfurization.

Coal cleaning is reasonably easy. Most coal cleaning processes take advantage of some difference, often density, between the carbonaceous portion of the coal and minerals associated with it, as described in Chapter 29. The process can add several dollars per ton to the cost of coal.

Large amounts of SO_x emissions are produced in the states located around the Ohio River Valley. These states have large numbers of power plants and heavy industries that burn coal. An approach to reducing the SO_x emissions in this region would be to switch these facilities to coals of lower sulfur content. Shipment of low-sulfur coal to the Midwest and northeast would roughly double the cost of fuel to these facilities. Not only would the fuel cost increase, but there also would be significant social and economic impacts on those states that produce high-sulfur coal. As a rule of thumb, the low-sulfur coal in the United States is located mostly in states west of the Mississippi, while coal in states east of the Mississippi tends to be high sulfur. Therefore, converting utilities and industries in the Ohio River Valley and elsewhere in the northeast to low-sulfur, western coal could cost many tens of thousands of jobs in such coal-producing states as Pennsylvania, Kentucky, Illinois, Indiana, and Ohio. The economic impact on those states would be well into the billions of dollars. Nevertheless, switching to different coal supply, of lower sulfur content, may be the least expensive (least expensive, that is, to the consumer of the coal) of the possible strategies for reducing SO_x emissions.

Flue gas desulfurization (FGD) is carried out in scrubbers. The principle of operation is that SO_x and its solutions in water are acidic, and therefore will react chemically with something that is basic or alkaline. The cheapest bases are lime (calcium hydroxide, $Ca(OH)_2$) and limestone (calcium carbonate, $CaCO_3$). We can illustrate one reaction inside a scrubber by presuming that the SO_x is in the form of SO_3, and that lime is the reagent being used to achieve desulfurization.

$$Ca(OH)_2 + SO_3 \rightarrow CaSO_4 + H_2O$$

The calcium sulfate product appears as an insoluble precipitate, usually in the form of gypsum. This material can be removed from the system as the so-called scrubber sludge. There is some interest in possible commercial applications of sludge, such as using it to manufacture gypsum wallboard. A diagram of the scrubber operation is shown in Figure 30.7.

It is very important to notice that a scrubber does *not* destroy pollution. What the scrubber does is to convert the pollution from a rather dilute, difficult to handle form—a small concentration of SO_x in flue gas—into a far more concentrated form (scrubber sludge) that is much easier to collect and handle.

Scrubbers work. They can remove more than 90% of the SO_x from flue gas. Since the mid-1970s, SO_x emissions from the United States and Canada have been reduced by about 30%, even though coal use was increasing.

A combination of coal cleaning and FGD allows a power plant to meet stringent SO_x emission regulations. However, several economic issues must be taken into account. First, scrubbers are not cheap. They can represent about a third of the capital cost of a power

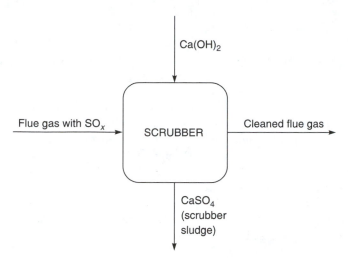

FIGURE 30.7 A scrubber removes SO$_x$ from flue gas by contacting the gas with a base, such as lime (calcium hydroxide). The sulfur is captured as the insoluble calcium sulfate (gypsum, or 'scrubber sludge').

plant, and the cost of the scrubber must be recovered somehow (in the form of the rate paid for electricity by the utility's customers). Second, scrubbers themselves require energy to operate. This energy has to come from somewhere, and the obvious source is the power plant itself. The use of some energy internally in the plant for scrubber operation necessarily reduces the net amount of energy available for sale to customers. In a plant without a scrubber, about 34% of the chemical energy in the coal is converted to electricity leaving the plant to the consumers. In an otherwise identical plant with a scrubber, the comparable figure might be about 30%. If the sludge cannot be marketed as a useful by-product, then there will be a disposal cost associated with handling the sludge. Cutting SO$_x$ emissions in half will cost well over a billion dollars per year. Most of this cost is likely to be passed along to the consumer in the form of electric rate increases.

Another approach to dealing with SO$_x$ emissions is the concept of 'emissions trading'. In this system, each company or utility that produces SO$_x$ would operate under a permit that specifies the maximum level of SO$_x$ that can be legally released. Exceeding the maximum could result in fines of up to $25,000 per day. Although one goal of the system is simply that each company achieve, and not exceed, its individual allowance level, some companies will actually perform better. Their SO$_x$ emissions will be below the legally permitted level. A power plant with emissions below its allowed level is assigned pollution 'credits.' Each credit conveys the right to emit one ton of SO$_x$. These credits can be sold to another plant that is unable to meet its emission allowance. There is thus a financial incentive for power producers to achieve significant reductions of emissions, because they can sell their extra credits to companies that are not meeting targets. Buying credits by the companies that cannot meet the SO$_x$ standards will allow them to continue operation, but few companies would be willing to tolerate for long the continuous expense of having to buy pollution credits. Thus these companies would have a financial incentive to work toward reducing their own emissions. These credits have been bought and sold in private transactions and in public auctions. There is even a commodity trading market in emission allowances on the Chicago Board of Trade. Prices have ranged up to $450 per credit.

Nitrogen Oxides

We've discussed previously some of the difficulties of dealing with NO_x reduction from power plants (Chapter 18). There are a number of reasons why NO_x has proven to be much harder to deal with than SO_x. Chemically, it is harder to remove nitrogen from the fuel than it is to remove sulfur. In the specific case of coal, much of the sulfur is present in various minerals (such as pyrite) that can be removed easily and cheaply. This isn't the case with nitrogen. Nitrogen is chemically bonded to the fuel molecules, and its removal would require a chemical process, rather than a relatively simple and inexpensive physical separation.

Second, a main reason why scrubbers work the way they do is that the product, calcium sulfate, is insoluble in water, so is relatively easy to collect and remove. The analogous nitrogen compound, calcium nitrate, remains soluble in water. In fact, the nitrate salts of all elements are readily soluble in water, so it would be much more difficult to collect and remove them.

Third, the formation of thermal NO_x is a persistent problem in high-temperature combustion equipment. Our customary approach to achieving high efficiencies is to make the value of T_{high} in the Carnot equation as high as possible, but this is exactly the situation that favors formation of thermal NO_x. This situation provides a difficult dilemma for combustion engineers who strive, on the one hand, for high system efficiencies, and, on the other hand, for low thermal NO_x production.

There is, as yet, no technology equivalent to FGD except 'flue gas denitrogenation,' which is in use on a similarly large scale in fossil-fuel-fired power plants. A variety of approaches to the problem are at various stages of research, development, or demonstration, in part under the aegis of the U.S. Department of Energy's Clean Coal Technology program. Selective catalytic nitrogen oxide reduction involves the reaction of ammonia with NO_x in the flue gases at relatively low temperatures. The reaction that takes place transforms nitrogen oxides, with about 80% conversion, into nitrogen and water vapor:

$$2NH_3 + NO + NO_2 \rightarrow 2N_2 + 3H_2O$$

Reburning is a technique aimed primarily at reducing formation of thermal NO_x. Coal is burned in the lower part of a boiler to provide 80–90% of the heat needed to generate the steam. Natural gas is then added in a 'reburn' region higher in the boiler. The constituents of the natural gas, such as methane, react with any nitrogen oxides produced by the coal combustion, and decompose them into nitrogen. A gas reburning system can reduce NO_x emissions by about 40%. Reburning can be combined with other techniques to reduce emissions even further. A related strategy is cofiring, in which coal and natural gas are burned simultaneously. Since there is essentially no sulfur in natural gas, SO_x emissions are reduced in direct proportion to the amount of gas that is cofired with the coal. NO_x emissions are also reduced.

The reduction of total NO_x emissions nationwide is particularly challenging because most of the sources are small, individually owned, and mobile. There are almost 150 million of them in the United States—cars and trucks. About half of the NO_x emissions in the United States comes from the transportation sector. It is chemically possible to reduce the NO_x emitted by cars and trucks by catalytic converters, which we will discuss in the next chapter. As the exhaust gases cool, there is a tendency for the thermal NO_x to decompose back to its constituent elements, nitrogen and oxygen. Normally this reaction proceeds slowly, but the appropriate catalyst can significantly increase its rate and decrease the amount of NO_x emitted. Catalytic converters reduce NO_x emissions by about 60% over the life of a vehicle.

Though SO_x emissions have been decreasing in the United States since the 1970s, NO_x has been increasing. In the past 50 years, NO_x emissions have doubled in the United States and tripled in Canada. Even though catalytic converters are doing a good job of reducing the NO_x emissions from each vehicle, the total number of vehicles in the United States is steadily rising. Since 1980, three years after the mandatory introduction of catalytic converters, NO_x emissions in North America have begun to level off, but the sheer number of new cars on the road every year keeps the total NO_x emissions high.

An alternative is to eliminate the internal combustion engine altogether, or at least to reduce the use of such engines. Not too many years ago, such a suggestion might have been considered a sacrilege by many people. Now, though, there is a small but growing interest in electric vehicles and hybrid vehicles that combine electric and internal combustion systems. Several major international car companies, such as Toyota and Honda, are marketing hybrid vehicles. These new cars offer the possibility of major reductions in NO_x emissions.

CITATION

1   Kennan, George F. *Memoirs 1950–1963*. Little, Brown: Boston, 1972; p. 257.

FOR FURTHER READING

Gebelein, Charles G. *Chemistry and Our World*. Brown: Dubuque, IA, 1997. An excellent introductory chemistry text with many real-world examples. Chapter 18, on air pollution is particularly relevant here.

Harte, John. 'Acid rain.' In: *The Energy–Environment Connection*. Hollander, Jack M. (Ed.). Island Press: Washington, 1992. A fine overview that covers the origins of acid rain, damage to organisms and ecosystems, and possible activities to address the problem.

McNeill, J.R. *Something New Under the Sun*. Norton: New York, 2000. Chapters 3 and 4 of this book discuss what we are doing to the atmosphere, including the problem of acid rain.

Pearce, Fred. *Acid Rain*. Penguin: Harmondsworth, UK, 1987. A discussion, easily read by persons of limited science background, on the problems of acid rain. The examples are primarily from British and European perspectives.

Tyson, Peter. *Acid Rain*. Chelsea House: New York, 1992. An overview of the sources of acid rain and environmental issues stemming from it. Parts are now somewhat dated, but it still makes a good overview for the reader with little scientific background.

31

VEHICLE EMISSIONS AND EMISSIONLESS VEHICLES

The exhaust from gasoline-engine automobiles and light trucks contributes to three of the most serious pollution problems in the world today: acid rain (Chapter 30), global warming (Chapter 32), and photochemical smog, which we discuss in this chapter. Photochemical smog first entered the public's consciousness in the 1940s, especially in Los Angeles. Smog can be so irritating to the eyes and respiratory tract that, during the Second World War, some people thought the smog was actually a poison gas attack by the Japanese. Within 20 years, smog was stunting tree growth 50 miles away from the city center. By 1976, Los Angeles' air had reached officially unhealthy levels three days out of four.

Air pollution from automobiles kills cells in the nose and causes burning in the eyes and throat. It damages the sensitive cilia (tiny air-cleansing hairs) that line the bronchiole tubes and may cause the lungs to age prematurely. It adds to the breathing problems of asthmatics and reduces lung function in healthy people.

This is not a problem confined to the United States. Air pollution from cars and light trucks has become a problem of serious proportions in at least half the cities of the world. Cities such as Athens, Budapest, and Mexico City now have to enforce emergency driving bans on hazy days to minimize health risk to their citizens. Car sales continue to grow in western Europe, Japan, and the United States. During the 1980s and early 1990s, greatest growth took place in the emerging industrialized nations of Asia, such as Korea and Taiwan. Rapid growth in sales of cars and light trucks is likely to occur in developing regions of Africa, Asia, and Latin America, and in the countries of the former Soviet Union and eastern Europe. This is of concern because most developing nations have few laws or regulations on automobile exhaust controls.

There are close to a billion cars and light trucks on the world's roads—more than double the number of vehicles that had been on the roads in the late 1980s. Part of the concern for the United States is that, though we have only about 4% of the world's population, we own half of all the cars in the world.

TAILPIPE EMISSIONS

Let's begin by making a survey of the possible emissions from cars. First, we review combustion reactions. When we discussed natural gas, we indicated three kinds of combustion conditions:

Complete combustion $CH_4 + 2O_2 \rightarrow CO_2 + 2H_2O$
Incomplete combustion $CH_4 + 1.5O_2 \rightarrow CO + 2H_2O$
Oxygen-starved combustion $CH_4 + O_2 \rightarrow C + 2H_2O$

These conditions also apply to the combustion of gasoline. Although gasoline is a mixture of many individual components, for simplicity we can represent gasoline as, say, C_7H_{16}. Then we have

Complete combustion $C_7H_{16} + 11O_2 \rightarrow 7CO_2 + 8H_2O$
Incomplete combustion $C_7H_{16} + 7.5O_2 \rightarrow 7CO + 8H_2O$
Oxygen-starved combustion $C_7H_{16} + 4O_2 \rightarrow 7C + 8H_2O$

Notice that in both cases there is a drop in the ratio of air (or oxygen) to fuel (Table 31.1) in going from complete to incomplete to oxygen-starved combustion.

Sulfur in the fuel is converted to SO_x. For gasoline-fueled vehicles, the sulfur content of the fuel is so low that, for all practical purposes, we can neglect automobiles as a potential source of SO_x emissions.

Nitrogen in fuel is converted to NO_x. We must remember that there are two sources of NO_x emissions. Fuel NO_x arises from nitrogen atoms chemically bonded to fuel molecules being converted to NO_x as the fuel burns. Thermal NO_x arises from nitrogen molecules (N_2) in the air being converted to NO_x as a result of the very high temperatures of combustion.

There is one more kind of emission. Suppose that some of the fuel does not burn at all. This kind of emission is called **unburned hydrocarbon**. Unburned hydrocarbon emissions have several sources. Some fuel molecules may actually sweep through the engine without being burned. Some gasoline evaporates from fuel tanks and fuel lines. Some is spilled on the ground (many of us do this occasionally at the self-service gasoline pumps of service stations) and evaporates.

These emissions are of concern for a number of reasons. First, carbon monoxide is lethal. Much of the 200 million tons of CO emitted per year is from cars. In the United States, about two-thirds of total CO emissions are from cars and light trucks. Exposure to carbon monoxide can lead to death, for example, when people accidentally operate their cars in closed garages (as when tuning up the engine at home). In some tragic cases, this is

TABLE 31.1 RATIO OF AIR OR OXYGEN TO FUEL FOR THE THREE KINDS OF COMBUSTION

	Air(Oxygen)/Fuel Ratio	
	Natural Gas	Gasoline
Complete combustion	2	11
Incomplete combustion	1.5	7.5
Oxygen-starved combustion	1	4

done deliberately as a form of relatively painless suicide. The risk arises from the fact that CO enters the bloodstream and disrupts the delivery of oxygen to the body's tissues by combining with hemoglobin, the molecule that normally carries oxygen. Because it combines more tightly with hemoglobin than does oxygen, it steadily reduces the oxygen-carrying capacity of the blood. At high concentrations, CO is deadly. At lower concentrations it can lead to angina, blurred vision, and impaired physical and mental coordination.

Carbon monoxide is a major pollutant in all cities. Many city dwellers—about 80 million Americans—suffer from exposure to elevated levels of CO. In 1970, the government set a standard specifying an area to be unhealthy if subjected to levels of over nine parts per million (ppm) during an eight-hour period at least twice a year, or if the level exceeded 35 ppm for two or more one-hour periods per year. Workers in New York City's garment district apparently experience a concentration of CO greater than 100 ppm every day.

Carbon monoxide from car exhausts may cause potentially fatal heart rhythm disturbances in otherwise healthy people. Joggers who exercise close to a busy highway could be affected by inhaling too much carbon monoxide. Since people sitting in nearby cars are affected, it is particularly hazardous to sit in traffic jams, when about 10% of the gas emitted from the exhaust is CO.

Unburned hydrocarbon emissions represent a significant waste of fuel. These losses amount to about 3,000 tons per *day* in the United States, just as vapor from filling station pumps or spills.

As we've seen (Chapter 30), NO_x contributes to acid rain formation. Cars create about one-third of the NO_x in the environment. Though significant NO_x is also produced in fossil-fuel-fired electricity-generating stations, worldwide, transportation is the largest source.

Nitrogen dioxide, NO_2, is a reddish-brown gas that is toxic when concentrated; the maximum safety limit for NO_2 is 5 ppm, but even at this very low level NO_2 causes eye and lung irritation. NO_2 causes the distinctive brownish haze typical of photochemical smog. NO_2 can irritate lungs and lower resistance to respiratory infections such as influenza.

Dealing with NO_x emissions is complicated by their being generated from natural, as well as human, sources. Natural sources include lightning discharges and production by bacteria. Outside of urban areas, large amounts are produced by burning of trees and other plant materials. Most NO_x eventually washes out of the atmosphere in acidic precipitation. Though we have seen (Chapter 30) the problems associated with acidic rain, it is nevertheless a way for green plants to get some of the nitrogen needed for their growth. The concern that focuses on NO_x emissions from automobiles is that they concentrate mostly in the urban areas where there are the most vehicles and, of course, the most people.

HOW SMOG FORMS

In addition to the concerns discussed above for carbon monoxide, NO_x, or unburned hydrocarbons considered singly, a significant air pollution problem occurs when these three substances react together to form smog. The complex chemical reactions involved in smog formation are promoted by the energy in sunlight. For that reason, this kind of smog is sometimes called **photochemical smog.** Smog formation is undesirable for a variety of reasons. Eye irritation occurs because some of the compounds produced during smog formation are chemically similar or identical to the components of tear gas. Respiratory tract irritation can occur even in otherwise healthy adults by exposure to smog. It is very harmful to persons with chronic respiratory problems, such as emphysema, and to infants whose respiratory tracts may not yet be fully developed. Smog is also aesthetically undesirable; thanks to the NO_x, it tends to have a dull grayish-brown color. Photochemical smog is typical in cities where sunshine is abundant and internal combustion engines exhaust large quantities of pollutants to the atmosphere.

The adjective *photochemical* is used to differentiate this form of air pollution from sulfuric acid smogs, which are quite different in origin. Sulfuric acid smogs are produced when high-sulfur-content coal is burned without emission control equipment. A droplet of water forms around a fly ash particle, and SO_x dissolves to form a solution of sulfuric acid. Sulfuric acid smogs have been implicated in major air-pollution disasters, such as the one in Donora, Pennsylvania in 1947. Because coal is only rarely used as a fuel for home heating any more, and because coal-fired electricity-generating plants have scrubbers, sulfuric acid smogs are virtually a thing of the past in most industrialized nations. Nowadays when the term 'smog' is used without a modifier, it can be taken to mean photochemical smog.

The chemical processes by which the three primary pollutants—carbon monoxide, unburned hydrocarbons, and NO_x, are converted into the secondary pollutants found in photochemical smog begins with the breakdown of nitrogen dioxide:

$$NO_2 \rightarrow NO + O$$

The energy for this process is provided by sunlight. Notice that one of the products of this reaction is not the O_2 molecule, familiar as a component of the atmosphere, but rather an oxygen atom, O. Oxygen atoms are extremely reactive.

If there were no other pollutants around, this breakdown of NO_2 would not be a concern. The oxygen atom would react with an oxygen molecule to form ozone:

$$O + O_2 \rightarrow O_3$$

There is little doubt that ground-level ozone concentrations have increased in recent years. Since O_3 is not a normal industrial pollution product, this increase is almost certainly due to photochemical pollution. The observed average atmospheric concentration of about 20 parts per billion (ppb) is below the safety limit of 80 ppb, but in photochemical smog, concentrations rise. The ozone would react with nitrogen monoxide to re-form the original nitrogen dioxide:

$$O_3 + NO \rightarrow NO_2 + O_2$$

The net effect, other than absorbing some energy from the sunlight, is that nothing much would have changed.

Ozone is quite toxic. Its safety limit is only 0.08 ppm. In the ozone layer of the stratosphere, O_3 protects against excessive ultraviolet radiation. At or near ground level, however, ozone is a serious health and environmental hazard. At concentrations common in many cities, some healthy adults and children experience coughing, painful breathing, and temporary reduction of lung function after an hour or two of exercise. The effects of long-term exposure to elevated levels of ozone are uncertain.

But of course there *are* other pollutants around, so that this innocuous scheme that absorbs sunlight and recycles other chemicals is not the one that actually occurs. Some of the reactive ozone molecules and oxygen atoms instead react with other species in the atmosphere, leading to smog formation. Air in an urban environment contains unburned hydrocarbons from car exhausts, evaporation, and spillage at filling stations. Atomic oxygen reacts with some of these hydrocarbons to form reactive, short-lived intermediate compounds. Those, in turn, react to form secondary pollutants such as aldehydes (e.g., formaldehyde⚛). As little as 0.2 ppm of nitrogen oxides and 1 ppm of reactive hydrocarbons are sufficient to initiate these reactions.

⚛ Pure formaldehyde is a gas. The compound is perhaps best known as its solution in water, usually called formalin. The water solution is used as a preservative for biological specimens and in embalming fluids. It is also used as a germicide and fungicide. The odor of formalin is probably unforgettable to those who have spent much time doing dissections in biology labs. Formaldehyde itself is extremely irritating to the mucous membranes of the respiratory system. It may also be a carcinogen.

A compound related to formaldehyde, acetaldehyde (which also has potentially severe effects on the respiratory system), undergoes a sequence of reactions with oxygen and NO_2, eventually forming peroxyacetylnitrate, $CH_3O_3NO_2$. Indeed, there is a family of related compounds, the peroxyacylnitrates, often called PANs. As pollutants, the PANs are much worse than either NO_x or O_3. The first step in their formation is the reaction of O_3 with a hydrocarbon from unburned gasoline to form an aldehyde, which then reacts with O_2 and NO_2. If we use a simple hydrocarbon such as ethane as an example, the reaction proceeds as

$$O_3 + CH_3CH_3 \rightarrow CH_3CHO + O_2$$

$$CH_3CHO + O_2 + NO_2 \rightarrow CH_3CO_3NO_2$$

One of the PANs, peroxybenzoylnitrate, irritates the eyes a hundred times worse than does formaldehyde. Concentrations as low as 0.02 ppm cause moderate to severe eye irritation. Peroxyacetylnitrate, the compound whose formation is illustrated in the equations directly above, is even worse.

CATALYTIC CONVERTERS

Cars and light trucks are the leading cause of photochemical smog. The rates at which the smog-forming reactions take place depend on several factors, including ambient temperature, the amount of available sunlight, and the presence of chemicals, such as unburned hydrocarbons and relative amounts of hydrocarbons and NO_x, in the air.

We have seen the case of electricity-generating plants, where we can add scrubbers and baghouses to help remove or reduce potential emissions. From an emissions perspective, power plants are called stationary sources. One virtue of a stationary source is that, as the name implies, it doesn't move around. It stays put, and we can add hardware to it to help minimize its impact on the environment. A second virtue is that, usually, we need not worry about the size, weight, or complexity (disregarding, for the moment, cost) of emission control devices on stationary sources. Automobiles and other vehicles are mobile sources. Here the size, weight, and complexity—and of course the cost—of emission control devices are a concern. The strategy is to destroy the emissions before they can exit the tailpipe. The potential emissions that are of special concern are the ones that we have been discussing here: carbon monoxide, unburned hydrocarbons, and thermal NO_x.

The way we try to destroy these potential emissions is in a device called a **catalytic converter**. The basic technology for the catalytic converter came into use in the 1970s and has improved steadily since then. Every American car made since 1975 has a catalytic converter that contains a catalyst in which the active species is platinum or palladium. The catalytic converter is designed to accelerate three chemical reactions:

$$CO + 0.5O_2 \rightarrow CO_2$$

$$C_7H_{16} + 11O_2 \rightarrow 7CO_2 + 8H_2O$$

$$2NO_2 \rightarrow N_2 + 2O_2$$

where, in this example, we've used C_7H_{16} to represent a component of unburned gasoline and NO_2 to be a component of NO_x. Notice that in two of these reactions we seek to *add* oxygen, and in one reaction we seek to *remove* oxygen. It is a very tricky business to do both (i.e., add and remove oxygen) with the same catalytic unit. How do we achieve this?

To answer this question, we first need to learn some of the jargon of the business. When the exact amount of oxygen needed for complete combustion is available, we call this condition **stoichiometric combustion**. If there is more air (oxygen) present than is needed for stoichiometric combustion, this condition is **fuel-lean** combustion. If there is not enough air (oxygen) available for stoichiometric combustion, this condition is **fuel-rich** combustion. We also need to recall that the amount of heat produced drops as we go from complete to incomplete to oxygen-starved combustion.

Now let's see what happens in our engines. At fuel-rich conditions, we don't have enough air (oxygen) for stoichiometric combustion. We should expect that emissions of carbon monoxide and unburned hydrocarbons will be high. In moving from fuel-rich to stoichiometric to fuel-lean combustion, more and more air (oxygen) becomes available, so we expect the carbon monoxide and unburned hydrocarbons to drop. As we move from fuel-lean to stoichiometric to fuel-rich conditions, the amount of heat produced decreases, so the temperature in the engine cylinder will be reduced, and thermal NO_x formation drops. We can summarize this as shown in Figure 31.1.

The catalytic converter works best when the concentrations of all three potential pollutants are high, around the point at which the curves in Figure 31.1 meet. The point where we get the highest simultaneous concentration of carbon monoxide, unburned hydrocarbons, and NO_x is right around stoichiometric combustion. In other words, the point of stoichiometric combustion in the engine is where the catalytic converter works best. It so happens that near-stoichiometric combustion is where your engine works best. Thus by making sure that you keep your engine tuned up, and running near the stoichiometric combustion condition, you help reduce emissions to the environment (because this is where the catalytic converter works best), get the best performance from the engine, and, likely, reduce the need for major maintenance on the engine.

There is little question that the catalytic converter is a major technological success. Fifteen years after the introduction of the catalytic converter, the average American car emitted 96% less carbon monoxide, 96% fewer hydrocarbons, and 76% less nitrogen oxides than did the cars of 1970, before catalytic converters were mandated. Unfortunately, vehicle exhaust pollution remains a serious and growing problem around the world. This is *not* a result of poor design or improper functioning of the catalytic converter. The problem, rather, is that there are a lot more of us, more people are able to buy cars, and those who own them are driving more. In other words, the increasing

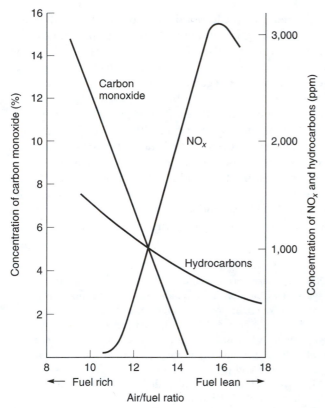

FIGURE 31.1 This graph shows how the levels of the tailpipe emissions vary as the combustion conditions in the engine change from fuel-rich to stoichiometric to fuel-lean.

number of cars on the roads around the world, and the greater use of these cars, has offset these emissions reductions *per car* achieved by the use of catalytic converters. For example, Americans drove 50% more miles in 1990 than they did in 1970, and most of that driving was in stop-and-go traffic, where more pollutants are emitted.

The next steps to reduction of vehicle emissions, after implementation of reformulated gasoline (discussed in the next section), are likely to be the mandatory removal of the oldest cars from the road. About 80% of all exhaust emissions are produced by 20% of the oldest vehicles. As older cars wear out and are replaced by newer models, emissions from vehicles should decrease, even with an increase in the number of vehicle-miles driven.

REFORMULATED GASOLINES

By the late 1980s, it became apparent that, even with converters, further reductions in emissions were needed. Although there remains some room for improvement of the catalytic converter, despite its acknowledged success in fighting pollution from cars, the present-day catalytic converter is approaching the limits of what we can do in the tailpipe of the car. The next step is actually to change the fuel—that is, reformulate the gasoline being burned.

The approach to **reformulated gasolines** is based on several observations. The lower the boiling point of a compound, the easier it is to evaporate. Thus we can reduce evaporation losses by eliminating or reducing the lowest boiling components of the gasoline—in other words, by shifting the boiling range upward. Aromatic compounds contribute to smoke and soot formation. They are also of concern as suspect carcinogens. Reformulated gasolines will essentially eliminate the presence of aromatics. Compounds containing an oxygen atom, generically called **oxygenates**, for example methyl tertiary-butyl ether, MTBE,⚛ are, in a sense, already partially combusted. Therefore, when they do burn in the engine their combustion temperature is lower. This helps reduce formation of thermal NO_x. Furthermore, oxygenates bring some oxygen in with the fuel, in addition to what comes in with the air, shifting the combustion to fuel-lean conditions. As a result, carbon monoxide and unburned hydrocarbon formation are reduced. As an added bonus, oxygenates have very high octane numbers, and their addition boosts the octane rating of gasolines (which would otherwise be reduced by removal of aromatics). The trick with oxygenates is finding ones that, unlike MTBE, do not pose a hazard. Thus the production of reformulated gasoline involves shifting the boiling range up, removing aromatics, and adding oxygenates.

⚛ For a time, in the 1980s and 90s, MTBE as a gasoline additive was considered to be splendid advance in the fight against air pollution from vehicles. Its future seemed so bright that there were concerns that there might not be enough production capacity to meet the expected demand. Then it was recognized that small quantities of MTBE will dissolve in water. If oxygenated gasoline leaks from storage tanks and gets into the water supply, some of the MTBE in the gasoline will dissolve in the water. Even small quantities of MTBE in water give it an offensive odor and taste. Concerns have also been raised about the possible toxicity of MTBE if it is ingested with drinking water. Though it appears now that the human health effects of MTBE may be less severe than once feared, nevertheless the use of MTBE may be restricted or eliminated. If so, MTBE will join a huge array of other technological developments of which it could be said, 'It seemed like a good idea at the time.'

LEVs, ULEVs, ZEVs, AND HYBRIDS

The catalytic converter was, and is, a superb technological advance in fighting air pollution caused by motor vehicles. The use of reformulated gasolines adds another step toward a cleaner environment. There still remain, though, areas where smog remains a problem. The catalytic converter represents a change to the conventional automobile. Reformulation changes the fuel. If these two steps together are not adequate to improve air quality to acceptable levels, then the last major step to be taken is to change the vehicle itself, specifically, to do away with the internal combustion altogether, or to combine a small gasoline engine with some other form of propulsion (usually electric motors). Protection of air quality by addressing emissions from motor vehicles has led to consideration of low-emission vehicles (LEVs), ultra-low-emission vehicles (ULEVs), and zero-emission vehicles (ZEVs). The strategies for designing such vehicles are generally those steps we've just mentioned: either relying entirely on electricity for propulsion, or on a combination—a hybrid—of gasoline and electric propulsion.

Ferdinand Porsche built the first hybrid vehicle, one that used both an electric motor and an internal combustion engine, in 1902. Hybrid vehicles can be thought of as being electric cars that are equipped with a small internal combustion engine in addition to the battery-driven (or fuel-cell-driven) electric motor. This gives the immediate advantage that they are not limited in driving range by the energy in the batteries. The range of the hybrid car is comparable to that of conventional gasoline-powered vehicles. Indeed, hybrid vehicles offer virtually all of the advantages of gasoline vehicles combined with most of the advantages of electric vehicles. A hybrid vehicle has electric batteries and an electric motor connected to the drive wheels, along with a gasoline-powered engine. The gasoline engine has sufficient horsepower for reasonable driving performance, and confers the added advantage that there is no need to 'plug in' the car to recharge the batteries. Instead, the batteries are kept charged by the generator connected to the engine. Fuel mileage is close to 70 miles per gallon; some anecdotal reports suggest that even better mileage can be gotten. A hybrid car is not a ZEV, because the gasoline engine will still produce some emissions. However, because the hybrid will run on electricity some of the time, it offers very low emissions compared to the conventional gasoline-fueled cars now on the road. Hybrid car technology has evolved to a point at which commercial models are available from some of the major car companies (Toyota and Honda being notable examples). Virtually all of the major companies (but apparently not, ironically, Porsche) are at least planning a hybrid. Now that hybrids are being used by the public, we will soon learn about their long-term durability and maintenance. Hybrids are comparatively complex because they have two drive systems, so may prove to be expensive to maintain.

The use of these vehicles can be required by some form of legislation at state or federal level. In the United States, California is the leader in setting legislative goals for introducing such cars, with the laudable intent of improving air quality, particularly in congested urban areas. However, genuine acceptance is more likely to come from the public's wanting to own such cars, not being required to do so. A significant shift away from our traditional gasoline-fueled internal combustion engine vehicles will occur when consumers decide that they really want hybrids or electric cars. What will that take? The gasoline-engine vehicle cannot be criticized on the basis of its performance; even small, inexpensive vehicles can easily reach speeds well above any legal limits. Today's vehicles also offer an extraordinary range of amenities and accessories. It's unlikely that hybrid or electric vehicles would be able to demonstrate clear superiority to gasoline-fueled vehicles either in speed or in comfort.

Consumer acceptance is a necessary criterion for the success of new vehicles, but not the only one. Modern gasoline-fueled vehicles are remarkably reliable. They usually require little major maintenance for many years, and can now go for long periods between routine tune-ups. We are used to that. New vehicles will have to approach that level of technical reliability. Similarly, most of us, except perhaps in small towns and rural areas, are used to having a gas station within close proximity, often open round-the-clock every day of the year. A few minutes spent at the gas pump suffices to refuel the vehicle. We are used to that, too. Ideally, new vehicles will have to be equally easy to refuel or recharge, and there will have to be an infrastructure in place to allow that to happen. A major switch in consumer preference for a new type of vehicle won't be triggered by vastly improved comfort or speed, and there may even be for a time some inconvenience in refueling or recharging. Likely, a wide acceptance of electric or hybrid vehicles may require a significant change in consumers' attitudes, a change perhaps brought about by concerns for the environment.

It is unrealistic to expect that there will be an instantaneous change-over from gasoline vehicles to electrics or hybrids. Rather there will likely be a long period of transition when all of these kinds of vehicles are on the road. The most likely 'transition car' from the gasoline vehicle to an all-electric vehicle is the hybrid.

Early in the 20th century, electric cars, using lead–acid batteries, were actually more popular than gasoline-powered vehicles. However, once car owners began to drive their newfangled vehicles over longer distances, it soon became apparent that gasoline offered an enormous advantage in energy density. The energy density is the amount of energy provided per volume, or per weight, of an energy source. Lead–acid batteries are heavy. (This can be verified by anyone who has ever changed the battery in his or her own car, or simply by feeling the weight of a battery in the auto parts section of a store.) Unfortunately, a lead–acid battery provides about 1% of the energy that can be obtained by burning an equal weight, or equal volume of gasoline. As a result, gasoline-fueled vehicles have a much greater range, and quickly established their dominance in transportation. Today, for every one electric vehicle on the road, there are about *eighty thousand* gasoline-fueled cars and light trucks.

Electric cars are perceived to have significant range and speed limitations. Both of these problems result from limitations in present battery technology. The most common batteries available for use in electric cars nowadays are the lead–acid storage batteries, similar to those used for starting gasoline-fueled cars. Some modifications in design would be made to adapt them to electric cars, but the basic principle of these batteries is at least a hundred years old. Lead–acid batteries in electric vehicles would result in driving ranges of 60–100 miles at top speeds of 50–60 miles per hour. To be sure, this limited range is not a problem for an 'around town' vehicle used for commuting to work, running errands, and other short trips. The average commuting distance is only about 12 miles, and 90% of all vehicle trips for any purpose are still under 60 miles in length. On the other hand, for those who must frequently drive long distances, or who must have some sort of emergency capability in a vehicle, electric cars could still serve as a second vehicle. With advanced battery technologies, now a subject of a significant national research effort, electric cars will be designed and built to accelerate rapidly and to cruise easily at interstate highway speeds.

Electric vehicles certainly offer tremendous advantages for the possible reduction of smog, global climate change (Chapter 32), and acid rain. They have no emissions of unburned hydrocarbons, NO_x, SO_x, carbon dioxide, carbon monoxide, or soot. Furthermore, an electric motor is quieter in operation than internal combustion engines, so electric cars would also help mitigate 'noise pollution' associated with vehicle traffic. However, when we consider an electric car as a zero-emission vehicle, it is important to remember that the adjective zero-emission refers to the *vehicle*. The electricity used to charge the batteries has to come from someplace. If that electricity is generated in a fossil-fuel-fired plant, there will be emissions to the environment from the plant itself. To be sure, controlling emissions from one large, stationary electricity-generating plant is much easier than trying to control emissions from millions of internal combustion engines driving around. Furthermore, the electricity-generating plant is likely to be in a relatively remote area, and not contributing its emissions to a congested downtown. It would be possible to have a zero-emission transportation system if the electricity used for charging the batteries came from such sources as solar (Chapter 36), wind (Chapter 35), hydro-electric, or nuclear plants. As we'll see below, electric cars can also be designed to make their own electricity onboard in fuel cells.

All storage batteries, regardless of their specific design, work on the same general principle. During charging, electrical energy is converted into chemical energy. When the battery is discharging (that is, providing electricity to us), the reverse process occurs, and chemical energy is converted back to electrical energy. One potentially great advantage is that it does not really matter where that original electricity came from. That is, the electricity could be generated in fossil-fuel, nuclear, solar, hydro- or wind plants, providing enormous flexibility for an overall energy economy. In contrast, the choices of fuels for internal combustion engine vehicles are very limited. However, in considering this advantage of electric vehicles operated by storage batteries, we must also take into account the demand for electricity that these cars would create. Replacing 10% of the transportation energy now provided by gasoline with electricity for storage batteries would increase the total load on our electricity-generating capacity by about 20%. Replacing *all* of our gasoline-derived energy with electricity would likely triple our electricity demand.

Additional advantages of electric cars come from the motors and drive systems. For one thing, electric motors are inherently more efficient than internal combustion engines. Second, when an electric car is stationary, the motor does not turn over, whereas in a conventional car the internal combustion engine is still running, consuming fuel, and producing emissions when the engine is idling. Third, electric propulsion systems are much simpler—in terms of the number of components—than those for internal combustion engine vehicles, so are also lighter and smaller. That means they are also less costly to produce and, with fewer parts, need fewer repairs. Periodic tune-ups are eliminated. Finally, electric cars can also be designed to take advantage of so-called regenerative braking. In this technology, when an electric car slows down, the braking energy is recaptured and returned as electricity to the battery. In an internal combustion engine car, the energy generated (often as heat) in braking is wasted.

The charging time, for completely discharged batteries in an electric car, is estimated to be about six hours. Therefore the most convenient period for the owner to recharge them would be overnight. Some commuters do have to put up with very long commutes, possibly close to the vehicle's driving range. For those situations, vehicles could be recharged during the day while the driver is at work, if shopping malls and places of employment offered 'plug-ins' for people to recharge their vehicles. The lengthy charging time would pose a serious limitation for using electric cars in long-distance travel. An intriguing concept to solve this problem is that the car owner would own the vehicle itself but not the batteries. When the batteries were getting low, the driver could pull into a 'battery station' where the entire set of discharged batteries would quickly be swapped for another complete set of charged batteries (no doubt for an appropriate fee).

For all their potential virtues, electric vehicles are not likely to eliminate gridlock, traffic accidents, urban sprawl, or any of the other impacts we discussed in Chapter 22. Taking care of those problems will not be solved by a change in vehicle technology. Rather, major societal changes, and changes in individual behavior will be needed. Such changes may not be made willingly.

It may prove that the future of electric and hybrid vehicles lies not with batteries, but with fuel cells. A fuel cell converts chemical energy directly into electrical energy, just like a battery does, and, also like a battery, has no potentially harmful emissions. The earliest concept of the fuel cell derives from the work of Humphry Davy in the 18th century. Little was done to convert Davy's original concept into a practical, commercial device until the advent of the space program. Fuel cells were quickly found to be useful for operating the electrical equipment of space vehicles. Fuel cells have been used to generate electricity for large communities (e.g., in sections of Tokyo) and to run buses. Today fuel cells are under intense research and development in numerous laboratories around the world. Many designs are being studied. Fundamentally, though, the idea is to conduct in the fuel cell a reaction chemically similar to combustion, but instead of releasing the energy of the reaction as heat, it is produced as electricity instead. One promising reaction is that of hydrogen:

$$2H_2 + O_2 \rightarrow 2H_2O$$

The combustion of hydrogen in air or oxygen can be spectacular (indeed involved in one of the most famous news events of the 20th century, the burning of the dirigible

Hindenberg). The same reaction takes place in the fuel cell, but now the reaction occurs by the relatively slow transfer of electrons. These electrons are transferred via a current in an external circuit, which can be made to do WORK. Fuel cells can be stacked together to increase both the voltage and the current. They have no moving parts, are quiet and environmentally clean. Fuel cells operating on hydrogen produce only water as an emission.

Fuel cells offer the advantage of the battery in an electric vehicle, but without its shortcomings. As long as fuel (for example, hydrogen and oxygen in the example above) is supplied to the cell, it will continue to operate. In a sense, we can think of a fuel cell as a kind of battery that will never go dead. Though hydrogen is thought to be the ideal fuel for a fuel cell, we'll see in Chapter 34 that there are some problems to be overcome in handling and storing hydrogen for vehicle use, particularly to give the vehicle a reasonable driving range. The alternative to storing hydrogen onboard, which is also under active investigation, is to make it as needed. The vehicle would be fueled with a material easier to handle and store such as gasoline, methanol, or methane. Then, as the vehicle is operating, the original fuel is 're-formed' to provide hydrogen. Using the simple molecule of methane as an example,

$$CH_4 + H_2O \rightarrow CO + 3H_2$$

The carbon monoxide by-product needs to be absorbed somehow, or destroyed by oxidizing it to carbon dioxide. However, a ready supply of hydrogen is available without needing to store it. The advantage of relying on gasoline is that the fueling infrastructure is already in place. Furthermore, being able to 'fill up' one's fuel cell vehicle at a conventional, familiar gas station may help greatly with consumer acceptance. Methanol has been considered, both as a source of hydrogen via re-forming and as a possible fuel itself for the fuel cells. We will discuss methanol in more detail in Chapter 34, and see that there are some potentially significant disadvantages to its handling and use. Critics of the onboard re-former approach say that it converts the car into a rolling chemical factory. Perhaps the future will see evolutionary changes such that the first successful fuel cell vehicles will be those with re-formers, and eventually successful designs will be developed for direct use of hydrogen.

FOR FURTHER READING

Aubrecht, Gordon. *Energy.* Merrill: Columbus, 1989; Chapter 21. This chapter discusses the automobile in various contexts, including air pollution, fuel economy, and comparisons with other forms of transportation. Though some of the data are now dated, the approaches and analyses used merit consideration.

Berger, John J. *Charging Ahead.* University of California: Berkeley, 1997; Chapter 23. The book provides an overview of various approaches to renewable energy, and prospects for their commercialization. The chapter cited here relates to electric vehicles, with plenty of examples current through the mid-1990s. Little scientific background is needed on the part of the reader.

Borowitz, Sidney. *Farewell Fossil Fuels.* Plenum: New York, 1999; Chapter 15. This chapter provides a useful discussion of batteries and fuel cells, mostly in the context of their applications in cars. Requires little or no previous scientific background.

Brower, Michael. *Cool Energy.* MIT Press: Cambridge, 1992; Chapter 8. This chapter actually deals with various ways of storing energy. Though it is not devoted entirely to vehicles, it does describe the use of batteries in electric cars.

Edelson, Edward. *Clean Air.* Chelsea House: New York, 1992; Chapter 5. This book provides a useful introduction to air-pollution issues for readers with little or no science background. This chapter discusses air pollution from automobiles.

Ford, Tim. 'Fuel-Cell Vehicles Offer Clean and Sustainable Mobility for the Future.' *Oil and Gas Journal* **1999**, *97*(50), 130–3. A good overview article on fuel cell vehicles.

Gebelein, Charles G. *Chemistry and Our World.* Brown: Dubuque, 1997; Chapter 18. A well-illustrated introductory college-level chemistry text, with plenty of real-world examples. This chapter discusses air pollution as related to energy use, including photochemical smog.

Hollander, Jack M.; Brown, Duncan. 'Air Pollution.' In: *The Energy–Environment Connection.* Hollander, Jack M. (Ed.). Island Press: Washington, 1992; Chapter 2. This chapter is not specifically devoted to air pollution from vehicles. However, it does provide useful information on various pollutants, such as CO and NO_x.

Kraushaar, J.J.; Ristinen, R.A. *Energy and Problems of a Technical Society.* Wiley: New York, 1988; Chapter 13. This chapter discusses pollution of the atmosphere, and includes useful discussions of photochemical smog, CO, NO_x, and catalytic converters.

McNeill, J.R. *Something New Under the Sun.* Norton: New York, 2000; Chapter 3. An excellent book, subtitled 'An environmental history of the twentieth-century world.' This chapter treats the urban history of the atmosphere, including the impacts of emissions from automobiles on urban air quality.

Moore, John W.; Stanitski, C.L.; Wood, James W.; Kotz, John C.; Joesten, Melvin D. *The Chemical World.* Saunders: Fort Worth, 1998; Chapter 14. This book is an introductory textbook for college-level chemistry courses. A good discussion is provided on the topics of chemical reactions in the atmosphere and urban air pollution.

Ristinen, Robert A.; Kraushaar, Jack J. *Energy and the Environment.* Wiley: New York, 1999; Chapter 8. This chapter provides a useful overview of hybrid vehicles, electric batteries, and fuel cells.

Rock, Maxine. *The Automobile and the Environment.* Chelsea House: New York, 1992. Written for persons having little previous science background, this book provides an overview of exactly what the title indicates, including discussions of pollution from automobiles and of electric cars.

Suzuki, Takashi. *The Romance of Engines.* Society of Automotive Engineers: Warrendale, 1997; Chapters 41, 42. A history of engine development, mostly automobile engines, and a treasure-trove of information for 'engine heads.' These two chapters discuss hybrids and the prospects for a hydrogen-fueled engine.

Zetsche, Dieter. 'The Automobile: Clean and Customized.' In: *Key Technologies for the 21st Century.* Freeman: New York, 1996; Chapter 5. A short article on possible automobile technologies for the new century.

32

THE GREENHOUSE EFFECT

The greenhouse effect is the single environmental issue related to energy use that could have by far the most drastic implications for our future.

OUR PLANET IS A GREENHOUSE

Let's begin by considering a growing plant. During the daylight hours, the plant is warmed by energy in sunlight. At night, when the heat of the sun is temporarily no longer available, the plant will cool. It does this for the reason we have already discussed (Chapter 11): heat will flow spontaneously from a warm object—in this case, the plant—to a cooler one—the air surrounding the plant. The plant loses heat by two mechanisms of heat transfer: convection, currents of warm air rising from the plant; and radiation, where in this case the plant likely radiates heat energy as infrared radiation. We can write an energy balance equation for this plant, just as we can for any other SYSTEM:

$$E_{IN} - E_{OUT} = E_{STORED}$$

For this system, E_{IN} is the energy provided by the sunlight, and E_{OUT} is the energy lost by convection and radiation of infrared light.

Now suppose we put the plant in a greenhouse (Figure 32.1). Once again, it will be warmed by the energy provided by sunlight. The sun emits energy in many regions of the electromagnetic spectrum. Much of the energy that we receive on Earth is in the 'visible' region of the spectrum. Ordinary glass is, obviously, transparent to visible light. (Glass would be a much less useful material to us if it weren't transparent.) The energy brought in by the sun's light helps to warm the contents of the greenhouse (Figure 32.2). Once again, when the daylight hours are over, the warmth of the sun is no longer provided and the plant will cool. In this case, though, two factors intervene to retard the cooling process. First, the roof of the greenhouse helps to trap convection currents of warm air, and keep

FIGURE 32.1 A plant growing in a greenhouse provides a simple model for helping to understand the planetary-scale greenhouse effect.

them inside the greenhouse. Second, and more importantly for our discussion, the glass helps trap infrared radiation.

As we've seen (Chapter 11), any object that is hotter than its surroundings will attempt to come to some equilibrium temperature by transferring heat to those surroundings. The region of the spectrum in which a hot object radiates depends upon its temperature. Some objects (an example being the heating coil on an electric stove set on 'high') radiate heat in the visible region. In fact, we can see this, and sometimes speak of objects as being 'red hot.' Objects that are only slightly warmer than their local environment do not radiate heat in the visible region of the spectrum. Instead, they radiate in the lower energy (longer wavelength) infrared region of the spectrum. We, for example, with body temperatures of 98.6°F, are only slightly warmer than our surroundings, so we radiate in the infrared. (This is the basis for the infrared-detecting gun sights that allow aiming of guns in the dark, as sometimes seen in tough guy, shoot-'em-up action movies.)

The essence of operation of a greenhouse is that, while ordinary glass is transparent in the visible, it is opaque in the infrared. Thus the radiated heat is trapped inside the greenhouse (Figure 32.3). In the case of a real greenhouse, there is a second mechanism by which heat is trapped inside: Heat transfer by convection (Chapter 11) is limited because the roof keeps currents of warm air from rising completely out of the greenhouse. We can write an energy balance equation to apply to this new SYSTEM (i.e., plant plus greenhouse):

$$E'_{IN} - E'_{OUT} = E'_{STORED}$$

In this equation the 'prime' signs ($'$) have no significance other than to remind us that we're writing the equation for a different system.

To understand the effect of the greenhouse, consider: First, the energy in is the same in both cases; it's the energy in the sunlight. But, E'_{OUT} is smaller than E_{OUT}, because some of the heat has been trapped in the greenhouse by its roof and by the glass being opaque to infrared radiation. If the energy going in is the same in both cases, but the energy going out has decreased for the greenhouse, then inevitably E'_{STORED} must be larger than E_{STORED}; that is, energy stored in the greenhouse is larger than the energy stored in the plant not in the greenhouse. (This is easy to verify by making up some simple numbers for the various terms in the two energy balance equations and doing the necessary arithmetic. For example, suppose the incoming solar radiation is 100, in arbi-

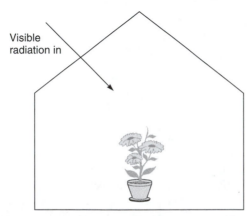

FIGURE 32.2 A significant source of ENERGY supplied to the greenhouse derives from sunlight coming through the glass walls and roof.

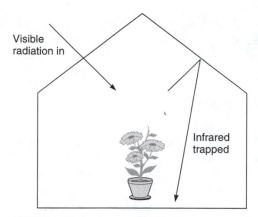

FIGURE 32.3 Some ENERGY is trapped inside the greenhouse in the form of infrared radiation, because glass is opaque to infrared. This trapping of energy reduces the amount that would otherwise be lost by being radiated into the environment.

trary energy units, and that the plant by itself loses 60 units, whereas with the greenhouse only 25 units are lost. Then the energy stored by the outside plant is 40 units, while the energy stored in the greenhouse is 75 units. More energy is stored in the greenhouse.)

The observable effect of the larger value of E_{STORED} for the greenhouse relative to the plant outside is that the greenhouse is warmer. More energy is stored as heat. We sometimes do greenhouse experiments ourselves. Almost everyone has had the experience of leaving a car parked in direct sunlight on a hot summer day with all the windows up. After several hours, the interior of the car becomes hot enough to be painful to touch, or to melt candies or plastics left inside. In this case, the car's roof and windows trap much of the heat provided by the sun (again by limiting heat loss via convection and infrared radiation), and the temperature inside soars. Of course, the ability of the greenhouse to maintain an artificially warm temperature inside is the reason why we build greenhouses in the first place—to provide an environment for growing fresh flowers, fruits, or vegetables in the winter, or for growing plants in regions of cool climates that normally require tropical or subtropical climates.

Now let's make a gigantic leap upward in scale, and consider how our planet works (Figure 32.4). Each day we receive energy from the sun. The energy reaching the surface of the planet is mostly in the visible and infrared regions of the spectrum. This radiation is absorbed by the Earth, warming the continents and oceans. The current average temperature of the planet, about 15°C, is much higher than the −270°C of outer space. Consequently, the Earth will try to spontaneously transfer heat to its cooler (in this case, frigid!) surroundings of outer space.

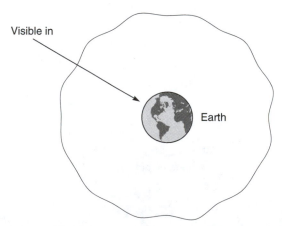

FIGURE 32.4 A significant source of ENERGY supplied to our planet derives from sunlight coming through the atmosphere.

Our atmosphere is like glass, in the sense that it is transparent to visible radiation. During the night, those objects that were heated during the day will cool off by radiating their heat to outer space, in the form of infrared radiation (Figure 32.5). Thus, our planet has achieved its own energy balance:

$$\text{Energy in } - \text{ Energy out} = \text{Energy stored}$$

where the amount of energy stored as heat maintains the average temperature that we enjoy. The principal components of our atmosphere, nitrogen and oxygen, are essentially transparent to the infrared radiation. However, the atmosphere also contains small amounts of other gases that absorb some infrared radiation (Figure 32.6). (As we will see below, without these gases, the average temperature of our planet would be substantially colder than it actually is.) The average temperature of the Earth results, in part, from the trapping of infrared radiation in the atmosphere by some of the atmosphere's components. Because these components—carbon dioxide and water vapor being examples—act analogously to the glass in the greenhouse, such gases are called **greenhouse gases**. By extension, this warming of Earth by trapping infrared is called the **greenhouse effect**.

What happens? If some of the infrared is trapped in the atmosphere, then the 'heat out' term in the heat balance equation drops:

Heat in	−	Heat out	=	Heat stored
If this stays the same ...		But this drops (because heat is trapped in the atmosphere) ...		Then this has to go up

If there were no greenhouse effect, then the average temperature of Earth would be a balance between the energy provided by the sun and that radiated back into outer space. Earth would have an average temperature of −18°C (0°F), so that almost all of the water on the planet would be frozen year-round. It is very questionable whether life as we know it could have evolved in such an 'ice world.' But even if it had, certainly our lives would be much different today if we had to cope with such temperatures. The greenhouse effect is essential in keeping our planet habitable for the species that have evolved here.

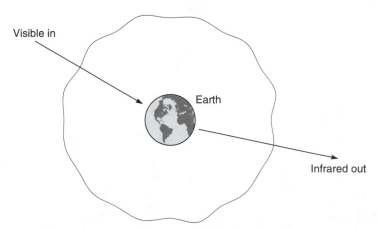

FIGURE 32.5 At night, that part of the planet not in direct sunlight will tend to cool by radiating infrared back into space, through the atmosphere.

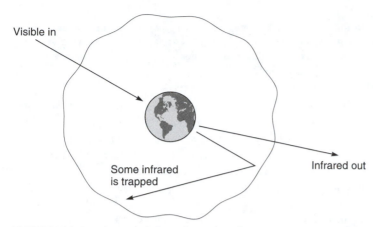

Visible in

Some infrared
is trapped

Infrared out

FIGURE 32.6   Some of the infrared radiation that might otherwise carry heat back into outer space is trapped in the planet's atmosphere by components of the atmosphere that absorb infrared radiation.

The composition of the atmosphere plays the crucial role in the Earth's enjoying an average temperature about 33°C warmer than would be expected by balancing solar radiation with infrared losses. Though water vapor and carbon dioxide represent only very small fractions of the total composition of the atmosphere (some 10,000–50,000 ppm and 350 ppm, respectively), they provide significant contributions to the greenhouse effect. The notion that the atmosphere might be involved in trapping some of the sun's heat originated with the French mathematician and physicist, Jean-Baptiste Joseph Fourier. About 200 years ago, he developed the comparison of the function of the atmosphere to that of the glass in a greenhouse. However, he did not identify which components of the atmosphere were responsible, nor how they acted. Sixty years after Fourier's original idea, John Tyndall, an English physicist, demonstrated that carbon dioxide and water vapor absorb heat, and calculated the effect that would result from these two compounds in the atmosphere. In principle, any molecule that can absorb infrared radiation is potentially a greenhouse gas. There are many such substances, in addition to the carbon dioxide and water. As examples, methane, the principal component of natural gas; nitrous oxide, one of the potential contributors to NO_x; ozone; and chlorofluorocarbons, formerly used in refrigeration and air conditioning, are among the other substances that help the atmosphere retain heat.

Fourier was one of the most brilliant intellects of his era, the late 18th and early 19th centuries. He made many original contributions to mathematical physics, including areas of heat conduction, harmonics in music, and prediction of tides. In addition, Fourier was keenly interested in ancient Egypt, and accompanied Napoleon on his ill-fated invasion of Egypt in 1798. Fourier also published books on Egyptology.

Tyndall was one of many outstanding experimental physicists in 19th-century England, and was a good friend of Michael Faraday. Perhaps Tyndall's most famous contribution is his explanation of why the sky is blue: Some of the sun's rays are scattered by molecules in the atmosphere. He also did experiments on the concept of spontaneous generation—

the notion that persisted for a long time in science that living organisms could arise spontaneously from nonliving matter, for example, that bacteria found on decaying food somehow originated 'spontaneously' from the food. Tyndall's experiments were the final nail in the coffin for this deservedly dead idea.

The atmosphere does not provide a mechanism only for warming. Other atmospheric effects can act to cool the planet. Besides acting as a greenhouse gas, water vapor also forms clouds, which help cool Earth by reflecting sunlight. Aerosols from volcanoes also absorb and reflect radiation, and usually act to cool the climate. When a major volcanic eruption occurs, most recently with Mount Pinatubo in the Philippines, unusually cool summers and harsher winters can persist for several years.

The composition of the atmosphere has a major role in determining the average global climate. As we've seen, without naturally occurring carbon dioxide and water vapor in the atmosphere, Earth would be a grim place to live. Any changes to the composition of the atmosphere could have potentially drastic effects on the environment and on life. We would refer to such impacts as **global climate change**. Current attention focuses on increases in the greenhouse gases.

We call the warming of the planet by some 30°C as a result of the carbon dioxide and water vapor that occur naturally in the atmosphere the **natural greenhouse effect**. It exists and occurs without human intervention. We know that carbon dioxide captures and returns the infrared radiation coming from the surface of the Earth. If the concentration of CO_2 in the atmosphere were to increase, the effect of this increase would reduce the value of E_{OUT} in Earth's energy balance. A century ago the Swedish chemist Svante Arrhenius℥ calculated that doubling the concentration of CO_2 would result in an increase of 5–6°C in the average temperature of the planet's surface. This would be global climate change on a grand scale; since the effect is one of heating, we also refer to it as **global warming** (Figure 32.7).

℥ Arrhenius made numerous contributions to the newly developing discipline of physical chemistry, most notably in the areas of the behavior of electrolyte solutions and in the theoretical treatment of rates of chemical reactions. Arrhenius received his doctoral degree from the University of Uppsala in Sweden, though it is said to have taken the personal intervention of Wilhelm Ostwald, one of the world's greatest chemists of that era, to persuade the Uppsala faculty that Arrhenius's work on the dissociation of electrolytes in solution did indeed have merit. In 1903, Arrhenius was awarded the Nobel Prize in chemistry.

In fact, as we discuss below, it appears that atmospheric CO_2 concentrations have been increasing steadily for over a century. The effect of increased atmospheric CO_2 is to trap even more infrared and thus to augment or enhance the amount of warming of the planet beyond that provided by the natural greenhouse effect. This additional heating by increased concentrations of CO_2 is then called the **enhanced greenhouse effect**. Since many, but not all, scientists believe that the increased CO_2 primarily results from human activities, it is also called the **anthropogenic greenhouse effect**.

Whether an enhanced greenhouse effect exists; if it does, whether it is or is not anthropogenic; and what, if anything we should do about it, lie at the center of a vigorous debate. Many believe that the increased CO_2 results directly from human use of energy, and especially from the combustion of fossil fuels. In the United States, fossil fuels provide 99% of our transportation energy and about 75% of the primary energy used to generate electricity. If the resolution of this debate results in a strong push to reduce anthropogenic

FIGURE 32.7   An example of what might be anticipated in a case of extreme global warming.

CO_2 emissions—either through legislation and regulation or through collective societal decisions—the enhanced greenhouse effect could be the greatest of all issues to impact how we use energy in the future, and from what sources we obtain it. The debate is complicated by the fact that it represents a problem in which the need for reliable scientific knowledge for its resolution exceeds the present-day ability of the scientific community to provide it.

AN ENHANCED GREENHOUSE EFFECT—THE EVIDENCE FOR GLOBAL CHANGE

The argument as developed to this point is that an increase in concentrations of greenhouse gases in the atmosphere would alter the energy balance of the planet, by trapping infrared radiation with the consequence of reducing the amount of energy radiated into outer space. Because the amount of energy provided to the planet remains constant, the effect of decreasing the E_{OUT} term in the energy balance is necessarily to increase the amount of energy stored. The observable effect would be an increase in average global temperatures (possibly accompanied by a host of other effects that are related to temperature). In this section we examine changes in atmospheric greenhouse gases and the possible effects of such changes.

Initially we focus on carbon dioxide, since it seems most closely linked to anthropogenic sources. Over the past 400,000 years, during which there were four major glacial cycles, atmospheric CO_2 concentrations averaged about 240 parts per million (ppm), with upward or downward swings of about 20%. (That period of Earth's history also witnessed four major cycles of glaciation and subsequent retreat of the glaciers—essentially four large swings in global climate—with this relatively small variation in CO_2.) For the period from A.D. 800 to about 1800, the CO_2 concentration in the atmosphere lay in the range 270–290 ppm. A rapid increase began in the first half of the 19th century, so that by 1900 the concentration had hit 295 ppm. Acceleration of the CO_2 concentration increase continued, leading to values of 310–315 ppm by the middle of the 20th century, and hitting 360–370 ppm by the dawn of the 21st century.

Atmospheric levels of CO_2 have been measured directly since 1957. The concentration records earlier than the 1950s consist of evidence from many different sources. However, analyses of ice cores serve as the primary data source because they provide a continuous record of past atmospheric composition. Based on these data, it can be said unequivocally that, for the last 18,000 years, atmospheric CO_2 concentrations fluctuated around 280 ppm, and that, in the relatively recent past, there has been an increase to a concentration of 360–370 ppm, with a current rate of increase of 1.5 ppm per year. Concentrations of CO_2 are now higher by at least 30% than the maximum concentrations that occurred prior to the rapid growth of CO_2 concentration in the past 150 years.

The modern rate of atmospheric CO_2 increase is accurately known from direct measurements of CO_2 in the atmosphere, as well as from analyses of air bubbles trapped in ice and firn. The increase in atmospheric concentrations of CO_2 measured at the observatory at Mauna Loa, Hawaii has been tracked since 1958. On average, the rate of increase was 0.83 ppm per year during the 1960s, 1.28 ppm per year in the 1970s, and 1.53 ppm per year in the 1980s. Not only is the absolute amount of CO_2 increasing, but the rate at which the concentration is increasing is also growing.

> Firn is a form of solid water that is essentially a transition between snow and glacial ice. Snow that lasts through a summer melting season becomes firn; the firn becomes glacial ice when it freezes to an extent that liquid water can no longer trickle through it.

We next must question whether this greatly increased CO_2 concentration, and its escalating rate of increase, is due to human activities. Three factors do appear to have a significant role in affecting atmospheric CO_2 concentration: growth in global population, in agricultural production, and in industrialization.

It is remarkable that the very rapid increase in atmospheric CO_2 has occurred since the industrial revolution of the 18th century. The industrial revolution was fueled, both figuratively and literally, with coal. The 20th century added its own contributions, especially in the great increase in petroleum consumption. We must remember that carbon dioxide is an inevitable product of the complete combustion of fossil fuels. Fossil fuels provide 99% of our transportation energy, 72% of our residential energy (by direct combustion, the rest of our residential energy coming from electricity), and 70% of our electricity. Industrial production today exceeds what it was at the turn of the last century (1900) by some 50 times. In the next half-century, it will likely grow by another factor of 5–10. Much of the growth in industrial production has depended on the combustion of fossil fuels, either as a direct source of energy or for electricity generation. The annual growth rate in energy consumption is typically 2–3%, much of which derives from coal. If present-day rates of fuel consumption continue unchanged, the atmospheric concentration of CO_2 is expected to be double the level that existed in 1860 sometime between 2030 and 2050.

Earth's population tripled in the 20th century and likely will double or triple again before reaching a plateau sometime toward the end of our century. Even if these additional people consume energy only at a subsistence level, they will, merely by existing on this planet, add to the increased use of energy. However, these people will, not unreasonably, aspire to something much more than a crude, impoverished subsistence-level existence. Increased energy use provides the key to improving their standard of living, with increased consumer goods, readily available transportation, abundant electricity, and all the other amenities of life that those of us fortunate enough to live in wealthy, industrialized nations often take for granted.

Furthermore, people have to eat. Conversion of forested land to agricultural use results in a redistribution of carbon from being locked up in plants or in the soil to being CO_2 in the atmosphere. This land use conversion represents another anthropogenic factor increasing atmospheric CO_2. Two trends drive this buildup: fossil fuel combustion, which accounts for about three-fourths of the addition, and deforestation, which nowadays is mainly a problem in the tropics, but a century ago occurred mostly in North America and the temperate parts of Asia. Deforestation accounts for about a quarter of the anthropogenic impact on atmospheric CO_2 (fossil fuel combustion accounts for virtually all the rest).

We have established that atmospheric CO_2 concentrations are increasing, and that there is at least a circumstantial link between the onset, and subsequent acceleration, of that increase and two human factors: the industrial revolution, with its heavy reliance on fossil fuel combustion; and the accelerating growth of the planet's population. Now we must inquire whether there is any evidence for global warming.

Drilling into the ice covering Greenland has provided about two miles' worth of ice cores dating back at least 110,000 years. A second project at the Russian Vostok Station in Antarctica has yielded over a mile of ice formed over a period of 160,000 years. The ratio of deuterium ($_1H^2$) to ordinary hydrogen ($_1H^1$) in the ice can be used to estimate the temperature at the time the snow fell. Higher temperatures tend to increase the deuterium/hydrogen ratio in the rain or snow. In addition, the bubbles of air trapped in the ice can be analyzed for carbon dioxide and other gases.

> Water molecules made of atoms of $_1H^1$ have a lower molecular mass than molecules of so-called heavy water, in which the hydrogen atoms are $_1H^2$. The water made with ordinary hydrogen has a molecular weight of 18, while that of heavy water is 20. Following a perfectly general rule for molecules that are chemically similar but different in molecular weight, the lighter ones evaporate more readily. Because of this effect, there is relatively more $_1H^1$ and less $_1H^2$ in water vapor that has evaporated into the atmosphere than in liquid water in the oceans. Rain or snow that condenses from atmospheric water vapor will also reflect this enrichment of $_1H^1$. However, as average global temperatures warm, more molecules of heavy water will evaporate from the liquid state, and consequently the isotope ratio, $_1H^2/_1H^1$, in rain or snow also varies with average temperature. This is the central concept in estimating ancient temperatures by analysis of the isotopic composition of ice.

Around the world, the average air temperature at Earth's surface has increased by 0.3–0.8°C over the past hundred years. Between 1890 and 1990 average surface temperatures increased 0.3–0.6°C. Nine of the ten hottest years on record occurred between 1987 and 1997, and the 1990s were the hottest decade in about six centuries. Changes this big and this fast are within the natural range of variation for swings in average global temperatures. However, they are rare in the temperature record for the past two million years, most likely have not occurred at any time within the past 10,000 years, and definitely are unprecedented for the past 600 years. Any anthropogenic changes in the energy balance of Earth, an increase in greenhouse gases or aerosols, will alter atmospheric and oceanic temperatures. As the atmosphere warms, changes in the amount of water vapor in the air, the amount of sea ice, and the extent of cloud cover will all work to amplify or reduce any warming that would lead to climate changes.

The increase in the global mean temperature of about 0.6–0.8°C since the start of the 20th century is associated with a stronger warming of daily minimum temperatures than with increasing daily maximums. Consequently, there has also been a reduction in spread of temperatures noted in a 24-hour period.

The frost-free season in the northeastern United States began days earlier in the mid-1990s than it did in the 1950s. For every country where the number of frost days has been examined, they have become fewer in number. This is consistent with the warming in average minimum temperature found for each country. As examples: during the period 1910–1998 there was a slight decrease in the number of days below freezing over the entire United States. In Australia and New Zealand, the frequency of days with temperatures below freezing decreased, with a simultaneous trend for warming in daily minimum temperatures. In northern and central Europe, the number of frost days has been decreasing since the 1930s, again associated with strong increases in winter minimum temperatures. In the past 30 years, the elevation at which the temperature is always below freezing has ascended almost 500 feet up the sides of mountains in the tropics.

In the 30 years from 1940 to 1970, average global temperatures actually dropped, by about 0.3°C. This change has been ascribed to a temporary increase in Earth's reflection of incoming sunlight back into space, caused by, among other things, an increase in planetary volcanic activity that put more aerosols and dust into the atmosphere. This drop in average temperature has been interpreted to mean that there is no anthropogenic or enhanced greenhouse effect. However, over the long term, that temperature decrease vanished comparatively quickly, and the short-lived 0.3°C dip has now been totally overwhelmed by a steady increase in temperature.

What do we think we know about the connection between carbon dioxide and global warming? Historically, for about 160,000 years, there appears to be a correlation between the concentration of carbon dioxide in the atmosphere and the average global temperature. Currently, the carbon dioxide content of the atmosphere is rising. Humankind is consuming fossil fuels—and hence producing carbon dioxide—at an unprecedented rate. The average global temperature is rising.

Certainly there were periods of global warming long before humans began burning fossil fuels. Furthermore, CO_2 is a relatively minor greenhouse constituent in the atmosphere; as we will see later in this section, there are other greenhouse gases besides CO_2. Generally, over the past 160,000 years, when the CO_2 concentration was high, average temperatures were high. Periods of high temperature have also been characterized by high atmospheric concentrations of methane. Such correlations do not necessarily *prove* that elevated atmospheric CO_2 and CH_4 caused the temperature increases. Presumably the converse could have taken place—in other words, high temperatures could have triggered natural phenomena that caused more CO_2 and CH_4 to enter the atmosphere. Also, it has been argued that the rise in global temperature since the middle of the 19th century may simply be the result of natural solar variations. Whether this temperature increase really is a consequence of the increased CO_2 concentration cannot, at present, be concluded with absolute certainty. Unquestionably the fact is that the rise in CO_2 at ever-increasing rates coincides with the onset of extensive use of fossil fuels, also at ever-increasing rates. This compelling circumstantial evidence implicates carbon dioxide from human-related sources as a cause of recent global warming. We must also recognize the importance of the difference between being able to detect a change and being able to attribute that change to some identifiable factor.

 It may be worth contemplating that two of the keenest intellects of the 19th century, Henry David Thoreau and Sherlock Holmes, held greatly different views on the reliability of circumstantial evidence. Thoreau indicated that circumstantial evidence could be very compelling, as in the case of finding a trout in one's milk (i.e., evidence that the milk had been heavily diluted with water before being sold). Holmes, on the other hand, warned that circumstantial evidence had a way of pointing in the opposite direction from the truth.

Roles for Other Greenhouse Gases

About half of global warming may be attributable to compounds other than carbon dioxide. Other greenhouse gases of concern include methane, nitrous oxide, ozone, and the chlorofluorocarbons (gases that formerly had been used extensively in refrigeration and air conditioning equipment). The effectiveness of these gases in contributing to global warming depends on their lifetime in the atmosphere, and their interactions with other gases and water vapor. These factors are combined into an indicator called global warming potential (GWP).

Methane, which has a GWP of 11, comes from a wide variety of sources. It is a major component of natural gas, so likely some methane has always leaked into the atmosphere naturally from porous or cracked rock in natural gas reservoirs. However, human use of natural gas has likely led to increased emissions from leaks in gas wells, pipelines, and storage. Methane has always been released by decaying plant matter. This again represents a natural source of methane, but, as with natural gas, any human activity that increases decay of plant material also leads to increased methane release. Two such anthropogenic sources are the decaying organic matter in landfills and decaying residue of cleared forests. Methane formed in the main New York City landfill is used for residential heating, but at most landfills it simply escapes into the atmosphere. Another major source of methane is cultivated rice paddies. Concentrations of methane now exceed previous high values by more than 250%. Methane has increased by 1.1% per year for the last 30 years.

 Since it is unlikely the world will ever run out of garbage, and since methane is an excellent fuel gas, some localities have started to harness the natural methane production in landfills and use the gas as a source of energy. This work appears promising (Chapter 34), but the methane from most of the world's landfills is simply allowed to escape into the atmosphere.

Another place where organic matter is broken down under conditions leading to the production of methane is in the digestive systems of cattle and sheep. There, bacteria break down the cellulose of plant material and form methane. A healthy adult cow, placidly chewing its way through life, can add about 18 cubic feet of methane to the air every day. The tiny little digestive tracts of termites operate on a similar kind of chemistry, converting cellulose to methane. An individual termite is hardly a match for a cow as a methane source, but the termites' contribution is in their sheer numbers: there's about a half-ton of termites for every man, woman, and child on the planet.

Over the past decade, atmospheric concentrations of nitrous oxide have shown a slow but steady rise, about 0.2–0.4% per year. Natural sources include formation in lightning and volcanic gases—basically, natural thermal NO_x. Major anthropogenic sources include artificial nitrogen fertilizers, discharge of sewage, and the burning of plant material (as when forests are cut and burned in land use conversion). In addition to its role in the greenhouse effect, nitrous oxide at high altitudes in the atmosphere contributes to stratospheric ozone depletion. Near the surface of the Earth, the reactions of nitrogen oxides and hydrocarbons lead to the production of ozone. Ozone can also act like a greenhouse gas, but its action depends very much upon the altitude at which it occurs. It appears to have its maximum warming effect at altitudes of about six miles (i.e., in the upper troposphere). Loss of ozone has a cooling effect in the stratosphere and may also promote slight cooling at the surface of the Earth.

Chlorofluorocarbons, already implicated in the destruction of stratospheric ozone, also absorb infrared radiation, extremely strongly in fact. Because these gases have been implicated in the destruction of the ozone layer in the stratosphere, their manufacture has been stopped and their use is being eliminated. But, because an enormous number of refrigerators and air conditioners using these chemicals are still in use, atmospheric concentrations of the chlorofluorocarbons will continue to rise for some time, until those units reach the end of their working lives and are replaced by newer models. Currently the atmospheric concentration of chlorofluorocarbons increases about 6% per year.

Atmospheric aerosols consist mainly of tiny particles of ammonium sulfate that form from sulfur dioxide released by natural or artificial sources. These particles promote global cooling by reflecting and scattering sunlight before it can reach Earth's surface. Also, aerosols can enhance the condensation of water droplets and, consequently, enhance cloud formation. Clouds also increase reflection of incoming sunlight back into space. Aerosols actually counter the effects of the greenhouse gases. A temporary drop in average global temperature that followed the eruption of Mount Pinatubo in 1991 was a consequence of the large volume of sulfur dioxide released by the volcano. We have seen (Chapter 30) that SO_x emissions are a major contributor to acid rain. That SO_x also contributes to aerosol effects that counter the action of greenhouse gases has led to arguments—apparently made in all seriousness—that acid rain could be good for us! Aerosols have a much shorter lifetime in the atmosphere than greenhouse gases.

Other Signs of Global Change

The effect of an anthropogenic greenhouse effect should have as its primary impact an increase in average global temperature. However, temperature has, in turn, many impacts on weather patterns and ecosystem changes. So, any temperature changes that occur will also be reflected in many related, albeit secondary, effects.

One impact of changing temperature is changes in the circulation patterns of the atmosphere and the oceans. These changed circulation patterns lead in turn to changes in weather patterns. In both the United States and elsewhere around the world, areas affected either by drought or by excessive wetness have increased. In the United States, the number of dry days in each winter increased steadily from 1973 to 1998. So did the frequency of dry spells lasting five days or more. Heatwaves also became increasingly prevalent. In June 1998, Amarillo, Texas experienced 13 days in a row with temperatures above 100°F. Brownsville, Texas spent 17 days with the *low* temperature never dropping below 80°F.

Property losses caused by weather-related catastrophes have grown steadily in the United States, from about a hundred million dollars a year in the 1950s to six billion dollars per year in the 1990s. Furthermore, the actual number of catastrophes grew from about ten per year in the 1950s to 35 in the 1990s. The 1990s, in addition to being the hottest decade on record, also experienced a record number of damaging storms.

A change in climate can affect many aspects of the health of an ecosystem, such as the abundance and distribution of various species. In addition, many biological processes undergo sudden shifts at some particular 'thresholds' either of temperature or of precipitation. The ability to tolerate frost or low levels of precipitation, as examples, can often determine the boundaries of the ranges of various plant and animal species. Changes in the number of days exceeding some particular temperature threshold for a certain species, or changes in the frequency of either droughts or extreme precipitation, can lead to changes in some species in an ecosystem. Over a long period of time, the cumulative effect of such changes could be drastic, in ultimately affecting distributions of many species.

For example, in the western parts of North America, the range of the Edith's checkerspot butterfly shifted northward by nearly 60 miles and upward on mountainsides by about 400 feet of altitude during the 20th century. These shifts in range turn out to correlate quite well with the shifts in constant average temperature, which in the same period moved northward by about 65 miles and upward about 350 feet. In the Mesa Verde Cloud Forest preserve in Costa Rica, about 20 species (out of an original total of 50) of amphibians have become extinct in the past two decades. During that time, three major periods of die-offs of frogs all occurred during unusually dry winters.

The increasing range of Edith's checkerspot results from the fact that warmer temperatures are, in effect, climbing up mountains. Unfortunately, in other parts of the world, mosquitoes, and the diseases they carry, are also moving upward. Especially in the tropics, a change in average temperature of just a few degrees can mean the difference between the likelihood of encountering freezing temperatures, or not. In the past 20 years, *Aedes aegypti* mosquitoes, which lived only at low altitudes because of the temperatures at which they can survive, have now been found above one mile of elevation in northern India and 1.3 miles in the Andes Mountains. These mosquitoes are carriers of dengue, and yellow fever may follow.✿ Dengue fever has already been reported at the one-mile mark in Taxco, Mexico.

✿ Both dengue fever and yellow fever are very serious diseases. Dengue fever, caused by a virus carried by *Aedes*, is characterized by muscle and joint pain, eruptions on the skin, enlargement of the lymph nodes, and a decrease of white blood cells. The victim is debilitated for weeks. Yellow fever is also a viral disease and has such effects as jaundice, slowed heartbeat, and tendency to bleeding.

In the 1990s, as a result of the succession of hot years, malaria outbreaks were reported in the United States as far north as New York (whereas 10 years earlier it had been confined to California). Malaria has now also been reported in Canada, in some of the countries of the former Soviet Union, and in Korea.

This spread of disease is a possible by-product of the disruption of ecosystems resulting from climate change. A healthy ecosystem has a balance of species; some will surely help keep down the population of disease-carrying organisms, such as mosquitoes, by eating them. When the ecological balance is upset, as could happen as a result of climate change, then there could be a decline in species that keep pests under control. As a result, diseases borne by mosquitoes or other carriers could spread further in range.

The effect of warm temperatures ascending mountains is also manifest in the fact that, in many parts of the world, glaciers at mountain summits are melting. Over the last 30 years, the elevation at which temperatures are always below freezing has ascended almost 500 feet on mountains in the tropics. Glaciers on tropical peaks are melting rapidly, taken as another sign of global warming.

POTENTIAL CONSEQUENCES OF AN ENHANCED GREENHOUSE EFFECT

An enhanced greenhouse effect, resulting from anthropogenic emissions of carbon dioxide and other greenhouse gases, could affect us in many ways. These include, but are certainly not limited to, rises in sea level, changes in agricultural productivity, and threats to human health. In this section we explore some of these possible consequences.

Increasing Average Global Temperatures

Different climate forecasting models predict different increases in average global temperature. This does not mean that modelers are incompetent or that the models are useless. Differences in predicted temperatures arise in part from using different initial assumptions (such as how rapidly atmospheric CO_2 concentrations will increase) and different degrees of sophistication or complexity built into the model. Most predictions of future temperatures, based on an assumed doubling of atmospheric CO_2, lie in the range of a warming of 1–6°C; many are in the narrower range of 2–5°C, with 'best estimates' often 2.5–3.5°C.

Part of the oral tradition of those involved in modeling complex physical phenomena (and the modeling of the global climate is as complex a modeling job as has ever been undertaken) says that, 'All models are wrong; some models are useful.' How do we know that we can put confidence in a model and that a reasonable likelihood exists of its results being accurate? Several tests of a model help build this confidence. First, the model can be tested to see if it can accurately simulate today's climate, or some features of the climate, such as the variations of seasons. Second, attention can be given to certain specific aspects of the model, such as its ability to predict the extent of cloudiness. Third, the model can be run 'backwards,' so to speak, to determine its accuracy in calculating ancient climates of Earth. Accurate results from all of these tests provide justification of the model's credibility. All of the major climate models undergo continual testing, fine-tuning, and retesting.

The oceans provide a huge thermal reservoir—they can soak up a great deal of heat without a large increase in water temperature. The high heat capacity of water is the physical basis of this behavior, and means that the oceans will warm slowly. That is, the anticipated warming predicted on the basis of a doubling of atmospheric CO_2 might be delayed for some years, or even decades, after the CO_2 in the atmosphere actually has doubled. Eventually, though, the full extent of warming will occur at the time when the oceans 'catch up' to the temperature increases.

If the climate models are accurate and *if* CO_2 concentrations double, New York City could expect about 50 days a year with temperatures above 90°F (instead of the 15 now typical for the city). In Dallas, the number of days per year with temperatures above 100°F would increase from the present 20 to 100—in other words, to nearly one-third of the

POTENTIAL CONSEQUENCES OF AN ENHANCED GREENHOUSE EFFECT

year. A government official in the first Bush administration, of the type whom former governor George Wallace of Alabama used to delight in calling 'pointy-headed bureaucrats,' alleged that a few degrees' climate warming would have no more effect than the increased temperature one would experience by getting off a flight bound for Boston in Washington instead. In other words, Boston would then have the climate that Washington does today. Among numerous issues that this glib response overlooks is this question: If a future Boston were to have the climate that Washington now does, what would Washington be like?

In the previous section we examined evidence that global change is occurring. The consequences have so far seemed relatively small. Most of us have taken these changes in our stride, if we've even bothered to notice them. The much greater warming forecast in our new century will be quite another thing, the proverbial 'whole new ballgame.'

Rising Sea Level

Doubling the CO_2 concentration in the atmosphere will raise sea level by about 8–30 inches. In the period after the last ice age, some 12,000 years ago, a 5°C temperature increase accompanied sea-level rises of some 300 feet, along with significant habitat changes for many species. Sea levels have already risen 5–6 inches over the last century, mainly due to the thermal expansion of all the water in the oceans as they warm.

An increase in sea level of about 15 feet potentially could inundate the Netherlands, and certainly would do so for many low-lying islands. Fifty million people of Bangladesh, half the total population, live on land within 15 feet of sea level. Even smaller sea-level increases—in the range 1.5–4.5 feet—would endanger such cities as New York, Venice, Bangkok, and Taipei. Across the world a large percentage of people live on, or very close to, sea coasts. Eighty percent of the population of Australia lives within 12 miles of the coast.

The ice cap at the North Pole appears to be melting, based on explorations in the summer of 2000. The formerly solid ice cap, which had long resisted passage by any ships other than specially equipped icebreakers, now may be turning to slush. The melting of the ice sheet over the North Pole will not contribute to a rise in sea level. (This can be verified by a simple experiment of filling a glass to the brim with water and ice cubes, and watching what happens to the water level as the ice melts.) However, if the Arctic Ocean were to be free of ice, it could be warmer by some 10°C in winter. This would surely have some effects on climate in the northern hemisphere.

In the Antarctic, the ice sheet is mostly on land. If this ice were to melt, which could potentially happen over only a few decades, the extra water added to the oceans just from melting of the Ross ice shelf would raise sea level by nearly 20 feet. Such a rise would drown most of the world's major seaports. If the sea were to rise by 30 feet, the United States would lose Louisiana, at least one-quarter of Florida, and parts of other states as far north as New Jersey. The impact there would be particularly severe, because much of the land that would be flooded is the heavily developed and highly populated area near New York City. Warming that was prolonged enough, and warm enough, could melt the entire Antarctic and Greenland ice sheets, bringing the sea level up another 200 feet.

The total melting of the Antarctic and Greenland ice sheets would likely take centuries, and the change likely would be slow enough to allow society to respond in some rational fashion. But—observations made on glaciers show that they can move surprisingly rapidly for relatively short periods, i.e., a few months to several years.

When this happens, the bottom of the glacier melts, and the resulting layer of water allows for even easier sliding. The sliding motion generates heat, and the heat in turn causes even more melting, more sliding motion, still more heat, in an escalating feedback process. If this were to happen to the Ross ice sheet, the predicted 20-foot rise in sea level would come very quickly, with disastrous results, because so little time would be available for any reasonable response.

The Shift of Prime Agricultural Regions

The consequences for agriculture of a doubling of atmospheric CO_2 are also likely to be significant. The regions of greatest agricultural productivity would probably change. The temperate climate zones ideal for farming will shift toward the poles. In some cases, this shift will cross national borders. For example, the combination of drought and high temperatures could reduce crop yields in the midwestern United States, but the growing range might then extend further into Canada. The boundary between permafrost (or tundra) and normal soil would move north by 60–120 miles per degree Celsius (roughly, per two degrees Fahrenheit) of temperature rise. Consequently, much more of Canada should be able to grow wheat while at the same time wheat production in the southern Great Plains will decrease. These changes could have enormous impacts on global geopolitics. What if Canada exported food to the United States, instead of vice versa? What if Siberia became warm enough to become a major food exporter to the rest of the world?

The species of plants grown would change as well. The corn belt in the United States would shift a hundred miles toward the southwest or northeast for each one degree Celsius change in average daily temperature during the growing season. Though the temperatures would permit growing corn in more northern latitudes, the soils there often are less suited for agriculture. Probably crop yields would decrease. Past experience has shown that world grain markets have responded by increasing grain prices for even a 1% change in the amount of grain coming to market. Price changes caused by drastic drops in the amount of corn, or other grains, being produced could be very severe.

Many species of trees take hundreds, even thousands, of years for substantial change of their growing range. Though individual trees are rooted to one spot, the species can migrate. Although individual species can migrate with slowly changing climate, whole habitats usually do not, so that the species composition of an ecosystem is dramatically altered. Habitats would change unpredictably, and some species would, in effect, be stranded in ecologically unsuitable territory. Trees and other plants provide the necessary habitat that various animals depend on for their survival. Rates of species extinction could be substantially increased if climate changes occur more rapidly than ecosystems are able to adapt. As climate changes, the climate tolerances of birds could cause them to change their ranges and migration patterns. Of course, birds can migrate in periods of days or weeks, while trees or other plants may take centuries to catch up. This too will alter the nature of ecosystems.

When the last ice age ended about 12,000 years ago and a period of warming set in, major ecological changes occurred worldwide. The global warming amounted to about 5°C. The time necessary for the ecosystems to adjust to this sustained warming was several thousand years. Now, however, the potential exists for this same level of warming to occur within a century. If it does, substantial change to plant and animal populations and their habitats are virtually inevitable. And, the more rapidly the climate changes, the more severe the ecological disruption.

The Spread of Deserts

If atmospheric CO_2 doubles, areas of several Mediterranean countries, parts of the United States, and some of the former Soviet Union countries will probably become deserts. Much of this land is presently good agricultural land. This increasing 'desertification' is already a severe problem in parts of Africa. It could also occur in the United States, in a way similar to—but far worse than—the 'dustbowl' of the 1930s. In fact, an example of how this might occur is in central Nebraska. Today, the Sand Hills region of Nebraska is home to over 700 species of plants, as well as an area of wetland that supports many kinds of migratory birds. Analyses of the geological history of the Sand Hills, combined with computer modeling of climate changes, suggest that a drop of about two to four inches in annual precipitation, combined with an increase in temperature rise of about 2–$5°C$ (4–$10°F$) would turn the Sand Hills back to desert. That is, an area of some 20,000 square miles in northern and central Nebraska would become a desert. (And this does not include changes that would certainly be occurring elsewhere on the Great Plains.)

If the midwest were to become drier as a result of climate change, the growing season would be shortened by about 10 days and droughts will become more frequent. Farms that depend on irrigation will be in deep trouble, because many of the water reservoirs in the midwest and southwest that would be useful for irrigation are already depleted. We have already reached a point at which more Colorado River water has been allocated for irrigation purposes than actually flows in most parts of the river! According to these allocations, people can, in principle, pull more water out of the Colorado than is actually in it. A $2°C$ temperature rise is predicted to cause a 40% decrease relative to the present-day flow. We can be sure that this situation would ignite fierce legal battles among the farmers, corporations, municipalities and states that depend on having that water.

Human Health Effects

An increase in average temperatures might increase the geographical range of disease-spreading insects. The result could be significant increases in malaria, yellow fever, and sleeping sickness. The insect carriers of these diseases are very sensitive to weather conditions. Cold limits mosquitoes to living in seasons and in geographic regions where temperatures stay above certain minimums. Freezing kills many eggs, larvae, and adults outright. *Anopheles* mosquitoes, which transmit the malaria parasite, cause sustained outbreaks of malaria only where temperatures routinely exceed $60°F$. *Aedes aegypti* mosquitoes, responsible for yellow fever and dengue fever, convey virus only in regions where temperatures seldom fall below $50°F$. Higher temperatures increase the rate at which the disease-causing organisms that live inside the mosquitoes reproduce and mature. As whole geographic regions experience warming, mosquitoes could expand into territories they formerly could not tolerate because of temperature limitations. As they do so, they would bring the diseases they carry with them. Furthermore, within their survivable range of temperatures, mosquitoes breed faster and bite more often as the air becomes warmer. Warmer nighttime and winter temperatures could enable them to cause more disease for longer periods in the areas they already infest.

Droughts that last longer and massive rainfalls have become more common as the average temperature has risen. Floods and droughts could undermine health in various ways. They could damage crops and make them vulnerable to infestation by pests and weeds, reducing food supplies and potentially contributing to malnutrition. Floods help trigger outbreaks of disease by creating breeding grounds for insects whose dried-out eggs will hatch in still water, promoting the emergence and spread of disease. The insects can make a further gain if climate change reduces the populations of predators that normally keep them in check.

The Possible Release of More Greenhouse Gases on a Warmer Earth

Global warming could possibly exacerbate the release of even more CO_2 or other greenhouse gases. The ability of gases to dissolve in liquids decreases as the temperature of the liquid increases. This is why, for instance, a warm carbonated beverage is more likely to foam when it's opened than would a cold one. Increasing the temperature of the oceans will decrease the solubility of CO_2.⚛ Also, substantial amounts of methane, which we've seen is a potent greenhouse gas, are currently contained in various environments such as mud on seafloors, bogs, and the permanently frozen soil of arctic regions (permafrost). As the temperature rises, these substances will also get warmer, making it more likely that they would release methane to the atmosphere. Global warming would also cause more water to evaporate. Since water vapor is a greenhouse gas, an increasing concentration of water vapor in the atmosphere could further add to the greenhouse effect.

⚛ There is, however, the possibility that more CO_2 could be absorbed, rather than released. An increase in ocean temperature could promote the growth of microscopic algae (phytoplankton) and hence increase CO_2 absorption by photosynthesis. However, if the ocean is warmer, water will not circulate as well as it does now, inhibiting the growth of phytoplankton. A warmer climate would mean that the tree line would move north, bringing with it added CO_2 absorbing capacity, again due to photosynthesis. But, related reductions in rainfall elsewhere might turn areas that nowadays have abundant vegetation into deserts, reducing CO_2 absorption.

Increased Plant Growth as a Benefit

There seems to be a complicated interaction between the roles of CO_2 and water in plant growth. Increased atmospheric CO_2 might not be entirely bad for plants. As CO_2 concentration increases, the rate at which water is lost by evaporation from the leaves falls. We know that CO_2 is essential for photosynthesis, but because the concentration of CO_2 in the air is relatively low, CO_2 molecules are difficult for a plant to catch. To give themselves the best opportunity for absorbing CO_2 molecules, plants keep small pores on the surfaces of their leaves open. But, at the same time these wide-open pores make it easy for water molecules inside the leaves to escape by evaporation. In other words, capturing a little CO_2 involves losing a lot of water. Plants can adapt to increasing CO_2 in one of two ways. One strategy is not to change the size of the pore openings. This allows more photosynthesis to occur (because there is more CO_2 to be captured) with the same amount of water loss. Alternatively, the second strategy is to close the pore openings somewhat. In this case, less water is lost by evaporation, and about the same amount of photosynthesis occurs. In dry conditions, increased CO_2 can essentially compensate for less water.

In greenhouses, both agricultural and wild plants show higher growth rates, between 20% and 40%, when the CO_2 concentration is doubled. The results vary greatly from one plant species to another, however. Plants that have low growth rates now, which is the case for many wild plants, don't seem to respond as well to increased CO_2 as do plants that normally have rapid growth rates. Many crop species fall into the latter category. Most of the studies done so far are somewhat artificial, in that they were done for relatively short periods and in greenhouses. Some studies under field conditions show few significant increases in growth with increased CO_2. Nevertheless, artificially increased CO_2 levels in indoor growing environments are already in commercial practice to accelerate the growth of crops of very high value. ⚛

⚛ Some of the details of this interesting work, already being used commercially in the United States and in the Netherlands, can be found in Chapter 3 of Michael Pollan's recent book, *The Botany of Desire* (Random House, 2001).

THE ROLE OF THE GLOBAL CARBON CYCLE

The global carbon cycle (Figure 32.8) allows us to track the removal of CO_2 from the atmosphere, and ways in which CO_2 is added to the atmosphere. Processes that add CO_2 to the atmosphere are called **sources**. They include the decay of organisms; fossil fuel combustion; decomposition of carbonate rocks, either by natural processes or in the manufacture of cement; and CO_2 coming out of solution from the ocean. Processes that remove CO_2 from the atmosphere are called **sinks**. They include photosynthesis, dissolving of CO_2 into the oceans, or the fixation of CO_2 into various carbonate forms. Changes in patterns of land use can act either as sources or sinks. If, for example, forest land is converted to agricultural uses, that change tends to increase atmospheric CO_2. On the other hand, growth of new forest would likely serve as a sink. Changes in the concentration of CO_2 in the atmosphere result from the relative balance between sources and sinks.

Patterns of changing land use could have a substantial effect on storage of carbon on land (in plants and soils), decreasing the potential store of carbon. In addition, deforestation releases one to two billion tons of carbon (as CO_2) to the atmosphere every year. Around the world, about 60,000 square miles of rainforest are cut down or burned every year. This is an area roughly equivalent to the size of Florida. Doing so does double harm. First, trees do a good job of removing CO_2 from the atmosphere; thus a significant carbon sink is removed. Second, regardless of whether the trees are burned to clear land quickly or are

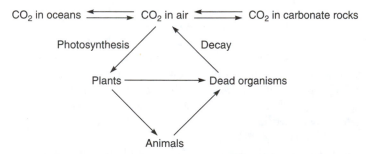

FIGURE 32.8   The global carbon cycle shows both sources for carbon dioxide, tending to increase its concentration in the atmosphere, and carbon dioxide sinks, that work to reducing its concentration.

simply left to decay, the carbon locked up in the now-dead trees will be returned to the atmosphere as CO_2. Perhaps the least damaging approach would involve using the trees as lumber (that is, so that they are not left to rot nor are burned), and to use the cleared land for agriculture, since the growing farm crops would absorb some CO_2. Even so, it may be that the ability of the land to absorb CO_2 drops by up to 80%. This reflects the fact that, if we consider equal areas, forests contain 10–20 times more carbon than cropland.

Until about 1950, the expansion of agriculture and forestry in the middle latitudes of the globe dominated carbon emissions from the land. Since then, though, the conversion of temperate forests to agricultural lands has decreased. In some places, forests are even making a comeback, growing on previously logged land and on abandoned agricultural land. As a result, since the 1950s, CO_2 emissions from the tropics, which have been increasing, are now greater than from temperate regions.

The dominant anthropogenic CO_2 sources are the use of fossil fuels and production of cement. As the generation of energy and the consumption of fossil fuels increased, so did the quantity of combustion products released to the atmosphere. Since 1860, the CO_2 concentration has increased from 290 ppm to 360 ppm, and the current rate of increase is about 1.5 ppm per year. Currently, fossil fuels containing five billion tons of carbon are burned annually. This represents four times the rate of release in 1950 and ten times the rate in 1900. During the 1990s, about a ton of carbon dioxide was produced each year for every human on the planet. The cumulative production of CO_2 from fossil fuel combustion since the beginning of the Industrial Revolution (i.e., since 1750) is about 280 billion tons of carbon.

The cement industry produces prodigious amounts of carbon dioxide from limestone. In the production of Portland cement, for example, limestone and clay are heated to temperatures of about 1450°C. In this process, the limestone decomposes:

$$CaCO_3 \quad \rightarrow \quad CaO \quad + \quad CO_2$$

limestone   lime

Subsequent reactions with the aluminum and silicon compounds in the clay produce a family of calcium silicate and calcium aluminate compounds. When water is added, the silicates and aluminates take on water molecules and the cement 'sets.'

The ocean contains more than 50 times as much carbon as the atmosphere. Most of the carbon in the ocean is in the form of dissolved bicarbonate and carbonate ions. The rest is organic, either as living organisms, or suspended or dissolved organic compounds. The water at the surface of the ocean is more acidic than the water in the deep ocean. As this relatively acidic water circulates, it will dissolve calcium carbonate. Doing so puts more CO_2 into the seawater. Over many years, perhaps centuries, much of this CO_2 will be released to the atmosphere.

However, the sea also takes up carbon dioxide in its surface layer. Some CO_2 is used by organisms living in the ocean, such as marine plants that live by photosynthesis and reef-building corals. The oceans act as a 'biological pump,' taking CO_2 from the atmosphere and into the ocean at a rate of about 10 billion tons per year. In the ocean, the plants and animals that took up the carbon convert it to carbonates in their skeletons, pass through their life cycles, and sooner or later die. Their skeletons sink to the ocean depths, where part of the carbonate is deposited in sediments, and the rest dissolves. Thus, the CO_2 taken out of the atmosphere slowly cycles into carbonate sediments or into the carbon reservoir of the deeper ocean. The deep ocean is thought to hold some 35 *trillion* tons of CO_2. The transfer of carbon between the surface and the very deep ocean is slow. As atmospheric CO_2 levels increase, the ocean responds by dissolving more CO_2 in its surface layer, and by mixing the CO_2-enriched surface waters downward through exchange with deeper waters. These deeper waters may have an enormous capacity for carbon. If so, the ocean should have the capability to absorb large quantities of CO_2 over many decades and perhaps even millennia.

On a time scale of 1,000–10,000 years, the ocean could absorb about 85% of the anthropogenic CO_2 that has been produced since the beginning of the Industrial Revolution. Some estimates suggest that the oceans could potentially absorb the CO_2 produced by burning the entire known amount of all the coal, oil, and natural gas available on Earth. Of course, nobody knows how the plants and animals living in the ocean would respond to such a massive uptake of CO_2. Certainly, changes in temperature or in the circulation patterns of the ocean and changes in the marine ecosystems would affect the rate of exchange of CO_2 between the atmosphere and the ocean, and thereby affect the concentration of CO_2 in the atmosphere.

Carbon on Earth is stored primarily in rocks and sediments. The oceans contain about 50 quadrillion tons of sedimentary carbonates, 38 trillion tons of dissolved inorganic carbon, and about 20 quadrillion tons of organic carbon. The atmosphere contains around 700 billion tons of carbon. Terrestrial vegetation contains some 500–900 billion tons of carbon, and the soil contains about twice that much. The reservoirs or sinks (the atmosphere, oceans, soil, and terrestrial biosphere) from which or into which carbon can easily be exchanged account for only a tiny fraction of the total carbon on Earth. But, it is only this tiny fraction that is available to play a role in biological, physical, and chemical processes at the Earth's surface. Anthropogenic carbon cycles among the atmosphere, oceans, and the terrestrial biosphere.

POSSIBLE POLICY OPTIONS FOR DEALING WITH AN ENHANCED GREENHOUSE EFFECT

What should, or could, we do if we are concerned that the greenhouse effect is real?

First, we can shift our fuel consumption to those fuels that produce more usable energy per unit weight of carbon dioxide emitted (specifically, use less coal and more natural gas). Some have argued that we must put an immediate moratorium on the use of fossil fuels, but there are no realistic short-term alternatives. Fuel switching from coal to natural gas not only cuts roughly in half the carbon dioxide produced per unit energy produced, but substantially reduces the sulfur dioxide emissions that contribute to acid rain.

Second, we can increase the efficiency of processes that consume fuel, to increase the amount of useful energy per unit carbon dioxide emitted. As examples, we can switch to cars that are more fuel efficient (more miles per gallon), increase the efficiency of electric power plant operation, and practice energy conservation at home. Energy efficiency and conservation not only reduce CO_2 emissions and other environmental impacts, but at the same time would reduce the dependency of countries such as the United States and Japan on imported energy supplies (Chapter 33), thereby reducing any balance-of-payments deficits, military budgets, and even threats to world security.

Third, we should consider seriously the 'nonfossil' sources of energy, even though such a switch cannot be made quickly. Hydropower can supply some additional electricity, but it appears to be nearly 'max-ed out,' since most available hydro sites are already in use, and environmental concerns about the consequences of dam-building make it unlikely that any more large-scale dams will be built in the United States. Solar energy (Chapter 36) does not yet seem ready to provide large-scale electricity generation. Certainly solar homes would be very important in those regions of ample sunlight. The use of biomass as an energy source (Chapter 34) is controversial, in the specific sense that advocates argue that biomass is 'CO$_2$-neutral,' but others contend that such a goal is impossible. In principle, the growing of new biomass would consume, by photosynthesis, as much carbon dioxide as was produced by the burning of an equal amount of biomass fuel for energy production, thus resulting in no *net* carbon dioxide emission. In practice, this perfect balance might be hard—indeed, impossible—to achieve. Also, very large areas of land would have to be set aside to grow the biomass crops to provide the energy. It would be important to use 'all electric' homes (eliminating the combustion of fuels for domestic heating) and electric transportation—electric streetcars, electric railway locomotives, and electric automobiles. Where would we get all the electricity needed for these electric homes and electric vehicles if we can't increase hydro, and solar is not ready for large-scale use? We'd need a serious reconsideration of nuclear fission as an energy source.

The 1997 Kyoto Protocol requires developed countries (industrialized nations referred to in the Protocol as 'Annex I Parties') to reduce emissions of greenhouse gases during the first 'commitment period,' 2008–2012, by an average of 5.2% below the levels of 1990. The nations of the European Union have allocated their 8% reduction commitment under the Protocol among their 15 member states. The United States had agreed to undertake a 7% reduction and Japan, 6%. The United States Senate has never ratified the Kyoto Protocol, and it appears that President Bush will not support its ratification or implementation. The developing countries (the so-called Group of 77 plus China, actually made up of over 130 nations) have no obligations under the Protocol to reduce emissions of greenhouse gases. To stabilize concentrations of anthropogenic CO$_2$ below 750 ppm, anthropogenic emissions will have to decline relative to today's levels. If anthropogenic emissions were to be held at 1990 levels, atmospheric concentrations of CO$_2$ will continue to increase through this century. Projections of future energy use indicate that CO$_2$ concentrations in the atmosphere will continue to rise in any reasonable scenario.

There is no short-term or even mid-term alternative to fossil fuels, at least on the scale that would be needed to reduce substantially anthropogenic CO$_2$ emissions. If large reductions in the source are not feasible, then the only alternative is to increase sinks, by finding ways to sequester CO$_2$. The United States Department of Energy has proposed that nations develop the potential to dispose of nearly four billion tons of CO$_2$ annually by the year 2025 and 15 billion tons by 2050. If these goals could be met, then the world could stabilize atmospheric concentrations of CO$_2$. Several sequestration options are under consideration, including the use of depleted oil formations.

One example is the work being done by Statoil, the Norwegian state oil company. Since 1996, Statoil and its partners have been producing natural gas that contains 9% CO$_2$, almost four times the amount allowed by exportation rules. Normally, gas companies that encounter CO$_2$-rich gas would separate the CO$_2$ and then vent it into the air. Instead, Statoil pumps the CO$_2$ down into a geologic formation a half-mile below the seafloor. If they had taken the standard approach, of venting the gas to the atmosphere, Statoil would single-handedly have boosted Norway's CO$_2$ emissions by 3%.

To be sure, there are those who consider that the greenhouse effect would be a good thing. We've seen that some evidence shows that plants grow faster in atmospheres of higher carbon dioxide concentration. That effect, coupled with the warmer average temperatures, might enhance crop yields and ease food shortages. A higher average temperature might actually reduce energy consumption for home heating (but not for home air conditioning!). From that perspective, the more industrial growth we have, likely coming from increased burning of fossil fuels, the better off (in terms of standard of living) we would be. We've also seen argument that the greenhouse effect is not an issue for worry, because it simply means that in the future Boston would have a climate like Washington does now, and, similarly, New York City will have one like Atlanta's. Nonetheless, we might ask whether the citizens of Boston and New York have been consulted, let alone those in Washington and Atlanta wondering what their cities might be like.

WHERE DO WE STAND?

Two points are well established: Carbon dioxide in the atmosphere contributes to an elevated global temperature, and the concentration of carbon dioxide in the atmosphere has been increasing over the past hundred or more years. There seems to be strong, but not unequivocal, evidence that the increase of carbon dioxide over the past century is a consequence of human activity. Certainly we have been burning fossil fuels at ever-increasing rates, and simultaneously working to destroy a major carbon sink, the Amazon rainforest. Many indicators support, or at least are consistent with, the notion that the average global temperature is rising and has been doing so for at least a century. It is, however, appropriate to question how accurate the temperature measurements were in, say, 1860. Also, it is worth questioning if temperature data obtained prior to global monitoring by satellites are truly representative of the planet as a whole. The connection between anthropogenic greenhouse gas emissions and the global warming is not definitively established, but evidence in favor of this connection is steadily growing. Whether the average global temperature will continue to increase as anthropogenic emissions of greenhouse gases increase, and if so, by how much, is still open to doubt.

Strong cases can now be made that surface temperatures are warmer than they were 150 years ago and that the burning of fossil fuels since the beginning of the Industrial Revolution has led to a significant increase in carbon dioxide. What we don't know, and what lies at the center of the debate, are the long-term consequences of continued carbon dioxide increases on Earth's climate, and ultimately upon our quality of life. It must be emphasized that there is no debate over scientific laws or the scientific method. Further, almost all would agree that the CO_2 concentration in the atmosphere is increasing. However, Earth's climate is the most complicated system scientists have ever attempted to understand, to model, and to predict. Uncertainties in the computer models result in small changes in the initial assumptions used for the model leading very different predictions of, say, the temperature increase 50 years into the future.

One side of the debate argues in favor of taking precautions. If we don't start now to curtail greenhouse gas emissions, it may be too late to prevent a catastrophe a half-century in the future. On the other side lies the argument that making policy, setting regulations, passing laws, or signing international protocols before we understand the problem (in fact, before it is unequivocally established that a problem exists), runs a high probability of failure. Some have faith that because science has in the past found solutions to major problems it will do so in the future. This position is sometimes phrased as, 'If we can put a man on the moon, surely we can . . . .'

FOR FURTHER READING

Alley, Richard B. *The Two-Mile Time Machine*. Princeton University Press: Princeton, 2000. A very readable account, by one of the world's foremost glaciologists, of how data taken from ice cores reveals information about climate change.

Arrhenius, S. 'On the Influence of Carbonic Acid in the Air upon the Temperature on the Ground.' *Philosophical Magazine* **1896**, *41*, 237. This is *the* seminal publication on the greenhouse effect. It should be accessible in large libraries, particularly at universities.

Aubrecht, Gordon J. *Energy*. Merrill: Columbus, 1989; Chapters 23, 24. The first of these two chapters provides a good discussion of what is meant by *climate*. The second then addresses some of the issues surrounding possible human effects on global climate.

Dyson, Freeman. 'The Greenhouse Effect: An Alternative View.' In: *The Faber Book of Science*. Carey, John (Ed.). Faber and Faber: London, 1995; pp. 492–494. This brief article suggests that our attention really needs to be focused on how the biosphere adapts to increased carbon dioxide concentrations.

Houghton, John. *Global Warming: The Complete Briefing*. Cambridge University: Cambridge, 1997. This book reviews the evidence that indicates global warming is occurring, what the likely impacts could be, and some policy recommendations. It is packed with data tables and graphs.

McNeill, J.R. *Something New Under the Sun*. Norton: New York, 2000; Chapter 4. An important book, providing a history of the 20th century with focus on environmental issues. This chapter presents a history of the atmosphere from regional and global perspectives, including global climate change.

Sarmiento, Jorge L.; Wofsy, Steven C. *A U.S. Carbon Cycle Science Plan*. U.S. Global Change Research Program: Washington, 1999. This is a publication on behalf of the federal agencies that jointly sponsored the Carbon and Climate Working Group. Though it seems highly questionable whether the administration of the second Bush would take action on any of the recommendations, this report contains much useful background information.

Schimel, D.; Enting, I.G.; Heimann, M.; Wigley, T.M.L.; Raynaud, D.; Alves, D.; Siegenthaler, U. 'CO_2 and the Carbon Cycle.' In: *The Carbon Cycle*. Wigley, T.M.L.; Schimel, D.S. (Eds.). Cambridge: Cambridge, 2000; Chapter 1.

Schneider, Stephen H. 'Global Climate Change.' In: *The Energy–Environment Connection*. Hollander, Jack M. (Ed.). Island Press: Washington, 1992; Chapter 4. A review, with a large number of references to additional reading, of how we might forecast changes in climate and verify the models, and how the ecosystem might respond.

Schneider, Stephen H. *Laboratory Earth*. Basic Books: New York, 1997. A well-written introduction to the field of Earth systems science for persons with little previous scientific background. Among the topics covered are climate change, and the modeling of human-induced climate change.

Schwartz, A. Truman; Bunce, Diane M.; Silberman, Robert G.; Stanitski, Conrad L.; Stratton, Wilmer J.; Zipp, Arden P. *Chemistry in Context*. WCB/McGraw-Hill: New York, 1997; Chapter 3. This chapter discusses the chemistry of global warming, with abundant illustrations and reasonably recent data.

In addition to these books or bound reports, global climate change has been the subject of a tidal wave of articles in popular magazines and scientific journals. Some recent ones that are particularly useful are listed below; they collectively represent a range of opinions, and most should be accessible in good libraries.

Brewer, Peter G.; Orr, Franklin M., Jr. 'CO_2: The Burning Issue.' *Chemistry and Industry* **2000** (17), 567–571.

Easterling, D.R.; Meehl, G.A.; Parmesan, C.; Changnon, S.A.; Karl, T.R.; Mearns, L.O. 'Climate Extremes: Observations, Modeling, and Impacts.' *Science* **2000**, *289*, 2068–2073.

Epstein, P.R. 'Is Global Warming Harmful to Health?' *Scientific American* **2000**, *283*(2), 50–57.

Holmes, Bob. 'Heads in the Clouds.' *New Scientist* **1999**, *162*(2185), 32–36.

Karl, Thomas R.; Ternberth, Kevin E. 'The human impact on climate.' *Scientific American: Earth from the Inside Out* **2000**, 64–69.

Malin, Clement B. 'Petroleum Industry Faces Challenge of Change in Confronting Global Warming.' *Oil and Gas Journal* **2000**, *98*(35), 58–63.

Monastersky, Richard. 'Sizzling June Fires Up Greenhouse Debate.' *Science News* **1998**, *154*, 52–53.

Monastersky, Richard. 'Good-bye to a Greenhouse Gas.' *Science News* **1999**, *155*, 392–394.

Park, Robert L. 'Voodoo Science and the Belief Gene.' *Skeptical Inquirer* **2000**, *24*(5), 24–29.

Sletto, Bjorn. 'Desert in Disguise.' *Earth* **1997**, *6*(1), 42–49.

33

FOSSIL ENERGY: RESERVES, RESOURCES, AND GEOPOLITICS

ENERGY RESERVES AND RESOURCES

There are certain ENERGY sources that offer the prospect of being truly inexhaustible. Water and wind are two. We will see that solar energy (Chapter 36) is another. Biomass (Chapter 34), and nuclear fusion (Chapter 37), if it ever proves practical, are two more. However, the common sources of energy that most of us rely on—fossil fuels for transportation, and fossil-fuel-generated electricity—are, despite our ordinary presumptions, exhaustible. Most of us take for granted an unlimited supply of electricity; if there is a limitation, it is in our ingenuity for plugging appliance upon appliance into the available outlets. Similarly, we presume the ability to purchase unlimited amounts of gasoline, anywhere, anytime. But, some day, in some sense of the term, we will run out. In this chapter we will focus on some of the questions of how much of a given supply of energy is available, and how long it is likely to last.

The concepts of reserves and resources that we discuss in this section apply to any energy source (and indeed to other natural substances as well, such as ores of metals). The **reserve** is the amount of a material that can be recovered economically with known technology. Reserves usually are established by detailed exploration. A **resource** is the entire amount of the material known *or estimated* to exist, regardless of the cost or technological developments needed to extract it. Put crudely, the reserve is what we're sure of, but the resource includes what we think is there. Notice that by this definition the amount of reserves is always less than the amount of the resource.

An analogy to illustrate the concepts of reserves and resources—and the various categories of each—is in one's personal finances. Your financial reserve would consist, first of all, of the amount of cash you have in hand, the amount of money in your checking and savings accounts at the bank, and whatever cash advance you might be able to obtain from your credit cards. Note that all of these quantities can—in principle—be determined quite accurately, and they represent money that you can count on having. This money we would say represents the **proved reserves**.

In addition to your proved reserves, you could raise money by selling some personal possessions, such as a car; or by liquidation of any investments (stocks or bonds) that you might have. This is a bit trickier, however, than emptying your purse or wallet and your bank account. First of all, we know that the stock and bond markets fluctuate daily, so that while you know that you can certainly realize some amount of money by liquidating your investments, you can't quite count on how much it will be on a particular day. Second, while used cars, antiques, and collectibles do have assessed values, if you were forced to sell your car in a hurry to raise money, you can't necessarily count on holding out till you receive the 'book value.' These situations represent cases where you know with reasonable certainty that you can obtain some money, but you can't determine in advance exactly how much. This money represents your **probable resources.**

Continuing on a money-raising campaign, there are loans that you might be able to get from banks, family members, or friends. You can likely assume that various sums of money would be available, but you have very little knowledge in advance of exactly how much you personally could obtain. This money counts as your **possible resource**. The amount of your possible resource is not very certain. Similarly, your ability to 'extract' the resource (money) from these sources is not certain, either. How many of your 'friends' would really give you an unsecured loan of money?

Hypothetical resources include things like winning a lottery. You know, of course, that some pot of money is available to be won. However, the jackpot amount fluctuates daily, as do your (exceptionally slim) odds of winning it. Finally, there are **speculative resources.** For example, some long-lost relative that you have never even heard of may suddenly die and leave all his or her money to you. We know that this does, on rare occasions, happen to people. But in your effort to raise money, you don't, first of all, even know that this source exists, and, secondly, have no idea of how much loot the old codger has anyway. These concepts are summarized in the diagram in Figure 33.1

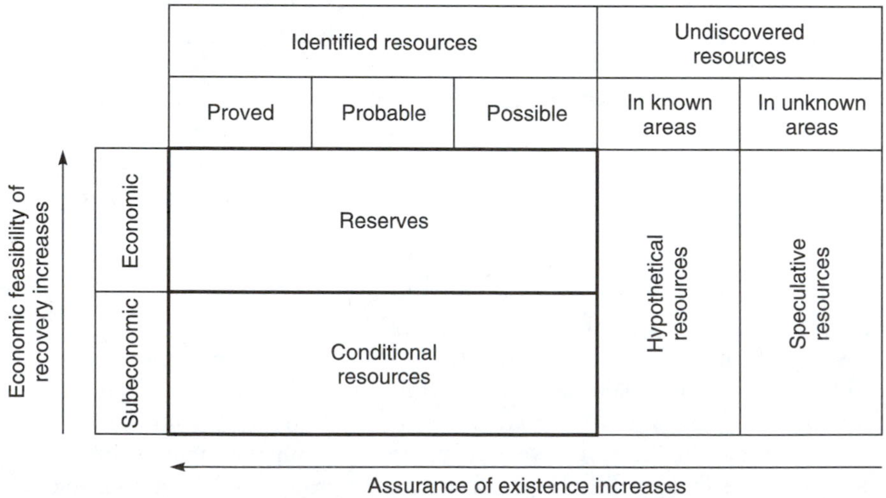

FIGURE 33.1 Reserves and resources are categorized according both to the certainty of their existence and to the economics of recovery.

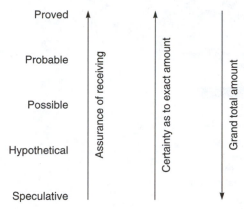

FIGURE 33.2 Relationships exist among the various categories of reserves or resources in terms of the total amount, the certainty of the exact amount, and the assurance of recovery.

To recap, note that there is pattern of variability among these classes of reserves and resources, that relate your assurance of receiving the money, your certainty as to the exact amount of each type that might be realized, and the grand total amount available (Figure 33.2). Your proved reserve might amount to, say, $1,000, but you can find out that amount to the exact penny, and you know with absolute certainty that it is indeed yours. Counting up all your money, right through the speculative resources, nothing says you can't claim to be with $50 billion. After all, long-lost Aunt Hildegarde might croak this very day, leaving you $49,999,999,000! (Then again, she might not, or she might be somebody else's aunt.)

The same situation exists when we discuss energy sources. The difference is that, with energy sources, we pay increased attention to the question of 'extraction,' that is, to a second dimension that incorporates the economic feasibility of extracting or recovering the material from the Earth. We paid little attention to this dimension in our example of trying to raise money, because presumably the 'technology' for 'extracting' money would simply involve phone calls, letters, or personal visits.

The amount of a particular energy source that is proven, probable, or possible, *and* economically recoverable constitutes the reserves. The amount proved, probable, or possible but 'subeconomic' (meaning that, with today's technology, it would be so expensive to extract them we couldn't afford to do it) represents conditional resources. The remaining amounts then represent the hypothetical resources and the speculative resources.

The amount of an energy source that is counted in any given category—as reserves, for example—changes over a period of time. In fact, sometimes the reserve of an energy source that we are using continuously will appear to increase. Several factors contribute to the fluctuation in the numerical value of the reserve.

First, the 'fineness' of the estimate is constantly being revised. For example, one method of exploring for coal involves drilling test holes into the Earth. In an early stage of exploration, drilling is done at very wide intervals, which sometimes may be several miles apart (Figure 33.3). In this example, we see that two test holes would have encountered a coal seam of some particular thickness. Perhaps the test holes are a mile apart. To estimate the amount of coal that has been discovered in this way, we might assume that the seam thickness is constant between the two drill holes. However, with further exploration and increased drilling at smaller and smaller spatial intervals, we might get much different pictures (Figure 33.4). The accuracy of the estimates of reserves is often related to the complexity of the subsurface geology. The reserves in a large deposit lying in a region of relatively simple geology often tend to be underestimated. In contrast, deposits in regions of complex geology tend to be overestimated.

A second reason for the fluctuation in reserves is that continued exploration for any energy source could discover entirely new fields. Particularly at times when energy is scarce and prices and profitability are high, an energy company might find that it pays off to mount a new campaign of exploration.

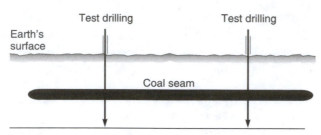

FIGURE 33.3 When exploring for a source of energy, such as coal, an initial reconnaissance might be made at intervals of a mile or more. The actual amount to be found in between the test drillings has to be estimated.

Third, improvements in the technology of extracting an energy supply from the Earth could shift subeconomic conditional resources into the reserves category. As improved technologies are brought to bear, resources once considered impractical to extract can become economic. For example, steady improvements are being made in drilling for oil and gas, even to the extent of being able to drill horizontal holes under the Earth's surface. Improved drilling technology may make feasible the recovery of oil or gas that once was inaccessible. Similarly, new mining technologies, perhaps involving robot miners, could make it feasible to extract coal from much thinner seams than are accessible with today's mining technology, or even to extract coal seams that are under water in flooded mines.

When we compare the availability of energy sources in the United States with those of other countries, it is important to recognize that the definition of *reserves* varies widely from region to region. In the United States, the Securities and Exchange Commission allows companies to call a reserve 'proved' only if—in the case of oil and gas—it lies near a producing well and there is 'reasonable certainty' that it can be recovered profitably at current prices, using existing extraction technology. However, governmental agencies or regulators in other countries do not enforce particular oil-reserve definitions. A particularly egregious example was the former Soviet Union, which routinely published wildly optimistic 'guesstimates' of its oil reserves. This practice seems to be continuing with agencies in the newly independent countries that formed from the breakup of the Soviet Union. Unfortunately, some analysts and economists have misinterpreted (i.e., believed in) these estimates of 'proved' reserves.

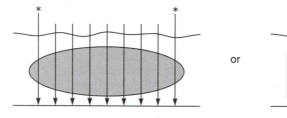

*indicates the original test drillings

FIGURE 33.4 When more detailed test drillings are made in an exploration program, drilling at smaller intervals than shown in Figure 33.3, it may prove that the amount actually found is much greater than, or more much less than, that originally estimated.

WE'RE HOOKED ON OIL

Most of this chapter focuses on petroleum. There are two reasons for this. First, a petroleum product—gasoline—is the one form of energy, other than electricity, that all of us use in our daily lives. Second, petroleum is the one energy source for which the United States has a significant dependence on imports.

> *Many commodities which are most useful to us are esteemed and desired but little. We cannot live without water, and yet in ordinary circumstances we set no value on it. Why is this? Simply because we usually have so much of it that its final degree of utility is reduced nearly to zero. We enjoy every day the almost infinite utility of water, but then we do not need to consume more than we have. Let the supply run short by drought, and we begin to feel the higher degrees of utility, of which we think but little at other times.*
> —Jevons[1]

Let's begin by reviewing our dependence on petroleum products. Automobiles and light trucks use gasoline. Heavy trucks, many buses, and many kinds of agricultural machinery (for example, tractors) use diesel engines, and hence diesel fuel. Most small ships also use diesel engines. Large ships rely on steam turbines, but the steam is generated using heat obtained from burning fuel oil. Airplanes, depending on the type of engine, use jet fuel or gasoline. In sum, about 99% of all the energy we use in transportation comes from petroleum. Or, looking from the other perspective, two-thirds of the oil that we use is consumed for transportation. Some electricity is generated by burning fuel oil, rather than coal, in boilers. Some homes are heated using fuel oil, LPG, or in some cases kerosene-fired space heaters. Most of the plastics and synthetic fabrics that are virtually ubiquitous in our daily lives are made from by-products from the petroleum industry (indeed, the chemicals used to make these synthetics are known as *petro*chemicals). The economic prosperity of the United States has been built on abundant cheap oil. Our modern industrial society is 'hooked' on petroleum.

As one example of our oil 'habit,' consider a typical supertanker. These ships are about a thousand feet long, or about as long as three football fields (including the end zones) laid end-to-end. A ship of this sort would put to sea from, say, the port of Ras Tanura in Saudi Arabia. It will be carrying two million barrels of crude oil. Coming from Saudi Arabia, it will be at sea for about five weeks. When the supertanker is unloaded, those two million barrels of oil, at sea for 35 days, will be enough to last the United States for just about *three hours.*

In mid-2001 the United States was consuming nearly 17 million barrels of petroleum per *day.* At that time, the world price of oil was about $24 per barrel. More than half of that oil was imported, much from the Middle East. The gap between domestic consumption and domestic production is rising steadily, as will be discussed below. The difference between the consumption and domestic production has to be made up from imports. Imports may, according to some forecasts, rise to be 70% or more of total American consumption. This already has several consequences.

First, we should consider the sheer cost of oil imports, and its impact on our balance of payments. With oil at $24 per barrel, we were spending, in midsummer 2001, about $210,000,000 *per day* on imported oil, or nearly $80 billion per year. In 2000, the balance-of-trade deficit for the United States was $245 billion. Essentially, one-third of our trade deficit is accounted for by imports of petroleum and petroleum products. A second consideration is the impact of oil on how we conduct and establish our foreign relations.

Comrade Stalin correctly said that whoever has oil predominates.

—Kirov[2]

The Persian Gulf War was fought for one reason—oil. Was our reluctance to intervene in the 'ethnic cleansing' of Bosnia related to the fact that Bosnia has no oil? How would we react as a nation if some foreign power, or if some major domestic unrest, threatened oil supplies we obtain from Venezuela, or from Mexico? Would we invade?

OIL ECONOMICS

Price increases for home heating oil and gasoline in 2000 and 2001 led to many concerns about the high cost of energy. Some people believed that these increases were a manifestation of price-gouging or 'obscene profits' of oil companies. In our concern for how much we shell out with each stop at a gas station, or for the size of the monthly check for home oil delivery, it is easy to lose sight of a remarkable fact: Oil is cheaper today, in real terms (that is, adjusted for inflation and fluctuations in the value of currency), than it was before the oil price shock of 1973. Oil is probably cheaper now than it ever has been in history. Americans pay the lowest oil prices in the world, but use the most energy per capita. Excepting only tap water, gasoline is by far the cheapest liquid we buy. Gasoline is cheaper than milk, than bottled water, or than soft drinks.

Indeed, not only is oil cheaper, in real terms, than ever before, the price of most commodities has been falling for the last 150 years. Iron ore is another example, with no direct relation to energy supply, of a material cheaper than it has ever been in history. These facts have led to one group of forecasters, sometimes called 'Simonists' in honor of the economist Herbert Simon, ⚛ who believe that prices of commodities will continue to fall forever, and that there will always be an abundant supply of, as examples, cheap oil, coal, and steel. A second group of forecasters, who are sometimes called 'Neo-Malthusians' for the early economist Thomas Malthus, ⚛ anticipate that sooner or later we will begin to run out of these materials, and prices will rise drastically.

⚛ Herbert Simon was a remarkable polymath who made many superb contributions to several fields, including, among others, economics (for which he won the Nobel Prize in 1978), the theory of organizations, psychology, and artificial intelligence. Simon argued that the price of basic commodities, in real dollars (that is, corrected for inflation and currency fluctuations), will steadily drop, as a result of such factors as continued improvements in technology and new discoveries (meaning both new reserves and new kinds of technologies). For about the past 150 years, Simon's position has been correct. As only one example, in the year 2000, iron ore, the basis of our steel industry, was cheaper than it ever had been in history (in constant dollars).

⚛ Thomas Malthus was a British economist of the late 18th and early 19th centuries whose best-known contribution was the idea that, sooner or later, a population will grow faster than its food supply, and therefore poverty (as well as the attendant social ills and possible outright starvation) are inevitable. For example, if unchecked, a population will grow geometrically (e.g., 2, 4, 8, 16 ...), whereas arable land brought into cultivation for food production will grow only arithmetically (e.g., 2, 4, 6, 8 ...). Since the amount of arable land is limited, then at some point there will be far more people than there is land and crop production to feed them. Malthus's original argument has been extended beyond food to any resource that is finite; at some point the demand for that resource (analogous to the need for food by an expanding population) exceeds its availability, resulting in economic and social problems.

Oil prices, in current costs (that is, not corrected for inflation or currency changes) have fluctuated a good bit in the past two decades. These price changes have direct consequences.

Up until the 'oil price shock' of 1979 (occasioned in part by the Islamic revolution in Iran), oil prices were dictated by the ministers of the Organization of Petroleum Exporting Countries (**OPEC**) under long-term contracts with the major oil companies. That arrangement, however, broke down during 1979–80, because some of the oil-producing nations found that they could sell their oil to traders at prices far above that set by OPEC. In the collapse of oil prices that soon followed, oil buyers became unwilling to purchase cargoes on which the price had been set when the supertankers were halfway around the world. The buyers feared that the value of the oil would drop during its lengthy ocean voyage. One result of this situation was that American companies instead bought oil from relatively nearby Venezuela and Mexico. And, thanks to that situation, Saudi oil prices dropped by about two-thirds between 1981 and 1985.

In 1985, Saudi Arabia made a major change by taking on the risk of price changes while the oil supertankers were in transit. Nowadays the value of the oil is calculated when it is unloaded at its destination. The price of the oil is determined by comparing it with prices of oil of similar quality sold on the open market. Until the ship arrives in the United States (or wherever its destination happens to be), nobody knows the price its load of oil will bring. Only when the oil is being pumped from the ship will the open market determine its value.

In the mid-1980s, oil prices were relatively high. One effect of those prices was to reduce demand for oil. As soon as the price of oil dropped, demand increased again. One reason for this rise in demand in the 1980s was that, as oil fell below $20 per barrel, it was used to replace natural gas in some applications, such as a fuel in power plants. In the late 1990s, crude oil inventories were very nearly at maximum capacity. As a result, the demand for new oil production fell, and prices plummeted as well.

When oil prices fell in 1999, to about half their previous level, there was not a corresponding upward swing in demand. This was ascribed to relatively mild winters in 1996/97 and 1997/98 (reducing demand for heating oil), and to the economic recession, facetiously called the 'Asian flu,' that hit many of the nations of Asia in the late 1990s. In addition, a decade's worth of improvements in using natural-gas-fired turbines to generate electricity made it undesirable to switch back to oil, even when the price of oil fell. In 1999, the price of oil was about $10 per barrel. Some optimists expected that it would soon be cut in half again, hitting the $5 mark, and that it would be 2010 before oil climbed back to $22 per barrel. (In fact, oil had already topped this price in 2000.)

When oil prices are around $10 per barrel, as they were in the late 1990s, every oil firm has to slash its exploration budget. With low-price oil, very few investments outside the Middle East make sense for a company. Some oil-producing countries are burdened with governments having little intellectual or moral enlightenment, and that rely heavily on income from oil sales to keep the domestic economies afloat. They may already be on the brink of economic collapse. Cheap oil prices would make it more likely that economic conditions in these countries will get worse, perhaps also affecting political freedom and other domestic policies.

For most of us, *the* petroleum product is gasoline. Though it is easy to complain about high gasoline prices when we stop at the gas station, gasoline is less costly (when adjusted for inflation) than it was 40 years ago. There are several reasons why this is so. One is the intense competition among oil companies that often prevents refiners and retailers from passing on the full impact of oil price hikes. A second reason is fuel taxes in the United States. In all industrialized nations but one (that lone exception being the United States), taxes make up so much of the retail price of gasoline that consumers can hardly notice any fluctuation in crude oil prices. In Europe, it is not uncommon for gasoline to be about $4 per gallon, and about 80% of the purchase price is tax. This $4 per gallon price of gasoline should also be interpreted in light of the fact that European nations import oil at the same price per barrel as does the United States. The high European gasoline taxes are used in many countries to support development of alternative energy sources and to support an excellent public transportation system. In America, fuel taxes are lower than in virtually any other industrialized nation, making up only about a third of the price.

We've seen (Chapter 32) that, as we begin the 21st century, the possibility of global climate change arises in part from increased concentrations of anthropogenic carbon dioxide in the atmosphere. Most of the world's industrialized nations agreed at the Kyoto summit in 1997 to binding targets to reduce emissions of greenhouse gases, including CO_2. To be sure, whether countries will hit these targets, and how they will accomplish that feat, are unclear. The possible effect on future oil demand and on the world oil industry is also unclear. On the one hand, it is possible that demand for oil in the industrialized world could be one casualty of the Kyoto accords, since natural gas produces less CO_2 per Btu of useful energy than does any petroleum product. But, on the other hand, cheap oil might encourage more emissions of carbon dioxide, by increasing the use of petroleum products.

Another aspect of the economics of oil is the recognition that oil, in addition to its current traded price of $24 per barrel has other, hidden costs associated with it. For example, the fact that the burning of fossil fuels may be having long-term effects on the environment has led some people to question whether the cost of cleaning up oil's environmental, political, and social costs also be taken into account in setting the 'true' price of oil. The immediate cost of the Persian Gulf War was about 90 billion dollars. As we include long-term costs of veterans' care and the continual sporadic air strikes on Iraq the cost is still mounting. Since that war was fought to assure our continued access to Middle Eastern oil, would it not be more appropriate to factor the 90 billion dollars into the price of oil, rather than to have paid it from the general treasury? Costs such as the continued military presence in the Persian Gulf and the cost of correcting environmental problems are sometimes known as external costs, or 'externalities.' When such externalities are then rolled into the cost of the oil itself, the process is referred to in 'bureaucrat-speak' as 'internalizing the externalities.'

Where are oil prices headed? Of course no one knows for certain. Global demand for oil is currently rising at more than 2% a year. Since 1985, energy use is up about 30% in Latin America, 40% in Africa, and 50% in Asia. One forecast suggests that worldwide demand for oil will increase 60% (to about 40 billion barrels a year) by 2020. On the other hand, oil demand may fall whenever the remarkably long period of economic growth in the United States eventually comes to an end. The Neo-Malthusian view is that we can expect a doubling or tripling of the price of oil in the next 10–15 years, followed by the onset of genuine physical shortages of oil. This view is generally consistent with the predicted increase in world oil demand by 60% by 2020. If that scenario comes to pass, the high gasoline prices that motorists began paying in the spring of 2000 could be an indicator of much more serious problems in oil production and delivery. The alternative, Simonist view is that oil will continue the near-150 year trend of commodity prices and its price will continue to drop, in real terms, into the foreseeable future.

LIFETIME ESTIMATES

When we have obtained some quantitative information about an energy source, we can then estimate its lifetime. The calculation itself is simple:

$$\text{Lifetime} = \frac{\text{Total amount available}}{\text{Amount used per year}}$$

As an example, the United States has a coal reserve of 182 billion tons. Our annual consumption rate is about one billion tons per year. Thus we have

$$\text{Lifetime} = \frac{182 \text{ billion tons reserve}}{1 \text{ billion tons used per year}}$$

$$= 182 \text{ years}$$

This calculation should seem quite straightforward, and it is, if we consider only the arithmetic. One has to be *extremely careful* in interpreting these lifetime calculations. Why?

First, what was the number that was used as the 'total amount available'? In our example above we used the economically recoverable reserves. Suppose that instead we used the estimated total resource. In that case,

$$\text{Lifetime} = \frac{2,925 \text{ billion tons resource}}{1 \text{ billion tons used per year}}$$

$$= 2,925 \text{ years}$$

In other words, we've now calculated a lifetime of the same material, coal, that's 16 times longer than our first calculation!

Second, when we use the rather conservative value of economically recoverable proved reserves, our calculation assumes that there will be no fluctuation in the amount of reserves. Historical data clearly show this assumption to be untrue. A prediction of the petroleum reserve having a lifetime of about 10 years has been made at least since the 1920s. Of course we have never run out of petroleum. We need to realize that the quantitative estimate of reserves changes with time, for reasons that we've previously discussed.

Third, how do we pick, or estimate, the amount to be used each year, especially if the lifetime is very long? Our examples above assume that the annual consumption rate of coal, a billion tons per year, would not change for at least a century. This assumption is also historically untrue, and could possibly lead us to develop a false sense of security. For example, over a period of about 60 years, from 1913 to 1973, the annual rate of coal consumption hovered around 500 million tons per year. It dropped somewhat during the Great Depression and increased during the Second World War, but for a six-decade period, the 500-million-ton number was reasonably accurate. After 1973, coal consumption doubled, to about one billion tons per year, over a period of about 15 years. Now, it appears to have leveled off again at the one billion ton figure. So—what number do we pick for annual rate of consumption? And how do we adjust the number for possible growth or shrinkage of consumption?

Whenever we see an estimate of the lifetime of some energy source, we should ask several questions. First, what do we mean by 'total amount'? Is it the economic, proved reserves, is it the total resource, or is it something in between? Second, how do we estimate the 'amount used per year'? Is it assumed to be constant, or is it adjusted for growth? If it's adjusted for growth, how was it adjusted? To return to our lifetime of 182 years for our coal supply, we should really express this as follows: The lifetime of America's *present economically recoverable coal reserve* is 182 years *at present annual rates of consumption.*

Are we going to run out of petroleum? Well, there is a straightforward answer: It depends. Using some recent figures for oil in the lifetime equation, we have

$$\text{Lifetime} = \frac{88,100 \text{ megatonnes}}{3,072 \text{ megatonnes per year}}$$

$$= 30 \text{ years (!)}$$

This simplistic calculation provides us the rather dire prediction that we will run out of oil within 30 years. Some forecasters have even suggested that the lifetime may be as small as 10 years; however, such forecasts have been made with regularity at least since 1920. (And in fact, in the 19th century there were similar forecasts that we would run out of coal within 10 years, also made on a fairly regular basis throughout that century.)

Needless to say, we have yet to run out of oil (or coal, or natural gas). So what is the problem? Why have these estimates been so wildly wrong? Are the people who have made them scoundrels or dolts? To address this issue, we must consider that the lifetime, estimated according to the above equation, is a ratio of two numbers, the reserve and the consumption rate. A change in *either* of these numbers will cause a change in the estimated lifetime.

With increased geological exploration, and more detailed knowledge, hypothetical resources become conditional resources, conditional resources become reserves, and possible or probable reserves become proved reserves. Improvements in extraction technology can be responsible for conditional resources becoming economical to recover, and hence adding to the reserve base. Thus more extensive exploration, improved extraction, or both will increase the reserves, and, as a mathematical consequence, increase the lifetime, assuming that there has been no change in consumption. On the other hand, changes in consumption rate will also impact lifetime. If the consumption rate drops, as might be caused, for example, by price increases or supply shortages, then the lifetime will increase. If the consumption rate goes up, lifetime drops. Both of these examples assume that there has been no change in reserves. In fact, both are likely to be changing at the same time, and for valid reasons. Thus the estimates of lifetimes are also changing continually. Both of these changes—in reserves and in consumption rates—are issues we will address in more detail later in this chapter.

So, is this calculation of a lifetime of 30 years for the world's oil reserve true? There are, of course, two possible answers. The optimistic answer is No, because predictions that the world is running out of oil, made at least since the 1920s, haven't come true yet. And, for all we know, there may be enormous quantities of oil yet to be found. Furthermore, we will not literally run out of oil, because as the supply declines, the price will rise, and thus reduce the rate of consumption. The pessimistic answer is Yes, mainly because since the 1970s the rate of discovery of new oil reserves is less than the rate of production and consumption of oil. We cannot realistically presume that current oil production could be constant for 30 years and then stop overnight. Such production could be maintained only if it were continually matched by new discovery, which, as we will see, is now far from the case. In all oil fields, production declines in the latter half of a field's life. The oil industry is already highly efficient and it seems unlikely that there is scope for endless further development of exploration and extraction technologies. In fact, much of the current technology development is aimed at increasing production rate, which accelerates depletion without adding to the reserves.

Here is an analogy that illustrates the pessimistic scenario. Suppose that you have a part-time job that pays $100 per week, and your expenses for living (rent, food, clothing, entertainment, etc.) amount to $75 per week. In essence, your rate of 'discovery' of reserves (money) is greater than your rate of consumption. As a result, your reserves will grow steadily, even though there is consumption on a regular basis (Table 33.1). In other words, after a hundred weeks, your reserve would have built up to $2,500. Now suppose you have the same part-time job, but your living expenses rise to $125 per week. This represents the case of the rate of consumption being greater than the rate of 'discovery.' What happens is shown in Table 33.2. The net effect of consumption being greater than discovery is that reserves dwindle and, no matter how large the reserves initially were, sooner or later you go broke. Another point that should be recognized from the previous table is that if the reserve initially is very large—as was the case with petroleum—then it's possible to 'coast' for a long time with consumption exceeding discovery before going broke. In this simple example of a fixed income and rate of expenditure, the initial reserve of $2,500 lasted for 99 weeks—nearly two years—even though 'consumption' always exceeded 'discovery.'

INCREASING THE PETROLEUM RESERVE

Recall the equation

$$\text{Lifetime} = \frac{\text{Total amount}}{\text{Amount used per year}}$$

We would like, of course, to have the value of the lifetime to be as large as possible. This is true not just for petroleum, but for any energy source. Mathematically, there are two ways to do this. One is to increase the numerator, the total amount available. The other is to decrease the denominator, the amount used per year.

What factors will increase the total oil reserve? One is to find more oil. Some scientists and oil industry analysts argue that we have not yet totally explored the world, and that improved satellite imaging, computer-based data analysis, and other techniques will lead to new finds of oil.

> *The cheerful doctrine that 'Something always turns up' is admissible only in a condition of profound ignorance, and we are now far past that condition—perhaps unhappily for ourselves. . . .*
> *It is, of course, true that not every acre of the earth's surface has been explored, but in a looser sense we do know what it contains, and that looser sense is enough. First, take*

TABLE 33.1   WHEN FINANCIAL INCOME EXCEEDS THE RATE OF SPENDING, FINANCIAL RESERVES INCREASE. WHEN THE RATE OF DISCOVERY OF NEW ENERGY SOURCES EXCEEDS THE RATE OF CONSUMING THEM, THOSE RESERVES INCREASE ALSO

Week	Discovery	Consumption	Reserve
1	100	75	25
2	100	75	50
3	100	75	75
100	100	75	2,500

TABLE 33.2 WHEN RATE OF CONSUMPTION EXCEEDS INCOME, THEN SOONER OR LATER WE HIT ZERO, REGARDLESS OF THE TOTAL AMOUNT THAT WAS ON HAND TO BEGIN WITH. WHEN THE RATE OF CONSUMING AN ENERGY SOURCE EXCEEDS THE RATE OF DISCOVERING NEW AMOUNTS (ADDING TO THE RESERVE) THE SAME EFFECT WILL HAPPEN

Week	Discovery	Consumption	Reserve
101	100	125	2,475
102	100	125	2,450
103	100	125	2,425
199	100	125	0!

an example near home. There are rich coal fields under us in parts of England, and it is not impossible that there are some more not yet discovered, but nobody would be likely to maintain that there is any chance whatever that there is 10 times as much undiscovered but mineable coal under us as that already known. In this loose sense we do know the amount of English coal.

—Darwin[3]

In addition it is thought that there may be vast quantities of so-called nonconventional oil available. There are three categories of nonconventional oil. First, it is already well established that there are huge deposits of heavy oils and bitumen⚛ in Canada, Venezuela, and Siberia (which can be produced only at very slow rates and with a prodigious effort). Second, there still exist prospects for enhancing production from oil reservoirs already discovered. Third, some possible contributions from very small fields, or oil reservoirs in very deep water might also add to the total. A small number of scientists believe that there may be vast pools of oil buried deep inside the Earth, formed not by the generally accepted kerogen maturation processes we discussed in Chapter 20, but rather from other geochemical pathways that may not even involve starting from living organisms.⚛

⚛ The word bitumen is used loosely to refer to a variety of extremely viscous, very high boiling hydrocarbons that have the approximate look and consistency of road tar. Potentially the most important bitumens are the so-called tar sands, which occur in many parts of the world and, collectively, contain immense quantities of hydrocarbons. In Canada alone the tar sand resource is estimated to be equivalent to about two *trillion* barrels of oil. The tar sands are deposits of bitumen with sand or sandstone.

⚛ Over the years, a few scientists have suggested that much, or all, of the world's petroleum does not come from the geochemical alteration of biologically derived organic matter in the relatively recent past, but has some other, nonbiological origin. One of the first to suggest such a concept was the Russian chemist Dimitri Mendeleev, one of the discoverers of the periodic classification of the elements. In recent years, the most visible proponent of such ideas has been Thomas Gold, who hypothesizes that the majority of the world's petroleum represents 'preplanetary' material that was entrapped in place when the Earth first coalesced many billions of years ago. The most recent argument in favor of this view is Gold's book *The Deep Hot Biosphere* (Springer-Verlag, 1999). A very substantial majority of geochemists reject the idea of a nonbiological origin of petroleum.

Certainly, improved technology is steadily being developed for recovering the oil we already know exists. In the 1960s, oil companies assumed that about 30% of the oil in a field was typically recoverable; now, however, they expect to recover 40–50%.

Finally, as the price of oil goes up, in response to its appearing to become more scarce, it becomes more profitable to explore for oil, and to extract some of the poorer quality oils that otherwise might have been left in the ground.

Anything can be done in the oil field. If you think there's oil somewhere, and have the dollars and desire to find out, it will be done. There is no such thing as 'I can't.'

— Bass[4]

LESSENING OUR DEPENDENCE ON PETROLEUM

What factors will decrease the amount of oil used per year? The best approach by far is through energy conservation. As the price of petroleum and its products rises, more and more people, and industries, take steps to curtail use. This includes, as examples, switching to more fuel-efficient cars, or using more insulation or turning thermostats back in the home to burn less heating oil.

Reducing oil consumption can be achieved with more efficient use of oil or by substituting alternative energy sources for oil. Together, these strategies offer the quickest and cheapest way of reducing dependence on petroleum, and should at the same time maximize competition and innovation. The Corporate Automobile Fuel Efficiency (CAFE) standards mandated in 1975 pushed cars from 13 miles per gallon (mpg) to 27.5 in 1986, saving five million barrels of oil a day.

In President Carter's term (1977–81) and the five years following it, oil imports from the Persian Gulf region fell by 87%. In the period from 1977 to 1985, the gross domestic product (Chapter 1) of the United States rose some 27%, while at the same time total oil imports dropped by 42%. That saving single-handedly took away one-eighth of OPEC's market. Worldwide, the oil market shrank by about a tenth; with OPEC's share drastically reduced, and OPEC's output being cut about in half. More fuel-efficient cars were the most important cause.

In 1985, President Reagan rolled back the **CAFE standards**. This lessening of car and light-truck efficiency standards doubled American oil imports from the Persian Gulf. The additional oil consumed as a result is equivalent to all the oil thought to be in the Arctic National Wildlife Refuge. If we had continued to conserve oil at the same rate as achieved in the 1976–85 period, or even if we had collectively bought new cars that got 5 mpg more than they did, the United States would no longer need Persian Gulf oil. For these reasons, it's argued that the biggest untapped oil field in America is hovering 18 inches off the ground—in the gas tanks of our vehicles. Unfortunately, federal policy in the 1980s discouraged energy efficiency. To be sure, concern for energy efficiency appeared needless after the 1986 oil price crash, which brought 10 years of cheap oil. At the same time, budget cuts in federally funded energy research and development slowed down technological innovation in energy productivity. Even as late as the Persian Gulf War in 1991, if the first President Bush had required that the average car get 32 mpg, that by itself would have displaced all Persian Gulf oil imports to the United States.

Aside from increasing the energy efficiency of devices (such as cars), it is important also to recognize the prodigious rate at which we simply waste the petroleum we do have. About 700 million gallons of oil end up in the ocean every year, from a variety of sources, all of which represent sloppy habits. Some of this oil is a result of used engine oil being poured down the drain. Some is from gasoline or lubricating oil that accumulates in streets or parking lots as a result of leaks, and is then carried away in storm runoff. Some is lost in leaks from offshore oil drilling. About 37 million gallons spill from oil tankers, and 15 million more from oil drilling rigs.

In addition, there are likely to be changes in the pattern of our energy use in the future. The use of electric cars is being mandated in California, and in other states as well. Hybrid vehicles (Chapter 31), capable of advertised fuel efficiencies of at least 70 mpg are now on sale in many parts of the country. Some fuels not derived from petroleum are being investigated, such as the so-called 'biodiesel,' which is derived directly from plants, and ethanol (Chapter 34). These changes, combined with increased energy efficiency, suggest a long-term lessening of our dependence on petroleum. Indeed, that seems to be the case.

Prior to the first oil price shock in 1973, oil consumption in the industrialized countries grew by more than two million barrels per day each year. Since the 1970s, however, the continuing decline in the growth of demand has been so dramatic that in 1995, consumption among the industrialized nations was only two million barrels per day higher than it was in 1975. Put in other terms, this change represents an average annual growth of only 0.1 million barrels per day. Oil consumption by industrialized nations in 1996 was less than it was at its peak in 1978, even though the GDP of these countries had grown by 42% during the same time.

In addition, the share of oil in the total consumption of primary energy has been falling steadily, in favor of increases in natural gas and in nuclear energy. This trend is particularly evident in Japan and Europe. In Japan, the share of oil in total energy consumption fell by more than 21 percentage points between 1973 and 1996, whereas the share of natural gas went up by more than 10 percentage points during the same period.

In the United States, the share of oil in meeting total energy demand was 40% in 1997. In comparison, in 1978, the year of peak oil consumption, oil's market share was nearly 50%. For each dollar of GDP, the United States used half the oil in 2000 than it did in 1975. In other words, the importance of oil to our overall economic well-being is substantially less than it was a quarter-century ago. Vehicle fuel efficiency reached an average of 21.5 mpg in 1997. However, the recent gains in fuel efficiency have slowed in the past few years, because many consumers have purchased sport utility vehicles or light trucks, vehicles that do not traditionally provide good gasoline mileage. Once again, it appears that CAFE was one of the best energy policies we ever had. However, even if vehicle efficiency stayed the same, demand for gasoline (and hence for petroleum) increased because most of us are driving more. Driving increased steadily from 9,000 miles per vehicle in 1980 to nearly 12,000 miles per vehicle by 1997.

OIL PRODUCTION

We can find more oil, that is, increase the reserve; we can diminish consumption by various measures. In addition, we need also to look at the status of oil production.

There are three points that need to be considered to project future oil production. First, we need to know how much oil has been extracted to date, that is, the cumulative production. Second, we need an estimate of reserves. The amount of reserves tells us the amount of oil that can be pumped out of the known oil fields before they have to be abandoned. Third, we need a good estimate of the amount of oil that still remains to be discovered and extracted from the Earth. The sum of these three numbers is the **ultimate recovery**, which is the total number of barrels of oil that will have been extracted by the time oil production comes to an end many decades or centuries in the future.

The rate at which any well—or any oil field, or any country—can produce oil always rises to a maximum and then, when about half the oil is gone, begins falling gradually back to zero. The oil geologist M. King Hubbert discovered that, in any large region, extraction of a finite resource rises along a bell-shaped curve that peaks when about half the resource is gone (Figure 33.5). This assumes that no legislative or regulatory restraints have been placed on extracting the oil, so that its extraction is essentially unhindered. (Indeed, the graph of extraction of *any* finite resource starts at zero, rises to a peak, and ends at zero.) Hubbert projected that oil production in the lower 48 states would reach a maximum around 1969, with a year or two leeway either way. In fact, oil production peaked in 1970, well within the prediction. The production of oil in the various nations of the former Soviet Union and the collective production of all producers outside the Middle East both follow the bell-shaped curves quite faithfully.

In all oil fields, peak production occurs at approximately the same time as midpoint of the total yield (Figure 33.6). (Again this relationship does not include those situations in which artificial legislative, regulatory, or other arrangements deliberately restrict production.) Remarkably, that prediction does not shift very much even if estimates are off somewhat, even by a few hundred billion barrels. Therefore, what we must pay particular attention to is not when all the oil is gone, but rather the point at which *half* is gone. That point will represent the peak of oil production. Likely, this peak point could occur between 2005 and 2015. Norway, for example, is likely to hit its peak in 2005.

The total world oil production, in the entire history of the oil industry, was 800 billion barrels by the end of 1997. It's estimated that the industry will be able to recover only about another trillion barrels of conventional oil. Thus the ultimate recovery is about 1.8 trillion barrels. Currently, the demand for oil worldwide is rising at about 2.5% a year.

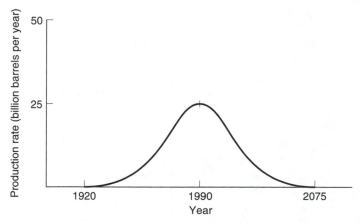

FIGURE 33.5 The Hubbert curve is a model for depletion of an energy source, or indeed of any finite resource. This Hubbert curve shows depletion of oil reserves.

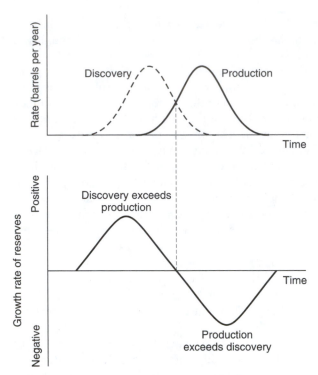

FIGURE 33.6 When production and depletion are compared, the important point is when half of the total amount of the energy source (here, oil) has been produced

This increase is driven largely by the expanding economies of the Far East. It would be even higher, but for the continuing economic struggles of Russia and the other countries of the former Soviet Union, as well as some of the Latin American nations. The world's midpoint of depletion will come when 900 billion barrels have been produced (that is, half the ultimate recovery) which is likely to be sometime in the next decade.

Once the approximately 900 billion barrels have been consumed, oil production is likely to begin to fall soon thereafter. Barring a major worldwide economic recession, and assuming that this half-way point in total recovery coincides with peak production, oil shortages could be expected to arrive in the early decades of this new century.

With respect to depletion of their domestic resources, the oil-producing countries fall into three broad groups. First, there are those countries that are already past their midpoint and, therefore, are in decline, examples being Germany and the United States. A second group comprises those countries that are now close to the midpoint. Examples in this group include Britain and Norway. The third group is those countries that are still at a very early stage of depletion. The principal members of this group are the Middle East countries (in particular, Iran, Iraq, Kuwait, Saudi Arabia, and the United Arab Emirates). For a while, at least, they will be able to make up the difference between the total demand for oil worldwide and the amount of oil that the other nations can supply. The Middle East countries can play this role only until they themselves reach the midpoint, after which they too will go into decline.

But—most of the oil that has been produced so far, as well as that most likely to be produced over the next 20 years is conventional oil. Much of it, especially that from the

Middle East, flows at high rates from giant fields. This oil is easy, cheap, and fast to produce. We should not lose sight of the fact that there is also nonconventional oil. As we've seen, this material includes heavy oils, bitumens, oil shales,⚛ and that amount of oil that could be obtained by enhanced recovery techniques.⚛ The Orinoco oil belt in Venezuela contains a trillion barrels of heavy oil. Tar sands and shale deposits in Canada and the former Soviet Union may contain the equivalent of more than 300 billion barrels of oil. Certainly, these nonconventional oils could help as production of conventional oil passes its midpoint. While it's not likely that nonconventional oil will affect the time of the peak in world oil production—at least not by very much—it could likely have a valuable contribution to total oil supplies during the time of the 'tail' (on Hubbert's bell-shaped curve) of depletion. Unfortunately, all of this nonconventional oil is both difficult and expensive to produce. Not only that, with today's technology it can be produced only at relatively slow rates. In fact, currently nonconventional oil amounts to about two million barrels a day, which might sound like an impressive amount of oil until we consider that it amounts only to 3% of total world oil production.

⚛ Oil shale is an organic-rich rock that contains no free oil. However, oil can be produced by heating the rock (a process usually called 'retorting') to drive oil out. Worldwide there are probably five to six times more recoverable reserves of oil locked up in oil shale than there are in conventional liquid petroleum. Unfortunately, producing oil from oil shale has formidable processing, economic, and environmental problems, and is not economically competitive with conventional petroleum. The last great flurry of interest in oil shale in the United States occurred in the late 1970s in response to the 'oil crisis' of the time.

⚛ The petroleum industry has available an array of advanced recovery or enhanced recovery methods aimed at producing additional oil from wells or stimulating production of oil from very small accumulations. These methods include, as examples, injecting steam into wells, flooding wells with chemicals to reduce the interactions of oil with the surrounding rocks, 'acidizing' to decompose carbonate rocks, or fracturing rocks surrounding the well.

World oil reserves have increased steadily over the past 20 years. Extrapolating that trend into the future would allow the conclusion that oil production will continue to rise unhindered for decades to come, still increasing well past 2020.⚛ However, almost 80% of the oil produced today flows from fields discovered prior to 1973, and the great majority of those fields are already in declining production. During the 1990s, oil companies discovered an average of seven billion barrels a year of new oil; but typical production worldwide is about 20 billion barrels a year—three times the rate of discovery. Now, in the early years of the 21st century the ratio is even worse—we are using about four times as much oil as we discover.

⚛ Mark Twain's essay on the length of the Mississippi River is the finest discussion of the perils of making long-range extrapolations from limited data that has ever been written. It should be required reading for anyone who uses or manipulates numerical measurements to determine trends. The essay appears in Chapter XVII of his *Life on the Mississippi*, available in many editions. In it, Twain 'proves,' on the basis of 20 years' worth of data, that, about a million years ago, the Mississippi must have been 1,300,000 miles long. As Twain puts it, 'There is something fascinating about science. One gets such wholesale returns of conjecture out of such a trifling investment of fact.'

Of course, advanced oil extraction technologies will buy a bit more time before production begins to fall. There is some relief to be had in the prospect of extracting more oil from existing fields by advanced technology and in the development of nonconventional oil. However, it is very difficult to quantify the amount and timing of such contributions. So, from an economic perspective, the timing of when the world will run completely out of oil is not really important; what truly matters is the timing of when production will begin to decline. Beyond that point, oil prices are almost sure to rise unless the world's demand for oil declines at the same time. Consequently, the question seems not to be whether, but when, worldwide oil production will start to decline, ushering in what might be called the 'permanent oil shock era.'

THE RACE BETWEEN DISCOVERY AND PRODUCTION

Oil must be found before it can be produced.

—Campbell[5]

In the mid-1970s there was a fundamental change in our pattern of petroleum use. The rate of finding new oil each year (that is, the quantity of oil discovered per year) became, for the first time ever, smaller than the annual rate of using oil. The rate of finding oil is analogous to depositing money into a bank account. Similarly, the rate of using oil is analogous to withdrawing money or writing checks. In essence, we are taking money out of the bank (petroleum out of the ground) faster than we are putting it into the bank (i.e., discovering more oil). This discussion is not unlike our previous discussion on energy balances (e.g., Chapter 4), and particularly the Calorie balance in our daily nutrition. Recall that we said then that withdrawing more than is being put in can only 'work' if we already have some money in the bank, or some Calories available in our body, to start with. For example, we need to have some initial amount of financial reserves already in the bank. We also said that, regardless of the original size of those reserves, if we persistently take out more than we put in, sooner or later we'll go broke. In *exactly the same way*, it doesn't matter how much oil was originally in the oil reserve—if we consume it at a rate faster than we find new oil, sooner or later we'll consume it all.

Why has the rate of discovery of new oil dropped? Several factors may have a role. First, there may indeed be less oil, and what oil does remain is more difficult to extract. Most areas of the world have already been explored, at least to some extent. Some believe, therefore, that the last ever major oil discovery, like Alaska's North Slope, has already been made, and that it is highly unlikely that ever again there will be another big oil discovery. Further developments in technology are needed to exploit the hard-to-extract reserves of oil. The technology needed to make subeconomic 'conditional resources' into economically recoverable reserves is developing only slowly. Finally, another factor may be policies of oil companies regarding exploration vs. acquisition. For most oil companies, it is easier to 'find' oil simply by buying another oil company, thereby obtaining that company's oil, than it is to mount a major campaign of exploration.

For example, consider that an oil company executive is charged with responsibility of increasing the reserves belonging to that company, and has been given a budget of a billion dollars. The first option available is to spend the billion dollars on exploratory drilling to find previously undiscovered oil. Suppose that a billion barrels of oil are discovered. This oil is added to the reserves of the company, in terms of ownership, but in addition, represents a billion barrels added to the total amount of known reserves worldwide. The probability of this exploration strategy being successful might be, say, 10%. A second option available to the executive is to spend the billion dollars in a merger or acquisition of another company that happens to have legal ownership of a billion barrels of reserves. Assuming that the acquisition takes place, the probability of success is 100%, and again in this example a billion barrels of oil have been added to the reserves owned by the company. However, in this case, no new oil has been discovered. Therefore the addition to the total amount of known reserves worldwide is zero. For the executive and for the corporation, the second option is the safer one. For the world's oil consumers, the first option is preferable, because, if successful, it does increase the amount of the world's oil reserves.

What would make it worth gambling on increased oil exploration? Increased oil prices, which would lead to higher sales prices for petroleum products and, ideally, higher profits for the corporation and a greater return to its stockholders. For example, suppose the odds of winning a bet (or, the odds of successfully finding a new oil reserve) are 10:1. No one would gamble a dollar if the winning stake was also only a dollar. A few people, perhaps, might be willing to wager a dollar at 10:1 odds if there were a chance of winning $10. Likely many people would wager a dollar to win a thousand dollars if there were one chance in 10 of winning. If the price of oil goes up, so does the value of new reserves. Thus, the greater the potential reward for gambling on exploration.

On the other hand, low oil prices depress exploration and drilling activity, especially in marginally producing areas such as the United States. The United States has, after all, been producing oil since 1859 and has been pretty thoroughly explored in terms of oil geology. When oil prices fall, the United States just can't compete. Those wells that produce only tiny quantities of oil per day (so-called stripper wells, which produce 10 barrels per day or less) are shut down, and the exploration rigs pack up and go back home.

By nationalizing their oil industries, and then by doing their best through OPEC to keep prices high in the 1970s and 80s, the Middle Eastern nations encouraged oil development elsewhere. With oil so profitable in those days, prospectors were motivated to search even remote, inhospitable parts of the world. Now, though, virtually the whole world has been mapped for oil. This brings us to the realization that indeed there may be no more new oil. Still, there is always the hope that huge deposits of oil may lie undetected in far-flung regions of the world. However, that seems rather unlikely. The extent of oil exploration has now been such that only locations in extremely deep water and the polar regions remain to be fully explored. Even in those cases, the prospect of finding oil seems reasonably well understood. In the years since 1950, the largest new oil discovery has been in the North Sea. The ultimate recovery from the North Sea is about 60 billion barrels, which is equivalent to less than three years of worldwide demand for oil. Because the world has now been so thoroughly explored, it seems unlikely that any new discovery of size comparable to the North Sea remains to be found.

Because of this, merely spending more money on oil exploration will not change the situation. Even after the price of oil hit all-time highs in the early 1980s, explorers armed with new technology for finding and recovering oil found few new fields. The rate of discovery of oil continued to decline. By some estimates, the oil industry has already found about 90% of all the oil that's likely ever to be found. Worldwide, the situation is this: Oil consumption is now about 24 billion barrels a year and is still going up, but we are finding less than six billion barrels, a rate of discovery that seems to be falling every year. Ninety percent of current world oil production comes from fields more than 20 years old; 70% comes from fields more than 30 years old.

Global primary energy demand is likely to increase by about 40% by 2010. Even with increased concern for anthropogenic carbon dioxide emissions (Chapter

32), it is likely that fossil fuels will, even then, supply 90% of the world's energy consumption. The Asian countries alone will account for nearly half of that increased demand.

In the United States, oil production hit a peak in 1970 at 10 million barrels per day. Production then steadily dropped off, to eight million barrels per day in 1976. Increased production in Alaska, thanks to the discovery of the North Slope oil, and increased drilling activity in the lower 48 states, thanks to high oil prices of the late 1970s and early 80s, led to a resurgence to nearly nine million barrels per day in 1985. However, by the close of the 20th century, total oil production in the United States was at its lowest level since 1950, having declined by one-third since 1985. Declining Alaskan North Slope production accounts for most of this downward trend.

WHAT DOES 'RUNNING OUT' REALLY MEAN?

> We are all going to run out of oil very soon. Our country sooner than others, to be sure, but all of us down to nothing, Venezuela and Arabia and China too, and it will be very interesting to see how we handle it.
>
> Bass[6]

What does it mean that we are going to run out of oil? Will we ever pump what is literally the last drop of oil out of the Earth? The answer to the second question is, of course not. We will never truly run 'out of' anything. As any energy source becomes increasingly scarce, its price will rise. This price increase will have several effects. It will become increasingly profitable to explore for additional sources. And, it will become increasingly profitable to invest in the research and development needed to develop new extraction technologies. Both of these activities add new reserves. Furthermore, at some point it becomes economically feasible to switch to a totally different source of energy. For example, it is thought that when the price of petroleum rises to about $35 per barrel (and appears likely to stay there for a long period of time), it will then be economically feasible to develop the production of synthetic liquid fuels from coal.

The world is not running out of oil. What we are likely to run out of, and soon if the Neo-Malthusians are correct, is the end of the era of abundant, cheap oil. From one economic perspective, resources such as oil are not truly finite in size, but merely finite at a certain price. In other words, if we want more oil, all we need do is drill more wells or develop some better extraction technology, or come up with some package of economic incentives to make it more profitable to produce oil. However, the reservoirs of conventional oil, especially the ever prolific ones such as in the Middle East, were formed only at certain periods in the geological history of the Earth, and formed under a limited set of geographical and geological circumstances. We can successfully find and extract oil now only if all the necessary geological conditions have already been met. Unfortunately, there is nothing we can do in the way of technology, or in providing economic incentives, that can have any effect whatsoever on the geological events that happened millions of years ago. Unfortunately oil is indeed a finite resource.

However, we must also face a second major concern, besides the issue of how much oil remains. That second concern is where the remaining oil is located—who's got it, who needs it, and what some of the implications are regarding the global distribution of oil. We now change our attention to aspects of the geopolitics of oil.

OIL IN THE UNITED STATES

As a rule of thumb, about half of the petroleum used in the United States is imported. That fraction fluctuates a bit from week to week, and appears to be increasing over time.

The leading source of imported oil in the recent past has been Saudi Arabia, from which we have been getting about 1.5 million barrels per day. Mexico has been running a close second, with about 1.3 million barrels per day. Together, these two sources supply about a third of our imported oil. They are followed very closely by Venezuela and Canada, each of which provide about 1.1–1.2 million barrels per day. Other countries that have been large suppliers of oil are Nigeria and Angola though at a quarter to half the level of countries such as Venezuela. Just as the exact fraction of oil that is imported fluctuates, so too do the amounts imported from each of these countries and their relative rankings as our sources of oil. However, as an approximate, but useful, estimate:

> HALF OF THE OIL WE USE IS IMPORTED, AND HALF OF THE IMPORTS COME FROM OPEC NATIONS.

Unfortunately, we in the United States have exploited our oil reserves longer, and certainly more fully, than any other nation. One result of this is that discovering and extracting another barrel of 'new' oil costs more in the United States than it does elsewhere. Furthermore, production from many wells in the United States, especially in the lower 48 states, has declined to relatively tiny amounts. For example, the average production per oil well in Saudi Arabia is 9,000 barrels a day. By comparison, in the United States, ours pump 15 (1/600th, or less than 0.2%). During the late 1980s we stopped producing two million barrels a day—as much as we imported from Kuwait at that time—because the price of oil fell so low that it was not profitable to produce domestic oil. Many stripper wells (those producing less than 10 barrels per day) were shut down because they were just not economical. However, if ways could be found to improve the economics of production from stripper wells, the total oil obtained from small amounts of production per well, but from a very large number of wells, could be a significant addition to our domestic oil supply.

This situation is further complicated by a possible future shortage of refinery capacity in the United States. In recent years, until about 2000, consumers have benefited from relatively low prices for gasoline, diesel fuel, and heating oil. However, the low prices of these commodities also discouraged oil companies from making the investment needed to construct and operate a new refinery. Depending on its capacity, a new refinery could cost about two billion dollars. No new refinery has been built in the United States since 1977.

As we've seen, the major oil discovery in the United States in the last half-century has been at Prudhoe Bay on Alaska's North Slope. The oil production activities in the Prudhoe Bay fields—pipelines, roads, drilling rigs, wells, waste treatment facilities, and airstrips, now cover about 800 square miles. The oil is brought south through the Trans-Alaska Pipeline System (sometimes known as TAPS). Alaska's North Slope is a very harsh environment, and the fact that oil can even be extracted and shipped from there is a tribute to the geologists, engineers, and workers whose efforts have made it possible. Oil almost anywhere else on Earth is more accessible and more reliably deliverable than from oil fields above the Arctic Circle. Importing oil from the world market—bought from many countries and brought into many different ports—may actually be better for America's energy security than is delivering oil to the lower 48 states through one single conduit, TAPS. The pipeline would be easy to disrupt by hostile military action or terrorists, and probably very difficult to repair.

The present focus is on the oil prospects in the Arctic National Wildlife Refuge (ANWR). Current estimates are that ANWR might contain about three billion barrels of oil. The prospect of exploring for, and then producing, this oil has ignited a fierce debate. Proponents argue that ANWR could provide one-tenth of our future domestic oil production. Opponents argue that ANWR is the last corner of North America that has the full range of arctic and subarctic ecosystems and that is still not touched (or ruined, as some would say) by human activities. If a decision were made to exploit the oil in ANWR, it's estimated that it would require 5–10 years before the oil was flowing in quantity to the lower 48. At the present rate of oil consumption, the three billion barrels of oil in ANWR are equivalent to six months' supply. Of course, we would not switch completely to arctic-refuge oil for six months, and then switch back again to some other source. Rather, oil from ANWR would add to the total of all the oil, domestic and imported, that we use. Nevertheless, the ANWR oil would likely be gone by 2020.

OIL IN THE MIDDLE EAST

The world's major petroleum reserves are in the Middle East. In terms of oil reserves, the top five countries in the world are Saudi Arabia, Iraq, United Arab Emirates, Kuwait, and Iran. Currently, sales of Iraqi oil are limited by United Nations' sanctions as a consequence of the Persian Gulf War. American relationships with Iran have been poisonous since 1979, though there is some hope for a 'thaw' as a result of the election of a comparatively moderate president in Iran. Some observers of foreign affairs believe that Saudi Arabia may be ripe for a fundamentalist Islamic revolution; if that happens, oil imports from there may be drastically curtailed.

Although the Middle East contains two-thirds of the world's oil reserves, it produces less than a third of the oil. If oil production were determined only by the cost of production and the quality of the oil—in other words, if there were no domestic or international political factors to reckon with—most of the world's oil would come from these countries. Undoubtedly, uncontrolled production by these countries would weaken oil prices and cause substantial reduction in investment and exploration and production activities of oil companies. Saudi Arabia could possibly supply most of the world single-handedly. Oil in the Middle East is cheap to extract. The actual production cost of the oil is about two dollars a barrel. We've seen that the biggest oil discovery in the recent past has been the North Sea. There, it costs about eight dollars a barrel to produce oil.

To illustrate the rich endowment of oil in the Middle East, we can consider the Ghawar field in Saudi Arabia. It was discovered in 1948. The field is about 150 miles long and, at its widest, is about 25 miles wide. The oil is contained in a layer of porous limestone that is about 100 feet thick. This single oil field in Saudi Arabia is estimated to contain 80 billion barrels of oil, an amount that represents more oil than is known to exist in any other single location anywhere, and that represents three times the remaining oil in the entire United States.

Though the Ghawar field contains an extraordinary quantity of oil, it's just one of many oil fields in Saudi Arabia. There are nearly 60 oil fields in Saudi Arabia, but oil is being produced from only seven. That production, from essentially one-eighth of all Saudi oil fields, is enough to satisfy all of the country's customers. All of the Saudi oil fields together contain about one-fourth of the world's oil reserves.

Not only are these oil fields in Saudi Arabia huge, they are also highly productive and, as we've seen, inexpensive to operate. The oil in the rocks under the desert is held there at high pressure. Because of that, once the reservoir rocks are penetrated by the drilling bits, the oil easily gushes to the surface without having to be pumped.

The oil travels by pipeline, still under pressure, to separator plants where the natural gas that is dissolved in the oil is released and separated. As the pressure is reduced, natural gas bubbles out of the oil, in the same way that carbon dioxide fizzes from a carbonated beverage when a can is opened. Any hydrogen sulfide in the oil is also removed. Then the oil is piped to ports at Yanbu, Ju'aymah or Ras Tanura.

There are some significant economic and geopolitical consequences of the fact that about half the world's remaining oil is in five countries in the Middle East. These five countries collectively are the only ones that can vary their oil output according to pre-vailing market conditions. And, when their market share is high, they can trigger a price crisis. This is exactly what happened in 1973. At that time these five Middle Eastern members of OPEC were able to hike prices not because oil was scarce (in fact, only about one-eighth of the world's reserves of conventional oil had been consumed to that point) but because they had managed to corner 36–38% of the market.

But—then along came the North Slope and the North Sea. Those new oil fields, coupled with a drop in demand for oil (occasioned at least in part by high prices) seriously reduced the 'clout' of these countries in the oil market. Prices collapsed. OPEC's share of the market was cut in half (to about 18%). From a peak of $283 billion in oil revenues obtained in 1980, OPEC's revenues fell to $77 billion in 1986, then rising again to $132 billion in 1995. If the 1995 revenues are adjusted for inflation and expressed in terms of 1980 dollars, then they would be less than $100 billion, about a third of what was amassed in 1980. Furthermore, non-OPEC oil production (and not counting the countries of the former Soviet Union and not counting the United States) tripled between 1976 and 1995. This increase in the supply of non-OPEC oil came at the expense of OPEC's market share. Today OPEC provides only a quarter of the world's oil requirements outside the former Soviet Union and the United States, compared with nearly 60% two decades earlier.

Because there are no new oil fields any more, certainly none comparable to a North Sea or a North Slope, the market share enjoyed by the Middle Eastern countries is now slowly increasing. It could hit 35% in the near future. This would be just about at the level these countries held in 1973 and, therefore, would provide them the potential of increas-ing prices. In fact, some analysts suggest that their share of the market could hit 50% in the 2010s. If that happens, the world could then see radical increases in oil prices. Such a time line could be delayed by the development of alternatives (such as biomass-derived fuels, Chapter 34) or by the emergence of stability in the Caspian Sea region that would allow those nations to become significant players in world oil markets.

There is no doubt though, that OPEC is able to flex its muscles. In what may be an indicator of things to come, OPEC lowered production quotas effective in the second quarter of 1999 for all members except Iraq. These countries reduced their output and allowed accumulated inventories of oil to be drawn down. In rather short time, worldwide oil prices jumped by 60% relative to the price at the end of 1998.

If OPEC, or the 'big five' Middle Eastern nations alone, do raise oil prices dras-tically, that action might help reduce demand for oil. The reduced demand would in turn likely depress production, possibly for as long as a decade. As a further result, actual physical shortages of oil would be delayed for a few years. Even so, by about 2010 they will be supplying half of the world's oil needs; and by 2015, they will be close to the midpoint of their depletion. Physical shortages of conventional oil would develop around 2020.

OIL IN THE CASPIAN REGION

A focus exclusively on the Middle East provides a gloomy picture for world oil sometime beyond 2020. There is, though, the potential for mitigating this problem with oil from a different region—around the Caspian Sea. However, many challenges, of domestic and international relations, of investment and infrastructure development, and of security, will have to be met and overcome first.

By 2015 to 2020, the Caspian Sea region could become the world's second largest source of oil, after the Middle East. Its three principal producers are Azerbaijan, Turkmenistan, and Kazakhstan (sometimes somewhat facetiously called 'the Stans'). These countries provide oil and natural gas to Europe via a pipeline system that runs across Russia.

That pipeline system represents the first problem. It's a jury-rigged, ramshackle, leaky system that, as it now exists, could not possibly handle the amount of oil and gas that would need to flow into world markets after about 2010. Much of the pipeline is by two Russian companies, Gazprom and Transneft. Russia has a substantial state investment in these companies. Despite (or perhaps because of) that state ownership, these companies do not have capital, in sound currency, sufficient to fix, let alone expand, the pipeline system. (This, however, has not stopped them from levying high toll charges on the oil and gas passing through the system.) Of course, the 'Stans' and those companies from outside the region that have invested heavily in the future development of the Caspian want to see the pipelines refurbished and extended. Unfortunately, this makes the problem all the knottier.

If the Caspian region does become a major player in world energy supply, the way the Middle East is today, then surely the United States and other energy-importing counties would have a strong interest in the stability and security of the region. It's estimated that we spend two dollars in the Department of Defense budget to protect every *one* dollar's worth of Persian Gulf oil coming into the United States. It seems rather unlikely that we would blithely allow countries like Russia, with grave economic problems and the potential to revert to an authoritarian, nationalistic government, or staunchly anti-American Iran to have some dominant role in the operation and security of the pipelines.

One possibility would be to find different routes to get the pipelines to consumers outside the Caspian region. Little would be served, from the perspective of the United States or Europe, by sending the Caspian oil northward into Russia. The only alternatives to the westward route that has the problems just mentioned would be to head to the east, or to the south. A route from Turkmenistan to Pakistan would have to pass through Afghanistan, a nation whose political future is still uncertain. A pipeline project heading through Iran would likely be impossible to finance, because the United States would bring economic sanctions with anyone doing business with Iran. Even if this problem were removed by a thaw in relations between the United States and Iran, any route heading south would terminate somewhere in the Middle East. Such a situation would mean that both the Middle East's own oil, and oil coming from the Caspian region, would have to pass through the Persian Gulf. The already high stakes for strategic dominance of the Persian Gulf would become all the greater. If Saudi Arabia were to undergo a fundamentalist Islamic revolution, or if there were a resurgence of Saddam Hussein's power in Iraq, much of the world's energy supply would be vulnerable.

If by some extraordinary stroke of good fortune in the tide of human affairs the problems just described could be resolved, there remains yet another: What is the Caspian, and who owns its resources? A decision on what exactly the Caspian is has a remarkable bearing on the answer to the second half of the question. For Russia and Iran, the Caspian is a lake. For Azerbaijan and the 'Stans,' it is an inland sea. At first sight, this dispute may seem to be an overly pedantic exercise in geographic nomenclature, but its outcome has important consequences. If the Caspian is indeed a lake, then its resources would have to be shared equally among all the surrounding countries, regardless of what they might be able to claim as their territorial waters. On the other hand, if the Caspian is a sea, then its resources (including, of course, oil) would be divided according to the extent of each country's territorial waters. These are determined by projecting the length of each nation's shoreline out into the Caspian. Since the lengths of the shorelines of the countries surrounding the Caspian are not equal, then neither would be the distribution of the shares of oil resources.

Given the difficulties of resolving these questions, it is difficult to see how the Caspian region will supply more than a small fraction of the world's oil anytime soon. Unless there are some very major changes in the politics and international affairs in the region in the near future, it seems unlikely that Caspian oil will have a major role in the world until well after 2010.

OIL IN CHINA AND THE PACIFIC RIM

It will be interesting to see how China will use its oil in world politics. There are some indications that China plans to rely on its enormous coal reserves for its domestic energy needs and use its oil for export. Strong Asian-Pacific economies, as in Japan and South Korea, and some of the rapidly developing economies, as in Singapore and Malaysia, are mostly in countries with little or no indigenous energy sources. Therefore, China has the opportunity to use its oil to affect the course of events in the Pacific Rim.

Some global petroleum forecasts suggest that by 2010 the deficit of petroleum (that is, the amount that's needed to be imported) in the Pacific Rim nations will be greater than the surplus amount in the Middle East. In addition, unrest is growing in some of the traditional large producers of petroleum. In Saudi Arabia, for example, there is widespread disaffection for the ruling family—and for the United States—outside the capital city. The situation seems ripe for another 'Iran-like' takeover by Islamic fundamentalists. If the Pacific Rim nations do indeed face a major petroleum shortage and, about the same time, an Islamic revolution in Saudi Arabia takes that petroleum supply out of the world market, there will be havoc with the world economy.

OIL IN EUROPE AND JAPAN

Europe and Japan have virtually no oil resources. Some oil from the North Sea is available in Europe, but virtually all of Japan's oil is imported. Despite that, both Europe and Japan have modern, stable industrial economies. What are the lessons of this that we in the United States might consider? Should we worry about the amount of oil we import, since Japan and many of the European countries have strong economies but yet import over 90% of their oil? After all, Germany and Japan import all their oil, though these countries are also very skilled at earning foreign exchange to pay for it. They rely on a global oil-trading and transport system so flexible that even the Persian Gulf War did not create lines at gas stations. A major difference between these countries and the United States is that, in Japan and virtually all European countries, gasoline is taxed very heavily. In some places it costs the equivalent of $4.50 per gallon. This high price encourages the use of small, highly fuel-efficient cars and a heavy reliance on public transportation. Indeed, in some countries the gasoline tax helps pay for the excellent public transportation systems. Many of the railway systems use electric locomotives, rather than diesel-engine locomotives. Some countries, particularly France, have made a significant investment in nuclear power. (It is fair to say that others, such as Germany, have plans to phase out nuclear power in the next few decades.) Many of these countries feel some obligation to try to help stabilize the nations of the former Soviet Union, helping to develop their oil industries and, by extension, their capabilities of exporting oil.

So far in this chapter we have focused almost exclusively on petroleum. Certainly one petroleum product, gasoline, is one of the most common energy sources on which all of us rely. And, because of the many geopolitical issues in the Middle East particularly, petroleum seems often in the news. However, there are alternatives, the two other major fossil fuels, natural gas and coal. We will end the chapter by considering each of these briefly.

NATURAL GAS

Recall that natural gas is a premium fuel, for several reasons. It has a high calorific value and instant on/off capability, and is clean burning and relatively easy to handle (Chapter 7). For these reasons, natural gas is an excellent fuel for domestic cooking and heating, and has become the preferred fuel for 'peak-load' power plants. These small plants, usually producing 50 MW of power or less, use natural-gas-fired combustion turbines to operate the generators. Current estimates are that the United States has abundant natural gas reserves. This, coupled with the fact that natural gas is a clean fuel, means there is a push to use natural gas in a variety of applications, including domestic heating, peak-load power plants, natural gas vehicles (such as city buses), and hybrid vehicles, some of which have both electric motors and natural gas engines.

In 1997, electric utilities turned to natural-gas-fired plants to pick up the slack created when the output from nuclear plants dropped. A year later demand for gas in the residential, commercial, and industrial sectors dropped, due to warmer than normal weather. Demand remained strong in the electric utility sector. The likely slowing of growth in nuclear power generation, coupled with the fact that no new nuclear plants are being built, will require the electric power industry to turn to other fuel sources in the future. Probably these will be coal for the large plants, and natural gas for the smaller ones. The natural gas industry can certainly expect stiff price competition from coal and possibly from fuel oil.

The major areas of the world possessing large reserves and resources of natural gas are the countries of the former Soviet Union (which collectively have about 37% of the total), Iran (with about 25%), the United States, Algeria, Saudi Arabia, Canada, and Mexico. The total in all the Middle Eastern countries amounts to about as much as is in the countries of the former Soviet Union.

If we examine the historical trends for lifetime estimates of natural gas, a story not unlike that for petroleum emerges. Table 33.3 shows some lifetime estimates made for natural gas at various times during the past half-century. And of course, the explanation is the same, the race between discovery and production. Present estimates suggest that production of gas is very likely to grow, and is not likely to peak at least until 2020, allowing it to be a potentially valuable substitute for conventional oil.

TABLE 33.3 ONE REASON FOR OPTIMISM ABOUT
FUTURE ENERGY SUPPLIES IS THAT LIFETIME
ESTIMATES, AS SHOWN HERE FOR NATURAL GAS, HAVE
INVARIABLY PROVEN WRONG

Year the Estimate Was Made	Estimated Lifetime (Years)	Predicted Year to Run Out of Natural Gas
1950	30	1980
1955	24	1979
1960	21	1981
1965	18	1983
1970	14	1984
1975	12	1987
1980	11	1991

Europe is not a major producer of natural gas. Only a few countries in Europe that produce oil from the North Sea also produce their own gas. But, because natural gas is a premium fuel, there is significant demand in Europe for it. Europe basically has two options.

The first is to depend on Russia and the other countries of the former Soviet Union. Nowadays the former Soviet Union is exporting natural gas to Europe via pipelines. This situation has some potentially positive outcomes. First, it provides a flow of hard currency into the former Soviet Union, which may help these countries to obtain consumer goods and the various services needed to rebuild their infrastructures (roads, bridges, railways, etc.). This could in turn lead to maintaining some political stability in these countries, which is of great interest and advantage to the more modernized countries of Western Europe. However, there is always the specter of some extremist figure like Vladimir Zhirinovsky coming to power in Russia, and stopping the exports of natural gas to Western Europe.

The second option is to depend on Algeria. This country exports gas to southern Europe (for example, to Italy) via pipelines on the seafloor of the Mediterranean. At the present time, Algeria has serious internal problems with Islamic fundamentalists seeking to disrupt and overthrow the current government. What might happen to natural gas exports if the Islamic fundamentalists come to power in Algeria is highly questionable.

One consideration affecting the world trade in natural gas is that the easiest way to transport it is via pipelines. Therefore, countries possessing abundant natural gas, but which are very distant from major gas markets may not have a major role in international trade in natural gas. This generally seems true of the countries in the Middle East, where there is little domestic demand for natural gas (because of the sparse population) and long distances to major markets elsewhere.

However, there is growing realization that natural gas does not need to be shipped as *gas*. Recent years have seen increased interest in so-called gas-to-liquids technology (GTL). The overall strategy is to convert natural gas into some liquid product, either near the gas wells, or at least at some site in the country where the gas is being produced. Then, the liquid product is shipped in tankers. Within this broad strategy, two approaches seem of most interest. One is the conversion of gas to liquid products resembling the paraffin fraction of petroleum. The other is to convert gas to methanol (Chapter 34). Of the two, the former is the more favored nowadays, because the product would be a substitute for petroleum, indeed much like the high-quality Pennsylvania crudes. In fact, in a GTL plant it would be possible to tailor the product to produce directly high yields of, say, diesel fuel, or some other desired product.

The only way that large quantities of natural gas can be shipped (as cargo, as distinct from transportation via a pipeline network) is to chill the gas to temperatures low enough to cause it to condense into a liquid (called liquefied natural gas, LNG). The temperatures required to produce and store LNG are about $-170°C$. For a time the United States imported LNG from Algeria, especially into the northeast, where there is a high demand and high population, but a long distance from the natural gas wells in the Gulf Coast. This practice raised some safety concerns. Suppose a leak allowed some of the LNG to vaporize. Mixtures of 5–15% of natural gas in air are explosive. What if an LNG tanker blew up in, say, Boston harbor? The loss of life and property damage would be enormous.

The United States has imported natural gas from Canada. As a rule, Canadian gas was more expensive than domestic gas from the Gulf Coast. This led to some awkward problems of pricing gas when one gas utility was selling both domestic and imported Canadian gas. For example, some situations developed in the upper Midwest where gas customers on one side of town had a relatively low gas price because their gas came up in a pipeline from the Gulf, while other customers elsewhere in the same town, served by the same utility, paid a much higher price, because their gas came across the border from Canada.

COAL

Only three areas of the world account for 90% of the world's total coal reserves: the countries of the former Soviet Union, the United States, and China. However, the total amount of coal in the world is so huge that even nations with a tiny percentage of the total world reserve are major coal exporting nations. This includes Australia (the world's number one coal exporting nation), South Africa, and Poland. Each of these three countries has, at best, about 1% of the total world coal reserve. Oddly, the United States is both an importer and an exporter of coal simultaneously. We export coal to the Pacific Rim and to Europe. We import coal from Colombia.

Today's proven coal reserve, divided by the present annual rate of coal consumption, suggests a lifetime of about 180 years for the world's coal. Regardless of what reserve base one uses, and what sort of predictions one makes about future growth rates, it seems reasonable that our domestic coal supplies could well represent a significant fraction of our energy needs for well over a century. This has led to the slogan 'Coal is Our Ace in the Hole,' meaning that we can always fall back on coal if other energy sources are not available.

Of course 'Coal is Our Ace in the Hole' has to be balanced against another slogan: 'Coal is a Dirty Fuel.' We have seen the considerable technology necessary to provide emission control devices on coal-fired power plants. Coal mining—especially strip mining—is very hard on the environment. Carbon dioxide emissions per unit energy produced are high.

Technology exists to convert coal to substitute natural gas and to synthetic crude oil. In principle, coal-derived gaseous and liquid fuels could replace natural gas and petroleum. However, this would involve a much higher cost for our energy. For example, the cost of synthetic crude oil from petroleum would be about $35 per barrel, compared with today's roughly $22. As a result, we have not an energy *di*lemma, but a 'tri-lemma:' Is the current satisfactory state of domestic natural gas supply real, and how long will it last?

How stable and secure are our sources of imported petroleum? What trade-offs will we, as a society, accept between the energy security provided by domestic coal reserves and the possible environmental problems from using it?

There is ample evidence that substitute or synthetic fuels derived from coal are technologically feasible. During the Second World War, Germany supplied most of its enormous liquid fuel needs by converting coal into liquid fuels. Of course, under the extremely harsh Nazi dictatorship, and the exigencies of fighting a two-front war, economic considerations of the cost of these coal-derived liquids was not an issue. For much of the second half of the 20th century, South Africa supplied almost all of its liquid fuel needs, along with most of its chemical products, from coal. During that time world revulsion at the apartheid regime made it extremely difficult for South Africa to purchase oil on open world markets. South Africa has negligible oil resources, but is abundantly endowed with coal. Again, international issues, in this case the embargo on oil sales to the apartheid regime, made the economic considerations of synthetic oil production from coal a secondary issue.

Europe has a centuries-long tradition of coal use. However, many of the European coal mines are by now very deep and, in many cases, in poor quality coal seams. The Europeans are shutting down many of their domestic mines. The energy options in Europe include the use of imported natural gas, importing coal (the coal-rich nations of the former Soviet Union and other East Bloc countries would love to get into the Western European coal market), and nuclear energy. Australia is a major supplier of coal to the Pacific Rim nations. In the future it may experience increased competition from Indonesia and Russia.

CITATIONS

1   William Stanley Jevons, quoted in Heilbroner, Robert. *Teachings from the Worldly Philosophy*. Norton: New York, 1996; p. 214.
2   Sergei Kirov, quoted in Knight, Amy. *Who Killed Kirov?* Hill and Wang: New York, 1999; p. 109.
3   Charles Darwin, quoted in Gregory, Kenneth. *The First Cuckoo*. Akadine Press: Pleasantville, NY, 1997; pp. 236–237. (The Charles Darwin quoted here is the grandson of the biologist who was one of the developers of the theory of evolution.)
4   Bass, Rick. *Oil Notes*. Houghton Mifflin: Boston, 1989; p. 62.
5   Campbell, Colin. 'How Secure Is Our Oil Supply?' *Science Spectra* **1998** (12), 18–24.
6   Bass, Rick. *Oil Notes*. Houghton Mifflin: Boston, 1989; p. 59.

FOR FURTHER READING

Anderson, Robert O. *Fundamentals of the Petroleum Industry*. University of Oklahoma: Norman, 1984. Though some of the information in this book is now dated, it is nevertheless an excellent overview of the industry for nonprofessionals. Various chapters touch on international issues, searching for oil, and modern oil extraction methods.

Bass, Rick. *Oil Notes*. Houghton Mifflin: Boston, 1989. An interesting look at the craft of being an oil geologist. This book is written in the form of a personal journal, mixing the oil business, geology, and the author's personal observations in a very readable manner.

Chernow, Ron. *Titan*. Random House: New York, 1998. This is the most recent biography of John D. Rockefeller, Sr., who created the biggest and most powerful monopoly in the history of the United States—Standard Oil.

Churchill, Winston. *The World Crisis*. Barnes and Noble: New York, 1993. Among his other accomplishments (and foibles), Churchill may have been the finest master of English rhetoric of the 20th century. This book is his memoir of the First World War. In Chapter VI, Churchill discusses the momentous decision to convert the British fleet from using coal as fuel to oil. This decision set off an enormous train of consequences relating to the geopolitics of oil.

Conaway, Charles. *The Petroleum Industry: A Nontechnical Guide*. PennWell: Tulsa, 1999. Well described by its title. Chapter 2, on petroleum origins and accumulation, and Chapter 3, on petroleum exploration, are particularly relevant.

Flavin, Christopher; Lenssen, Nicholas. *Power Surge*. Norton: New York, 1994. This book focuses primarily on future sources of energy. Chapter 2 provides the authors' view of the potential for another 'oil shock.'

Levorsen, A.I. *Geology of Petroleum*, Freeman: San Francisco, 1954. This is probably *the* classic text on its subject. Though unfortunately long out-of-print, it is accessible through libraries or used-book stores. The book presumes some prior background in chemistry and geology on the part of the reader.

McPhee, John. *Annals of the Former World*. Farrar, Straus and Giroux: New York, 1998. The best book on geology ever written for the nonscientist or nonspecialist reader, by our best writer of nonfiction. Much information on oil geology, including resource issues. Highly recommended.

In addition to these books, the articles listed below represent a few of the numerous publications on the availability, economics, and geopolitics of oil published in the past few years. They represent a cross-section of opinions. Most should be accessible through a good library.

Anon. 'Cheap Oil: the Next Shock?' *The Economist* **1999**, *350*(8109), 23–25.

Beck, Robert J. 'OPEC Discipline, Inventory Cuts Key to Oil Prices in Second Half.' *Oil and Gas Journal* **1999**, *97*(30), 49–64.

Campbell, Colin. 'How Secure is Our Oil Supply?' *Science Spectra* **1998** (12), 18–24.

Campbell, Colin J.; Laherrère, Jean H. 'The End of Cheap Oil.' *Scientific American* **1998**, *278*(3), 78–83.

Chalabi, Fadhil J. 'OPEC's Obituary.' *Foreign Policy* **1997** (109), 126–140.

Cook, William J.; Sheets, Kenneth; Black, Robert F.; Pasternak, Douglas. 'Hostage to oil.' *U.S. News & World Report* **1990**, *109*(14), 56–64.

Hersh, Seymour M. 'The Price of Oil.' *The New Yorker* **2001**, 48–65.

Lovins, Amory; Lovins, Hunter. 'The False Promise of Alaskan Oil.' *Foreign Affairs* **2001**, *80*(4), 72–85.

Morse, Edward L. 'A Case for Oil.' *Index on Censorship* **1997**, *26*(4), 134–141.

Motavalli, Jim. 'Running on Empty.' *E Magazine* **2000**, *XI*(4), 34–39.

Nixon, Will. 'Energy for the Next Century.' *E Magazine* **1991**, *II*(3), 30–39.

Olcott, Martha Brill. 'The Caspian's False Promise.' *Foreign Policy* **1998** (111), 95–113.

Pearce, Fred. 'Dry Future.' *New Scientist* **1999**, *163*(2194), 49.

Peterson, F.M.; Johnson, J.E. 'What Are the Consequences of Low Oil Prices?' *Hydrocarbon Processing* **1999**, *78*(2), 53–55.

Roy, Olivier. 'Crude Manoeuvres.' *Index on Censorship* **1997**, *26*(4), 144–152.

34

RENEWABLE ENERGY FROM BIOMASS

Throughout most of the world today, especially the industrialized nations, there is an enormous dependence on fossil fuels. In the United States, we rely on fossil fuels for 99% of our transportation energy, 70–75% of electricity generation, and 70–75% of industrial and residential heating. As we move forward in the 21st century, there are two major factors that are likely to put tremendous pressure on fossil fuel use.

The first is continued concern about global climate change, or the greenhouse effect (Chapter 32). Increases in atmospheric carbon dioxide concentrations are sometimes simplistically attributed entirely to anthropogenic CO_2 from fossil fuel combustion. The facts that CO_2 is not the only greenhouse gas, that human activities are not the only source of CO_2 emissions, and fossil fuel combustion is not the only human activity that generates CO_2 somehow get lost. Regardless, concerns about global climate change are very likely to lead to increased interest in energy sources and technologies that do not result in a net increase in atmospheric CO_2 concentrations.

The second issue is the concern about reserves, resources, and geopolitics (Chapter 33). The potentially short lifetime of remaining oil reserves make it prudent to consider other sources of energy that are not likely to be depleted. Energy sources that are not subject to depletion are called **renewables**. Examples of renewables include wind, water, solar, and biomass. The term 'renewables' implies that this source of ENERGY is continuously being renewed, either through natural events, as in wind energy or solar energy, or through human intervention, such as the energy crops discussed later in this chapter. Such energy sources that are harvested but then regrown are also known as **sustainable energy sources**. Furthermore, renewables do not rely on localized concentrations in seams or reservoirs, as do coal, oil, or natural gas, but rather can be utilized almost anywhere. Therefore renewables are not nearly so likely to be impacted by concerns of geopolitics and possible embargoes as are the fossil fuels.

Let's take stock of some of the various large-scale energy sources we have discussed so far.

(1) *Hydroelectricity*. Without doubt, the actual generation of hydroelectricity is very clean, since there are no emissions of any sort. However, in the United States at least the number of potential natural sites for hydroelectric generation is very limited. In years past, we have created artificial sites (that is, dams) for hydro plants. The dams themselves have significant environmental problems, and in recent years there has been a growing tendency to dismantle existing dams. Barring a very radical shift in public acceptance for dams and their associated environmental problems, there is unlikely to be any substantial increase in hydroelectricity, and hydro generation may even decline in the future.

(2) *Fossil fuels*. Though still the mainstay of our energy consumption, fossil fuel combustion has been directly linked to such environmental problems as acid rain and both kinds of smog. Like it or not, *all* fossil fuel combustion produces carbon dioxide, which increases the anthropogenic contribution to atmospheric CO_2 concentrations, linked to global climate change. And, with oil in particular there are concerns both for the lifetime of oil and, in the United States, an increasing dependence on imports to meet our oil needs.

(3) *Nuclear energy*. Currently, nuclear energy has a negative public image, due largely to the Three Mile Island and Chernobyl accidents. Nuclear energy proponents now like to make the point that nuclear energy is the only 'emissions free' large-scale source of electricity (somehow overlooking hydro). This is indeed true if we consider emissions in the sense of CO_2, SO_x, NO_x, and 'RO$_x$.' There remains, however, the knotty and unsolved problem of nuclear waste disposal. Allied with this problem is the issue of nuclear weapons proliferation involving by-product $_{94}Pu^{239}$ (Chapter 28).

If we wished to dwell on the negative aspects, this is a pretty dismal list of the shortcomings of hydro, fossil, and nuclear. Suppose that we could draw up a set of specifications for an 'ideal' energy source. We might say that we want something that: (1) is essentially limitless, either through some characteristic of its own or because we can take steps to renew it (this means that we need not worry about lifetime estimates); (2) produces little or no pollution, such as SO_x or NO_x, and no radioactive waste (this eliminates, or at least reduces, many environmental concerns); (3) produces no *net* CO_2 increase in the atmosphere (this eliminates one anthropogenic contribution to atmospheric carbon dioxide concentrations); (4) is a domestic resource (this eliminates concern about dependence on foreign sources, with the associated economic and geopolitical issues). A variety of energy sources meet, or are thought to meet, these criteria. Indeed, it is the renewables that come closest.

We have discussed the use of water as an energy source and, particularly, hydroelectricity previously (Chapters 8 and 15). The modern applications of wind energy are discussed in Chapter 35, and solar energy, in Chapter 36. The other major renewable energy resource is biomass, the subject of the remainder of this chapter.

INTRODUCTION TO BIOMASS

Most materials that contain carbon, or carbon and hydrogen, will burn, and could, at least in principle, be used as fuel.⊗ We use the term **biomass** to refer to any type of animal or plant material—that is, materials produced by life processes—that can be converted into energy or used directly as energy sources. Biomass stands in distinction to the fossil fuels, which also derive from plant material, but in which case some very long geological process is needed to form the fuel. Broadly, biomass falls into three categories: wastes (e.g., straw, bagasse, garbage), standing forests (e.g., firewood), and energy crops (e.g., corn grown to produce ethanol). In principle, biomass should be inexhaustible, renewable in the literal sense of the word, provided that we grow a new plant to replace each one harvested for energy. In principle (though not necessarily in practice, as we will see later), biomass should have no net effect on atmospheric CO_2, because removal of CO_2 by photosynthesis in the growing 'replacement' plant should exactly balance CO_2 production when the biomass is consumed. When biomass material is used for energy production it is sometimes called a biofuel.

⚛ Just because something *can be* used as a fuel does not mean that it necessarily *will be*. We've discussed back in Chapter 3 that it's possible, in principle, to heat your home by burning sugar in the furnace. Not many of us would care to do that. More than two centuries ago, Lavosier demonstrated that diamonds, a form of pure carbon, can be made to burn. Imagine the cost of operating a pulverized-diamond-fired boiler!

With a renewed interest in biomass in the early 21st century, it may be easy to lose sight of the fact that, for all but a small fraction of human history, biomass has been our predominant source of fuel. This was true well into the mid-19th century. The most important biomass source was wood. Dried plants, oils extracted from plant parts, dried animal dung, and even animal fat also contributed to domestic heating, lighting, and cooking needs until the latter part of the 19th century. The rise of coal as an important fuel for stationary steam engines and steam locomotives through the 19th century, and then the increasing importance of petroleum products as transportation fuels in the 20th century were responsible for displacing biomass as an important energy source.

Nowadays, biomass supplies about 15% of the world's primary energy. At least two billion people depend almost exclusively on biomass for their energy supplies. However, most of these people live in developing nations where the per capita energy consumption is low; all together, their total energy demand is still relatively small. Biomass supplies over 35% of the energy in developing nations but has the potential of supplying even more. (This would require an increased effort on two fronts: improving biomass production, and doing a better job of utilizing agricultural wastes.) Only about 5% of the energy used in the United States is produced by biomass-fueled systems. About half of that comes from firewood used for space heating or for cooking. Most of the rest comes from ethanol (made from corn) used as a vehicle fuel, and from the burning of municipal solid wastes. These sources together make biomass the largest source of renewable energy used in the United States, other than hydroelectricity. In comparison to the United States, some European nations (Denmark, Austria, Sweden, and Finland) obtain about 10% of their energy from biomass. Some nations far surpass the average of 35% for developing countries; India for example, obtains more than half of its energy from biomass, and Tanzania, virtually all.

The solid biomass (e.g., wood) standing in forests is thought to be comparable with the world's proven natural gas and oil reserves. The energy in this amount of biomass is roughly 50 times greater than the world's primary energy consumption. If we included, in addition to the wood in forests, the biomass of grasslands and crops (though we must recognize that these are nowadays used for food production!) the total for all solid biomass is about 200 times the global energy consumption.

Biomass fuels utilize the chemical energy fixed by photosynthesis and stored within plants. Consider the carbon cycle in nature (Figure 34.1). The key process is photo-

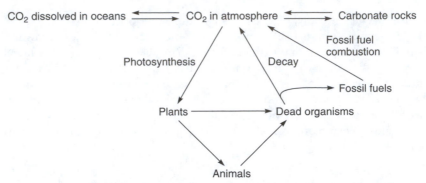

FIGURE 34.1 Biomass energy begins with the important step of photosynthesis in the global carbon cycle.

synthesis—the combining of carbon dioxide with water to produce sugars and oxygen, facilitated by light.

The energy of sunlight facilitates the transformation illustrated in Figure 34.2. Since the sugars are in a STATE of relatively high chemical potential energy, at some time later we can allow the reverse transformation to occur (Figure 34.3). The chemical potential energy created by the plant during photosynthesis can be released to create heat for domestic cooking and heating, to produce industrial process heat (this is a common use of biomass in the paper industry), or can be converted to other energy forms: electricity, or gaseous or liquid fuels. Therefore,

BIOMASS IS A WAY OF STORING SOLAR ENERGY FOR LATER USE.

To create the chemical potential energy stored in biomass requires sunlight and water (to drive the process of photosynthesis), nutrients for the growing plants, and land on which to grow the plants. Unfortunately, the maximum possible photosynthetic conversion of sunlight to chemical energy via photosynthesis is only about 7%. Just as real heat engines never actually achieve the ideal Carnot efficiency, neither do plants achieve this level of maximum conversion. Sugar cane is about 2% efficient in converting available solar energy to biomass, and corn about 1% efficient. More likely, though, only about 0.3% of the solar energy of falling on land is stored as sugars and other compounds in land plants. Plants growing in water, such as algae, come much closer to the ideal maximum.

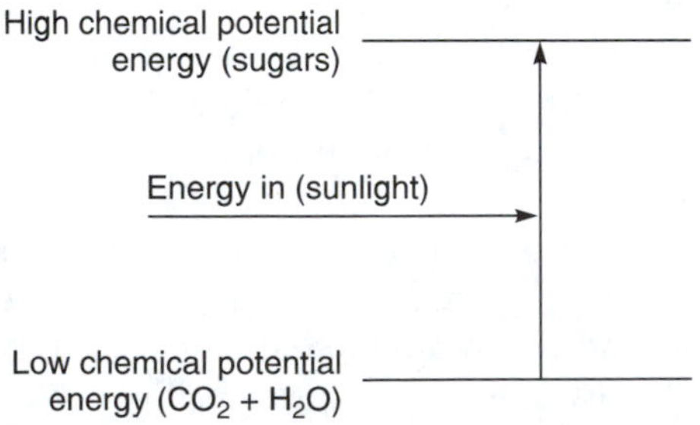

There are a number of attractive features associated with biomass as an energy source:

(1) Biomass is an indigenous resource in most parts of the world. Some kinds of plants grow almost everywhere. If you can grow your own energy, then there is no reason to be fearful of, or held hostage to, countries owning large supplies of a scarce resource like oil.

High chemical potential energy (sugars)

Energy in (sunlight)

Low chemical potential energy ($CO_2 + H_2O$)

FIGURE 34.2 Energy from sunlight is the input needed for the 'reverse energy diagram' converting carbon dioxide and water to simple sugars.

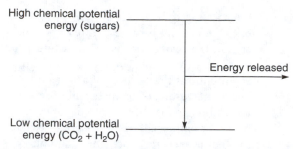

FIGURE 34.3 When biomass is used as an ENERGY source, the heat liberated comes from the conversion of the plant matter (a sugar molecule as a simple example) back to carbon dioxide and water. Thus biomass can be considered as a way of storing solar energy for utilization later.

(2) Biomass is renewable. Agricultural fields produce every year, and trees can be regrown over longer periods, say, 30–100 years.

(3) Biomass provides a way of conveniently storing energy, which is not true of other renewables. As an example, we can cut and stack wood. But we can't store wind. We will see later (Chapter 36) that one of the major challenges in solar energy is the fact that the sun doesn't shine at night; we can't store sunshine.

(4) Biomass captures CO_2 during growth. We have said (Chapter 20) that the formation of fossil fuels represents a 'detour' in the global carbon cycle. The problem that we now face, with increasing CO_2 concentrations in the atmosphere and potential global climate change, is that we are extracting and burning fossil fuels—generating anthropogenic CO_2—at a rate faster than photosynthesis by living plants can catch up and pull that CO_2 back out of the atmosphere. In a sense, the global carbon cycle is now out of balance because one side of the cycle is running much faster than the other. Atmospheric CO_2 is increasing. Biomass utilization represents a way of short-circuiting the global carbon cycle (Figure 34.4) so that in principle (though perhaps not in practice) if we grow exactly as many new plants as are harvested for energy, then we will remove CO_2 from the atmosphere by photosynthesis as rapidly as we produce it by combustion. The net effect should be that the CO_2 level should be constant. That is, there is no *net* change in atmospheric CO_2 concentration.

(5) Biomass is extremely flexible or versatile. It includes solid fuels (wood, charcoal, forest and crop residues, industrial and municipal wastes), liquid fuels (methanol, ethanol, and vegetable oils), and at least one gaseous fuel (biogas).

(6) Production of biomass is one of the few methods of obtaining energy sources that does not arouse much public opposition, at least so far.

Is Biomass CO_2-Neutral?

Let's consider in a bit more detail the issue of biomass utilization being 'CO_2-neutral,' that is, producing no net change in atmospheric CO_2 concentration. As we've seen, biomass utilization involves burning directly plants or products derived from plants, and hence short-circuiting the global carbon cycle. The basis of that 'short circuit' is that, in principle,

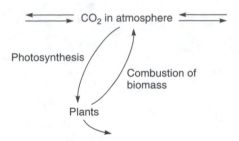

FIGURE 34.4 Biomass represents a short circuit in the global carbon cycle, reducing the cycle simply to photosynthesis followed by burning of the biomass. In principle, though probably not in practice, just as much carbon dioxide would be removed from the atmosphere by photosynthesis as is put into it from the burning of the biomass.

biomass-based fuels need not be a net source of CO_2 emissions because CO_2 released during combustion would be cycled back into plant materials by photosynthesis. The crucial assumption is that growth of a new crop following harvesting of biomass will offset all of the carbon released from the harvesting and use of the biomass. If trees have been harvested for biomass energy, then the assumption is that if we plant enough trees to replace the ones just cut down, the new growth will absorb an amount of CO_2 equivalent to that produced by using the trees as biomass energy. In either case—the regrowth of crops or the planting of new trees, there should be no net CO_2 accumulation in the atmosphere.

However, it is very questionable whether this ideal would ever be achieved in practice. It *might* be achieved if there was no loss of carbon compounds from the soil (the loss of which would also produce CO_2) and if the regrown biomass, of whatever kind, contained the same amount of carbon as the harvested biomass. If the new plants—those intended to replace the ones just harvested—are of a different kind, this ideal may not be met. An extreme case might be the cutting of an old, mature forest and its replacement by a new forest of fast-growing trees that could be harvested every few years. The new forest will not contain as much carbon as did the old one. To be sure, it is also possible to gain, rather than lose, in carbon uptake. If this same fast-growing forest were planted to replace grasses harvested for biomass, in this case the forest would contain much more carbon.

Furthermore, it is important to recognize that energy is also required to cultivate, harvest, and process the biomass, as well as to transport it. Very likely the farm machinery and transportation rely on devices burning gasoline or diesel fuel. If electricity is used in the processing, it likely came from generating stations that burn fossil fuels. The CO_2 emissions from these sources must be taken into account in considering whether the biomass energy system is truly CO_2-neutral.

In addition to considering these additional CO_2 sources, we must also consider a factor of time. There will be some interval of time between the release of CO_2 into the atmosphere when the biomass is burned, and the removal of a corresponding amount of CO_2 when replacement plants are grown. If we are considering plants that are harvested annually, then this delay time may be only a matter of a few months. With rapidly growing trees, the delay may be a few years. With the more traditional trees that we normally associate with forests, the delay time may be several decades.

Energy Crops

Today, much of the biomass that is used as an energy source is actually a by-product of other crops, or other operations. Examples include bagasse, the waste material left from sugar cane (Figure 34.5) processing, and paper mill wastes, which satisfy more than half the energy needs of the pulp and paper industry. To go beyond this—to make biomass a significant contributor to our energy economy—will require production of biomass on a much larger scale. The need for larger-scale production leads to the concept of 'energy

FIGURE 34.5 Sugar cane is an excellent example of an energy crop. The sugars can be fermented to produce ethanol, and the residue (bagasse) can be burned as a fuel.

crops.' Two kinds of energy crops have so far received the greatest consideration for their future potential in the United States: so-called short-rotation woody crops, and herbaceous crops. Short-rotation woody crops include poplar, willow, and, in warm climates, eucalyptus. Switchgrass and sorghum are examples of herbaceous energy crops. The best choice of an energy crop will depend on the specific local conditions (prevailing climate and quality of soil, as examples) and the expected use of the crop.

If biomass is to become an important energy source, the plants selected as energy crops should be efficient in producing an energy-rich product and producing it quickly. Broad-leaf plants have short growing seasons and are efficient at photosynthesis. These kinds of plants would be preferred. Several types of herbaceous plant are also being considered, specifically rapidly growing annual or perennial grasses that can be harvested with existing farm equipment. Examples of such plants include sorghum, switchgrass, and Napier grass.⊛ Ideally, energy crops would be grown much like ordinary food crops. If evaluated in terms of potential profit, sorghum may prove to be the best energy crop. However, large-scale cultivation could increase soil erosion. The perennial varieties offer the advantages that their roots help hold soil in place and, once the perennials have established themselves, they are highly resistant to weeds.

⊛ Sorghum is the name for various kinds of cornlike grasses that include, as examples, milo and kafir. Varieties of sorghum have been used in Africa and Asia since prehistory, where to this day they provide food for millions of people. In the United States, sorghum varieties are mainly used as cattle fodder. Sorghums generally have good drought resistance. Switchgrass is a perennial grass that can grow up to six feet high. It is currently grown in the American and Canadian midwest for livestock fodder, and has been considered as a potential feedstock for ethanol production. One of the more ambitious schemes to have been floated suggested planting 35 million acres of switchgrass on the

Canadian prairies, and using it to produce ethanol to fulfill Canada's gasoline needs. Napier grass, sometimes also called elephant grass or Uganda grass, is a tropical forage grass of very high yield, so high in fact that in some countries it is considered a weed. Some clumps can grow to over 20 feet tall. It is a native of tropical Africa, but now occurs also in South America, the Philippines, Hawaii, and in some of the southern states.

Plants chosen as energy crops should ideally require little or no fertilizer, because considerable energy is used in producing fertilizers. The plants should be ones with good resistance to insect attack and plant diseases. Like fertilizers, pesticides and herbicides require large amounts of energy to produce. The plants should need little water, because in many farming areas of the world water is often in short supply.

In addition to the direct advantage of providing biomass as a source of energy, these crops offer other advantages as well. Lands not suited for food crops, or that have deteriorated to a point of no longer being suitable for food crops, could possibly be returned to productivity. Several hundred million acres might be made productive again. Energy crops could reduce soil erosion and water runoff from land. Rural economic development could be promoted by planting energy crops. Cultivating energy crops could be a valuable source of income for farmers. The energy crops could provide a steady supplemental income, or allow some unused land to be farmed without requiring much investment in new equipment.

Of course there are also perceived disadvantages to the energy crop concept. Energy crops may require large quantities of water. The ideal plants for energy crops—ones that are both fast-growing and hardy—could become nuisances if they began growing outside the farm. They could conceivably displace native plant species, upsetting the local ecosystem and having negative impacts on the animal life as well. The growth of energy crops could affect biodiversity via the destruction of habitats if, as examples, forests are more intensely harvested or land set aside for multi-crop farming is converted to monoculture energy crops. Though individual energy crops could ideally be grown with less pesticide, herbicide, and fertilizer than food crops, large-scale energy farming could still lead to increases in chemical use (and the energy demand to produce these chemicals) because more land would come under cultivation. There is also an issue of energy crops competing with, or displacing, food crops. In many developed countries there is already competition for the use of land, water, and nutrients among various forms of food production, including animal farming. If a significant new market for energy crops were to develop, farmers could choose to grow energy crops on their prime land to achieve the highest possible yields and profits. Their doing so could result in some displacement of conventional food production.

If energy crops do succeed and become an important contributor to our energy needs, farmers will need to consider their business in a different light. Ever since the time in prehistory when the first farmer produced more food than was needed just for himself and his family, and managed to sell or barter the excess, farmers have been in the business of selling to people or corporations who were purchasers of food products. If energy crops succeed, farmers will now be selling to customers such as electric utilities and fuel processing companies, a change that will represent the proverbial 'whole new ballgame.'

A related economic issue is the question of whether a specific crop is of more value as food or as fuel. As a crop is growing, the farmer would need to decide whether it would be better to sell that crop as a food supply, or to sell it as a fuel. Should corn, for example, be grown for food or as a source of fuel (ethanol)? In addition, what about crop surpluses? They can be stored against future shortages, or they could be burned as a source of low-cost biomass. However, if the latter course were chosen, we might then ask why land should be devoted to growing surplus crops in the first place. Perhaps that same 'extra' land—that is, the land producing the surplus—could be put to better use in growing a better-quality energy crop.

Aquatic Energy Crops

In an energy economy that relies heavily on biomass, the amount of energy that can be produced will ultimately be limited by the amount of available land on which the biomass can be grown. Estimates of land requirements for biomass to make a significant contribution vary a good bit, depending on the initial assumptions. Regardless, it is clear that large amounts of land would be needed. As one example, the production of liquid transportation fuels from biomass in amounts sufficient to meet the needs in the United States would likely require converting an area equal to about two-thirds of our present cropland to energy production. (Of course, energy crops may not need to be grown on prime cropland.) The concern about the availability of land as a constraint, and about the displacement of food production for energy production may be lessened by considering aquatic plants as energy crops.

Algae, seaweed, and other aquatic plants can be intensively grown in certain areas of the sea or in inland lakes or ponds. Doing so would eliminate the competition of energy crops for land with conventional agriculture. Marine algae, such as kelp, have been considered as potential renewable resources. The water hyacinth, which nowadays is considered to be a weed, has been studied as a feedstock for producing gaseous fuels. Aquatic weeds are a hazard in some waterways, especially in tropical or subtropical regions, and often have to be harvested anyway, just to keep the waterway open. The energy value is then a useful by-product.

Possible Disadvantages of Biomass

Like every other energy source, biomass has both advantages and disadvantages. Sometimes the advantages or disadvantages can be more perceived than real. Nevertheless, several concerns have been raised about biomass energy, particularly if it were to be used on a scale much larger than at present.

One issue is the efficiency of use of solar energy. As an approximation, biomass can be considered to convert 1% of the incoming solar energy into plant material through photosynthesis. In comparison, the efficiencies may be 10–20% for conversion of solar energy to electricity via photovoltaic cells, and 30% for a generating plant that uses solar energy to generate steam. (Both of these technologies are discussed in Chapter 36.)

Biomass is not an emission-free energy system. We have already touched on the issue of whether biomass is really 'CO_2-neutral.' The burning of biomass will produce substances that could lead to air pollution if emitted to the environment. These substances include carbon monoxide, nitrogen oxides, and especially particulates such as soot and ash. The amounts of these substances produced varies with the specific technology being used. For example, some biomass processes produce relatively little NO_x, especially thermal NO_x, because the combustion temperatures are lower than in fossil-fuel-fired processes. Actual emissions to the environment will also depend on what, if any, emission-control equipment is used. Emissions from biomass-fueled electricity-generating plants are likely to be similar to emissions from coal-fired power plants, with the notable exception that biomass units would produce very little SO_x. It is likely that particulate emissions would be the most serious problem.

The issue of the amount of land required is another concern. Barring the extensive adoption of energy crops grown in water, the production of biomass inevitably requires land, and the amount of available land has to be considered in assessing the ultimate potential of biomass fuels on a global scale. Also needing consideration is the trade-off between the use of land for food crops or its use for energy crops. Some estimates of land requirements are very large indeed. For example, a small electricity station using solid biomass as the fuel to raise steam might require 1,500 acres of land (almost six square miles) to grow the biomass to generate one megawatt of power. There may not be enough arable land in Europe—even if it were *all* used for energy crops and not food—to supply Europe's energy needs entirely from biomass. Taking the favorable assumption that we could grow all energy crops worldwide as fast as sugarcane grows in Brazil, meeting the world's energy needs with biomass would require that two-thirds of the land now used for food crops be diverted to energy crops.

At least some, and perhaps most, processes for using biomass or converting it into other fuel forms appear to need an investment of energy that is greater than the energy released when the biomass is burned. In such cases, we would likely be better off not using energy to grow, harvest, transport, and process biomass, but rather to use that same energy directly.

WOOD

Wood was probably the first fuel to be used by humans. Until the 19th century, it was the predominant fuel in the world. As recently as 1800, about 95% of all energy used in the world came from wood. Wood was finally replaced by coal as the major energy source in 1880. Even in relatively recent times, wood has supplied a large amount of energy for certain countries. In the Soviet Union in the 1920s, energy was scarce because the bitter and brutal civil war had paralyzed much of the coal and oil industry. As a result, about 85% of the energy used by the young Soviet Union came from wood, until coal mines were reopened. During the last year of the Second World War, most of Japan's industry had been virtually destroyed. Energy imports were cut off. At the time, about half of the energy used in Japan came from wood. Wood is still popular in the United States as a supplemental fuel, for example in a wood burner used to heat a basement or garage, and in fireplaces. In many developing countries, people even today rely very heavily on wood as an energy source for domestic heating and cooking. For example, cooking with wood accounts for roughly 60% of all energy consumed in the African Sahel (i.e., the semiarid region south of the Sahara). We've seen (Chapter 7) that wood is not a very desirable energy source with respect to natural gas or oil. However, for most poor people in developing countries, wood is the only choice available.

The heating value of wood, pound for pound, is not as good as competing fuels. The major components of wood are cellulose and lignin.⚛⚛ The molecules of these materials contain abundant numbers of oxygen atoms. During the conversion of cellulose and lignin into kerogen, and then the kerogen into coal, many of these oxygen atoms are removed by the 'cooking' inside the Earth. As a result, coal contains much less chemically combined oxygen than does wood. When either fuel is burned, the carbon atoms are ultimately converted to carbon dioxide. But, when carbon atoms are bonded to oxygen atoms already present in the fuel, those carbon atoms are, in a sense, already partially oxidized. As a result, less energy is liberated from the fuel containing the greater amount of oxygen. Though there is variation among the heating values of different kinds of coals and different kinds of wood, as a rule-of-thumb, wood has half the heating value of coal, and one-fourth the heating value of natural gas. With a poor-quality wood, this fraction would drop to an eighth (looked at from a different angle, replacing natural gas by wood would require burning four to eight times as much wood as gas). The actual heating value of wood depends on the kind of tree from which it came, and whether it was allowed to dry before burning. A good grade of dry wood will provide about 6,000 Btu per pound.

Cellulose is one of the most abundant organic compounds on Earth. At least a billion tons a year of cellulose (not counting the many other components of plants) are produced each year via photosynthesis. Its molecular structure consists of several thousand glucose units linked together. The complete hydrolysis of cellulose would produce the simple sugar glucose. Cellulose is one of the major structural materials in plants; as examples, various kinds of wood contain 30–40% cellulose, and cotton, about 90%.

Lignin is also a major constituent of most plants, but much different in composition from cellulose. It is composed of three chemical 'building blocks,' not one as in the case of cellulose, and, rather than being simple sugars, the building blocks of lignin are aromatic compounds that can be considered derivatives of benzene. The properties of a particular lignin depend on the relative proportions of its three basic constituents. Lignin occurs in all woody plants and accounts for a quarter to a third of the (dry) mass of such plants. Lignin provides much of the structural rigidity of such plants. It is much more resistant to chemical degradation than is cellulose; through a series of geochemical transformations it eventually transforms into a major component of coal.

Uses of Wood as an Energy Source

Certainly the most straightforward way of using wood as an energy source is to burn it. We've mentioned the continuing reliance on wood for domestic heating and cooking by many of the people—perhaps as many as 90%—in developing countries. Unfortunately, the small-scale burning of wood in such applications is very inefficient. At best, only 10%, possibly closer to 5%, of the heat liberated from the wood is used effectively in a traditional open fire. Much good could be done in designing stoves, simple enough and cheap enough so that people would use them. An enclosed fire in a stove that regulates the air intake and directs the hot combustion gases into a chimney would very likely increase the efficiency by at least a factor of four. The same heating and cooking could be achieved with one-fourth to one-fifth the wood consumption of an open fire. This saving would be important from an environmental perspective (e.g., as it impacts deforestation, discussed below), and an economic one, because in some poor countries 40% of household expenditures are for wood.

On a commercial, rather than domestic, scale, the direct combustion of wood is also the cheapest and most straightforward option. Wood combustion could be used simply as a source of heat, or to generate steam. In areas of the world where lumbering is a major industry, electricity-generating plants burn a wood residue called 'hog fuel' to produce energy. The pulp and paper industry is a leader in so-called cogeneration processes, which provide the simultaneous production of heat and electricity for industrial use. If it's burned in well-designed and well-run combustion equipment, wood will produce less

NO_x emissions than coal or fuel oil. Wood has a low sulfur content, so will also produce less SO_x emissions.

A second approach to wood utilization involves pyrolysis, the breaking down of the wood by heat. If wood is heated to about 250°C in the absence of air (so that it won't catch on fire), the moisture and volatile components are driven off by the heat, leaving a solid product that has a high carbon content. We recognize this material as charcoal. The charcoal is a good solid fuel. Also, the charcoal can be transported and handled more economically than an equivalent amount of wood. In some countries, charcoal is the most widely used household fuel in urban areas. The gaseous and liquid products that are driven off during the pyrolysis also have value. At one time, for example, methanol (discussed in the next section) was recovered from wood in this way. If the gaseous and liquid products are used as well as the charcoal, the energy obtained is equivalent to an overall efficiency up to 80%.

A third approach to wood utilization is gasification, the conversion of wood into a gaseous fuel. Wood is reacted with steam and oxygen to produce a gaseous fuel that is mainly carbon monoxide and hydrogen. This fuel is burned in a gas turbine, similar to a turbojet engine. The turbine is connected to an electricity generator. The hot combustion gases leaving the turbine are used to produce hot water that is used for heating nearby homes. A pioneering plant in Värnano, Sweden produced electricity plus hot water with 80% efficiency, and no SO_x emissions. During the years of the Great Depression and then the Second World War, small gasifiers were used on motor vehicles to convert wood or charcoal to gaseous fuel for the engine. These units required considerable maintenance, but nonetheless did work.⚛

⚛ During the 1930s and again during the Second World War there were efforts in a number of countries to convert automobiles to run on a gaseous fuel produced onboard from wood or charcoal (or coal). Many of these vehicle conversions were done by private individuals with home workshop equipment. The gas producers and accessory equipment take up considerable room in the vehicle, and require careful maintenance. Nevertheless, they do function, and it is entirely possible to use wood as an automobile fuel, via its conversion to a gas. Details are provided in the book *Producer Gas for Motor Vehicles* (Cash and Cash; see Further Reading list).

Problems with the Use of Wood

The rapid population increase in developing nations has caused extensive cutting of forests for fuel. As the forest is cut back from the population center, more human time and effort must go into foraging for fuel. In some cases this may involve 5–10 hours of work per day, every day. Usually it is the women who have this job. Their labor is lost to other, potentially more productive, activities such as child rearing, farm work, or work in the home. As more effort goes into finding fuel, less goes into raising crops, tending animals, or light manufacturing, establishing a vicious cycle of increasing poverty and lowered standard of living. There may be several possible solutions to this problem. In the short term, it would be useful to devise ways of burning existing wood more efficiently, the issue of stoves vs. open fires. In the medium-term, perhaps ways can be found for utilizing indigenous energy sources other than wood. For example, it may prove that even a poor grade of coal might be better to use than wood, especially if use of the coal would reduce forest cutting. It may be possible also to exploit other available energy sources, such as solar. The long-term solution requires population control. Unfortunately, in many countries this solution will be very controversial due to religious or cultural beliefs.

Large-scale wood use leads to deforestation. A forest is not just trees or wood—it is an entire ecosystem. As forests are cut, animal habitat is destroyed along with the trees. One effect may be to reduce the habitat of birds that are useful in keeping insects in check. In addition, trees are very helpful for holding soil in place—absorbing moisture plus holding soil. The loss of trees leads to increasing water runoff and soil erosion. The soil erosion leads in turn both to the loss of valuable topsoil and to water pollution from suspended soil and mud washed into rivers or lakes. Mudslides can occur. The desperate need for energy combined with exploding populations is causing people to destroy their own environment. Twenty years ago wood was being cut in Africa at a rate of about five million acres per year. About 90% of this wood was burned as fuel. Deforestation and desertification is widespread and increasing in Africa. In those past 20 years, the southern edge of the Sahara desert has moved nearly 70 miles further south.

Where wood is heavily used as an energy source, the destruction of forests often leads to their replacement by open fields or, where 'slash and burn' agriculture is practiced, by scorched earth. The transformation from forest land to open fields or scorched earth causes a change in the fraction of sunlight that is reflected back from the surface (the **albedo**). A change in the albedo can in turn lead to climate change. The destruction of the Amazon rainforest, which is probably the most awful ecological catastrophe now in progress, is due largely to slash and burn agriculture. If only deforestation in the tropics is considered, in the last two million years the albedo is estimated to have changed by 0.001 units (0.1 watt per square meter). In the last quarter-century, the change has been 0.00035 units. So over a period of 25 years, humans have caused a change of about a third of the value that took two million years in nature; at this rate, some seven centuries from now the albedo change would be 0.01 unit, enough to change the climate by 2°C. (This statement should also be considered in light of Mark Twain's eloquent discussion of the length of the Mississippi River—note on p. 517.)

Wood requires both time and, especially, energy to be harvested. When highly mechanized wood-harvesting equipment is used, the energy expended in cutting the wood is likely greater than the energy available from burning it. That is, if we consider the energy consumed in gasoline for chainsaws, diesel fuel for operating the logging equipment, gasoline or diesel fuel used to transport the wood to the point of use, and energy consumed by the workers themselves, this sum is likely greater than the energy liberated when the wood is burned. In other words, the total energy available to society has actually decreased!

What happens when we burn wood? The good news is that low combustion temperatures lead to low NO_x emissions. SO_x emissions are low, because wood normally has a low sulfur content. Wood fires can produce unburned particulates, both of ash and of soot. In addition, the slow heating of wood in the fire can result in 'creosote' tars accumulating in the chimney (Chapter 6). When the weather problem known as a thermal inversion❀ occurs, large smoke clouds from wood fires can build up in the atmosphere and cause severe local air pollution problems.

❀ A thermal inversion is an atmospheric phenomenon in which a layer of warm air moves on top of a layer of cooler air that is close to the ground. This event is especially bad in regions with significant air pollution, because the upper, warmer air layer acts as a 'cap' that prevents the ground-level air, laden with pollutants, from rising and dispersing. This causes the air pollutants to collect near ground level (where, of course, we live). If the thermal inversion persists for several days, ground-level pollutant levels can become high enough to cause serious problems. Eventually, high winds or rain will break up the layer of warm air and allow the pollutants to disperse.

Is Wood a 'Renewable'?

Although wood is classified as a 'renewable,' in fact it may take 20 to more than 100 years to grow a tree to replace a mature tree cut down. Without very strict forest management in place for 10 years or more we can't regrow trees as fast as we cut them. Wood is a renewable resource if we assume that a new tree is planted for each one cut down and that we measure in units of decades. It is obviously not true that wood is a renewable resource in the short term, because a seedling one or two years old is not equivalent to a mature tree.

For this reason, the least desirable way is to use trees that have been growing a long time. Fortunately, there are practices that allow trees to be used and renewed on relatively short time frames. Some tree species, such as poplars, eucalyptus, silver maple, and black locust, can be used as energy crops by the practice known as coppicing. In this approach, year-old trees—willow being a good example—are cut down near the ground. This causes the cut stumps to sprout about a half-dozen new shoots, each of which grows rapidly over the next three to five years. Depending on the species used and how rapidly trees grow in the chosen location, the tree shoots would be harvested every two to eight years. The harvested coppice wood can be dried, chipped, and used as a fuel. Since new shoots grow from the cut stumps, replanting is required only once for every three or four times that the coppice wood is harvested.

Likely, competition from the forest-products industry will tend to limit the amount of wood harvested as an energy crop from existing commercial forests. Environmental concerns about the destruction of old-growth forest ecosystems and the apparent devastation caused by clear-cutting may limit, or even eliminate, the opening up of forests not presently being commercially exploited for the harvesting of energy crops.

Agroforestry, a combination of food and wood production on the same land, offers a way to increase yields of food, firewood, and fodder for cattle. The experience of some African nations such as Kenya and Nigeria has shown that mixing corn and leucaena trees can increase corn production anywhere from 40% to 80% relative to that from a field growing only corn. At the same time, the yield of wood is about two tons per acre.

Leucaena trees are natives of tropical America that have now been introduced to most of the tropics. They typically grow to heights of about 30 feet, and sometimes twice that in selected locations. Currently these trees are used as shade plants for coffee, cacao, and rubber crops.

METHANOL

Alcohols as Liquid Fuels for Vehicles

Since the dawn of practical internal combustion engines in the late 19th century, liquid fuels have been the overwhelming choice as the energy source for these engines and the vehicles that use them. Among the liquid fuels, petroleum derivatives, gasoline and diesel fuel, have dominated.

Why liquid fuels? Simply because neither of the alternative choices is as suitable. First, solid fuels, such as finely pulverized coal, are not well suited for use in Otto or diesel engines. One problem is that such fuels invariably produce ash. Ash could build up in the cylinders, erode cylinder walls, pistons, and exhaust valves, or add to air pollution unless some sort of ash-collection device could be added to the vehicle. Gases, on the other hand, can be superb fuels for internal combustion engines. However, unless the gas can be liquefied or compressed to high pressure, it is difficult to carry enough gaseous fuel on board a small vehicle to give it much practical driving range. On large vehicles that travel short distances, and that have access to specialized fueling facilities (an excellent example being city buses), compressed natural gas is proving to be useful fuel choice. Liquid fuels provide, on the one hand, clean-burning characteristics similar to gases, and, on the other, a density close to that of solid fuels, to allow an ample fuel supply to be carried on the vehicle.

The 21st century may see petroleum derivatives displaced, at least in part, from their present near-total domination of the liquid fuel market. Several factors may have a role in this displacement: concern for long-term supplies of petroleum or our dependence on imported sources (Chapter 33), concern about urban air quality (Chapter 31), or concern about anthropogenic CO_2 emissions (Chapter 32). If we were still to rely on liquid fuels for transportation energy, then likely replacements for petroleum are two alcohol fuels: methanol and ethanol. Both of these fuels can be made from biomass.

Why these alcohol fuels? Together they offer a number of potential advantages as replacements for petroleum. First, they can be burned in existing spark-ignition engines (Otto-type engines) with minimal modification. Cars and light trucks already on the road could be adapted to these fuels. Car-makers could continue to use existing production facilities and methods to build engines, eliminating the enormous capital investment needed to tool up to make an entirely new kind of engine. Though not so good as petroleum, the alcohol fuels have sufficient energy density (i.e., Btus produced per pound or gallon of fuel burned) that fill-ups are not needed so often as with uncompressed gaseous fuels. The need derives primarily from the need for transportation fuels. Their use would not add undue weight to the vehicle, as heavy tanks for compressed gas might. The fact that they are liquids conveys yet another advantage: An enormous infrastructure already exists for the handling, storage, and transporting of liquid fuels (i.e., gasoline). A conversion from petroleum to alcohol fuels could be effected with minimal change in the transportation-fuel infrastructure. Finally, we already have the know-how to make both methanol and ethanol in large quantities at reasonable cost.

Producing methanol and ethanol could become an important future role for biomass. Improvements in the conversion of biomass to these liquids, and the possible introduction of less expensive energy crops would be the two key developments that could open this very large market for these fuels. Since the alcohols can be burned in existing combustion systems with only slight modification they could be an energy source providing a relatively smooth transition between our current energy economy that is dominated by fossil fuels and an energy future that relies on nearly pollution-free energy sources such as solar energy and wind-generated electricity.

Production of Methanol

In the past, methanol was produced from wood. For that reason, it is still sometimes known by the common name 'wood alcohol.' Hardwoods—beech, hickory, maple, or birch—were the favored source. Heating the wood in the absence of air vaporized a whole array of substances that produced a gas useful for illumination, a tar containing dozens of useful chemicals, and a liquid called pyroligneous acid. (The solid portion of the wood remained behind as charcoal.) Methanol can be recovered from the pyroligneous acid.

Today most methanol is made from natural gas via a three-step process. First, the natural gas is reacted with steam to produce a mixture of carbon monoxide and hydrogen:

$$CH_4 + H_2O \rightarrow CO + 3H_2$$

The process is called steam re-forming. (This is a very different meaning of the term 're-forming' than the catalytic re-forming process discussed in Chapter 21.) The mixture of carbon monoxide and hydrogen is called synthesis gas. Synthesis gas is an exceptionally versatile material, used, as the name implies, for synthesis of other substances, of which methanol is one. For the synthesis of methanol, the proportion of hydrogen and carbon monoxide needs to be 'shifted' from a 3:1 ratio to 2:1. That shifting occurs by reacting the gas with carbon dioxide, a process that reduces the amount of hydrogen and increases the amount of carbon dioxide:

$$H_2 + CO_2 \rightarrow H_2O + CO$$

In the third step, the 'shifted' synthesis gas reacts in the presence of a catalyst to produce methanol:

$$CO + 2H_2O \rightarrow CH_3OH$$

Virtually all the methanol now used in the United States comes from this process.

The importance of this process is that its long-term use is not dependent on the availability of cheap natural gas. If for any reason natural gas became in short supply, or prices became unacceptably high, synthesis gas can be made from other sources. In fact, virtually any carbonaceous material can be converted to synthesis gas.⚛ Biomass is a candidate for future production of methanol. (So is coal.⚛)

⚛ In the sometimes quirky terminology of fuel processing and utilization, the process of converting a solid fuel, such as biomass, to synthesis gas is called 'gasification' and not steam re-forming, even though the biomass would be reacted with steam. The exact ratio of hydrogen to carbon monoxide in the synthesis gas depends heavily on the relative amounts of hydrogen and carbon in the material that is gasified or steam re-formed. If there is proportionately too much hydrogen and not enough carbon monoxide, the composition is shifted by reaction with $CO_2 : H_2 + CO_2 \rightarrow H_2O + CO$. On the other hand, if there is proportionately too little hydrogen and too much carbon monoxide, the composition can be shifted in the other direction by reacting with steam: $CO + H_2O \rightarrow CO_2 + H_2$. The versatility and the beauty of synthesis gas is twofold: synthesis gas can, in principle, be made from *any* carbonaceous material; and the composition of the synthesis gas can be shifted to any desired proportion of hydrogen and carbon monoxide.

⚛ Coal can be converted to synthetic liquid fuels in several ways. The two main strategies are direct liquefaction, in which the coal is reacted with hydrogen, usually in the presence of a catalyst and a solvent; and indirect liquefaction, in which the coal is first converted to synthesis gas and then the synthesis gas in turn is converted to liquid products. During the Second World War much of the liquid fuel supply used for the German war effort was produced from coal by these two processes, which led to a campaign of bombing synthetic-fuel plants by the American and British air forces. In the apartheid era of South Africa, virtually all of the liquid fuels used in South Africa were produced by the indirect liquefaction of coal, because the racial policies at that time resulted in many nations refusing to sell oil to South Africa. The production of synthetic liquid fuels from coal has never been economically feasible in an open market (i.e., one not constrained by war or embargoes).

Advantages of Methanol

Generally a standard spark-ignition automobile engine can be modified to run on methanol. (Some of the modifications are required because methanol can cause deterioration of rubber and plastic parts.) Indeed, methanol is already used successfully as a fuel in racing cars. Methanol has a very high octane number, about 110, so that it can be used in engines of very high compression ratio. Cars designed specifically to run on methanol would gain a 10–20% benefit in thermal efficiency relative to similar cars designed for gasoline, because of the higher compression ratio. A high compression ratio also means that a methanol-fueled car could have higher acceleration than comparable gasoline-fueled cars (which is why methanol is favored in racing cars). Methanol also works well in blends with gasoline. A blend of 85% methanol and 15% gasoline (called M85) can be used in current car engines with very minor modifications. It is also possible to devise a 'flexible fuel vehicle' that could run on gasoline or M85.

Use of methanol would potentially reduce air pollution associated with motor vehicles. A methanol–air mixture has a lower flame temperature than a gasoline–air mixture, therefore leading to much less thermal NO_x production. Methanol contains no sulfur, so there would be no SO_x emissions. Methanol burns well at fuel-lean conditions, helping to reduce CO emissions, and yet does not produce soot at fuel-rich conditions. Methanol evaporates more slowly than gasoline, helping to reduce evaporative emissions of unburned hydrocarbons.

We already know how to make methanol in large-tonnage quantities. Estimates of the costs of large-scale methanol production for vehicle fuel vary; it is thought that it would be reasonably cost-competitive with gasoline.

Methanol can also be used in fuel cells, making it a possible fuel for hybrid vehicles (Chapter 31) or in electric vehicles that use fuel cells as the source of electricity. Whether methanol would see use in these applications depends on whether effective methanol fuel cells can be developed, and on public acceptance of hybrid or fuel-cell electric vehicles.

Disadvantages of Methanol

The heating value of methanol, on a volumetric basis, is about 76,000 Btu/gal; gasoline is about 132,000 Btu/gal. To move a car a given distance (e.g., one mile) requires a certain amount of WORK. Methanol provides only about 58% of the ENERGY as a comparable volume of gasoline. That means we will get 58% of the mileage, the miles per gallon. For example, a car that gets 25 miles per gallon with gasoline might get 14–15 on methanol. (This comparison assumes two cars that are otherwise identical except for the fuel, and are driven under comparable conditions.) This has two consequences. First, either the driver would have to put up with stopping at a 'methanol station' somewhat more frequently than at a gas station, or else the fuel tank on the methanol vehicle would have to be larger. Second, a person driving the same number of miles would have to purchase substantially more (almost twice as much) methanol as gasoline. To be truly cost-competitive, methanol would have to sell at about half the cost per gallon as gasoline.

The second major concern with methanol is safety. If methanol is accidentally ingested, blindness or death can occur. Drinking about an ounce of methanol can cause death; drinking half that amount can lead to permanent blindness. Other health effects at lower levels of exposure include severe headaches, visual impairment, and convulsions. Though few might ever drink methanol, the same problems can be caused by inhalation of the vapors or, insidiously, by absorption through the skin.⊛ Being splashed with methanol as a result of an accidental spillage could result in its being absorbed through the skin. In addition to the direct health effects, methanol burns with a nearly colorless flame. It might be hard to see an accidental methanol flame, unless something has been deliberately added to the methanol to make the flame luminous. (One such additive could be a small amount of gasoline.)

> ⊛ Sometimes people in advanced stages of alcoholism, especially those who are really down-and-out, will drink methanol by inadvertence. These folks, in their desperation for a drink, may find some wood alcohol and assume that 'alcohol is alcohol.' Doing so can literally be a fatal mistake. A small amount of wood alcohol in the system can cause incurable blindness or death. Sad to say, such events still occur, judging by occasional reports in the news media.

A vehicle set up to burn gasoline would need some adjustment of the fuel injection system because of the different air:fuel ratio needed to burn methanol. Comparing methanol with octane, for example,

$$2CH_3OH + 3O_2 \rightarrow 2CO_2 + 4H_2O$$
$$2C_8H_{18} + 25O_2 \rightarrow 16CO_2 + 18H_2O$$

illustrates the difference in the proportions of oxygen and fuel molecules needed for complete combustion.

Unlike gasoline, methanol is soluble in water. If water is accidentally mixed with gasoline (in a storage tank, for example), a layer of gasoline will float on a layer of water. It would be possible to separate the gasoline from the water. Methanol would form a solution with the water. Thus it could be possible to have water blending into a methanol fuel (much more severely than with gasoline) causing problems for running the engine.

The fact that methanol turns to vapor less easily than gasoline is, as we've seen, an advantage in terms of potential reduction of emissions of unburned hydrocarbons. At the same time, though, this makes starting a methanol-fueled engine more difficult in cold weather.

Methanol contaminating the engine oil can severely degrade the lubricating characteristics of the oil. In extreme cases, this could result in serious maintenance problems for the engine. Neither this problem, nor the attack on rubber or plastic mentioned previously, is of much concern when methanol is used in small amounts in a blend with gasoline. For M85 fuel or pure methanol (M100) these problems would need to be addressed.

Finally, although methanol can be produced cheaply today, in large amounts, we must remember that today's methanol production uses cheap natural gas as the starting material. It may prove difficult for biomass to compete with natural gas as a feedstock for methanol synthesis without substantial tax incentives. For one thing, using biomass would require replacing a relatively simple steam re-forming unit to make synthesis gas with a more complex, and therefore more expensive, biomass gasifier. Some current estimates suggest that biomass-derived methanol would be twice as expensive as that from natural gas.

ETHANOL

Ethanol is also called ethyl alcohol or grain alcohol. A very large family of alcohols exists, but when the word *alcohol* is used by itself (as in the term 'alcoholic beverages,' for example), almost always it is ethanol that is meant. It is related to methanol as ethane is to methane. A major source of ethanol is the fermentation of carbohydrates in plants, such as corn and other grains—hence its common name of grain alcohol. The commercial product denatured alcohol is also mainly ethanol, deliberately adulterated with various compounds that make it undrinkable, and therefore immune from the labyrinth of regulations and taxes associated with beverage alcohol.

Since ethanol is a readily combustible liquid, it deserves consideration as a liquid fuel for motor vehicles. In the early years of the 20th century, ethanol actually competed with gasoline as an automotive fuel. Henry Ford built a small number of 'fuel-flexible' Model Ts that could run on ethanol or gasoline. During the First World War, Ford was a strong proponent of increased alcohol fuel production, and attempted to convince President Wilson that American cars should be converted to run on ethanol instead of gasoline.

> It beats me how anyone can really think that sanctions will bring down the Smith regime in Rhodesia in a matter of months.
>
> During the war we completely blockaded Madagascar for two whole years because of their pro-Vichy loyalties, yet when our troops arrived there in 1942 they found life going on quite serenely.
>
> The cars were running more or less happily on rum, of which Madagascar is a large producer.
>
> —Anderson[1]

Production of Ethanol

The interest in ethanol comes from the fact that it can be made from plant materials grown every year—that is, it could be renewable—plus the fact that it can also be produced from waste materials such as sugar beet or sugar cane pulp and potato peelings, as well as from surplus crops. **Fermentation** is the process that converts plant materials to ethanol. The process of fermentation goes back at least 10,000 years. It is likely the second-oldest chemical process to be exploited by humans (the oldest being combustion). And, like combustion, fermentation was probably discovered independently by many early prehistoric societies. Beer was prescribed as a medicine in Sumer at least 4,000 years ago. The oldest set of laws still preserved—those written by Hammurabi in Babylon sometime around 1770 B.C.—included regulations for drinking establishments.

In the fermentation process, starches or simple sugars from biomass are decomposed by yeasts or other microorganisms (more specifically, by enzymes produced by these microorganisms) to produce ethanol. Lavosier was the first to show that a sugar molecule will produce two molecules of ethanol and two of carbon dioxide:

$$C_6H_{12}O_6 \rightarrow 2C_2H_5OH + 2CO_2$$

About 85% of the chemical potential energy in the molecule of sugar is retained in the two molecules of ethanol.

Sources of starch for fermentation are mainly grain crops but also include root plants, such as potatoes. Sugar sources include cane and beet sugar, sorghum, and artichokes. The ethanol business in the United States nowadays is based on corn, whereas other grains and sugarcane are used elsewhere. Corn is the most important material for ethanol production in the United States for two reasons. First, the high starch content of corn makes fermentation relatively easy. Second, in most years so much corn is grown that millions of tons are left over as surplus, so can be purchased by the ethanol industry at a low price. Fuel ethanol production in the United States is about 1.5 billion gallons per year. It uses 4% of the total corn crop, and saves petroleum equivalent to about 60,000 barrels per day. The fermentation process for ethanol production is illustrated in Figure 34.6.

We've seen that concerns about the land area required for growing biomass could be allayed by relying on aquatic plants. Duckweed represents an aquatic plant source for ethanol.⊗ Duckweed grows rapidly. It can grow in brackish (i.e., slightly salty) or polluted water. Duckweed plants are small enough that they can be pumped through pipes, and do not need to be chopped up prior to the fermentation step.

⊗ Duckweeds are floating aquatic plants of various kinds that range from the so-called giant duckweed that grows into thick, dense masses to another variety believed to be the smallest flowering plant on Earth. One of the duckweeds has the intriguing name of mud midget. Generally the duckweeds are considered to be invasive plants in aquatic ecosystems. Harvesting and using the duckweeds would likely be beneficial, not only for the use of the biomass, but also for reducing their impact on the local ecosystem. Because even the giant duckweed is, individually, a relatively small plant, duckweeds would not have to be chopped up before being fed to a biomass conversion or combustion system.

There are other routes to ethanol besides fermentation. Into the 1980s the major source of ethanol used ethylene. Enormous tonnages of ethylene are produced every year by the petroleum industry. Ethylene is a very versatile chemical used in the production of a wide array of synthetic materials, the ubiquitous polyethylene being just one. In the presence of a catalyst (e.g., phosphoric acid) ethylene reacts with water to produce ethanol:

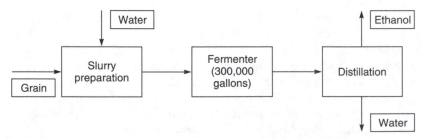

FIGURE 34.6 This schematic diagram illustrates the principal steps in the ethanol production process.

$$C_2H_4 + H_2O \rightarrow C_2H_5OH$$

The technology for this process is well known. However, this ethanol-production process ultimately depends on petroleum as the source of ethylene. If petroleum were to become in short supply, so too would ethylene. There would be no advantage in relying on an ethylene-to-ethanol process to address petroleum shortages. An alternative route to ethanol from biomass uses bacteria for direct production of high yields of ethanol. As we'll see later in this section, the distillation step for recovering ethanol from the fermentation mixture requires a great deal of energy. A process that would convert biomass directly to ethanol, bypassing the distillation step, would be very attractive. In addition, it might be possible to develop strains of bacteria that would convert other forms of biomass, such as wood, to ethanol.

Advantages of Ethanol

A potential long-term advantage of ethanol lies in the versatility of the fermentation process. Virtually any biomass material that provides starches or sugars could, in principle, be fermented to ethanol. Fuel ethanol production would not be reliant upon, nor hostage to, a single type of energy crop. Not only could energy crops such as grain grown specifically for ethanol production be used, but also grain wastes (that might, e.g., have otherwise rotted in storage), or some food wastes, such as potato peels.

It's been known for a century that ethanol can be a suitable fuel for automobile engines. Henry Ford had many character defects, but idiocy was not one of them. Like its chemical sibling methanol, ethanol can be made in high tonnage by known technology. It has a high octane number (about 110), which makes it well suited to high-compression-ratio spark ignition engines (but not diesels). Ethanol burns cleanly, with reduced thermal NO_x emissions and no fuel NO_x or SO_x. A 90:10 blend of gasoline and ethanol, sometimes called **gasohol** (discussed further below) can be run in standard automobile engines with no modifications.

If we consider the sequence of chemical reactions involved in the fermentation of sugar to ethanol, the burning of ethanol as a fuel, and then the photosynthesis in a plant grown to replace the one harvested for ethanol production, we have:

Fermentation:	$C_6H_{12}O_6 \rightarrow 2C_2H_5OH + 2CO_2$
Combustion:	$2C_2H_5OH + 6O_2 \rightarrow 4CO_2 + 6H_2O$
Photosynthesis:	$6CO_2 + 6H_2O \rightarrow C_6H_{12}O_6 + 6O_2$
Net:	$C_6H_{12}O_6 \rightarrow C_6H_{12}O_6$

The overall process simply involves a sugar molecule from one plant being regenerated in the next year's planting, with no net CO_2 production provided that we replace each plant harvested with a new one. In principle (though not very likely in practice), use of ethanol as a fuel should be CO_2-neutral.

Disadvantages of Ethanol

As a fuel, ethanol has a heating value of 101,000 Btu/gal. Thus it has about three-fourths of the energy of gasoline on a volumetric basis. This is not so bad as methanol. It means, though, that a car getting 24 miles per gallon with gasoline would get 18 mpg with ethanol (assuming again otherwise comparable cars under otherwise comparable driving conditions). Ethanol shares with methanol the issues of water solubility and a need to modify the fuel system to accommodate a different fuel-to-air ratio.

The raising of corn for ethanol production is very energy-intensive—plowing, fertilizing, harvesting, and transporting the corn all require energy. In addition, the distillation step as currently practiced also requires large amounts of energy (as heat), which is generally supplied from petroleum or natural gas. Furthermore, the fertilizers used in corn production, often in massive amounts, are made from petroleum-derived chemicals. These facts lead to a serious concern that the energy derived by burning fuel ethanol may be less than the energy needed to make it in the first place. In other words, producing fuel ethanol might cost us energy, rather than providing us with energy.

If we focus only on the distillation step itself, it is questionable whether even that one step breaks even in terms of energy use. If we stack the deck, so to speak, with the most favorable assumptions—that the energy for distillation would come from biomass or solar, that the residue that did not ferment (the so-called distiller's grain) would be salvaged as animal feed, and that the ethanol would be blended into a gasohol, then making a gallon of ethanol might save us a gallon of gasoline. But, if we consider using pure ethanol as a fuel, providing distillation heat from petroleum, and not recovering the distiller's grain, then producing a gallon of ethanol *costs* us the net energy equivalent of a quarter- to half-gallon of gasoline. A breakthrough in chemical engineering that would replace the distillation step with some other operation for recovering the ethanol from the fermentation mixture would be helpful, as would further progress in alternative biomass-based processes that don't even require distillation.

Since it is likely that ethanol production consumes more energy than it provides in the ethanol, certainly we cannot claim that ethanol is truly a CO_2-neutral energy source. The farm machinery and transportation equipment all emit CO_2, as does the generation of the heat needed for distillation. Substituting biomass or solar energy for fossil energy wherever possible in the ethanol production system would help to bring ethanol closer to being truly CO_2-neutral. Nowadays it's estimated that CO_2 emissions from the entire ethanol production cycle are only slightly less than if we had simply used gasoline in the first place.

As is the case with virtually any biomass energy source, we must address the questions of whether there is enough biomass and enough land. It's estimated that approximately 10% of the corn crop would be required to provide 1% of our gasoline needs. This is not a good ratio. *All* of the corn grown in the United States would supply ethanol equivalent to only 10% of our gasoline. Growing crops for energy takes land or crops away from food production. With a large percentage of the world's population malnourished and in some cases even starving, does it make sense to divert lots of agricultural land to the production of 'energy crops'? Though it's important to remember that many other biomass sources can be converted to ethanol, corn is the dominant feedstock in the United States. The growing of corn contributes to erosion of agricultural land; topsoil washed away eventually contributes to water pollution.

If much corn were diverted to ethanol production, the price of corn as a food would likely rise. Furthermore, since much of our corn is used as animal feed, it is also likely that the price of meat would rise. Fortunately, in a somewhat bizarre sense, we usually produce substantial crop surpluses, thanks in part to government subsidies that actually encourage overplanting. Much of this surplus grain rots in storage. If we're going to have crop surpluses and wastage anyway, perhaps it would be better to convert the surplus corn (and other crops) to ethanol. Doing so might allow replacing some 3–5% of our current gasoline consumption with ethanol before corn and meat prices would be affected.

A final issue is the true cost of ethanol. The current price of fuel ethanol is $1.10–1.20 per gallon. Much of that cost, some 40%, is the cost of the corn itself. This is another reason for considering using surplus grains that would otherwise rot, or completely different feedstocks. Much of the remainder of the cost comes from chopping up the corn and fermenting it. In the United States, ethanol is economically competitive because of a federal subsidy of about 54¢ per gallon. If this subsidy were to be removed, then ethanol could be competitive only with major breakthroughs in finding lower-cost production methods, and in finding a source that is cheaper than corn. Even if the entire corn plant could be converted to ethanol (which is certainly not the case now), it is unlikely that ethanol could compete economically with petroleum because of the substantial energy requirements in the farming process. Ethanol produced from sugarcane would be economically competitive only if the price of oil rose to about $30 per barrel.

Ethanol in the United States and Brazil

Beginning in the 1970s, in response to the OPEC oil embargoes, the United States developed gasohol, a blend of 10% ethanol and 90% gasoline. The ethanol came from agricultural products or agricultural wastes. The idea was simply to extend supplies of gasoline. In principle, no engine modifications are needed to run on gasohol. Anecdotal reports on gasohol experience were highly controversial. Some drivers swore by it; others swore at it, and claimed it caused harm to the engine and fuel system if used for a long time. In the 1970s and early 80s, some car companies voided warranties if alcohol fuels were used. The negative reports probably stemmed from two issues: Alcohol can soften or dissolve some rubber or plastic parts in the engine or fuel system that are impervious to gasoline. Ethanol absorbs water, which can lead to increased corrosion of fuel system components, and poor combustion performance of the fuel. Gasohol is cost-competitive only because of the 54¢ per gallon subsidy on ethanol.

At about the same time that the United States began making gasohol, Brazil—one of the world's largest sugar producers—found a world situation where sugar was very cheap and oil, which Brazil had to import, was very expensive. Brazil undertook a major national program, one far larger than in the United States, on alcohol fuels. Two blends were developed: 22% ethanol:78% gasoline, and 98% ethanol:2% gasoline. (The reason for using the odd 98:2 ethanol:gasoline blend, and not pure ethanol, was to keep people from drinking it.) Eventually about 60% of Brazil's liquid fuel needs came from ethanol.

By 1990, about four million vehicles were running on the 98:2 blend, and another six to eight million on the 22:78 blend. The program was a great success while sugar was cheap and oil expensive. When oil prices dropped and sugar escalated, the program was a financial disaster. By the late 1990s, Brazil was making nearly four billion gallons of ethanol per year from sugar cane. The amount of energy in this ethanol is equivalent to consumption of about 140,000 barrels of petroleum per day. Brazil has learned to make careful use of the biomass, recovering the unfermented material to make methane as a fuel gas, various chemicals and fertilizers, and animal feed. Nevertheless, the Brazilian alcohol industry, like the American, relies on subsidies to be profitable.

PLANT OILS

A trip down the cooking-oil aisle of any well-stocked supermarket will show that there are numerous kinds of oils that derive from seeds, nuts, or other plant parts. Olive oil, peanut oil, corn oil, palm oil, and sunflower oil represent some of the examples. That is, many kinds of plants produce seeds, nuts, or other parts that can be sources of oils (Figure 34.7). The technology for extracting these oils is well known and very old, going back at least to the time of Vitruvius. Olives crushed in waterwheel-operated mills provided a source of olive oil to the ancient Romans. Over the centuries, millions of home cooks have made empirical observations that attest to the fact that these oils will indeed catch on fire and burn.

Since it is easily demonstrable that these oils will burn, from time to time there has been interest in their use as fuels. As early as 1900, a demonstration diesel engine was operated in Paris on peanut oil. In 1912, Rudolf Diesel predicted that plant oils would be as important as petroleum. The Japanese navy used soybean oil as a fuel in the battleship *Yamoto*, the largest and most powerful battleship ever built. The oil 'crises' of the 1970s renewed interest in alternative sources of liquid fuels for motor vehicles. Generally, the plant oils have been— are—better suited to operation of diesel engines than spark ignition engines. Nowadays oils from rapeseed (Figure 34.8) and soybeans provide a small fraction of the diesel fuel used in Europe and the United States, respectively.

Many potential advantages accrue from using plant oils as liquid fuels. Like other forms of biomass, plant oils are, in principle, renewable. We need only grow a new crop of plants to replace the ones harvested for the oil to assure a continuous supply. And, like other forms of biomass, plant oils would, in principle, be CO_2-neutral. We might again call this into question if we were to include the energy expended (and hence the CO_2 produced) in farming, harvesting transportation, and processing. The by-product of plant oil production is a high-protein material called 'meal' that can be used as cattle feed.

As is the case with the alcohol fuels, plant oil fuels produce no SO_x or fuel NO_x emissions. Thermal NO_x emissions are relatively low. However, the production of plant oils requires much less energy than production of ethanol from corn. In particular, no

FIGURE 34.7 The olives from these trees can be processed to olive oil, which can be burned as a fuel or further processed into biodiesel.

FIGURE 34.8 Rapeseed oil is another example of a plant oil that is another energy source.

distillation step is needed, and, as we've seen, that step alone is very energy intensive. If the seeds or nuts were crushed in a water-operated mill, the CO_2 emission associated with production could be quite low. Another potential advantage of plant oils, relative to ethanol, is that their production does not require a distiller's license or involvement of the federal bureaucracy.

Compared to conventional petroleum-derived diesel fuel, plant oils have a higher viscosity and a lower cetane number (about 33). However, an even more serious problem with the direct use of plant oils is the very short operating life of a diesel engine before a major overhaul is needed. It's not unreasonable to expect 10,000 hours of operation from an engine running conventional diesel fuel before a major engine overhaul or rebuilding is needed. In contrast, some reports suggest that the comparable figure is 600 hours for a diesel engine running on sunflower oil and 100 hours on linseed oil.

Linseed oil, made by pressing oil from the linseed, has been used for generations as an additive to varnish, paint, and putty to help these substances dry into a hard, protective material. In addition, the oil itself is often rubbed into fine woodwork (such as furniture or the wooden stocks of sporting firearms) to provide a waterproof, hard finish to the wood. There is little wonder that such a material would quickly turn to 'gunk' in an engine.

These problems with the direct use of plant oils in diesel engines have generally been solved by reacting plant oils with methanol. This is a genuine chemical reaction, not a physical mixing. The generic process is

Plant oils + Methanol → Methyl esters + Glycerine

Using a slightly simplified composition of olive oil as an example,

$$(C_{17}H_{33}CO_2)_3C_3H_5 + 3CH_3OH \rightarrow 3C_{17}H_{33}CO_2CH_3 + C_3H_5(OH)_3$$

The chemical process is called **transesterification**; ethanol and butanol (C_4H_9OH) have been suggested to be used instead of the methanol. If the methanol (or other alcohol) were to be derived from biomass itself, then this would be an 'all-biomass' process.

The molecules in these transesterified plant oils (that is, the methyl esters) are roughly the size of the molecules in diesel fuel. The methyl esters are sometimes called 'biodiesel.' Biodiesel has viscosity comparable to conventional petroleum-derived diesel, and cetane numbers of about 50 (compared to about 33 for the raw oil). It appears to provide a much longer engine operating life than raw plant oils. Like regular diesel fuel, it is a safer product to handle and store than is gasoline (or ethanol). Biodiesel has to get very hot before its vapors will sustain an open flame. Biodiesel does mix with lubricating oils, so may cause some lubricity and engine wear problems.

More exotic sources of fuel are those plants that produce a sap that can be used as a vehicle fuel. The jojoba plant produces an oil now used in cosmetics, hair oil, and transmission fluid. Jojoba oil is very similar to the natural oil produced by the sperm whale. The gopher plant (*Euphorbia lathyrus*) grows wild in northern California. Its sap is an emulsion of hydrocarbons in water, and it can produce about 25 barrels of crude oil per acre. It offers the advantage of being able to grow well in both wet and dry regions. Guayule has a sap like rubber. A tree in the Amazon supposedly produces pure diesel oil.⚛ A hundred trees could produce 25 barrels of diesel fuel per year from one acre of land. It's also possible that aquatic plants and algae could produce a variety of fuel oil substitutes.

⚛ The jojoba, or goat nut, is a perennial evergreen shrub that grows in the Sonoran Desert of Arizona, California, and Mexico. Native Americans used jojoba oil for cooking and for various medical applications. Commercially, jojoba wax has an extraordinary range of applications, from massage oils to printer's ink, from cosmetics to transmission fluids. The gopher plant is a member of the euphorbia genus of latex-producing plants; other popular names include petroleum plant and milk bush. It grows in arid subtropical and tropical regions, and was the subject of considerable interest as a source of a possible substitute for petroleum during the oil shortages of the 1970s. Guayule is a shrub, seldom more than three feet tall, native to Mexico and the southwestern part of Texas. It is currently a source of natural rubber and of latex, the latter being especially valuable for such medical products as surgical balloons and gloves.

A resource we will probably never run out of is used French fry oil. In many communities, waste cooking oil is currently sent to landfills. Instead, it would be advantageous to be able to use it (or its transesterified product) as a fuel for trucks (operated by food processing companies, for example). Current estimates suggest a cost of about one dollar per gallon. An advantage of this material is that, in the event of a spill, it soon biodegrades and does not contaminate the environment.

Different plants would likely be grown for their oils in different parts of the world. The local climate and political and economic factors would be important in selecting the appropriate plant-oil crop for a given location. Nowadays soybean oil is dominant in the United States. An estimated cost for biodiesel from soybean oil would be about $1.35/gallon. If the entire soybean crop in the United States were diverted to the production of biodiesel, this would supply about 15% of total diesel fuel demand. In the southern states, winter rapeseed could yield large quantities of fuel oil at a competitive price, because it can be grown in the off-season when food crops would not be in production.

The yield of plant oils (in terms of useful Btus produced per acre of land) is much lower than that obtainable from the total quantity of biomass that could be grown on the same amount of land. However, substantially less processing is required to generate a useful transportation fuel, especially in comparison to ethanol production from corn.

Very likely, woody and herbaceous energy crops would be grown for conversion to methanol or ethanol for use in spark-ignition-engine automobiles, while other crops would be raised for their oils that can be substituted for diesel and fuel oil. An intriguing process that combines several ideas is shown in Figure 34.9. Three useful energy products are produced: charcoal, methanol, and biodiesel. Both the protein-rich animal meal and glycerine would be valuable by-products of this process. (Glycerine has a wide array of uses in commercial products ranging from cosmetics to dynamite.)

MUNICIPAL SOLID WASTE

One of the concerns driving the interest in renewables is that of finding energy sources that we are unlikely to run out of, so that issues such as lifetime of reserves and geopolitics of energy supply (Chapter 33) can be minimized or even eliminated. One possible energy source that we will never run short of is waste. In affluent, industrialized nations like the United States or Britain, each person produces, on average, a ton of solid waste per year.

Many of the materials that we routinely throw away are potentially combustible and could be used as fuels: paper (e.g., newspapers and magazines), cardboard, plastics, discarded textiles, and rubber tires are examples. In some communities grass clippings and leaf rakings are also mingled with the household wastes. This aggregate of waste material has been given various names, such as municipal solid waste, eco-fuel, or refuse-derived fuel. In addition to what we each produce as garbage every year, other sorts of waste are also potential fuels, including industrial wastes, agricultural wastes, and forestry residues.

Roughly 60% of municipal solid waste consists of combustible materials of the sort listed above. Across the United States, no more than about 10% of municipal solid waste is converted to energy. Perhaps another 10% is recycled, and the rest is disposed of by adding it to landfills. As we'll see in the next section, waste buried in a landfill could still produce usable energy, in the form of methane. However, methane is also a potent greenhouse gas; if it is formed in landfills and not collected or burned, it could contribute to global climate change.

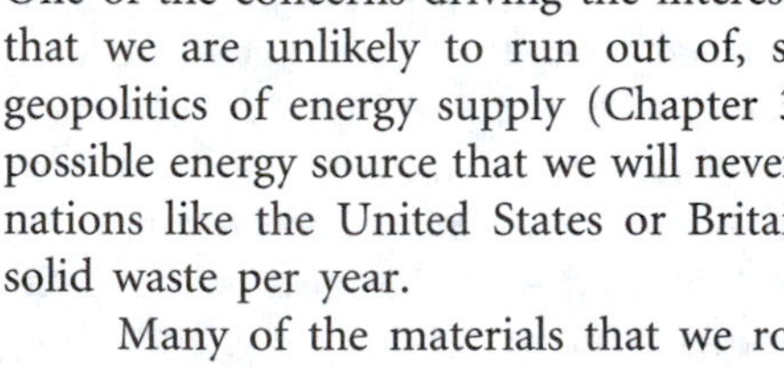

Wood ———► Methanol + Charcoal (solid fuel)

+

Plants ———► Plant oils + Cattle food

Biodiesel

FIGURE 34.9 This concept of a biomass conversion scheme illustrates the versatility of biomass as an energy source, producing methanol, charcoal, and biodiesel, as well as a by-product that is useful as cattle feed.

Using municipal solid waste as an energy source accomplishes several useful things. First, this does provide us with some useful energy that otherwise might literally have been thrown away. Second, it extends the useful life of landfills by reducing significantly the volume of material to be put into the landfill. In many areas, landfills are at or reaching their capacity. Communities need to take on the extra expense of hauling garbage to distant landfills. Public opposition, cost of acquiring the land, and various environmental regulations make it difficult to open new landfills. And, with these concerns in mind, burning municipal solid waste instead of putting it in a landfill can also provide economic benefits. It may actually cost less to burn the waste than to put it in a landfill. In addition, it might be possible to make money, or receive some form of credit, for the energy produced from burning the waste, as opposed to the certainty of incurring a cost for adding it to a landfill.

Municipal solid waste can be burned in incinerators designed to run on this kind of fuel. Incinerators reduce the total volume of waste to about 10–12% of the original volume burned, helping the problem mentioned above, reducing the volume of material to be put in landfills. It can also be burned in fluidized-bed combustors (FBCs), in which air is blown through a bed of solid material such as limestone or sand. The particles of the solid are suspended in the air stream, such that the bed looks, and feels, like a fluid. FBCs can be designed to have a wide tolerance for fuel specifications, and, in comparison with other forms of combustion equipment, are relatively easy to 'turn down' to respond to changes in demand for energy.

While much of what we throw away is combustible, some of it isn't. Glass and metals won't burn. The use of municipal solid waste as a fuel is improved if the non-combustible material is removed beforehand. A separation of noncombustible from combustible material needs to be made. This could be done where the waste is generated—in the home—if the majority of householders cooperate with such a program. Since many communities now have recycling programs that encourage householders to separate glass, metal, plastics, and paper anyway, it may be easy to effect this separation in the home. The alternative would be to do the separation after the collected waste has been taken to the incinerator location. Making the separation there might prove to be a messy and labor-intensive job. It could, though, provide a possible source of employment and income for unskilled workers.

Burning municipal solid waste leads to some problems in pollution control that are not shared by other forms of biomass or by the fossil fuels. Some of the plastics can produce potentially harmful emissions when they burn. Poly(vinyl chloride) has an incredible range of uses, being another one of the plastics that is almost ubiquitous in our society. Examples of its uses include PVC pipe, garden hose, plastic films (such as food wrap), artificial leather, and plastic toys. Burning poly(vinyl chloride) could form hydrogen chloride, which dissolves in water to become hydrochloric acid. It could also form other organic compounds containing chlorine, many of which are health hazards. Urethane foams and acrylic fibers can produce hydrogen cyanide, which is toxic. Some of the heavy metals that might be contained in noncombustibles not completely separated from the waste can also be health hazards if released to the environment. Mercury and lead are examples. These chlorides, cyanides, and heavy metals need to be contained in the incineration process or captured before they escape to the environment.

Municipal solid waste is not the only waste source that is a potentially useful fuel. Agricultural wastes, such as stalks or leaves of plants, and related food processing wastes, such as potato peelings, are potential sources of energy. Hawaii offers an excellent example; there, about 10% of the electricity comes from using bagasse (the residue from processing sugar cane) as a fuel to raise steam in electricity-generating plants. Residues from the logging industry, such as tree stumps and branches, are usually left behind after the logging operations and could also be used as fuel.

METHANE FROM BIOLOGICAL SOURCES

We've seen that municipal solid waste can be thought of as being a renewable, because garbage is one resource we're likely never to run short of, nor to stop producing. The alternative to the direct burning of municipal solid waste is to take advantage of the fact that many biological materials, one example being animal manure, undergo anaerobic digestion to produce methane. This process can occur naturally, as waste organic matter occurs naturally in a landfill, or by deliberate intent in equipment designed for the purpose. The reaction is facilitated by anaerobic bacteria (i.e., bacteria that do not use oxygen for their life processes). Taking a simple sugar molecule as an example,

$$C_6H_{12}O_6 \rightarrow 3CH_4 + 3CO_2$$

The gaseous fuel produced by the anaerobic digestion of wastes is sometimes called **biogas**. The important component is methane, which is an excellent fuel.

Many kinds of biomass could be (and are) converted to biogas. These include animal waste (e.g., manure from farms or feedlots); the organic portion of municipal solid wastes (garbage, specifically the organic portion); agricultural and food processing residues, such as bagasse, sawdust, and scrap wood; sewage, and various kinds of industrial wastes.

The technology is basically very simple. It uses a large tank (the digestor) from which air can be excluded. Usually a mixer in the tank assures that the digesting material is slowly stirred. The necessary anaerobic bacteria may be in the original waste being digested, such as animal manure, or they may be intentionally introduced. The digester operates at a working temperature in the range 95–140°F. The gaseous products—the biogas—bubble to the surface and can be collected for use. Many feedlots, put bluntly, stink. They are concentrated sources not only of what is known by the somewhat more polite euphemism 'olfactory pollution,' and of water pollution from storm water runoff carrying wastes into the water system.

In addition to producing the desired biogas as an energy source, other advantages derive from the digestion of wastes. The process can be used as a waste treatment operation to reduce pollution hazards and odors. Biogas production can be of some benefit to the public health. The digestion process can kill bacteria and parasites that were present in the waste and that could be threats to the health of humans or other animals. The solid by-product that remains after digestion is a fertilizer rich in nitrogen that can be returned to the land.

The anaerobic digestion process would normally proceed to yield one molecule of CO_2 with every molecule of CH_4. Because CO_2 is noncombustible, biogas of this composition would have a much lower heating value than natural gas (which contains more than 90% methane) unless the CO_2 is removed. Doing so might ultimately release this CO_2 into the atmosphere unless special provisions were made to capture and sequester it. Continued research on biogas production has yielded ways of modifying the process so that an excess of methane, relative to CO_2, can be produced. The methane concentration in the gas could be over 70%. More than 85% of the chemical potential energy of the original waste material being digested is retained in the biogas.

Organic waste sent to a landfill will eventually produce methane when it becomes buried deeply enough to allow anaerobic processes to occur. Methane leaking out of landfills has several potential problems. First, methane is a substantially more potent greenhouse gas than carbon dioxide (on a molecule-for-molecule basis, methane is about 11 times better at absorbing infrared). Second, methane is flammable and, in some concentrations, methane/air mixtures are explosive. A potential safety hazard exists.

Digesting wastes rather than consigning them to a landfill, or, at the least, deliberately tapping into the landfill to capture the so-called 'landfill gas' can therefore provide several advantages.

Biogas production from digestion of wastes is becoming an attractive technology for some developing nations where there are very high population densities (which translate directly into a large, unending supply of human waste, and a large energy demand). In China, over seven million small digestion systems have been installed, mostly family-sized units, to help meet the cooking, and heating of small rural communities. At least 25 million people now cook with biogas. There are also about 10,000 biogas plants producing gaseous fuel for farms or light industries. The United States now has over 120 biogas plants, the majority of which are used to collect the gases produced in landfills. Disney World, a site ideal for collection of high concentrations of human waste, has an experimental digestion unit to produce a gas about 95% CH_4. Britain now has over 50 biogas-producing sites in operation.

Biogas from anaerobic digestion of biomass has been used successfully in two applications: energy self-sufficiency on farms, providing energy for farm vehicles, electricity, and heat; and energy sufficiency for running a municipal waste system—vehicles and heating. When the biogas has been available to operate collection vehicles, its use essentially eliminates the energy cost of collecting garbage. Biogas could be used directly for heating or in internal combustion engines. Biogas production offers an economical route to on-site electricity production wherever sufficient wastes are available and there is an identified demand for the electricity. Potentially there is no reason why methane separated from biogas generated on a large scale could not supplement natural gas in pipelines. So far it has not been used that way.

Of course, burning the methane in biogas still produces CO_2. But, given a choice, the CO_2 produced is a less harmful greenhouse gas than CH_4. (If biogas were to be produced on a large scale, leakage of methane from the digesters would have to be minimized because methane is a potent greenhouse gas.) There are no SO_x emissions from burning biogas.

HYDROGEN

Almost any rocket launch provides a spectacular demonstration of the prospects of hydrogen as a fuel. Hydrogen currently is the fuel of choice for large rockets; the National Aeronautics and Space Administration (NASA) is the country's largest single consumer of hydrogen. Though few, if any, people envision a day of rocket-propelled cars, hydrogen nevertheless offers some attractive possibilities as a source of energy for motor vehicles.

Hydrogen can be produced from several sources. One is the steam re-forming or gasification of fossil fuels. Here, methane offers an attractive choice because it contains the highest proportion of hydrogen by weight (25%) of any chemical compound. As we've seen, the steam re-forming of methane produces synthesis gas. 'Shifting' the synthesis gas by reacting with more steam converts the carbon monoxide to dioxide:

$$CH_4 + H_2O \rightarrow CO + 3H_2$$
$$CO + H_2O \rightarrow CO_2 + H_2$$

Removing or sequestering the CO_2 leaves essentially pure hydrogen. This is presently the most popular route to hydrogen. Gasifying fuel oil or coal as the starting material will get us to the same point. However, because oil and coal contain proportionately more carbon and less hydrogen than does methane, more CO_2 will be produced requiring removal (or will be emitted to the environment). Biomass can replace the fossil fuels in this application. Biomass gasification and the steam re-forming of methane from anaerobic digestion of biomass will both provide opportunities for production of hydrogen. Some bacteria can produce hydrogen directly from biomass under anaerobic conditions (that is, without the intervening production of methane and its subsequent conversion to hydrogen). Presently the yields of hydrogen are low, but as research progresses it may prove possible to make hydrogen directly from biomass just as methane is made from wastes nowadays.

Perhaps the ultimate hydrogen resource is water. Splitting the water molecule produces hydrogen directly:

$$2H_2O \rightarrow 2H_2 + O_2$$

Electricity is usually the energy source for this process; the splitting apart of a molecule by electricity is called **electrolysis**. About three-fourths of Earth's surface is covered by water. Immense quantities are available. This fact alone suggests that we need not be concerned about running out of hydrogen. If the planet ever runs out of water, we will have a lot more pressing worries than how to drive down to the local discount emporium. But, the horrific notion of running out of water would not come to pass, because the combustion of the hydrogen regenerates water:

$$2H_2 + O \rightarrow 2H_2O$$

On a volumetric basis, liquid hydrogen has a heat of combustion of 36,500 Btu/gal. This does not compare well with other fuels, even with methanol. This is due to the very low density of hydrogen. When compared on a weight basis, however, hydrogen looks exceptionally good, providing 52,000 Btu/lb. This is more than double the heating value per pound of methane, quadruple that of coal, and 10 times better than wood.

Hydrogen provides several potentially significant advantages. First, on a pound-for-pound basis it has a very high energy content. Second, its combustion produces only water as a product, and can be considered to be nonpolluting. (Actually, in high-temperature combustion systems, the possibility of making thermal NO_x still must be taken into account.) The burning of hydrogen produces no CO_2. If the hydrogen were made by electrolysis of water, and the electricity were made from hydro-, nuclear, or solar plants, then there would be no CO_2 emission at all. If water is the source of the hydrogen, then it is perpetually renewable, since water is produced again during the combustion.

Of course, some problems exist. The first is how to make hydrogen cheaply enough to be cost-competitive with present petroleum-derived fuels. The electrolysis of water would require very inexpensive electricity, probably from hydro plants. The gasification or steam re-forming of fossil fuels is known technology, indeed in commercial practice, but adds both cost and energy, raising the question of why we wouldn't use the fossil fuel directly in the first place. The same is true of biomass. Some chemical reactions will split water. For example, steam will react with red-hot scrap iron:

$$H_2O + Fe \rightarrow FeO + H_2$$

There is much interest in finding a catalyst that will help break the water molecule apart by chemical reactions but at much lower temperatures. On the other hand, water will break apart without a catalyst at extremely high temperatures, several thousands of kelvins. It might be possible to obtain such temperatures by focusing solar energy onto a tiny volume.

Unfortunately, the breaking of the water molecule will require a certain amount of energy, regardless of how it's done. When the hydrogen burns with oxygen, that same (in principle) amount of energy will be released again. If that could be achieved, we would be able to 'break even' in terms of energy, eventually recovering the same energy from burning hydrogen that was used to make it. In that sense, we could regard hydrogen as an 'energy carrier.' That is, we would invest a certain amount of energy at, say, a hydroelectric plant to produce a quantity of hydrogen. Later, at some different location, our investment would be repaid when that energy is liberated by the burning of the hydrogen. We must also consider that some energy, regardless of how little, is going to be required simply to move the hydrogen to its point of use. And a far more formidable problem is that no process of any kind will work with 100% efficiency. Even if we merely electrolyzed water and burned the hydrogen right on the spot (disregarding what strange motivation might lead us to do this) we would never recover all of the energy expended to make the hydrogen. Even more so than the example of ethanol from corn, we will lose more energy making the hydrogen than is 'paid back' when we later use it as a fuel.

A third issue is to how to handle and store the hydrogen once we've made it. Under ordinary conditions, hydrogen is a gas of very low density. To store a sufficient quantity of hydrogen to provide any reasonable amount of energy (in a car, for example), the hydrogen would have to be liquefied or stored as a compressed gas at very high pressures. To liquefy hydrogen at atmospheric pressure requires that we cool it to $-423°F$ (20 K). 'Slush hydrogen' is a mixture of solid and liquid hydrogen—a hydrogen frappe—held at about $-435°F$. This material can be handled and stored by an organization having the financial wherewithal and technical resources of NASA, but is hardly suitable for the average homeowner commuting to work or driving to the supermarket. Storing compressed gaseous hydrogen requires very strong high-pressure containers, likely to be very heavy and adding substantially to the weight of the vehicle. Since H_2 is the smallest known stable molecule, it has a knack of worming its way through very small pores or holes, and can easily leak. The best approach to handling and storing hydrogen seems to be finding some material that will adsorb and hold large quantities of hydrogen, but from which the hydrogen can easily be removed (desorbed) as fuel is needed. Very likely the best material for this application will turn out to be one of the forms of carbon, an element of which most of the forms are themselves of low density.

These concerns do not imply that there is no hope for hydrogen. Hydrogen is already being transported by rail, truck, and pipeline, and stored in liquid or gaseous forms with a superb record of safety. Hydrogen-fueled city buses have been demonstrated successfully in Chicago and Vancouver. These buses used compressed hydrogen to operate fuel cells. Since buses are bigger and heavier than cars to begin with, they can more easily accommodate the weight of the storage tanks and of the fuel cells.

The fuel cells may hold the key to hydrogen's future as a large-scale energy resource. Hydrogen is an excellent fuel for fuel cells. Continuing engineering development of fuel cells is bringing them closer to the market for both stationary uses, in the on-site generation of electricity, and transportation uses, in hybrid or electric vehicles. The next few years are likely to see increasing applications in both markets. An important stimulus for the development of the small fuel cells for vehicle use is California's requirement that 2% of new cars sold in 2003 be zero-emission vehicles. As improvements in fuel cell technology move ahead together with improved systems for purifying and storing hydrogen made from fossil fuels or biomass, interest in hydrogen as an energy source will continue to evolve. Nonetheless, the small-vehicle market will require small, inexpensive, efficient fuel cells; safe and simple ways to store hydrogen onboard; and a hydrogen refueling infrastructure—hydrogen stations.

TWO SUMMARY POINTS

Let's summarize two major points of all the foregoing discussion:

(1) Is biomass 'renewable' energy? Well, yes and no. It depends on the time scale we use. Grains, animal wastes, and garbage are renewable on a scale of a year or less. However, wood definitely is not. We cannot cut down a tree and replace it with another tree the same size the next year. Unless there is some forest management cycle over about 40 years, wood is not really renewable. In addition, at least for some forms of biomass fuels, such as ethanol from corn, it is likely that more energy is expended in growing and processing the biomass than is recovered when the fuel is burned.

(2) Regardless of how we use biomass as an energy source, burning wood, ethanol, methane, or any other biomass form all put CO_2 into the atmosphere. So too do the farm vehicles, transportation system, and processing operations. Therefore if the global climate change induced by anthropogenic CO_2 is considered to be a problem, the use of biomass does not help. In fact, cutting down trees exacerbates the problem, because trees are an excellent way of removing CO_2 from air. Thus if the global climate change is a concern, we still need to rely on hydro, nuclear, and solar.

CITATION

1   Anderson, F.C., quoted in Gregory, Kenneth. *The Second Cuckoo*. Akadine Press: Pleasantville, NY, 1997; p. 199.

FOR FURTHER READING

Aubrecht, Gordon. *Energy*. Merrill: Columbus, 1989; Chapter 16. This chapter discusses aspects of biomass energy. Topics include biomass fuel and conversion of biomass to other materials.

Berger, John J. *Charging Ahead*. University of California: Berkeley, 1997; Chapters 18 and 19. These two chapters together constitute a discussion of 'bioenergy,' including some of the aspects of biomass-based businesses.

Borowitz, Sidney. *Farewell Fossil Fuels*. Plenum: New York, 1999; Chapter 11. As the title implies, the book discusses alternatives to the use of fossil fuels. Chapter 11 provides a brief overview of biomass energy.

Brower, Michael. *Cool Energy*. MIT Press: Cambridge, 1992; Chapter 5. The focus of this book is renewable energy, in many forms, including solar and wind. Chapter 5 treats biomass.

Cash, John D.; Cash, Martin G. *Producer Gas for Motor Vehicles*. Lindsay: Bradley, IL, 1997. This is a reprint of a book originally published in 1942, now available in an inexpensive paperback edition. It describes ways of converting wood into a gaseous fuel, and even how to run an automobile on such a fuel.

Dunn, Seth. 'Decarbonizing the Energy Economy.' In: *State of the World 2001*. Starke, Linda (Ed.). Norton: New York, 2001; Chapter 5.

Flavin, Christopher; Lenssen, Nicholas. *Power Surge*. Norton: New York, 1994. Chapter 9 discusses approaches to biomass energy for the future. Most of the book is devoted to various aspects of future use of energy, including wind and solar as well as biomass.

Johansson, Thomas B.; Kelly, Henry; Reddy, Amulya K.N.; Williams, Robert H. *Renewable Energy*. Island Press: Washington, 1993. This book is a collection of chapters, by various authors, on most aspects of renewables, including hydrogen, biomass, and biogas.

Kheshgi, Haroon S.; Prince, Roger C.; Marland, Gregg. 'The Potential of Biomass Fuels in the Context of Global Climate Change: Focus on Transportation Fuels.' In: *Annual Review of Energy and the Environment*. Socolow, Robert H.; Anderson, Dennis; Harte, John (Eds.). Annual Reviews: Palo Alto, 2000; Vol. 25, pp. 199–244.

Lenssen, Nicholas. 'Providing Energy in Developing Countries.' In: *State of the World 1993*. Starke, Linda (Ed.). Norton: New York, 1993, Chapter 6.

Moore, John W.; Stanitski, Conrad L.; Wood, James L.; Kotz, John C.; Joesten, Melvin D. *The Chemical World*. Saunders: Fort Worth, 1998; Chapter 11. This is an introductory college-level chemistry textbook. Chapter 11 includes a useful introduction to the molecular structures and chemical properties of the simple alcohols, such as methanol and ethanol.

Pasztor, Janos; Kristoferson, Lars. 'Biomass Energy.' In: *The Energy–Environment Connection*. Hollander, Jack M. (Ed.). Island Press: Washington, 1992; Chapter 7. This discussion focuses largely on environmental issues associated with the use of biomass.

Riley, Robert Q. *Alternative Cars in the 21st Century*. Society of Automotive Engineers: Warrendale, PA, 1994. Chapter 4 describes alternative fuels for automobiles, including methanol, ethanol, natural gas, and hydrogen.

Röbbelen, Gergard; Downey, R. Keith; Ashri, Amram. *Oil Crops of the World*. McGraw-Hill: New York, 1989. This is a fine overview of all sorts of plant-derived oils, including use of olive oil, jojoba, and sunflower oil. The focus, however, is on all uses of these plant oils, and not just their use in energy production.

Shepherd, W.; Shepherd, D.W. *Energy Studies*. Imperial College Press: London, 1998; Chapter 13. This chapter is a good overview of biomass energy, including the various approaches for converting biomass into liquid or gaseous fuels, and the production of energy from waste materials.

Sørensen, Bent. *Renewable Energy*. Academic Press: San Diego, 2000. A review of most forms of renewables, including environmental impact and economic issues.

Tillman, David A. *Wood as an Energy Resource*. Academic Press: New York, 1978. Chapter 5 provides information on new combustion systems for wood, as well as conversion of wood to other fuel forms.

Wagner, Rudolf. *A Handbook of Chemical Technology*. Lindsay Publications: Bradley, IL, 1988. This reprint of a book originally published in 1872 is a cornucopia of information for anyone interested in the history of industrial chemistry. Division VIII includes information on making methanol ('wood spirit') from wood.

Wright, F.B. *Distillation of Alcohol and De-naturing*. Lindsay: Bradley, IL, 1994. Another old-timer rescued from obscurity and reissued as an inexpensive paperback by Lindsay Publications. This book introduces the fermentation of a variety of biomass sources to produce ethanol, and then the many considerations needed to distill the fermented materials to produce alcohol in high purity.

35

ELECTRICITY FROM WIND

The last chapter introduced the concept of renewables, and the ideas that the push for replacing fossil fuels with renewables stems from two issues: (1) limiting emissions of greenhouse gases, particularly carbon dioxide; and (2) not being hostage to resource depletion or geopolitics. Among the various categories of renewables, wind is thought to have great promise as a future energy resource. It shares the same advantages of other renewables—it is essentially inexhaustible, it is a domestic energy resource, and its use produces no pollution, hazardous by-products, nor any significant danger to the public.

WHERE THE WIND COMES FROM

Winds are generated by the uneven heating of the Earth and its atmosphere by the sun. The Earth is generally warmest at the equator. As the air there is heated, it expands. (Remember Charles and Gay-Lussac—increasing the temperature of a gas increases its volume.) The same mass of air now occupies a larger volume, which is another way of saying that the density of that portion of air has decreased. Because the density has decreased, the air will rise. It is this expansion by heating and consequent reduction of density that is the origin of the bit of folk wisdom that 'hot air rises.'

Two effects follow from the rising of the heated air. First, cooler air from elsewhere will flow into the region, to replace the heated air that has ascended into the atmosphere. This flow of air is felt by us as a wind. The second effect is that the heated air that has risen into the atmosphere will travel at high altitudes until it cools off. As it cools, its volume decreases (Charles and Gay-Lussac again) and it becomes more dense. In doing so, it sinks to lower elevations.

Essentially, a gigantic 'conveyor belt' of air movement has been established. Air masses that benefit by being warmed by the sun rise, and are replaced by cooler air flowing into the region. At the same time, the warmed air travels away from the equator, cools, and sinks, replacing some of the cooler air that had moved toward the equator. These large circulation patterns, the 'conveyor belt' model, owe their origin to the differential heating of Earth by the sun.

WIND ENERGY IS AN INDIRECT FORM OF SOLAR ENERGY.

This model focuses on a global scale, and explains how atmospheric circulation can distribute energy, originally obtained from the sun, toward the poles. The other global-scale phenomenon that contributes to the pattern of winds is the rotation of the Earth. The rotation adds the effect of stirring or swirling into the pattern of air transport established by the 'conveyor belt.'

On a local scale, this simplistic model becomes much more complex. Each component of the local landscape—soil, rocks, plants, and water—all have different abilities to absorb solar energy or to reflect or reradiate it back into the atmosphere. Each, then, experiences a different degree of heating. Furthermore, different landscapes will also have different features of topography: Some places may be flat, others hilly or mountainous. Some may incorporate large bodies of water. Urban landscapes may be marked by numerous tall buildings. As a result, the frequency, direction, and speed of winds can vary greatly from one location to another.

For centuries, sailors were well aware of these variations, and amassed an enormous body of empirical knowledge about the distribution, frequency, and speed of winds around the world.

WIND ENERGY IN THE RECENT PAST

We've seen in Chapter 9 how wind was an important energy source in the Middle Ages and Renaissance. Although the rise of the steam engine and the Industrial Revolution that the steam engine helped foster diminished the role of wind as a significant source of energy, wind never left the picture entirely.

> One of the great curiosities in Zealand, the flourishing Holland colony in Ottawa County, Michigan, is the great, awkward and unmanageable concern called the Windmill. This is a monstrous wooden pile in the form of an octagon tower. The mill is moved by the force of the wind striking against four winding slats, covered with canvas. They were sawing, or attempting to saw, while I was there. Occasionally, with a fair wind, the saws would strike a few minutes quite lively, then draw a few slower strokes and then entirely stop, perhaps for half an hour. An enterprising individual is now putting up a steam sawmill, which will do a better business.
> —Scientific American[1]

Discussing the potential application of wind energy in the 21st century is by no means a suggestion that we should somehow change our technologies and patterns of energy use back to those of medieval times.

Windmills continued to be important throughout the 19th century and well into the 20th. They were important sources of energy throughout much of the rural United States at least into the 1940s. There does not seem to be an accurate census of how many windmills were in use, but estimates range from many hundreds of thousands up to about six million. Through much of the 19th century, these windmills provided ENERGY for crucial chores such as pumping water for livestock or for irrigation. It's not uncommon to see a windmill in the background of scenes from Westerns in the movies or on television. Even now it's still possible to see the occasional windmill when driving through rural areas where farming or ranching is important (Figure 35.1).

In the late 19th century the technology of generating electricity had been worked out by Gramme and other pioneering electrical engineers (Chapter 15). Soon thereafter, windmills were operating small generators to produce electricity either for recharging batteries or for operating small electrical devices on the farm. For example, batteries recharged by the wind made it possible for farmers and ranchers living far from electricity distribution lines to get news and entertainment by radio.

The use of windmills to produce electricity dates from the late 1880s in the United States. The first such installations were erected in Massachusetts, along the coast. In 1890, a mansion in Cleveland may have been the first major dwelling in the United States to be lighted by electricity; the generator was operated by a windmill mounted on a nearby tower that was 60 feet (five to six stories) tall. A half-century later there were probably hundreds of thousands of so-called 'home light plants' operating in the United States, mostly in the windy Great Plains.

These applications of wind energy in the United States were killed off by the steady spread of electrical grids that brought electricity from central generating plants to all parts of the country. The process was hastened by the rural electrification programs of the 1930s. That supply of electricity, coming via distribution lines from plants tens or hundreds of miles away, was ample for all the many applications of electricity around the farm or ranch. The small wind-operated generators fell into disuse.

In Europe, Denmark has been the leader in applications of wind technology. At the same time—roughly around 1890—that wind was being used for small-scale electricity generation in the rural United States, the Danish engineer Poul la Cour was developing similar devices. Indeed, la Cour is the founding father of the wind-to-electricity technology that we will discuss later in this chapter. By the time of the First

FIGURE 35.1 Windmills, an important source of mechanical WORK in the Middle Ages, never really disappeared. They are still very useful in rural areas, e.g., for pumping water.

World War several hundred wind-operated generators were in use in Denmark, mostly along the coast.

The easy availability of cheap oil after the Second World War was a major factor in eliminating European interest in the use of wind to produce electricity. However, that period of cheap oil lasted only about a quarter-century, until the oil price shocks of the 1970s. The Danish response to the rapid run-up in oil prices in the early 1970s provided the impetus for the revitalization of the wind industry. The Danish government took several steps to encourage a shift back to the use of wind. These steps included a 30% subsidy on the purchase of wind machines, and a requirement that electricity utilities purchase the electricity that these devices produced.

WIND TURBINES

Gramme and his contemporaries were responsible for the engineering development of Faraday's scientific discovery—that a moving magnetic field generates an electric current—into the practical generators that are still used to produce virtually all of the electricity that we use today. Only the small fraction that we derive from batteries (Chapter 13) and the tiny fraction from solar cells (Chapter 36) do not come from modern variants of the 'Gramme machine.' Since Gramme's generator involves rotating an electric conductor in a magnetic field, the essence of electricity generation is that of finding ways to turn the generator as cheaply and reliably as we can. We need to find a reliable source of rotary mechanical WORK to operate the generator. This WORK is provided to the generator by a turbine, a device designed to capture some of the kinetic energy of a moving fluid and convert that ENERGY into rotating mechanical WORK.

As we've seen, the fluid that passes through the turbine is the so-called working fluid. We have already met several working fluids: water in hydroelectric plants, steam in fossil-fuel or nuclear plants, and combustion gases in jet engines. Since wind represents a fluid (air) in motion, it could, in principle, be used as the working fluid in a turbine. We use the term **wind turbine** to refer to machines that use the ENERGY of wind specifically to supply WORK to an electricity generator (Figure 35.2). Nowadays the term *windmill* is usually restricted to mean devices that use the wind for other applications of mechanical WORK, such as pumping water or making flour.✥

> ✥ Actually neither the wind turbine nor the windmill represents the most powerful of wind-operated devices. That honor goes to the large sailing ships of the 18th and 19th centuries, the best of which could extract as much as 10,000 horsepower from the wind.

The focus of interest for applications of wind energy in the 21st century is entirely on the wind turbine. There seems to be no interest in reviving windmills.

Installations using windmills and waterwheels could be configured such that the axis of the wheel was either vertical or horizontal. The same is true with wind turbines. In a horizontal axis machine, the turbine blades look somewhat like the propeller on an airplane. The axle on which the turbine blades are mounted is connected, usually via a gearbox, to the axle of the generator. The generator, gearbox, and any necessary components of the electrical system or control systems are all mounted in a nacelle, which sites on top of the turbine tower, directly behind the turbine blades. The electricity that is generated is conducted down to ground level by a cable.

FIGURE 35.2 The future of wind energy lies in the wind turbine, which uses the kinetic energy of wind to operate an electricity generator.

The vertical axis wind turbine represents the alternative design. These devices look like enormously large eggbeaters, having two aluminum blades attached to a central, vertical mast. The vertical axis wind turbines have several design advantages relative to the horizontal axis type: The gearbox and generator are on the ground, making maintenance much easier; wind coming from any direction can be used without having to turn the unit into the wind; and the blades are much lighter in weight than those on horizontal axis machines.

Two other considerations of design of wind turbines are the generating capacity of an individual unit, and whether they would be constant- or variable-speed devices. After considerable testing and development work in the 1980s and 90s, the industry seems presently to have adopted wind turbines in two general size ranges: small machines that would be used by individual farms or households that are not connected to the electricity distribution grid, and larger machines generating electricity for distribution by an electric utility. The smaller units have generating capacities typically in the range 0.5 to about 50 kW, whereas the larger machines, intended to feed electricity into a grid, would be in the range 50 to about 600 kW.

The occurrence of gusts, during which the wind speed could double and then perhaps drop to even lower than original value in a matter of seconds, has to be taken into account in the design of wind turbines. The constant-speed wind turbines must be designed and built to endure higher mechanical loads on their drive trains and gearboxes as wind speed increases. A variable-speed machine can operate at a range of wind speeds while still delivering constant output. Fluctuating mechanical loads imposed by the winds are more easily withstood by blades and drive train. The drive train can be less rugged, and thus made lighter and less expensively. A further advantage of the variable-speed wind turbines is their ability to deliver electricity over a wider range of wind speeds. A disadvantage of variable-speed devices is an increased cost the necessary electrical control system and so-called 'power conditioning' equipment that are needed to provide electricity output of a constant quality even while the speed of the turbine itself is changing. Of course, in areas that have fairly constant wind speeds there would be ample reason to stay with the simpler constant-speed devices.

Current commercial designs of wind turbines can operate over a wide range of wind speeds, from a little under 10 miles per hour to about 60. Rapid strides have been made in improving their reliability. Some of the pioneering designs of the 1980s were out of service more than half the time. Nowadays the 'down time' for a wind turbine is in the range 2–5%. The approximate power delivered by the wind is given by the equation

$$P = \frac{d^2 v^3}{2,000}$$

where P is the power in kilowatts; d the diameter, in meters, of the circle swept out by the rotors; and v, the wind speed in meters per second. The theoretical maximum efficiency of converting wind ENERGY to WORK, the **Betz limit**, is 16/27ths, essentially 59%. Just as the Carnot efficiency tells us the theoretical maximum efficiency of a heat engine, and the real efficiencies of actual operating engines are lower than the ideal Carnot efficiency, in the same way the efficiency of actual wind turbines is some 50–70% of the Betz limit. The conversion of mechanical WORK to electrical ENERGY in the generator is quite high (typical of most generators), above 90%.

Though the future of wind as an energy source seems to be in the development of wind farms (discussed below), each with hundreds of wind turbines having individual capacities in the several hundred kilowatt range, some remarkable small designs have been developed that are tributes to human ingenuity. The Savonius rotor is a low-tech device that seems to take its design inspiration from the Islamic vertical mill (Chapter 9). At first glance, a sketch of the Savonius rotor might give the impression that the device is nothing more than an oil drum sawn in half lengthwise and opened outwards to make two 'cups' capable of catching the wind. In this the first impression is exactly correct, for that is in fact the easiest way to make a Savonius rotor. A virtue of this device is that it can be constructed for very little money by do-it-yourselfers with common tools. Though the Savonius rotor is not nearly so efficient as the more sophisticated, large commercial devices, it can be built and installed in impoverished villages and used, for example, to operate water pumps. Another remarkable design is a collapsible, portable wind turbine used by Mongolian nomads. The spead of blades is only about six feet. The turbine can be taken down, packed up, and moved to the next camp. These small portable units have generating capacities of about 50–250 watts.

THE POTENTIAL CONTRIBUTION OF WIND AS A MODERN ENERGY SOURCE

As is true of any energy resource, the estimates of the extent to which we can count on wind to meet future energy needs are dependent on the assumptions used in making the estimates. And as is also true of any other energy resource, there is considerable variation in the magnitudes of the estimates. Nevertheless, the consensus seems to be that the potential for wind to meet future electricity generation needs is enormous.

If we start with the wind energy theoretically available, ignoring for the moment engineering issues of building practical wind turbine devices to capture all that energy, and ignoring societal issues of where wind turbine installations might actually be located, then the potential is extraordinary. The total amount of kinetic energy dissipated by winds across the United States is some 40 times larger than our annual energy consumption. In other words, we could—in theory—satisfy *all* our energy needs by extracting just one-fortieth of the energy from the wind.

Of course, there are limitations imposed because we should assume reliance on today's wind turbine machines. As noted above, their efficiencies are limited to some 50–70% of the Betz limit. Also, wind turbine towers are at most a few hundred feet tall; winds at higher elevations from the ground are not utilized. Nonetheless, if we restrict ourselves to current technology, there is sufficient wind energy in the 'lower 48' (that is, not including Alaska and Hawaii) to provide about 1.3 times the total electricity production (for 1998) in the United States. This is a conservative estimate, because it assumes— almost certainly incorrectly—that there will be no further developments or improvements in wind turbine technology.

Furthermore, it is not likely that every possible site for wind-to-electricity energy conversion will be developed. There are several reasons for this limitation. Some wind sites may be so remote that it would be impractical, or not cost effective, to connect them to a distribution grid. Other sites may be in environmentally sensitive areas where society might decide to limit development. (Indeed, land-use issues may most likely be among the most contentious issues for wind energy development.) Still others might be in highly populated urban areas where the costs of acquiring the land, razing existing buildings, and developing the site might be prohibitive. If we accepted no limitations at all on land use (probably a very unrealistic assumption), then, with these other factors taken into account, the prime spots for wind energy in the United States could provide about 55% of our electricity needs. If we assume on the other hand that only open range land would be developed, then this figure drops to about 20%. Then, if we superimpose estimates for the limitations imposed by current technology—such as the efficiencies of conversion and height limits on wind turbine towers—then the upper and lower limits on the portion of our electricity demand that could be supplied by wind are 40% at the highest and about 10% at the lowest.

Being able to supply some 10–40% of our electricity needs with wind might seem like a rather significant reduction from the grandiose idea that wind could provide 40 times more total energy than we use. However, if wind were to supply 10% of our electricity needs, that alone is equivalent to half the electricity supplied by nuclear plants. If 40% were to be supplied by wind, that is equivalent to about two-thirds of the electricity from coal-fired plants.

Worldwide, the total wind flow is ample to supply several hundred times the world's total annual energy consumption. Once again, the same issues discussed above would apply to estimating how much of this potential might actually be tapped, and are even more difficult to take into account. For example, some countries have very strict limitations on land use, while others have essentially none. It seems reasonable to expect that over the next several decades about 10% of total worldwide electricity production could be supplied by wind.

THE PRIME LOCATIONS FOR WIND ENERGY DEVELOPMENT

The area of the United States that appears to be of highest potential for wind energy development is the northern Great Plains: the Dakotas, Montana, and Wyoming. The conversion of North and South Dakota into an enormous wind energy plantation might be able to supply up to 80% of the nation's electricity needs. That estimate includes protecting the environmentally sensitive areas. Unfortunately, these states have some of the lowest population densities of anywhere in the country—in other words, the customers for all this electricity live someplace else. Significant investment might be needed in improving the transmission systems to get the electricity to other parts of the country, coupled with contentious right-of-way issues.

In the more populous east, the ridge-and-valley topography of Pennsylvania has some of the best wind resources. In fact, these winds have been exploited for decades by glider and sailplane enthusiasts, for millennia by migrating raptors.

Some of the best locations are over oceans, but generally near to shore. This raises the possibility of offshore generation of electricity from ocean winds. Such offshore generation might eliminate, or at least reduce, some potentially contentious issues of transmission rights-of-way and alternative uses for the 'land' (or, in this case, ocean) surface.

Before the first kilowatt-hour of electricity can be generated and sold, much work needs to be done. First, sites with appropriate wind potential need to be identified and assessed. Then, land leases have to be negotiated and land-use permits obtained. Agreements have to be set up for the transmission of the electricity and its purchase. Turbines must be designed, manufactured, and installed, along with an array of electricity collection equipment, substations, and interconnection facilities. All of this hardware has to be installed on site, tested, and maintained.

In choosing a site for a wind turbine installation, having wind that blows reasonably constantly throughout the day and throughout the year is usually much more important than finding a place that has high peak wind speeds but has them only occasionally. Detailed data on wind speeds, directions, and duration over a number of years are needed to help make a proper selection of a site. More than just the nature of the wind is important in selecting a site. Obstacles such as hills, various kinds of vegetation, and buildings or other structures can greatly alter the wind profiles, and even create regions of strong turbulence, such that a simple average wind-velocity distribution will not give an adequate description of what might be encountered at the site.

The feasibility of developing a particular site depends not only on the average wind speed and current land use (as in issues of the kinds of vegetation or structures on the land), but also on more complicated factors such as daily and seasonal variations in wind speed. Wind never blows at a steady speed for 24 hours a day throughout the year. Wind is almost always variable. The POWER generated depends critically on the speed; as shown above, the electric power depends on the cube of wind velocity—doubling the wind speed increases power by a factor of eight. As a result, an accurate assessment of the pattern of winds at a prospective site is vital to success. Questions such as whether there is a windy season, and if so, when; the strength of the strongest gusts; and how long 'calms' might last are all important parts of this assessment.

To be sure, wind speed and direction generally follow daily and seasonal patterns that are remarkably predictable despite some occasional wide fluctuations. Had this not been true, the ability of sailors to travel and explore the oceans and to use them for trade routes would have been seriously hampered. Often a database of several years' worth of wind speed measurements will provide a fairly accurate assessment of the amount of energy that a wind turbine site can be expected to deliver at any given time of day or year.

Assuming that such data are available to allow making reasonable predictions of wind energy, then a more important, and sometimes more complicated, issue emerges: how well the wind electricity generation correlates with the timing of needs for electricity by the utilities. If a wind installation can provide electricity at times when demand is highest, so much the better; if wind generation is highest when demand is lowest, the system is 'out of synch' and may not be so worthwhile an investment.

ELECTRICITY FROM WIND: THE CURRENT SCENE

The commercial generation of electricity from wind nowadays relies on the wind farm concept. A **wind farm** is a location containing large numbers—from dozens to hundreds—of wind turbines, each with a generating capacity of 100–500 kW (Figure 35.3).

Wind farms are now in operation in many countries around the world. There is still a market for the smaller machines that are used individually to supply electricity for a single user, such as a ranch, rather than for a utility.

Though we have seen that the greatest wind potential in the United States is in the upper Great Plains, it is California that is, by far, the leading state in exploiting wind energy. Indeed the state of California is well ahead of many nations in the use of wind.

FIGURE 35.3 For large-scale electricity production, many wind turbines would be clustered together into wind farms.

Three factors all came together in the late 1970s or early 80s to give California its current dominance in wind energy. One was a report commissioned by the state government that identified three strong windy sites in California as possibly suitable for the establishment of a wind energy industry. A second was enactment, at both the state and federal levels, of tax credits for the development of energy resources not using fossil fuels. The third was federal legislation, the Public Utility Regulatory Policies Act, that required utilities to buy electricity generated by renewable energy sources at approximately the cost of their own generated power. These three factors together led to a rush to develop wind energy in California beginning in the early 1980s.

From 1981 to 1986, more than 15,000 wind turbines with a total peak capacity of 1,300 MW were installed in California, especially in the regions around the Tehachapi and Altamont passes. In 1985 alone about 400 MW of capacity were installed, equivalent to a medium-sized power plant. As things turned out, some of the early wind turbine designs were absolute disasters. In some cases, a blade would snap off, usually followed rather quickly by the turbine shedding its other blades. Nevertheless, the designers and engineers persevered, with dramatic improvements. As California's wind farms developed, the cost of wind-generated electricity dropped from 50 cents per kilowatt hour in 1980, an exorbitant price at that time, to about five to seven cents in the mid-1990s. That cost is quite competitive with electricity from most other sources.

California's wind farms now supply about 2% of the total electricity demand in the state. Though the percentage figure is small, it is applied against a very large total energy demand. California is one of the most populous, energy-intensive, and wealthy states. Energy demand is high.

By the mid-1990s the total capacity of wind turbines in the United States was 1,745 MW. On a nationwide scale, this was less than 2% of the total electricity-generating capacity from renewable energy sources, and less than 0.3% of the total electricity-generating capacity of the nation. Nevertheless, this wind-generated electricity is equivalent to the capacity of three to five medium-sized power plants. The United States produces about 40% of the world's wind electricity, with most coming from California. However, the rest of the world is catching up quickly. In 1985, we had 95% of the world's installed generating capacity, so in about 15 years that 95% figure has been cut by more than half.

The continuing modest demand for small units to be installed in remote or rural locations should not be neglected. About 6,000 such machines were put in use across the United States during the 1980s.

In Europe, the current leader in wind technology is Denmark. In fact, Denmark deserves the credit for taking the lead, worldwide, in revitalizing wind technology in the 1970s. In response to the two oil price shocks of that decade, Denmark made a conscious, national decision to switch to renewables, most notably biomass and wind. Currently, Denmark has nearly 4,000 wind turbines, which supply 3% (500 MW) of Danish electricity demand.❁ The national energy plan calls for reaching 10% by 2005.

❁ In reading any kind of statistics, and certainly ones on energy, it is always important to distinguish between percentages and absolute numbers. On a *percentage* basis, Denmark meets about twice as much of its electricity demand from wind as does California—roughly 3% vs. 1.5%. However, the *absolute* electricity generation from wind in California more than triples that of Denmark—about 1,600 MW vs. 500 MW, respectively. The principal reason, which in no way diminishes the tremendous Danish contributions to wind energy development, is that there are only about one-third as many people in Denmark as there are in the Los Angeles–Anaheim–Riverside–Oxnard metropolitan sprawl alone.

Europe as a whole pulled ahead of the United States in the mid-1990s. Total installed wind-turbine generating capacity is now about 2,300 MW in Europe versus about 1,770 MW for the United States.

Worldwide, the wind industry is steadily growing. Total annual sales passed the billion-dollar mark in the early 1990s and hit $1.5 billion by 1995. More than 20,000 wind turbines are now in operation around the world.

THE ADVANTAGES OF WIND AS AN ENERGY SOURCE FOR ELECTRICITY GENERATION

Wind is a very environmentally friendly energy resource. It is essentially pollution free. There is no air or water pollution, no acid precipitation nor smog. There is no pollution of groundwater and drinking water supplies. There are no SO_x emissions to attack and degrade plants. There is no radioactive waste, fly ash, or scrubber sludge. There are no carbon dioxide emissions, nor emissions of any other greenhouse gases, so there is no impact on global climate change. About the only imaginable effect on the environment is a slight reduction in the strength of the wind as it passes through a wind farm.

A wind farm poses no significant threat to the safety of the public. There will not be an 'Altamont' or a 'Tehachapi' disaster as we now speak of Three Mile Island or Chernobyl. Overall, it is hard to imagine an energy source more benign to the environment than wind technology.

Each individual wind turbine requires a relatively short time to build. In fact, they could be built on an assembly line. If that were done, it would be possible to produce enough wind turbines to generate a thousand megawatts in a year. In comparison, the construction time to build a thousand-megawatt fossil-fuel power plant would be several years and, for a nuclear plant, about 10 years. Furthermore, the wind turbines are modular. Each wind farm contains dozens, or perhaps hundreds. Thus once a site for a wind farm has been selected, new turbines can be added quickly and relatively inexpensively as electricity demand increases. In comparison, fossil or nuclear plants would usually be built in increments of several hundred megawatts at a time, at a cost well into the hundreds of millions of dollars.⌬

⌬ There is one significant, and increasingly important, exception to this. Natural gas can be used to fire a *gas* turbine connected to a generator in a design similar to that of an aviation gas turbine (jet) engine. Small, natural-gas-fired gas turbine plants can be built relatively quickly and at much lower cost than the several-hundred-megawatt behemoths. Nevertheless, even these plants probably do not offer the potential flexibility of expanding a wind farm a few turbines at a time.

Although the total area occupied by a wind farm may be large, the only actual use of the ground surface is the 'footprint' of each wind turbine. Over the area of an entire wind farm, the turbines themselves use no more than about 5% of the total land. The remaining land can be put to other uses, including farming (as J.S. Bach once pointed out, sheep may safely graze), ranching, forestry, or recreation. Because the turbines themselves occupy only a small fraction of the land area, wind energy development is ideally suited to farming areas. In Europe, farmers plant crops right up to the base of the turbine towers. In California, cows graze close to the turbine towers. The leasing of land for wind turbines provides benefits to landowners through extra income and increased land values.

Wind also has a profound advantage over solar, which is perceived to be one of the other environmentally benign renewable energy technologies (Chapter 36). Solar suffers the undeniable disadvantage that the sun doesn't shine at night. On the other hand, wind can blow day or night, and on sunny or cloudy days. In fact, winds are often at their strongest and most reliable during the coldest and darkest nights of winter, and this is precisely the time when most of us have our greatest demand for electricity.

POTENTIAL DISADVANTAGES OF WIND AS AN ENERGY SOURCE

The very first issue, perhaps so obvious that it would be easy to overlook, is that we want to put wind turbines only in areas that are windy. As we've seen, some of the regions of the United States that are ideal for wind energy development, such as the Dakotas, are far from major population and industrial centers. There could possibly be significant investment required for transmission lines to get the electricity from where it is generated to where it is needed.

But even when a suitably windy site is located, we must still recognize that, in most places, the wind does not blow all the time, nor does it usually blow steadily. This intermittent or episodic nature of the wind causes a variety of problems.

On a very short time scale, minutes or even seconds, there are gusts of wind that bring sudden, sharp changes in wind speed. On a time scale of hours, there are lulls and freshenings; an example that we all experience is the way the wind picks up, dies down, and picks up again as a sequence of rain storms pass through the area. On the time scale of a day, there are also variations such as a drop in wind speed at sunset. Still more variations occur on scales ranging from several days to months to years. A wind gust can put serious mechanical strains on an individual turbine, but the effects on a wind farm are usually mitigated by having numerous turbines spread across the area of the farm. Local weather variations can be compensated for by having several wind farms, located 50–100 miles apart.

To take into account long periods of calm, a wind energy system would need some sort of back-up generating capacity. First, there would be need for some back-up that could come on line in an hour or so. Second, to compensate for the entire wind farm being becalmed for a long period of time, back-up capacity that essentially duplicates the whole system would be needed. In addition, an energy storage system might be needed in those areas where there is a mismatch between the times of peak generation from the wind turbines and times of peak demand by the customers. (These issues might be addressed, in favorable cases, by having multiple wind farms located some distance apart. If numerous, scattered wind farms were linked into a single grid, a drop in wind speed in one area might be compensated by an increase in wind somewhere else.) The costs of these back-up and energy storage systems would have to be factored into the total cost of the wind installation.

Nevertheless, nuclear, fossil-fuel, and hydro plants can operate 24 hours per day, and can generate electricity on demand. Wind farms are necessarily dependent on wind; for some wind farms, the average output might be only one-fifth of the rated capacity. In this specific respect, these other energy sources have a significant advantage relative to wind.

Because of these concerns about the intermittent nature of wind, there seems to be a general rule of thumb that no more than about 20% of a nation's electricity supply should come from wind. If the percentage is much higher than that, there is a risk of occasional electricity supply shortages.

The issue of land use is seen, by different camps, of course, to be either an advantage of wind, or a significant disadvantage. Electricity production using wind turbines requires a great deal of space. Because of the turbulence created by the rotating blades, the turbines have to be placed between 300–1,000 feet apart. A reasonable-sized wind farm will have at least a hundred wind turbines. As we've seen, the turbines themselves occupy less than 5% of the area of the wind farm, so that the rest of the space can be used for such activities as farming or ranching. The concerns arise when the total area set aside for the wind farm must be weighed against other competing uses of the same land. Perhaps the most contentious problem that might arise is setting aside for a wind farm land that might otherwise have been used for housing developments. Concerns would also surely arise if roads for access to the wind turbines had to be cut through forests, and more so if the forest itself were to be cleared to make way for the wind farm.

Noise is another contentious issue relating to wind farms. There's no getting around the fact that wind turbines make noise. The 'silent turbine' has yet to be designed. As a result, it is likely that regulations might be established to limit the noise production from a wind farm, and to determine how close turbines may be sited to occupied buildings. Again, Denmark seems to be leading the way; the Danes have a regulation that the noise level at the house nearest to any turbine must be less than 45 decibels, which is about the level of a conversation.

The decibel is a unit of sound intensity. Like the pH scale we've met in Chapter 30, it is a logarithmic scale. The faintest detectable (by humans) sound is arbitrarily assigned a value of one. The number of decibels between any two sound intensity levels is taken to be 10 times the logarithm of the ratio of the two intensities. A faint whisper or very light rustling of leaves is about 10 decibels; ordinary conversation about 40; light to moderate traffic, 60; and an aircraft engine, about 120. The aircraft engine is 10^{12} times more intense than the faintest detectable sound (that is, 1,000,000,000,000 times!).

A second problem, besides loss of turbine blades, has surfaced at the Altamont Pass, California wind farm—bird kills. Birds can be killed by flying into the spinning blades of a turbine. At Altamont Pass, about 300 raptors are killed per year, including birds as large as golden eagles. No utility would want to face public relations nightmare of being accused of wiping out endangered birds. However, bird kills are certainly not unique to wind energy systems. Birds are killed or injured by flying into transmission and distribution lines coming from other kinds of electricity-generating plants too. Furthermore, not all wind farms have experienced the problem found at Altamont Pass. The wind farm at Altamont Pass is particularly likely to experience bird kills because it happens to lie on a bird migration route, and the land around the wind turbines has a large, active population of the small mammals that are favored prey of the raptors.

A final argument raised against wind farms is 'visual pollution.' Though these installations produce no tangible pollution in the form of emissions to the air or water, many people simply object to the sight of tens or hundreds of wind turbines, and their associated transmission lines, on the landscape. A wind farm of modest output might cover, say, three square miles of land and incorporate 50–60 wind turbines, each well over a hundred feet tall. That wind farm would produce perhaps 25 MW of power, whereas a new coal or nuclear plant would be at least 1,000 MW. To help assure that the turbines capture as much wind as possible, it is good to locate the wind farm on a hilltop. Thus a situation develops in which people for many miles around wind up looking at dozens and dozens of turbines. Unlike many other kinds of utility or industrial installations—even junkyards—a screen of tall, thick trees cannot be used to hide a wind farm from view, because it would defeat the whole purpose, by screening the wind, too.

> ⚛ There is much truth in the cliché that beauty lies in the eye of the beholder. Though cries of 'visual pollution!' have been raised against proposed wind farms because people find them unsightly or downright ugly, it is fair to say that others find some of the turbine designs visually attractive. This seems especially true for the vertical-axis 'eggbeater' design that is thought to be attractive in a futuristic sense, perhaps an emblem of a future of clean, renewable energy.

We've seen that some offshore winds can be especially powerful, and suggest the potential of locating wind farms on the ocean surface. These offshore wind farms could potentially negate many of the concerns relating to land use, noise, and visual pollution.

COMPARISON OF WIND ENERGY WITH OTHER SOURCES OF ELECTRICITY

Cost is always of concern for any energy system. At least two measures of cost must be considered. The first is the cost to build and install the generating system. The second is the actual cost of the electricity that is eventually delivered to the consumers.

Great strides were made in the 1980s in improving wind technology, strides which had a definite impact on the economics. At the beginning of the 1980s, the cost of a wind farm was about $3,000/kW of generating capacity. By the end of the decade, that figure had dropped to $1,000–1,200/kW. Some estimates now indicate that the cost is under $1,000/kW. To put these figures into perspective, a new coal-fired plant with the necessary emissions control equipment would cost about $1,300/kW, and a nuclear plant, assuming that all the necessary permits were granted and the lawsuits settled, would be about $2,000/kW.

However, on some wind farm sites, the electricity actually generated might be no more than about 25% of the maximum possible. A typical fossil fuel plant, in comparison, can produce up to 75% of its maximum possible annual output—in other words, two or three times as much electricity per unit of generating capacity as the wind plant. So, to replace a 500 MW coal-fired plant a utility would need to install over 1,000 MW of wind farm capacity. (One might argue that this issue is counterbalanced by the fact that the 'fuel'—the wind—for the wind farm is free, certainly not true of the coal plant.)

Just as the cost of building a wind farm dropped significantly, so too has the cost of the electricity. In the early 1980s the cost of electricity ranged up to 50 cents per kilowatt-hour; by 1990, the cost was down to five to eight cents. This cost is very competitive with electricity from other sources. The National Energy Policy Act of 1992 contained a 1.5 cents per kilowatt-hour production tax credit for wind-generated electricity, which further helped the cost of wind-generated electricity.

The characteristics of the local wind resource are an important factor in determining the value of the electricity produced. The value of electricity may be reduced if winds are at their peak when daily electricity demand is lowest, or may increase if wind generation is high at the same times that demand is high.

The major costs in the wind power industry are in capital equipment, since the turbines do not need 'fuel.' The wind-generated electricity will be considerably more expensive if the site chosen for the wind farm is located far from existing transmission lines, so that they too would have to be included in the cost of the facility. The maintenance costs on modern wind turbines are extremely low. If some way could be devised to account for the environmental and social damages caused by fossil fuel or nuclear plants, these additional costs would enhance the economic advantages of wind.

Availability represents the percentage of time the unit is actually operating (that is, not including scheduled maintenance shutdowns and unexpected outages). For example, if a plant operates for 7,000 hours in a year its availability would be—given 8,760 hours in a year—80%. Currently the availability is about 90% for wind, which is equivalent to fossil or nuclear plants. Optimistic projections are eventual availabilities of 95–96%. Achieving these availabilites will require more rugged designs, better materials of construction, and more careful attention to maintenance.

The **capacity factor** is the amount of electricity actually generated (e.g., over the course of a year) divided by the maximum generation at the peak capacity of the plant. As an example, if a plant had the maximum capacity of 500 million kWh per year and actually did generate 250 million kWh, its capacity factor would be 50%. The capacity factors for wind farms range from about 25% to 50%. In comparison, nuclear plants have capacity factors of 70% and coal, 75–80%.

On balance, wind is an essentially pollution-free method of generating electricity that appears now to be very cost-competitive with conventional sources. It seems reasonable, therefore, that

> WIND IS THE RENEWABLE ENERGY RESOURCE WITH FEWEST TECHNICAL, ECONOMIC, OR ENVIRONMENTAL HURDLES TO MAKING A SIGNIFICANT CONTRIBUTION TO OUR ELECTRICITY SUPPLY.

CITATION

1   From the August, 1849 issue of *Scientific American* magazine, reprinted in the August, 1999 issue (volume 281, number 2, p. 10).

FOR FURTHER READING

Berger, John J. *Charging Ahead*. University of California Press: Berkeley, 1997. Subtitled 'The business of renewable energy and what it means for America,' this book covers, in engaging style, most aspects of renewable energy. Part III consists of several chapters discussing wind energy. Connoisseurs of disasters will enjoy the stories of wind turbines disintegrating.

Borowitz, Sidney. *Farewell Fossil Fuels*. Plenum: New York, 1999. Chapter 12 of this book deals with wind energy. Despite the title, the book treats not only renewables, but also coal, petroleum, and nuclear energy.

Brower, Michael. *Cool Energy*. MIT Press: Cambridge, 1992. An easily readable book on renewable energy resources. Chapter 4 deals with wind energy.

DeBlieu, Jan. *Wind*. Houghton Mifflin: Boston, 1998. This is a broad-ranging book about wind itself, not specifically focused on energy. Chapter 9 does discuss wind farms, including personal visits and interviews by the author.

Powell, F.E. *Windmills and Wind Motors*. Lindsay: Bradley, IL, 1985. This inexpensive paperback is a reprint of a book originally published in 1910. It would be of use to those interested in the recent history of windmill technology, and to anyone wanting to build either a model windmill or an actual working device.

Ramage, Janet. *Energy: A Guidebook*. Oxford: Oxford, 1997. This is an excellent introductory book on energy technology, requiring only modest mathematics and science background. Most of the examples relate to energy use in Britain. Chapter 10 discusses wind energy.

Ristinen, Robert A.; Kraushaar, Jack J. *Energy and the Environment*. Wiley: New York, 1999. A relatively recent paperback book the cover of which is graced with wind turbines. The relevant discussion will be found in Chapter 5.

Sørensen, Bent. *Renewable Energy*. Academic Press: San Diego, 2000. This book is a very extensive compendium of information on most forms of renewable energy. It is mainly aimed at professionals in the field. Section 3.2 covers wind energy.

Walker, John F.; Jenkins, Nicholas. *Wind Energy Technology*. Wiley: Chichester, 1997. This handy paperback book is probably the best reasonably short introduction to the field. Most of the examples pertain to the British energy scene, but are useful for American readers as well.

ENERGY FROM THE SUN

If we were to draw up a list of criteria for the ideal energy source, what would we want? Probably our major considerations would be these: It should be environmentally friendly; that is, have little or no harm to the environment. Second, it should be inexhaustible, or least renewable, so we need not worry about issues of reserves, resources, and lifetimes. Perhaps we would also say that it should be domestic, so we need not worry about geopolitics and possible import restrictions.

Most of the world continues to rely heavily on energy from fossil fuels. This field has seen the development of technical solutions to SO_x, NO_x, and particulate emissions, so that even the fuel that potentially would be most difficult to burn cleanly—coal—can be used with relatively small impact on the environment. However, there is no practical solution for the CO_2 emissions, which would be a serious problem if the greenhouse effect and resulting global climate change (Chapter 32) are real. If the United States faces up to the challenge of meeting the CO_2 emission reductions of the 1997 Kyoto accords, a potentially significant reduction in fossil fuel use may be required. Nuclear energy solves the CO_2 emission problem for large-scale electricity generation—there aren't any CO_2 emissions. However, several issues related to nuclear energy concern society, including the possibility of reactor accidents, the continuing problem of waste disposal, and nuclear weapons proliferation. Hydroelectricity has many advantages (such as no 'fuel' costs and the fact that it is essentially pollution-free). However, while the actual electricity generation itself—water flowing through a turbine connected to a generator—is a very clean energy source, the impact on ecosystems of dams constructed to provide artificial sites for hydroelectricity is now of such concern that some dams are deliberately being dismantled. Whether a new, large dam for hydroelectricity generation would ever be built in the United States is questionable. It's not yet clear how much we can depend on wind energy, which certainly shares many of the advantages of hydro.

Fortunately, yet another energy source meets all three of the criteria we set out above: solar energy. Solar energy shares many of the advantages of hydropower: there is no cost for the 'fuel,' it is essentially inexhaustible, and there is virtually no pollution. In addition, solar energy has one additional advantage: in principle, it can be used in any area of abundant sunshine, whereas hydropower is restricted to sites where there are water sources (rivers, waterfalls, or dams). It can be used by any country or any region, eliminating many of the geopolitical issues concerned with unevenly distributed resources such as petroleum.

We will defer a discussion of where the sun gets its energy to the next chapter. For the moment we consider only that the sun's energy derives from the conversion of mass to energy according to Einstein's equation,

$$E = mc^2$$

(Chapter 26). The mass loss associated with the nuclear processes in the sun amounts to four million tons per *second*. Since c^2 is itself a huge number, the energy production per second in the sun is so large that we have no way of translating it into our normal experience. We must appreciate that the sun radiates this energy into all directions in space (Figure 36.1). Since the Earth is a tiny 'dot' 93 million miles away from the sun, virtually all of the energy produced by the sun misses the Earth. In fact, the fraction of the sun's energy hitting the Earth is only one two-billionth (1/2,000,000,000th!) of the sun's total output.

This fraction is so tiny it would not seem worth bothering with. And yet, almost all the energy that we use derives in some fashion from the sun. Growing plants require sunlight to provide the energy for photosynthesis, the conversion of carbon dioxide and water into the chemicals needed for the plant's life processes. Our use of food (Chapter 3), and of firewood or other plant products for energy (Chapter 6), depends directly on the sunlight-driven photosynthesis process. The fossil fuels derive from the remains of once-living organisms preserved in the Earth's crust (Chapter 20); the organisms preserved as fossil fuels once required sunlight and photosynthesis. The water driving waterwheels (Chapter 8) or hydroelectric turbines (Chapter 15) is supplied by the Earth's hydrological cycle, which is also driven by the energy in sunlight. Wind results from differences in the amount of heating of Earth's atmosphere by the sun (Chapter 9). Only the energy provided by nuclear fission reactors, and the comparatively tiny amount of energy derived from chemical processes in batteries do not derive, directly or indirectly, from the energy of the sun. Furthermore, if we could collect completely this tiny fraction of the sun's total

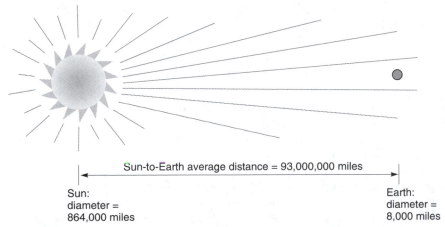

Sun-to-Earth average distance = 93,000,000 miles

Sun:
diameter =
864,000 miles

Earth:
diameter =
8,000 miles

FIGURE 36.1 From the perspective of the sun, the Earth is just a tiny dot 93 million miles away. Only a tiny fraction of the sun's energy arrives on Earth.

energy that falls on Earth, and convert it into useful ENERGY or WORK with 100% efficiency, we could provide energy equivalent to the current annual energy consumption of the whole world in only 40 minutes. Of course, we haven't quite learned how to do that yet.

ENERGY FROM THE SUN TO EARTH

In the course of an hour, the energy radiated by the sun into all directions of space is equivalent to the amount of heat that would be released by burning 400,000,000,000,000,000,000,000 tons of coal. Our little dot, our Earth, receives only 1/2,000,000,000th of this amount of radiation into the upper atmosphere. Not all of this energy actually makes its way to Earth's surface. One-third is reflected back directly into space, and the atmosphere absorbs another quarter (about 25%). Clouds, the ice caps, and the oceans cause the reflection of solar radiation back to space. The energy absorbed in the atmosphere helps drive the hydrological cycle and weather patterns. The remaining energy that actually falls on Earth's surface provides warmth, evaporates water into the hydrological cycle, and provides the energy for photosynthesis and plant growth, giving us crops and firewood.

Not all parts of Earth's surface receive the same amounts of solar radiation, nor does the solar radiation at one particular spot remain constant through the year. As we observe literally every day, solar radiation is at a maximum at some time during the daylight hours but is zero at night. (We will see later in this chapter that the lack of solar radiation at nighttime represents a significant complication in the use of solar energy.) Further, solar radiation is less in winter than in summer. This derives from two factors: In winter the sun appears lower in the sky than in the summer so that the POWER provided by the sun is less; and in winter the available daylight hours are fewer. A third complication comes from weather patterns: Even in a given season at a given location on Earth, the available solar radiation can vary from day to day because of changes in the prevailing cloud cover.

The effects of latitude and local weather patterns result in considerable differences in available solar radiation in a particular country. In the United States, for example, the average amount of sunlight received in a year varies by about a factor of two from New England and the Northeast to the desert Southwest states. In Great Britain the mean annual solar power received changes significantly through the course of the year, with a distribution month-by-month that peaks in June. Further south, the nations in northern Africa, such as Egypt and Sudan, receive about three times as much solar energy as Britain, with a month-by-month distribution that shows much less variation.

The promise and challenge of solar energy derive from two issues. One is that it is a clean, environmentally friendly form of energy that will last forever (at least in terms of the longevity of the human species) and that is inexhaustible. The other is that even after the tiny fraction of the sun's energy is received by Earth, and further filtered by reflection and absorption, the amount that does hit Earth's surface could potentially provide enormous quantities of energy throughout the world. Sunshine delivers energy as heat and light to Earth at a rate of some 15,000 times the entire energy consumption of the world. In the course of three weeks, Earth receives as much energy from sunlight as is stored in our entire reserves of coal, oil, and natural gas. The sunlight falling on the United States in the course of a year represents an amount of energy equivalent to 500 times the current annual energy consumption. On average, the solar energy falling received on an acre of ground in the 'lower 48' of the United States every day is equivalent to the energy of 10 barrels of oil.

The total amount of solar energy falling on the Earth is more than enough for present human needs and amply sufficient to provide any anticipated future demand. Similarly, for any of the other industrialized countries of the world, average annual radiation is several hundred times the total energy needs. For those developing nations, where per capita energy use (and standard of living) is still low, the development of solar energy could potentially provide an opportunity to improve the standard of living to that of the wealthier nations with relatively little environmental disruption.

Though solar energy offers considerable promise, there remain significant problems on the road to its large-scale commercial use. First, solar energy is not a concentrated form of energy, in contrast to, say, the energy in a flame or in current of electricity. Considerable space and ingenuity are required to capture large amounts of solar energy. We must inevitably cope with the fact that the sun does not shine at night. Neither solar energy nor its two direct products, heat and electricity, can be stored easily. Because of the effects of latitude, local weather conditions, and terrain, different parts of the world, and of individual countries, receive different amounts of sunlight. In some parts of the world, such as the northern portion of the United States, Canada, and much of Britain and Europe, solar radiation is strongest in the summer, when the energy demand is relatively low, and weakest in winter, when there are significant energy demands for heating and lighting. (To be sure, there are other places where the amount of incoming solar radiation matches better the energy demand—for example, in southern portion of the United States, electricity demand in summer exceeds that in winter because of the widespread use of air-conditioning and refrigeration.)

By 2025, the worldwide demand for fuel is likely to increase by about 30%, and that for electricity will more than triple. Even with more efficient production and use of our present energy sources, and with vigorous measures for energy conservation, it seems inevitable that new sources of energy will be required to meet the world's demand. Solar energy, if developed for large-scale use, could provide a significant portion, perhaps more than half, of the electricity demand and could displace some of the requirement for fuel.

Solar energy will have two major categories of application: providing space heat and hot water for buildings, and generating electricity. Each of these two major categories has within it two broad strategies. Approaches to building heat and hot water generation are divided into passive and active methods. So-called indirect and direct strategies are proposed for electricity generation. Each is discussed in the sections that follow.

PASSIVE SOLAR HEATING OF BUILDINGS

There are two types of solar energy systems used to heat and cool houses, passive and active. A **passive solar energy system** is one designed to use the entire building—walls, floors, windows, and roof—to collect, channel, and circulate heated or cooled air. In its simplest form, a passive solar system uses no mechanical devices or other moving parts (such as pumps or fans) to distribute heat. Passive solar takes advantage of the easily observed fact that all rooms with a window facing the sun are heated when sunlight shines in. In the United States, Canada, and northern Europe, this heating provided by solar energy can contribute some 10–20% of the annual heating requirements of the house. A passive solar structure is built and landscaped so that it becomes, in effect, one large solar collector. The passive solar approach operates by providing for direct admission of solar radiation, retaining heat by thermal insulation and heat storage in building components, and distributing heat input distributed by natural heat-transfer processes of conduction, convection, and radiation. A well-constructed passive solar building offers advantages of simplicity, reliability, durability, and economy. A quarter-million passive solar homes have been built in the United States.

In the classical Greek drama *Prometheus Bound*, Prometheus, the god of fire, tells how he first found the people of Earth, lacking 'the knowledge of houses turned to face toward the sun.' The playwright, Aeschylus, was not referring to his own people, the ancient Greeks, who had an appreciation for solar energy. The Greeks were apparently the first to make extensive use of passive solar design. Their houses were constructed of adobe walls some 18 inches thick on the northern side to keep out cold north winds. The southern side, in contrast, faced a portico supported by wooden pillars that led to an open courtyard. This design permitted sunlight to enter the main living areas of the home through the portico. The solar radiation was trapped in the earthen floor and thick adobe walls. Both the floor and walls provided a useful form of heat storage and allowed for a gentle release of solar warmth throughout the night. These buildings first appeared in Greece over 2,000 years ago. Their design spread to the nations of classical Islam, where they became common throughout the past millennium.

In addition to his writings on waterwheels (Chapter 8), the Roman architect Vitruvius also provided guidelines for orienting houses, public buildings, temples, and even whole cities to the sun. Many Roman remains in various parts of Europe and the Mediterranean countries demonstrate that the Romans used solar water heating. Many Roman baths obtained water from man-made channels open to the sun and lined with grooved black slate so that water absorbed heat from the sun-warmed slate as it ran through the grooves.

To provide the simplicity and reliability that are potential advantages of passive solar, a building requires both intelligent design and the right materials and components. The most important building concept is the orientation of the house to the south to take advantage of the sun during the winter. (Throughout this discussion we assume that the building is in the Northern Hemisphere.) Once the building is properly oriented, two things are necessary for a passive solar system. One is windows, to let the solar energy in, and which can help trap heat by preventing it from being reflected back to the atmosphere. The second is some so-called thermal mass that serves to collect the heat, store it, and pass it on to the air space in the building. Both of these key components of the passive solar system are part of the structure of the building, justifying the notion that a well-designed passive solar building is, in effect, one large solar collector. Some other relatively simple measures of building design and construction, such as providing for natural ventilation and sensible landscaping, with overhead shading for summer cooling, can reduce a typical building's annual heating and cooling costs by about 15–25% while at the same time adding little or nothing to its construction cost. (However, the cost of retrofitting an existing building for passive solar features, especially if it is not properly oriented toward the south, is usually prohibitive.)

On sunny days a south-facing window gains more energy from sunlight than it loses through infrared radiation from the surfaces in the room, and the larger the window, the warmer the room. Most of the windows in a passive solar building should face south, or south and east, and should be large to maximize the amount of solar radiation admitted to the building. The north wall should have only few, small windows, or none at all, to minimize any energy losses through the windows. Double- or even triple-paned windows can be used to help hold heat during the winter.

While large windows are desirable to maximize the amount of solar radiation admitted, when the sun is not shining, or at night, the windows also allow energy back out, and can actually cool the room. One solution to this problem uses windows that reflect radiation back into the room. Glass itself traps infrared better than it does visible light, but special coatings on the glass can enhance this effect. Thick, well-insulated drapes or blinds can also be closed to help reduce heat loss at night or on cloudy days.

As sunlight passes through the windows, it should strike surfaces that absorb the energy and can later radiate it back into the rooms as heat, causing a greenhouse effect. Thick walls and floors made of tile, stone, concrete, adobe, or brick serve to store heat. They can be a foot or more in thickness. Stone or brick fireplaces also provide this effect. After dark, the absorbed heat absorbed in these thick floors and walls is radiated back into the building space. To help further with heat retention, wall studs are two by six, two by eight, or even two by ten inches to provide space for more insulation than could be obtained with the customary two-by-four-inch studding.

An alternative to relying simply on direct solar heating is to use a specially designed interior wall for thermal storage. This type of structure was developed by Felix Trombe in the 1950s and is often called a **Trombe wall**. A large concrete wall with blackened outer surface is located immediately inside the south-facing double-glazed window. The incoming radiation heats the air in the narrow cavity between the wall and the window but is mostly absorbed by the wall. Heat absorbed by the wall causes the interior wall temperature to rise and to transfer heat energy into the house indirectly by conduction, convection, and radiation. The air between the glazing and the wall rises as it is warmed and is distributed throughout the building. If the wall is vented at top and bottom, the heated air can be circulated around the house either by natural convection currents or, if need be, by a fan. At night, heat losses by radiation back out through the window can be greatly reduced by use of screens or blinds across the window. Trombe walls have disadvantages of high construction costs and reducing the available space within the room.

As an example of successful passive solar design, 'Advanced House,' in Toronto, has a sun space that warms air by sunlight before it circulates through the house. The wall adjoining the sun space is brick, which has a high heat capacity. At night, the heat stored in this wall during the day partly warms the house. Orienting the sun space toward the south, to collect the maximum amount of sunlight, increases its effectiveness. The windows consist of triple panes coated with a material that does not affect the transmission of visible light but does block loss of heat. Walls and ceilings have heavy insulation to inhibit the loss of heat. Circulation of air within the house prevents heat from collecting near the ceilings (which would make the living space nearer the floor cold). Currently the construction cost for such homes is greater than that for conventional ones. But, the money saved in substantially reduced heating costs could pay for the additional construction investment within 10 years.

Without doubt, one of the biggest problems with any solar building design is heat storage. When the sun is shining, it provides more heat than a building can comfortably use. On the other hand, some method has to be used to store heat for use later, on cloudy days and at night. Put in other terms, the problem for solar space heating is that when you need the heat, you often haven't got the sun. Unfortunately, technology for long-term storage of heat—in support of any kind of energy system—does not yet exist. Many solar homes use a second, back-up energy system in combination with the solar design.

Heating is not the only possibility with a passive solar system. Natural air conditioning can also be obtained. With appropriate building design, the heat of the sun induces convection currents that draw cool air into the building and reduce the inside temperature. Islamic architecture has used this principle for centuries. Many Islamic buildings have a 'chimney' that draws up hot air and brings air into the building past the north-facing surfaces that remain cool throughout the day. A modern version of this concept uses a Trombe wall to create the air movement. The hot air is vented to the outside, while the incoming air is cooled by north-facing, heavy masonry surfaces.

ACTIVE COLLECTION OF SOLAR ENERGY FOR BUILDING HEAT AND HOT WATER

An **active solar energy system** does not rely simply on the design of the building itself, but rather employs solar collectors, usually mounted on the roof, that capture the energy of sunlight to heat either air or a liquid that is pumped to a heat storage unit. An active solar heating system, by its design, achieves three things: the collection or trapping of solar energy; transferring heat from the collector to a fluid that distributes heat where it can be used; and, usually, some means of storing the collected heat for use later when the sun is not shining. Most active solar systems require pumps or blowers, so cannot operate independently of other energy sources (usually electricity).

Scientists and inventors have always been intrigued with the possibility of finding ways to capture and concentrate the energy of sunlight. The first recorded application of solar energy is the story of Archimedes, who, in 213 B.C. allegedly used an array of mirrors to focus sunlight on the wooden ships of a Roman invasion fleet, setting them afire and forestalling, for a time, the Roman conquest of Syracuse. Likely the story is apocryphal, though it is certain that Archimedes did write a book called *On Burning Mirrors*. Many authors of modern times question whether the technology for constructing and focusing a large array of mirrors was available to Archimedes. However, some 2,000 years later the French naturalist George Buffon used an array of 168 mirrors, each six inches on a side, to ignite a woodpile nearly 200 feet away, suggesting that it just might have been possible for Archimedes to set fire to a wooden ship if indeed he had the ability to make and focus large arrays of mirrors.

Early chemists used specially shaped mirrors as a source of heat in the laboratory. A 16th-century process for making perfume involved placing flowers in a flask of water and using concentrated solar heat to extract the fragrant essences from the flowers into the water. At roughly the same time, Leonardo da Vinci proposed a large mirror (a *really* large mirror—some four miles across!) that would generate boiling water for a dyeing factory. Like many of Leonardo's other inventions, this too was many years ahead of its time, and it is unlikely that technology was available at the time to build such a device. Toward the end of the 18th century, Antoine Lavosier experimented with solar furnaces in a search for a powerful source of heat. With a 52-inch diameter lens, he was able to melt platinum, which means that he likely achieved temperatures of about 1,780°C (3,240°F).

The first solar water heaters were simple metal tanks painted black and tilted toward the sun. In 1891, Clarence Kemp patented the 'Climax': a combination of metal heating tanks to collect heat from sunlight with enclosing insulating boxes that helped retain the heat. The Climax became the first commercial solar water heater, sold in models from 32- to 700-gallon sizes. The competing 'Day and Night,' invented by William Bailey in 1909, offered as a selling point its ability, as its name implied, to supply hot water during both the day and night. This unit consisted of two parts, a solar heater and a water storage tank that was placed next to the kitchen stove to keep the water warm at night.

Nowadays the most common means of collecting the sun's rays for heating or producing hot water is with a **flat-plate solar collector**. The concept dates to the pioneering work done during the middle of the 18th century by the Swiss scientist Nicolas de Saussure. His design, which still forms the fundamental basis of the modern flat-plate collector, consisted of glass plates spaced above a blackened surface enclosed in an insulated box. This early design achieved temperatures as high as 300°F.

There are many different designs of flat-plate collectors, but all are similar in fundamentals (Figure 36.2). Any modern flat-plate collector has four main parts: an absorbing surface that faces outward to receive the sunlight, a system of pipes that carry the heated fluid, a sheet of glass on the front, and insulation on the sides and back.

The absorbing surface is a flat metal plate, painted black to absorb heat. The collector plate absorbs solar energy, becoming warm in the process. The absorbed heat in turn warms a circulating fluid. Good thermal contact between plate and fluid is important.

Pipes run into the box and are attached to the black plate so that heat travels by conduction through the metal and into the heat-transfer fluid. Although the heat-transfer fluid can in principle be either a liquid or gas, liquids are more efficient. In a conventional flat-plate collector used in homes, water is by far the best heat-transfer liquid because of its high heat capacity, low cost, and relative ease of use. In active solar systems used in climates where freezing temperatures can be encountered, antifreeze can be added to the water.

The glass cover functions like the glass in a greenhouse. It allows visible light from the sun to enter the collector, but traps the infrared heat emissions from the metal plate, raising the efficiency of the collector. In addition, the sheet of glass also reduces loss of heat by convection currents from the collector plate.

Heat losses can also occur by conduction through any material in contact with the collector. Because of this, good insulation around the sides and back is important.

Normally, several flat-plate collectors will be connected in a series. The heated water is pumped either to a storage tank or to a traditional hot-water heating system using radiators or baseboard units.

A flat-plate collector's efficiency depends on a number of factors, foremost the amount of solar energy hitting it, which in turn is affected by the collector's orientation

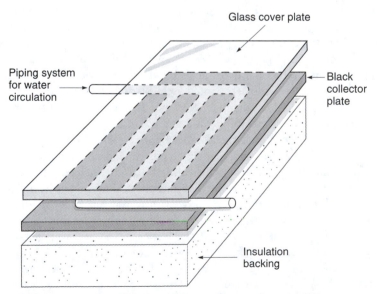

FIGURE 36.2 The flat-plate solar collector is a simple, easy-to-build device for heating water.

and tilt. Collectors should be mounted on a roof with a southern exposure and, if possible, at an optimum angle determined by the latitude of the building. A collector facing directly into the sun will receive the most insolation. Collectors can be designed that are motorized and able to tilt, so that during the course of the day a control system can position the collector so that it is always facing into the sun. However, such a movable collector system would be expensive, due to its complexity, so is not a good candidate for use in a simple home system.

Incoming solar energy is also affected by the prevailing weather—whether the day is clear and sunny, hazy, or with heavy cloud cover—and by the season of the year, which determines the height of the sun above the horizon. The sun's height affects the strength of the solar radiation when it reaches the collector surface.

Apart from the weather, the energy collected will depend on a number of factors. The design and materials of construction determine the fraction of incoming solar energy that will be transmitted by the glass, and the fraction of that transmitted energy that will be absorbed by the collector plate.

The solar collector must be plumbed into the building's water supply or heating system. In northern climates, where the water in the collector contains antifreeze, it must be kept separate from the domestic water supply, because antifreeze is poisonous. As the morning sun begins to warm the collector, the fluid will eventually reach a point at which it is hotter than the water in a 'preheat' water tank. Then the circulating pump is switched on. The hot fluid from the collector is sent through a heat exchanger to heat the water in the preheat tank. This heated water can be used directly as a source of heat, or stored in a hot water tank to provide a source of hot water.

In most applications, heat from the solar collector is usually transferred to an insulated storage medium such as a water tank and distributed through the building as needed. A well-insulated 1,500-gallon tank can retain its heat for four to six days. If a flat-plate collector system is used to heat a large building, problems may arise because the collector may not generate a high enough temperature throughout the entire building, due to heat losses from the hot water as it circulates throughout the building. If higher temperatures are required than can be achieved by flat-plate collectors, the energy in the sunlight could be concentrated by mirrors or lenses, but, like the mechanized, tilting collectors, these designs are likely to be prohibitively expensive for simple domestic use.

In most homes, the hot water heater is one of the most costly appliances, in terms of the electricity or fuel (usually natural gas) needed to keep it operating. The energy (and hence cost) savings that might be achieved with a solar collector can vary considerably, depending on, for example, the local climate and how careful or wasteful the household is in its energy use. Rough estimates for a home in the temperate regions of the Northern Hemisphere occupied by a family that is careful in its use of energy are that a solar collector might achieve a 40% reduction in energy costs needed for hot water. In this situation, a typical domestic requirement might be one square yard of collector area, plus about 10 gallons of water storage for each person. A supplementary storage tank would need to be of at least 30–50 gallons capacity.

Active solar energy is a technology in widespread and growing use around the world. Israel probably leads the world in the practical application of solar energy. About a million solar hot water heaters have been sold and installed in the past 50 years, providing the energy for generating more than 80% of the hot water used in Israel. At least two-thirds of homes obtain hot water from solar collectors. Israeli law now requires all new residential buildings smaller than 10 stories to use solar water heating. Australia has also made a serious commitment to solar energy. In some parts of Australia, solar hot water heaters are required by law, as are solar-heated distillation units for producing drinkable water from salt water. In some places in Australia, the market penetration for solar equipment now exceeds 30%. Japan has at least three million small domestic solar water heating installations, with roof-mounted storage tanks and inclined flat-plate collectors. One-sixth of all houses in Greece are estimated to use solar systems. Further south, in Cyprus, more than 90% of homes now feature with solar water heaters. Even in Great Britain, a nation stereotyped for its miserable weather, there are an estimated 40,000 solar collectors, with annual sales of solar hot water systems now exceeding a thousand units. In the 1980s, solar water heaters became mandatory for new houses in parts of California.

Active solar energy systems share some of the same problems or limitations we have already discussed for passive systems. The most obvious is that solar energy varies in at least three ways: day-to-day variation, or even variation during the course of a day, because of changing weather; a seasonal variation from changes in the sun's position in the sky; and the inevitable problem that solar energy is available only during daylight hours. Because of this variable and intermittent nature of solar energy, it is generally prudent to have a back-up energy source, as well as provisions for storing some of the solar energy. As we've seen, the successful long-term storage of heat is a scientific and engineering problem that has not yet been solved. Fortunately, for many domestic hot water systems or building heat systems, it is usually only necessary for the heat energy to be stored for a few hours.

Unfortunately, commercial solar water heating systems are expensive to install in many countries, particularly when they are added to an existing building as a retrofit, and not incorporated into the design of a new building under construction. Costs vary, depending on, for example, the complexity of the design (i.e., whether it is to be used only to provide hot water, or for hot water and heat), materials used, and local labor costs. An installation for a typical home would likely cost several thousand dollars at least. Thus payback time—the time needed to save an amount of money using present sources of energy equal to the installation costs plus the operating costs—becomes of concern. Again, many factors contribute to the overall payback time, including the local climate, and the local cost of the energy source (natural gas or electricity) being displaced by solar energy. Further, it is common that the entire installation cost of a new active solar energy system would have to be paid 'up front.' For most of us, doing so requires borrowing money at commercial interest rates. This cost is also a factor in determining payback time. As a rule of thumb, a payback time of less than 10 years is desirable; anything longer indicates that the installation of the solar energy system is prohibitively expensive. Obviously regions with sunny climates, such as many of the southern parts of the United States or the Mediterranean countries, are more favorable for installation of such systems.

INDIRECT CONVERSION OF SOLAR ENERGY TO ELECTRICITY

The principal alternative to using solar energy for building heat or hot water is its use as the primary energy source for generating electricity. Solar energy provides a source of heat. This heat could, in principle, be used to raise a head of steam for a steam turbine/generator set. In this approach, the sun's energy is converted, via steam, into electricity. Since the energy in sunlight is not converted directly into electricity, this approach is sometimes referred to as **indirect conversion**. Since the solar energy is used to produce heat, such plants are also sometimes called solar thermal power stations. All of the principles are the same as those for a fossil fuel or nuclear plant. A source of heat is used to produce the steam that drives a turbine. The only significant difference is the source of heat: the combustion of a fuel, the fissioning of uranium, or capturing the heat in sunlight.

To provide steam to a turbine requires significantly higher temperatures in the fluid than can be achieved using the simple flat-plate solar collector. To realize these high temperatures—certainly above 100°C and preferably well above that—from the heat in solar energy requires the use of devices that not only collect the solar energy but also concentrate it. As we will see, these devices are, therefore, more complex and sophisticated than those used only for building heat or hot water generation. Further, the maximum amount of energy can be collected if some provision exists to allow the collectors to track the position of the sun in the sky. A tracking system also adds to the complexity of the design.

One of the principal strategies for indirect solar uses a type of collector known as a linear focusing collector. This kind of collector is more commonly known as the parabolic trough (Figure 36.3). It relies on a fundamental principle of the parabola: Regardless of what point on the surface of the parabola is struck by a ray of the sun, the ray is *always* reflected to one particular spot, the 'focus' of the parabola (Figure 36.4). As the name implies, the parabolic trough is constructed from a reflecting material having a parabolic shape. A pipe, usually painted black to help absorb heat, runs the length of the trough at the exact focus of the parabolic shape, thereby collecting all of the sun's rays that strike anywhere in the interior of the trough. The pipe is filled with a heat-transfer fluid, usually water or an oil. The heated fluid, which could be at temperatures up to 500°C, is pumped along the pipe. The more sophisticated designs feature sensors that determine the position of the sun and relay this information to motors that keep the collectors steadily tracking the sun.

The collection area of a single parabolic trough unit may be no more than a few square yards. To generate sufficient high-temperature steam to operate an electricity-generating plant requires hundreds or even thousands of the individual trough modules. Because each module has its own heat collection pipe, this system is sometimes referred to as a distributed collector system.

FIGURE 36.3 The parabolic trough solar collector provides much higher temperatures and can be used, e.g., to produce a head of steam to operate a turbine for electricity generation.

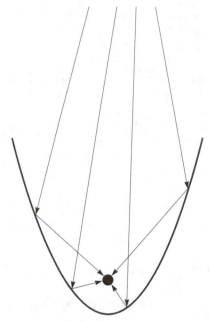

FIGURE 36.4 The principle on which the parabolic trough collector functions is that a ray of sunlight striking *anywhere* inside the parabola will be reflected to the same spot.

A system of distributed parabolic trough collectors lends itself easily to a wide variety of applications that require relatively low amounts of energy or power. This system can be used to pump water for irrigation, to produce steam for industrial purposes, to generate electricity in small- and medium-sized installations, and to supply heat for residential use. However, the large number of individual modules that might be needed limits the distributed collector approach mainly to smaller installations. On the other hand, parabolic trough modules are well suited to 'low-tech' assembly-line construction. They can potentially be produced in a factory and shipped to the installation site, where they could be connected together, as many as needed to meet the energy requirement. This approach may offer much simpler construction, and possibly lower construction costs, than a custom-designed electricity-generating station erected on site.

Nine indirect solar installations in the Mojave Desert of California can generate a total of 354 MW. The largest consists of about 900 parabolic mirror troughs, each about a city block long. The maximum temperature achieved by the focused sunlight is 390°C. A small natural-gas-fired system is used to further raise the temperatures and to supplement the solar energy system in cloudy weather. In 1984, an Israeli company, Luz International Ltd., installed the first such plant, in Harper Valley, California, based on a smaller, experimental plant operating in Israel. The cost of the initial installation was about $6,000 per kilowatt of generating capacity; the cost of the electricity produced was 26.5 cents per kilowatt-hour. Both costs are enormous compared to a standard generating plant. The only way to make a profit with such costs is with the aid of government subsidies intended to stimulate the use of solar and other renewable energy resources. Each of the eight subsequent Luz-type plants was more efficient than its predecessor. The last cost about $3,000 per kilowatt to install and produced electricity costing about nine cents per kilowatt-hour. Though the construction cost remains quite high compared to conventional fossil or nuclear plants, the electricity cost approaches (albeit from the high side) that of more conventionally generated electricity.

The Luz system has some difficulties besides the financial ones. Any such plant requires large areas of land with abundant sunshine. Neither requirement is always available in sufficient quantity in highly industrialized countries. The cost of land in developed, populous nations can be very high. Many parts of the United States, Canada, Britain, and Europe have a significant proportion of cloudy days throughout the year. And, as is true of every approach to using solar energy, we must always be content with the fact that the sun does not shine at night. Some system for energy storage has to be provided, or some sort of back-up energy source must be available.

The alternative to the distributed collector system uses a system of mirrors to reflect solar energy onto a single, central collector that provides enough heat to raise steam. The steam drives a steam turbine just like any other electric power plant. A conceptual sketch is given in Figure 36.5.

This system, sometimes known as the solar power tower, uses a heat-exchange fluid, usually water, held in a receiver composed of a series of blackened pipes, or in a tank, located on the top of a tall tower. The fluid is heated by solar energy reflected from thousands of individual mirrors, called heliostats (Figure 36.6). The heat-exchange fluid in the central receiver absorbs the heat from the focused solar radiation. A 100 MW solar electric facility would require about 24,000 heliostats spread out over an area of 1.8 square miles if it were located in sunny region such as the desert in central California. This central receiver approach is considered better suited for large-scale applications than is the distributed collector strategy.

Because the collector receives the reflected radiation from a large array of mirrors, each unit of surface area on the collector receives many times the power it would receive if it simply faced the sun directly. This provides a substantial concentration of energy and power relative to that obtained by spreading out a larger number of individual parabolic trough collector modules. This concentration of the total energy at the collector means that it operates at much higher temperatures, some 400–500°C. The result is a solar-to-electric efficiency of nearly 40%. This efficiency exceeds all but the most modern of fossil fuel plants.

The first large central receiver system was built in Odeillo, in southern France. Another was added at nearby Tavgasonne. They were 1 MW and 2 MW plants, respectively. These plants are very small compared with fossil or nuclear installations, where the generating capacity is in the hundreds or even thousands of megawatts. The plant at Odeillo uses 63 tracking heliostats covering an area of 30,000 square feet (about three-quarters of an acre).

In the United States, the Solar One plant was built at Barstow in the Mojave Desert. This location averages 300 cloudless days per year. Solar One used 1,818 heliostats to reflect the solar radiation onto the boiler at the top of the 'power tower.' The combined reflecting area of these mirrors was about 18 acres. In the tower, a heat transfer fluid was

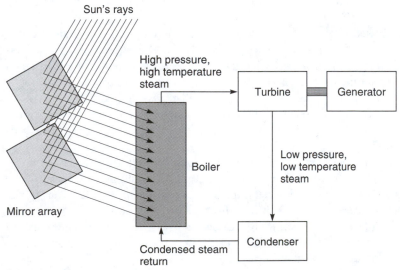

FIGURE 36.5 Mirrors (heliostats) can be used to reflect the sun's rays onto a tank of water, heating it enough to generate steam for a turbine.

FIGURE 36.6 An entire field of heliostats is required to capture enough solar energy to operate a small electricity-generating plant.

heated to 500°C and carried the heat to the steam generators of the power plant. The peak output of Solar One was 10 MW, again very small in comparison with standard fossil or nuclear technology. Solar One cost $142 million to build; the heliostat field accounted for almost half of the total cost. Solar One was closed in 1988 after operating successfully for six years. One disadvantage of the plant, aside from economics, was that steam pressure dropped quickly when clouds blocked the sun. The loss of steam pressure would force the plant to shut down, and it took some time to restart once sunlight returned.

In Tennant Creek, Australia, over 20 dish-shaped mirrors, each with an area of 4,300 square feet, superheat steam to 1500°C. This steam is then fed to the usual turbine for operating an electricity generator.

Solar power towers can only be feasible when located in regions of abundant sunny days, such as deserts. Further, the experience gained at Solar One showed that the principal drawback was interruption of electricity generation on a cloudy day, or even when clouds pass by on a sunny day. The start-up time after an interruption was considerable. Although the concept has been shown to be a technical success, it seems that the cost of electricity produced from such plants is much too high to be competitive with more conventional sources, and has to fall before the solar power tower concept can be economical.

One interesting proposed solution to these problems is to place a solar power tower next to an existing fossil fuel plant. The conventional plants would use solar heat where possible to operate the normal steam generators, but on days when too little solar power was available, the fossil fuel could be used as the back-up. An alternative concept would be give up on the idea of using large, centrally located electricity-generating plants. Instead one might try to use the solar in small, widely distributed, plants that might have greater chance of being economical immediately.

Despite these potential problems, solar thermal power plants using steam turbines currently seem most likely to be the first large-scale solar electricity producers. Compared to conventional fossil or nuclear plants, they offer several advantages. The solar energy itself is free. The only potential pollution problem comes from disposal of the waste heat, though it must be said that solar energy is no different from other sources of energy in this respect. The most troublesome feature of solar electricity plants is the fluctuation in available solar energy resulting from clouds and nighttime. Therefore, as with any sort of solar system, an adequate means of energy storage is needed.

PHOTOVOLTAICS—THE DIRECT CONVERSION OF SOLAR ENERGY TO ELECTRICITY

One potential important application of the direct use of solar energy is to replace the use of firewood with a cookstove that uses the direct rays of the sun. This is not a significant problem in the United States, but it is an important one in other parts of the world. There are hundreds of millions of people in the developing world where cooking is their most important energy use. The widespread deforestation in those countries has caused great hardship and ecological damage. Many of these countries have lots of sunshine and are good candidates for solar cookstoves.

So far, every method of generating electricity we have discussed is based on using the kinetic energy of a fluid—water in hydro plants and steam in all others—to drive a turbine. The approach derives from Michael Faraday's original experiment showing that a moving magnetic field generates an electric current, followed by the successful engineering development by Zenobè Gramme and his contemporaries. Fossil, nuclear, and indirect solar plants all use large and technologically complicated devices basically to boil water. But why not just throw all that stuff—boiler, turbine, generator, condenser, cooling tower—away and convert the energy of sunlight directly into electricity? This approach is called **direct solar electricity generation**, or **photovoltaics**.

In general, the photovoltaic effect occurs when light strikes two dissimilar materials and produces electrical potential energy. As we will see below, certain combinations of materials respond to the energy in visible sunlight. In combination, they make photovoltaic cell, or solar cell. Such cells can be thought of as being 'solar batteries.' Unlike primary or secondary cells that rely on chemical reactions to produce the electrical potential, a photovoltaic cell will not 'go dead,' so long as sunlight continues to strike it.

The conversion of light directly to electricity, by the photovoltaic effect, was discovered by the Henri Becquerel's father, Edmond, in 1839. An explanation of the physical basis of the photoelectric effect— how the wavelengths of light determine their ability to knock electrons out of the atoms—was developed by Albert Einstein. This research, not the theory of relativity, garnered Einstein his Nobel Prize in physics. Development of the first practical photovoltaic cell occurred in 1954, at Bell Laboratories.

The key material in most common photovoltaic cells is silicon. Silicon has four electrons in its outermost 'shell' of electrons. In a solid piece of silicon, the atoms are arranged in a regular, three-dimensional array, called a lattice. Each silicon atom is surrounded by four neighboring silicon atoms. The chemical bond between any pair of silicon atoms is formed by each of the atoms contributing one of its electrons to form a pair (Figure 36.7). The electrons in silicon are tightly held in these electron-pair bonds. For any substance to conduct an electric current, some of its electrons must be loosely held, so as to allow electrons to move when the material is subjected to an electrical potential. Since electrons are tightly held in silicon, under normal circumstances silicon is not a good conductor of electricity.

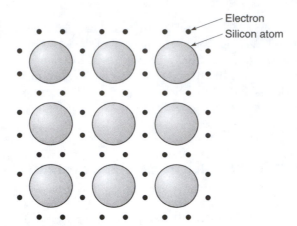

FIGURE 36.7 In pure silicon, electrons are strongly held in chemical bonds between pairs of silicon atoms.

The electrical conductivity, that is, the current-carrying facility, of any material depends at the atomic scale, on the number of electrons that can move freely within the material under the influence of an electrical potential. For any material that acts as an insulator (or, from the other perspective, a nonconductor) electrons are tightly bound to the constituent atoms of the material and therefore cannot move. In contrast, materials that are conductors have large numbers of electrons that are free to move under the influence of relatively small potential differences. Stephen Gray developed the broad classifications of materials as insulators or conductors more than 200 years ago (Chapter 12). We have to recognize a third category, materials that have relatively few electrons capable of moving under the influence of an electrical potential, and which, therefore, sustain only small currents for a given potential difference. Such materials conduct electricity better than does an insulator, but yet not so good as a true conductor. They are called **semiconductors**.

Suppose that we were to take a piece of silicon and replace some of the silicon atoms with, for example, arsenic. Adding a small quantity—equivalent to about one atom in a million—of a second element to pure silicon is called **doping**. Arsenic has five electrons in its outer shell. If relatively few of the silicon atoms are replaced, the material retains the crystalline nature of silicon, but now has some 'extra' electrons, thanks to the inclusion of some atoms of arsenic (Figure 36.8). If sufficient energy is provided, these 'extra' electrons are able to move under the influence of a potential difference. In other words, we have created a semiconductor by doping silicon with arsenic. Because this specific doping process has provided the material with 'extra' electrons, and the electrons have a negative charge, we call this material an **n-type semiconductor**, where the 'n' reminds us that the semiconducting properties are due to these extra electrons.

Instead, we could take another piece of silicon, and replace some of the atoms with, for example, gallium (which has only three electrons in its outer shell). Again we would instead have a material that retains the crystalline nature of silicon, but which, in this case, has some 'missing' electrons, because each atom of gallium contributes three, rather than four, electrons. An electron being 'missing' is equivalent to there being a 'hole' among the electrons in the material (Figure 36.9).

Since the electron has a negative charge, a 'hole' conceptually would have a positive charge. This material, gallium-doped silicon, is also a semiconductor. In this case, the

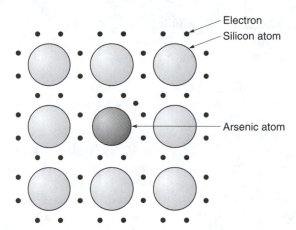

FIGURE 36.8 When silicon is doped with an element like arsenic, 'extra' electrons are contributed to the structure, which are not tightly held in localized chemical bonds.

shortage of electrons means that it is the 'holes', not the 'extra' electrons that migrate when exposed to an electrical potential. Since a 'hole' can be considered equivalent to a positive charge we call a material such as gallium-doped silicon a **p-type semiconductor** ('p' for positive, since if an electron is negative, a 'hole' would be positive).

Silicon is the second most common element in the Earth's crust. We need never worry about running out of silicon. The relatively high cost of solar cells, a topic we will revisit below, is certainly not due to scarcity of the necessary raw material, and reserves tend to be listed as unlimited. Rather, the cost comes from the requirement for purity, and in the difficulty of fabrication. The first challenge is to produce *extremely* pure silicon. Fewer than one impurity atom in a million can be allowed. This requirement demands rigorous purification and vigilant steps to ensure that the silicon stays pure. The second

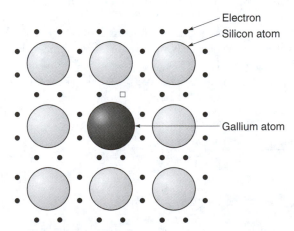

FIGURE 36.9 When silicon is doped with an element like gallium, there are not enough electrons to provide two-electron bonds between pairs of atoms, creating electron 'holes' in the structure (indicated by the empty 'box').

factor comes from the difficulties in melting and subsequently recrystallizing the extremely pure silicon very slowly and carefully to produce an almost perfect crystalline array. This requirement for very high-quality single silicon crystals poses great difficulties in the manufacturing cost and drives up the price.

Suppose now that we put the two different types of semiconductors together, forming an **n–p junction**. Some of the 'extra' electrons in the n-type semiconductor could cross the junction boundary, because they would be attracted to the positive holes. Similarly, some of the holes might migrate across the boundary to the n-type material. We might expect that if we connected an n-type and a p-type semiconductor, the 'extra' electrons ought to flow from the n-type into the 'holes' in the p-type. The separation of the charges creates a difference in electrical potential energy (voltage) across the junction. However, electrons do not flow easily in a semiconductor (as implied by its name) as compared to a regular electrical conductor like copper. Instead, to get electrons to flow in a semiconductor, it is necessary to supply some external energy to provide a 'kick' to get them moving. The energy needed to 'kick' electrons across an n-type semi-conductor into a p-type turns out to be about the amount of energy available in sunlight (Figure 36.10). Thus by joining an n-type and a p-type semiconductor, connecting it into an electrical circuit, and allowing sunlight to fall on it, we can achieve direct conversion of solar energy (sunlight) to electricity: The electrons liberated or kicked loose by the energy in sunlight will move under the influence of this potential difference to create a current. This current can be brought outside the cell by connecting the n-type semiconductor to the p-type with a wire (i.e., a good conductor) and be used to do some WORK.

A photovoltaic cell typically includes sheets of n- and p-type silicon in close contact. As we've seen, the n-type semiconductor is rich in electrons and the p-type is rich in holes. The tendency for electrons to diffuse from the n-region into the p-region, and for the holes to move in the other direction generates a voltage or potential difference at the junction. This voltage difference accelerates the electrons released when sunlight strikes the cell. If the two layers are connected by a wire or other conductor, electrons will flow through the external circuit from the n-semiconductor where their concentration is higher to the p-semiconductor where it is lower. The result is a direct current of electricity. As long as the cell is exposed to light, the current will continue to flow, with the only energy source being the solar radiation. A typical photovoltaic unit, or module, consists of 20–40 of these two-layered cells.

The present generation of photovoltaic cells only converts one-sixth of the incoming solar energy falling on the cells into electricity. Solar cells now under development can convert sunlight to electricity with an efficiency of 30–35%.

The power output of a solar cell varies with the intensity of light. If the light intensity is cut in half, then the current will also drop by half. The voltage will decrease

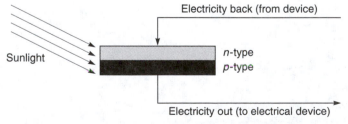

FIGURE 36.10 An n–p junction semiconductor provides the key for direct production of electricity from the energy in sunlight—the solar cell

by only a few percent. In bright sunlight, a cell 10 cm square will give an output of about 0.5 V and 3 A, about 1.5 W of power. Voltage also depends on the temperature of the cell, and increases slightly (by about 0.5% per degree) above 25°C. With current technology, it would take an area of six square miles covered with photovoltaic cells to replace a 2,000 MW power station. The first significant use of solar cells was to provide electricity for NASA spacecraft. There, cost was irrelevant. The intensity of solar radiation in outer space is so high that the low efficiency of the cells was not a major limitation of their use.

Modules of photovoltaic cells are connected to form panels that are assembled into groups called **arrays**. The actual current and voltage produced by the array depends on how the modules and panels are wired together. The array is connected to a secondary-cell battery that has two functions. First, the battery stores electricity, so that electricity will be available when the sun is not shining. Second, the battery and its control system are set up to send electricity into the external circuit at an even rate. This is vital, because the current from the array could fluctuate from one moment to the next as clouds pass overhead. With present technology, an array with an area of about one square yard could provide some tens of watts of power, enough for several small electric lights, or for operating portable radios. With about two square yards of cells, sufficient electrical energy is produced to drive a small electric motor, as, for example, in a water pump.

A typical photovoltaic installation for a home would require three main components: first, of course, the array of solar cell modules; second, a set of batteries to provide both a steady output of electricity and a means of energy storage; and, third, since most common household devices use alternating, rather than direct current, a device to transform the direct current output of the batteries into alternating current. Assuming a power output for a single cell of about a half-kilowatt per square foot, the array of modules would need to cover 600–900 square feet to generate enough electricity for a single family home. Assuming that the array was mounted on the roof, an area some 20 × 30 feet to 20 × 45 feet would be needed. Normally the array would be mounted to allow for air circulate underneath. Such an installation also has to take into consideration the amount of sunlight available in the specific geographic location, the availability of electricity from a utility (as, for example, for a back-up), and a realistic estimate of the amount of electricity required by the family.

Thirty houses in Gardner, Massachusetts have been supplied with arrays of solar cells. Each solar panel covers about one-fourth of the roof area and supplies 2.2 kW, enough to provide energy for heating, lighting, and the many other electrical appliances. The initial cost was $10,000 per house, but their lifetime is estimated at 30 years, with little or no maintenance in the meantime. Based on estimated annual natural gas and electricity costs for a typical house in that area, about $1,500, the payback time would be about seven years just for the cost of the solar panels. Allowing for some interest costs on the initial installation price, the payback would occur in about 10 years.

The alternative to building-by-building installations of solar panels is to set up centralized photovoltaic generating stations. A 350 kW solar voltaic system is in operation supplying two villages outside Riyadh, Saudi Arabia. In the United States, the largest photovoltaic installation, the Carrisa Plains station of Areo Solar, near San Luis Obispo, California, generates about 7 MW at peak capacity. The annual energy production is about 13 million kilowatt-hours. This would be an energy supply sufficient for a town with a population of about 10,000.

A big plant, that could supply energy for a city of a quarter-million people, would need to generate 200 MW of power. This plant would have to cover approximately a square mile of land. At currently attainable levels of operating efficiency, all the electricity needs of the United States could be supplied by a photovoltaic station having an area of a little over 7,000 square miles. This is roughly equivalent to paving over the entire state of New Jersey or Massachusetts, or Rhode Island, Connecticut and Delaware combined, or about one-tenth the land area of South Dakota or Nebraska.

Though steady progress continues in both single-building applications, as well as small-scale central generating plants, the major developments in photovoltaics have mainly focused on smaller applications in remarkably large markets. The first major consumer application was in photography, where solar cells were used as the exposure meters in cameras. When first introduced to the market, they were sometimes called 'electric eye' cameras. Solar calculators, wristwatches, and clocks are now familiar to most people. These solar cells may be small in size individually, but collectively they account for more than a quarter of the worldwide annual output of photovoltaics. Sales of cells and modules for such consumer goods are increasing steadily. At least a hundred million of these devices have already been sold worldwide.

The largest market for photovoltaic cells is currently in providing on-site generation of electricity to operate equipment or appliances in communities far from utility grids. These applications include remote communications systems, such as telephone or microwave relay stations; water pumps for drinking water, livestock, or irrigation; navigational aids such as warning buoys or remote lighthouses; and roadside call boxes. Before photovoltaics were available, these devices needed to use either secondary batteries, which needed to be recharged, or small generators operated by diesel engines (the engines needing occasional refueling).

In many parts of the world, solar cell arrays provide energy for street lighting. During daylight hours, the photovoltaic electricity charges a battery. After dark the battery operates the lighting until dawn. For example, highway traffic lights in rural Alaska, far from power lines, are supplied in this fashion.

Providing electrical energy for villages in less-developed countries represents another very beneficial application of photovoltaics. More than two million villages worldwide lack electricity for their water supply, refrigeration, lighting, or other basic needs of daily life. In many cases, the cost of extending utility grids to these remote locations would be prohibitive. In the most impoverished of regions, the utility grids may not even exist in the first place. Such on-site photovoltaic systems could meet much of the need for electricity for basic human activities at lower cost than trying to expand utility grids or build new, centralized generating facilities. This application of photovoltaics represents a market with very large future potential and chances for growth. Furthermore, the availability of photovoltaic refrigerators to keep vital medicines or vaccines intact could literally save lives in remote, impoverished regions.

Photovoltaics have also been tried as energy sources for transportation. In 1980, the *Solar Challenger*, a specially built aircraft weighing 185 lb, carried a pilot safely using only solar energy. The longest flight was 18 miles. Solar cars continue to be developed, with sufficient interest to hold long-distance solar car races in several countries.

Both direct and indirect solar have many advantages: no cost for the primary energy (as opposed to purchasing coal or uranium), an inexhaustible resource, no pollution or emissions from the generation process, minimal maintenance, and no by-product harmful to humans or hazardous waste disposal. In fact, direct solar has no moving parts at all (unless the solar cells are mounted on some movable array to 'track' the position of the sun in the sky). Photovoltaic systems do not use water for either cooling or steam generation; so are well suited to remote or arid regions. Photovoltaics can operate on any scale, from small hand-held consumer devices to full-scale electricity plants generating at least 7 MW. Indirect solar is not applicable for small devices, but it too has been tested at the 10 MW scale. Neglecting the issues of manufacturing inexpensive photovoltaic cells, there is really only one fundamental problem with solar energy—the sun doesn't shine at night.

Developing countries all have problems that arise from the very restricted spread of electricity distribution systems, the absence of rapid communications, and the acute shortage of skilled maintenance engineers. In most villages and many towns, electricity generated by central power stations is unavailable and, at the same time, poor distribution and lack of maintenance makes diesel generation both expensive and unreliable. Such countries, so long as they are provided with plentiful sunshine, could avoid building (and having to pay for!) large central generating stations and transmission networks by installing solar arrays to supply electricity where and when it is needed for a town, a village, a factory, or household. Photovoltaic generators could not only produce electricity more cheaply when compared to the total investment needed in new generating stations and transmission grids, but are also more reliable in terms of maintenance and upkeep.

Terrestrial or Earth-bound solar power stations are only viable in regions of intense sunlight, such as the Middle East, and countries with a Mediterranean climate, like the southern parts of the United States and Mexico. Otherwise, it is hard to see how an electric utility company can rely entirely on solar, unless a clever method can be found to store some of the energy generated during the daytime.

One answer to the storage question usually means the use of rechargeable batteries. Another approach involves a hybrid solar energy system, in which a low-cost fossil fuel system is added to a photovoltaic system to compensate for variations in sunlight, have been suggested. A combined solar–natural gas hybrid would be one of the most environmentally benign ways of using fossil fuels to generate electricity. A method now being tried in Japan is to use some of the electricity generated by direct solar during the daytime to operate pumps that pump water to large ponds on the top of a mountain. At night, the water is then allowed to flow back down hill and generate hydroelectricity.

The reason that we perceive that the sun does not shine at night arises from the fact that the part of the planet we happen to occupy rotates out of the path of the sunlight. Of course the sun does not really 'set' or 'go out.' Since sunlight is always available in outer space, we can get around the problem of the sun not shining at night by moving the photovoltaic generating station into space. The conceptual Satellite Solar Power System would consist of an orbiting satellite with huge thin wings, some $7\frac{1}{2}$ miles across, covered with photovoltaic cells. Such a satellite would generate 5,000 MW of power, which then needs to be transmitted to Earth. This could be done handily if the satellite were parked in a synchronous orbit so that it could focus a beam of microwaves nearly continuously onto a receiver at a fixed location on Earth. However, it has been pointed out that one of the principal applications of microwaves is to cook things, and concern has been expressed about the possibility of the beam wandering off target and microwaving the local population, plants, and animals—finally realizing science fiction's 'death ray from outer space.' Such a satellite-mounted photovoltaic station could cost about 50 billion dollars—roughly ten times greater than a coal-fired plant of similar capacity.

The cost of solar electricity depends on three principal factors: The first is the efficiency of the solar cells—the fraction of energy in the sunlight striking the cells that is converted to electrical energy. The higher the efficiency, the fewer the cells required to produce a given amount of electricity, or the more electricity can be generated with a given number of cells. The second factor is the cost to produce the cells themselves. The third is the overall expenditure required to install, operate, and maintain a solar energy facility. The situation is complicated by the fact that the more efficient cells are also the more expensive. Thus it may prove better, for overall cost considerations, to use a large number of cheap, but low-efficiency cells.

The cost of producing electricity dropped from three dollars per kilowatt-hour in 1974 to 30 cents per kilowatt-hour in 1990. Clearly, this represents a tremendous improvement. But, to be competitive with electricity generated in conventional fossil fuel plants, the cost of electricity delivered to consumers must now be around six to eight cents per kilowatt-hour. Some forecasts suggest that photovoltaic electricity may drop to 10 cents per kilowatt-hour, bringing it close to being competitive with conventional sources, within the next decade.

Not only is the cost of producing electricity dropping, but so is the price of the cells themselves. This price is generally expressed as dollars per peak watt, where a **peak watt** is defined as one watt of power generated when the solar radiation power is 1,000 W per square meter. Forty years ago the cost of a solar cell was $1,000 per peak watt. The major impetus to develop photovoltaic cells came with the growth of the American space program in the 1960s, when a lightweight, energy source was needed for satellites. Prices of solar cells plummeted in the 1970s, from $300 per peak watt in 1973 down to $10 in 1983. By 1986, the cost was down to $5 per peak watt, and in the $2–3 range by the early 1990s.

Quite possibly the greatest short-term impact of photovoltaics could be in developing countries. At least two billion people in such countries lack electricity. The cost of photovoltaic devices to replace the kerosene lamps and stoves is still beyond the resources of these very poor people. Private foundations or public programs, such as via the United Nations could possibly assist with funds for installing these photovoltaic devices.

ADVANTAGES OF SOLAR AS AN ENERGY SOURCE

As we have seen, solar energy has many attractive features. It is available anywhere on Earth and can be collected easily with portable devices. This advantage minimizes the problems of transporting fuel from where it is produced to where it is used to generate energy and of transporting power from where it is generated to where it is ultimately used. Solar energy is nonpolluting and clean. It produces no air or water pollution. There is no problem of disposing of hazardous wastes. No mining or drilling is needed. The most important advantage of solar energy is that it is virtually inexhaustible. The energy source—the sun—cannot be owned and monopolized, as fossil fuels have become because of the unequal geographic distribution of known reserves.

Of course, solar energy also has its drawbacks. Sunlight is a rather dilute energy source. The large amount of land required for certain solar applications, such as central station power plants, may pose a problem in some areas, especially where wildlife protection is a concern. To produce solar energy equal to the total United States' consumption of energy, at least 0.5% of the country's landmass would have to be used for collection. This mass would amount to about 18,000 square miles, or one-sixth the area of the state of Arizona. A second major disadvantage is the inconsistency of direct sunlight as an energy source. Not only night, but clouds and seasonal changes affect the rate at which direct sunlight can be collected. Erratic sunlight creates the need for methods of storing the energy collected until it is needed. Back-up systems using other energy sources would be needed to supply solar homes or offices during prolonged periods of bad weather when stored energy might be depleted. One possible problem with the large-scale use of indirect solar plants is the generation of thermal pollution, just as is true with any technology using steam turbines. Indirect solar plants also require cooling water, which may be costly or scarce in desert areas.

Ultimately, cost is the key. When the overall cost has been brought down to six to eight cents per kilowatt-hour, solar energy has the potential to come into widespread use in the generation of electricity. Another potential barrier, though, resides in the legal system. A significant problem for further development of solar energy lies in the language of present-day building codes. Of the more than 10,000 such codes nationwide, very few provide for the installation of solar collectors. If provisions for solar energy systems are not in the code, individuals are powerless to install them. Furthermore, only about six states have dealt with the issue of the legal right to access to sunlight. Without such legal protection, the potential buyer of a solar collector system must think twice about committing money if a cranky neighbor's whims, the growth of trees, or the construction of a tall building nearby block the homeowner's access to sunlight. We must recognize that, to reach solar collectors, the sunlight often has to pass through air space not owned or controlled by the solar collector owner. It would seem rather prudent on the part of someone intending to install solar equipment to ensure that the necessary intervening airspace would not be subsequently blocked by the actions of other people. Is access to the sun a legal right?

FOR FURTHER READING

Aubrecht, Gordon J. *Energy*. Merrill: Columbus, 1989. Chapter 15 of this useful introductory text on energy science deals with solar energy.

Berger, John J. *Charging Ahead*. University of California: Berkeley, 1997. A major section of this book, some dozen chapters, is devoted to various aspects of solar energy. Well-written and intended for readers with little previous scientific background; includes interviews with, and anecdotes of, some of the key people in the development of a solar industry.

Borowitz, Sidney. *Farewell Fossil Fuels*. Plenum: New York, 1999. The principal focus of this book is to examine alternative energy sources to lessen (or indeed eliminate) dependence on fossil fuels. Chapters 9 and 10 discuss the utilization of solar energy.

Brooke, Bob. *Solar Energy*. Chelsea House: New York, 1992. This book reviews most aspects of solar energy, written for readers with little previous science background.

Brower, Michael. *Cool Energy*. MIT Press: Cambridge, 1992. This book discusses the use of various forms of renewable energy, including environmental implications and economic issues. Chapter 3 reviews solar energy.

Hill, Robert; O'Keefe, Phil; Snape, Colin. *The Future of Energy Use*. Earthscan: London, 1995; Chapter 7. A portion of this chapter is devoted to photovoltaics, and information of the potential for solar energy, mainly in Britain.

Hoagland, William. 'Solar Energy.' In: *Key Technologies for the 21st Century*. Freeman: New York, 1996; Chapter 15. The book is a compilation of articles from *Scientific American* magazine. This chapter is devoted to solar energy and some of its possible applications in the near-term future.

Johansson, Thomas B.; Kelly, Henry; Reddy, Amulya K.N.; Williams, Robert H. *Renewable Energy*. Island Press: Washington, 1993. This book is a collection of chapters by various experts in their respective fields. Seven chapters are devoted to various aspects of solar energy. Intended for readers with some technical background.

Ramage, Janet. *Energy: A Guidebook*. Oxford: Oxford, 1997; Chapter 11. A fine book reviewing many kinds of energy, including fossil, nuclear, and others. This chapter discusses solar energy. Available as an inexpensive paperback.

Schwartz, A. Truman; Bunce, Diane M.; Silberman, Robert G.; Stanitski, Conrad L.; Stratton, Wilmer J.; Zipp, Arden P. *Chemistry in Context*, WCB/McGraw-Hill: New York, 1997; Chapter 9. A well-illustrated and useful introductory text on chemistry. This chapter has a good section on photovoltaics.

Shepherd, W.; Shepherd, D.W. *Energy Studies*. Imperial College Press: London, 1998; Chapters 11, 12. The first of these two chapters discusses solar energy for space heating and water heating. The second discusses photovoltaics.

Smith, Howard Bud. *Energy*. Goodheart-Willcox: Tinley Park, 1993; Chapter 10. An introductory text on energy, abundantly illustrated, for readers with little prior background in science. This chapter discusses solar energy.

Sørensen, Bent. *Renewable Energy*. Academic Press: San Diego, 2000. This book treats many aspects of renewables. Much useful information on solar is scattered throughout the book.

NUCLEAR FUSION: BRINGING THE SUN TO EARTH

The light and warmth provided by the sun are one of the most obvious aspects of our daily existence. Humans, and perhaps our prehuman ancestors, must have speculated about the nature of the brilliant, glowing ball that appears to signal the start of day, traverses the sky, and then disappears again. Many early societies must also have speculated about the changing of the seasons—why the sun appears lower in the sky, and for a shorter period of time, each day until a turning point is reached and the sun begins gradually to regain its former vigor, but only until the next turning point, when the cycle starts all over again. The apparent movement of the sun across the sky is an illusion created by the rotation of the Earth; that part of the planet we happen to be standing on gradually rotates out of the path of the sun's rays. The progression of the seasons derives from the revolution of the Earth around the sun, and the Earth's axis of rotation being tilted somewhat relative to the plane of Earth's orbit. During the course of the year, our part of the planet receives greater or lesser amounts of sunlight—so is warmed to greater or lesser degrees—and the tilt of Earth's axis relative to the plane of its orbit creates the appearance of the sun rising higher or lower in the sky. These explanations of the sun's behavior have been known for centuries, and derive from astronomical observations and the geometry of Earth's orbit. It took far longer to develop the understanding of the source of the sun's energy.

WHERE THE SUN GETS ITS ENERGY

It must have seemed to early humans that something in the sky is on fire. On Earth, fire provides heat and light; this thing in the sky that gives heat and light to our whole world must itself be some kind of fire. For millennia the nature of the sun, the source of its 'fire,' and its curious course across the sky provided the raw material for colorful myths and religious beliefs. With the dawn of modern science in the 17th century the early scientists too began to speculate on what this energy source must be. They realized that the sun produces a prodigious quantity of energy, warming and lighting a comparative speck of dirt orbiting some 93 million miles away, decade after decade, century upon century, millennium to millennium. Nothing in scientific experience of phenomena on Earth seemed capable of accounting for this unimaginably vast outpouring of energy. The 19th century was fueled by coal. It should not be surprising that one of the more inspired guesses from that era was that the sun must be some gigantic lump of coal, blazing away up in the sky. (Lest we be too quick or too harsh in criticizing those early scientists with the hindsight of our much greater knowledge at the start of the 21st century, it's good to remember that our knowledge of nuclear processes, dating from the discovery of Becquerel, is little over a hundred years old, and even into the early years of the 20th century some eminent scientists adamantly refused to accept that matter is made of atoms.)

The first step toward understanding came from a determination of the sun's composition. In the 1850s, Robert Bunsen and his colleague Gustav Kirchhoff invented a device, the spectroscope, that could determine the identities of the elements based on the kinds of light they emit when heated to incandescence. Bunsen has become much better known for his invention of the handy gas-fired heater—the Bunsen burner—used by generations of chemists and chemistry students. Kirchhoff provided the fundamental mathematical analysis of the characteristic light emission from glowing objects. The Bunsen-Kirchhoff spectroscope was a major advance in analytical chemistry, and its use quickly led to the discovery of several previously unknown chemical elements. But since the spectroscope required only the light from a glowing object, and not necessarily the object itself, astronomers could use it to study the light from stars. There was no need to obtain a sample of the star's material and bring it back to Earth for analysis—the star's light would suffice. During the solar eclipse of 1868 the French astronomer Pierre Janssen noticed some unusual features in the spectrum of the sun. In England, the team of Joseph Lockyer, an astronomer, and Edward Frankland, a chemist, realized that Janssen had observed the spectral 'fingerprint' of a chemical element unknown (at the time) on Earth. Frankland proposed the name helium, from the Greek word for sun, *helios*, for this new element.

Spectroscopic chemical analysis of the sun revealed that the dominant component, by far, in the sun is hydrogen. Helium was second in abundance. The Earth's solid crust consists of compounds of silicon, aluminum, iron, and calcium in combination with oxygen. Hydrogen and helium are gases at all but extremely low temperatures. In the 1920s, the English astronomer Arthur Eddington recognized that, if the sun were gaseous throughout, as we might infer from its composition, it could be stable only if its interior temperature was extremely high, some millions of kelvins. At the extreme temperatures and pressures in the interior of the sun, atomic nuclei could actually be forced together. As we have seen, radioactive decay is a spontaneous change. Radioactivity occurs at the temperatures and pressures we ordinarily encounter in everyday life. Under these everyday conditions, no way exists for atomic nuclei to be forced together, because of the natural electrical repulsion between two nuclei of the same charge. However, under the extreme conditions of temperature and pressure in the interior of the sun, nuclear reactions that could not take place at ordinary conditions would become spontaneous. Any nuclear process that is made to proceed at extreme temperatures is called a **thermonuclear reaction**.

In 1938, Hans Bethe, a German-born physicist who was one of the many brilliant scientists that emigrated to the United States in the 1930s to escape the Nazis, recognized that the process must be one of **nuclear fusion**—the joining together of small nuclei, such as hydrogen, to produce larger ones. Bethe worked out a sequence of processes, the net result of which would be the conversion of $_1H^1$ to $_2He^4$. Bethe proposed an entire series of reactions, in which, as examples, hydrogen reacted with carbon to build up nitrogen and then oxygen (the third-most abundant element in the sun). The oxygen nucleus broke apart to form helium and carbon, helium being a net product of the process and the carbon being available to begin a new cycle. Thus the net effect of the reactions converts hydrogen to helium. Subsequently other scientists have refined Bethe's model, and added other possible reactions, but it was Bethe's work, building on Eddington's insights, that provided the first real understanding of the source of the sun's energy.

We can consider a very simple hypothetical fusion reaction represented by

$$_1H^2 + {_1}H^2 \rightarrow {_2}He^4$$

(Notice that this equation conforms to the rules for balancing nuclear equations that were introduced in Chapter 26, but, in contrast to all of the equations of radioactive decay or fission processes, here the product nucleus is *larger than* the reacting nuclei.) As with radioactivity or fission processes, there are two ways of considering the energy release from this fusion reaction.

First, we can once again use the curve of binding energy, Figure 37.1. Because the curve of binding energy is so much steeper on the left-hand side, then we can expect a substantial release of energy in a fusion process, much more than in fission. Although the change in the number of nucleons—the horizontal step across the curve—is smaller than most radioactive processes and very much smaller than in fission, the very steep slope of the curve means that the energy release—the vertical leap upward—is extremely large compared to radioactivity or fission.

Alternatively, we can apply Einstein's equation for the equivalence of mass and energy. To make a very scrupulous accounting of the equation shown above, the mass of the two $_1H^2$ nuclei is 4.02820. The mass number of the $_2He^4$ produced in the reaction is 4.00280. The loss of mass is 0.0254, which amounts to 0.63%. In contrast, in the fission of uranium via the process

$$_{92}U^{235} + {_0}n^1 \rightarrow \text{etc.}$$

the mass loss is about 0.056%. The mass loss in fusion is proportionately much greater than in fission, by about a factor of 10. In other words, on a weight-for-weight basis, over 10 times as much energy is available in nuclear fusion as in nuclear fission via Einstein's equation, $E = mc^2$.

The energy released by such fusion of hydrogen to helium in the sun is so enormously large that, while scientists can calculate the quantities of energy produced and

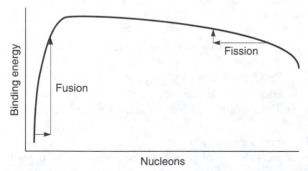

FIGURE 37.1   The curve of binding energy illustrates the much larger energy release from fusion compared to fission.

hydrogen consumed, they are almost impossible to contemplate. We know from Einstein's equation that the energy of the sun must be obtained at the expense of its mass. The sun converts about 650 million tons of $_1H^1$ to $_2He^4$ every second. The corresponding mass loss, m, is 4,600,000 tons per *second*! Yet despite this seemingly prodigious conversion of mass to energy, the sun's mass is so large that it has already been radiating energy for some five billion years, and astronomers believe that it will continue to do so for about five billion years more.

We have seen (Chapter 36) that only two primary sources of energy are available to us: solar and nuclear. Although our direct use of the sun's energy is a quite small proportion of total energy consumption, water, wind, food, firewood, biomass, and the fossil fuels can all be traced ultimately to solar energy. Now we see that solar energy derives from thermonuclear processes—nuclear fusion—in the sun. So, in the last analysis, all of the energy that we use in any form derives from processes involving atomic nuclei.

HARNESSING THE ENERGY OF NUCLEAR FUSION

Although nuclear fusion reactions require enormous temperatures, on the order of tens of millions of kelvins, to be initiated, they are spontaneous reactions under such conditions. Exactly like every other spontaneous process, we can in principle insert some device into this system that will allow us to capture some of the energy released from fusion as useful work. As we will see, the main problem lies in creating the temperatures required to ignite the spontaneous thermonuclear reactions.

A half-century ago, only one device was known that could generate, on Earth, the necessary temperature to start a thermonuclear reaction—an atomic bomb. Recognizing the potential for 10 times as much energy to be released from fusion as from fission, and having available a 'match' to light the nuclear fire (the atomic bomb), scientists of the late 1940s and early 1950s envisioned a bomb that would make use of the energy released from fusion. At first it was called the 'super' (as in super-bomb), but has become known as a thermonuclear bomb, or, more familiarly, the hydrogen bomb or H-bomb.

Even before the first demonstrated explosion of the hydrogen bomb by the United States in 1952, nuclear scientists recognized that such a weapon would be of unprecedented destructive power. Many hesitated to work on the 'super' or even actively protested its development. Perhaps the most prominent person in the United States to raise doubts of the wisdom and morality of building the 'super' was J. Robert Oppenheimer, who had been the civilian leader of the Manhattan Project that led to the development of the atomic (fission) bomb. In the swirl of anticommunist paranoia of early 1950s America, Oppenheimer was publicly disgraced by having his security clearance and access to secret information revoked. His career was effectively ruined. Among the scientists on the other side of the argument, pushing for rapid development of the 'super' and vigorously condemning Oppenheimer was Edward Teller. In the United States, Teller became known as the 'father of the hydrogen bomb' and the role model for the title character in the film *Doctor Strangelove*.

When we turn from methods of blowing each other up to finding ways of harnessing the energy release from fusion, several issues arise: What will we use as the fuel? How will we achieve the temperatures needed to ignite it? How can we find a container to hold the equivalent of a miniature sun, or a controlled hydrogen bomb? How can we capture useful energy from the process?

Fuel

To achieve nuclear fusion, two nuclei must be smashed together. To bring this about requires significant energy, because normally two objects with similar electric charges (in this case, the positive charges due to the protons in the nucleus) will normally repel each other. The nuclei must have sufficient kinetic energy to overcome the electrical repulsion. The greater the positive charge on the nucleus, that is, the more protons in the nucleus, the greater will be the electrical repulsion to be overcome. For this reason, it would be best for the fuel to consist of nuclei having the smallest possible number of protons—namely, one. The element of atomic number 1 is hydrogen, so hydrogen is the fuel of choice for nuclear fusion.

The simplest approach would be to use $_1H^1$ as the hydrogen fusion, since that is the most common form of hydrogen. $_1H^1$ is also the principal fuel for generating the sun's energy. Unfortunately, though, the temperatures required to ignite a $_1H^1$ fusion reaction are much too high to be achieved in any sort of fusion reactor that is likely to operate on Earth.

Deuterium, $_1H^2$, will undergo fusion at a temperature lower than that needed for $_1H^1$ fusion. The even heavier isotope, tritium ($_1H^3$) will undergo fusion at still lower temperature. However, $_1H^3$ is unstable and would be difficult to collect in reasonable quantities. That leaves $_1H^2$ as the best possible fuel.

To consider the fusion reactions of deuterium, this hydrogen isotope can react in one of two ways:

$$_1H^2 + _1H^2 \rightarrow _2He^3 + _0n^1$$
$$_1H^2 + _1H^2 \rightarrow _1H^3 + _1H^1$$

The two nuclei of deuterium have about equal probability of reacting in these two ways. In the second reaction, the tritium that is formed as a product reacts quickly with another deuterium nucleus:

$$_1H^3 + _1H^2 \rightarrow _2He^4 + _0n^1$$

So, we can then represent the overall fusion process as the net sum of these three individual reactions, that is,

$$5_1H^2 \rightarrow _2He^3 + _2He^4 + _1H^1 + 2_0n^1$$

Notice that each of the three individual reactions produces a pair of particles. For example, two deuterium nuclei produce a $_2He^3$ nucleus and a neutron rather than simply fusing to produce a single $_2He^4$ (which would still balance the nuclear equation). As a rule, fusion requires creation of at least two products, which share the released energy. Much of the energy liberated in the fusion is actually carried off by the neutrons.

The total energy that would be produced by the complete fusion of one gram of $_1H^2$ is 226 *billion* (!) Btus. However, deuterium is not the common isotope of hydrogen, and in fact only one of every 6,500 hydrogen atoms occurring in nature is the deuterium isotope. The deuterium must be extracted from the water and separated from the ordinary isotope (that is, $_1H^1$). The technology for doing this is now well known. While it may seem at first sight that only one atom out of 6,500 is the desired isotope, there is still enough deuterium available so that the complete fusion of the deuterium that could be obtained by processing one liter of water would still yield about 10 million Btus of energy.

The equations shown above indicate that any tritium formed in the process will also undergo fusion by reacting with another deuterium nucleus. It is also possible to begin the fusion reaction using a mixture of deuterium and tritium as the fuel, rather than only deuterium. This mixture will ignite and begin to fuse at a lower temperature than if pure deuterium were used. A disadvantage of tritium is that it is radioactive and does not occur in nature. Fusion reactors that rely on tritium as part of the fuel must use a radioactive material. Tritium is a β-emitter with a half-life of 12 years. It can be made in various nuclear processes such as

$$_5B^{10} + _0n^1 \rightarrow _1H^3 + _2He^4 + _2He^4$$

or

$$_3Li^6 + _0n^1 \rightarrow _1H^3 + _2He^4$$

Ignition

Most experimental fusion reactors are based on the fusion of the deuterium with the tritium nucleus. This requires very severe conditions. At the normal temperatures of chemical reactions or in chemical engineering practice, which are seldom above 1,000°C, deuterium and tritium atoms exist with their nuclei and electrons just as any other atom. When these atoms approach each other, they reach a point at which their approach is so close that the electron clouds around each atom repel each other. (Like charges repel!)

Fusion is much harder to initiate than fission, because it requires that two nuclei be brought extremely close together. To overcome the electron repulsion between atoms we can strip away the electrons. This is done by heating the deuterium–tritium mixture to very high temperatures, generating a **plasma**. Temperatures of about 40 million kelvins are required to effect this. A plasma is a cloud of ionized gas. In the plasma, the nuclei can approach each other without the interference of the electron clouds. But again, because each nucleus has a positive charge, the nuclei will repel each other as they approach. The challenge is to give each nucleus enough kinetic energy to overcome the repulsion energy and let the nuclei slam into each other. The speeds needed to achieve sufficient kinetic energies require in turn extremely high temperatures, which heretofore have been found only in the center of the sun or other stars. For deuterium fusion, generating the kinetic energy required to overcome the repulsive energy of the nuclei is done by heating the plasma to temperatures on the order of 100 million kelvins. In the case of a deuterium–tritium mixture, the necessary temperatures are still above 50 million kelvins.

This leads straightaway to the first engineering problem: How do we heat the plasma to 100 million kelvins? There are at least three approaches. One is to use very high electric currents, analogous to the resistance heating of the filaments in light bulbs or in the heating elements on electric stoves. High-temperature plasmas do conduct electricity; the currents used can be millions of ampères. To achieve even more intense heating, radiofrequency systems similar to those in microwave ovens can be used. A second approach uses a battery of extremely powerful lasers. This so-called inertial system uses a mixture of deuterium and tritium frozen into small pellets less than one millimeter in diameter. One of these is placed in the middle of a vacuum chamber and hit from all sides by intense laser beams. The laser bombardment takes place in a small fraction of a second, heating the fusion fuel rapidly without expanding it. The outer layer of the pellet vaporizes almost instantly and expands outward with great force. This very rapid vaporization with its outward force also creates a force pushing against the remaining interior of the pellet, causing it to implode. The great density and heat of the imploding pellet ignites the fusion reaction. A third method is to inject a beam of $_1H^2$ or $_1H^3$ nuclei into the plasma. (In a hydrogen bomb, the high temperatures necessary for fusion are achieved by first detonating an atomic bomb, which is an integral part of the hydrogen bomb.)

It may not seem so at first sight, but thermonuclear fusion has much in common with the more familiar processes of chemical combustion. In both cases, a high temperature is necessary to start the reaction, and the energy released by the reaction maintains a high enough temperature to spread the fire. In an ordinary chemical reaction, some of the energy stored in chemical bonds is released to us when the atoms of the reacting molecules recombine into new, more tightly bound product molecules. In thermonuclear fusion reactions, some of the energy stored as nuclear binding energy is released to us as the nuclei recombine into new, more tightly bound nuclei. The significant differences between these chemical and nuclear processes are the temperatures at which they occur and the amounts of energy released for a given mass of reactants.

Reactors

We might say that operating a fusion reactor is an approach to controlling and exploiting the energy of a hydrogen bomb. This analogy isn't exact, because there's no way of having a fusion explosion in a reactor. But, the key point is that, just as a hydrogen bomb is much more powerful than an atomic bomb, the engineering problems to achieve controlled nuclear fusion are much more formidable than to build and operate fission power reactor.

The second major problem of achieving controlled nuclear fusion, once ignition has been achieved, is how to contain the tremendous temperatures of the plasma. (This is analogous to an ancient joke about two chemists. The first claims to be trying to invent an acid so powerful it will dissolve any known substance. The second immediately asks, 'Oh yeah? So what are you going to keep it in?') A hundred million kelvins is not only way above the melting point of any known material, it's also way above the boiling point of any material. In fact, five *thousand* kelvins is about enough to melt or vaporize any known substance. We might expect that a plasma reaction, once started, could burn a hole through anything.

As we've noted, high-temperature plasmas conduct electricity. Therefore, they can be confined and controlled by electric and magnetic forces. Indeed, the plasma can be confined simply by a properly shaped magnetic field. This approach to containing a plasma undergoing fusion relies on the observation first made by Ørsted: A moving electric current generates a magnetic field. Since the plasma has an electric charge, when it is set in motion it will generate a magnetic field (Figure 37.2). If we enclose the magnetic field in an even stronger magnetic field of the same polarity (like poles repel!) then this strong, external magnetic field will tend to compress the plasma. The compression of the plasma by the applied external field serves two purposes: First, it pushes the plasma away from the walls of the reactor vessel. Thus the materials the reactor is made of do not have to do the impossible job of sustaining a temperature of millions of degrees. This process is called **magnetic confinement**. Second, by forcing the plasma into a smaller volume, the density of nuclei (that is, the number of nuclei per unit volume of the plasma) is substantially increased. This in turn increases the likelihood that two nuclei will collide and react. Magnetic confinement is one of three major strategies being pursued to achieve practical, controlled fusion reactions.

In practice, most experimental fusion reactors are, in effect, a hollow doughnut made of copper (Figure 37.3). Very strong electromagnets are mounted on the outside for magnetic containment. This type of reactor is called a **tokamak** (from the Russian: *tok*

Magnetic field generated by moving plasma

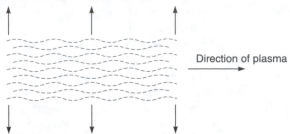

Direction of plasma

FIGURE 37.2 A plasma, which is a collection of electrically charged particles in motion, generates its own magnetic field just as Ørsted demonstrated for a simple Voltaic pile.

means current, and *mak*, or *mag*, is short for magnetic). The fundamental concepts of the tokamak reactor were developed in the early 1950s by Igor Tamm and Andrei Sakharov, then at Moscow University.

The main magnetic field of the tokamak forms a 'magnetic doughnut.' The magnetic field is formed inside a doughnut-shaped chamber by electromagnets around the outside of the chamber. Plasma nuclei inside the tokamak spiral around the magnetic field lines and don't touch the chamber walls.

The use of a magnetic field gets away from the possible problem of finding a material to confine the extreme temperatures of the plasma. A magnetic field is not made of any material substance, can exist at any temperature, and can exert forces on moving electrical charges. If so-called magnetic walls could be designed with appropriate strength the magnetic field alone could contain the plasma and produce the necessary fusion temperatures. The plasma will not melt or burn a hole through the walls of the reactor in part because it is at an extremely low density. The density of nuclei in the plasma is only about 1/10,000th of the normal density of the atmosphere. The total amount of heat in such an exceptionally low-density material is very small. Furthermore, even if the plasma did contact the walls of the reactor, such contact would only slow the ions and, consequently, cool the plasma. Cooling due to contact with the container wall would be likely to stop the reactions completely. (However, this also adds an inherent level of safety to the fusion reactor, because if the slightest thing goes wrong with magnetic confinement, the plasma will contact the walls, cool, and shut down the fusion reactions.)

FIGURE 37.3 The tokamak reactor is a doughnut-shaped device in which the plasma is confined and the fusion reactions can occur.

A system somewhat related to the tokamak is the magnetic mirror system, which uses open-ended reactors rather than the 'magnetic doughnut.' This approach is based on the fact that a plasma cannot pass through a magnetic field. The reactor is set up such that the external magnetic field is strongest at the open ends. This strong magnetic field at the ends of the reactor is called the 'magnetic mirror.' It turns, or reflects, the plasma back toward the center of the reactor where the magnetic field is weaker. The plasma will reflect back and forth between the magnetic mirrors at the reactor ends.

A second major approach to fusion reactor design relies on inertial confinement. As discussed above, intense pulses of laser light heat and compress a tiny sphere of deuterium and tritium. Though the laser pulses last only a few billionths of a second, they are nevertheless sufficient to vaporize the surface of the fuel particle. As the surface expands outward with explosive force, the core simultaneously experiences enormous inward forces that cause it to implode. The implosion compresses the fuel to a density more than 20 times that of lead. The intense compression caused by this implosion causes the temperature to rise to that needed to ignite fusion. The successive ignition of a series of such fuel pellets dropped steadily into the laser beams in a fusion power plant might produce a steady stream of electrical power. The success of this technique clearly requires precise timing, of the dropping of the pellets, along with the development of more efficient lasers.

Energy Capture

Finally, if the fusion reaction can be ignited, and then confined, successfully, it is still necessary for us to find some way of withdrawing or capturing a portion of the energy release in the form of useful energy for us. At least two methods of capturing some of the energy in a useful form are under consideration. One is the direct conversion of the current of charged particles from the reaction into electricity. A second is capturing the neutrons liberated from the fusion reaction as heat in a liquid lithium blanket surrounding the reactor. Heat is then converted to electricity by raising steam to operate a conventional steam turbine and generator.

Withdrawing power from a fusion reactor relies on Faraday's principle that a conductor flowing or moving in a magnetic field generates an electric current. In this instance the plasma is the conductor; allowing it to flow through a magnetic field generates an electric current (Figure 37.4). This approach is called **magnetohydrodynamics**.

As we have seen in the discussion of fusion reactions, most of the energy of nuclear fusion is in the kinetic energy of the product neutrons. If these neutrons could be stopped and captured, their kinetic energy would be converted into heat. One scheme would surround the reactor with a blanket of lithium metal to capture the neutrons. Doing so serves a double purpose. First, the neutrons release their kinetic energy as heat, which can be used to raise steam. Second, as the lithium captures neutrons, it 'breeds' tritium via the reaction

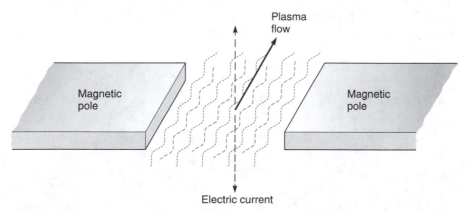

FIGURE 37.4 The flowing, reacting plasma in a fusion reactor can be used to generate an electric current.

$$_0n^1 + {_3}Li^6 \rightarrow {_2}He^4 + {_1}H^3$$

Unfortunately, the tritium and the high-energy neutrons from the fusion reactions make the reactor structure radioactive. However, the quantity of radioactive material in a fusion reactor would be about a hundred times less than in a fission reactor. The tritium would be recycled into the reactor as fuel. Furthermore, tritium has a half-life of 12.6 years, compared with the very long half-lives of some highly radioactive fission products (e.g., 24,400 years for the very dangerous plutonium).

So far, the energy output of most experimental fusion reactions is much smaller than the energy needed to heat and confine the plasma. Even at 100 million kelvins, more energy must be put into the plasma than is given off by fusion. At about 350 million kelvins, the fusion reactions finally reach a point at which they can produce as much energy as is put into generating the plasma. This point is called **break-even**. So far, break-even in experimental reactors has been achieved for no more than a fraction of a second. No one has ever built, let alone operated, a practical fusion power reactor that has achieved break-even for sustained periods of time. As we will see, there are profound reasons to hope that a breakthrough will be achieved and the large-scale development of fusion energy will become practical.

COMPARING FUSION AND FISSION AS SOURCES OF NUCLEAR ENERGY

Let's review very briefly the principal concept of 'atomic energy,' generating electricity from nuclear fission processes (Figure 37.5). We've identified several concerns with this approach. The prospect exists to have a serious accident, releasing radioactive material to the environment, as at Three Mile Island or Chernobyl. Even with fuel reprocessing, fission reactors still produce some waste material that has to be stored safely. The total supply of uranium is limited. The United States has an estimated one-third of the Western world's resources of uranium ore. However, the uranium content of the ore itself may only be about 0.2%.

The possible benefit of nuclear energy is that the amount of energy per pound of uranium undergoing fission is enormous, compared to fossil fuels; larger by at least a factor of a million. Thus it could be of great interest to find an alternative way to release energy from some sort of nuclear process, while at the same time overcoming the concerns we now have with fission processes.

One advantage of fusion is that deuterium is extremely abundant (in water) but uranium or thorium, for use in fission reactors, are rare metals. Together, uranium and

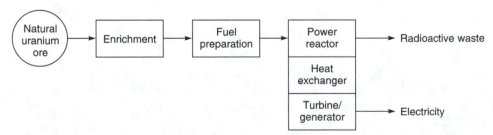

FIGURE 37.5 The many steps in securing ENERGY from nuclear fission make fission a much more complex process than fusion. The challenge is to develop an operating, commercial-scale fusion reactor

thorium account for about 10–12 parts per million of the Earth's crust. To be sure, the energy that could be liberated from the fission of these elements represents about 10 times as much energy as could be obtained from the Earth's total supply of coal, oil, and gas. However, only a small part of Earth's supply of these fissionable fuels can be extracted from the crust with reasonable ease. Furthermore, just like the fossil fuels, the reserves of uranium and thorium are not evenly distributed around the world. Fusion, on the other hand, relies primarily on deuterium, which is available in water all over the world.

In contrast to fission reaction processes, there is no net production of radioactive material. Tritium is radioactive, but would be captured and recycled into the reactor, so it would be retained in the system. Indeed, it's an advantage that tritium is produced in the reaction, because its half-life is so short (12.6 years) that it does not occur in nature. If it were to be used as a fusion fuel, and were not actually made during fusion, it would have to be produced anyway, via some other nuclear process. The high-energy neutrons produced in fusion would be dangerous if emitted to the environment, but in an operating fusion reactor would likely be captured in the blanket of lithium. The neutrons may make some of the internal parts of the reactor radioactive. Nevertheless, a fusion reactor likely will produce relatively little radioactive waste, substantially less than a fission reactor.

If we consider a fusion reaction based on deuterium alone, the reactions that take place are

$$_1H^2 + {}_1H^2 \rightarrow {}_2He^3 + {}_0n^1$$
$$_1H^2 + {}_1H^2 \rightarrow {}_1H^3 + {}_1H^1$$
$$_1H^3 + {}_1H^2 \rightarrow {}_2He^4 + {}_0n^1$$

The net reaction (that is, adding up these three reactions and eliminating species that occur on both sides of the equation) is

$$5_1H^2 \rightarrow {}_2He^3 + {}_2He^4 + {}_1H^1 + 2_0n^1$$

The only fusion products are helium and ordinary hydrogen (i.e., $_1H^1$), which are stable and safe. There is no problem with production of radioactive waste, or of fissionable materials that terrorists or rogue nations could use to produce atomic weapons.

Overall, then, fusion offers numerous advantages relative to fission as a source of nuclear energy. Extracting deuterium from water is not as disruptive to the environment, nor nearly so dangerous, as mining uranium, enriching it, and processing it into fuel for fission reactors. Fusion reactors do not contain a large inventory of radioactive material, in contrast to the numerous fuel assemblies in a fission reactor. The danger of a reactor accident is virtually eliminated. If any part of the system fails, the plasma will cool very rapidly and the fusion reaction will stop almost instantly. Besides, fusion reactors cannot become supercritical and get out of control because, unlike fission, there is no critical mass for fusion. There is no need to rely on human intervention (such as scramming a fusion reactor) or computer-based control systems taking over. Any component failure means the reaction stops. There is no air pollution, nor any CO_2 emissions, because there is no combustion; furthermore, the only product is helium. Storage of radioactive waste is much less of a problem with a fusion reactor. Much or all of the tritium will be recycled, and the only other likely radioactive material might be some internal parts of the reactor. The half-life of any tritium that would be regarded as waste is a good bit shorter than many fission by-products. There are no fissionable by-products (such as plutonium) that could be used to manufacture weapons, so nuclear weapons proliferation is not an issue with fusion. The problem of thermal pollution, characteristic of conventional steam-turbine plants, could be avoided if the option of direct generation of electricity by magnetohydrodynamics is chosen.

THE PROMISE AND THE FRUSTRATION OF FUSION ENERGY

If fusion reactors could ever achieve break-even and operate in that way for some time, so that they would be reliable, long-term suppliers of electricity, the promise that is held out is truly fabulous. The estimates of the amount of energy that could be obtained from fusion, and comparisons with other energy sources, necessarily depend on the assumptions made in doing the calculations, such as the reserves of various fuels and possible future rates of energy consumption. Because of that, there is some variation among projections of the benefits that could be obtained from fusion energy. Regardless of whether estimates agree in exact detail, there is no doubt that, if nuclear fusion could be made to succeed as a commercially viable energy source, the energy requirements of humankind would be solved for the foreseeable future.

Let's start with a quart of water. By fusion of the $_1H^2$ contained in that water, we would obtain as much energy as we would get by burning 75 gallons of gasoline. Five gallons of water contain one gram of deuterium, which when fused releases as much energy as 2,500 gallons of gasoline. Complete fusion of the deuterium in a cubic meter of water would yield energy equivalent to 2,000 barrels of oil. Each cubic kilometer of ocean contains enough deuterium to be roughly equivalent to all of Earth's oil reserves. Each ton of deuterium produces 60 million times more energy than a ton of coal.

If we took the very parochial view that we would use deuterium fusion only to supply the United States' energy needs, it would take us 600 billion years, at present rates of energy consumption, to use up the deuterium in the oceans. Since the universe is only some 10 billion years old, there's nothing to worry about in terms of there being enough deuterium to meet our energy needs. Obviously, this notion is pretty fanciful. Suppose we add two constraints: first, that we would be able to use only 1% of the deuterium in the oceans, and, second, that everyone in the whole world would be able to use energy at the present rate at which it's consumed in the United States. Then the best we can hope for is energy to last the planet for the next 300 million years. We've seen (Chapter 35) that, depending upon the values used for total reserve or resource and annual consumption rates, it would appear that fossil fuels have lifetimes of, at best, some few centuries. At present rates of energy consumption, and assuming access to all the deuterium, the people of the world could be using fusion energy for the next 50 billion years. Long before we come close to running out of deuterium, the sun will have lived through its natural life cycle and have extinguished all life on Earth—perhaps even the Earth itself.

But when considering this fantastic promise of fusion energy, we must also keep in mind the challenges that have to be met. Basically, there are three. The first is to create and heat the plasma at temperatures over a hundred million kelvins. The second challenge is to keep enough plasma away from the container walls for long enough to permit abundant reactions to occur. Indeed, the most serious problem facing fusion is that of achieving a net production of energy. As we've seen, a fusion reactor needs energy for generating the plasma and for operating the electromagnets used for magnetic confinement. A commercially viable fusion reactor should produce enough electricity to satisfy these needs and have some left over to sell to the public. So far, experimental fusion reactors are only able to satisfy their own energy needs for a few seconds at best. The third challenge is to design a practical, safe, and economic fusion reactor. Magnetic confinement fusion reactors are now close to break-even. Capital costs for commercial fusion reactors are difficult to estimate, but may be within a factor of two of the costs of other power plants. The cost of fuel would be cheap. The lasers and other technologies needed for inertial confinement fusion are very complex and troublesome, so it remains very questionable whether inertial confinement can compete with magnetic confinement, and in fact whether inertial confinement would ever be commercially viable as a source of energy. Optimists and proponents of fusion believe that the technology may be commercially available within the next 25–50 years. Others are not so sure, and sometimes paraphrase an old joke to say that 'fusion is the energy source of the future, and it always will be.'

FOR FURTHER READING

Asimov, Isaac. *The History of Physics*. Walker: New York, 1966. This is a very fine introduction to physics (at least through the early 1960s), well written for the reader with little previous science background. Chapter 41 provides information on nuclear fusion.

Aubrecht, Gordon J. *Energy*. Merrill: Columbus, 1989. Chapters 6 and 14 of this text discuss aspects of nuclear fusion.

Bloomfield, Louis A. *How Things Work*. Wiley: New York, 1997. This is a fine introductory textbook on physics, centered around examples of how real-world devices function. Chapter 18 provides information relevant to nuclear fusion.

Furth, Howard P. 'Fusion.' In: *Key Technologies for the 21st Century*. Freeman: New York, 1996. This relatively inexpensive paperback is a compilation of articles from *Scientific American* magazine. Chapter 16 talks about the future prospects for fusion energy.

Hewitt, Paul G. *Conceptual Physics*. Addison-Wesley: Reading, 1998. This is an introductory text on physics, well illustrated with a remarkable number of conceptual diagrams and cartoons. Chapter 33 provides information on fusion.

Shepherd, W.; Shepherd, D.W. *Energy Studies*. Imperial College Press: London, 1998; Chapter 8. This chapter covers most aspects of nuclear energy, with a concluding section on fusion. The information includes research and development on fusion reactors through the mid-1990s.

Smith, Howard Bud. *Energy*. Goodheart-Willcox: Tinley Park, 1993. Chapter 12 of this introductory text on energy provides a discussion of fusion. The book is aimed at readers with little or no prior technical background.

GLOSSARY

This glossary provides brief definitions or examples of the terms identified in the text in **boldface** type. The number appearing in square brackets is a reference to the chapter in which that term is introduced and explained.

Absolute zero [11] The lowest possible temperature; the temperature at which molecular motion vanishes and an object would have no heat energy (i.e., be at the zero point of thermal potential energy).

Acid neutralizing capacity [30] The ability of a lake to resist a change in pH when acids are added to it.

Acid rain [30] Rainfall that has a pH less than 5.6, the pH of a solution of carbon dioxide in water.

Actinides [28] The family of radioactive chemical elements with atomic numbers from 89 to 103. The best-known of the actinides are uranium and plutonium.

Active solar energy system [36] A system for capturing solar energy for space heating or producing hot water that relies on devices (such as a **flat-plate solar collector**) specifically added to the building for capturing the solar energy.

Adiabatic expansion [24] The expansion of a gas accompanied by a flow of heat into the system, so that the temperature of the gas does not remain constant.

Albedo [34] The fraction of radiation reflected from a surface, often used to refer to the reflection of sunlight from the Earth's surface.

Alkylation [21] In petroleum refining, a process by which small molecules (of four or fewer carbon atoms) are combined to produce molecules boiling in the gasoline range. The product of alkylation is also of very high octane number.

Ampère or **amp** (unit) [12] The unit of electric current. The symbol for amp is A.

Anthropogenic [18] Caused by, or originating from, human activity.

Anthropogenic greenhouse effect [32] Climate change (usually assumed to be warming) that is caused by carbon dioxide or other infrared-absorbing gases introduced to the atmosphere by human activities, such as fossil fuel combustion.

Array [36] In solar energy, a group of multiple photovoltaic cells or solar collectors.

Atomic number [25] The number of protons in the **nucleus**. The atomic number unequivocally determines the chemical identity of an atom.

Autoignition temperature [21] The temperature at which a **fuel** (e.g., gasoline) and air mixture will ignite by itself with no other energy source, such as a spark plug or match.

Availability [35] The percentage of time that a device is actually operating, not including scheduled shutdowns for maintenance.

Becquerel (unit) [28] A unit expressing the **radioactivity** of a substance, defined as one radioactive decomposition per second. The symbol for a becquerel is Bq.

Beneficiation [29] In general, improving the chemical composition or physical properties of an energy resource; the term is usually restricted to coal. The goal is usually to reduce the amount of potential pollution-forming materials prior to combustion, to improve the calorific value of the fuel, or to do both. Examples could include reducing the amount of ash-forming minerals or reducing the amount of sulfur.

Betz limit [35] The theoretical maximum limit, 59%, of conversion of energy to work in a wind-operated device.

Biogas [34] A mixture of methane and carbon dioxide produced by anaerobic bacteria acting on organic matter, particularly material deposited in landfill.

Biomass [34] Any of several sources of energy, such as wood or plant oils, obtained from recently living organisms.

Blend stock [21] Any of several petroleum refinery streams that are subsequently blended together to produce the various grades of gasoline—87, 89, 93 octane—on the market.

Boiler [17] A device for converting water into steam.

Boiling water reactor [27] A type of nuclear reactor in which water is boiled inside the reactor core, and the resulting steam is then used directly in **turbines**.

Break-even [37] In nuclear fusion technology, the point at which the energy produced from a fusion reaction is equal to the energy needed to cause it to occur.

British thermal unit [3] The amount of heat required to raise the temperature of one pound of water by one degree Fahrenheit. The symbol for British thermal unit is Btu.

CAFE standards [33] An acronym for corporate automotive fuel efficiency.

Caloric [11] An invisible and weightless fluid that was supposed (incorrectly) to account for the transfer of heat from one object to another.

Calorie [3] The amount of heat required to raise the temperature of one gram of water by one degree Celsius. The symbol for calorie is cal.

Calorimeter [3] A device used to measure the heat absorbed or released from a process, usually a chemical reaction.

Capacity factor [35] The ratio of the average actual use of a system to its designed capacity for producing, e.g., energy or work.

Carnot efficiency [11] A measure of the ability of a **heat engine** to convert energy to work, determined as the ratio of the difference in temperature between the hot and cold sides of the engine to the hot-side temperature (when the temperatures are expressed in kelvins).

Catalyst [21] A substance that increases the rate of a chemical reaction without itself undergoing permanent change in the reaction.

Catalytic converter [31] A device installed in the exhaust system of automobiles to destroy the carbon monoxide, nitrogen oxides, and unburned hydrocarbons coming from the engine, before they can be emitted to the environment.

Catalytic cracking [21] A process in which high-boiling petroleum fractions are reacted in the presence of a catalyst to increase both the yield and octane number of gasoline.

Catalytic reforming [21] A process in which gasoline or naphtha is reacted in the presence of a catalyst to enhance the octane number.

Cetane number [24] A standard of performance of diesel fuel, based on a comparison of the fuel being tested with a mixture of cetane (assigned a value of 100) and methyl-naphthalene (assigned a value of 0). The cetane number of the fuel is equal to the percentage of cetane in the cetane/methylnaphthalene blend.

Chain reaction [26] In general, any process in which a product from one step starts the next step. In nuclear processes, a chain reaction uses **neutrons** produced in the **fission** of one nucleus then to cause fission to occur in another nucleus.

China syndrome [28] The notion that, in the event of a catastrophic nuclear reactor accident, the hot, radioactive contents would burn through the pressure vessel and the containment structure, and keep going until they had burned a hole completely through the Earth and emerged in China. The term is based on the popular, but false, concept that China is on the exact opposite side of the Earth from the United States.

Coal cleaning [18] Any of several processes that are intended to reduce the amount of sulfur and/or noncombustible, ash-forming minerals in coal before it is burned.

Compression ratio [21] In internal combustion engines, the ratio of cylinder volume when the piston is at the bottom of its stroke to the cylinder volume when the piston is at the top of its stroke.

Condenser [18] A device for converting steam—such as that exiting a turbine—back to water.

Condensing turbine [16] A steam **turbine** in which a condenser helps to remove or draw off the low-temperature, low-pressure steam leaving the turbine by condensing that steam back to liquid water.

Conduction (heat) [11] One of three possible mechanisms for the transfer of heat, conduction occurs via an atom-to-atom (or molecule-to-molecule) transmission of kinetic energy from the hot end of a substance to the cooler end. The atoms or molecules at the hot end have a higher kinetic energy than those at the cool end; this kinetic energy is transferred through the body as the high-kinetic-energy atoms or molecules jostle neighbors of lower kinetic energy.

Conductor (electricity) [12] A substance, most commonly a metal, that allows electric current to flow through it easily.

Containment structure [27] A building manufactured of reinforced concrete to house a nuclear reactor. Its principal function—implied by its name—is to contain any radio-active materials in the case of a severe accident to the pressure vessel.

Control rod [27] Devices made from good neutron-absorbing materials, such as boron, and used to regulate the rate of **fission** in a nuclear reactor.

Convection [11] One of three possible mechanisms for the transfer of heat. Convection occurs via the motion of heated bodies of liquids or gases. The common saying that 'hot air rises' is in fact a manifestation of convective heat transfer.

Coolant [27] A substance used to remove heat from the **core** of a nuclear reactor. In some reactor designs, ordinary water is used as the coolant.

Cooling tower [18] A structure that is used to lower the temperature of warm water leaving a condenser before the water is returned to the environment. Water is sprayed into the cooling tower and is cooled by a current of air rising through the tower.

Core [27] The portion of a nuclear reactor that contains the fuel rods, control rods, and **moderator**, and through which coolant circulates.

Cracking [21] Any process for reducing the size, and hence boiling temperatures, of molecules. Often used to enhance the yield of gasoline in a refinery.

Curie (unit) [28] A unit of **radioactivity** equal to 3.7×10^{10} radioactive disintegrations per second. Symbols for the curie are c or Ci.

Current (electric) [12] The amount of electric charge flowing per unit time.

Curve of binding energy [26] A graphical representation of the average binding energy per **nucleon** as a function of the number of nucleons. The greater the binding energy, the more stable is the **nucleus**.

Deforestation [6] An environmental problem caused by the virtually complete cutting of trees in an area, often the result of cutting trees for firewood.

Delayed neutron [27] A delayed **neutron** is one emitted from an energetically excited nucleus *after* a **fission** event has occurred, in contrast to a neutron produced in the fission itself.

Direct solar energy generation [36] The conversion of solar energy to electricity, without intervening steps, using photovoltaic solar cells.

Dissolved gas [20] Hydrocarbons that are normally gaseous (i.e., the components of natural gas) that are dissolved in the liquid hydrocarbons in a petroleum reservoir. The term 'solution gas' is also used.

Doping [36] Treatment of a very pure substance, such as highly purified silicon, with small and controlled amounts of another element to achieve a desired alteration of properties, as in the manufacture of semiconductors.

Dynamo [13] An electricity generator.

ECCS [27] The emergency core cooling system, a safety feature of nuclear reactors that, in case of severe problems, is designed to flood the **core** with massive quantities of either cold water or a solution of neutron-absorbing material in water.

Ecology [6] The branch of science that studies the relationships between organisms and their environment.

Ecosystem [6] An ecological community consisting of all the species that live within an area and the environment of that area, all considered as a single system.

Efficiency [11] Most generally, the ratio of work obtained from a system to the amount of energy expended. In heat engines, often expressed as the **Carnot efficiency.**

Electric current [12] See **current (electric).**

Electrolysis [34] The splitting apart of molecules by means of an electric current. This is one process considered for the production of hydrogen from water.

Electromagnetic radiation [11] Energy that is emitted or absorbed in the form of radiation that travels at the speed of light and is characterized by various wavelengths, ranging from radio waves (very long wavelengths) through infrared radiation, visible light, to ultraviolet light (short wavelengths), X-rays and γ-rays (very short wavelengths). What we normally think of as 'light' (i.e., visible light) represents only a tiny fraction of the entire range of electromagnetic radiation.

Electromotive force [12] The difference in electrical potential energy between two points. The electromotive force of a battery or other source of electricity is the amount of work done on an electric charge as it moves through the battery (or other source). Often abbreviated as emf or e.m.f.

Electron [25] One of the fundamental components of atoms, assigned an electric charge of -1. Electrons have about $1/1,800$th of the mass of the neutron or proton, so their mass number is generally taken to be 0.

Energy [2] The ability or capacity to do work.

Energy balance [4] A relationship stating that, for any system under consideration, the difference between the energy put into the system and the energy removed from the system must somehow be stored in the system. If more energy is removed than was put in, then some energy must have been removed from storage.

Energy intensity [1] The ratio of energy consumed by a country to its total output of goods and services, i.e., the ratio of energy consumption to gross domestic product for any country.

Enhanced greenhouse effect [32] The trapping of heat in the Earth's atmosphere, and consequent rise of global temperatures, as a result of infrared-absorbing gases introduced into the atmosphere by human activity.

Exothermic [17] Any process, usually a chemical reaction, that produces or liberates heat.

Extraction [29] The removal or recovery of an energy resource from the Earth as, e.g., coal mining. The term applies in a more general sense to removal or recovery of anything, such as an ore, from the Earth.

Fermentation [34] A process in which sugar molecules are converted into ethanol. Fermentation is facilitated by enzymes present in yeasts or bacteria.

Fire-tube boiler [17] A type of boiler equipped with a number of tubes through which the hot gases produced by burning the fuel are passed. This increases the available surface area for heat transfer.

Fission [26] The splitting of an atomic **nucleus** approximately in half.

Fission products [26] The nuclei produced as a result of fission. Often (though not always), these products are radioactive.

Flashing [28] The rapid, sometimes nearly instantaneous, conversion of a liquid to its vapor, as, e.g., the flashing of water to steam. Flashing usually occurs when heat is added quickly to the liquid or when the pressure on the liquid is quickly reduced.

Flat-plate solar collector [36] A device used to absorb solar energy for heating water; the hot water produced could be used for space heating or as domestic hot water. As the name implies, it consists of a flat plate, usually painted in a dark color to help absorb solar radiation, and a system of pipes mounted on the plate. Water circulating through the pipes is heated and then distributed through the building.

Flue-gas desulfurization [18] Any process by which sulfur oxides are removed from the gaseous products of fuel combustion before they would be emitted to the environment. Usually this process is carried out in a scrubber by reacting the sulfur oxides with lime.

Fly ash [18] Small particles of the noncombustible ash produced from burning coal, which are carried out of the boiler with the gaseous products of combustion.

Free radicals [28] Chemical species containing an unpaired electron. Commonly, free radicals are the intermediate stages of a chemical reaction and are highly reactive. In contrast to ions, free radicals do not have an electric charge.

Fuel [5] Any substance that is consumed, usually by combustion, to liberate energy, usually in the form of heat.

Fuel assembly [27] A combination of a number of **fuel rods**, usually with some piece(s) of structural material to hold the rods in place. An entire fuel assembly is loaded into, or removed from, a nuclear reactor core rather than adding or removing individual fuel rods.

Fuel-lean [31] A combustion condition, usually in an internal combustion engine, in which the air-to-fuel ratio is greater than that needed for complete combustion of the fuel.

Fuel NO_x [18] Nitrogen oxides that are produced by the oxidation of nitrogen atoms chemically combined with molecules of the fuel.

Fuel-rich [31] A combustion condition, usually in an internal combustion engine, in which the air-to-fuel ratio is less than that needed for complete combustion of the fuel.

Fuel rod [27] A device for containing the fuel used in a nuclear reactor, it consists of a long, thin tube made of a metal capable of withstanding extreme temperatures (such as zirconium) into which pellets of enriched uranium oxide are loaded.

Furnace [17] A device in which a fuel is burned to liberate heat.

Gas cap gas [20] Gaseous hydrocarbons (i.e., natural gas) occurring as a free gaseous phase with liquid hydrocarbons in a petroleum reservoir. Also sometimes called associated gas.

Gasohol [34] A fuel composed of nominally 90% gasoline and 10% ethanol.

Global carbon cycle [20] A model that accounts for the fate of carbon in biological and geological processes as well as human activities. The global carbon cycle is often used for (but not restricted to) developing an understanding of the effect of various natural processes or human activities on the concentration of carbon dioxide in the atmosphere.

Global climate change [32] A change in the climate of the planet, potentially (but not necessarily) resulting from changes in the composition of the atmosphere. When the term is used in its most general sense, the climate change could be one of warming or cooling.

Global warming [32] **Global climate change** that results in an increase in the average temperature of the planet. Sometimes used synonymously with the term **greenhouse effect**.

Gravitational potential energy [8] Energy that a body or object has by virtue of its position in a gravitational field; on Earth the zero point of gravitational potential energy is taken to be the center of the Earth, so the gravitational potential energy of an object is conveniently indicated by its height above the Earth's surface (or, strictly speaking, distance from the Earth's center).

Greenhouse effect [32] An effect on Earth's climate caused by trapping heat in the atmosphere, so named by analogy to the trapping of heat in a greenhouse by the glass windows.

Greenhouse gas [32] A gas that absorbs some infrared radiation that otherwise would be lost to outer space, thereby trapping some heat in the atmosphere. Two (but not the only) greenhouse gases are carbon dioxide and methane.

Gross domestic product [1] The total output of goods and services in a country. Abbreviated as GDP.

Half-life [28] The time required for one-half of the original amount of a substance to decay or be destroyed. This term is usually applied to radioactive substances, but sometimes appears in the media in other contexts, e.g., the half-life of information or knowledge.

Heat capacity [11] For any object or system, the amount of heat required to raise its temperature by one degree. The heat capacity can be expressed also as the ratio of the heat supplied to the corresponding temperature change caused by that heat. The heat capacity per unit mass is known as the **specific heat** and is a characteristic of the material constituting the object or system.

Heat engine [11] A device that extracts useful work from a spontaneous flow of heat.

Horsepower [10] A unit of **power** equal to 746 watts or 33,000 foot-pounds per minute. Originally derived from the power that could be exerted by a horse in pulling or lifting a load.

Hydroelectricity [15] Electricity produced in systems in which water (flowing over a waterfall or a dam) is used as the working fluid in the **turbine** used to operate the generator.

Hydrolysis [3] Any chemical process in which a molecule is broken down by the action of water, such as in the conversion of starches to simple sugars.

Hypothetical resource [33] An amount of material not yet discovered, but estimated to exist in areas known to contain other deposits or reservoirs of that material.

Indirect conversion [36] A process for producing electricity from solar energy in which the solar energy is used to produce steam, and the steam in turn operates a turbine connected to a generator.

Insulator [12] A material through which an electric current does not readily flow, such as plastics, rubber, or wood.

Ion [25] A chemical species that has an electric charge, produced when an atom gains or loses electrons.

Isothermal expansion [24] The expansion of a gas that occurs at a constant temperature.

Isotope [25] Atoms of the same chemical element (i.e., having the same atomic number) but having different **mass numbers**.

Kerogen [20] Any of several organic substances produced by the chemical recombination of products of the anaerobic decay of organic matter. On subsequent heating inside the Earth's crust, kerogens transform into the fossil fuels.

Liquefied petroleum gases [20] Propane and butane sold in liquid form for use as a fuel. These gases can be liquefied at normal temperatures by moderately increased pressure (125 and 30 pounds per square inch, respectively), and are sold in cylinders or canisters under these pressures. Often known as LPG, LP gas, or bottle gas. Propane is often used as an industrial fuel, sometimes as a vehicle fuel (as in propane-fueled forklift trucks or even taxis), and is a good domestic fuel in regions where natural gas is not available. Liquefied butane is the fuel inside disposable cigarette lighters.

Magnetic confinement [37] A technique that uses very powerful, externally applied magnetic fields to confine the **plasma** in a nuclear fusion reactor.

Magnetohydrodynamics [37] The interaction of an electrically conducting fluid, such as a **plasma** or ionized gas, with a magnetic field; commonly abbreviated as MHD.

Mass number [25] The sum of the numbers of **protons** and **neutrons** in an atomic **nucleus**.

Mechanical equivalent of heat [11] The amount of mechanical work that is equivalent to one unit of heat energy; 4.186 joules per calorie, or 778 foot-pounds per Btu.

Meltdown [28] A serious and potentially catastrophic event in which the temperature in the core of a nuclear reactor becomes so high that the core melts.

Mined land reclamation [29] A set of procedures, usually mandated by legislation or regulation, intended to restore the topography and ecosystem of land on which mining has occurred to the conditions that existed before the mining took place.

Mobile source [30] A source of air pollution that is capable of moving, and usually does move, from place to place. Motor vehicles are examples of mobile sources.

Moderator [27] A substance used in the core of a nuclear reactor to adjust (or moderate) the kinetic energy of the neutrons produced from nuclear fission, so that their energy will be in the appropriate range to cause another fission, and continue the chain reaction.

Motor fuel alkylate [21] The product of **alkylation**, usually having octane numbers of about 100.

Multiplication factor [27] Represents the ratio of the number of **neutrons** present in a nuclear reactor as a result of one particular 'generation' of fission reactions to that number present before those reactions occurred. For example, if 1,000 neutrons were involved in fission reactions that collectively produced 2,000 neutrons, the multiplication factor would be 2.

Natural gasoline [20] A mixture of hydrocarbons that are liquids at ordinary temperatures, but exist largely or completely in the gaseous state at the temperature of an oil reservoir. When gas from the reservoir is cooled from reservoir temperature to normal temperatures, these compounds condense back to the liquid state, the liquid being called natural gasoline.

Natural greenhouse effect [32] The warming of the Earth caused by trapping of radiation by components of the atmosphere—principally carbon dioxide and water vapor—that occur naturally, rather than being in the atmosphere as a result of human activity.

Neutron [25] One of the fundamental components of the atomic nucleus, assigned an electric charge of 0 and a mass number of 1.

n–p Junction [36] A material containing both **n-type semiconductors** and **p-type semiconductors**; or the junction between the two types.

n-Type semiconductor [36] A **semiconductor** that has been produced by treating an element with small amounts of another element that has more valence electrons; e.g., a semiconductor produced by treating silicon (four valence electrons) with arsenic (five valence electrons).

Nuclear fusion [37] A process in which two small nuclei are combined to produce a larger one.

Nucleon [25] A generic term referring to any component of an atomic nucleus, i.e., protons or neutrons.

Nucleus [25] The central structure in an atom, consisting of neutrons and protons, and accounting for more than 99.9% of the mass of the atom.

Nutritional calorie [3] A measure of energy in food, equal to 1,000 calories or one kilocalorie.

OPEC [33] The Organization of Petroleum Exporting Countries, a cartel that seeks to set world petroleum prices and to regulate oil production among its members.

Oxygenates [31] Any of several families of chemical compounds containing an atom of oxygen, used in gasoline to reduce the tendency to evaporation and reduce the production of carbon monoxide and nitrogen oxides during combustion.

Particulate emissions [18] Fine particles of ash and/or soot produced during combustion of a fuel and released to the environment.

Passive solar energy system [36] A system for capturing solar energy for space heating in buildings that relies on the design of the building itself, and not on having to use specific devices for capturing the solar energy.

Peak watt [36] In photovoltaics, the generation of one watt of electrical power when the solar energy is 1,000 watts per square meter.

pH [30] A logarithmic scale used for measuring the acidity or basicity of a solution in water. The pH is given by the negative logarithm of the concentration of hydrogen ions; for example, a solution with a hydrogen-ion concentration of 10^{-5} would have a pH of 5. The pH of chemically pure water is 7; solutions with pH values below 7 are acidic, and, above 7, basic.

Photochemical smog [31] An air pollution problem resulting from chemical reactions among carbon monoxide, nitrogen oxides, and unburned hydrocarbons from vehicle exhausts, facilitated by sunlight.

Photosynthesis [3] A process, facilitated by sunlight, in which carbon dioxide and water are used by living plants to produce sugars and other chemical compounds needed for the plant's life processes.

Photovoltaics [36] The conversion of solar energy to electricity.

Plasma [37] A highly, or completely, ionized gas, such as produced in nuclear fusion reactors.

Possible resource [33] An amount of resource that may exist but for which no data or exploration actually confirms its existence.

Power [2] The rate of using **energy** or doing **work**.

Pressure-tube reactor [27] A nuclear reactor design that relies on placing fuel rods inside tubes that contain coolant moving through at high pressure, the tubes being surrounded by moderator at low pressure.

Pressure vessel [27] A very thick, steel vessel used to hold the core of a nuclear reactor.

Pressurized water reactor [27] A type of nuclear reactor in which the water coolant is maintained under high pressure so that it does not boil; the high-temperature liquid water is used as a heat source to produce steam in a separate unit (the heat exchanger or steam generator). In contrast to the simpler boiling water reactor design, the steam actually sent to the turbines has never been exposed to the reactor core.

Primary cell [13] Commonly called a 'battery,' this device produces electricity by reactions of chemicals sealed inside and which are not reversible; i.e., the battery is not rechargeable.

Probable resource [33] An amount of resource based on some body of limited evidence, data, or exploration, but not commercially exploited.

Prompt neutron [27] A neutron released during the fission process.

Proton [25] One of the fundamental components of the atomic nucleus, assigned an electric charge of +1 and a mass number of 1.

Proved reserve [33] The amount that can be recovered with reasonable certainty from reservoirs whose existence and limits have been established, and recovered using existing economic technology.

p-Type semiconductor [36] A **semiconductor** that has been produced by treating an element with small amounts of another element that has fewer valence electrons; e.g., a semiconductor produced by treating silicon (four valence electrons) with indium (three valence electrons).

Pulverized coal firing [17] A method of burning coal in which finely ground coal is suspended in a stream of air and blown into a boiler. This method is intended to maximize the rate at which the coal can be burned, and hence maximize the rate at which heat is released.

Rad (unit) [28] A unit that expresses the absorbed dose of radiation, equivalent to 100 ergs per gram (or 0.01 joules per kilogram).

Radioactivity [25] The spontaneous emission of radiation from atoms.

Reformulated gasoline [31] A type of gasoline developed to help combat air pollution from automobiles. The principal changes involved in reformulation of gasoline include changing the boiling range, to reduce vapor pressure and hence production of unburned hydrocarbons by evaporation; adding oxygenates to reduce formation of carbon monoxide and NO_x, and also to enhance octane number; and reducing aromatics, to reduce soot formation.

Rem (unit) [28] The röntgen-equivalent-man, the dosage of radiation equal to the amount that produces the same damage in humans as one **röntgen** (see below) of high-energy X-rays.

Renewables [34] Any of several possible energy sources—such as wind, biomass, or solar—that in principle are inexhaustible and the use of which is generally accompanied by little or no pollution.

Reserve [33] The amount of a material that can be recovered by known, economical technology, and usually established by detailed geological exploration.

Reservoir [20] An underground rock formation in which oil or natural gas accumulates.

Resistance [12] The opposition to flow of an electric current.

Resistivity [12] A characteristic property of a material that takes into account the resistance to the flow of an electric current, the cross-sectional area of current flow, and the length of the path through which the current must flow.

Resource [33] The entire amount of a material known or estimated to exist.

Röntgen (unit) [28] A dosage of exposure to ionizing X-rays or γ-rays. The symbol usually used is R; in older publications r is sometimes used.

Scramming [27] The insertion of all control rods into the core of a nuclear reactor, which is done in emergencies to shut down the reactor.

Scrubber [18] A device for performing flue gas desulfurization, usually by contacting the sulfur-laden flue gases with lime or limestone.

Secondary cell [13] Also called a rechargeable battery or storage battery. A device that produces electricity via chemical reactions in which the chemicals can be restored by recharging, i.e., sending an externally supplied electric current through the cell.

Semiconductor [36] A material of which the electrical conductivity is intermediate between that of a good conductor and that of an insulator.

Sink [32] In the global carbon cycle, anything that removes carbon dioxide from the atmosphere. An example of a very important sink is photosynthesis by growing plants.

Sour [20] Natural gas or oil containing hydrogen sulfide, or, generally, sulfur compounds of any sort.

Source [32] In the global carbon cycle, anything that adds carbon dioxide to the atmosphere. For example, the combustion of fossil fuels is a CO_2 source.

Specific heat [3] A characteristic property of a substance indicating the amount of heat required to raise its temperature by one degree per unit mass of the substance.

Speculative resources [33] An amount of material not yet discovered, and estimated to exist in areas not yet known to contain other deposits or reservoirs of that material.

Spent [27] In reference to nuclear fuel or fuel rods, the condition in which the amount of remaining fissionable U^{235} is too low to sustain a chain reaction any longer.

State [5] A particular condition of a **system**, usually specified by observations of one or more of its properties.

Stationary source [30] A source of air pollution that is incapable of moving. Electricity-generating plants are an example of a stationary source.

Steam-tube boiler [17] A boiler in which water and steam are contained in a (usually large) number of tubes, this design maximizing the surface area available for transferring heat to the water or steam.

Stoichiometric combustion [31] A combustion condition in which exactly enough air is available for complete combustion, i.e., conversion of all the carbon in the fuel to carbon dioxide and all the hydrogen in the fuel to water.

Storage battery [13] See **secondary cell**.

Straight-run gasoline [21] Gasoline produced directly by distillation of petroleum.

Superconductor [12] A material in which the electrical resistivity is essentially zero.

Sustainable energy source [34] An energy source that can be regrown after being harvested. An example would be sunflowers harvested to produce oil, with a new crop grown the next year.

Sweet [20] Natural gas or oil containing little or no hydrogen sulfide or other sulfur-containing compounds.

Sweetening [20] A process by which hydrogen sulfide is removed from sour natural gas.

System [4] That part of the universe that we take or set aside for our observation or study.

Thermal cracking [21] Any of several processes relying on heat to break large petroleum molecules into smaller ones. This increases the gasoline yield, but nowadays is mainly used to reduce the viscosity of crude oil or high-boiling petroleum fractions.

Thermal NO_x [18] Oxides of nitrogen produced by the oxidation of nitrogen molecules in the air, facilitated by the extremely high temperatures in some combustion systems.

Thermonuclear reaction [37] A **nuclear fusion** process brought about by extremely high temperatures.

Tokamak [37] A type of **nuclear fusion** reactor in which the plasma is confined inside a doughnut-shaped (toroidal) vessel.

Transesterification [34] In the context of biomass energy, a reaction of plant oils with methanol to produce methyl esters that are superior to the original plant oils as fuels for diesel engines.

Transmutation [26] The conversion of one chemical element into another.

Trombe wall [36] An architectural device, usually installed inside a bank of windows, which is used for absorbing solar radiation during daylight hours and radiating the absorbed energy as heat to the interior of the building at night.

Turbine [15] A device that extracts rotary mechanical work from the kinetic energy of a high-pressure moving fluid.

Ultimate recovery [33] The total amount of a material, such as oil, that will have been removed from the Earth when production eventually comes to an end.

Unburned hydrocarbons [31] Gasoline components emitted to the air, either by evaporation from fuel tanks and fuel lines, or by being swept through the engine without burning.

Uranium enrichment [27] Any of several processes for increasing the proportion of fissionable U^{235} to nonfissionable U^{238}. The products are weapons-grade uranium (about 90% U^{235}), reactor-grade uranium (about 3% U^{235}), and residual depleted uranium (essentially 100% U^{238}).

Utilization [29] Liberation of the energy from an energy source in a useful and controlled fashion. Usually this term refers to burning a fuel to liberate its energy as heat.

Vapor lock [21] Interruption of the flow of gasoline in the fuel system of an automobile, caused by evaporation of the gasoline to form bubbles of vapor in the fuel lines.

Vapor pressure [20] The pressure exerted by the molecules of a vapor in equilibrium with a liquid. Vapor pressure increases with increasing temperature; when the vapor pressure becomes equal to the prevailing atmospheric pressure, the liquid boils. The higher the vapor pressure of a liquid at a given temperature, the greater its tendency to evaporate.

Volt (unit) [12] The unit of electrical potential energy.

Water-tube boiler [17] Another term for **steam-tube boiler**.

Water wall [18] One or more of the walls of a water-tube or steam-tube boiler, made almost completely from the tubes themselves.

Watt (unit) [12] A unit of power, normally used in electrical systems.

Wind farm [35] A collection of **wind turbines** in one location, used to generate electricity.

Wind turbine [35] A device in which the kinetic energy of air (i.e., wind) is used to operate an electricity generator.

Work [2] Moving an object into or out of position, especially against a force or resistance.

Working fluid [15] The high-pressure fluid, commonly water or steam, used to operate a **turbine**.

BIBLIOGRAPHY: RESOURCES FOR LEARNING MORE

A THREE-FOOT SHELF OF BOOKS

The list below represents a selection of books that bear directly on subjects covered in throughout this book. Most are still in print. All are either introductory textbooks or are written for the lay reader who has little previous background in science. The entire collection would fill only a single, three-foot-long shelf, with just enough space left over for a good dictionary. The half-dozen of these books that are particularly noteworthy are denoted by a (☆) symbol.

The Physical Science Background

☆ Asimov, Isaac. *The History of Physics*. Walker: New York, 1966.

Bloomfield, Louis. *How Things Work: The Physics of Everyday Life*. Wiley: New York, 1997.

Gebelein, Charles G. *Chemistry and Our World*. Brown: Dubuque, 1997.

Glashow, Sheldon L. *From Alchemy to Quarks: The Study of Physics as a Liberal Art*. Brooks/Cole: Pacific Grove, 1994.

Hewitt, Paul G. *Conceptual Physics*. Addison-Wesley: Reading, 1998.

☆ McPhee, John. *Annals of the Former World*. Farrar, Straus, Giroux: New York, 1998.

Schwartz, A. Truman; Bunce, Diane M.; Silberman, Robert G.; Stanitski, Conrad L.; Stratton, Wilmer J.; Zipp, Arden P. *Chemistry in Context: Applying Chemistry to Society*. McGraw-Hill: New York, 1997.

Historical Development of Technology

☆ Cardwell, Donald. *The Norton History of Technology*. Norton: New York, 1995.

Derry, T.K.; Williams, Trevor I. *A Short History of Technology: From the Earliest Times to A.D. 1900.* Dover: New York, 1993.

Josephson, Paul R. *Red Atom: Russia's Nuclear Program from Stalin to Today.* Freeman: New York, 2000.

Kirby, Richard Shelton; Withington, Sidney; Darling, Arthur Burr; Kilgour, Frederick Gridley. *Engineering in History.* Dover: New York, 1990.

Romer, Alfred (Ed.). *The Discovery of Radioactivity and Transmutation.* Dover: New York, 1964.

Strandh, Sigvard. *The History of the Machine.* Dorset: New York, 1979.

Usher, Abbott Payson. *A History of Mechanical Inventions.* Dover: New York, 1988.

Energy Science and Technology

Aubrecht, Gordon. *Energy.* Merrill: Columbus, 1989 (second edition published by Prentice-Hall: Englewood Cliffs, 1995).

Berger, John J. *Charging Ahead: The Business of Renewable Energy and What It Means for America.* University of California: Berkeley, 1997.

Cassedy, Edward S.; Grossman, Peter Z. *Introduction to Energy: Resources, Technology and Society.* Cambridge University: Cambridge, 1998.

☆   Faraday, Michael. *The Chemical History of a Candle.* Collier: New York, 1962 (various editions of this book are available).

Flavin, Christopher; Lenssen, Nicholas. *Power Surge: Guide to the Coming Energy Revolution.* Norton: New York, 1994.

Hill, Robert; O'Keefe, Phil; Snape, Colin. *The Future of Energy Use.* Earthscan: London, 1995.

☆   Ramage, Janet. *Energy: A Guidebook.* Oxford University: Oxford, 1997.

Ristinen, Robert A.; Kraushaar, Jack J. *Energy and the Environment.* Wiley: New York, 1999.

Rossotti, Hazel. *Fire.* Oxford University: Oxford, 1993.

Fuels

Berger, Bill D.; Anderson, Kenneth E. *Modern Petroleum: A Basic Primer of the Industry.* PennWell: Tulsa, 1992.

Johansson, Thomas B.; Kelly, Henry; Reddy, Amulya K.N.; Williams, Robert H. (Eds.). *Renewable Energy: Sources for Fuels and Electricity.* Island Press: Washington, 1993.

Lorenzetti, Maureen Shields. *Alternative Motor Fuels: A Nontechnical Guide.* PennWell: Tulsa, 1996.

Röbbelen, G.; Downey, R.K.; Ashri, A. *Oil Crops of the World.* McGraw-Hill: New York, 1989.

Schobert, Harold H. *Coal: The Energy Source of the Past and Future.* American Chemical Society: Washington, 1987.

Tillman, David A. *Wood as an Energy Resource.* Academic Press: New York, 1978.

Environmental Issues

Christianson, Gale E. *Greenhouse.* Walker: New York, 1999.

Goldemberg, José. *Energy. Environment, and Development.* Earthscan: London, 1996.

Hollander, Jack M. (Ed.). *The Energy–Environment Connection.* Island Press: Washington, 1992.

Houghton, John. *Global Warming: The Complete Briefing.* Cambridge University: Cambridge, 1997.

☆   McNeill, J.R. *Something New Under the Sun: An Environmental History of the Twentieth-Century World.* Norton: New York, 2000.

Schneider, Stephen H. *Laboratory Earth.* Basic Books: New York, 1997.

REFERENCE SOURCES

Two specialized dictionaries that are useful references for the reader exploring energy science and technology are the *McGraw-Hill Dictionary of Scientific and Technical Terms* (McGraw-Hill: New York, 1994), and *A Dictionary of Mining, Mineral, and Related Terms* (U.S. Department of the Interior, 1968). The latter seems unfortunately to be out of print, though sometimes can be encountered in used-book stores, and is a cornucopia for those wanting to find out about drill bortz, gin pits, or a rollman. Of course, anyone wanting to learn more about anything is well-served by starting with the *Encyclopaedia Britannica*.

OTHER RESOURCES

The magazine *American Heritage of Invention and Technology* is an excellent source of articles on the development of various forms of technology (covering a much wider field than only energy technology) and their societal implications. The articles are generally written for the layperson.

    Lindsay Publications, Bradley, Illinois, reprints textbooks and monographs from the late 19th and early 20th century. Many relate to some aspect of energy science or engineering. Most are available as inexpensive paperbacks. They are excellent sources for readers interested in the history of technology or for adventurous do-it-yourselfers. Collectively, the various reprinted books offered by this firm cover a wide range of technologies, some of which are seriously weird.

INDEX

Aborigines, 41
Absolute zero, 133–134, 138, 139
    concept, 134
    definition, 134
Abul-Hasan Al-Hasudi, 102
AC system, 191–193, 201, 277
Acetaldehyde, 467
Acid mine drainage, 428–429
Acid neutralizing capacity, 449
Acid rain, 7, 441–461
    causes of, 444–448
    definition, 443
    destruction of terrestrial plants, 453–455
    effects on aquatic ecosystems, 450–452
    effects on buildings and statues, 448–449
    environmental consequences, 448–456
    geographical factors, 456–457
    human health effects, 455–456
    'natural,' 443–444
Acidification of natural waters, 449–450
Acids, 442
Actinides, 413
Adair, Gene, 193
Adiabatic expansion, 346
Adipose tissue, 34
Aeneas, 40
Aerosols, 448–449, 488
Africa, energy crisis, 56
Agricultural wastes, 553
Agriculture, 1
    shift of prime regions, 492
Agroforestry, 58, 544
Air, 18

components, 76
    in combustion process, 75–76
Air brake, 191
Air pollution, 246, 306, 308, 434, 437, 438, 463, 543
Airline crashes, 314
Airplanes, 271, 272, 328
    diesel engine, 350–351
    gas turbine engine for, 276
ALARA principle, 411
Alaskan pipeline, 431
Alaska's North Slope, 521
Albedo, 543
Alcohol, 272
Alcohol fuels, 544–545, 553
Aldehydes, 309
Algae, 539
Alkylation, 302
Alley, Richard B., 499
Allowed direction, 173
Alloys, 42
al-Mas'ûdî, 102
α-decay, 372
α-particles, 362, 372, 375–377, 403–405, 408, 410
Alternating current (AC), 191–193, 201, 277
Alternative fuels, 277–278
Aluminum, 456
Amber, 153
American Civil War, 95
Ammonia, 133, 438
Ammonium sulfate, 488
Ampère (unit), 160
Ampère, André Marie, 160, 175

Amphibians, 451–452
Amphibious digger, 264, 265
Amphibolos, 264
Anaerobic digestion, 559–560
Analytical engine, 189
Anasazi people, 30
Andaman Islands, 39
Anderson, F.C., 549, 563
Anderson, Kenneth E., 295, 309, 439
Anderson, Robert O., 295, 309, 439, 529
Animal energy, 28
Animal muscles, 27
Animal wastes, 55, 438
Animals
    and work, 15
    domestication, 14–15
    for transportation, 257–259
Anthracosis, 428
Anthropogenic greenhouse effect, 482
Anthropogenic SO_x, 247
Antiknock agents, 305
Antimony sulfide, 78
Antipater of Thessalonica, 89, 99
Antipollution legislation, 431
Aphrodite, 40
Aquatic energy crops, 539
Aquatic systems
    effects of acid rain, 450–452
    remediation, 452
Arc light, 181, 182, 183, 184, 186
Archimedes, 587
Arctic National Wildlife Refuge (ANWR), 522
Aristotle, 9–10
Armature, 192
Armengaud, Marcel, 332
Aromatic hydrocarbons, 288–289, 291, 339
    removal, 308
Arrays, 599
Arrhenius, Svante, 482, 500
Arsenic, 597
Ash, 251–252
Ash hopper, 244
Ash residue, 243
Ashes, 60, 72
Asimov, Isaac, 16, 150, 167, 228, 365, 384, 401,
    617
Atmosphere, 29, 447
Atmospheric engine, 115–119, 130
Atmospheric pollution, 61
Atmospheric pressure, 114, 116
Atomic bomb, 394, 399, 403, 406, 413–414, 608,
    611
Atomic energy, 355–366, 383, 614
    see also Nuclear energy
Atomic number, 364, 365, 369, 375, 377
Atomic structure, 362–365, 373
Atomic weight, 364
Atoms, 355–367
Aubrecht, Gordon, 16, 167, 385, 401, 439, 474, 500,
    563, 603, 617

Auto Club, The, 321
Auto touring, 322
Autoignition temperature, 300
Automobile, 264–266, 270
    development and impact, 311–325
    early impact, 271
    mass ownership, 322
    prehistory, 268
    time line of major developments, 312
    transforming the American scene, 322–325
    use effects, 317–325
Automobile battery, 174
Automobile emissions, 463–475
Automobile manufacturing, 314–317
Average global temperature increases, 490–491

B vitamins, 30
Babbage, Charles, 189
Babcock and Wilcox Company, 240
Background radiation, 409–410
Backlog, use of term, 53
Bacon, Francis, 141
Baghdad, 93
Bailey, William, 587
Baldwin, Neil, 193
Banû Mûsà brothers, 102
Barbegal, 91
Barber, John, 331
Baritz, Loren, 325
Barium, 377, 378
Barium cyanoplatinate, 357
Barnacles, 430
Bases, 442
Bass, Rick, 513, 520, 529
Batteries, 172–174, 470–474, 601
Battery, forerunner of, 172
Becquerel, Edmund, 595
Becquerel, Henri, 358–362, 365, 381, 384
Becquerel (unit), 404
Belisarius, 86
Bell, Henry, 262
Benson, Richard, 132, 150
Bentele, Max, 341
Bentonite clay, 303
Benz, Karl Friedrich, 269, 314
Benz(a)anthracene, 61
Benzene, 288–289, 291
Benzo(a)pyrene, 61
Benzo(b)fluoranthene, 61
Benzo(e)pyrene, 61
Benzo(j)fluorothene, 61
Berger, Bill D., 295, 309, 439
Berger, John J., 474, 563, 579, 603
Berman, Marshall, 325
Bernal, J.D., 126, 150
Berry, R. Stephen, 16, 99
β-decay, 372, 376, 377
β-particles, 362, 377, 404–405, 408
β-radiation, 376

Bethe, Hans, 606–607
Bicycles, 313–314
Billington, David P., 193
Biogas, 559–560
Biological pump, 496
Biomass, 531–564
    attractive features, 534–535
    categories, 533
    CO_2-neutral, 535–536, 539
    conversion, 557
    definition, 532
    disadvantages of, 539–540
    emissions, 539
    gasification, 561
    mechanism, 533–534
    overview, 532–540
    solid, 533
Bitumen, 512, 517
Black lung disease, 428
Black, Newton Henry, 78
Blake, William, 94, 99
Blast furnace, 95
Blend stocks, 307
Blending, 307
Bleviss, Deborah L., 309
Bloomfield, L.A., 151, 167, 617
Boats, 259–260
Bodanis, David, 385
Boeing 707 airliner, 332
Bofors, 276, 328
Bogotá, Colombia, 203
Bohr, Niels, 378, 379, 387
Boiler explosions, 124, 263
Boilers, 116, 229–240, 398
    design improvements, 242–243
    full system, 245
    function, 230, 232–236
    heat loss, 243
    heat transfer surface, 233
    increase in requirements, 241
    inspection, 263
    simple design, 232
    types, 233–236
Boiling water reactor (BWR), 394–395
Booth, J.C., 327
Borowitz, Sidney, 474, 563, 579, 603
Boulton and Watt, 268
Boulton, Matthew, 124, 126
Boyle, Robert, 141, 219–223
Boyle's law, 220–223, 345
Brazil, 58, 553
Breast wheel, 86
Briggs, Asa, 99, 126
British thermal unit (Btu), 20
British Thomson-Houston Co., 330
Bronowski, Jacob, 47, 167, 376, 384
Brϕnsted (or Brϕnsted–Lowry) acid, 445
Brooke, Bob, 603
Brooks, Laura, 110
Brower, Michael, 209, 474, 563, 579, 603

Brown, Duncan, 475
Brown's Ferry nuclear plant, 413
Brunel, Isambard Kingdom, 262–263
Bruno, Leonard C., 126
Buchanan, R.A., 127
Buffering capacity, 449
Buffon, George, 587
Bunsen, Robert, 606
Burke, J., 167
Burkett, Allan R., 228
Burner cans, 334
Burners, 53–54, 243, 244, 333
Burning
    process, 25, 42–47
    wood, 59
    see also Combustion; Fire
Buses, 264–266
Bush, President, 514
Butane, 286, 288, 292
Butanol, 556
Butler, Samuel, 31

CAFE, 513, 514
Calcium carbonate, 444, 458
Calcium compounds, 408
Calcium hydroxide, 458
Calculators, 6
Calculus, 11
Calder Hall nuclear power station, 399
Calder, N., 371, 384
Calley, John, 119
Caloric theory of heat, 140–142
Calorie (unit), 20, 34, 35
Calorimeter, 18–19
Calvin, Melvin, 29
Campbell, Colin, 518, 529
Campbell, J.L., 63
Cancer of the scrotum, 72
Caneva, Kenneth L., 31
Canterbury Tales, 92
Capacity factor, 579
Car. See Automobile
Carbohydrates, 23–24, 34
Carbon, 17, 18, 22, 77, 78, 409
Carbon atom, 45, 287–288, 290
Carbon cycle, 281–283
Carbon dioxide, 23, 25, 29, 44–46, 62, 179, 400–401,
    432–434, 438, 443, 482–486, 490, 491,
    493–495, 499, 508, 531, 535–536, 551, 552,
    554, 559, 560, 563
Carbon monoxide, 76–77, 306, 308, 432–433, 437,
    464–466, 468, 474
Carbon Petroleum Dubbs, 303
Carbonic acid, 23, 443
Carcinogens, 61, 72
Cardano, Girolamo, 114
Cardwell, Donald, 127, 151, 167, 180, 193, 228, 341,
    401
Caret, Robert L. et al., 31

Carey, John, 73, 78, 107, 110, 193, 278, 360, 365, 374, 375, 384
Carnot cycle, 224
Carnot efficiency, 138, 224–226, 396
Carnot equation, 338, 394
Carnot, Nicholas-Leonard Sadi, 135–136, 143, 148, 150, 224, 274
Carson River, 95
Carter, President, 513
Cartledge, Bryan, 425
Cash, John D., 563
Cash, Martin G., 563
Caspian region, oil supplies, 524–525
Cassedy, Edward S., 8, 401, 425
Cast iron stoves, 71
Catalyst, 303
Catalytic converter, 467–469
Catalytic cracking, 303–304
Catalytic reforming, 304–305
Cataract, 447
Cathode-ray tube, 356
Cauciescu, Nicolai, 3
Cayley, George, 268, 271, 272
Celestine III, Pope, 103
Cellulose, 59, 540–541
Celsius scale, 134
Central heating
    history, 65–66
    public baths, 66
Ceramics, 42
Cesium, 408
Cetane number, 351–352, 555
Chadwick, James, 385, 406
Chain reaction, 381–384, 389, 392, 394
Chapelon, André, 278
Char, 59
Charles, Jacques Alexandre Cæsar, 130, 133, 219, 220, 223
Charles' Law, 133, 139, 334, 345
Charlotte Dundas, 262
Chaucer, Geoffrey, 92, 93, 99
Chelyabinsk-40 facility, 412
Chemical analyses, 17
Chemical bond, 596
Chemical potential energy, 84, 132, 139, 173, 174, 433
Cheney, Margaret, 194
Chernobyl nuclear reactor accident, 391, 397, 412, 419–424, 614
Chernow, Ron, 529
Chess-playing Turk, 213–214
Chimney cleaning, 72
Chimney sweep, 72, 73
China, 65, 89, 102
    oil supplies, 525
China syndrome, 412
Chlorofluorocarbons, 481, 488
Chrysene, 61
Chrysler, 276
Churchill, Winston, 275, 530

Claremont, 262
Clay, 42
Clay minerals, 303
Clean Air Act, 1990, 308, 436, 457
Clean air acts, 247, 248
Clean Coal Technology Program, 250
Climax heating system, 587
Cloud point, 339
Coal, 7, 71–74, 281, 546–547
    chemical potential energy, 119
    cleaning, 431–432
    constituents, 244
    costs, 254
    energy release, 23–25
    exploring, 503–504
    low-sulfur, 7, 436
    molecular composition, 244
    partial digestion, 25
    structure, 244
    sulfur content, 445
Coal beneficiation, 431–432
Coal cleaning, 247, 436, 458
Coal combustion, by-products, 244–247
Coal dust, 428
Coal-fired power plant, 241–255
    overall layout, 252–254
Coal firing, 230, 236–239
Coal gas, 268
Coal mines, 428
    draining, 117
Coal reserves, 509, 510, 528–529
Coalbrookdale, 95
Cobalt, 413
Cogeneration process, 541
Cohen, I. Bernard, 63
Cohn, Steven Mark, 401
Coleridge, Samuel Taylor, 123
Collier, John G., 401
Collier Trophy, 351
Colorado River, 206, 208
Combustion, 237, 551
    chemistry, 75–77
    complete, 76
    energy release, 47
    for home comfort, 65–79
    incomplete, 76
    methane, 45, 46
    oxygen-starved, 77, 78
    wood, problems associated with, 60–61
Combustion chamber, 238, 331
Combustion conditions, 464, 469
Combustion gases, 243
Combustion performance, 298–299
Combustion process, 433
    air in, 75–76
Combustion products, 441
Combution, methane, 46
Compact disks, 6
Complete combustion, 76
    burners, 53–54

Compressed air, 182
Compressed natural gas, 278
Compression ignition, 343
Compression ignition engine, 346
Compression ratio, 298, 299, 347, 352
Compression stroke, 346, 347
Conant, James Bryant, 78
Conaway, Charles, 295, 530
Concorde airliner, 330
Condenser, 253–254, 398
Condensing, 118
Condensing turbine, 215
Conduction, 148–149, 230, 231, 235
Conductors, 163
Constantinople, 66
Consumption rate, 512
Containment structure, 391
Control rods, 389, 390, 392, 394, 396, 398
Convection, 149, 230, 231, 235
Cooke, A.F., 240
Cooking of food, 41
Cooking ovens, 41
Cooking vessels, 42
Coolants, 391, 396
Cooling towers, 398, 434
Cooling water, 253
Copepods, 451, 452
Cord, measure of wood, 51, 54
Corn, 552, 553
Corporate Automobile Fuel Efficiency (CAFE)
     standards, 513, 514
Cotton industry, 122
Cracking, 302–304
Crayfish, 452
Creosote, 61, 72
Cross, Gary, 279
Crude oil, 287, 290, 429
     analysis, 292
     categories, 291–292
     chemical compounds, 293
     inventories, 507
Crutzen, Paul J., 255
Cube taps, 3
Culm banks, 432
Cultures, two, 7
Cummins, Lyle, 354
Curie, Irène, 406
Curie, Marie, 361, 367, 374, 405
Curie, Pierre, 374
Curve of binding energy, 369–370, 374, 380, 607
Cyanides, 431
Cybernetics, 122
Cyclic paraffins, 288, 291
Cycloalkanes, 288
Cylinder, 116

Dahlin, Dennis, 63, 79
Daimler, Gottlieb, 269, 270, 314
Dalén, Nils, 331, 332, 341

Dams
     environmental factors, 200, 203–209
     pro and con, 204–209
     safety concerns, 204–205
Darling, Arthur, 100, 127, 180
Darwin, Charles, 512, 529
Davenport, Thomas, 189
Davies, Robertson, 74, 78, 79
Davy, Sir Humphry, 144, 172, 183, 230
Davy, M.J.B., 340
DC system, 191–193, 201
De architectura, 90
DeBlieu, Jan, 110, 579
de Camp, L. Sprague, 31, 99, 110
de Caus, Solomon, 213
Decibel (unit), 577
de Duve, Christian, 84, 99, 173, 180
Deforestation, 58, 543
Degen, Paula, 279
de Hautefeuille, Jean, 268
DeHaviland Comet, 332
de Laval, Carl Gustaf Patrik, 214, 228
de Laval turbine, 214–215, 332
Delayed neutrons, 392–393
della Porta, Andrea, 213
de Lorenzi, Otto, 240
Dengue fever, 489
Denmark, wind technology, 567–568, 574, 577
Deptford Power Station, 190
Derry, T.K., 31, 47, 99, 110, 127
de Saussure, Nicholas, 587
Desertification, 58
Deserts, spread of, 493
Desulfurization, 437
Deuterium, 390, 609–610, 614–615
Developing countries, 601
Developing nations. See Third World
Dibenz(a,h)anthracene, 61
Dibenzo(a,l)pyrene, 61
Dickens, Charles, 11, 16, 23, 31
Diesel engine, 264, 274–275, 277, 337, 343–354
     advantages, 352
     airplane, 350–351
     commercial applications, 348
     early development, 344–345
     marine applications, 349
     operation, 345–348
     railways, 349–350
     road transportation, 350
Diesel fuel, 1, 289, 304, 343–354, 556
     cetane number, 351
     properties, 351
     quality issues, 351
Diesel locomotive, 275, 349
Diesel Motor Co. of America, 349
Diesel, Rudolf, 274, 343, 348, 352, 353
Diesel-electric locomotive, 275, 349–350
Diesel-engine generator, 419–420
Digestion, limitations, 25–29
Direct current (DC), 191–193, 201

Direct solar electricity generation, 595
Disease and climate change, 489–490, 493
Disordered energy, 38
Dissolved gas, 285
Distillation, 293
    straight-run gasoline from, 301
Distillation tower, 293–295
Distribution system, 166
Dixon, Jeremiah, 172
Domesday Book, 91, 92
Domestic heating, 65–79
    fuels, 71–75
Doping, 596–597
Doyle, A. Conan, 185, 384
Dreadnought, 219
Drexel, Morgan and Company, 187
Driftwood, legislation, 50–51
Drive-in movies, 323
Drive-in restaurant, 323, 324
Dubbs, P.C., 303
Duckweed, 550
Du Fay, C.F., 156, 170
Dunn, Seth, 563
Duquesne Light Company Colfax station, 227
Dynamo, 177, 186
    see also Generator
Dyson, Freeman, 500

ECCS. *See* Emergency core cooling system
Ecology, 58
Economic development, 6
Ecosystem destruction, 58
Eddington, Arthur, 606
Edelson, Edward, 439, 475
Edison Electric Light Company, 187
Edison Electric Tube Company, 187
Edison General Electric Company, 188
Edison Lamp Works, 187
Edison Machine Works, 187
Edison, Thomas, 185–187, 191, 193, 201
Edith's checkerspot butterfly, 489
Efficiency, 224, 227, 232, 239, 253, 338, 347, 352,
    438, 593
    concept, 130
    definition, 129
    formulas for, 138–139
    notion of, 129
    of heat engine, 134, 138
    of steam engine, 122
    quantifying, 134–140
    thermal, 129–151
Einstein, Albert, 373, 375, 595
Einstein's equation, 582, 607–608
Eisenhower administration, 321
Electric car, 270, 470–472, 514
Electric chair, 191
Electric charge, 156, 158, 160, 169, 173, 367,
    407–408
Electric current, 159–160, 171, 175

definition, 159
Electric eye camera, 600
Electric fan, 188
Electric fluid, 155, 159, 169
Electric grid, 202
Electric iron, 188
Electric lighting, 183–184
    domestic, 184
Electric locomotive, 190, 276
Electric motor, 7, 189–191
Electric power, 165–167
Electric power plant, 7
Electric rates, 188, 189
Electric relays, 184
Electric street railway, 277
Electrical conductivity, 596
Electrical energy, 177
Electrical equivalent of heat, 146
Electrical fluid, 153
Electrical generator. *See* Generator
Electrical outlet, 3
Electrical potential, 173
Electrical potential difference, 170
Electrical potential energy, 174
Electrical system, 159
    energy diagram, 160
Electricity, 153–168
    alternatives, 4
    availability, 3
    consumption, 2
    development, 5
    early investigators, 153–159
    from jet engines, 338
    from wind. *See* Wind energy
    generation, 169–180
    impacts on society, 181–194
    limitations, 3
    prehistory, 153
    transportation, 276–277
    uses, 2
Electricity-generating plants, 226
    coal-fired, 241–255
    costs, 254–255
Electricity-generating stations, 188
Electricity-generation, early application, 181–182
Electrochemical cell, 172
Electromagnetic radiation, 149, 356
Electromotive force, 160–164
Electron holes, 597–598
Electrons, 362–364, 372, 595–596, 597
Electroscopic force, 160–164
Electrostatic precipitator, 251
Electrostatic system of units, 408
Elephant grass, 538
Eling, Egidius, 331–332
Eliot, George, 144, 150
Ellis, C.D., 385
Emergency core cooling system (ECCS), 398–399,
    418
Emissionless vehicles, 463–475

Emissions, 435–438, 445–448, 450, 456, 457–461, 508
   biomass, 539
   see also Vehicle emissions and specific types
Energy
   access to, 6
   and food, 17, 20–23, 29–30, 34, 35
   and heat, 43
   and money (analogy), 33–34, 38
   and power, 26
   changes in patterns of use, 514
   conversion to work, 36–37
   definition, 144
   dependence on, 4
   developments, 4–5
   domestic uses, 2–3
   in fuels, 21–23
   human use of, 14–15
   origins, 17
   reliance on, 4
   transfer, 11
   ubiquitous, 2
   work and power, 10–13
   see also specific sources of energy
Energy balance, 33–38
   definition, 35
Energy consumption, 4
   and income growth, 5
Energy conversion, 7, 109, 146, 147
Energy crisis
   Africa, 56
   and wood, 49–52
   Third World, 59
   US, 56
Energy crops, 536–539, 552
   aquatic, 539
   disadvantages, 538
   plants chosen as, 538
Energy density, 472
Energy diagram, 46, 131, 148, 178, 179, 195, 239, 371, 373, 433
   electrical system, 160
   generic, 46, 84, 85
   steam turbine, 226
   thermal system, 132
   waterwheel, 85
Energy extraction, 84
Energy intensity, definition, 5
Energy release, 22, 23, 25–26
   combustion, 47
Energy reserves, 501–504
Energy resources, 501–504
   lifetime estimates of, 509–511
Energy sources, 511, 531–532
   categories, 55
   commercial, 55
   new, 55
   traditional, 55, 56
   see also specific sources of energy
Energy storage, 34

Energy supply, extracting, 504
Energy transformations in nuclear power plant, 397
Energy use
   and national well-being, 4–6
   overview, 1–4
Engels, Friedrich, 130, 150, 189
Engine, duty, 122
Engine knock, 299–301, 308
Enhanced greenhouse effect, 482–490
   policy options, 497–499
Enhanced recovery methods, 517
Enriched uranium, 394, 397
Environment, 94, 427–440
   strategies for addressing, 435–438
   and water power, 98–99
Environmental control, 255
Environmental effects, 61, 244–247
Environmental problems, firewood, 58
Environmental protection, 254
Environmental regulations, 247
Enzymes, 23–25
Eternal flame, 40
Ethane, 286
Ethanol, 277, 278, 308, 309, 436, 545, 549, 556
   advantages, 551
   CO_2-neutral, 551, 552
   cost of, 553
   disadvantages, 551–553
   production, 549–551
Ethell, Jeffrey L., 279, 341
Ethyl alcohol. *See* Ethanol
Ethyl gasoline, 306
Europe, oil supplies, 525–526
Evans, Oliver, 264
Exhaust emissions, 306, 463–475
Exhaust stroke, 347
Exothermic reaction, 237
Exploration, 503–504, 519
External combustion engine, 267
Exxon Valdez, 429, 430

Fahrenheit scale, 134
Fan jet engine, 335
Faraday, Michael, 23, 31, 150, 175, 181, 185, 288, 595
Fats, 23–24
Federal Energy Regulatory Commission, 202
Fermentation, 39, 549, 551, 553
Fermi, Enrico, 376, 377, 393
Ferris, George Jr., 95
Ferris wheel, 95
Field magnet, 192
Filaments, 184, 185, 187
Financial reserves, 511
Fire, 14, 18, 25, 39–47
   creation, 77–78
   discovery, 39–40
   early uses, 40–42
   heat from, 41
   humankind's first contact, 40

Fire (*cont.*)
    importance to primitive humans, 40
    see also Burning
Fire and light, 41
Fire-cover bell, 68
Fire piston, 343, 344
Fire-setting, 42
Fire-tube boiler, 233–234
Firewood, 49–64
    availability, 57
    environmental problems, 58
    gathering, 57
    ideal source, 54
    maximum yields, 55
    over-use, 51
    standard measure of quantity, 51
Firn, 484
First World War, 124
Fission. *See* Nuclear fission
Fitch, John, 262
Flash boilers, 266
Flashing, 422
Flat-plate solar collector, 587–589
Flavin, Christopher, 530, 563
Floating mill, 86
Flue-gas desulfurization (FGD), 248, 437, 458
Fluidized-bed combustion, 432, 558
Fluorescence, 358–359
Fly ash, 252
Food
    chemical components, 23–24
    cooking of, 41
    and energy, 17, 20–23, 29–30, 34, 35
    major components, 17
Food and fuels, 17–18
Food consumption, 22
Food preservation, 41–42
Food processing, 1
Foot-pound (unit), 13–14
Forbidden direction, 173
Ford, Henry, 266, 270–271, 297, 315–316
Ford Model T, 270–271, 278, 298, 315–316, 323
Ford, Tim, 475
Forests, 533
Formaldehyde, 467
Forth Banks generating station, 216
Fossil fuels, 281–283, 531, 532
    characteristics, 281
    definition, 281
    environmental consequences, 427–440
    extraction, 427–431
    types, 281
Four-stroke engines, 298
Four-stroke gasoline engine, 271
Four-stroke reciprocating internal combustion
    engine, 331
Fourier, Jean-Baptiste, 481
Fourneyron, Bénoit, 199
Fowler, John M., 255, 425, 439
France, 91

Francis, James, 199
Francis turbine, 199
Franklin, Benjamin, 51, 68–70, 154–159, 169–170
Franklin stove, 51–53, 70–71
Free radicals, 405
Freeman, Castle Jr., 64
French Revolution, 175
Frisch, Otto, 378, 379, 384
Fuel assembly, 388
Fuel cells, 474, 562
Fuel injection, 346, 347
Fuel-lean combustion, 468
Fuel oil, 74, 304
Fuel-rich combustion, 468
Fuel rods, 388, 390, 413, 418–419
Fuel utilization, 432–435
Fuels, 43
    alternatives, 4, 277–278
    chemically complex, 25
    chemically simple, 25
    definition, 281
    domestic heating, 65, 71–75
    energy in, 21–23
    major components, 18
Fuels and foods, 17–18
Fuller, Edmund, 99
Fulton, Robert, 262, 263, 271, 311
Furnace, function, 230
Furth, Howard P., 617
Fusion reactor. *See* Nuclear fusion; Nuclear reactor
Fussell, Paul, 209

Gallium, 596, 597
Galperin, Anne L., 401, 425
Galvani, Luigi, 169–170
γ-decay, 372
γ-rays, 362, 404–405, 408
Gas cap gas, 286
Gas chromatography, 292
Gas drilling, 504
Gas engine, 267
Gas-fired appliances, 75
Gas heaters, 74
Gas turbine, 329, 338
Gas turbine engine, 328
    for airplanes, 276
    for automobiles, 337
    for road vehicles, 336
Gases, behavior of, 219–223
Gaslight, 186–187
Gasohol, 308, 553
Gasoline, 1, 266, 269, 270, 273, 276, 294, 297–309,
    328, 507–508
    availability, 3
    components, 300
    costs, 3–4
    future directions, 308–309
    high-octane, 304
    lead-free, 6

market demand, 297
production, 301–309
rationing, 3
uses, 2
yield, 301
see also Reformulated gasolines
Gasoline engine, 278
Gay-Lussac, Joseph, 130, 133, 150, 219, 220, 223
Gears, 121
    development, 91
    invention, 90–91
Gebelein, Charles G., 30, 31, 228, 255, 439, 461, 475
General Electric Company, 188
General Motors Acceptance Corporation, 322
Generator, 175–179, 181, 183, 189, 195
    see also Dynamo
Generic energy diagram, 46, 84, 85
Geological exploration, 510
Geopolitics, 531
Gies, Francis, 99, 110
Gies, Joseph, 99, 110
Gilbert, William, 153
Gimpel, Jean, 110
Gladstone, William, 177
Glashow, Sheldon, 363, 365, 385
Glashow, S.L., 151, 167
Glasstone, Samuel, 385
Global carbon cycle, 281–283, 495–497
Global climate change, 204, 482, 531
    articles in popular magazines and scientific
        journals, 500
    signs of, 488–490
Global warming, 58, 482, 483, 486
Global warming potential (GWP), 487
Gloster Aircraft Co., 329
Gloster Meteor jet fighter, 332
Goethe, Johann Wolfgang von, 204, 209
Gogol, Nikolai, 50
Gold and Silver Telegraph Company, 185
Gold, Thomas, 512–513
Goldemberg, J., 8, 63, 439
Goodyear blimp, 350
Gopher plant, 556
Gorky, Maxim, 429, 439
Gould, Stepen Jay, 38
Governor of an engine, 122
Graedel, Thomas E., 255
Grain alcohol. See Ethanol
Gram–centimeter–second system, 408
Gramme, Zénobe, 181, 183, 189, 195, 567, 568
Grand Canyon, 207
Grand Coulee Dam, 203, 204
Grapes of Wrath, The, 322
Graphite, 393–394
Graphite moderator, 391, 412
Graphite–water reactor, 397
Gratzer, Walter, 191, 193, 354
Gravitational energy system, 148

Gravitational potential energy, 84, 97, 124, 132, 137, 138
    definition, 82
Gravitational system, 159
Gray, Stephen, 155–156, 161
Great Depression, 5, 204, 208, 275, 351, 509, 542
Great Eastern Railway, 261
Great Western, 262–263
Great Western Steamship Company, 262
Greece, 88, 93
Greeks, 101
Greenhouse effect, 438, 477–500, 531
    anthropogenic, 482
    enhanced, 482–490
    natural, 482
    overview, 477–483
    present situation, 499
Greenhouse gases, 508
    other than carbon dioxide, 487–488
    release of, 494
Gross domestic product (GDP), 5–6
Grossman, Peter Z., 8, 401, 425
Grübler, Arnulf, 8
Guayule, 556
Guibert, J.C., 341, 354
Guillerme, André E., 100
Gunpowder engine, 268, 272
Gunpowder-fueled pumps, 268
Gunston, Bill, 341
Guthrie, Woodie, 204

Häfele, Wolf, 425
Hahn, Otto, 377, 378, 384
Haldane, J.B.S., 383, 384
Half-life, 404, 405, 408
Han Dynasty, 65
Hart, Ivor B., 127
Harte John, 461
Hawke, David Freeman, 100
Hawkins, N., 127
Hawks, Ellison, 180, 228, 240, 354
Hazen, M.H., 47, 63, 79
Hazen, R.M., 47, 63, 365, 401
Health effects of radiation, 403–407
Hearth, 41
Heat, 129–151
    Caloric theory of, 140–142
    definition, 140, 141, 145–146
    electrical equivalent of, 146
    as energy, 230
    and energy, 43
    from fire, 41
    as form of motion, 230–231
    generation, 37
    mechanical equivalent of, 145
    and work, equivalence, 144–147
Heat capacity, definition, 149–150
Heat engine, 136, 224, 299
    analysis, 143

Heat engine (*cont.*)
    definition, 132
    efficiency, 134, 138
    steam turbine as, 224–227
Heat exchanger, 395, 396
Heat flow, 131, 148–150
Heat transfer, 135, 148–149, 229–240
    mechanisms, 230–232
Heavy hydrogen, 390
Heavy oils, 517
Heavy water, 391
Hectare (unit), 207
Heinkel, Ernst, 276, 329, 340
Heinkel He 178, 276
Heinrich, Bernd, 31
Helicopter, 336
Heliostat, 593–594
Helium, 370, 606
Helmholtz, Herman von, 146–147
Hemispheres of Magdeburg, 114
Henry, Charles C., 277
Herbaceous crops, 537
Hero of Alexandria, 113, 212
Hestia, 40
Hetch-Hetchy Dam project, 207
Hewitt, Geoffrey F., 401
Hewitt, Paul G., 363, 366, 385, 617
Hickory, 60
Higginson, Francis, 50, 63
High-pressure steam, 260, 264
Highsmith, Phillip, 16
Hill, Donald, 100, 110
Hill, Robert, 8, 64, 603
Hill, Robert *et al.*, 425, 439
Hills, Richard L., 110, 127
Hoagland, William, 603
Hog fuel, 541
Hohenemser, Christoph *et al.*, 425
Hollander, Jack, 439, 475
Holmes, Sherlock, 377, 487
Homewood, Brian, 209
Hooke, Robert, 220
Hoover Dam, 202, 204, 208
Hoover, Herbert, 351
Horizontal waterwheel, 88–89
Horseless carriage, 259, 264, 267, 270, 312
Horsepower, 27–28, 95, 97, 123
Horseshoe Falls, 201
Hotton, Peter, 79
Houdry, Eugene, 303
Houghton, John, 500
Houwink, R., 31
Hubbert curve, 515
Hubbert, M. King, 515
Hugo, Victor, 87
Human energy, 17–32, 35
    for transportation, 313
Human health effects of enhanced greenhouse effect,
        493–494
Human muscles, 28

Human power, limitations, 28
Human use of energy, 14–15
Hünecke, Klaus, 341
Hunt, J.L., 63
Hunter, Christine, 79
Huygens, Christiaan, 115
Hybrid vehicles, 470–474, 514
Hydraulic power, 182
Hydrocarbons, 284, 287, 289, 300, 306, 308
    see also Unburned hydrocarbons
Hydrochloric acid, 442
Hydrocracking, 304
Hydroelectric power, 199–204
    advantages, 200
    economics, 200
    Niagara Falls, 200
    prospective sites, 202
    statistics, 199–200
    transmission lines, 201
Hydroelectricity, 197, 532
    potential for increasing generating capacity, 203
Hydrogen, 17, 18, 364, 370, 606, 609
Hydrogen atoms, 45, 288
Hydrogen bomb, 611
Hydrogen fuel, 560–562
Hydrogen ion concentration, 441, 446
Hydrogen isotopes, 365, 390, 609–610
Hydrogen sulfide, 286, 289
Hydrologic cycle, 97
Hydrolysis, 23–24
Hydronium ion, 442
Hydropower. *See* Hydroelectric power
Hypothetical resources, 502, 510

Ibsen, Henrik, 457
Ignition promoters, 352
Implosion, first recorded, 115
Impulse turbine, 197, 198
Impulse wheel, 196
Incandescent light, 184, 185, 187
Income and energy, 5
Incomplete combustion, 76
Indeno(1,2,3-cd)pyrene, 61
India, 88, 90
Indicated work, 121
Indicator, 121
Industrial growth, 318
Industrial Revolution, 4, 94, 95
Infrared radiation, 480, 481
Internal combustion engine, 5, 116, 267–269, 274,
        278
Ion, 364
Ion exchange, 449
Iran, 92
Iraq, 4, 92
Iron making, 95
Isisu Dam, 208
Isle of Man, 97
Isobutane, 288

Isomerization, 304
Isothermal expansion, 346
Isotope, 365
Isotope separation, 387
Israel, Paul, 194
Istanbul, 66
Itaipu plant, 199

Jakab, Peter L., 279
James Bay, 208
James, Peter, 79
Japan, 514
    oil supplies, 525–526
Jean, Gimpel, 99, 100
Jefferson, Thomas, 154, 167
Jehl, F., 184, 193
Jenkins, Nicholas, 580
Jet, use of term, 329
Jet aircraft, 330
    time line of major events, 340
Jet airliner, 332
Jet car, 336–338
Jet engine, 276, 304, 327–341
    development, 329
    electricity from, 338
    history, 328–338
Jet fuel, 289, 294, 304, 327–341
    history, 327
    kerosene as, 338–340
    production, 339–340
    treated with specialized chemicals, 339
Jevons, William Stanley, 3, 8, 505, 529
Johansson, Thomas B. et al., 563, 603
John Fitch, 262
Johnson, Mark, 438–439
Joint Regulatory Agency, 263
Jojoba, 556
Joliot, Frédéric, 406
Joliot-Curie, Irène, 405
Josephson, Paul R., 401, 425
Joule, James Prescott, 144–147
Joyce, James, 240
Junkers Jumo 004 engine, 329

Kalfus, Ken, 426
Kelvin (unit), 134, 138
Kelvin, Lord. See Thompson, Sir William (Lord
    Kelvin)
Kelvin scale, 134
Kemmler, William, 191
Kemp, Clarence, 587
Kennan, George F., 461
Kennedy, John, 40
Kerogen, 283–285
Kerosene, 74, 267, 276, 294, 297, 327–328
    jet fuel, 338–340
Kheshgi, Haroon S. et al., 564
Kier, Samuel M., 327
Kier's Rock Oil, 327

Kilgour, Frederick Gridley, 100, 127, 180
Kilowatt-hour (kWh) (unit), 165
Kinetic energy, 81–82, 93, 98, 101, 109–110, 133,
    147, 177, 182
    conversion, 132
    definition, 82
Kirby, Richard Shelton, 100, 127, 180
Kirby, Richard Shelton et al., 279
Kirchhoff, Gustav, 606
Kirov, Sergei, 506, 529
Kite-in-a-thunderstorm experiment, 157
Kraushaar, Jack, 255, 401, 426, 440, 475, 580
Kristoferson, Lars, 564
Kyoto Protocol, 498
Kyoto summit, 508

la Cour, Poul, 567
Laidler, Keith J., 127
Lakoff, George, 438–439
Landels, J.G., 99
Landfills, 252, 487, 559
Langley, Samuel Pierpont, 272
Lavoisier, Antoine Laurent, 43, 44, 587
Lead, 306–307
Lead isotope, 409
Leaded gasoline, 305–306
League of Women Voters, 426
Lebon, Philippe, 268
Le Corbusier, 320
Lee, Edmund, 108
Legislation, driftwood, 50–51
Leibniz, Gottfried, 10, 11
Lemale, Charles, 332
Lenoir, Etienne, 268
Lenoir, Joseph Etienne, 268
Lenssen, Nicholas, 530, 563, 564
Leon, George deLucenay, 168
Leonardo da Vinci, 105–107, 113, 115–116, 196,
    587
Leucaena tree, 544
Leukemia, 405, 407
Levorsen, A.I., 530
Lewis acids, 445
Leyden jar, 157, 158
Lichen, 454
Lifetime estimates
    energy resources, 509–511
    natural gas, 527
Lifetime of oil reserve, 511
Light, 357
    and fire, 41
Light bulb, 165–166, 184, 185, 187
Light water, 390
Light-water-moderated reactor, 394
Lighters, 343
Lighthouses, 183
Lightning, 40, 154, 157, 169
Lignin, 59, 284, 540–541
Linear focusing collector, 591

Linseed oil, 555
Liquefied petroleum gas (LPG), 74, 278, 286, 292, 293
Liverpool and Manchester Railway, 260
Locke, John, 141, 150
Lockheed F-80 Shooting Star jet fighter, 332
Locomotives, 7, 190, 260–261, 311, 349–350
Logs, 53
Lombe, John and Thomas, 94
London Underground Railway, 276
Low-emission vehicles (LEVs), 470–474
Low-energy society, 28
Lowrey, Grosvenor, 187
Lubricating oils, 267
Luminescence, 356
Lung cancer, 410
Luria, Salvador E., 38
Lusitania, 218–219
Luz International Ltd., 592
Lymphoid tissue, 405–406
Lyons, John W., 47
Lysine, 30

McAdam, John Loudon, 258–259
Macaulay, David, 100, 209
McDonald, Alan, 8
McFarland, E.L, 63
McGregor, Robert Kuhn, 63
Mach, Ernst, 362, 365
McNeill, J.R., 255, 309, 439, 461, 475, 500
McPhee, John, 530
McShane, Clay, 279
Magie, W.F., 365
Magnetic compass, 260
Magnetic confinement, 611
Magnetic doughnut, 612
Magnetic field, 175–178, 192
Magnetohydrodynamics, 613
Malthus, Thomas, 506–507
Managed woodlot, 54
Manhattan Project, 608
Manly, Charles, 273
Mann, Thomas, 23, 31, 366
Margolis, Jonathan, 439
Mariotte, Edmé, 220
Martin, Thomas, 194
Marx, Karl, 189
Mass number, 364, 365, 375
Matches, 77–78, 343
    safety, 77–78
    strike-anywhere, 77
Matthiessen, Peter, 31
Mauretania, 218
Maxim, Hiram, 272
Maxwell, James Clark, 222
Maybach, Wilhelm, 269, 314
Mechanical energy, 145
Mechanical equivalent of heat, 145
Mechanical solutions, 29

Mechanical system, 146
Mechanical work, 93
Medawar, Peter, 190, 193, 435, 439
Meitner, Lise, 377, 378
Mekong River, 208
Meltdown, 412, 418
Mendeleev, Dimitri, 512
Mercury, 451
Messerschmidt Me 262 Stormbird, 276, 329, 330, 332
Metal fatigue, 332
Metals, 42
Methane, 43–44, 46, 75, 76, 78, 179, 286, 486, 487
    biological sources, 559–560
    combustion, 45, 46
Methanol, 309, 436, 544–549, 555
    advantages, 547
    disadvantages, 547–549
    production, 545–547
Methyl tertiary butyl ether (MTBE), 306–309, 470
Metric system, 20
Middle East, oil supplies, 522–523
Midgley, Thomas, 305
Migration, 58–59
Mill, John Stuart, 130, 150
Mill tailings, 414–415
Miller's Tale, The, 92
Mined land reclamation, 427–428
    recontouring and fertilizing and seeding (or replanting), 428
Mini-hydro plants, 199
Missing mass, 373
Mobile sources, 447
Moderator, 390–391
Moissan, Ferdinand Frédéric Henri, 362
Money and energy (analogy), 33–34, 38
Money and spending (analogy), 11
Moore, John W. *et al.*, 475, 564
Morgan, Alfred P., 168
Morgan, Hugh, 341
Morgan, J.P., 187
Morus, Iwan Rhys, 194
Motor courts, 323
Motor fuel alkylate, 302
MTBE. *See* Methyl tertiary butyl ether
Mulford, Clarence E., 205, 209
Multiple phase currents, 192
Multiple-unit system, 191
Multiplication constant, 392
Multiplication factor, 392
Municipal solid waste, 557–558
Murdoch, William, 260, 268
Mutations in genetic material, 407

n-type semiconductor, 596
n–p junction, 598
Nakic'enovic', Nebojosa, 8
Nanjing, 89
Naphtha, 294

Naphthenes, 288, 289
Napier grass, 537
Napoleon Bonaparte, 171
Narmada project, 208
National Aeronautics and Space Administration
    (NASA), 560
National Research Council, 401, 439
National Wild and Scenic Rivers Act 1968, 209
Natural gas, 25, 43–44, 74, 277, 281, 284–287, 292,
    436
    calorific value, 286
    domestic heating, 74–75
    domestic use, 286–287
    extraction, 429
    lifetime estimates, 527
    reserves, 526–528
Natural gasoline, 286
Natural greenhouse effect, 482
Negative charge, 156
Neon, 133
Neutron, 363, 365, 370, 372, 387, 392
New England, 95
New York, 95
Newcomb, T.P., 279
Newcomen engine, 119, 120, 122, 130, 227
Newcomen, Thomas, 118–119, 123, 311
Newton, Isaac, 141, 220
Niagara Falls, 193, 200, 207
NIMBY syndrome, 417
Nitrogen, 17, 18, 76, 244, 289, 363, 433
Nitrogen oxides. See NO$_x$
Noddack, I., 377, 384
Nomadic peoples, 58–59
Nonconventional oil, 517
Noria, 88
Norman, Andrew et al., 354
Norse wheel, 89
Notre Dame of Paris, 87
NO$_x$, 7, 61, 244, 247, 250–251, 433, 434, 437–438,
    441, 445, 447, 448, 450, 456, 466, 468, 481
    as cause of acid rain, 444–448
    emissions, 460–461, 464, 465, 542
Nuclear binding energy, 367–372, 381
    curve of, 369–370, 374, 380, 607
    definition, 368
Nuclear bomb, 415
Nuclear car, 322
Nuclear controversy, 403–426
Nuclear energy, 383, 394, 438, 532
    economics, 399–401
    future, 424–425
    history, 399
    time line of major events, 400
    see also Atomic energy
Nuclear fission, 376–381, 387–393, 408
    comparison with nuclear fusion, 607–608,
        614–616
    first applications, 394
    steam cycle in, 397–398
Nuclear fusion, 605–617

comparison with nuclear fission, 607–608,
    614–616
definition, 606
energy capture, 613–614
engineering problems, 610
fuel for, 609–610
harnessing, 608–614
ignition, 610–611
overview, 605–608
promise and frustration of, 616–617
reactors, 611–613
Nuclear magnetic resonance (NMR) spectroscopy,
    163
Nuclear potential energy, 371
Nuclear power plants, 6, 387–402
    decommissioning, 425
    energy transformations in, 397
    radiation, 399
Nuclear processes, 371, 373
Nuclear reactions, 182, 373
Nuclear reactor
    accidents, 399–401, 412, 417–424
    components, 387–391
    core processes, 392–393
    emissions, 411–412
    first construction, 393–394
    safety issues, 411–413
    safety systems, 398–399
Nuclear Regulatory Commission, 411, 418
Nuclear structure, 372
Nuclear waste, 413–417
Nuclear Waste Policy Act 1982, 417
Nuclear weapons, 415
Nucleon, 364, 368, 371
Nucleus, 362, 363, 368, 369, 371, 372
Nutritional calorie, 20
Nye, David E., 127

Oak, 60
O'Brian, Patrick, 157, 167, 279
Octane enhancer, 305–307
Octane number, 300–301, 304, 305, 308
Odén, Svante, 444
Ohain, Hans-Joachim von, 329, 331
Ohm (unit), 161
Ohm, George Simon, 160
Oil burners, 75
Oil City, Pennsylvania, 430
Oil consumption, 505
Oil crises, 554
Oil demand, 508
Oil discovery vs. oil production, 518–520
Oil drilling, 504
Oil economics, 506–508
Oil exploration, 519
Oil extraction technologies, 518
Oil fields, 429
Oil firing for steamships, 275
Oil fuels, 276

Oil imports, 505
Oil industry
    environmental problems, 429
    see also Petroleum industry
Oil lamps, 297
Oil pipelines, 430–431
Oil price shock, 507, 514
Oil prices, 507–508, 519
Oil production, 514–518
    vs. oil discovery, 518–520
Oil products. *See* Petroleum products
Oil refineries, waste materials, 431
Oil refining, environmental problems, 431
Oil reserves, 510–511, 515, 517
    exhaustion, 520
Oil reservoir, 284–285
Oil shales, 517
Oil shortages, 508
Oil spills, 429–431
Oil supplies
    Caspian region, 524–525
    China, 525
    Europe, 525–526
    Japan, 525–526
    Middle East, 522–523
    Pacific Rim, 525
    United States, 521–522
Oil wells, 267
O'Keefe, Phil, 8, 64, 603
Old–deep crudes, 291, 292
Old–shallow crudes, 291
Olive oil, 554, 556
OPEC, 507, 513, 519, 521, 523
Oppenheimer, J. Robert, 608
Orata, Caius Sergius, 65
Organic waste, 559–560
Organization of Petroleum Exporting Countries. *See* OPEC
Orinoco oil belt, 517
Ørsted, Hans Christian, 174–176, 181
Otto cycle, 268–271
Otto-cycle engines and their performance, 298–299
Otto–Langen engine, 269
Otto, Nikolaus, 268–271, 298, 314, 331
Outlet strips, 3
Overhead cables, 277
Overpopulation, 58
Overshot waterwheel, 85–86, 97, 196
    construction, 86–87
Ox, 28
Oxides, 244
Oxygen, 17, 18, 24, 44, 46, 76, 77, 179, 289, 466, 467, 468
Oxygen atoms, 45
Oxygen-starved combustion, 77, 78
Oxygenated gasolines, 308
Oxygenates, 470
Ozone, 466, 467, 481
Ozone layer, 447

p-type semiconductor, 597
Pacific Rim, oil supplies, 525
Packard Motor Car Company, 350, 351
Paper-making, 93
Papin, Denis, 113, 115–116, 123, 126, 261–262, 268
Parabolic trough solar collector, 591–592
Paraffin oil, 267
Paraffins, 287, 288, 302
Parallel wiring, 188
Parsons, Charles (later Sir Charles), 215, 229
Parsons turbine, 215–218, 330–331
Parsons turbine/generator set. *See* Steam turbine/generator set
Particulate emissions, 251–252
Pasachoff, Naomi, 385
Pasztor, Janos, 564
Peale, Charles, 69
Pearce, Fred, 461
Pearl Street Electricity Generating Station, New York, 185, 187, 188, 191
Peat moss, 283
Peierls, Rudolf E., 401
Peking Man, 39
Pelton, Lester, 198
Pelton turbine, 197
Pelton wheel, 198
Penaud, Alphonse, 273
Pennington, John, 272
Pennsylvania, 95
Pennsylvania crude, 291, 327
Pennsylvania Fire-Place, 51
Pennsylvania Rock Oil Company, 267, 327
Pentane, 287, 289
Peroxyacetylnitrates (PANs), 467
Persian Gulf War, 4, 506, 513–514
Perspiration, 35–36
Peru, 93
Petrolea, 261
Petroleum, 4, 7, 261, 266–267, 277, 281–295, 311
    age–depth classification, 287–292
    chemical composition, 287
    dependence on imports, 505
    lessening the dependence on, 513–514
    partial digestion, 25
Petroleum derivatives, 545
Petroleum extraction, 429
Petroleum industry, origin, 328
Petroleum products, 57, 292–295, 508
    dependence on, 505
    domestic heating, 74
    linear paraffin molecules in, 302
    see also specific products
Petroleum refining, 292–295, 301–309
    costs, 307–308
    essence of, 293
Petroleum reserves, increasing, 511–513
Petroski, Henry, 100, 127, 209
pH scale, 441–443
pH values, distribution, 446

Phosphorescence, 360
Phosphorus, 77, 78
    Red phosphorus, 78
Photochemical smog. *See* Smog
Photons, 30
Photosynthesis, 29–30, 62, 533–534, 551
Photovoltaic cells, 595–602
Photovoltaics, 595–602
    applications, 600
    components, 599
    transportation, 600
Pine, 60
Pipelines, 430–431
Piperidine, 174
Piston, 116
Piston-in-cylinder design, 118
Plankton, 446
Plant growth and enhanced greenhouse effect,
        494–495
Plant oils, 553–557
    CO_2-neutral, 554
Plants, 29–30, 62
    effect of acid rain, 453–455
Platinum, 304
Platinum-based catalytic converter, 306
Plato, 174
Platt, Harold L., 194
Plows, 28
Plutonium, 413–415
Poe, Edgar Allan, 214
Pohl, Frederick, 209
Poirier, Jean-Pierre, 47
Poliomyelitis, 434
Pollution, 60–61, 244, 248, 252–253, 427, 435–436,
        463
    see also specific types and pollutants
Polonium, 409
Pompey, 89
Poncelet undershot waterwheel, 85, 86
Porsche, Ferdinand, 470–472
Positive charge, 156
Possible resource, 502
Post mills, 103–105
    improvements, 107
Potassium chlorate, 77, 78
Potassium uranyl sulfate, 359
Potential, 161
Potential energy, 82–83, 147
    see also Chemical potential energy; Gravitational
        potential energy; Thermal potential energy
Pottery, 42
Pour point, 339
Powell, F.E., 579
Power
    concept, 11–12
    definition, 12, 165
    and energy, 26
    formula, 118
    relative amounts, 27
    work and energy, 10–13

Power grid, 202
Power, Henry, 220
Power Jets Ltd, 329
Power plants, 182
Power stroke, 348
Premium fuel, 526
Pressure cooker, 113
Pressure-tube reactor (RBMK), 397
Pressure vessel, 391
Pressurized water reactor (PWR), 395–396
Primary cells, 173, 174
Primary energy, 514
Primary pollutants, 436
Pringle, Sir John, 158
Probable resources, 502
Prometheus Bound, 585
Prompt neutrons, 392–393
Propane, 286, 292
Propeller-jet turbine, 328
Properties of matter, 223
Proteins, 23–24
Proton, 363, 365, 367, 369, 370, 372
Proved reserves, 501, 504
Public baths, central heating, 66
Pullman, Bernard, 366
Pulp and paper industry, 541
Pulverized coal, 238–239, 243
Pulverized coal-fired boiler, 239
Pumping station, 95
Purrington, R.D., 16, 151
Pynchon, T., 172, 180
Pyramids, construction, 28
Pyrolysis, 542
Pyroscaphe, 262

Quantum mechanics, 363
Quixote, Don, 105, 123

Rad (unit), 407
Radiation, 230–232, 235, 362
    background, 409–410
    health effects of, 403–407
    nuclear power plants, 399
    types, 403–404
Radiation damage, 410
Radiation dose, measuring, 407
Radiation exposure, 406, 407
Radiation exposure threshold, 410–411
Radiation sickness, 406
Radiation therapy, 413
Radiation units, 407–408
Radioactive by-products, 377
Radioactive decay, 372, 405, 406
Radioactive isotopes, 409, 413
Radioactive material, 411
Radioactive waste, 413–417
Radioactivity, 361, 362, 367
    definition, 367, 368
Radium, 375, 377

Radon, 409
Rafts, 259
Railroads, 260–261, 314
    electrification, 277–278
    see also Locomotives
Ramage, Janet, 168, 180, 209, 426, 580, 603
Ramelli, Agostino, 100, 110
Ramie, 92
Rankine, William, 140, 150
Rapeseed oil, 555
Reaction turbine, 197, 198, 215
Reciprocating motion, 121
Refining. *See* Petroleum refining
Reformulated gasolines, 308, 469–470
Refrigerator, 178, 188
Renewable woodlot, 54
Renewables, 49, 54–55, 531, 544, 559, 563
    see also Biomass and specific sources
Reserves, 531
    categorization, 502
    definition, 501, 504
    fluctuation, 503
    relationships, 503
Resistance, 160–164
    definition, 160
Resistance heating, 166
Resistivity, 162
Resources, 531
    categorization, 502
    definition, 501
    relationships, 503
Reverse energy diagram, 173, 178, 179, 195, 196
Revolutionary War, 52, 142, 158
Reynolds, Francis D., 187, 193, 325
Rhine River, 87
Riley, C.J., 279
Riley, Robert Q., 564
Röistinen, Robert A., 255, 401, 426, 440, 475, 580
Road building, 258–259, 313, 321
Road transportation, diesel engine, 350
Röbbelen, Gergard *et al.*, 564
Rock, Maxine, 475
Rocket, 261
Rolls-Royce, 330
Romans, 101
Rome, 65, 93
Romer, Alfred, 366, 385
Röntgen (unit), 407
Röntgen equivalent man (rem), 408
Röntgen equivalent physical (rep), 408
Röntgen, Wilhelm Konrad, 355, 356, 357, 358, 365
Roosevelt, Theodore, 219
Rose, Joshua, 127
Rose, Mark H., 194
Rose, Paul Lawrence, 401
Rossotti, Hazel, 47, 79
Rotating machinery, 122
Routledge, Robert, 194
Rover Jet 1, 276
Royce Hailey's Pig Stand, 323

Rühmkorff induction coil, 357
Rumford, Count Benjamin Thompson, 69, 141–142, 150, 230
Rural Electrification Program, 327
Rutherford, Ernest, 375, 376, 383, 384, 385

Safety matches, 77–78
Sailing ships, 259–260
Sakharov, Andrei, 612
Salt cedar, 206–207
Salvadori, Mario, 32
Sarmiento, Jorge L., 500
Satellite Solar Power System, 601
Saunders, Robin, 209
Savery, Thomas, 116–117, 123, 126
Schimel, D. *et al.*, 500
Schneider, Steven H., 500
Schwartz, A. Truman *et al.*, 255, 500, 603
Scientific knowledge, 9
Scramming, 398
Scrubber, 248, 255, 436, 458, 459
Sculpins, 451
Sea level rise, 491–492
Seaborg, Glenn T., 402, 426
Seaweed, 539
Second World War, 3, 5, 17, 276
Secondary cells, 173, 174
Secondary pollutants, 436, 467
Seeger, Pete, 204
Seine River, 87
Selandia, 275
Selden, George, 315
Semiconductors, 596
Series wiring, 183–184, 188
Sevenair, John P., 228
Shepherd, D.W., 402, 564, 604, 617
Shepherd, W., 402, 564, 604, 617
Ship propulsion, steam turbines, 218–219
Ships, 261–264
Shopping mall, 323, 324
Short-rotation woody crops, 537
Siemens, Ernst Werner von, 183, 277
Siemens, Sir William, 193
Silicon, 595–602
Silliman, Benjamin Jr., 266–267, 327
Sime, Ruth Lewis, 385
Simon, Herbert, 506
Sinclair, Upton, 295
Singer, Joseph G., 240
Singer, Sam, 38
Sinks, 495
Skin cancer, 447
Slag heaps, 432
Slavery, 28
Smeaton, John, 119
Smith, Howard Bud, 402, 604, 617
Smog, 246, 247, 463, 465–467
    formation, 465–467
    reduction of, 472

Smoke, 61, 543
Smoke eater, 69
Smoke emissions, 69
Smoke formation, 339
Smoke particles, 61
Smoke pollution, 60–61
Smoke problem in home heating, 67–69
Smoking technique, 41
Snape, Colin, 8, 64, 603
Snow, C.P., 7, 376, 384
Social solutions, 28
Softwoods, 60
Soil erosion, 58
Solar Challenger, 600
Solar collectors, 603
Solar energy, 30, 55, 62, 97–98, 282, 438, 534,
    581–604
    advantages, 581–582, 600, 602–603
    cost, 601–603
    direct conversion to electricity, 595–602
    indirect conversion to electricity, 590–595
    major categories of application, 584
    overview, 605–608
    problems for large-scale commercial use, 584
    radiation to earth, 583–584
    storage, 601
    wind energy as indirect form, 566
Solar heating
    active system, 587–590
    passive system, 584–587
Solar One plant, 593–594
Solar power stations, 601
Solar power tower, 593–594
Solid-fuel diesel, 344–345
Soot, 69, 72, 77, 78, 289, 432–433
Sørensen, Bent, 564, 580, 604
Sorghum, 537
Sour gas, 286
Sources, definition, 495–497
SO$_x$, 7, 60, 61, 244, 246–250, 433–437, 441, 445–448,
    450, 456, 488
    anthropogenic, 247
    as cause of acid rain, 444–448
    emissions, 458–459, 461, 464
Spark plug, 299
Specific heat, 19
Spectroscope, 606
Speculative resources, 502
Spent fuel rods, 389, 414–415
Sphere of Aeolus, 213
Spontaneous change, 84
Spurr, R.T., 279
Stagecoach travel, 259
Standards of living, 6
Stanley 'Steamer,' 270–271
Starch, 24
Static electricity, 169
Stationary engines, 349–351
Stationary sources, 447
Steam

application, 311
energy for the Industrial Revolution, 125–126
prehistory, 113
for transportation, 260–266
Steam automobiles, 312
Steam carriage, 124, 259, 264, 265
Steam cycle in nuclear fission, 397–398
Steam digester, 113
Steam dredge, 264
Steam engine, 4, 108, 109, 113–127, 130, 135, 140,
    183, 189, 212, 226, 264, 266, 267, 312
    analysis, 136
    derivation, 113
    developments by James Watt, 119–124
    efficiency of, 122
    first successful, 118
    improved efficiency, 122
    inefficiency, 125
Steam engines, disadvantages, 212
Steam jacket, 122
Steam locomotive, 7, 190, 260, 311
Steam-propelled vehicles, 266, 312
Steam rate, 229, 233
Steam-tube boiler, 234
Steam turbine, 211–228, 264, 266
    energy diagram, 226
    as heat engines, 224–227
    prehistory, 212–214
    ship propulsion, 218–219
Steam turbine/generator set, 196–197, 216, 227–228
Steamboat, 262
Steamboats, 261–264, 311
Steamships, oil firing for, 275
Steerable rudder, 260
Steinbeck, John, 322
Stephenson, George, 260–261, 311
Stevin, Simon, 107
Stilgoe, J.R., 63
Stinson-Detroiter plane, 350
Stock ticker, 185
Stockton and Darlington Railway, 260
Stoff, Joshua, 279
Stoichiometric combustion, 44, 468
Stoker-fired fire-tube boiler, 237
Stoker-fired water-tube boiler, 235
Stoner, Carol Hupping, 64
Stoves, 69–71
Strabo, 89
Straight-run gasoline, 301
Strandh, Sigvard, 32, 100, 111, 127, 151, 180, 210,
    228, 279
Strassman, Fritz, 377
Street lighting, 600
Street, Robert, 268
Strip mining, 427
Strontium, 408
Subsidence, 429
Suburbs development, 318–320
Sugar cane, 92, 537
Sugar molecules, 29

Sugar(s), 24, 62
Sulfur, 17, 18, 244, 289, 290, 432–433
Sulfur compounds, 289, 339, 428, 437–438
Sulfur oxides. *See* SO_x
Sulfur removal, 339
Sulfuric acid, 428, 445, 446, 448–449
    alkylation, 302
Sun-and-planet gears, 121
Superconductor, 163–164
Superheating, 235
Supertankers, 505, 507
Suspension firing, 238
Sustainable energy, 531
Sustained-yield management, 54
Suzuki, Takashi, 279, 341, 354, 475
Sweetening, 286, 292
Swift, Jonathan, 158
Switchgrass, 537
Synthesis gas, 546, 560
Synthetic materials, 7
System, 43–45, 84
    concept, 35
    spontaneous change of state, 46
    states, 46
Szilard, Leo, 383–384
Szostak, Rick, 279

Tailings, 414–415
Tailpipe emissions, 464–465, 469
Tamarisks, 206–207
Tamm, Igor, 612
Tara, 91
Teller, Edward, 608
Temperature, 20, 130–132, 146, 148
    definition, 131
    lowest possible, 134
Temperature difference, 136, 137
Temperature scales, 134
Tennant Creek, Australia, 594
Tenner, Edward, 63, 79, 440
Tennessee Valley Authority, 204
Terminology, 9–13
Tesla, Nikola, 192, 193, 201
Test drillings, 504
Tetraethyllead, 305, 306
Textile industry, 94
Thales, 153
Theophrastus, 67
Theory of Radioactive Transformation, 374
Thermal cracking, 302–303
Thermal efficiency, 129–151
Thermal energy, 134, 136, 146, 148
    conversion, 132, 139
    definition, 145–146
    systems, 148
Thermal equilibrium, 148
Thermal inversion, 543
Thermal pollution, 226
Thermal potential energy, 130–132, 148

    definition, 131
Thermal potential energy diagram, 131
Thermal system, energy diagram, 132
Thermal underwear, 142
Thermodynamics, 135, 140, 143
Thermonuclear fusion, 611
Thermonuclear reaction, 606, 608
Third World, 57, 200
    energy crisis, 59
    energy needs, 55
    wood, 55–57
Thompson, Benjamin, *See* Rumford, Count
Thompson, Sir William (Lord Kelvin), 137
Thomson, J. Arthur, 210
Thomson, James, 199
Thomson, J.J., 358, 365
Thomson-Houston Company, 188
Thoreau, Henry David, 9, 16, 487
Thorium, 373, 374, 614–615
Thorpe, Nick, 79
Three Gorges Dam, 208
Three Mile Island nuclear reactor accident, 412,
    417–419, 614
Tiber River, 87
Tillman, David A., 64, 564
Tobacco, 61
Tokamak reactor, 611–612
Torrey Canyon, 429
Total energy demand, 514
Towneley, Richard, 220
Trans-Alaska Pipeline System (TAPS), 521
Transesterification, 556
Transmission lines, 179, 182, 188
    hydroelectric plant, 201
    interconnection, 202
Transmutation, 373–376
Transportation, 257–295
    earliest vehicles, 258
    early sources, 260
    electricity for, 276–277
    energy demands, 277–278
    historical change, 311
    humans and other animals, 257–259
    photovoltaics, 600
    pollutants, 436
    steam for, 260–266
Treadmill, 122
Trees. *See* Agroforestry; Firewood
Trefil, James, 50, 63, 151, 365, 401
Trevithick engine, 130
Trevithick, Richard, 124, 260, 264, 311
2,2,4-Trimethylpentane, 300
Trip hammer, 89
Tritium, 390, 609–610
Trombe, Felix, 586
Trombe wall, 586–587
Trucks, 264–266, 270, 278
Tuberculosis, 94
Tungsten, 187
Turbine, 196

Turbinia, 218
Turbofan engine, 335–336
Turbojet engine, 333–335
Turboprop engine, 329, 335, 336
Turboshaft engine, 336
Turret mill, 105–108
Twain, Mark, 424, 425, 517
Two Cultures and the Scientific Revolution, 7
Tyndall, John, 481
Tyson, Peter, 461

Uganda grass, 538
Ultra-low-emission vehicles (ULEVs), 470–474
'Umar I, Caliph, 102
Unburned hydrocarbons, 433, 437, 464–468
Underground mining, 428
Undershot waterwheel, 84–85, 196
United Nations Convention on the Non-Navigable
        Uses of Transboundary Waterways, 208
United States
    central heating, 66
    energy crises, 56
    ethanol, 553
    oil supplies, 521–522
United States Naval Proving Grounds, 216
Units, 20
Uranium, 362, 374, 377–380, 382, 383, 387, 392–394,
        408, 413–415, 614–615
Uranium decay, 373
Uranium enrichment, 387
Uranium hexafluoride, 387
Uranium oxide, 393–394
Usher, Abbott Payson, 100, 111, 210, 228

Vacuum, 114
Vacuum Oil Co., 303
Vapor lock, 307
Vapor pressure, 294
Vehicle emissions, 463–475
Venezuela, 4, 517
Ventura, 291
Verberg, Carol J., 440
Verne, Jules, 194
Vestal Virgins, 40
Vitamins, 30
Vitruvius, 90–91, 100, 103, 585
Voight, Henry, 262
Volatiles, 59, 60
Volcanoes, 40
Volta, Alessandro, 170–172, 183
Voltage, 167
Voltaic cells, 171, 172
Voltaic pile, 171, 172, 174, 183, 189
von Bayer, Hans Christian, 151
von Guericke, Otto, 114–115, 154
von Kempelen, Wolfgang, 213–214, 331

Wagner, Rudolf, 564
Walker, John F., 580

Walzer, Peter, 309
Wanamaker department store, 181, 182
Washing machines, 188
Waste heat, 36–38, 434
Waste materials, oil refineries, 431
Waste products, 252
Water, 44–46, 62, 179
    infinite utility, 3
    as source of energy, 99
Water mills, 92, 94, 96, 125
Water power, 97–98
    and environment, 98–99
Water-tube boiler, 234–236, 243
Water turbine, 195–199
Water vapor, 25, 97
Water wall, 243, 244
Waterfall, 82, 134, 135, 136
    efficiency, 147
    waterwheel operated by, 83
Watermill, limitations, 109
Waterwheels, 15, 81–100, 124, 137, 196
    ancient world, 87–90
    application, 88
    basic concept, 81
    basic design, 84
    early modern world, 94–97
    energy diagram, 85
    medieval, 91–94
    oldest known, 89
    operated by waterfall, 83
    types, 84–87
    in US, 125
Watt (unit), 165
Watt, James, 119–124, 126, 213, 227, 260, 271, 311
Webbert, C., 278
Weinberg, Alvin, 416, 425
Wendt, Gerald, 426
Wescott, Lynanne, 279
Westinghouse, George, 191
    Company, 188, 191, 201
Whale oil, 297
Wheeler, John, 378, 379, 387
White, Michael, 111
Whittle, Frank, 276, 329–330, 341
William the Conqueror, 91
Williams, Trevor, 31, 47, 99, 110, 127
Wind
    generation, 565–566
    origin, 109–110
Wind direction, 103–104, 109
Wind energy, 101–111, 438, 565–580
    advantages for electricity generation, 575–576
    comparison with other energy sources, 578–579
    current scene, 573–575
    early modern age, 108–109
    future, 570
    indirect form of solar energy, 566
    limited applications, 108
    potential application, 566
    potential contribution, 570–571

Wind energy (*cont.*)
    potential disadvantages, 576
    prime locations for development, 571–573
    recent uses, 566–567
Wind farms, 573
    visual pollution, 577
Wind propulsion, 259–260
Wind technology, Denmark, 567–568, 574, 577
Wind turbines, 568–570
    advantages, 575
    bird kills, 577
    choice of site, 572
    footprint, 575
Windmills, 15, 566–567
    advantages, 104, 108
    Dutch, 107
    fantail, 108–109
    importance in industrial economy, 104–105
    Islamic world, 101–103
    limitations, 109
    state-of-the-art, 108
    see also Post mills; Turret mill
Windscale plant, 412
Windscale reactor, 399
Withington, Sidney, 100, 127, 180
Wofsy, Steven, 500
Wood, 261, 436, 540–544
    burning, 59
    combustion, 543
    components, 540–541
    constraints concerning use, 56
    cutting, drying, and transporting, 55
    direct combustion, 541
    and energy crises, 49–52
    as energy source, 54, 62, 541–542
    as fuel, 49, 66
    gasification, 542
    hardening, 41–42
    harvesting, 543
    heat produced per pound, 60
    as heat source, 62
    heating value, 540–541
    as only source of engery, 53
    problems with use of, 542–544
    problems associated with combustion of, 60–61
    renewable, 544
    Third World, 55–57
    use for heating in the home, 53
    use in industrialized world, 52–55
    see also Firewood
Wood, Margaret, 79
Wood smoke, 61
Wordsworth, Dorothy, 126
Wordsworth, William, 123
Work, 224, 226
    concept, 144
    definition, 10, 13, 144
    and energy, 147
    energy, and power, 10–13
    energy conversion to, 36–37
    examples, 10, 13–14
    and heat equivalence, 144–147
    implications, 13
    waterwheel, 81
Working fluid, 196, 212
    criteria, 211
Wright, F.B., 564
Wright, Wilbur and Orville, 271–274, 278, 328, 334
Wristwatches, 6

X-rays
    discovery, 355–358
    overexposure, 403

Yangtze River, 208
Yergin, Daniel, 296
Yevtushenko, Yevgeny, 205, 209
Young, James, 266
Young, Thomas, 11
Young–deep crudes, 291
Young–shallow crudes, 291

Zeolites, 304, 449
Zero-emission vehicles (ZEVs), 470–474
Zetsche, Dieter, 475
Zirconium metal cladding, 418–419